POWERS OF TEN

Power	Prefix	Symbol
10^{18}	Exa	E
10^{15}	Peta	P
10^{12}	Tera	T
10^{9}	Giga	G
10^{6}	Mega	M
10^{3}	kilo	k
10^{2}	hecto	h
10^{1}	deka	da
10^{-1}	deci	d
10^{-2}	centi	c
10^{-3}	milli	m
10^{-6}	micro	μ
10^{-9}	nano	n
10^{-12}	pico	p
10^{-15}	femto	f
10^{-18}	atto	a

THE GREEK ALPHABET

Upper Case	Lower Case	Name	Upper Case	Lower Case	Name
A	α	Alpha	N	ν	Nu
B	β	Beta	Ξ	ξ	Xi
Γ	γ	Gamma	O	o	Omicron
Δ	δ	Delta	Π	π	Pi
E	ε	Epsilon	P	ρ	Rho
Z	ζ	Zeta	Σ	σ,s	Sigma
H	η	Eta	T	τ	Tau
Θ	θ	Theta	Υ	υ	Upsilon
I	ι	Iota	Φ	ϕ	Phi
K	κ	Kappa	X	χ	Chi
Λ	λ	Lambda	Ψ	ψ	Psi
M	μ	Mu	Ω	ω	Omega

ELEMENTS OF Electromagnetics

SECOND EDITION

Matthew N. O. Sadiku

Temple University

Philadelphia, PA

New York Oxford

OXFORD UNIVERSITY PRESS

Oxford University Press

Oxford New York
Athens Auckland Bangkok Bombay Calcutta Cape Town
Dar es Salaam Delhi Florence Hong Kong Istanbul Karachi
Kuala Lumpur Madras Madrid Melbourne Mexico City
Nairobi Paris Singapore Taipei Tokyo Toronto

and associated companies in

Berlin Ibadan

Published by Oxford University Press, Inc.,
198 Madison Avenue, New York, New York 10016

Printed in the United States of America on acid-free paper

0-19-510368-8

9 8 7 6 5

To my wife Christianah Yemisi
and our daughters Ann Tomi
and Joyce Bolu

PREFACE

The basic objective of the first edition has been retained in this revision: to present EM concepts in a clearer and more interesting manner than earlier texts. This objective is achieved in the following ways:

1. In order to avoid complicating matters by covering EM and mathematical concepts simultaneously, vector analysis is covered at the beginning of the text and applied gradually. This approach avoids breaking in repeatedly with more background on vector analysis, thereby creating discontinuity in the flow of thought. It also separates mathematical theorems from physical concepts and makes it easier for the student to grasp the generality of those theorems.
2. Each chapter starts with a brief introduction that serves as a guide to the whole chapter and also links the chapter to the rest of the book. The introduction helps the students see the need for the chapter and how it relates to the previous chapter. Key points are emphasized to draw the reader's attention. A brief summary of the major concepts is discussed toward the end of the chapter.
3. Each chapter includes a reasonable amount of solved examples. Since the examples are part of the text, they are clearly explained without asking the reader to fill in missing steps. Thoroughly worked out examples give students confidence to solve problems themselves and to learn to apply concepts, which is an integral part of engineering education. Each illustrative example is followed by a problem in the form of a Practice Exercise, with the answer provided.
4. At the end of each chapter are ten review questions in the form of multiple-choice objective items. It has been found that open-ended questions, although intended to be thought-provoking, are ignored by most students. Objective review questions with answers immediately following them provide encouragement for students to do the problems and gain immediate feedback.

   A large number of problems are provided and are presented in the same order as the material in the main text. Problems of intermediate difficulty are identified by a single asterisk; the most difficult problems are marked with a double asterisk. Enough problems are provided to allow the instructor to choose some as examples and assign some as homework problems. Answers to odd-numbered problems are provided in Appendix E.
5. Since most practical applications involve time-varying fields, five chapters are devoted to such fields. However, static fields are given proper emphasis

because they are special cases of dynamic fields. Ignorance of electrostatics is no longer acceptable because there are large industries, such as copier and computer peripheral manufacturing, that rely on a clear understanding of electrostatics.

6. The last chapter covers numerical methods with practical applications and computer programs. This chapter is of paramount importance because most practical problems are only solvable using numerical techniques.
7. Important formulas are boxed to help students identify essential formulas. Over 130 illustrative examples and 300 figures are given in the text. Some additional learning aids such as basic mathematical formulas and identities are included in the Appendix. Another guide is a special note to students, which follows this preface.

Although this book is intended to be self-explanatory and useful for self-instruction, the personal contact that is always needed in teaching is not forgotten. The actual choice of course topics, as well as emphasis, depends on the preference of the individual instructor. For example, the instructor who feels that too much space is devoted to vector analysis or static fields may skip some of the materials; however, the students may use them as reference. Also, having covered Chapters 1 to 3, it is possible to explore Chapters 9 to 14. Instructors who disagree with the vector-calculus-first approach may proceed with Chapters 1 and 2, then skip to Chapter 4 and refer to Chapter 3 as needed. Enough material is covered for two-semester courses. If the text is to be covered in one semester, some sections may be skipped, explained briefly, or assigned as homework. Sections marked with the dagger sign (†) may be in this category.

A suggested schedule for a four-hour semester coverage is as follows:

Suggested Schedule

Chapter	Title	Approximate number of hours
1	Vector Algebra	2
2	Coordinate Systems and Transformation	2
3	Vector Calculus	4
4	Electrostatic Fields	6
5	Electric Fields in Material Space	4
6	Electrostatic Boundary-Value Problems	5
7	Magnetostatic Fields	4
8	Magnetic Forces, Materials, and Devices	6
9	Maxwell's Equations	4
10	Electromagnetic Wave Propagation	5
11	Transmission Lines	5
12	Waveguides	4
13	Antennas	5
14	Numerical Methods	(6)
	Exams	4
	TOTAL	60

Acknowledgments

Sir Isaac Newton once said, ''If I have seen farther than other men, it is because I have stood on the shoulders of giants.'' I am most grateful to many authors whose works I have drawn from. They are the giants on whose shoulders I stood while working on this project. I would like to acknowledge colleagues who have used the first edition and made many helpful suggestions for improvement. Many students have also identified errors and confusing statements. I owe special thanks to the reviewers, whose comments have helped to improve the contents of the text. These reviewers include K. D. Bennett, Lafayette College; John E. Mulholland, Villanova University; R. Bollini, Southern Illinois University at Edwardsville; Kenneth Carpenter, Kansas State University; Stuart M. Wentsworth, Auburn University; H. Grebel, New Jersey Institute of Technology; and Robert J. Coleman, University of North Carolina-Charlotte.

I would like to express my gratitude to my student Raymond Garcia for helping in various ways. Special thanks are due to Jerry Sagliocca and Jiening Ao for critically going over the entire manuscript. The secretarial support of Michelle Ayers is gratefully acknowledged. I am greatly indebted to Dr. Charles Alexander, dean of the College of Engineering and Architecture at Temple University, for his support, enthusiasm, and effort made to ensure that the first edition was successful. Special thanks are due to Dr. Brian Butz, chairman of the Department of Electrical Engineering, for his constant support and encouragement. To my wife, Chris, and our daughters, Ann and Joyce, I owe special thanks for their unfailing support and understanding. To them again I wish to dedicate this edition.

Comments, suggestions, and corrections are always welcome.

Matthew N. O. Sadiku
December 1993

A NOTE TO THE STUDENT

Electromagnetic theory is generally regarded by most students as one of the most difficult courses in physics or the electrical engineering curriculum. But this misconception may be proved wrong if you take some precautions. From experience, the following ideas are provided to help you perform to the best of your ability with the aid of this textbook:

1. Pay particular attention to Part I on *Vector Analysis,* the mathematical tool for this course. Without a clear understanding of this section, you may have problems with the rest of the book.
2. Do not attempt to memorize too many formulas. Memorize only the basic ones, which are usually boxed, and try to derive others from these. Try to understand how formulas are related. Obviously, there is nothing like a general formula for solving all problems. Each formula has some limitations due to the assumptions made in obtaining it. Be aware of those assumptions and use the formula accordingly.
3. Try to identify the key words or terms in a given definition or law. Knowing the meaning of these key words is essential for proper application of the definition or law.
4. Attempt to solve as many problems as you can. Practice is the best way to gain skill. The best way to understand the formulas and assimilate the material is by solving problems. It is recommended that you solve at least the problems in the Practice Exercise immediately following each illustrative example. Sketch a diagram illustrating the problem before attempting to solve it mathematically. Sketching the diagram not only makes the problem easier to solve, it also helps you understand the problem by simplifying and organizing your thinking process. Note that unless otherwise stated, all distances are in meters. For example $(2, -1, 5)$ actually means $(2\text{ m}, -1\text{ m}, 5\text{ m})$.

A list of the powers of ten and Greek letters commonly used throughout this text is provided in the tables located on the inside cover. Important formulas in calculus, vectors, and complex analysis are provided in Appendix A. Appendix C contains part of the computer programs developed and applied in Chapter 14. A list of symbols used throughout the text is given in Appendix D. Answers to odd-numbered problems are in Appendix E.

CONTENTS

PART 1

Vector Analysis

CHAPTER

Vector Algebra

1

Manhood, not scholarship, is the first aim of education.

—ERNEST T. SETAN

1.1 INTRODUCTION

Electromagnetics (EM) is a branch of physics or electrical engineering in which electric and magnetic phenomena are studied. It may be regarded as the study of the interactions between electric charges at rest and in motion. It entails the analysis, synthesis, physical interpretation, and application of electric and magnetic fields.

EM principles find applications in various allied disciplines such as microwaves, antennas, electric machines, satellite communications, bioelectromagnetics, plasmas, nuclear research, fiber optics, electromagnetic interference and compatibility, electromechanical energy conversion, radar meteorology, and remote sensing.[1,2] In physical medicine, for example, EM power, either in the form of shortwaves or microwaves, is used to heat deep tissues and to stimulate certain physiological responses in order to relieve certain pathological conditions. EM fields are used in induction heaters for melting, forging, annealing, surface hardening, and soldering operations. Dielectric heating equipment uses shortwaves to join or seal thin sheets of plastic materials. EM energy offers many new and exciting possibilities in agriculture. It is used, for example, to change vegetable taste by reducing acidity.

EM devices include transformers, electric relays, radio/TV, telephone, electric motors, transmission lines, waveguides, antennas, radars, and lasers. The design of these devices requires thorough knowledge of the laws and principles of EM.

[1]For numerous applications of electrostatics, see J. M. Crowley, *Fundamentals of Applied Electrostatics*. New York: John Wiley & Sons, 1986.

[2]For other areas of applications of EM, see, for example, D. Teplitz, ed., *Electromagnetism: Paths to Research*. New York: Plenum Press, 1982.

†1.2 A PREVIEW OF THE BOOK

It is worthwhile to have a broad view of the path we shall take in our journey through this book. This book is logically divided into four major parts.

In Part I, consisting of chapters 1 to 3, vector analysis is thoroughly reviewed. The time spent on this part should be regarded as valuable because vector analysis serves as the mathematical foundation for this course. If the foundation is not well laid, it will surely affect the whole structure.

Part II, comprising chapters 4 to 6, deals mainly with static charges producing electrostatic fields. The basic laws, Gauss's and Coulomb's, governing electrostatic phenomena are presented and applied to different problems. Part III, comprised of chapters 7 and 8, has to do with charges moving with uniform speed (or DC currents) producing magnetostatic fields. Biot-Savart's and Ampere's laws are used in analyzing such fields.

The last section, Part IV, consists of chapters 9 to 14 where charges can be accelerated to produce time-varying fields. Maxwell's equations, in their general form, are put together and are used in deriving governing equations for transmission lines, rectangular waveguides, resonant cavities, and antennas. The last chapter is on numerical methods commonly used in EM.

The subject of electromagnetic phenomena in this book can be summarized in Maxwell's equations:

$$\nabla \cdot \mathbf{D} = \rho_v \tag{1.1}$$

$$\nabla \cdot \mathbf{B} = 0 \tag{1.2}$$

$$\nabla \times \mathbf{E} = -\frac{\partial \mathbf{B}}{\partial t} \tag{1.3}$$

$$\nabla \times \mathbf{H} = \mathbf{J} + \frac{\partial \mathbf{D}}{\partial t} \tag{1.4}$$

where ∇ = the vector differential operator
$\mathbf{D}$ = the electric flux density
$\mathbf{B}$ = the magnetic flux density
$\mathbf{E}$ = the electric field intensity
$\mathbf{H}$ = the magnetic field intensity
ρ_v = the volume charge density
and $\mathbf{J}$ = the current density.

Maxwell based these equations on previously known results, both experimental and theoretical. A quick look at these equations shows that we shall be dealing with vector quantities. It is consequently logical that we spend some time in Part I examining the

†Indicates sections that may be skipped, explained briefly, or assigned as homework if the text is covered in one semester.

mathematical tools required for this course. The derivation of eqs. (1.1) to (1.4) for time-invariant conditions and the physical significance of the quantities **D**, **B**, **E**, **H**, **J**, and ρ_v will be our aim in parts II and III. In Part IV, we shall reexamine the equations for time-varying situations and apply them in our study of practical EM devices.

1.3 SCALARS AND VECTORS

Vector analysis is a mathematical tool with which electromagnetic (EM) concepts are most conveniently expressed and best comprehended. We must first learn its rules and techniques before we can confidently apply it. Since most students taking this course have little exposure to vector analysis, considerable attention is given to it in this and the next two chapters.[3] This chapter introduces the basic concepts of vector algebra in Cartesian coordinates only. The next chapter builds on this and extends to other coordinate systems.

A quantity is said to be a *scalar* if it has only magnitude. Quantities such as time, mass, distance, temperature, entropy, electric potential, and population are scalars. A quantity is called a *vector* if it has both magnitude and direction. Vector quantities include velocity, force, displacement, and electric field intensity. Another class of physical quantities is called *tensors,* of which scalars and vectors are special cases. For most of the time, we shall be concerned with scalars and vectors.[4]

To distinguish between a scalar and a vector it is customary to represent a vector by a letter with an arrow on top of it, such as $\vec{A}$ and $\vec{B}$, or by a letter in boldface type such as **A** and **B**. A scalar is represented simply by a letter—e.g., A, B, U, and V.

EM theory is essentially a study of some particular fields. We define a *field* as a function that specifies a particular quantity everywhere in a region. If the quantity is scalar (or vector), the field is said to be a scalar (or vector) field. Examples of scalar fields are temperature distribution in a building, sound intensity in a theater, electric potential in a region, and refractive index of a stratified medium. The gravitational force on a body in space and the velocity of raindrops in the atmosphere are examples of vector fields.

1.4 UNIT VECTOR

A vector **A** has both magnitude and direction. The *magnitude* of **A** is a scalar written as A or $|\mathbf{A}|$. A *unit vector* $\mathbf{a}_A$ along **A** is defined as a vector whose magnitude is unity (i.e., 1) and its direction is along **A**, that is,

....

[3]The reader who feels no need for review of vector algebra can skip to the next chapter.

[4]For an elementary treatment of tensors, see, for example, A. I. Borisenko and I. E. Tarapor, *Vector and Tensor Analysis with Application*. Englewood Cliffs, NJ: Prentice-Hall, 1968.

$$\mathbf{a}_A = \frac{\mathbf{A}}{|\mathbf{A}|} = \frac{\mathbf{A}}{A} \qquad [1.5]$$

Note that $|\mathbf{a}_A| = 1$. Thus we may write **A** as

$$\mathbf{A} = A\mathbf{a}_A \qquad [1.6]$$

which completely specifies **A** in terms of its magnitude A and its direction $\mathbf{a}_A$.

A vector **A** in Cartesian (or rectangular) coordinates may be represented as

$$(A_x, A_y, A_z) \qquad \text{or} \qquad A_x\mathbf{a}_x + A_y\mathbf{a}_y + A_z\mathbf{a}_z \qquad [1.7]$$

where A_x, A_y, and A_z are called the *components* of **A** in the x, y, and z directions respectively; $\mathbf{a}_x$, $\mathbf{a}_y$, and $\mathbf{a}_z$ are unit vectors in the x, y, and z directions respectively. For example, $\mathbf{a}_x$ is a dimensionless vector of magnitude one in the direction of the increase of the x-axis. The unit vectors $\mathbf{a}_x$, $\mathbf{a}_y$, and $\mathbf{a}_z$ are illustrated in Figure 1.1(a), and the components of **A** along the coordinate axes are shown in Figure 1.1(b). The magnitude of vector **A** is given by

$$A = \sqrt{A_x^2 + A_y^2 + A_z^2} \qquad [1.8]$$

and the unit vector along **A** is given by

$$\mathbf{a}_A = \frac{A_x\mathbf{a}_x + A_y\mathbf{a}_y + A_z\mathbf{a}_z}{\sqrt{A_x^2 + A_y^2 + A_z^2}} \qquad [1.9]$$

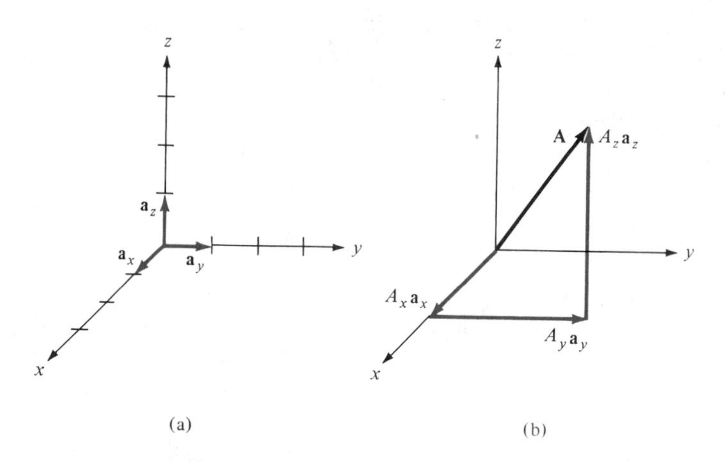

Figure 1.1 (a) Unit vectors $\mathbf{a}_x$, $\mathbf{a}_y$, and $\mathbf{a}_z$, (b) components of **A** along $\mathbf{a}_x$, $\mathbf{a}_y$, and $\mathbf{a}_z$.

†1.5 VECTOR ADDITION AND SUBTRACTION

Two vectors **A** and **B** can be added together to give another vector **C**; that is,

$$\mathbf{C} = \mathbf{A} + \mathbf{B} \tag{1.10}$$

The vector addition is carried out component by component. Thus, if $\mathbf{A} = (A_x, A_y, A_z)$ and $\mathbf{B} = (B_x, B_y, B_z)$,

$$\mathbf{C} = (A_x + B_x)\mathbf{a}_x + (A_y + B_y)\mathbf{a}_y + (A_z + B_z)\mathbf{a}_z \tag{1.11}$$

Vector subtraction is similarly carried out as

$$\begin{aligned} \mathbf{D} &= \mathbf{A} - \mathbf{B} = \mathbf{A} + (-\mathbf{B}) \\ &= (A_x - B_x)\mathbf{a}_x + (A_y - B_y)\mathbf{a}_y + (A_z - B_z)\mathbf{a}_z \end{aligned} \tag{1.12}$$

Graphically, vector addition and subtraction are obtained by either the parallelogram rule or the head-to-tail rule as portrayed in Figures 1.2 and 1.3, respectively.

The three basic laws of algebra obeyed by any given vectors **A**, **B**, and **C**, are summarized as follows:

Law	**Addition**	**Multiplication**
Commutative	$\mathbf{A} + \mathbf{B} = \mathbf{B} + \mathbf{A}$	$k\mathbf{A} = \mathbf{A}k$
Associative	$\mathbf{A} + (\mathbf{B} + \mathbf{C}) = (\mathbf{A} + \mathbf{B}) + \mathbf{C}$	$k(\ell\mathbf{A}) = (k\ell)\mathbf{A}$
Distributive	$k(\mathbf{A} + \mathbf{B}) = k\mathbf{A} + k\mathbf{B}$	

where k and ℓ are scalars. Multiplication of a vector with another vector will be discussed in Section 1.7.

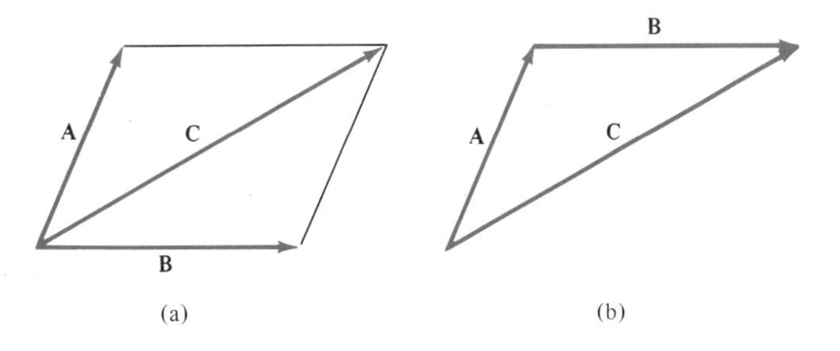

Figure 1.2 Vector addition $\mathbf{C} = \mathbf{A} + \mathbf{B}$: **(a)** parallelogram rule, **(b)** head-to-tail rule.

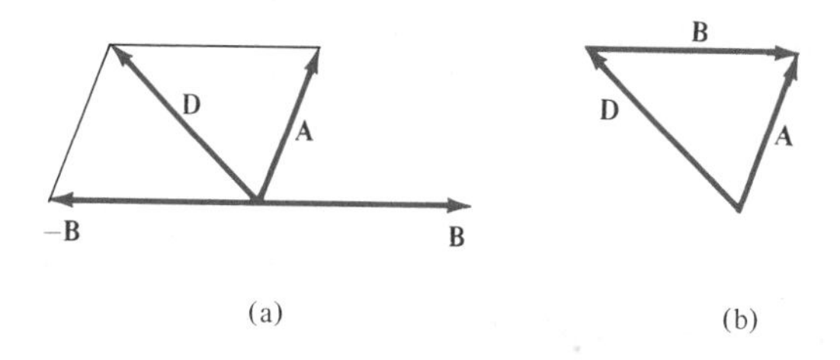

Figure 1.3 Vector subtraction $\mathbf{D} = \mathbf{A} - \mathbf{B}$: **(a)** parallelogram rule, **(b)** head-to-tail rule.

1.6 POSITION AND DISTANCE VECTORS

A point P in Cartesian coordinates may be represented by (x, y, z). The *position vector* $\mathbf{r}_P$ (or *radius vector*) of point P is defined as the directed distance from the origin O to P; that is,

$$\mathbf{r}_P = OP = x\mathbf{a}_x + y\mathbf{a}_y + z\mathbf{a}_z \tag{1.13}$$

Point (3, 4, 5), for example, and its position vector $3\mathbf{a}_x + 4\mathbf{a}_y + 5\mathbf{a}_z$ are shown in Figure 1.4.

If two points P and Q are given by (x_P, y_P, z_P) and (x_Q, y_Q, z_Q), the *distance vector* (or *separation vector*) is the displacement from P to Q as shown in Figure 1.5; that is,

$$\begin{aligned} \mathbf{r}_{PQ} &= r_Q - r_P \\ &= (x_Q - x_P)\mathbf{a}_x + (y_Q - y_P)\mathbf{a}_y + (z_Q - z_P)\mathbf{a}_z \end{aligned} \tag{1.14}$$

The difference between a point P and a vector $\mathbf{A}$ should be noted. Though both P and $\mathbf{A}$ may be represented in the same manner as (x, y, z) and (A_x, A_y, A_z) respectively,

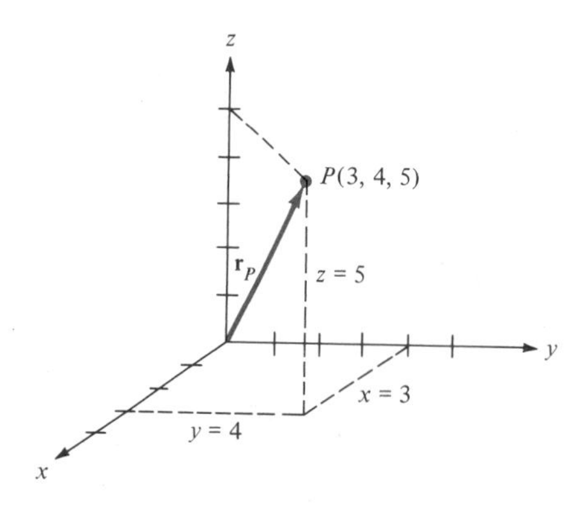

Figure 1.4 Illustration of position vector $\mathbf{r}_p = 3\mathbf{a}_x + 4\mathbf{a}_y + 5\mathbf{a}_z$.

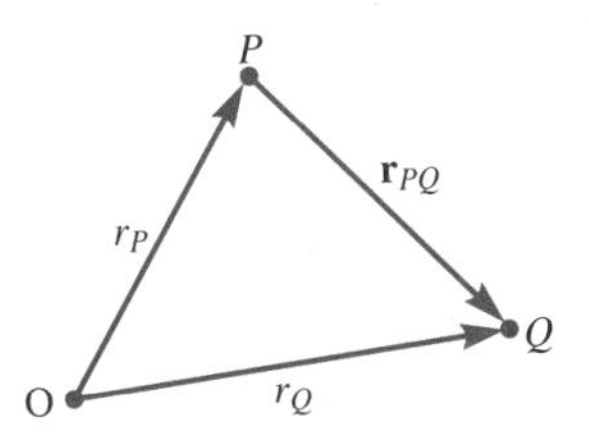

Figure 1.5 Distance vector $\mathbf{r}_{PQ}$.

the point P is not a vector; only its position vector $\mathbf{r}_p$ is a vector. Vector **A** may depend on point P, however. For example, if $\mathbf{A} = 2xy\mathbf{a}_x + y^2\mathbf{a}_y - xz^2\mathbf{a}_z$ and P is $(2, -1, 4)$, then **A** at P would be $-4\mathbf{a}_x + \mathbf{a}_y - 32\mathbf{a}_z$. A vector field is said to be *constant* or *uniform* if it does not depend on space variables x, y, and z. For example, vector $\mathbf{B} = 3\mathbf{a}_x - 2\mathbf{a}_y + 10\mathbf{a}_z$ is a uniform vector while vector $\mathbf{A} = 2xy\mathbf{a}_x + y^2\mathbf{a}_y - xz^2\mathbf{a}_z$ is not uniform because **B** is the same everywhere whereas **A** varies from point to point.

EXAMPLE 1.1

If $\mathbf{A} = 10\mathbf{a}_x - 4\mathbf{a}_y + 6\mathbf{a}_z$ and $\mathbf{B} = 2\mathbf{a}_x + \mathbf{a}_y$, find: (a) the component of **A** along $\mathbf{a}_y$, (b) the magnitude of $3\mathbf{A} - \mathbf{B}$, (c) a unit vector along $\mathbf{A} + 2\mathbf{B}$.

SOLUTION

(a) The component of **A** along $\mathbf{a}_y$ is $A_y = -4$.

(b)
$$\begin{aligned} 3\mathbf{A} - \mathbf{B} &= 3\,(10, -4, 6) - (2, 1, 0) \\ &= (30, -12, 18) - (2, 1, 0) \\ &= (28, -13, 18) \end{aligned}$$

Hence

$$\begin{aligned} |3\mathbf{A} - \mathbf{B}| &= \sqrt{28^2 + (-13)^2 + (18)^2} = \sqrt{1277} \\ &= 35.74 \end{aligned}$$

(c) Let $\mathbf{C} = \mathbf{A} + 2\mathbf{B} = (10, -4, 6) + (4, 2, 0) = (14, -2, 6)$.

A unit vector along **C** is

$$\mathbf{a}_c = \frac{\mathbf{C}}{|\mathbf{C}|} = \frac{(14, -2, 6)}{\sqrt{14^2 + (-2)^2 + 6^2}}$$

or

$$\mathbf{a}_c = 0.9113\mathbf{a}_x - 0.1302\mathbf{a}_y + 0.3906\mathbf{a}_z$$

Note that $|\mathbf{a}_c| = 1$ as expected. ■

PRACTICE EXERCISE 1.1

Given vectors $\mathbf{A} = \mathbf{a}_x + 3\mathbf{a}_z$ and $\mathbf{B} = 5\mathbf{a}_x + 2\mathbf{a}_y - 6\mathbf{a}_z$, determine

(a) $|\mathbf{A} + \mathbf{B}|$

(b) $5\mathbf{A} - \mathbf{B}$

(c) The component of $\mathbf{A}$ along $\mathbf{a}_y$

(d) A unit vector parallel to $3\mathbf{A} + \mathbf{B}$

ANSWER (a) 7, (b) (0, −2, 21), (c) 0, (d) ± (0.9117, 0.2279, 0.3419).

EXAMPLE 1.2

Points P and Q are located at (0, 2, 4) and (−3, 1, 5). Calculate

(a) The position vector P

(b) The distance vector from P to Q

(c) The distance between P and Q

(d) A vector parallel to PQ with magnitude of 10

SOLUTION (a) $\mathbf{r}_p = 0\mathbf{a}_x + 2\mathbf{a}_y + 4\mathbf{a}_z = 2\mathbf{a}_y + 4\mathbf{a}_z$

(b) $\mathbf{r}_{PQ} = \mathbf{r}_Q - \mathbf{r}_P = (-3, 1, 5) - (0, 2, 4) = (-3, -1, 1)$
or $\mathbf{r}_{PQ} = -3\mathbf{a}_x - \mathbf{a}_y + \mathbf{a}_z$

(c) Since $\mathbf{r}_{PQ}$ is the distance vector from P to Q, the distance between P and Q is the magnitude of this vector; that is,

$$d = |\mathbf{r}_{PQ}| = \sqrt{9 + 1 + 1} = 3.317$$

Alternatively:

$$d = \sqrt{(x_Q - x_P)^2 + (y_Q - y_P)^2 + (z_Q - z_P)^2} = \sqrt{9 + 1 + 1} = 3.317$$

(d) Let the required vector be $\mathbf{A}$, then

$$\mathbf{A} = A\mathbf{a}_A$$

where $A = 10$ is the magnitude of $\mathbf{A}$. Since $\mathbf{A}$ is parallel to PQ, it must have the same unit vector as $\mathbf{r}_{PQ}$ or $\mathbf{r}_{QP}$. Hence,

$$\mathbf{a}_A = \pm\frac{\mathbf{r}_{PQ}}{|\mathbf{r}_{PQ}|} = \pm\frac{(-3, -1, 1)}{3.317}$$

and

$$\mathbf{A} = \pm\frac{10\,(-3, -1, -1)}{3.317} = \pm(-9.045\,\mathbf{a}_x - 3.015\,\mathbf{a}_y + 3.015\,\mathbf{a}_z)$$

■

PRACTICE EXERCISE 1.2

Given points $P(1, -3, 5)$, $Q(2, 4, 6)$, and $R(0, 3, 8)$, find: (a) the distance vector $\mathbf{r}_{QR}$, (b) the distance between Q and R, (c) the angle between QP and QR, and (d) the area of the triangle PQR.

ANSWER (a) $-2\mathbf{a}_x - \mathbf{a}_y + 2\mathbf{a}_z$, (b) 3, (c) 70.93°, (d) 10.12.

EXAMPLE 1.3

A river flows southeast at 10 km/hr and a boat flows upon it with its bow pointed in the direction of travel. A man walks upon the deck at 2 km/hr in a direction to the right and perpendicular to the direction of the boat's movement. Find the velocity of the man with respect to the earth.

SOLUTION Consider Figure 1.6 as illustrating the problem. The velocity of the boat is

$$\mathbf{u}_b = 10\,(\cos 45°\ \mathbf{a}_x - \sin 45°\ \mathbf{a}_y)$$
$$= 7.071\ \mathbf{a}_x - 7.071\ \mathbf{a}_y \text{ km/hr}$$

The velocity of the man with respect to the boat (relative velocity) is

$$\mathbf{u}_m = 2\,(-\cos 45°\ \mathbf{a}_x - \sin 45°\ \mathbf{a}_y)$$
$$= -1.414\ \mathbf{a}_x - 1.414\ \mathbf{a}_y \text{ km/hr}$$

Thus the absolute velocity of the man is

$$\mathbf{u}_{ab} = \mathbf{u}_m + \mathbf{u}_b = 5.657\ \mathbf{a}_x - 8.485\ \mathbf{a}_y$$
$$|\mathbf{u}_{ab}| = 10.2\ \underline{/-56.3°}$$

that is, 10.2 km/hr at 56.3° south of east. ■

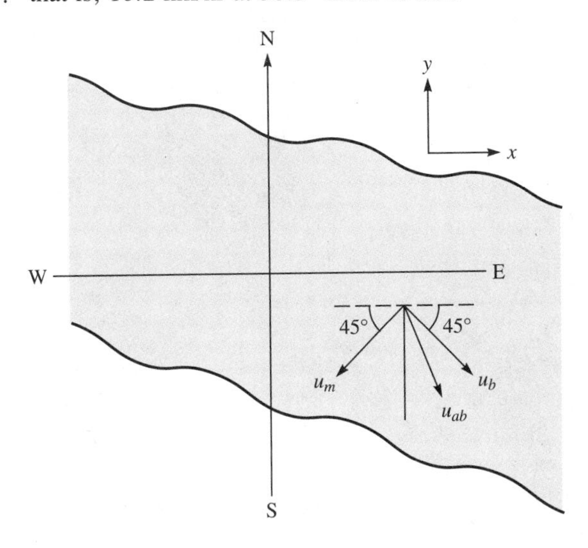

Figure 1.6 For Example 1.3.

PRACTICE EXERCISE 1.3

An airplane has a ground speed of 350 km/hr in the direction due west. If there is a wind blowing northwest at 40 km/hr, calculate the true air speed and heading of the airplane.

ANSWER 379.3 km/hr, 4.275° north of west.

EXAMPLE 1.4

Sketch the vector field

$$\mathbf{F} = x\mathbf{a}_x + y\mathbf{a}_y$$

SOLUTION For every point (x, y), $\mathbf{F}$ is evaluated. For example, at (1, 1), $\mathbf{F} = \mathbf{a}_x + \mathbf{a}_y$; at (1, −1), $\mathbf{F} = \mathbf{a}_x - \mathbf{a}_y$; at (−1, 1), $\mathbf{F} = -\mathbf{a}_x + \mathbf{a}_y$; at (−1, −1), $\mathbf{F} = -\mathbf{a}_x - \mathbf{a}_y$; on line $x = 0$, $\mathbf{F} = y\mathbf{a}_y$; on line $y = 0$, $\mathbf{F} = x\mathbf{a}_x$; and so on. By plotting a few points, the nature of the sketch is made evident as shown in Figure 1.7. Note that each arrow in the figure is in the radial direction and has a length equal to the distance from the origin. By convention, an arrow is drawn with its tail at the point where the vector is evaluated.[5] ■

[5]For a method of sketching three-dimensional vector fields, see A. J. B. Fuller and M. L. X. dos Santos, "New method for the display of three-dimensional vector fields," *IEE Proc.*, Pt. A, vol. 127, no. 7, Sept. 1980, pp. 435–442.

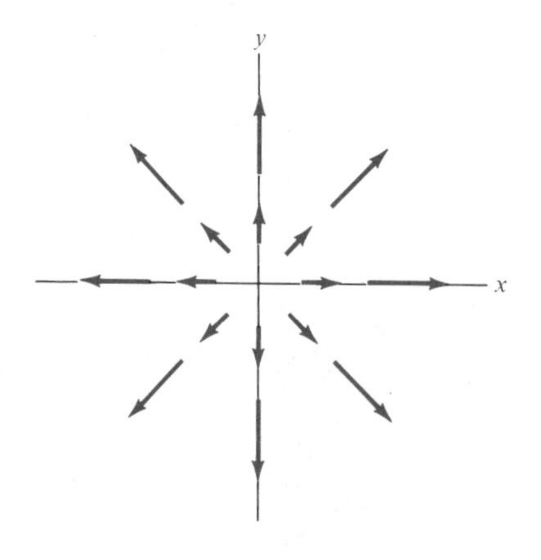

Figure 1.7 For Example 1.4.

PRACTICE EXERCISE 1.4

Sketch the vector field

$$\mathbf{G} = \frac{-y\mathbf{a}_x + x\mathbf{a}_y}{[x^2 + y^2]^{1/2}}$$

ANSWER See Figure 1.8.

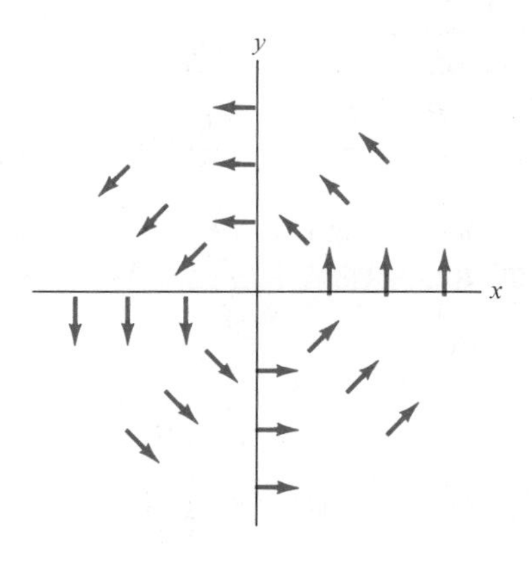

Figure 1.8 For Practice Exercise 1.4.

1.7 VECTOR MULTIPLICATION

When two vectors **A** and **B** are multiplied, the result is either a scalar or a vector depending on how they are multiplied. Thus there are two types of vector multiplication:

1. Scalar (or dot) product: $\mathbf{A} \cdot \mathbf{B}$
2. Vector (or cross) product: $\mathbf{A} \times \mathbf{B}$.

Multiplication of three vectors **A**, **B**, and **C** can result in either:

3. Scalar triple product: $\mathbf{A} \cdot (\mathbf{B} \times \mathbf{C})$

or

4. Vector triple product: $\mathbf{A} \times (\mathbf{B} \times \mathbf{C})$.

A. Dot Product

The dot product of two vectors **A** and **B**, written as $\mathbf{A} \cdot \mathbf{B}$, is defined geometrically as the product of the magnitude of **B** and the projection of **A** onto **B** (or vice versa). Thus:

$$\boxed{\mathbf{A} \cdot \mathbf{B} = AB \cos \theta_{AB}} \quad [1.15]$$

where θ_{AB} is the *smaller* angle between **A** and **B**. The result of $\mathbf{A} \cdot \mathbf{B}$ is called either the *scalar product* because it is scalar, or the *dot product* due to the dot sign. If $\mathbf{A} = (A_x, A_y, A_z)$ and $\mathbf{B} = (B_x, B_y, B_z)$, then

$$\boxed{\mathbf{A} \cdot \mathbf{B} = A_xB_x + A_yB_y + A_zB_z} \quad [1.16]$$

which is obtained by multiplying **A** and **B** component by component.

Note that dot product obeys the following:

(i) *Commutative law:*

$$\mathbf{A} \cdot \mathbf{B} = \mathbf{B} \cdot \mathbf{A} \quad [1.17]$$

(ii) *Distributive law:*

$$\mathbf{A} \cdot (\mathbf{B} + \mathbf{C}) = \mathbf{A} \cdot \mathbf{B} + \mathbf{A} \cdot \mathbf{C} \quad [1.18]$$

(iii)

$$\mathbf{A} \cdot \mathbf{A} = |\mathbf{A}|^2 = A^2 \quad [1.19]$$

Also note that

$$\mathbf{a}_x \cdot \mathbf{a}_y = \mathbf{a}_y \cdot \mathbf{a}_z = \mathbf{a}_z \cdot \mathbf{a}_x = 0 \quad [1.20a]$$

$$\mathbf{a}_x \cdot \mathbf{a}_x = \mathbf{a}_y \cdot \mathbf{a}_y = \mathbf{a}_z \cdot \mathbf{a}_z = 1 \quad [1.20b]$$

It is easy to prove the identities in eqs. (1.17) to (1.20) by applying eq. (1.15) or (1.16).

B. Cross Product

The cross product of two vectors **A** and **B**, written as $\mathbf{A} \times \mathbf{B}$, as defined as

$$\boxed{\mathbf{A} \times \mathbf{B} = AB \sin \theta_{AB}\mathbf{a}_n} \quad [1.21]$$

where $\mathbf{a}_n$ is a unit vector normal to the plane containing **A** and **B**. The direction of $\mathbf{a}_n$ is taken as the direction of the right thumb when the fingers of the right hand rotate from **A** to **B** as shown in Figure 1.9(a). Alternatively, the direction of $\mathbf{a}_n$ is taken as that of the advance of a right-handed screw as **A** is turned into **B** as shown in Figure 1.9(b).

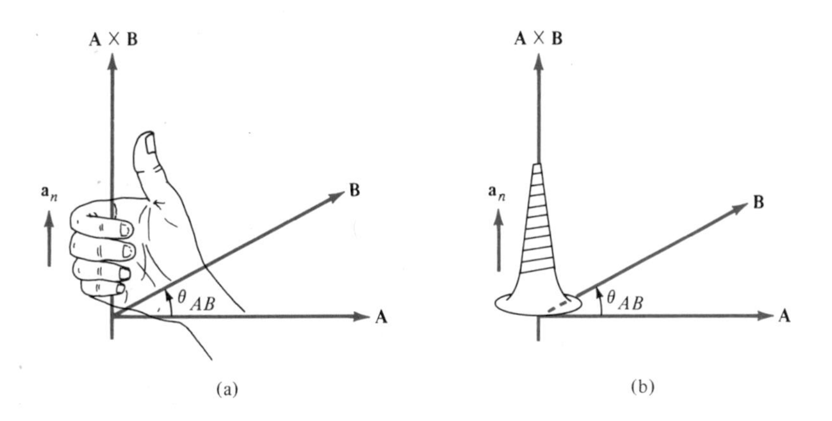

Figure 1.9 Direction of $\mathbf{A} \times \mathbf{B}$ and $\mathbf{a}_n$ using **(a)** right-hand rule, **(b)** right-handed screw rule.

The vector multiplication of eq. (1.21) is called *cross product* due to the cross sign; it is also called *vector product* because the result is a vector. If $\mathbf{A} = (A_x, A_y, A_z)$ and $\mathbf{B} = (B_x, B_y, B_z)$, then

$$\boxed{\mathbf{A} \times \mathbf{B} = \begin{vmatrix} \mathbf{a}_x & \mathbf{a}_y & \mathbf{a}_z \\ A_x & A_y & A_z \\ B_x & B_y & B_z \end{vmatrix}} \tag{1.22a}$$

$$= (A_yB_z - A_zB_y)\mathbf{a}_x + (A_zB_x - A_xB_z)\mathbf{a}_y + (A_xB_y - A_yB_x)\mathbf{a}_z \tag{1.22b}$$

which is obtained by ''crossing'' terms in cyclic permutation, hence the name cross product.

Note that the cross product has the following basic properties:

(i) It is not commutative:

$$\mathbf{A} \times \mathbf{B} \neq \mathbf{B} \times \mathbf{A} \tag{1.23a}$$

It is anticommutative:

$$\mathbf{A} \times \mathbf{B} = -\mathbf{B} \times \mathbf{A} \tag{1.23b}$$

(ii) It is not associative:

$$\mathbf{A} \times (\mathbf{B} \times \mathbf{C}) \neq (\mathbf{A} \times \mathbf{B}) \times \mathbf{C} \tag{1.24}$$

(iii) It is distributive:

$$\mathbf{A} \times (\mathbf{B} + \mathbf{C}) = \mathbf{A} \times \mathbf{B} + \mathbf{A} \times \mathbf{C} \tag{1.25}$$

(iv)

$$\mathbf{A} \times \mathbf{A} = 0 \tag{1.26}$$

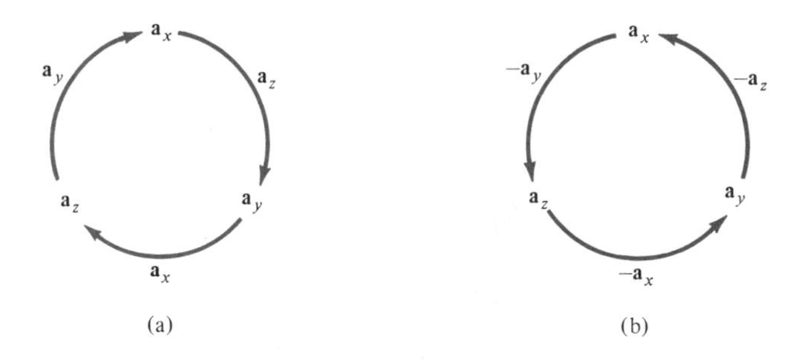

Figure 1.10 Cross product using cyclic permutation: **(a)** moving clockwise leads to positive results; **(b)** moving counterclockwise leads to negative results.

Also note that

$$\mathbf{a}_x \times \mathbf{a}_y = \mathbf{a}_z$$
$$\mathbf{a}_y \times \mathbf{a}_z = \mathbf{a}_x \qquad [1.27]$$
$$\mathbf{a}_z \times \mathbf{a}_x = \mathbf{a}_y$$

which are obtained in cyclic permutation and illustrated in Figure 1.10. The identities in eqs. (1.25) to (1.27) are easily verified using eq. (1.21) or (1.22). It should be noted that in obtaining $\mathbf{a}_n$, we have used the right-hand or right-handed screw rule because we want to be consistent with our coordinate system illustrated in Figure 1.1, which is right-handed. A right-handed coordinate system is one in which the right-hand rule is satisfied: that is, $\mathbf{a}_x \times \mathbf{a}_y = \mathbf{a}_z$ is obeyed. In a left-handed system, we follow the left-hand or left-handed screw rule and $\mathbf{a}_x \times \mathbf{a}_y = -\mathbf{a}_z$ is satisfied. Throughout this book, we shall stick to right-handed coordinate systems.

Just as multiplication of two vectors gives a scalar or vector result, multiplication of three vectors **A**, **B**, and **C** gives a scalar or vector result depending on how the vectors are multiplied. Thus we have scalar or vector triple product.

C. Scalar Triple Product

Given three vectors **A**, **B**, and **C**, we define the scalar triple product as

$$\boxed{\mathbf{A} \cdot (\mathbf{B} \times \mathbf{C}) = \mathbf{B} \cdot (\mathbf{C} \times \mathbf{A}) = \mathbf{C} \cdot (\mathbf{A} \times \mathbf{B})} \qquad [1.28]$$

obtained in cyclic permutation. If $\mathbf{A} = (A_x, A_y, A_z)$, $\mathbf{B} = (B_x, B_y, B_z)$, and $\mathbf{C} = (C_x, C_y, C_z)$, then $\mathbf{A} \cdot (\mathbf{B} \times \mathbf{C})$ is the volume of a parallelepiped having **A**, **B**, and **C** as edges and is easily obtained by finding the determinant of the 3×3 matrix formed by **A**, **B**, and **C**; that is,

$$\mathbf{A} \cdot (\mathbf{B} \times \mathbf{C}) = \begin{vmatrix} A_x & A_y & A_z \\ B_x & B_y & B_z \\ C_x & C_y & C_z \end{vmatrix} \tag{1.29}$$

Since the result of this vector multiplication is scalar, eq. (1.28) or (1.29) is called the *scalar triple product*.

D. Vector Triple Product

For vectors **A**, **B**, and **C**, we define the vector triple product as

$$\boxed{\mathbf{A} \times (\mathbf{B} \times \mathbf{C}) = \mathbf{B}(\mathbf{A} \cdot \mathbf{C}) - \mathbf{C}(\mathbf{A} \cdot \mathbf{B})} \tag{1.30}$$

obtained using the "bac-cab" rule. It should be noted that

$$(\mathbf{A} \cdot \mathbf{B})\mathbf{C} \neq \mathbf{A}(\mathbf{B} \cdot \mathbf{C}) \tag{1.31}$$

but

$$(\mathbf{A} \cdot \mathbf{B})\mathbf{C} = \mathbf{C}(\mathbf{A} \cdot \mathbf{B}). \tag{1.32}$$

1.8 COMPONENTS OF A VECTOR

A direct application of vector product is its use in determining the projection (or component) of a vector in a given direction. The projection can be scalar or vector. Given a vector **A**, we define the *scalar component* A_B of **A** along vector **B** as (see Figure 1.11a)

$$A_B = A \cos \theta_{AB} = |\mathbf{A}||\mathbf{a}_B| \cos \theta_{AB}$$

or

$$\boxed{A_B = \mathbf{A} \cdot \mathbf{a}_B} \tag{1.33}$$

The *vector component* $\mathbf{A}_B$ of **A** along **B** is simply the scalar component in eq. (1.33) multiplied by a unit vector along **B**; that is,

$$\boxed{\mathbf{A}_B = A_B \mathbf{a}_B = (\mathbf{A} \cdot \mathbf{a}_B)\mathbf{a}_B} \tag{1.34}$$

Both the scalar and vector components of **A** are illustrated in Figure 1.11. Notice from Figure 1.11(b) that the vector can be resolved into two orthogonal components: one component $\mathbf{A}_B$ parallel to **B**, another $(\mathbf{A} - \mathbf{A}_B)$ perpendicular to **B**. In fact, our Cartesian representation of a vector is essentially resolving the vector into three mutually orthogonal components as in Figure 1.1(b).

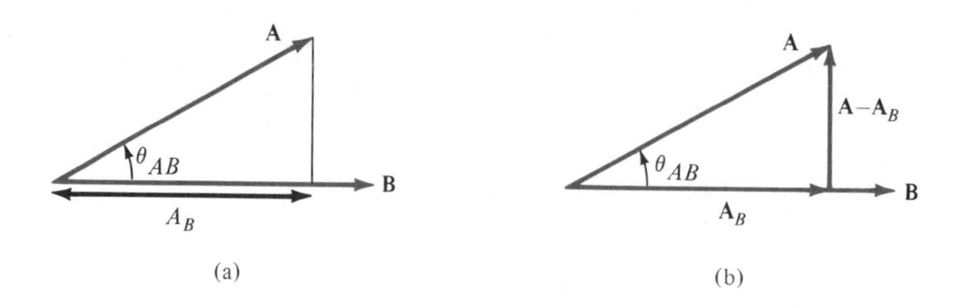

Figure 1.11 Components of **A** along **B**: **(a)** scalar component A_B, **(b)** vector component $\mathbf{A}_B$.

We have considered addition, subtraction, and multiplication of vectors. However, division of vectors **A/B** has not been considered because it is undefined except when **A** and **B** are parallel so that $\mathbf{A} = k\mathbf{B}$, where k is a constant. Differentiation and integration of vectors will be considered in Chapter 3.

EXAMPLE 1.5

Given vectors $\mathbf{A} = 3\mathbf{a}_x + 4\mathbf{a}_y + \mathbf{a}_z$ and $\mathbf{B} = 2\mathbf{a}_y - 5\mathbf{a}_z$, find the angle between **A** and **B**.

SOLUTION

The angle θ_{AB} can be found by using either dot product or cross product.

$$\mathbf{A} \cdot \mathbf{B} = (3, 4, 1) \cdot (0, 2, -5)$$
$$= 0 + 8 - 5 = 3$$
$$|\mathbf{A}| = \sqrt{3^2 + 4^2 + 1^2} = \sqrt{26}$$
$$|\mathbf{B}| = \sqrt{0^2 + 2^2 + (-5)^2} = \sqrt{29}$$
$$\cos\theta_{AB} = \frac{\mathbf{A} \cdot \mathbf{B}}{|\mathbf{A}||\mathbf{B}|} = \frac{3}{\sqrt{(26)(29)}} = 0.1092$$
$$\theta_{AB} = \cos^{-1}0.1092 = 83.73°$$

Alternatively:

$$\mathbf{A} \times \mathbf{B} = \begin{vmatrix} \mathbf{a}_x & \mathbf{a}_y & \mathbf{a}_z \\ 3 & 4 & 1 \\ 0 & 2 & -5 \end{vmatrix}$$
$$= (-20 - 2)\mathbf{a}_x + (0 + 15)\mathbf{a}_y + (6 - 0)\mathbf{a}_z$$
$$= (-22, 15, 6)$$
$$|\mathbf{A} \times \mathbf{B}| = \sqrt{(-22)^2 + 15^2 + 6^2} = \sqrt{745}$$
$$\sin\theta_{AB} = \frac{|\mathbf{A} \times \mathbf{B}|}{|\mathbf{A}||\mathbf{B}|} = \frac{\sqrt{745}}{\sqrt{(26)(29)}} = 0.994$$
$$\theta_{AB} = \cos^{-1}0.994 = 83.73°$$

■

PRACTICE EXERCISE 1.5

If $\mathbf{A} = \mathbf{a}_x + 3\mathbf{a}_z$ and $\mathbf{B} = 5\mathbf{a}_x + 2\mathbf{a}_y - 6\mathbf{a}_z$, find θ_{AB}.

ANSWER 120.6°.

EXAMPLE 1.6

Three field quantities are given by

$$\mathbf{P} = 2\mathbf{a}_x - \mathbf{a}_z$$

$$\mathbf{Q} = 2\mathbf{a}_x - \mathbf{a}_y + 2\mathbf{a}_z$$

$$\mathbf{R} = 2\mathbf{a}_x - 3\mathbf{a}_y + \mathbf{a}_z$$

Determine

(a) $(\mathbf{P} + \mathbf{Q}) \times (\mathbf{P} - \mathbf{Q})$
(b) $\mathbf{Q} \cdot \mathbf{R} \times \mathbf{P}$
(c) $\mathbf{P} \cdot \mathbf{Q} \times \mathbf{R}$
(d) $\sin \theta_{QR}$
(e) $\mathbf{P} \times (\mathbf{Q} \times \mathbf{R})$
(f) A unit vector perpendicular to both $\mathbf{Q}$ and $\mathbf{R}$
(g) The component of $\mathbf{P}$ along $\mathbf{Q}$

SOLUTION

(a)
$$\begin{aligned}(\mathbf{P} + \mathbf{Q}) \times (\mathbf{P} - \mathbf{Q}) &= \mathbf{P} \times (\mathbf{P} - \mathbf{Q}) + \mathbf{Q} \times (\mathbf{P} - \mathbf{Q}) \\ &= \mathbf{P} \times \mathbf{P} - \mathbf{P} \times \mathbf{Q} + \mathbf{Q} \times \mathbf{P} - \mathbf{Q} \times \mathbf{Q} \\ &= 0 + \mathbf{Q} \times \mathbf{P} + \mathbf{Q} \times \mathbf{P} - 0 \\ &= 2\mathbf{Q} \times \mathbf{P} \\ &= 2\begin{vmatrix} \mathbf{a}_x & \mathbf{a}_y & \mathbf{a}_z \\ 2 & -1 & 2 \\ 2 & 0 & -1 \end{vmatrix} \\ &= 2(1 - 0)\mathbf{a}_x + 2(4 + 2)\mathbf{a}_y + 2(0 + 2)\mathbf{a}_z \\ &= 2\mathbf{a}_x + 12\mathbf{a}_y + 4\mathbf{a}_z\end{aligned}$$

(b) The only way $\mathbf{Q} \cdot \mathbf{R} \times \mathbf{P}$ makes sense is

$$\begin{aligned}\mathbf{Q} \cdot (\mathbf{R} \times \mathbf{P}) &= (2, -1, 2) \cdot \begin{vmatrix} \mathbf{a}_x & \mathbf{a}_y & \mathbf{a}_z \\ 2 & -3 & 1 \\ 2 & 0 & -1 \end{vmatrix} \\ &= (2, -1, 2) \cdot (3, 4, 6) \\ &= 6 - 4 + 12 = 14.\end{aligned}$$

Alternatively:

$$\mathbf{Q} \cdot (\mathbf{R} \times \mathbf{P}) = \begin{vmatrix} 2 & -1 & 2 \\ 2 & -3 & 1 \\ 2 & 0 & -1 \end{vmatrix}$$

To find the determinant of a 3 × 3 matrix, we repeat the first two rows and cross multiply; when the cross multiplication is from right to left, the result should be negated as shown below. This technique of finding a determinant applies only to a 3 × 3 matrix. Hence

$$\mathbf{Q} \cdot (\mathbf{R} \times \mathbf{P}) = \begin{vmatrix} 2 & -1 & 2 \\ 2 & -3 & 1 \\ 2 & 0 & -1 \\ 2 & -1 & 2 \\ 2 & -3 & 1 \end{vmatrix}$$

$$= +6 + 0 - 2 + 12 - 0 - 2$$

$$= 14$$

as obtained before.

(c) From eq. (1.28)

$$\mathbf{P} \cdot (\mathbf{Q} \times \mathbf{R}) = \mathbf{Q} \cdot (\mathbf{R} \times \mathbf{P}) = 14$$

or

$$\mathbf{P} \cdot (\mathbf{Q} \times \mathbf{R}) = (2, 0, -1) \cdot (5, 2, -4)$$
$$= 10 + 0 + 4$$
$$= 14$$

(d)

$$\sin \theta_{QR} = \frac{|\mathbf{Q} \times \mathbf{R}|}{|\mathbf{Q}||\mathbf{R}|} = \frac{|(5, 2, -4)|}{|(2, -1, 2)||(2, -3, 1)|}$$
$$= \frac{\sqrt{45}}{3\sqrt{14}} = \frac{\sqrt{5}}{\sqrt{14}} = 0.5976$$

(e)

$$\mathbf{P} \times (\mathbf{Q} \times \mathbf{R}) = (2, 0, -1) \times (5, 2, -4)$$
$$= (2, 3, 4)$$

Using the bac-cab rule,

$$\mathbf{P} \times (\mathbf{Q} \times \mathbf{R}) = \mathbf{Q}(\mathbf{P} \cdot \mathbf{R}) - \mathbf{R}(\mathbf{P} \cdot \mathbf{Q})$$
$$= (2, -1, 2)\,(4 + 0 - 1) - (2, -3, 1)\,(4 + 0 - 2)$$
$$= (2, 3, 4)$$

(f) A unit vector perpendicular to both **Q** and **R** is given by

$$\mathbf{a} = \frac{\pm \mathbf{Q} \times \mathbf{R}}{|\mathbf{Q} \times \mathbf{R}|} = \frac{\pm (5, 2, -4)}{\sqrt{45}}$$

$$= \pm (0.745, 0.298, -0.596)$$

Note that $|\mathbf{a}| = 1$, $\mathbf{a} \cdot \mathbf{Q} = 0 = \mathbf{a} \cdot \mathbf{R}$. Any of these can be used to check **a**.

(g) The component of **P** along **Q** is

$$\mathbf{P}_Q = |\mathbf{P}| \cos \theta_{PQ} \, \mathbf{a}_Q$$

$$= (\mathbf{P} \cdot \mathbf{a}_Q)\mathbf{a}_Q = \frac{(\mathbf{P} \cdot \mathbf{Q})\mathbf{Q}}{|\mathbf{Q}|^2}$$

$$= \frac{(4 + 0 - 2)(2, -1, 2)}{(4 + 1 + 4)} = \frac{2}{9}(2, -1,2)$$

$$= 0.4444\mathbf{a}_x - 0.2222\mathbf{a}_y + 0.4444\mathbf{a}_z.$$

■

PRACTICE EXERCISE 1.6

Let $\mathbf{E} = 3\mathbf{a}_y + 4\mathbf{a}_z$ and $\mathbf{F} = 4\mathbf{a}_x - 10\mathbf{a}_y + 5\mathbf{a}_z$.

(a) Find the component of **E** along **F**.

(b) Determine a unit vector perpendicular to both **E** and **F**.

ANSWER (a) $(-0.2837, 0.7092, -0.3546)$, (b) $\pm (0.9398, 0.2734, -0.205)$.

EXAMPLE 1.7

Derive the cosine formula

$$a^2 = b^2 + c^2 - 2bc \cos A$$

and the sine formula

$$\frac{\sin A}{a} = \frac{\sin B}{b} = \frac{\sin C}{c}$$

using dot product and cross product respectively.

SOLUTION Consider a triangle as shown in Figure 1.12. From the figure, we notice that

$$\mathbf{a} + \mathbf{b} + \mathbf{c} = 0$$

that is,

$$\mathbf{b} + \mathbf{c} = -\mathbf{a}$$

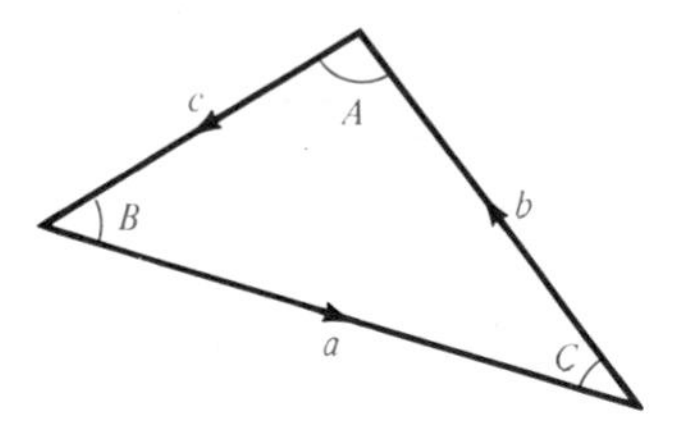

Figure 1.12 For Example 1.7.

Hence,

$$a^2 = \mathbf{a} \cdot \mathbf{a} = (\mathbf{b} + \mathbf{c}) \cdot (\mathbf{b} + \mathbf{c})$$
$$= \mathbf{b} \cdot \mathbf{b} + \mathbf{c} \cdot \mathbf{c} + 2\mathbf{b} \cdot \mathbf{c}$$
$$a^2 = b^2 + c^2 - 2bc \cos A$$

where A is the angle between **b** and **c**.

The area of a triangle is half of the product of its height and base. Hence,

$$\left|\frac{1}{2}\mathbf{a} \times \mathbf{b}\right| = \left|\frac{1}{2}\mathbf{b} \times \mathbf{c}\right| = \left|\frac{1}{2}\mathbf{c} \times \mathbf{a}\right|$$

$$ab \sin C = bc \sin A = ca \sin B$$

Dividing through by abc gives

$$\frac{\sin A}{a} = \frac{\sin B}{b} = \frac{\sin C}{c}$$

■

PRACTICE EXERCISE 1.7

Show that vectors $\mathbf{a} = (4, 0, -1)$, $\mathbf{b} = (1, 3, 4)$, and $\mathbf{c} = (-5, -3, -3)$ form the sides of a triangle. Is this a right angle triangle? Calculate the area of the triangle.

ANSWER Yes, 10.5.

EXAMPLE 1.8

Show that points $P_1(5, 2, -4)$, $P_2(1, 1, 2)$, and $P_3(-3, 0, 8)$ all lie on a straight line. Determine the shortest distance between the line and point $P_4(3, -1, 0)$.

SOLUTION The distance vector $\mathbf{r}_{P_1P_2}$ is given by

$$\mathbf{r}_{P_1P_2} = \mathbf{r}_{P_2} - \mathbf{r}_{P_1} = (1, 1, 2) - (5, 2, -4)$$
$$= (-4, -1, 6)$$

Similarly,

$$\mathbf{r}_{P_1P_3} = \mathbf{r}_{P_3} - \mathbf{r}_{P_1} = (-3, 0, 8) - (5, 2, -4)$$
$$= (-8, -2, 12)$$
$$\mathbf{r}_{P_1P_4} = \mathbf{r}_{P_4} - \mathbf{r}_{P_1} = (3, -1, 0) - (5, 2, -4)$$
$$= (-2, -3, 4)$$
$$\mathbf{r}_{P_1P_2} \times \mathbf{r}_{P_1P_3} = \begin{vmatrix} \mathbf{a}_x & \mathbf{a}_y & \mathbf{a}_z \\ -4 & -1 & 6 \\ -8 & -2 & 12 \end{vmatrix}$$
$$= (0, 0, 0)$$

showing that the angle between $\mathbf{r}_{P_1P_2}$ and $\mathbf{r}_{P_1P_3}$ is zero ($\sin\theta = 0$). This implies that P_1, P_2, and P_3 lie on a straight line.

Alternatively, the vector equation of the straight line is easily determined from Figure 1.13(a). For any point P on the line joining P_1 and P_2

$$\mathbf{r}_{P_1P} = \lambda\, \mathbf{r}_{P_1P_2}$$

where λ is a constant. Hence the position vector $\mathbf{r}_P$ of the point P must satisfy

$$\mathbf{r}_P - \mathbf{r}_{P_1} = \lambda(\mathbf{r}_{P_2} - \mathbf{r}_{P_1})$$

that is,

$$\mathbf{r}_P = \mathbf{r}_{P_1} + \lambda(\mathbf{r}_{P_2} - \mathbf{r}_{P_1})$$
$$= (5, 2, -4) - \lambda(4, 1, -6)$$
$$\mathbf{r}_P = (5 - 4\lambda, 2 - \lambda, -4 + 6\lambda)$$

This is the vector equation of the straight line joining P_1 and P_2. If P_3 is on this line, the position vector of P_3 must satisfy the equation; $\mathbf{r}_3$ does satisfy the equation when $\lambda = 2$.

The shortest distance between the line and point $P_4(3, -1, 0)$ is the perpendicular distance from the point to the line. From Figure 1.13(b), it is clear that

$$d = r_{P_1P_4} \sin\theta = |\mathbf{r}_{P_1P_4} \times \mathbf{a}_{P_1P_2}|$$
$$= \frac{|(-2, -3, 4) \times (-4, -1, 6)|}{|(-4, -1, 6)|}$$
$$= \frac{\sqrt{312}}{\sqrt{53}} = 2.426$$

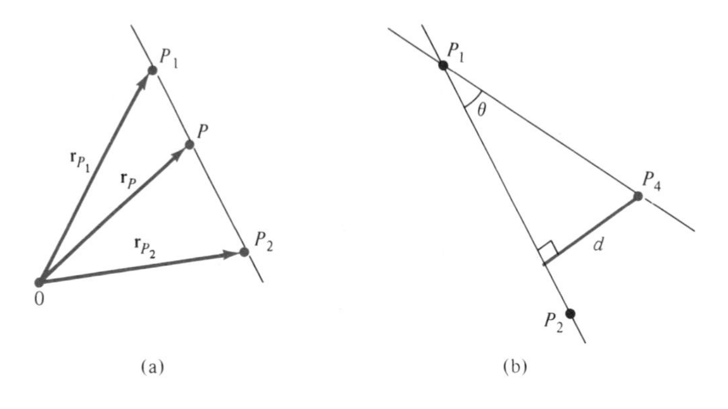

Figure 1.13 For Example 1.8.

Any point on the line may be used as a reference point. Thus, instead of using P_1 as a reference point, we could use P_3 so that

$$d = |\mathbf{r}_{P_3P_4}| \sin \theta' = |\mathbf{r}_{P_3P_4} \times \mathbf{a}_{P_3P_1}| \qquad \blacksquare$$

PRACTICE EXERCISE 1.8

If P_1 is $(1, 2, -3)$ and P_2 is $(-4, 0, 5)$, find

(a) The distance P_1P_2

(b) The vector equation of the line P_1P_2

(c) The shortest distance between the line P_1P_2 and point $P_3(7, -1, 2)$

ANSWER (a) 9.644, (b) $(1 - 5\lambda)\mathbf{a}_x + 2(1 - \lambda)\mathbf{a}_y + (8\lambda - 3)\mathbf{a}_z$, (c) 8.2.

SUMMARY

1. A field is a function that specifies a quantity in space. For example, $\mathbf{A}(x, y, z)$ is a vector field whereas $V(x, y, z)$ is a scalar field.
2. A vector $\mathbf{A}$ is uniquely specified by its magnitude and a unit vector along it, that is, $\mathbf{A} = A\mathbf{a}_A$.
3. Multiplying two vectors $\mathbf{A}$ and $\mathbf{B}$ results in either a scalar $\mathbf{A} \cdot \mathbf{B} = AB \cos \theta_{AB}$ or a vector $\mathbf{A} \times \mathbf{B} = AB \sin \theta_{AB}\, \mathbf{a}_n$. Multiplying three vectors $\mathbf{A}$, $\mathbf{B}$, and $\mathbf{C}$ yields a scalar $\mathbf{A} \cdot (\mathbf{B} \times \mathbf{C})$ or a vector $\mathbf{A} \times (\mathbf{B} \times \mathbf{C})$.
4. The scalar projection (or component) of vector $\mathbf{A}$ onto $\mathbf{B}$ is $A_B = \mathbf{A} \cdot \mathbf{a}_B$ whereas vector projection of $\mathbf{A}$ onto $\mathbf{B}$ is $\mathbf{A}_B = A_B\mathbf{a}_B$.

REVIEW QUESTIONS

1.1 Identify which of the following quantities is not a vector: (a) force, (b) momentum, (c) acceleration, (d) work, (e) weight.

1.2 Which of the following is not a scalar field?

(a) Displacement of a mosquito in space

(b) Light intensity in a drawing room

(c) Temperature distribution in your classroom

(d) Atmospheric pressure in a given region

(e) Humidity of a city

1.3 The rectangular coordinate systems shown in Figure 1.14 are right-handed except:

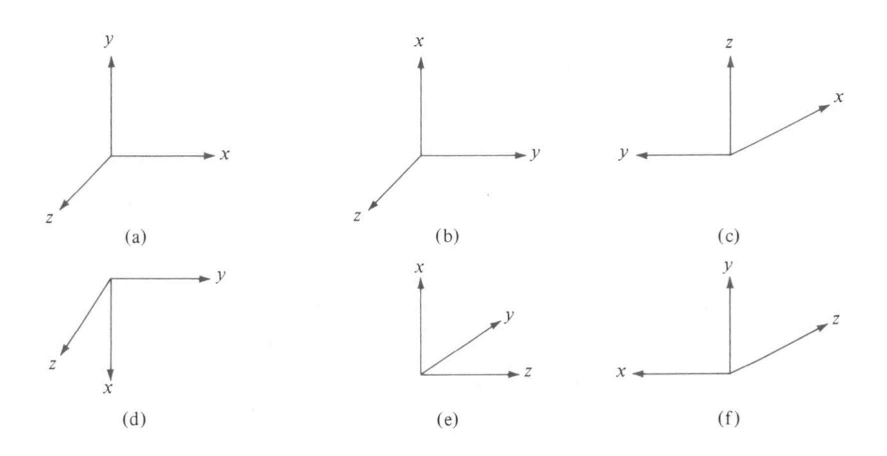

Figure 1.14 For Review Question 1.3.

1.4 Which of these is correct?

(a) $\mathbf{A} \times \mathbf{A} = |\mathbf{A}|^2$

(b) $\mathbf{A} \times \mathbf{B} + \mathbf{B} \times \mathbf{A} = 0$

(c) $\mathbf{A} \cdot \mathbf{B} \cdot \mathbf{C} = \mathbf{B} \cdot \mathbf{C} \cdot \mathbf{A}$

(d) $\mathbf{a}_x \cdot \mathbf{a}_y = \mathbf{a}_z$

(e) $\mathbf{a}_k = \mathbf{a}_x - \mathbf{a}_y$
where $\mathbf{a}_k$ is a unit vector.

1.5 Which of the following identities is not valid?

(a) $\mathbf{a}\ (\mathbf{b} + \mathbf{c}) = \mathbf{ab} + \mathbf{bc}$

(b) $\mathbf{a} \times (\mathbf{b} + \mathbf{c}) = \mathbf{a} \times \mathbf{b} + \mathbf{a} \times \mathbf{c}$

(c) $\mathbf{a} \cdot \mathbf{b} = \mathbf{b} \cdot \mathbf{a}$

(d) $\mathbf{c} \cdot (\mathbf{a} \times \mathbf{b}) = -\mathbf{b} \cdot (\mathbf{a} \times \mathbf{c})$

(e) $\mathbf{a}_A \cdot \mathbf{a}_B = \cos \theta_{AB}$

1.6 If $\mathbf{A} = 2\mathbf{a}_x + 5\mathbf{a}_y - 6\mathbf{a}_z$, then $\mathbf{A} \cdot \mathbf{a}_x + \mathbf{A} \cdot \mathbf{a}_y + \mathbf{A} \cdot \mathbf{a}_z$ is

(a) -60

(b) -6

(c) 1

(d) 2

1.7 Let $\mathbf{F} = 2\mathbf{a}_x - 6\mathbf{a}_y + 10\mathbf{a}_z$ and $\mathbf{G} = \mathbf{a}_x + G_y\mathbf{a}_y + 5\mathbf{a}_z$. If $\mathbf{F}$ and $\mathbf{G}$ have the same unit vector, G_y is

(a) -6

(b) -3

(c) 0

(d) 6

1.8 Given that $\mathbf{A} = \mathbf{a}_x + \alpha\mathbf{a}_y + \mathbf{a}_z$ and $\mathbf{B} = \alpha\mathbf{a}_x + \mathbf{a}_y + \mathbf{a}_z$, if $\mathbf{A}$ and $\mathbf{B}$ are normal to each other, α is

(a) -2

(b) $-1/2$

(c) 0

(d) 1

(e) 2

1.9 The component of $6\mathbf{a}_x + 2\mathbf{a}_y - 3\mathbf{a}_z$ along $3\mathbf{a}_x - 4\mathbf{a}_y$ is

(a) $-12\mathbf{a}_x - 9\mathbf{a}_y - 3\mathbf{a}_z$

(b) $30\mathbf{a}_x - 40\mathbf{a}_y$

(c) 10/7

(d) 2

(e) 10

1.10 Given $\mathbf{A} = -6\mathbf{a}_x + 3\mathbf{a}_y + 2\mathbf{a}_z$, the projection of $\mathbf{A}$ along $\mathbf{a}_y$ is

(a) -12

(b) -4

(c) 3

(d) 7

(e) 12

Answers: 1.1d, 1.2a, 1.3b,e, 1.4b, 1.5a, 1.6c, 1.7b, 1.8b, 1.9d, 1.10c.

PROBLEMS

1.1 Let $\mathbf{A} = 2\mathbf{a}_x + 5\mathbf{a}_y - 3\mathbf{a}_z$, $\mathbf{B} = 3\mathbf{a}_x - 4\mathbf{a}_y$, and $\mathbf{C} = \mathbf{a}_x + \mathbf{a}_y + \mathbf{a}_z$. (a) Determine $\mathbf{A} + 2\mathbf{B}$. (b) Calculate $|\mathbf{A} - 5\mathbf{C}|$. (c) For what values of k is $|k\mathbf{B}| = 2$? (d) Find $(\mathbf{A} \times \mathbf{B})/(\mathbf{A} \cdot \mathbf{B})$.

1.2 For vectors $\mathbf{A} = 2\mathbf{a}_x + 5\mathbf{a}_y + 4\mathbf{a}_z$, $\mathbf{B} = 3\mathbf{a}_x + \mathbf{a}_y + 5\mathbf{a}_z$, and $\mathbf{C} = \mathbf{a}_x - 6\mathbf{a}_z$, calculate:

(a) $\mathbf{A} \cdot \mathbf{B}$

(b) $\mathbf{A} \times \mathbf{B}$

(c) $\mathbf{A} \cdot (\mathbf{B} - \mathbf{C})$

(d) $\mathbf{A} \times (\mathbf{B} + \mathbf{C})$

1.3 Three field quantities are given by $\mathbf{P} = \mathbf{a}_x - 5\mathbf{a}_y + 3\mathbf{a}_z$, $\mathbf{Q} = 3\mathbf{a}_x + 2\mathbf{a}_y + 4\mathbf{a}_z$, and $\mathbf{R} = \mathbf{a}_x - \mathbf{a}_y$. Determine:

(a) $\mathbf{P} \cdot \mathbf{Q}$

(b) $\cos \theta_{PQ}$

(c) $\mathbf{Q} \times \mathbf{R}$

(d) $\sin \theta_{QR}$

(e) $\mathbf{P} \cdot (\mathbf{Q} \times \mathbf{R})$

(f) $\mathbf{Q} \cdot \mathbf{R} \times \mathbf{P}$

(g) $\mathbf{P} \times (\mathbf{Q} \times \mathbf{R})$

(h) $(\mathbf{P} \times \mathbf{Q}) \times \mathbf{R}$

1.4 If $\mathbf{A} = \alpha\mathbf{a}_x + 2\mathbf{a}_y + 10\mathbf{a}_z$ and $\mathbf{B} = 4\alpha\mathbf{a}_x + 8\mathbf{a}_y - 2\alpha\mathbf{a}_z$, for what values of α are $\mathbf{A}$ and $\mathbf{B}$ perpendicular?

1.5 For $\mathbf{U} = U_x\mathbf{a}_x + 5\mathbf{a}_y - \mathbf{a}_z$, $\mathbf{V} = 2\mathbf{a}_x - V_y\mathbf{a}_y + 3\mathbf{a}_z$, and $\mathbf{W} = 6\mathbf{a}_x + \mathbf{a}_y + W_z\mathbf{a}_z$, obtain U_x, V_y, and W_z such that **U**, **V**, and **W** are mutually orthogonal.

1.6 Given vectors $\mathbf{T} = 2\mathbf{a}_x - 6\mathbf{a}_y + 3\mathbf{a}_z$ and $\mathbf{S} = \mathbf{a}_x + 2\mathbf{a}_y + \mathbf{a}_z$, find: (a) the scalar projection of **T** on **S**, (b) the vector projection of **S** on **T**, (c) the smaller angle between **T** and **S**.

1.7 If $\mathbf{A} = 2\mathbf{a}_x + 3\mathbf{a}_y - 4\mathbf{a}_z$ and $\mathbf{B} = -6\mathbf{a}_x - 4\mathbf{a}_y + \mathbf{a}_z$, find the scalar and vector components of $\mathbf{A} \times \mathbf{B}$ along the direction of $\mathbf{C} = \mathbf{a}_x - \mathbf{a}_y + \mathbf{a}_z$.

1.8 Calculate the angles that vector $\mathbf{H} = 3\mathbf{a}_x + 5\mathbf{a}_y - 8\mathbf{a}_z$ makes with the x-, y-, and z-axes.

1.9 Find the triple scalar product of **P**, **Q**, and **R** given that

$$\mathbf{P} = 2\mathbf{a}_x - \mathbf{a}_y + \mathbf{a}_z$$

$$\mathbf{Q} = \mathbf{a}_x + \mathbf{a}_y + \mathbf{a}_z$$

and

$$\mathbf{R} = 2\mathbf{a}_x + 3\mathbf{a}_z.$$

1.10 Verify that:

(a) $\mathbf{A} \cdot (\mathbf{A} \times \mathbf{B}) = 0 = \mathbf{B} \cdot (\mathbf{A} \times \mathbf{B})$

(b) $(\mathbf{A} \cdot \mathbf{B})^2 + |\mathbf{A} \times \mathbf{B}|^2 = (AB)^2$

(c) If $\mathbf{A} = (A_x, A_y, A_z)$, then $\mathbf{A} = (\mathbf{A} \cdot \mathbf{a}_x)\mathbf{a}_x + (\mathbf{A} \cdot \mathbf{a}_y)\mathbf{a}_y + (\mathbf{A} \cdot \mathbf{a}_z)\mathbf{a}_z$.

1.11 Points $P_1(1, 2, 3)$, $P_2(-5, 2, 0)$, and $P_3(2, 7, -3)$ form a triangle in space. Calculate the area of the triangle.

1.12 The vertices of a triangle are located at $(4, 1, -3)$, $(-2, 5, 4)$, and $(0, 1, 6)$. Find the three angles of the triangle.

***1.13** If **r** is the position vector of the point (x, y, z) and **A** is a constant vector, show that:

(a) $(\mathbf{r} - \mathbf{A}) \cdot \mathbf{A} = 0$ is the equation of a constant plane

(b) $(\mathbf{r} - \mathbf{A}) \cdot \mathbf{r} = 0$ is the equation of a sphere

(c) Also show that the result of part (a) is of the form $Ax + By + Cz + D = 0$ where $D = -(A^2 + B^2 + C^2)$, and that of part (b) is of the form $x^2 + y^2 + z^2 = r^2$.

....

*Single-asterisks indicate problems of intermediate difficulty.

***1.14** (a) Prove that $\mathbf{P} = \cos\theta_1\, \mathbf{a}_x + \sin\theta_1\, \mathbf{a}_y$ and $\mathbf{Q} = \cos\theta_2 \mathbf{a}_x + \sin\theta_2\, \mathbf{a}_y$ are unit vectors in the xy-plane respectively making angles θ_1 and θ_2 with the x-axis.

(b) By means of dot product, obtain the formula for $\cos(\theta_2 - \theta_1)$. By similarly formulating **P** and **Q**, obtain the formula for $\cos(\theta_2 + \theta_1)$.

(c) If θ is the angle between **P** and **Q**, find $\frac{1}{2}|\mathbf{P} - \mathbf{Q}|$ in terms of θ.

***1.15** Let the position vectors of three points be represented by **A**, **B**, and **C**.

(a) Show that the shortest distance to the origin from the plane formed by the three points is given by

$$d = \frac{\mathbf{A} \cdot [(\mathbf{A} - \mathbf{B}) \times (\mathbf{A} - \mathbf{C})]}{|(\mathbf{A} - \mathbf{B}) \times (\mathbf{A} - \mathbf{C})|}$$

(b) Show that vector $(\mathbf{A} \times \mathbf{B}) + (\mathbf{B} \times \mathbf{C}) + (\mathbf{C} \times \mathbf{A})$ is perpendicular to the plane.

(c) Determine d if the points (2, 1, 0), (−1, 4, 3), and (2, 3, −5) form a plane.

1.16 Consider a rigid body rotating with a constant angular velocity ω radians per second about a fixed axis through O as in Figure 1.15. Let **r** be the distance vector from O to P, the position of a particle in the body. The velocity **u** of the body at P is $|\mathbf{u}| = d\omega = |\mathbf{r}| \sin\theta\, |\boldsymbol{\omega}|$ or $\mathbf{u} = \boldsymbol{\omega} \times \mathbf{r}$. If the rigid body is rotating with 3 radians per second about an axis parallel to $\mathbf{a}_x - 2\mathbf{a}_y + 2\mathbf{a}_z$ and passing through point (2, −3, 1), determine the velocity of the body at (1, 3, 4).

1.17 Decompose vector $10\mathbf{a}_x - 5\mathbf{a}_y + 4\mathbf{a}_z$ into vectors parallel and perpendicular to vector $3\mathbf{a}_x + \mathbf{a}_y$.

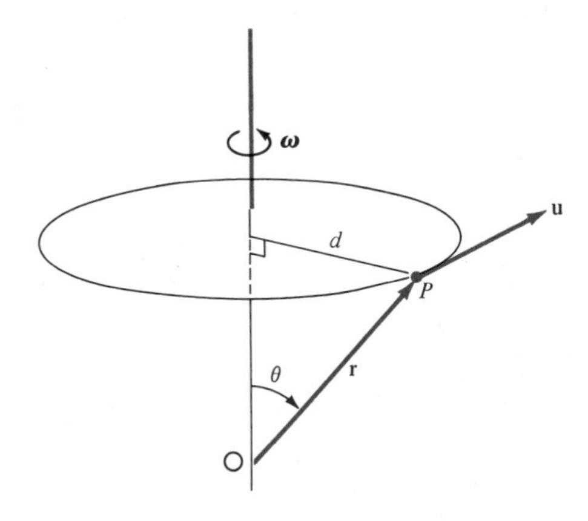

Figure 1.15 For Problem 1.16.

1.18 Given $\mathbf{A} = x^2y\mathbf{a}_x - yz\mathbf{a}_y + yz^2\mathbf{a}_z$, determine:

(a) The magnitude of $\mathbf{A}$ at point $T(2, -1, 3)$

(b) The distance vector from T to S if S is 5.6 units away from T and in the same direction as $\mathbf{A}$ at T

(c) The position vector of S

1.19 Let $\mathbf{Q} = (2x - y)\mathbf{a}_x + (4y + z)\mathbf{a}_y + (4x - 2z)\mathbf{a}_z$.

(a) Determine a unit vector in the direction of $\mathbf{Q}$ at $P(1, 2, 1)$.

(b) Find the component of $\mathbf{Q}$ at P in the direction of PT where T is point $(5, 3, -4)$.

(c) Where is $\mathbf{Q}$ the same as the unit vector of $\mathbf{a}_x + 11\mathbf{a}_y + 10\mathbf{a}_z$?

1.20 $\mathbf{E}$ and $\mathbf{F}$ are vector fields given by $\mathbf{E} = 2x\mathbf{a}_x + \mathbf{a}_y + yz\mathbf{a}_z$ and $\mathbf{F} = xy\mathbf{a}_x - y^2\mathbf{a}_y + xyz\mathbf{a}_z$. Determine:

(a) $|\mathbf{E}|$ at $(1, 2, 3)$

(b) The component of $\mathbf{E}$ along $\mathbf{F}$ at $(1, 2, 3)$

(c) A vector perpendicular to both $\mathbf{E}$ and $\mathbf{F}$ at $(0, 1, -3)$ whose magnitude is unity

***1.21** In Figure 1.11(b), the component of $\mathbf{A}$ perpendicular to $\mathbf{B}$ and in the plane containing $\mathbf{A}$ and $\mathbf{B}$ is given by $\mathbf{A} - \mathbf{A}_B$. Show that this component can be obtained from the triple product $\mathbf{a}_B \times (\mathbf{A} \times \mathbf{a}_B)$ so that $\mathbf{A} = (\mathbf{A} \cdot \mathbf{a}_B)\mathbf{a}_B + \mathbf{a}_B \times (\mathbf{A} \times \mathbf{a}_B)$.

CHAPTER

Coordinate Systems and Transformation

2

The heights by great men reached and kept were not attained by sudden flight, but they, while their companions slept, were toiling upward in the night.[1]

—HENRY W. LONGFELLOW

2.1 INTRODUCTION

In general, the physical quantities we shall be dealing with in EM are functions of space and time. In order to describe the spatial variations of the quantities, we must be able to define all points uniquely in space in a suitable manner. This requires using an appropriate coordinate system.

A point or vector can be represented in any curvilinear coordinate system, which may be orthogonal or nonorthogonal. An orthogonal system is one in which the coordinates are mutually perpendicular. Nonorthogonal systems are hard to work with and they are of little or no practical use. Examples of orthogonal coordinate systems include the Cartesian (or rectangular), the circular cylindrical, the spherical, the elliptic cylindrical, the parabolic cylindrical, the conical, the prolate spheroidal, the oblate spheroidal, and the ellipsoidal.[2] A considerable amount of work and time may be saved by choosing a coordinate system that best fits a given problem. A hard problem in one coordinate system may turn out to be easy in another system.

In this text, we shall restrict ourselves to the three best-known coordinate systems: the Cartesian, the circular cylindrical, and the spherical. Although we have considered the Cartesian system in Chapter 1, we shall consider it in detail in this chapter. We should bear in mind that the concepts covered in Chapter 1 and demonstrated in

....

[1]Quotations at the beginning of chapters 2 to 14 are taken with permission from A. C. and D. G. Remley, eds., *Leaves of Gold*. Fort Worth, TX: Brownlow Publishing Co., Inc., 1948.

[2]For an introductory treatment of these coordinate systems, see M. R. Spigel, *Mathematical Handbook of Formulas and Tables*. New York: McGraw-Hill, 1968, pp. 124–130.

Cartesian coordinates are equally applicable to other systems of coordinates. For example, the procedure for finding dot or cross product of two vectors in a cylindrical system is the same as that used in the Cartesian system in Chapter 1.

Sometimes, it is necessary to transform points and vectors from one coordinate system to another. The techniques for doing this will be presented and illustrated with examples.

2.2 CARTESIAN COORDINATES (x, y, z)

As mentioned in Chapter 1, a point P can be represented as (x, y, z) as illustrated in Figure 1.1. The ranges of the coordinate variables x, y, and z are

$$\begin{gathered} -\infty < x < \infty \\ -\infty < y < \infty \\ -\infty < z < \infty \end{gathered} \tag{2.1}$$

A vector $\mathbf{A}$ in Cartesian (otherwise known as rectangular) coordinates can be written as

$$(A_x, A_y, A_z) \qquad \text{or} \qquad A_x\mathbf{a}_x + A_y\mathbf{a}_y + A_z\mathbf{a}_z \tag{2.2}$$

where $\mathbf{a}_x$, $\mathbf{a}_y$, and $\mathbf{a}_z$ are unit vectors along the x-, y-, and z-directions as shown in Figure 1.1.

2.3 CIRCULAR CYLINDRICAL COORDINATES (ρ, ϕ, Z)

The circular cylindrical coordinate system is very convenient whenever we are dealing with problems having cylindrical symmetry.

A point P in cylindrical coordinates is represented as (ρ, ϕ, z) and is as shown in Figure 2.1. Observe Figure 2.1 closely and note how we define each space variable: ρ is the radius of the cylinder passing through P or the radial distance from the z-axis; ϕ, called the *azimuthal* angle, is measured from the x-axis in the xy-plane; and z is the same as in the Cartesian system. The ranges of the variables are

$$\begin{gathered} 0 \leq \rho < \infty \\ 0 \leq \phi < 2\pi \\ -\infty < z < \infty \end{gathered} \tag{2.3}$$

A vector $\mathbf{A}$ in cylindrical coordinates can be written as

$$(A_\rho, A_\phi, A_z) \qquad \text{or} \qquad A_\rho\mathbf{a}_\rho + A_\phi\mathbf{a}_\phi + A_z\mathbf{a}_z \tag{2.4}$$

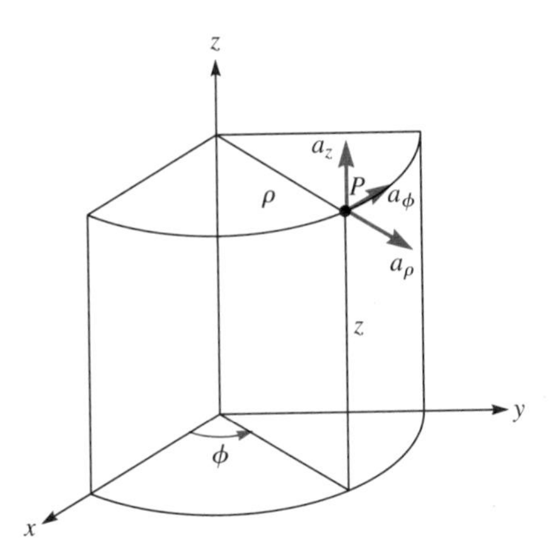

Figure 2.1 Point P and unit vectors in the cylindrical coordinate system.

where $\mathbf{a}_\rho$, $\mathbf{a}_\phi$, and $\mathbf{a}_z$ are unit vectors in the ρ-, ϕ-, and z-directions as illustrated in Figure 2.1. Note that $\mathbf{a}_\phi$ is not in degrees; it assumes the unit vector of $\mathbf{A}$. For example, if a force of 10 N acts on a particle in a circular motion, the force may be represented as $\mathbf{F} = 10\mathbf{a}_\phi$ N. In this case, $\mathbf{a}_\phi$ is in newtons.

The magnitude of $\mathbf{A}$ is

$$|\mathbf{A}| = (A_\rho^2 + A_\phi^2 + A_z^2)^{1/2} \quad [2.5]$$

Notice that the unit vectors $\mathbf{a}_\rho$, $\mathbf{a}_\phi$, and $\mathbf{a}_z$ are mutually perpendicular because our coordinate system is orthogonal; $\mathbf{a}_\rho$ points in the direction of increasing ρ, $\mathbf{a}_\phi$ in the direction of increasing ϕ, and $\mathbf{a}_z$ in the positive z-direction. Thus,

$$\mathbf{a}_\rho \cdot \mathbf{a}_\rho = \mathbf{a}_\phi \cdot \mathbf{a}_\phi = \mathbf{a}_z \cdot \mathbf{a}_z = 1 \quad [2.6a]$$

$$\mathbf{a}_\rho \cdot \mathbf{a}_\phi = \mathbf{a}_\phi \cdot \mathbf{a}_z = \mathbf{a}_z \cdot \mathbf{a}_\rho = 0 \quad [2.6b]$$

$$\mathbf{a}_\rho \times \mathbf{a}_\phi = \mathbf{a}_z \quad [2.6c]$$

$$\mathbf{a}_\phi \times \mathbf{a}_z = \mathbf{a}_\rho \quad [2.6d]$$

$$\mathbf{a}_z \times \mathbf{a}_\rho = \mathbf{a}_\phi \quad [2.6e]$$

where eqs. (2.6c) to (2.6e) are obtained in cyclic permutation (see Figure 1.10).

The relationships between the variables (x, y, z) of the Cartesian coordinate system and those of the cylindrical system (ρ, ϕ, z) are easily obtained from Figure 2.2 as

$$\boxed{\rho = \sqrt{x^2 + y^2}, \qquad \phi = \tan^{-1}\frac{y}{x}, \qquad z = z} \quad [2.7]$$

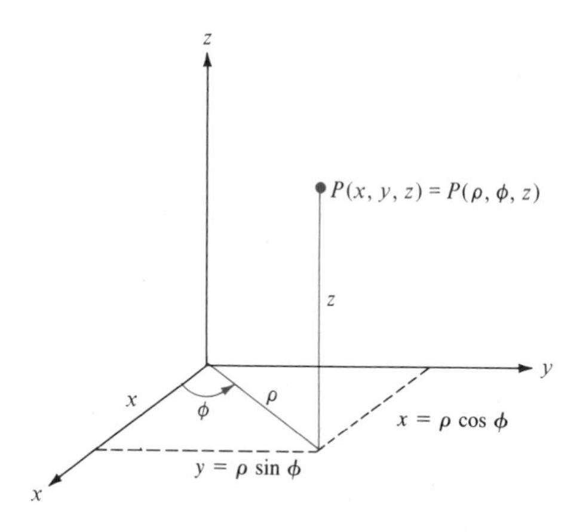

Figure 2.2 Relationship between (x, y, z) and (ρ, ϕ, z).

or

$$\boxed{x = \rho \cos \phi, \qquad y = \rho \sin \phi, \qquad z = z} \tag{2.8}$$

Whereas eq. (2.7) is for transforming a point from Cartesian (x, y, z) to cylindrical (ρ, ϕ, z) coordinates, eq. (2.8) is for $(\rho, \phi, z) \rightarrow (x, y, z)$ transformation.

The relations between $(\mathbf{a}_x, \mathbf{a}_y, \mathbf{a}_z)$ and $(\mathbf{a}_\rho, \mathbf{a}_\phi, \mathbf{a}_z)$ are obtained geometrically from Figure 2.3:

$$\begin{aligned} \mathbf{a}_x &= \cos \phi \, \mathbf{a}_\rho - \sin \phi \, \mathbf{a}_\phi \\ \mathbf{a}_y &= \sin \phi \, \mathbf{a}_\rho + \cos \phi \, \mathbf{a}_\phi \\ \mathbf{a}_z &= \mathbf{a}_z \end{aligned} \tag{2.9}$$

or

$$\begin{aligned} \mathbf{a}_\rho &= \cos \phi \, \mathbf{a}_x + \sin \phi \, \mathbf{a}_y \\ \mathbf{a}_\phi &= -\sin \phi \, \mathbf{a}_x + \cos \phi \, \mathbf{a}_y \\ \mathbf{a}_z &= \mathbf{a}_z \end{aligned} \tag{2.10}$$

Finally, the relationships between (A_x, A_y, A_z) and (A_ρ, A_ϕ, A_z) are obtained by simply substituting eq. (2.9) into eq. (2.2) and collecting terms. Thus

$$\mathbf{A} = (A_x \cos \phi + A_y \sin \phi)\mathbf{a}_\rho + (-A_x \sin \phi + A_y \cos \phi)\mathbf{a}_\phi + A_z \mathbf{a}_z \tag{2.11}$$

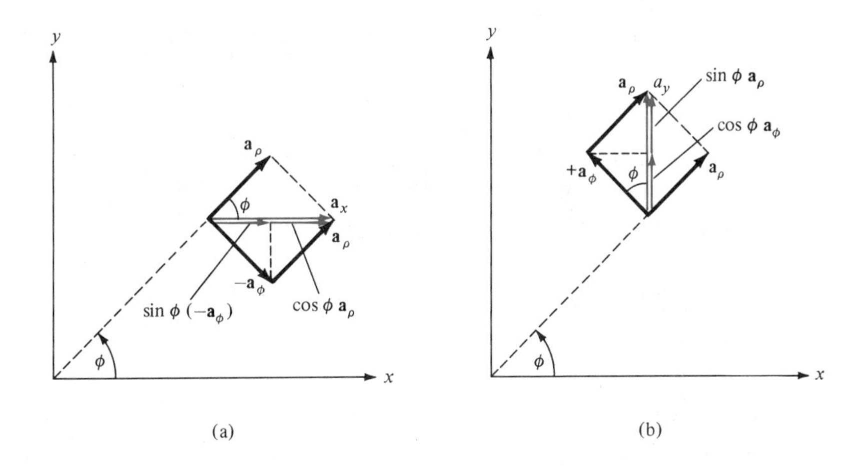

Figure 2.3 Unit vector transformation: **(a)** cylindrical components of $\mathbf{a}_x$, **(b)** cylindrical components of $\mathbf{a}_y$.

or

$$\begin{aligned} A_\rho &= A_x \cos\phi + A_y \sin\phi \\ A_\phi &= -A_x \sin\phi + A_y \cos\phi \\ A_z &= A_z \end{aligned} \qquad [2.12]$$

In matrix form, we have the transformation of vector **A** from (A_x, A_y, A_z) to (A_ρ, A_ϕ, A_z) as

$$\begin{bmatrix} A_\rho \\ A_\phi \\ A_z \end{bmatrix} = \begin{bmatrix} \cos\phi & \sin\phi & 0 \\ -\sin\phi & \cos\phi & 0 \\ 0 & 0 & 1 \end{bmatrix} \begin{bmatrix} A_x \\ A_y \\ A_z \end{bmatrix} \qquad [2.13]$$

The inverse of the transformation $(A_\rho, A_\phi, A_z) \rightarrow (A_x, A_y, A_z)$ is obtained as

$$\begin{bmatrix} A_x \\ A_y \\ A_z \end{bmatrix} = \begin{bmatrix} \cos\phi & \sin\phi & 0 \\ -\sin\phi & \cos\phi & 0 \\ 0 & 0 & 1 \end{bmatrix}^{-1} \begin{bmatrix} A_\rho \\ A_\phi \\ A_z \end{bmatrix} \qquad [2.14]$$

or directly from eqs. (2.5) and (2.10). Thus

$$\begin{bmatrix} A_x \\ A_y \\ A_z \end{bmatrix} = \begin{bmatrix} \cos\phi & -\sin\phi & 0 \\ \sin\phi & \cos\phi & 0 \\ 0 & 0 & 1 \end{bmatrix} \begin{bmatrix} A_\rho \\ A_\phi \\ A_z \end{bmatrix} \qquad [2.15]$$

An alternative way of obtaining eq. (2.14) or (2.15) is using the dot product. For example:

$$\begin{bmatrix} A_x \\ A_y \\ A_z \end{bmatrix} = \begin{bmatrix} \mathbf{a}_x \cdot \mathbf{a}_\rho & \mathbf{a}_x \cdot \mathbf{a}_\phi & \mathbf{a}_x \cdot \mathbf{a}_z \\ \mathbf{a}_y \cdot \mathbf{a}_\rho & \mathbf{a}_y \cdot \mathbf{a}_\phi & \mathbf{a}_y \cdot \mathbf{a}_z \\ \mathbf{a}_z \cdot \mathbf{a}_\rho & \mathbf{a}_z \cdot \mathbf{a}_\phi & \mathbf{a}_z \cdot \mathbf{a}_z \end{bmatrix} \begin{bmatrix} A_\rho \\ A_\phi \\ A_z \end{bmatrix} \quad [2.16]$$

The derivation of this is left as an exercise.

2.4 SPHERICAL COORDINATES (r, θ, ϕ)

The spherical coordinate system is most appropriate when dealing with problems having a degree of spherical symmetry. A point P can be represented as (r, θ, ϕ) and is illustrated in Figure 2.4. From Figure 2.4, we notice that r is defined as the distance from the origin to point P or the radius of a sphere centered at the origin and passing through P; θ (called the *colatitude*) is the angle between the z-axis and the position vector of P; and ϕ is measured from the x-axis (the same azimuthal angle in cylindrical coordinates). According to these definitions, the ranges of the variables are

$$\begin{gathered} 0 \leq r < \infty \\ 0 \leq \theta \leq \pi \\ 0 \leq \phi < 2\pi \end{gathered} \quad [2.17]$$

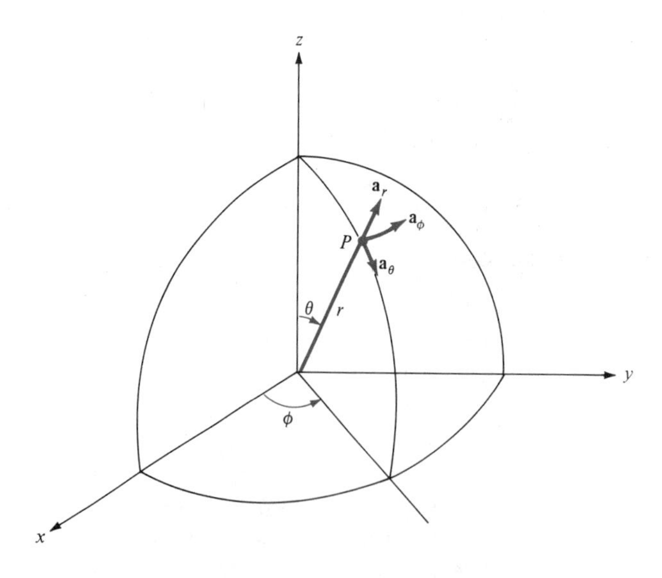

Figure 2.4 Point P and unit vectors in spherical coordinates.

A vector **A** in spherical coordinates may be written as

$$(A_r, A_\theta, A_\phi) \qquad \text{or} \qquad A_r\mathbf{a}_r + A_\theta\mathbf{a}_\theta + A_\phi\mathbf{a}_\phi \tag{2.18}$$

where $\mathbf{a}_r$, $\mathbf{a}_\theta$, and $\mathbf{a}_\phi$ are unit vectors along the r-, θ-, and ϕ-directions. The magnitude of **A** is

$$|\mathbf{A}| = (A_r^2 + A_\theta^2 + A_\theta^2)^{1/2} \tag{2.19}$$

The unit vectors $\mathbf{a}_r$, $\mathbf{a}_\theta$, and $\mathbf{a}_\phi$ are mutually orthogonal: $\mathbf{a}_r$ being directed along the radius or in the direction of increasing r, $\mathbf{a}_\theta$ in the direction of increasing θ, and $\mathbf{a}_\phi$ in the direction of increasing ϕ. Thus,

$$\begin{aligned}
\mathbf{a}_r \cdot \mathbf{a}_r &= \mathbf{a}_\theta \cdot \mathbf{a}_\theta = \mathbf{a}_\phi \cdot \mathbf{a}_\phi = 1 \\
\mathbf{a}_r \cdot \mathbf{a}_\theta &= \mathbf{a}_\theta \cdot \mathbf{a}_\phi = \mathbf{a}_\phi \cdot \mathbf{a}_r = 0 \\
\mathbf{a}_r \times \mathbf{a}_\theta &= \mathbf{a}_\phi \\
\mathbf{a}_\theta \times \mathbf{a}_\phi &= \mathbf{a}_r \\
\mathbf{a}_\phi \times \mathbf{a}_r &= \mathbf{a}_\theta
\end{aligned} \tag{2.20}$$

The space variables (x, y, z) in Cartesian coordinates can be related to variables (r, θ, ϕ) of a spherical coordinate system. From Figure 2.5 it is easy to notice that

$$r = \sqrt{x^2 + y^2 + z^2}, \qquad \theta = \tan^{-1}\frac{\sqrt{x^2 + y^2}}{z}, \qquad \phi = \tan^{-1}\frac{y}{x} \tag{2.21}$$

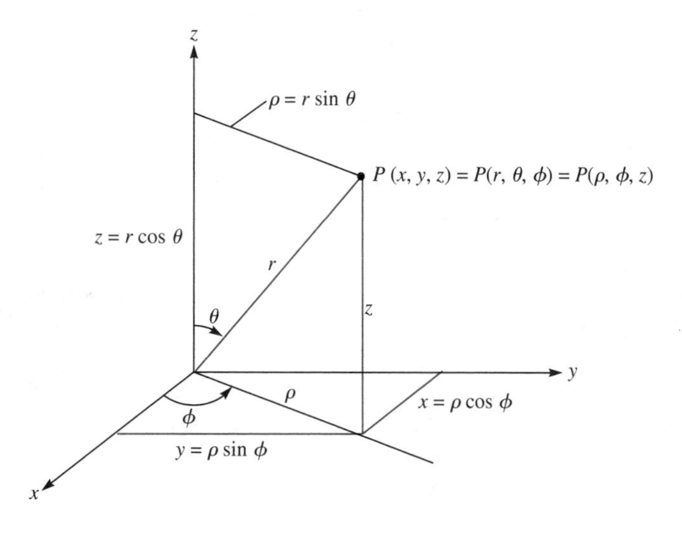

Figure 2.5 Relationships between space variables (x, y, z), (r, θ, ϕ), and (ρ, ϕ, z).

or

$$\boxed{x = r \sin \theta \cos \phi, \qquad y = r \sin \theta \sin \phi, \qquad z = r \cos \theta} \tag{2.22}$$

In eq. (2.21), we have $(x, y, z) \rightarrow (r, \theta, \phi)$ point transformation and in eq. (2.22), it is $(r, \theta, \phi) \rightarrow (x, y, z)$ point transformation.

The unit vectors $\mathbf{a}_x$, $\mathbf{a}_y$, $\mathbf{a}_z$ and $\mathbf{a}_r$, $\mathbf{a}_\theta$, $\mathbf{a}_\phi$ are related as follows:

$$\begin{aligned}
\mathbf{a}_x &= \sin \theta \cos \phi \; \mathbf{a}_r + \cos \theta \cos \phi \; \mathbf{a}_\theta - \sin \phi \; \mathbf{a}_\phi \\
\mathbf{a}_y &= \sin \theta \sin \phi \; \mathbf{a}_r + \cos \theta \sin \phi \; \mathbf{a}_\theta + \cos \phi \; \mathbf{a}_\phi \\
\mathbf{a}_z &= \cos \theta \; \mathbf{a}_r - \sin \theta \; \mathbf{a}_\theta
\end{aligned} \tag{2.23}$$

or

$$\begin{aligned}
\mathbf{a}_r &= \sin \theta \cos \phi \; \mathbf{a}_x + \sin \theta \sin \phi \; \mathbf{a}_y + \cos \theta \; \mathbf{a}_z \\
\mathbf{a}_\theta &= \cos \theta \cos \phi \; \mathbf{a}_x + \cos \theta \sin \phi \; \mathbf{a}_y - \sin \theta \; \mathbf{a}_z \\
\mathbf{a}_\phi &= -\sin \phi \; \mathbf{a}_x + \cos \phi \; \mathbf{a}_y
\end{aligned} \tag{2.24}$$

The components of vector $\mathbf{A} = (A_x, A_y, A_z)$ and $\mathbf{A} = (A_r, A_\theta, A_\phi)$ are related by substituting eq. (2.23) into eq. (2.2) and collecting terms. Thus,

$$\begin{aligned}
\mathbf{A} = {} & (A_x \sin \theta \cos \phi + A_y \sin \theta \sin \phi + A_z \cos \theta)\mathbf{a}_r + (A_x \cos \theta \cos \phi \\
& + A_y \cos \theta \sin \phi - A_z \sin \theta)\mathbf{a}_\theta + (-A_x \sin \phi + A_y \cos \phi)\mathbf{a}_\phi
\end{aligned} \tag{2.25}$$

and from this, we obtain

$$\begin{aligned}
A_r &= A_x \sin \theta \cos \phi + A_y \sin \theta \sin \phi + A_z \cos \theta \\
A_\theta &= A_x \cos \theta \cos \phi + A_y \cos \theta \sin \phi - A_z \sin \theta \\
A_\phi &= -A_x \sin \phi + A_y \cos \phi
\end{aligned} \tag{2.26}$$

In matrix form, the $(A_x, A_y, A_z) \rightarrow (A_r, A_\theta, A_\phi)$ vector transformation is performed according to

$$\boxed{\begin{bmatrix} A_r \\ A_\theta \\ A_\phi \end{bmatrix} = \begin{bmatrix} \sin \theta \cos \phi & \sin \theta \sin \phi & \cos \theta \\ \cos \theta \cos \phi & \cos \theta \sin \phi & -\sin \theta \\ -\sin \phi & \cos \phi & 0 \end{bmatrix} \begin{bmatrix} A_x \\ A_y \\ A_z \end{bmatrix}} \tag{2.27}$$

The inverse transformation $(A_r, A_\theta, A_\phi) \rightarrow (A_x, A_y, A_z)$ is similarly obtained, or we obtain it from eq. (2.23). Thus,

$$\begin{bmatrix} A_x \\ A_y \\ A_z \end{bmatrix} = \begin{bmatrix} \sin\theta\cos\phi & \cos\theta\cos\phi & -\sin\phi \\ \sin\theta\sin\phi & \cos\theta\sin\phi & \cos\phi \\ \cos\theta & -\sin\theta & 0 \end{bmatrix} \begin{bmatrix} A_r \\ A_\theta \\ A_\phi \end{bmatrix} \quad [2.28]$$

Alternatively, we may obtain eqs. (2.27) and 2.28) using the dot product. For example,

$$\begin{bmatrix} A_r \\ A_\theta \\ A_\phi \end{bmatrix} = \begin{bmatrix} \mathbf{a}_r \cdot \mathbf{a}_x & \mathbf{a}_r \cdot \mathbf{a}_y & \mathbf{a}_r \cdot \mathbf{a}_z \\ \mathbf{a}_\theta \cdot \mathbf{a}_x & \mathbf{a}_\theta \cdot \mathbf{a}_y & \mathbf{a}_\theta \cdot \mathbf{a}_z \\ \mathbf{a}_\phi \cdot \mathbf{a}_x & \mathbf{a}_\phi \cdot \mathbf{a}_y & \mathbf{a}_\phi \cdot \mathbf{a}_z \end{bmatrix} \begin{bmatrix} A_x \\ A_y \\ A_z \end{bmatrix} \quad [2.29]$$

For the sake of completeness, it may be instructive to obtain the point or vector transformation relationships between cylindrical and spherical coordinates using Figures 2.5 and 2.6 (where ϕ is held constant since it is common to both systems). This will be left as an exercise (see Problem 2.6). Note that in point or vector transformation the point or vector has not changed; it is only expressed differently. Thus, for example, the magnitude of a vector will remain the same after the transformation and this may serve as a way of checking the result of the transformation.

The distance between two points is usually necessary in EM theory. The distance d btween two points with position vectors $\mathbf{r}_1$ and $\mathbf{r}_2$ is generally given by

$$d = |\mathbf{r}_2 - \mathbf{r}_1| \quad [2.30]$$

or

$$d^2 = (x_2 - x_1)^2 + (y_2 - y_1)^2 + (z_2 - z_1)^2 \text{ (Cartesian)} \quad [2.31]$$

$$d^2 = \rho_2^2 + \rho_1^2 - 2\rho_1\rho_2\cos(\phi_2 - \phi_1) + (z_2 - z_1)^2 \text{ (cylindrical)} \quad [2.32]$$

$$d^2 = r_2^2 + r_1^2 - 2r_1r_2\cos\theta_2\cos\theta_1 - 2r_1r_2\sin\theta_2\sin\theta_1\cos(\phi_2 - \phi_1) \text{ (spherical)} \quad [2.33]$$

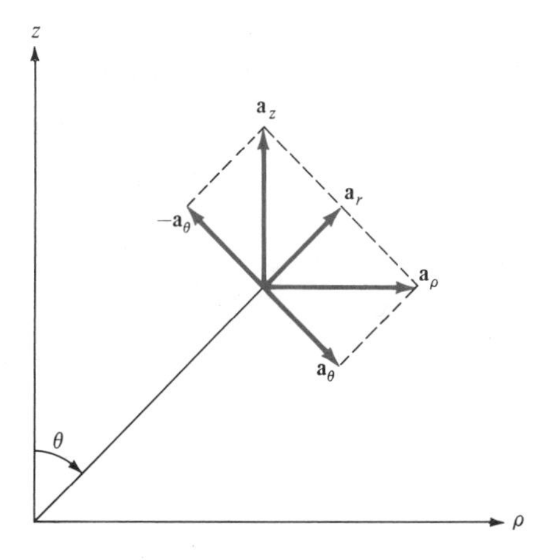

Figure 2.6 Unit vector transformations for cylindrical and spherical coordinates.

EXAMPLE 2.1

Given point $P(-2, 6, 3)$ and vector $\mathbf{A} = y\mathbf{a}_x + (x + z)\mathbf{a}_y$, express P and $\mathbf{A}$ in cylindrical and spherical coordinates. Evaluate $\mathbf{A}$ at P in the Cartesian, cylindrical, and spherical systems.

SOLUTION At point P: $x = -2$, $y = 6$, $z = 3$. Hence,

$$\rho = \sqrt{x^2 + y^2} = \sqrt{4 + 36} = 6.32$$

$$\phi = \tan^{-1}\frac{y}{x} = \tan^{-1}\frac{6}{-2} = 108.43°$$

$$z = 3$$

$$r = \sqrt{x^2 + y^2 + z^2} = \sqrt{4 + 36 + 9} = 7$$

$$\theta = \tan^{-1}\frac{\sqrt{x^2 + y^2}}{z} = \tan^{-1}\frac{\sqrt{40}}{3} = 64.62°$$

Thus,

$$P(-2, 6, 3) = P(6.32, 108.43°, 3) = P(7, 64.62°, 108.43°)$$

In the Cartesian system, $\mathbf{A}$ at P is

$$\mathbf{A} = 6\mathbf{a}_x + \mathbf{a}_y$$

For vector $\mathbf{A}$, $A_x = y$, $A_y = x + z$, $A_z = 0$. Hence, in the cylindrical system

$$\begin{bmatrix} A_\rho \\ A_\phi \\ A_z \end{bmatrix} = \begin{bmatrix} \cos\phi & \sin\phi & 0 \\ -\sin\phi & \cos\phi & 0 \\ 0 & 0 & 1 \end{bmatrix} \begin{bmatrix} y \\ x + z \\ 0 \end{bmatrix}$$

or

$$A_\rho = y\cos\phi + (x + z)\sin\phi$$

$$A_\phi = -y\sin\phi + (x + z)\cos\phi$$

$$A_z = 0$$

But $x = \rho\cos\phi$, $y = \rho\sin\phi$, and substituting these yields

$$\mathbf{A} = (A_\rho, A_\phi, A_z) = [\rho\cos\phi\sin\phi + (\rho\cos\phi + z)\sin\phi]\mathbf{a}_\rho + [-\rho\sin^2\phi + (\rho\cos\phi + z)\cos\phi]\mathbf{a}_\phi$$

At P

$$\rho = \sqrt{40}, \qquad \tan\phi = \frac{6}{-2}$$

Hence,

$$\cos\phi = \frac{-2}{\sqrt{40}}, \qquad \sin\phi = \frac{6}{\sqrt{40}}$$

$$\begin{aligned}
\mathbf{A} &= \left[\sqrt{40}\cdot\frac{-2}{\sqrt{40}}\cdot\frac{6}{\sqrt{40}} + \left(\sqrt{40}\cdot\frac{-2}{\sqrt{40}} + 3\right)\cdot\frac{6}{\sqrt{40}}\right]\mathbf{a}_\rho \\
&\quad + \left[-\sqrt{40}\cdot\frac{36}{40} + \left(\sqrt{40}\cdot\frac{-2}{\sqrt{40}} + 3\right)\cdot\frac{-2}{\sqrt{40}}\right]\mathbf{a}_\phi \\
&= \frac{-6}{\sqrt{40}}\mathbf{a}_\rho - \frac{38}{\sqrt{40}}\mathbf{a}_\phi = -0.9487\mathbf{a}_\rho - 6.008\mathbf{a}_\phi
\end{aligned}$$

Similarly, in the spherical system

$$\begin{bmatrix} A_r \\ A_\theta \\ A_\phi \end{bmatrix} = \begin{bmatrix} \sin\theta\cos\phi & \sin\theta\sin\phi & \cos\theta \\ \cos\theta\cos\phi & \cos\theta\sin\phi & -\sin\theta \\ -\sin\phi & \cos\phi & 0 \end{bmatrix} \begin{bmatrix} y \\ x + z \\ 0 \end{bmatrix}$$

or

$$A_r = y\sin\theta\cos\phi + (x + z)\sin\theta\sin\phi$$

$$A_\theta = y\cos\theta\cos\phi + (x + z)\cos\theta\sin\phi$$

$$A_\phi = -y\sin\phi + (x + z)\cos\phi$$

But $x = r\sin\theta\cos\phi$, $y = r\sin\theta\sin\phi$, and $z = r\cos\theta$. Substituting these yields

$$\begin{aligned}
\mathbf{A} &= (A_r, A_\theta, A_\phi) \\
&= r[\sin^2\theta\cos\phi\sin\phi + (\sin\theta\cos\phi + \cos\theta)\sin\theta\sin\phi]\mathbf{a}_r \\
&\quad + r[\sin\theta\cos\theta\sin\phi\cos\phi + (\sin\theta\cos\phi + \cos\theta)\cos\theta\sin\phi]\mathbf{a}_\theta \\
&\quad + r[-\sin\theta\sin^2\phi + (\sin\theta\cos\phi + \cos\theta)\cos\phi]\mathbf{a}_\phi
\end{aligned}$$

At P

$$r = 7, \qquad \tan\phi = \frac{6}{-2}, \qquad \tan\theta = \frac{\sqrt{40}}{3}$$

Hence,

$$\cos\phi = \frac{-2}{\sqrt{40}}, \qquad \sin\phi = \frac{6}{\sqrt{40}}, \qquad \cos\theta = \frac{3}{7}, \qquad \sin\theta = \frac{\sqrt{40}}{7}$$

$$\mathbf{A} = 7 \cdot \left[\frac{40}{49} \cdot \frac{-2}{\sqrt{40}} \cdot \frac{6}{\sqrt{40}} + \left(\frac{\sqrt{40}}{7} \cdot \frac{-2}{\sqrt{40}} + \frac{3}{7}\right) \cdot \frac{\sqrt{40}}{7} \cdot \frac{6}{\sqrt{40}}\right] \mathbf{a}_r$$

$$+ 7 \cdot \left[\frac{\sqrt{40}}{7} \cdot \frac{3}{7} \cdot \frac{6}{\sqrt{40}} \cdot \frac{-2}{\sqrt{40}} + \left(\frac{\sqrt{40}}{7} \cdot \frac{-2}{\sqrt{40}} + \frac{3}{7}\right) \cdot \frac{3}{7} \cdot \frac{6}{\sqrt{40}}\right] \mathbf{a}_\theta$$

$$+ 7 \cdot \left[\frac{-\sqrt{40}}{7} \cdot \frac{36}{40} + \left(\frac{\sqrt{40}}{7} \cdot \frac{-2}{\sqrt{40}} + \frac{3}{7}\right) \cdot \frac{-2}{\sqrt{40}}\right] \mathbf{a}_\phi$$

$$= \frac{-6}{7} \mathbf{a}_r - \frac{18}{7\sqrt{40}} \mathbf{a}_\theta - \frac{38}{\sqrt{40}} \mathbf{a}_\phi$$

$$= -0.8571\mathbf{a}_r - 0.4066\mathbf{a}_\theta - 6.008\mathbf{a}_\phi$$

Note that $|\mathbf{A}|$ is the same in the three systems; that is,

$$|\mathbf{A}(x, y, z)| = |\mathbf{A}(\rho, \phi, z)| = |\mathbf{A}(r, \theta, \phi)| = 6.083$$ ■

PRACTICE EXERCISE 2.1

(a) Convert points $P(1, 3, 5)$, $T(0, -4, 3)$, and $S(-3, -4, -10)$ from Cartesian to cylindrical and spherical coordinates.

(b) Transform vector

$$\mathbf{Q} = \frac{\sqrt{x^2 + y^2}\mathbf{a}_x}{\sqrt{x^2 + y^2 + z^2}} - \frac{yz\ \mathbf{a}_z}{\sqrt{x^2 + y^2 + z^2}}$$

to cylindrical and spherical coordinates.

(c) Evaluate $\mathbf{Q}$ at T in the three coordinate systems.

ANSWER

(a) $P(3.162, 71.56°, 5)$, $P(5.916, 32.31°, 71.56°)$, $T(4, 270°, 3)$, $T(5, 53.13°, 270°)$, $S(5, 233.1°, -10)$, $S(11.18, 153.43°, 233.1°)$

(b) $\dfrac{\rho}{\sqrt{\rho^2 + z^2}}(\cos\phi\ \mathbf{a}_\rho - \sin\phi\ \mathbf{a}_\phi - z\sin\phi\ \mathbf{a}_z)$, $\sin\theta\,(\sin\theta\cos\phi - r\cos^2\theta\sin\phi)\mathbf{a}_r + \sin\theta\cos\theta\,(\cos\phi + r\sin\theta\sin\phi)\mathbf{a}_\theta - \sin\theta\sin\phi\ \mathbf{a}_\phi$

(c) $0.8\mathbf{a}_x + 2.4\mathbf{a}_z$, $0.8\mathbf{a}_\phi + 2.4\mathbf{a}_z$, $1.44\mathbf{a}_r - 1.92\mathbf{a}_\theta + 0.8\mathbf{a}_\phi$.

EXAMPLE 2.2

Express vector

$$\mathbf{B} = \frac{10}{r}\,\mathbf{a}_r + r\cos\theta\,\mathbf{a}_\theta + \mathbf{a}_\phi$$

in Cartesian and cylindrical coordinates. Find **B** $(-3, 4, 0)$ and **B** $(5, \pi/2, -2)$.

SOLUTION Using eq. (2.28):

$$\begin{bmatrix} B_x \\ B_y \\ B_z \end{bmatrix} = \begin{bmatrix} \sin\theta\cos\phi & \cos\theta\cos\phi & -\sin\phi \\ \sin\theta\sin\phi & \cos\theta\sin\phi & \cos\phi \\ \cos\theta & -\sin\theta & 0 \end{bmatrix} \begin{bmatrix} \dfrac{10}{r} \\ r\cos\theta \\ 1 \end{bmatrix}$$

or

$$B_x = \frac{10}{r}\sin\theta\cos\phi + r\cos^2\theta\cos\phi - \sin\phi$$

$$B_y = \frac{10}{r}\sin\theta\sin\phi + r\cos^2\theta\sin\phi + \cos\phi$$

$$B_z = \frac{10}{r}\cos\theta - r\cos\theta\sin\theta.$$

But $r = \sqrt{x^2 + y^2 + z^2}$, $\theta = \tan^{-1}\dfrac{\sqrt{x^2 + y^2}}{z}$, and $\phi = \tan^{-1}\dfrac{y}{x}$

Hence,

$$\sin\theta = \frac{\sqrt{x^2 + y^2}}{\sqrt{x^2 + y^2 + z^2}}, \qquad \cos\theta = \frac{z}{\sqrt{x^2 + y^2 + z^2}}$$

$$\sin\phi = \frac{y}{\sqrt{x^2 + y^2}}, \qquad \cos\phi = \frac{x}{\sqrt{x^2 + y^2}}$$

Substituting all these gives

$$B_x = \frac{10\sqrt{x^2 + y^2}}{(x^2 + y^2 + z^2)} \cdot \frac{x}{\sqrt{x^2 + y^2}} + \frac{\sqrt{x^2 + y^2 + z^2}}{(x^2 + y^2 + z^2)} \cdot \frac{z^2 x}{\sqrt{x^2 + y^2}} - \frac{y}{\sqrt{x^2 + y^2}}$$

$$= \frac{10x}{x^2 + y^2 + z^2} + \frac{xz^2}{\sqrt{(x^2 + y^2)(x^2 + y^2 + z^2)}} - \frac{y}{\sqrt{(x^2 + y^2)}}$$

$$B_y = \frac{10\sqrt{x^2 + y^2}}{(x^2 + y^2 + z^2)} \cdot \frac{y}{\sqrt{x^2 + y^2}} + \frac{\sqrt{x^2 + y^2 + z^2}}{x^2 + y^2 + z^2} \cdot \frac{z^2 y}{\sqrt{x^2 + y^2}} - \frac{y}{\sqrt{x^2 + y^2}}$$

$$= \frac{10y}{x^2 + y^2 + z^2} + \frac{yz^2}{\sqrt{(x^2 + y^2)(x^2 + y^2 + z^2)}} + \frac{x}{\sqrt{x^2 + y^2}}$$

$$B_z = \frac{10z}{x^2 + y^2 + z^2} - \frac{z\sqrt{x^2 + y^2}}{\sqrt{x^2 + y^2 + z^2}}$$

$$\mathbf{B} = B_x \mathbf{a}_x + B_y \mathbf{a}_y + B_z \mathbf{a}_z$$

where B_x, B_y, and B_z are as given above.

At $(-3, 4, 0)$, $x = -3$, $y = 4$, and $z = 0$, so

$$B_x = -\frac{30}{25} + 0 - \frac{4}{5} = -2$$

$$B_y = \frac{40}{25} + 0 - \frac{3}{5} = 1$$

$$B_z = 0 - 0 = 0$$

Thus,

$$\mathbf{B} = -2\mathbf{a}_x + \mathbf{a}_y$$

For spherical to cylindrical vector transformation (see Problem 2.6),

$$\begin{bmatrix} B_\rho \\ B_\phi \\ B_z \end{bmatrix} = \begin{bmatrix} \sin\theta & \cos\theta & 0 \\ 0 & 0 & 1 \\ \cos\theta & -\sin\theta & 0 \end{bmatrix} \begin{bmatrix} \dfrac{10}{r} \\ r\cos\theta \\ 1 \end{bmatrix}$$

or

$$B_\rho = \frac{10}{r}\sin\theta + r\cos^2\theta$$

$$B_\phi = 1$$

$$B_z = \frac{10}{r}\cos\theta - r\sin\theta\cos\theta$$

But $r = \sqrt{\rho^2 + z^2}$ and $\theta = \tan^{-1}\dfrac{\rho}{z}$

Thus,

$$\sin\theta = \frac{\rho}{\sqrt{\rho^2 + z^2}}, \qquad \cos\theta = \frac{z}{\sqrt{\rho^2 + z^2}}$$

$$B_\rho = \frac{10\rho}{\rho^2 + z^2} + \sqrt{\rho^2 + z^2} \cdot \frac{z^2}{\rho^2 + z^2}$$

$$B_z = \frac{10z}{\rho^2 + z^2} - \sqrt{\rho^2 + z^2} \cdot \frac{\rho z}{\rho^2 + z^2}$$

Hence,

$$\mathbf{B} = \left(\frac{10\rho}{\rho^2 + z^2} + \frac{z^2}{\sqrt{\rho^2 + z^2}}\right)\mathbf{a}_\rho + \mathbf{a}_\phi + \left(\frac{10z}{\rho^2 + z^2} - \frac{\rho z}{\sqrt{\rho^2 + z^2}}\right)\mathbf{a}_z$$

At $(5, \pi/2, -2)$, $\rho = 5$, $\phi = \pi/2$, and $z = -2$, so

$$\begin{aligned}\mathbf{B} &= \left(\frac{50}{29} + \frac{4}{\sqrt{29}}\right)\mathbf{a}_\rho + \mathbf{a}_\phi + \left(\frac{-20}{29} + \frac{10}{\sqrt{29}}\right)\mathbf{a}_z \\ &= 2.467\mathbf{a}_\rho + \mathbf{a}_\phi + 1.167\mathbf{a}_z\end{aligned}$$

Note that at $(-3, 4, 0)$,

$$|\mathbf{B}(x, y, z)| = |\mathbf{B}(\rho, \phi, z)| = |\mathbf{B}(r, \theta, \phi)| = 2.907$$

This may be used to check the correctness of the result whenever possible. ■

PRACTICE EXERCISE 2.2

Express the following vectors in Cartesian coordinates:

(a) $\mathbf{A} = \rho z \sin\phi\, \mathbf{a}_\rho + 3\rho\cos\phi\, \mathbf{a}_\phi + \rho\cos\phi\sin\phi\, \mathbf{a}_z$

(b) $\mathbf{B} = r^2\, \mathbf{a}_r + \sin\theta\, \mathbf{a}_\phi$

ANSWER

(a) $\mathbf{A} = \dfrac{1}{\sqrt{x^2 + y^2}}[(xyz - 3xy)\,\mathbf{a}_x + (zy^2 + 3x^2)\,\mathbf{a}_y + xy\,\mathbf{a}_z]$

(b) $\mathbf{B} = \dfrac{1}{\sqrt{x^2 + y^2 + z^2}}\{[x(x^2 + y^2 + z^2) - y]\mathbf{a}_x$
$+ [y(x^2 + y^2 + z^2) + x]\mathbf{a}_y + z(x^2 + y^2 + z^2)\mathbf{a}_z\}$

†2.5 CONSTANT-COORDINATE SURFACES

Surfaces in Cartesian, cylindrical, or spherical coordinate systems are easily generated by keeping one of the coordinate variables constant and allowing the other two to vary. In the Cartesian system, if we keep x constant and allow y and z to vary, an infinite plane is generated. Thus we could have infinite planes

$$\begin{aligned} x &= \text{constant} \\ y &= \text{constant} \\ z &= \text{constant} \end{aligned} \tag{2.34}$$

which are perpendicular to the x-, y-, and z-axes respectively as shown in Figure 2.7. The intersection of two planes is a line. For example,

$$x = \text{constant}, \qquad y = \text{constant} \tag{2.35}$$

is the line RPQ parallel to the z-axis. The intersection of three planes is a point. For example,

$$x = \text{constant}, \qquad y = \text{constant}, \qquad z = \text{constant} \tag{2.36}$$

is the point $P(x, y, z)$. Thus we may define point P as the intersection of three orthogonal infinite planes. If P is $(1, -5, 3)$, then P is the intersection of planes $x = 1$, $y = -5$, and $z = 3$.

Orthogonal surfaces in cylindrical coordinates can likewise be generated. The surfaces

$$\begin{aligned} \rho &= \text{constant} \\ \phi &= \text{constant} \\ z &= \text{constant} \end{aligned} \tag{2.37}$$

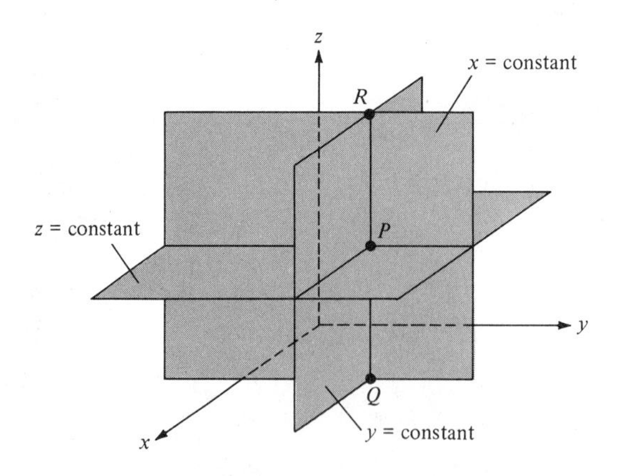

Figure 2.7 Constant x, y, and z surfaces.

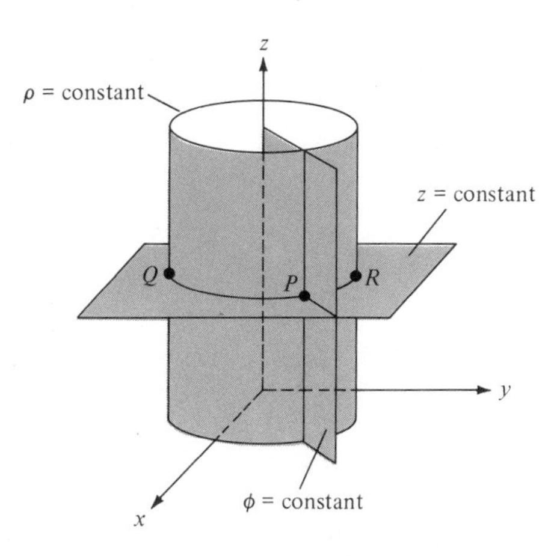

Figure 2.8 Constant ρ, ϕ, and z surfaces.

are illustrated in Figure 2.8, where it is easy to observe that ρ = constant is a circular cylinder, ϕ = constant is a semi-infinite plane with its edge along the z-axis, and z = constant is the same infinite plane as in a Cartesian system. Where two surfaces meet is either a line or a circle. Thus,

$$z = \text{constant}, \qquad \rho = \text{constant} \tag{2.38}$$

is a circle QPR of radius ρ, whereas z = constant, ϕ = constant is a semi-infinite line. A point is an intersection of the three surfaces in eq. (2.37). Thus,

$$\rho = 2, \qquad \phi = 60^\circ, \qquad z = 5 \tag{2.39}$$

is the point $P(2, 60^\circ, 5)$.

The orthogonal nature of the spherical coordinate system is evident by considering the three surfaces

$$\begin{aligned} r &= \text{constant} \\ \theta &= \text{constant} \\ \phi &= \text{constant} \end{aligned} \tag{2.40}$$

which are shown in Figure 2.9, where we notice that r = constant is a sphere with its center at the origin; θ = constant is a circular cone with the z-axis as its axis and the origin as its vertex; ϕ = constant is the semi-infinite plane as in a cylindrical system. A line is formed by the intersection of two surfaces. For example:

$$r = \text{constant}, \qquad \phi = \text{constant} \tag{2.41}$$

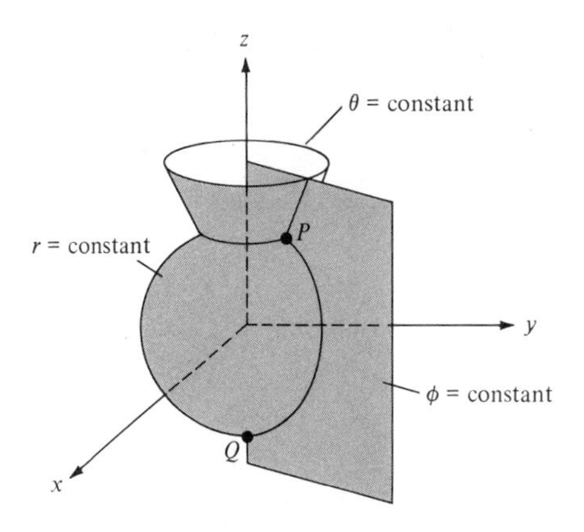

Figure 2.9 Constant r, θ, and ϕ surfaces.

is a semicircle passing through Q and P. The intersection of three surfaces gives a point. Thus,

$$r = 5, \qquad \theta = 30°, \quad \phi = 60° \tag{2.42}$$

is the point $P(5, 30°, 60°)$. We notice that in general, a point in three-dimensional space can be identified as the intersection of three mutually orthogonal surfaces. Also, a unit normal vector to the surface $n =$ constant is $\pm\, \mathbf{a}_n$, where n is x, y, z, ρ, ϕ, r, or θ. For example, to plane $x = 5$, a unit normal vector is $\pm\, \mathbf{a}_x$ and to plane $\phi = 20°$, a unit normal vector is $\mathbf{a}_\phi$.

EXAMPLE 2.3

Two uniform vector fields are given by $\mathbf{E} = -5\mathbf{a}_\rho + 10\mathbf{a}_\phi + 3\mathbf{a}_z$ and $\mathbf{F} = \mathbf{a}_\rho + 2\mathbf{a}_\phi - 6\mathbf{a}_z$. Calculate

(a) $|\mathbf{E} \times \mathbf{F}|$

(b) The vector component of $\mathbf{E}$ at $P(5, \pi/2, 3)$ parallel to the line $x = 2$, $z = 3$

(c) The angle $\mathbf{E}$ makes with the surface $z = 3$ at P

SOLUTION (a)

$$\mathbf{E} \times \mathbf{F} = \begin{vmatrix} \mathbf{a}_\rho & \mathbf{a}_\phi & \mathbf{a}_z \\ -5 & 10 & 3 \\ 1 & 2 & -6 \end{vmatrix}$$

$$= (-60 - 6)\mathbf{a}_\rho + (3 - 30)\mathbf{a}_\phi + (-10 - 10)\mathbf{a}_z$$

$$= (-66, -27, -20)$$

$$|\mathbf{E} \times \mathbf{F}| = \sqrt{66^2 + 27^2 + 20^2} = 74.06$$

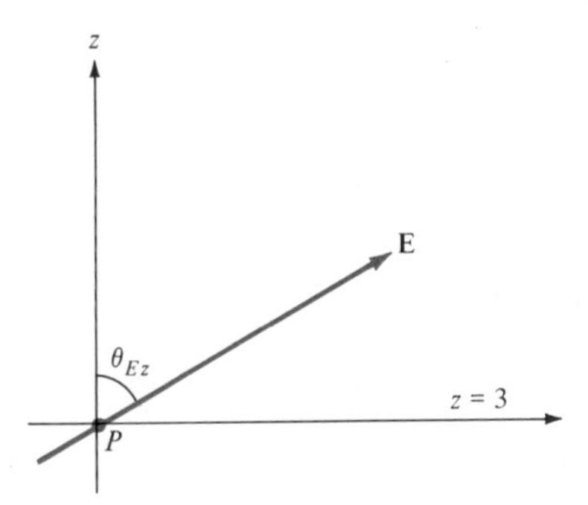

Figure 2.10 For Example 2.3(c).

(b) Line $x = 2$, $z = 3$ is parallel to the y-axis, so the component of **E** parallel to the given line is

$$(\mathbf{E} \cdot \mathbf{a}_y)\mathbf{a}_y$$

But at $P(5, \pi/2, 3)$

$$\mathbf{a}_y = \sin\phi\ \mathbf{a}_\rho + \cos\phi\ \mathbf{a}_\phi$$
$$= \sin\pi/2\ \mathbf{a}_\rho + \cos\pi/2\ \mathbf{a}_\phi = \mathbf{a}_\rho$$

Therefore,

$$(\mathbf{E} \cdot \mathbf{a}_y)\mathbf{a}_y = (\mathbf{E} \cdot \mathbf{a}_\rho)\mathbf{a}_\rho = -5\mathbf{a}_\rho \qquad (\text{or } -5\mathbf{a}_y)$$

(c) Utilizing the fact that the z-axis is normal to the surface $z = 3$, the angle between the z-axis and **E**, as shown in Figure 2.10, can be found using the dot product:

$$\mathbf{E} \cdot \mathbf{a}_z = |\mathbf{E}|\ (1) \cos\theta_{Ez}$$

$$\cos\theta_{Ez} = \frac{3}{\sqrt{134}} = 0.2592 \rightarrow \theta_{Ez} = 74.98°$$

Hence, the angle between $z = 3$ and **E** is

$$90° - \theta_{Ez} = 15.02°$$

■

PRACTICE EXERCISE 2.3

Given the vector field

$$\mathbf{H} = \rho z \cos\phi\ \mathbf{a}_\rho + e^{-z}\sin\frac{\phi}{2}\ \mathbf{a}_\phi + \rho^2\mathbf{a}_z$$

At point $(1, \pi/3, 0)$, find

(a) $\mathbf{H} \cdot \mathbf{a}_x$

(b) $\mathbf{H} \times \mathbf{a}_\theta$

(c) The vector component of **H** normal to surface $\rho = 1$

(d) The scalar component of **H** tangential to the plane $z = 0$

ANSWER (a) -0.433, (b) $-0.5\ \mathbf{a}_\rho$, (c) $0\ \mathbf{a}_\rho$, (d) 0.5.

EXAMPLE 2.4

Given a vector field

$$\mathbf{D} = r \sin \phi\ \mathbf{a}_r - \frac{1}{r} \sin \theta \cos \phi\ \mathbf{a}_\theta + r^2\mathbf{a}_\phi$$

determine

(a) **D** at $P(10, 150°, 330°)$

(b) The component of **D** tangential to the spherical surface $r = 10$ at P

(c) A unit vector at P perpendicular to **D** and tangential to the cone $\theta = 150°$

SOLUTION (a) At P, $r = 10$, $\theta = 150°$, and $\phi = 330°$. Hence,

$$\mathbf{D} = 10 \sin 330°\ \mathbf{a}_r - \frac{1}{10} \sin 150° \cos 330°\ \mathbf{a}_\phi + 100\ \mathbf{a}_\theta = (-5, 0.043, 100)$$

(b) Any vector **D** can always be resolved into two orthogonal components:

$$\mathbf{D} = \mathbf{D}_t + \mathbf{D}_n$$

where D_t is tangential to a given surface and $\mathbf{D}_n$ is normal to it. In our case,

$$\mathbf{D}_n = r \sin \phi\ \mathbf{a}_r = -5\mathbf{a}_r$$

Hence,

$$\mathbf{D}_t = \mathbf{D} - \mathbf{D}_n = 0.043\mathbf{a}_\theta + 100\mathbf{a}_\phi$$

(c) A vector at P perpendicular to **D** and tangential to the cone $\theta = 150°$ is the same as the vector perpendicular to both **D** and $\mathbf{a}_\theta$. Hence,

$$\mathbf{D} \times \mathbf{a}_\theta = \begin{vmatrix} \mathbf{a}_r & \mathbf{a}_\theta & \mathbf{a}_\phi \\ -5 & 0.043 & 100 \\ 0 & 1 & 0 \end{vmatrix}$$

$$= -100\mathbf{a}_r - 5\mathbf{a}_\phi$$

A unit vector along this is

$$\mathbf{a} = \frac{-100\mathbf{a}_r - 5\mathbf{a}_\phi}{\sqrt{100^2 + 5^2}} = -0.9988\mathbf{a}_r - 0.0499\mathbf{a}_\phi$$

■

PRACTICE EXERCISE 2.4

If $\mathbf{A} = 3\mathbf{a}_r + 2\mathbf{a}_\theta - 6\mathbf{a}_\phi$ and $\mathbf{B} = 4\mathbf{a}_r + 3\mathbf{a}_\phi$, determine

(a) $\mathbf{A} \cdot \mathbf{B}$

(b) $|\mathbf{A} \times \mathbf{B}|$

(c) The vector component of $\mathbf{A}$ along $\mathbf{a}_z$ at $(1, \pi/3, 5\pi/4)$

ANSWER (a) -6, (b) 34.48, (c) $-0.116\mathbf{a}_r + 0.201\mathbf{a}_\theta$.

SUMMARY

1. The three common coordinate systems we shall use throughout the text are the Cartesian (or rectangular), the circular cylindrical, and the spherical.
2. A point P is represented as $P(x, y, z)$, $P(\rho, \phi, z)$, and $P(r, \theta, \phi)$ in the Cartesian, cylindrical, and spherical systems respectively. A vector field $\mathbf{A}$ is represented as (A_x, A_y, A_z) or $A_x\mathbf{a}_x + A_y\mathbf{a}_y + A_z\mathbf{a}_z$ in the Cartesian system, as (A_ρ, A_ϕ, A_z) or $A_\rho\mathbf{a}_\rho + A_\phi\mathbf{a}_\phi + A_z\mathbf{a}_z$ in the cylindrical system, and as (A_r, A_θ, A_ϕ) or $A_r\mathbf{a}_r + A_\theta\mathbf{a}_\theta + A_\phi\mathbf{a}_\phi$ in the spherical system. It is preferable that mathematical operations (addition, subtraction, product, etc.) be performed in the same coordinate system. Thus, point and vector transformations should be performed whenever necessary.
3. Fixing one space variable defines a surface; fixing two defines a line; fixing three defines a point.
4. A unit normal vector to surface n = constant is $\pm\,\mathbf{a}_n$.

REVIEW QUESTIONS

2.1 The ranges of θ and ϕ as given by eq. (2.17) are not the only possible ones. The following are all alternative ranges of θ and ϕ, except

(a) $0 \leq \theta < 2\pi,\ 0 \leq \phi \leq \pi$

(b) $0 \leq \theta < 2\pi,\ 0 \leq \phi < 2\pi$

(c) $-\pi \leq \theta \leq \pi,\ 0 \leq \phi \leq \pi$

(d) $-\pi/2 \leq \theta \leq \pi/2,\ 0 \leq \phi < 2\pi$

(e) $0 \leq \theta \leq \pi,\ -\pi \leq \phi < \pi$

(f) $-\pi \leq \theta < \pi,\ -\pi \leq \phi < \pi$

2.2 At Cartesian point $(-3, 4, -1)$, which of these is incorrect?

(a) $\rho = -5$

(b) $r = \sqrt{26}$

(c) $\theta = \tan^{-1} \dfrac{5}{-1}$

(d) $\phi = \tan^{-1} \dfrac{4}{-3}$

2.3 Which of these is not valid at point $(0, 4, 0)$?

(a) $\mathbf{a}_\phi = -\mathbf{a}_x$

(b) $\mathbf{a}_\theta = -\mathbf{a}_z$

(c) $\mathbf{a}_r = 4\mathbf{a}_y$

(d) $\mathbf{a}_\rho = \mathbf{a}_y$

2.4 A unit normal vector to the cylindrical surface $\rho = 4$ is:

(a) $\mathbf{a}_\rho$

(b) $\mathbf{a}_\phi$

(c) $\mathbf{a}_z$

(d) none of the above

2.5 If every point in space, $\mathbf{a}_\phi \cdot \mathbf{a}_\theta = 1$.

(a) True (b) False

2.6 If $\mathbf{H} = 4\mathbf{a}_\rho - 3\mathbf{a}_\phi + 5\mathbf{a}_z$, at $(1, \pi/2, 0)$ the component of $\mathbf{H}$ parallel to surface $\rho = 1$ is

(a) $4\mathbf{a}_\rho$

(b) $5\mathbf{a}_z$

(c) $-3\mathbf{a}_\phi$

(d) $-3\mathbf{a}_\phi + 5\mathbf{a}_z$

(e) $5\mathbf{a}_\phi + 3\mathbf{a}_z$

2.7 Given $\mathbf{G} = 20\mathbf{a}_r + 50\mathbf{a}_\theta + 40\mathbf{a}_\phi$, at $(1, \pi/2, \pi/6)$ the component of $\mathbf{G}$ perpendicular to surface $\theta = \pi/2$ is

(a) $20\mathbf{a}_r$

(b) $50\mathbf{a}_\theta$

(c) $40\mathbf{a}_\phi$

(d) $20\mathbf{a}_r + 40\mathbf{a}_\theta$

(e) $-40\mathbf{a}_r + 20\mathbf{a}_\phi$

2.8 Where surfaces $\rho = 2$ and $z = 1$ intersect is

(a) an infinite plane

(b) a semi-infinite plane

(c) a circle

(d) a cylinder

(e) a cone

2.9 Match the items in the left list with those in the right list. Each answer can be used once, more than once, or not at all.

(a) $\theta = \pi/4$	(i) infinite plane
(b) $\phi = 2\pi/3$	(ii) semi-infinite plane
(c) $x = -10$	(iii) circle
(d) $r = 1, \theta = \pi/3, \phi = \pi/2$	(iv) semicircle
(e) $\rho = 5$	(v) straight line
(f) $\rho = 3, \phi = 5\pi/3$	(vi) cone
(g) $\rho = 10, z = 1$	(vii) cylinder
(h) $r = 4, \phi = \pi/6$	(viii) sphere
(i) $r = 5, \theta = \pi/3$	(ix) cube

2.10 A wedge is described by $z = 0$, $30° < \phi < 60°$. Which of the following is incorrect:

(a) The wedge lies in the $x - y$ plane.

(b) It is infinitely long

(c) On the wedge, $0 < \rho < \infty$

(d) A unit normal to the wedge is $\pm\mathbf{a}_z$

(e) The wedge includes neither the x-axis nor the y-axis

Answers: 2.1b,f, 2.2a, 2.3c, 2.4a, 2.5b, 2.6d, 2.7b, 2.8c, 2.9a-(vi), b-(ii), c-(i), d-(x), e-(vii), f-(v), g-(iii), h-(iv), i-(iii), 2.10b.

PROBLEMS

2.1 Convert the following points to Cartesian coordinates:

(a) $P_1(5, 120°, 0)$

(b) $P_2(1, 30°, -10)$

(c) $P_3(10, 3\pi/4, \pi/2)$

(d) $P_4(3, 30°, 240°)$

2.2 Express the following points in cylindrical and spherical coordinates:

(a) $P(1, -4, -3)$

(b) $Q(3, 0, 5)$

(c) $R(-2, 6, 0)$

2.3 Transform the following vectors to cylindrical and spherical coordinates:

(a) $\mathbf{P} = (y + z)\mathbf{a}_x$

(b) $\mathbf{Q} = y\mathbf{a}_x + xz\mathbf{a}_y + (x + y)\mathbf{a}_z$

(c) $\mathbf{T} = \left[\dfrac{x^2}{x^2 + y^2} - y^2\right]\mathbf{a}_x + \left[\dfrac{xy}{x^2 + y^2} + xy\right]\mathbf{a}_y + \mathbf{a}_z$

(d) $\mathbf{S} = \dfrac{y}{x^2 + y^2}\mathbf{a}_x - \dfrac{x}{x^2 + y^2}\mathbf{a}_y + 10\mathbf{a}_z$

2.4 Express the following vectors in the Cartesian system:

(a) $\mathbf{A} = (\rho^2 z^2 \cos^2\phi \sin\phi + \rho z \sin^2\phi)\mathbf{a}_\rho + (\rho z \sin\phi \cos\phi - \rho^2 z^2 \cos\phi \sin^2\phi)\mathbf{a}_\phi + \rho^2 \sin\phi \cos\phi \, \mathbf{a}_z$

(b) $\mathbf{B} = 6r^2 \sin\theta \cos\phi \, \mathbf{a}_r + 4r \cos\theta \sin\phi \, \mathbf{a}_\theta + r^3\mathbf{a}_\phi$

2.5 Prove the following:

(a) $\mathbf{a}_x \cdot \mathbf{a}_\rho = \cos\phi$

$\mathbf{a}_x \cdot \mathbf{a}_\phi = -\sin\phi$

$\mathbf{a}_y \cdot \mathbf{a}_\rho = \sin\phi$

$\mathbf{a}_y \cdot \mathbf{a}_\phi = \cos\phi$

(b) The 3×3 matrix in eq. (2.15) is the same as

$$\begin{bmatrix} \mathbf{a}_x \cdot \mathbf{a}_\rho & \mathbf{a}_x \cdot \mathbf{a}_\phi & \mathbf{a}_x \cdot \mathbf{a}_z \\ \mathbf{a}_y \cdot \mathbf{a}_\rho & \mathbf{a}_y \cdot \mathbf{a}_\phi & \mathbf{a}_y \cdot \mathbf{a}_z \\ \mathbf{a}_z \cdot \mathbf{a}_\rho & \mathbf{a}_z \cdot \mathbf{a}_\phi & \mathbf{a}_z \cdot \mathbf{a}_z \end{bmatrix}$$

(c) $\mathbf{a}_x \cdot \mathbf{a}_r = \sin\theta \cos\phi$

$\mathbf{a}_x \cdot \mathbf{a}_\theta = \cos\theta \cos\phi$

$\mathbf{a}_y \cdot \mathbf{a}_r = \sin\theta \sin\phi$

$\mathbf{a}_y \cdot \mathbf{a}_\theta = \cos\theta \sin\phi$

$\mathbf{a}_z \cdot \mathbf{a}_r = \cos\theta$

$\mathbf{a}_z \cdot \mathbf{a}_\theta = -\sin\theta$

2.6 (a) Show that point transformation between cylindrical and spherical coordinates is obtained using

$$r = \sqrt{\rho^2 + z^2}, \qquad \theta = \tan^{-1}\frac{\rho}{z}, \qquad \phi = \phi$$

or

$$\rho = r\sin\theta, \qquad z = r\cos\theta, \qquad \phi = \phi$$

(b) Show that vector transformation between cylindrical and spherical coordinates is obtained using

$$\begin{bmatrix} A_r \\ A_\theta \\ A_\phi \end{bmatrix} = \begin{bmatrix} \sin\theta & 0 & \cos\theta \\ \cos\theta & 0 & -\sin\theta \\ 0 & 1 & 0 \end{bmatrix} \begin{bmatrix} A_\rho \\ A_\phi \\ A_z \end{bmatrix}$$

or

$$\begin{bmatrix} A_\rho \\ A_\phi \\ A_z \end{bmatrix} = \begin{bmatrix} \sin\theta & \cos\theta & 0 \\ 0 & 0 & 1 \\ \cos\theta & -\sin\theta & 0 \end{bmatrix} \begin{bmatrix} A_r \\ A_\theta \\ A_\phi \end{bmatrix}$$

(*Hint:* Make use of Figures 2.5 and 2.6.)

2.7 In practice Exercise 2.2, express **A** in spherical and **B** in cylindrical coordinates. Evaluate **A** at $(10, \pi/2, 3\pi/4)$ and **B** at $(2, \pi/6, 1)$.

2.8 The transformation $(A_\rho, A_\phi, A_z) \rightarrow (A_x, A_y, A_z)$ in eq. (2.15) is not complete. Complete it by expressing $\cos\phi$ and $\sin\phi$ in terms of x, y, and z. Do the same thing to the transformation $(A_r, A_\theta, A_\phi) \rightarrow (A_x, A_y, A_z)$ in eq. (2.28).

2.9 Calculate the distance between the following pairs of points:

(a) $(2, 1, 5)$ and $(6, -1, 2)$

(b) $(3, \pi/2, -1)$ and $(5, 3\pi/2, 5)$

(c) $(10, \pi/4, 3\pi/4)$ and $(5, \pi/6, 7\pi/4)$.

2.10 (a) Express the point $(8, -15, 12)$ in spherical coordinates.

(b) Transform vector $\mathbf{F} = 2xy\mathbf{a}_x - x^2\mathbf{a}_y$ into cylindrical coordinates.

2.11 Describe the intersection of the following surfaces:

(a) $x = 2$, $y = 5$

(b) $x = 2$, $y = -1$, $z = 10$

(c) $r = 10$, $\theta = 30°$

(d) $\rho = 5$, $\phi = 40°$

(e) $\phi = 60°$, $z = 10$

(f) $r = 5$, $\phi = 90°$

2.12 At point $T(2, 3, -4)$, express $\mathbf{a}_z$ in the spherical system and $\mathbf{a}_r$ in the rectangular system.

***2.13** Given vectors $\mathbf{A} = 2\mathbf{a}_x + 4\mathbf{a}_y + 10\mathbf{a}_z$ and $\mathbf{B} = -5\mathbf{a}_\rho + \mathbf{a}_\phi - 3\mathbf{a}_z$, find

(a) $\mathbf{A} + \mathbf{B}$ at $P(0, 2, -5)$

(b) The angle between $\mathbf{A}$ and $\mathbf{B}$ at P

(c) The scalar component of $\mathbf{A}$ along $\mathbf{B}$ at P

2.14 Given that $\mathbf{G} = (x + y^2)\mathbf{a}_x + xz\mathbf{a}_y + (z^2 + zy)\mathbf{a}_z$, find the vector component of $\mathbf{G}$ along $\mathbf{a}_\phi$ at point $P(8, 30°, 60°)$. Your answer should be left in the Cartesian system.

***2.15** If $\mathbf{J} = r \sin\theta \cos\phi \, \mathbf{a}_r - \cos 2\theta \sin\phi \, \mathbf{a}_\theta + \tan\frac{\theta}{2} \ln r \, \mathbf{a}_\phi$ at $T(2, \pi/2, 3\pi/2)$, determine the vector component of $\mathbf{J}$ that is

(a) Parallel to $\mathbf{a}_z$

(b) Normal to surface $\phi = 3\pi/2$

(c) Tangential to the spherical surface $r = 2$

(d) Parallel to the line $y = -2$, $z = 0$

***2.16** Let $\mathbf{A} = \rho \cos\phi \, \mathbf{a}_\rho + z \sin\phi \, \mathbf{a}_\phi - \rho z^2 \mathbf{a}_z$ and $\mathbf{B} = r \sin\theta \, \mathbf{a}_r + r^2 \cos\theta \sin\phi \, \mathbf{a}_\theta$. Find

(a) $\mathbf{A}$ and $\mathbf{B}$ at $P(3, -2, 6)$

(b) The component (in spherical coordinates) of $\mathbf{A}$ along $\mathbf{B}$ at P

(c) A unit vector (in cylindrical coordinates) perpendicular to both $\mathbf{A}$ and $\mathbf{B}$ at P

***2.17** If the components of a vector **F** in space are defined as

$$F_x = F \cos \alpha \qquad F_y = F \cos \beta \qquad F_z = F \cos \gamma$$

(a) Express α, β, and γ in terms of θ and ϕ.

(b) Show that $\cos^2\alpha + \cos^2\beta + \cos^2\gamma = 1$.

***2.18** From Problem 2.17, we notice that another way of defining a point P in space is $(r, \alpha, \beta, \gamma)$ where the variables are portrayed in Figure 2.11. Using this definition, find $(r, \alpha, \beta, \gamma)$ for the following points:

(a) $(-2, 3, 6)$

(b) $(4, 30°, -3)$

(c) $(3, 30°, 60°)$

(*Hint:* r is the spherical r, $0 \leq \alpha, \beta, \gamma < 2\pi$.)

2.19 A vector field is expressed in "mixed" coordinate variables as

$$\mathbf{J} = \frac{3xz}{r^2}\,\mathbf{a}_x + \frac{3y \cos \theta}{r}\,\mathbf{a}_y + \left[2 - \frac{3y^2}{r^2} - \frac{3x^2}{r^2}\right]\mathbf{a}_z$$

Express **J** completely in spherical coordinates.

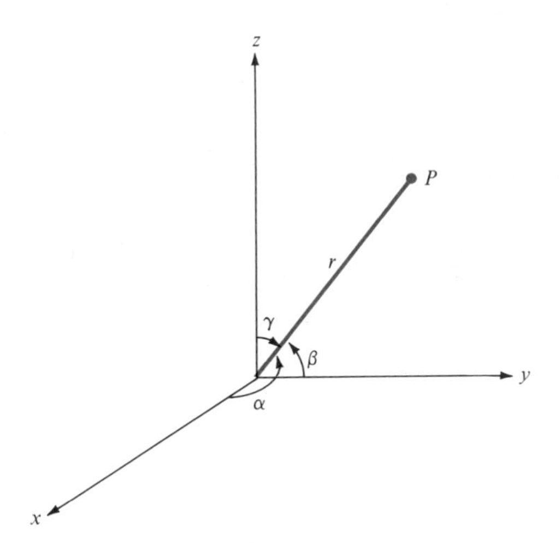

Figure 2.11 For Problem 2.18.

CHAPTER

Vector Calculus

3

No human investigation can claim to be scientific if it doesn't pass the test of mathematical proof.

—LEONARDO DA VINCI

3.1 INTRODUCTION

Chapter 1 is mainly on vector addition, subtraction, and multiplication in Cartesian coordinates, and Chapter 2 extends all these to other coordinate systems. This chapter deals with integration and differentiation of vectors.

The concepts introduced in this chapter provide a convenient language for expressing certain fundamental ideas in electromagnetics or mathematics in general. A student may feel uneasy about these concepts at first—not seeing ''what good'' they are. Such a student is advised to concentrate simply on learning the mathematical techniques and to wait for their applications in subsequent chapters.

3.2 DIFFERENTIAL LENGTH, AREA, AND VOLUME

Differential elements in length, area, and volume are useful in vector calculus. They are defined in the Cartesian, cylindrical, and spherical coordinate systems.

A. Cartesian Coordinates

From Figure 3.1, we notice that

(1) Differential displacement is given by

$$d\mathbf{l} = dx\,\mathbf{a}_x + dy\,\mathbf{a}_y + dz\,\mathbf{a}_z \tag{3.1}$$

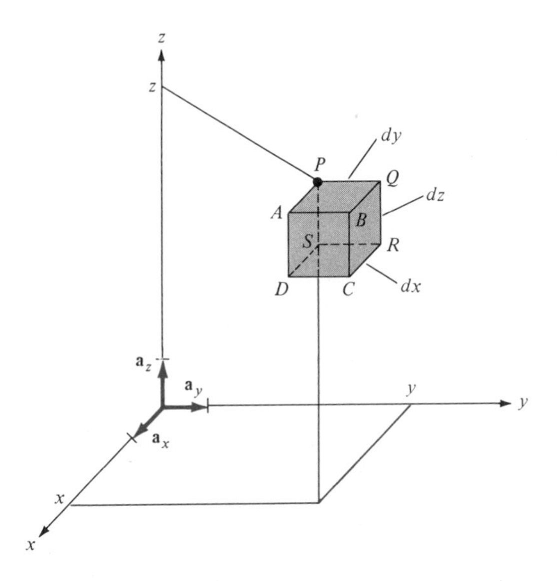

Figure 3.1 Differential elements in the right-handed Cartesian coordinate system.

(2) Differential normal area is given by

$$d\mathbf{S} = \begin{array}{l} dy\, dz\, \mathbf{a}_x \\ dx\, dz\, \mathbf{a}_y \\ dx\, dy\, \mathbf{a}_z \end{array} \tag{3.2}$$

and illustrated in Figure 3.2.

(3) Differential volume is given by

$$dv = dx\, dy\, dz \tag{3.3}$$

These differential elements are very important as they will be referred to again and again throughout the book. The student is encouraged not to memorize them, however, but to learn to derive them from Figure 3.1. Notice from eqs. (3.1) to (3.3) that $d\mathbf{l}$ and $d\mathbf{S}$ are vectors whereas dv is a scalar. Observe from Figure 3.1 that if we move from point P to Q (or Q *to* P), for example, $d\mathbf{l} = dy\, \mathbf{a}_y$ because we are moving in the y-direction and if we move from Q to S (or S to Q), $d\mathbf{l} = dy\, \mathbf{a}_y + dz\, \mathbf{a}_z$ because we

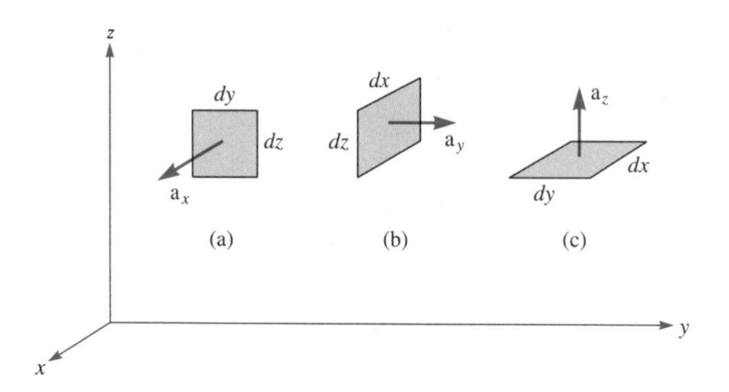

Figure 3.2 Differential normal areas in Cartesian coordinates: **(a)** $d\mathbf{S} = dy\, dz\, \mathbf{a}_x$, **(b)** $d\mathbf{S} = dx\, dz\, \mathbf{a}_y$, **(c)** $d\mathbf{S} = dx\, dy\, \mathbf{a}_z$.

have to move dy along y, dz along z, and $dx = 0$ (no movement along x). Similarly, to move from D to Q would mean that $d\mathbf{l} = dx\, \mathbf{a}_x + dy\, \mathbf{a}_y + dz\, \mathbf{a}_z$.

The way $d\mathbf{S}$ is defined is important. The differential surface (or area) element $d\mathbf{S}$ may generally be defined as

$$d\mathbf{S} = dS\, \mathbf{a}_n \qquad [3.4]$$

where dS is the area of the surface element and $\mathbf{a}_n$ is a unit vector normal to the surface dS (and directed away from the volume if dS is part of the surface describing a volume). If we consider surface $ABCD$ in Figure 3.1, for example, $d\mathbf{S} = dy\, dz\, \mathbf{a}_x$ whereas for surface $PQRS$, $d\mathbf{S} = -dy\, dz\, \mathbf{a}_x$ because $\mathbf{a}_n = -\mathbf{a}_x$ is normal to $PQRS$.

What we have to remember at all times about differential elements is $d\mathbf{l}$ and how to get $d\mathbf{S}$ and dv from it. Once $d\mathbf{l}$ is remembered, $d\mathbf{S}$ and dv can easily be found. For example, $d\mathbf{S}$ along $\mathbf{a}_x$ can be obtained from $d\mathbf{l}$ in eq. (3.1) by multiplying the components of $d\mathbf{l}$ along $\mathbf{a}_y$ and $\mathbf{a}_z$; that is, $dy\, dz\, \mathbf{a}_x$. Similarly, $d\mathbf{S}$ along $\mathbf{a}_z$ is the product of the components of $d\mathbf{l}$ along $\mathbf{a}_x$ and $\mathbf{a}_y$; that is $dx\, dy\, \mathbf{a}_z$. Also, dv can be obtained from $d\mathbf{l}$ as the product of the three components of $d\mathbf{l}$; that is, $dx\, dy\, dz$. The idea developed here for Cartesian coordinates will now be extended to other coordinate systems.

B. Cylindrical Coordinates

Notice from Figure 3.3 that in cylindrical coordinates, differential elements can be found as follows:

(1) Differential displacement is given by

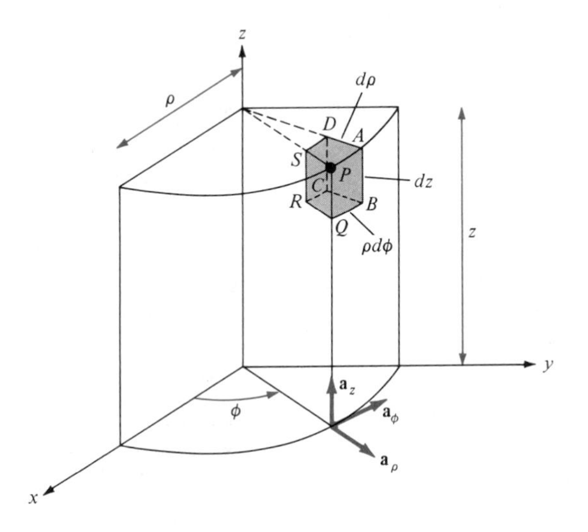

Figure 3.3 Differential elements in cylindrical coordinates.

$$d\mathbf{l} = d\rho\, \mathbf{a}_\rho + \rho\, d\phi\, \mathbf{a}_\phi + dz\, \mathbf{a}_z \qquad [3.5]$$

arc?

(2) Differential normal area is given by

$$d\mathbf{S} = \begin{array}{l} \rho\, d\phi\, dz\, \mathbf{a}_\rho \\ d\rho\, dz\, \mathbf{a}_\phi \\ \rho\, d\phi\, d\rho\, \mathbf{a}_z \end{array} \qquad [3.6]$$

and illustrated in Figure 3.4.

(3) Differential volume is given by

$$dv = \rho\, d\rho\, d\phi\, dz \qquad [3.7]$$

As mentioned in the previous section on Cartesian coordinates, we only need to remember $d\mathbf{l}$; $d\mathbf{S}$ and dv can easily be obtained from $d\mathbf{l}$. For example, $d\mathbf{S}$ along $\mathbf{a}_z$ is the product of the components of $d\mathbf{l}$ along $\mathbf{a}_\rho$ and $\mathbf{a}_\phi$; that is, $d\rho\, \rho\, d\phi\, \mathbf{a}_z$. Also dv is the product of the three components of $d\mathbf{l}$; that is, $d\rho\, \rho\, d\phi\, dz$.

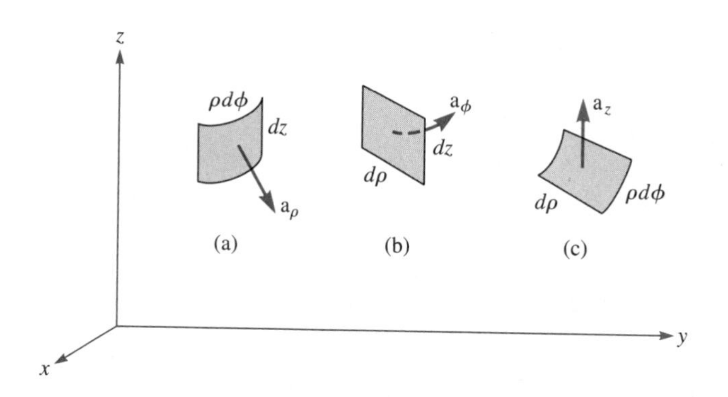

Figure 3.4 Differential normal areas in cylindrical coordinates: **(a)** $d\mathbf{S} = \rho\, d\phi\, dz\, \mathbf{a}_\rho$, **(b)** $d\mathbf{S} = d\rho\, dz\, \mathbf{a}_\phi$, **(c)** $d\mathbf{S} = \rho\, d\phi\, d\rho \mathbf{a}_z$.

C. Spherical Coordinates

From Figure 3.5, we notice that in spherical coordinates,

(1) The differential displacement is

$$d\mathbf{l} = dr\, \mathbf{a}_r + r\, d\theta\, \mathbf{a}_\theta + r \sin\theta\, d\phi\, \mathbf{a}_\phi \tag{3.8}$$

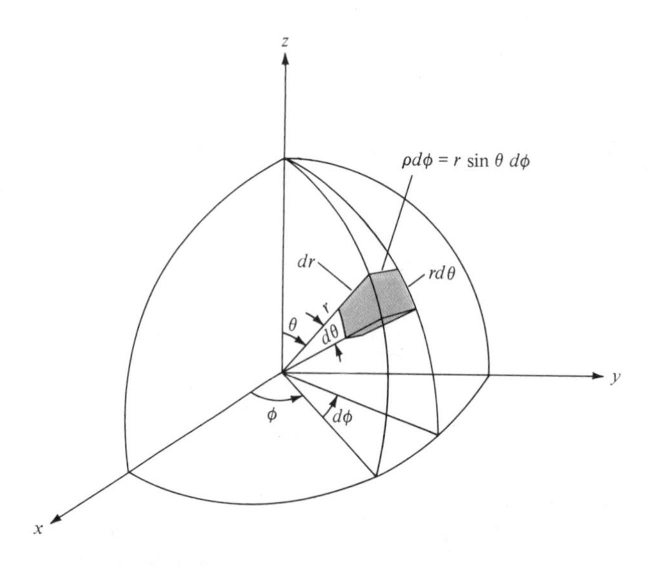

Figure 3.5 Differential elements in the spherical coordinate system.

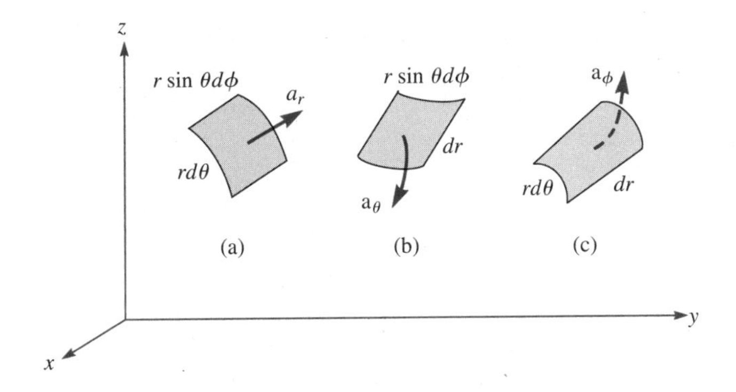

Figure 3.6 Differential normal areas in spherical coordinates: **(a)** $d\mathbf{S} = r^2 \sin\theta \, d\theta \, d\phi \, \mathbf{a}_r$, **(b)** $d\mathbf{S} = r \sin\theta \, dr \, d\phi \, \mathbf{a}_\theta$, **(c)** $d\mathbf{S} = r \, dr \, d\theta \, \mathbf{a}_\phi$.

(2) The differential normal area is

$$\begin{aligned} d\mathbf{S} = & \; r^2 \sin\theta \, d\theta \, d\phi \, \mathbf{a}_r \\ & \; r \sin\theta \, dr \, d\phi \, \mathbf{a}_\theta \\ & \; r \, dr \, d\theta \, \mathbf{a}_\phi \end{aligned} \qquad [3.9]$$

and illustrated in Figure 3.6.

(3) The differential volume is

$$dv = r^2 \sin\theta \, dr \, d\theta \, d\phi \qquad [3.10]$$

Again, we need to remember only $d\mathbf{l}$ from which $d\mathbf{S}$ and dv are easily obtained. For example, $d\mathbf{S}$ along $\mathbf{a}_\theta$ is obtained as the product of the components of $d\mathbf{l}$ along $\mathbf{a}_r$ and $\mathbf{a}_\phi$; that is, $dr \cdot r \sin\theta \, d\phi$; dv is the product of the three components of $d\mathbf{l}$; that is, $dr \cdot r \, d\theta \cdot r \sin\theta \, d\phi$.

EXAMPLE 3.1

Consider the object shown in Figure 3.7. Calculate

(a) The distance *BC*

(b) The distance *CD*

(c) The surface area *ABCD*

(d) The surface area *ABO*

(e) The surface area *AOFD*

(f) The volume *ABDCFO*

SOLUTION Although points A, B, C, and D are given in Cartesian coordinates, it is obvious that the object has cylindrical symmetry. Hence, we solve the problem in cylindrical coordinates. The points are transformed from Cartesian to cylindrical coordinates as follows:

$$A(5, 0, 0) \to A(5, 0°, 0)$$

$$B(0, 5, 0) \to B\left(5, \frac{\pi}{2}, 0\right)$$

$$C(0, 5, 10) \to C\left(5, \frac{\pi}{2}, 10\right)$$

$$D(5, 0, 10) \to D(5, 0°, 10)$$

(a) Along BC, $dl = dz$; hence,

$$BC = \int dl = \int_0^{10} dz = 10$$

(b) Along CD, $dl = \rho\, d\phi$ and $\rho = 5$, so

$$CD = \int_0^{\pi/2} \rho\, d\phi = 5\,\phi \Big|_0^{\pi/2} = 2.5\pi$$

(c) For $ABCD$, $dS = \rho\, d\phi\, dz$, $\rho = 5$. Hence,

$$\text{area } ABCD = \int dS = \int_{\phi=0}^{\pi/2} \int_{z=0}^{10} \rho\, d\phi\, dz = 5 \int_0^{\pi/2} d\phi \int_0^{10} dz \Big|_{\rho=5} = 25\pi$$

(d) For ABO, $dS = \rho\, d\phi\, d\rho$ and $z = 0$, so

$$\text{area } ABO = \int_{\phi=0}^{\pi/2} \int_{\rho=0}^{5} \rho\, d\phi\, d\rho = \int_0^{\pi/2} d\phi \int_0^5 \rho\, d\rho = 6.25\pi$$

(e) For $AOFD$, $dS = d\rho\, dz$ and $\phi = 0°$, so

$$\text{area } AOFD = \int_{\rho=0}^{5} \int_{z=0}^{10} d\rho\, dz = 50$$

(f) For volume $ABDCFO$, $dv = \rho\, d\phi\, dz\, d\rho$. Hence,

$$v = \int dv = \int_{\rho=0}^{5} \int_{\phi=0}^{\pi/2} \int_{z=0}^{10} \rho\, d\phi\, dz\, d\rho = \int_0^{10} dz \int_0^{\pi/2} d\phi \int_0^5 \rho\, d\rho = 62.5\pi$$

■

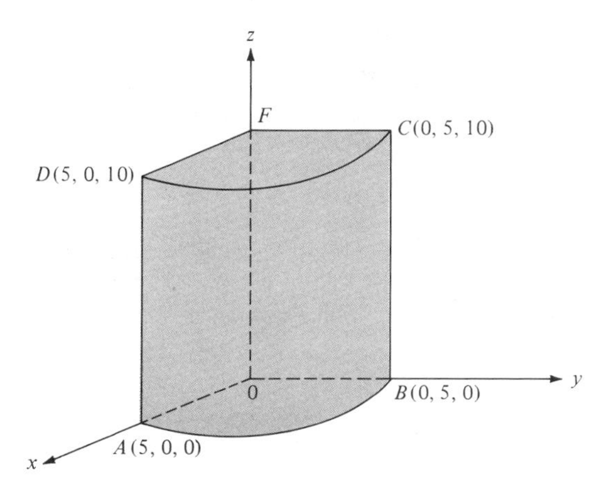

Figure 3.7 For Example 3.1.

PRACTICE EXERCISE 3.1

Refer to Figure 3.26; disregard the differential lengths and imagine that the object is part of a spherical shell. It may be described as $3 \leq r \leq 5$, $60° \leq \theta \leq 90°$, $45° \leq \phi \leq 60°$ where surface $r = 3$ is the same as *AEHD*, surface $\theta = 60°$ is *AEFB*, and surface $\phi = 45°$ is *ABCD*. Calculate

(a) The distance *DH*

(b) The distance *FG*

(c) The surface area *AEHD*

(d) The surface area *ABDC*

(e) The volume of the object

ANSWER (a) 0.7854, (b) 2.618, (c) 1.179, (d) 4.189, (e) 4.276.

3.3 LINE, SURFACE, AND VOLUME INTEGRALS

The familiar concept of integration will now be extended to cases when the integrand involves a vector. By a line we mean the path along a curve in space. We shall use terms such as *line, curve,* and *contour* interchangeably. Given a vector field **A** and a curve L, we define the integral

$$\int_L \mathbf{A} \cdot d\mathbf{l} = \int_a^b |\mathbf{A}| \cos\theta \, dl \qquad [3.11]$$

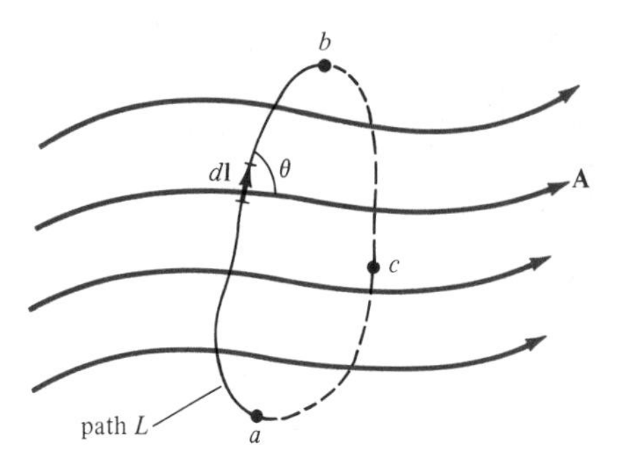

Figure 3.8 Path of integration of vector field **A**.

as the *line integral* of **A** around L (see Figure 3.8). It is the integral of the tangential component of **A** along the curve L. If the path of integration is a closed curve such as *abca* in Figure 3.1, eq. (3.11) becomes a closed contour integral

$$\oint_L \mathbf{A} \cdot d\mathbf{l} \tag{3.12}$$

which is called the *circulation* of **A** around L.

Given a vector field **A**, continuous in a region containing the smooth surface S, we define the *surface integral* or the *flux* of **A** through S (see Fig. 3.9) as

$$\Psi = \int_S |\mathbf{A}| \cos\theta \, dS = \int_S \mathbf{A} \cdot \mathbf{a}_n \, dS$$

or simply

$$\Psi = \int_S \mathbf{A} \cdot d\mathbf{S} \tag{3.13}$$

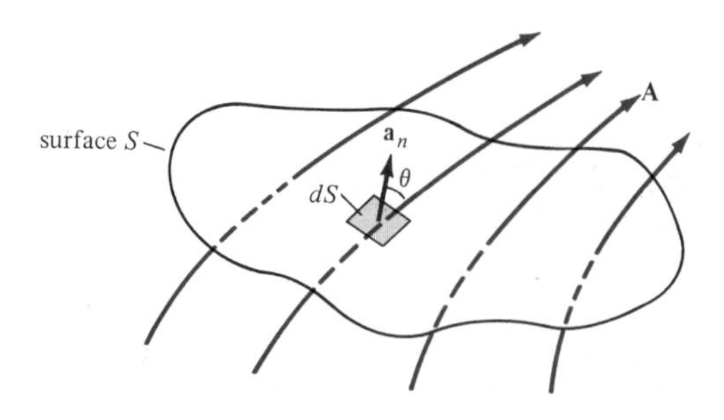

Figure 3.9 The flux of a vector field **A** through surface S.

where, at any point on S, $\mathbf{a}_n$ is the unit normal to S. For a closed surface (defining a volume), eq. (3.13) becomes

$$\Psi = \oint_S \mathbf{A} \cdot d\mathbf{S} \qquad [3.14]$$

which is referred to as the *net outward flux of* **A** from S. Notice that a closed path defines an open surface whereas a closed surface defines a volume (see Figures 3.11 and 3.13).

We define the integral

$$\int_v \rho_v \, dv \qquad [3.15]$$

as the *volume integral* of the scalar ρ_v over the volume v. The physical meaning of a line, surface, or volume integral depends on the nature of the physical quantity represented by **A** or ρ_v. Note that $d\mathbf{l}$, $d\mathbf{S}$, and dv are all as defined in Section 3.2.

EXAMPLE

3.2

Given that $\mathbf{F} = x^2\mathbf{a}_x - xz\mathbf{a}_y - y^2\mathbf{a}_z$, calculate the circulation of **F** around the (closed) path shown in Figure 3.10.

SOLUTION

The circulation of **F** around path L is given by

$$\oint_L \mathbf{F} \cdot d\mathbf{l} = \left(\int_1 + \int_2 + \int_3 + \int_4\right) \mathbf{F} \cdot d\mathbf{l}$$

where the path is broken into segments numbered 1 to 4 as shown in Figure 3.10.

For segment 1, $y = 0 = z$

$$\mathbf{F} = x^2\mathbf{a}_x, \qquad d\mathbf{l} = dx\,\mathbf{a}_x$$

Notice that $d\mathbf{l}$ is always taken as along $+\mathbf{a}_x$ so that the direction on 1 is taken care of by the limits of integration. Thus,

$$\int_1 \mathbf{F} \cdot d\mathbf{l} = \int_1^0 x^2 dx = \frac{x^3}{3}\bigg|_1^0 = -\frac{1}{3}$$

For segment 2, $x = 0 = z$, $\mathbf{F} = -y^2\mathbf{a}_z$, $d\mathbf{l} = dy\,\mathbf{a}_y$, $\mathbf{F} \cdot d\mathbf{l} = 0$. Hence,

$$\int_2 \mathbf{F} \cdot d\mathbf{l} = 0$$

For segment 3, $y = 1$, $\mathbf{F} = x^2\mathbf{a}_x - xz\mathbf{a}_y - \mathbf{a}_z$, and $d\mathbf{l} = dx\,\mathbf{a}_x + dz\,\mathbf{a}_z$, so

$$\int_3 \mathbf{F} \cdot d\mathbf{l} = \int (x^2 dx - dz)$$

But on 3, $z = x$; that is, $dx = dz$. Hence,

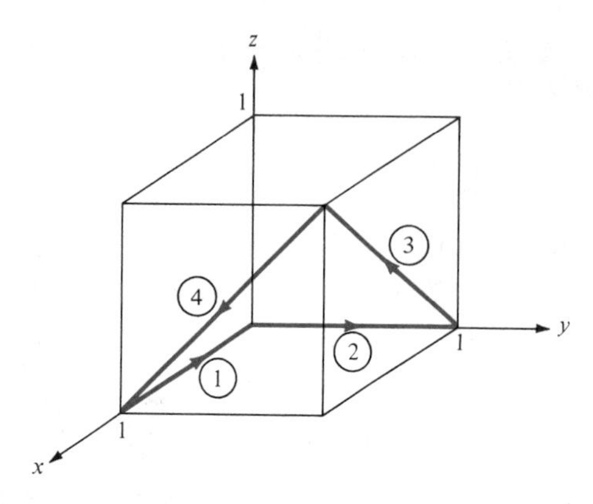

Figure 3.10 For Example 3.2.

$$\int_3 \mathbf{F} \cdot d\mathbf{l} = \int_0^1 (x^2 - 1)\, dx = \frac{x^3}{3} - x \bigg|_0^1 = -\frac{2}{3}$$

For segment 4, $x = 1$, so $\mathbf{F} = \mathbf{a}_x - z\mathbf{a}_y - y^2\mathbf{a}_z$, and $d\mathbf{l} = dy\, \mathbf{a}_y + dz\, \mathbf{a}_z$. Hence,

$$\int_4 \mathbf{F} \cdot d\mathbf{l} = \int (-z\, dy - y^2 dz)$$

But on 4, $z = y$; that is, $dz = dy$, so

$$\int_4 \mathbf{F} \cdot d\mathbf{l} = \int_1^0 (-y - y^2)\, dy = -\frac{y^2}{2} - \frac{y^3}{3} \bigg|_1^0 = \frac{5}{6}$$

By putting all these together, we obtain

$$\oint_L \mathbf{F} \cdot d\mathbf{l} = -\frac{1}{3} + 0 - \frac{2}{3} + \frac{5}{6} = -\frac{1}{6}$$

■

PRACTICE EXERCISE 3.2

Calculate the circulation of

$$\mathbf{A} = \rho \cos \phi\, \mathbf{a}_\rho + z \sin \phi\, \mathbf{a}_z$$

around the edge L of the wedge defined by $0 \leq \rho \leq 2$, $0 \leq \phi \leq 60°$, $z = 0$ and shown in Figure 3.11.

ANSWER 1.

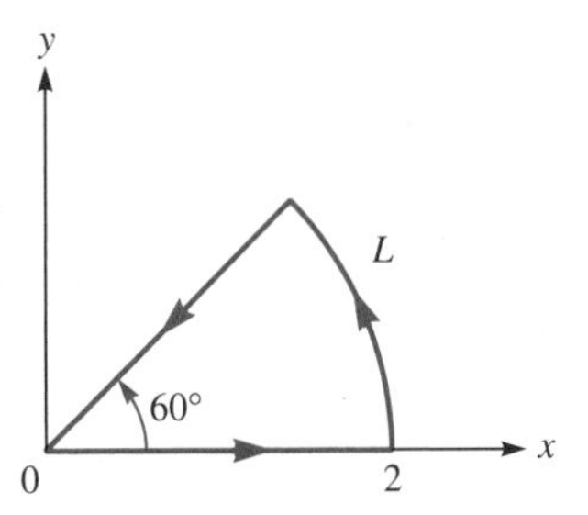

Figure 3.11 For Practice Exercise 3.2.

3.4 DEL OPERATOR

The del operator, written ∇, is the vector differential operator. In Cartesian coordinates,

$$\nabla = \frac{\partial}{\partial x}\mathbf{a}_x + \frac{\partial}{\partial y}\mathbf{a}_y + \frac{\partial}{\partial z}\mathbf{a}_z \tag{3.16}$$

This vector differential operator, otherwise known as the *gradient operator,* is not a vector in itself, but when it operates on a scalar function, for example, a vector ensues. The operator is useful in defining

1. The gradient of a scalar V, written as ∇V
2. The divergence of a vector $\mathbf{A}$, written as $\nabla \cdot \mathbf{A}$
3. The curl of a vector $\mathbf{A}$, written as $\nabla \times \mathbf{A}$
4. The Laplacian of a scalar V, written as $\nabla^2 V$

Each of these will be defined in detail in the subsequent sections. Before we do that, it is appropriate to obtain expressions for the del operator ∇ in cylindrical and spherical coordinates. This is easily done by using eqs. (2.7) and (2.21) along with the transformation formulas of sections 2.3 and 2.4.

To obtain ∇ in terms of ρ, ϕ, and z, we recall from eq. (2.7) that[1]

$$\rho = \sqrt{x^2 + y^2}, \qquad \tan\phi = \frac{y}{x}$$

[1] A more general way of deriving ∇, $\nabla \cdot \mathbf{A}$, $\nabla \times \mathbf{A}$, ∇V, and $\nabla^2 V$ is using the curvilinear coordinates. See, for example, M. R. Spiegel, *Vector Analysis and an Introduction to Tensor Analysis.* New York: McGraw-Hill, 1959, pp. 135–165.

Hence

$$\frac{\partial}{\partial x} = \cos\phi \frac{\partial}{\partial \rho} - \frac{\sin\phi}{\rho}\frac{\partial}{\partial \phi} \tag{3.17}$$

$$\frac{\partial}{\partial y} = \sin\phi \frac{\partial}{\partial \rho} + \frac{\cos\phi}{\rho}\frac{\partial}{\partial \phi} \tag{3.18}$$

Substituting eqs. (3.17) and (3.18) into eq. (3.16) and making use of eq. (2.9), we obtain ∇ in cylindrical coordinates as

$$\boxed{\nabla = \mathbf{a}_\rho \frac{\partial}{\partial \rho} + \mathbf{a}_\phi \frac{1}{\rho}\frac{\partial}{\partial \phi} + \mathbf{a}_z \frac{\partial}{\partial z}} \tag{3.19}$$

Similarly, to obtain ∇ in terms of r, θ, and ϕ, we use

$$r = \sqrt{x^2 + y^2 + z^2}, \qquad \tan\theta = \frac{\sqrt{x^2 + y^2}}{z}, \qquad \tan\phi = \frac{y}{x}$$

to obtain

$$\frac{\partial}{\partial x} = \sin\theta\cos\phi \frac{\partial}{\partial r} + \frac{\cos\theta\cos\phi}{r}\frac{\partial}{\partial \theta} - \frac{\sin\phi}{\rho}\frac{\partial}{\partial \phi} \tag{3.20}$$

$$\frac{\partial}{\partial y} = \sin\theta\sin\phi \frac{\partial}{\partial r} + \frac{\cos\theta\sin\phi}{r}\frac{\partial}{\partial \theta} + \frac{\cos\phi}{\rho}\frac{\partial}{\partial \phi} \tag{3.21}$$

$$\frac{\partial}{\partial z} = \cos\theta \frac{\partial}{\partial r} - \frac{\sin\theta}{r}\frac{\partial}{\partial \theta} \tag{3.22}$$

Substituting eqs. (3.20) to (3.22) into eq. (3.16) and using eq. (2.23) results in ∇ in spherical coordinates:

$$\boxed{\nabla = \mathbf{a}_r \frac{\partial}{\partial r} + \mathbf{a}_\theta \frac{1}{r}\frac{\partial}{\partial \theta} + \mathbf{a}_\phi \frac{1}{r\sin\theta}\frac{\partial}{\partial \phi}} \tag{3.23}$$

Notice that in eqs. (3.19) and (3.23), the unit vectors are placed to the right of the differential operators because the unit vectors depend on the angles.

3.5 GRADIENT OF A SCALAR

By definition, the gradient of a scalar field V is a vector that represents both the magnitude and the direction of the maximum space rate of increase of V. A mathematical expression for the gradient can be obtained by evaluating the difference in the field dV between points P_1 and P_2 of Figure 3.12 where V_1, V_2, and V_3 are contours on which V is constant. From calculus,

$$\begin{aligned} dV &= \frac{\partial V}{\partial x} dx + \frac{\partial V}{\partial y} dy + \frac{\partial V}{\partial z} dz \\ &= \left(\frac{\partial V}{\partial x} \mathbf{a}_x + \frac{\partial V}{\partial y} \mathbf{a}_y + \frac{\partial V}{\partial z} \mathbf{a}_z \right) \cdot (dx\, \mathbf{a}_x + dy\, \mathbf{a}_y + dz\, \mathbf{a}_z) \end{aligned} \qquad [3.24]$$

For convenience, let

$$\mathbf{G} = \frac{\partial V}{\partial x} \mathbf{a}_x + \frac{\partial V}{\partial y} \mathbf{a}_y + \frac{\partial V}{\partial z} \mathbf{a}_z \qquad [3.25]$$

Then

$$dV = \mathbf{G} \cdot d\mathbf{l} = G \cos \theta \, dl$$

or

$$\frac{dV}{dl} = G \cos \theta \qquad [3.26]$$

where $d\mathbf{l}$ is the differential displacement from P_1 to P_2 and θ is the angle between $\mathbf{G}$ and $d\mathbf{l}$. From eq. (3.26), we notice that dV/dl is a maximum when $\theta = 0$, that is, when $d\mathbf{l}$ is in the direction of $\mathbf{G}$. Hence,

$$\left. \frac{dV}{dl} \right|_{\max} = \frac{dV}{dn} = G \qquad [3.27]$$

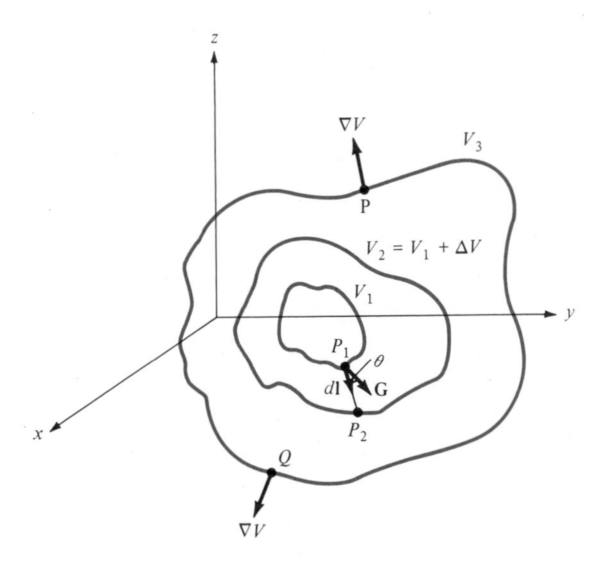

Figure 3.12 Gradient of a scalar.

where dV/dn is the normal derivative. Thus G has its magnitude and direction as those of the maximum rate of change of V. By definition, $\mathbf{G}$ is the gradient of V. Therefore:

$$\text{grad } V = \nabla V = \frac{\partial V}{\partial x}\mathbf{a}_x + \frac{\partial V}{\partial y}\mathbf{a}_y + \frac{\partial V}{\partial z}\mathbf{a}_z \tag{3.28}$$

By using eq. (3.28) in conjunction with eqs. (3.16), (3.19), and (3.23), the gradient of V can be expressed in Cartesian, cylindrical, and spherical coordinates. For Cartesian coordinates

$$\boxed{\nabla V = \frac{\partial V}{\partial x}\mathbf{a}_x + \frac{\partial V}{\partial y}\mathbf{a}_y + \frac{\partial V}{\partial z}\mathbf{a}_z}$$

for cylindrical coordinates,

$$\boxed{\nabla V = \frac{\partial V}{\partial \rho}\mathbf{a}_\rho + \frac{1}{\rho}\frac{\partial V}{\partial \phi}\mathbf{a}_\phi + \frac{\partial V}{\partial z}\mathbf{a}_z} \tag{3.29}$$

and for spherical coordinates,

$$\boxed{\nabla V = \frac{\partial V}{\partial r}\mathbf{a}_r + \frac{1}{r}\frac{\partial V}{\partial \theta}\mathbf{a}_\theta + \frac{1}{r\sin\theta}\frac{\partial V}{\partial \phi}\mathbf{a}_\phi} \tag{3.30}$$

The following computation formulas on gradient, which are easily proved, should be noted:

(a) $\nabla(V + U) = \nabla V + \nabla U$ **[3.31a]**

(b) $\nabla(VU) = V\nabla U + U\nabla V$ **[3.31b]**

(c) $\nabla\left[\frac{V}{U}\right] = \frac{U\nabla V - V\nabla U}{U^2}$ **[3.31c]**

(d) $\nabla V^n = nV^{n-1}\nabla V$ **[3.31d]**

where U and V are scalars and n is an integer.

Also take note of the following fundamental properties of the gradient of a scalar field V:

1. The magnitude of ∇V equals the maximum rate of change in V per unit distance.
2. ∇V points in the direction of the maximum rate of change in V.

3. ∇V at any point is perpendicular to the constant V surface which passes through that point (see points P and Q in Figure 3.12).
4. The projection (or component) of ∇V in the direction of a unit vector **a** is $\nabla V \cdot \mathbf{a}$ and is called the *directional derivative* of V along **a**. This is the rate of change of V in the direction of **a**. For example, dV/dl in eq. (3.26) is the directional derivative of V along P_1P_2 in Figure 3.12. Thus the gradient of a scalar function V provides us with both the direction in which V changes most rapidly and the magnitude of the maximum directional derivative of V.
5. If $\mathbf{A} = \nabla V$, V is said to be the scalar potential of **A**.

EXAMPLE 3.3

Find the gradient of the following scalar fields:

(a) $V = e^{-z}\sin 2x \cosh y$

(b) $U = \rho^2 z \cos 2\phi$

(c) $W = 10r \sin^2\theta \cos \phi$

SOLUTION

(a)
$$\nabla V = \frac{\partial V}{\partial x}\mathbf{a}_x + \frac{\partial V}{\partial y}\mathbf{a}_y + \frac{\partial V}{\partial z}\mathbf{a}_z$$
$$= 2e^{-z}\cos 2x \cosh y\, \mathbf{a}_x + e^{-z}\sin 2x \sinh y\, \mathbf{a}_y - e^{-z}\sin 2x \cosh y\, \mathbf{a}_z$$

(b)
$$\nabla U = \frac{\partial U}{\partial \rho}\mathbf{a}_\rho + \frac{1}{\rho}\frac{\partial U}{\partial \phi}\mathbf{a}_\phi + \frac{\partial U}{\partial z}\mathbf{a}_z$$
$$= 2\rho z \cos 2\phi\, \mathbf{a}_\rho - 2\rho z \sin 2\phi\, \mathbf{a}_\phi + \rho^2 \cos 2\phi\, \mathbf{a}_z$$

(c)
$$\nabla W = \frac{\partial W}{\partial r}\mathbf{a}_r + \frac{1}{r}\frac{\partial W}{\partial \theta}\mathbf{a}_\theta + \frac{1}{r \sin\theta}\frac{\partial W}{\partial \phi}\mathbf{a}_\phi$$
$$= 10 \sin^2\theta \cos \phi\, \mathbf{a}_r + 10 \sin 2\theta \cos \phi\, \mathbf{a}_\theta - 10 \sin \theta \sin \phi\, \mathbf{a}_\phi$$

■

PRACTICE EXERCISE 3.3

Determine the gradient of the following scalar fields:

(a) $U = x^2 y + xyz$

(b) $V = \rho z \sin \phi + z^2 \cos^2\phi + \rho^2$

(c) $f = \cos \theta \sin \phi \ln r + r^2\phi$

ANSWER

(a) $y(2x + z)\mathbf{a}_x + x(x + z)\mathbf{a}_y + xy\mathbf{a}_z$

(b) $(z \sin \phi + 2\rho)\mathbf{a}_\rho + (z \cos \phi - \frac{z^2}{\rho} \sin 2\phi)\mathbf{a}_\phi + (\rho \sin \phi + 2z \cos^2\phi)\mathbf{a}_z$

(c) $\left(\frac{\cos \theta \sin \phi}{r} + 2r\phi\right) \mathbf{a}_r - \frac{\sin \theta \sin \phi}{r} \ln r \, \mathbf{a}_\theta +$

$\left(\frac{\cot \theta}{r} \cos \phi \ln r + r \operatorname{cosec} \theta\right) \mathbf{a}_\phi$

EXAMPLE 3.4

Given $W = x^2y^2 + xyz$, compute ∇W and the direction derivative dW/dl in the direction $3\mathbf{a}_x + 4\mathbf{a}_y + 12\mathbf{a}_z$ at $(2, -1, 0)$.

SOLUTION

$$\nabla W = \frac{\partial W}{\partial x} \mathbf{a}_x + \frac{\partial W}{\partial y} \mathbf{a}_y + \frac{\partial W}{\partial z} \mathbf{a}_z$$

$$= (2xy^2 + yz)\mathbf{a}_x + (2x^2y + xz)\mathbf{a}_y + (xy)\mathbf{a}_z$$

At $(2, -1, 0)$: $\nabla W = 4\mathbf{a}_x - 8\mathbf{a}_y - 2\mathbf{a}_z$

Hence,

$$\frac{dW}{dl} = \nabla W \cdot \mathbf{a}_l = (4, -8, -2) \cdot \frac{(3, 4, 12)}{13} = -\frac{44}{13}$$ ■

PRACTICE EXERCISE 3.4

Given $\Phi = xy + yz + xz$, find gradient Φ at point (1, 2, 3) and the directional derivative of Φ at the same point in the direction toward point (3, 4, 4).

ANSWER $5\mathbf{a}_x + 4\mathbf{a}_y + 3\mathbf{a}_z$, 7.

EXAMPLE 3.5

Find the angle at which line $x = y = 2z$ intersects the ellipsoid $x^2 + y^2 + 2z^2 = 10$.

SOLUTION

Let the line and the ellipsoid meet at angle ψ as shown in Figure 3.13. The line $x = y = 2z$ can be represented by

$$\mathbf{r}(\lambda) = 2\lambda\mathbf{a}_x + 2\lambda\mathbf{a}_y + \lambda\mathbf{a}_z$$

where λ is a parameter. Where the line and the ellipsoid meet,

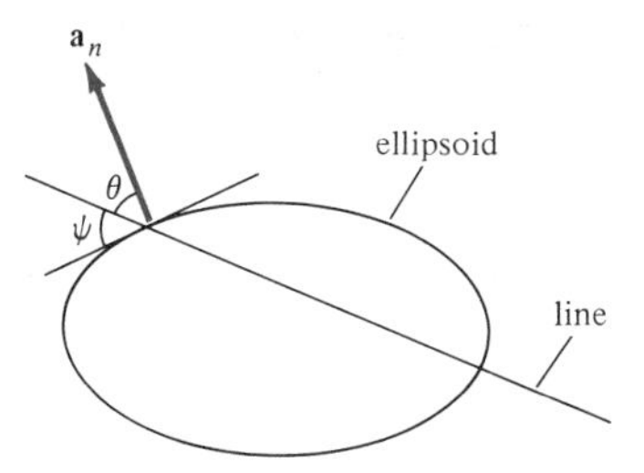

Figure 3.13 For Example 3.5; plane of intersection of a line with an ellipsoid.

$$(2\lambda)^2 + (2\lambda)^2 + 2\lambda^2 = 10 \quad \rightarrow \quad \lambda = \pm 1$$

Taking $\lambda = 1$ (for the moment), the point of intersection is $(x, y, z) = (2, 2, 1)$. At this point, $\mathbf{r} = 2\mathbf{a}_x + 2\mathbf{a}_y + \mathbf{a}_z$.

The surface of the ellipsoid is defined by

$$f(x, y, z) = x^2 + y^2 + 2z^2 - 10$$

The gradient of f is

$$\nabla f = 2x\,\mathbf{a}_x + 2y\,\mathbf{a}_y + 4z\,\mathbf{a}_z$$

At (2, 2, 1), $\nabla f = 4\mathbf{a}_x + 4\mathbf{a}_y + 4\mathbf{a}_z$. Hence, a unit vector normal to the ellipsoid at the point of intersection is

$$\mathbf{a}_n = \pm \frac{\nabla f}{|\nabla f|} = \pm \frac{\mathbf{a}_x + \mathbf{a}_y + \mathbf{a}_z}{\sqrt{3}}$$

Taking the positive sign (for the moment), the angle between $\mathbf{a}_n$ and $\mathbf{r}$ is given by

$$\cos\theta = \frac{\mathbf{a}_n \cdot \mathbf{r}}{|\mathbf{a}_n \cdot \mathbf{r}|} = \frac{2 + 2 + 1}{\sqrt{3}\sqrt{9}} = \frac{5}{3\sqrt{3}} = \sin\psi$$

Hence, $\psi = 74.21°$. Because we had choices of + or − for λ and $\mathbf{a}_n$, there are actually four possible angles, given by $\sin\psi = \pm 5/(3\sqrt{3})$. ■

PRACTICE EXERCISE 3.5

Calculate the angle between the normals to the surfaces $x^2y + z = 3$ and $x \log z - y^2 = -4$ at the point of intersection $(-1, 2, 1)$.

ANSWER 73.4°.

3.6 DIVERGENCE OF A VECTOR AND DIVERGENCE THEOREM

From Section 3.3, we have noticed that the net outflow of the flux of a vector field **A** from a closed surface S is obtained from the integral $\oint \mathbf{A} \cdot d\mathbf{S}$. We now define the divergence of **A** as the net outward flow of flux per unit volume over a closed incremental surface. In other words, the divergence of **A** at a given point P is the *outward* flux per unit volume as the volume shrinks about P. Hence,

$$\text{div } \mathbf{A} = \nabla \cdot \mathbf{A} = \lim_{\Delta v \to 0} \frac{\oint_S \mathbf{A} \cdot d\mathbf{S}}{\Delta v} \quad [3.32]$$

where Δv is the volume enclosed by the closed surface S in which P is located. Physically, we may regard the divergence of the vector field **A** at a given point as a measure of how much the field diverges or emanates from that point. Figure 3.14(a) shows that the divergence of a vector field at point P is positive because the vector diverges (or spreads out) at P. In Figure 3.14(b) a vector field has negative divergence (or convergence) at P, and in Figure 3.14(c) a vector field has zero divergence at P.

We can obtain an expression for $\nabla \cdot \mathbf{A}$ in Cartesian coordinates from the definition in eq. (3.32). Suppose we wish to evaluate the divergence of a vector field **A** at point $P(x_o, y_o, z_o)$; we let the point be enclosed by a differential volume as in Figure 3.15. The surface integral in eq. (3.32) is obtained from

$$\oint_S \mathbf{A} \cdot d\mathbf{S} = \left(\int_{\text{front}} + \int_{\text{back}} + \int_{\text{left}} + \int_{\text{right}} + \int_{\text{top}} + \int_{\text{bottom}} \right) \mathbf{A} \cdot d\mathbf{S} \quad [3.33]$$

A three-dimensional Taylor series expansion of A_x about P is

$$A_x(x, y, z) = A_x(x_o, y_o, z_o) + (x - x_o) \left.\frac{\partial A_x}{\partial x}\right|_P + (y - y_o) \left.\frac{\partial A_x}{\partial y}\right|_P + (z - z_o) \left.\frac{\partial A_x}{\partial z}\right|_P + \text{higher-order terms.} \quad [3.34]$$

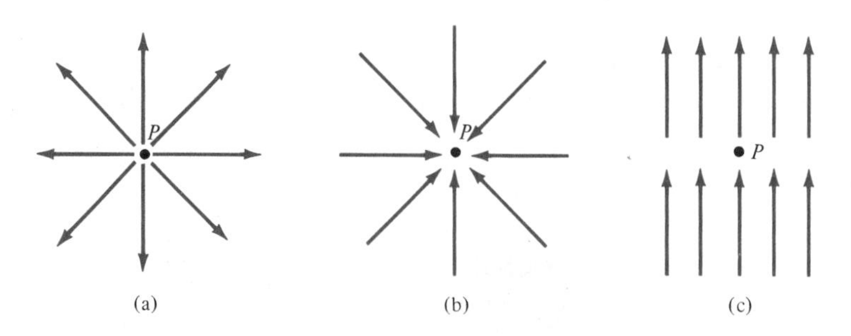

Figure 3.14 Illustration of the divergence of a vector field at P: **(a)** positive divergence, **(b)** negative divergence, **(c)** zero divergence.

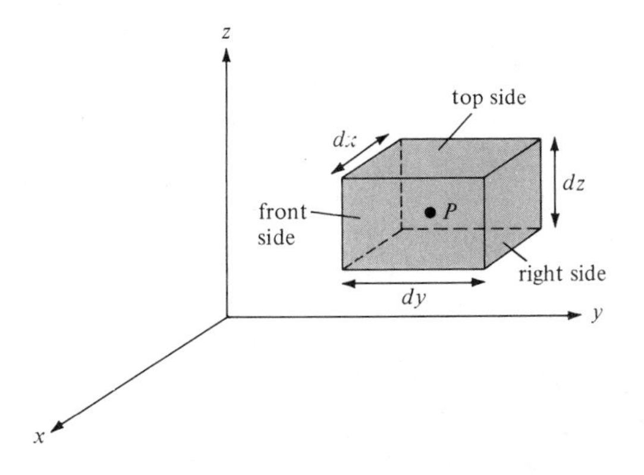

Figure 3.15 Evaluation of $\nabla \cdot \mathbf{A}$ at point $P(x_o, y_o, z_o)$.

For the front side, $x = x_o + dx/2$ and $d\mathbf{S} = dy\, dz\, \mathbf{a}_x$. Then,

$$\int_{\text{front}} \mathbf{A} \cdot d\mathbf{S} = dy\, dz \left[A_x(x_o, y_o, z_o) + \frac{dx}{2} \frac{\partial A_x}{\partial x} \bigg|_P \right] + \text{higher-order terms}$$

For the back side, $x = x_o - dx/2$, $d\mathbf{S} = dy\, dz(-\mathbf{a}_x)$. Then,

$$\int_{\text{back}} \mathbf{A} \cdot d\mathbf{S} = -dy\, dz \left[A_x(x_o, y_o, z_o) - \frac{dx}{2} \frac{\partial A_x}{\partial x} \bigg|_P \right] + \text{higher-order terms}$$

Hence,

$$\int_{\text{front}} \mathbf{A} \cdot d\mathbf{S} + \int_{\text{back}} \mathbf{A} \cdot d\mathbf{S} = dx\, dy\, dz \frac{\partial A_x}{\partial x} \bigg|_P + \text{higher-order terms} \tag{3.35}$$

By taking similar steps, we obtain

$$\int_{\text{left}} \mathbf{A} \cdot d\mathbf{S} + \int_{\text{right}} \mathbf{A} \cdot d\mathbf{S} = dx\, dy\, dz \frac{\partial A_y}{\partial y} \bigg|_P + \text{higher-order terms} \tag{3.36}$$

and

$$\int_{\text{top}} \mathbf{A} \cdot d\mathbf{S} + \int_{\text{bottom}} \mathbf{A} \cdot d\mathbf{S} = dx\, dy\, dz \frac{\partial A_z}{\partial z} \bigg|_P + \text{higher-order terms} \tag{3.37}$$

Substituting eqs. (3.35) to (3.37) into eq. (3.33) and noting that $\Delta v = dx\, dy\, dz$, we get

$$\lim_{\Delta v \to 0} \frac{\oint_S \mathbf{A} \cdot d\mathbf{S}}{\Delta v} = \left(\frac{\partial A_x}{\partial x} + \frac{\partial A_y}{\partial y} + \frac{\partial A_z}{\partial z} \right) \bigg|_{\text{at } P} \tag{3.38}$$

because the higher-order terms will vanish as $\Delta v \to 0$. Thus, the divergence of **A** at point $P(x_o, y_o, z_o)$ in a Cartesian system is given by

$$\nabla \cdot \mathbf{A} = \frac{\partial A_x}{\partial x} + \frac{\partial A_y}{\partial y} + \frac{\partial A_z}{\partial z} \tag{3.39}$$

Similar expressions for $\nabla \cdot \mathbf{A}$ in other coordinate systems can be obtained directly from eq. (3.32) or by transforming eq. (3.39) into the appropriate coordinate system. In cylindrical coordinates, substituting eqs. (2.15), (3.17), and (3.18) into eq. (3.39) yields

$$\nabla \cdot \mathbf{A} = \frac{1}{\rho}\frac{\partial}{\partial \rho}(\rho A_\rho) + \frac{1}{\rho}\frac{\partial A_\phi}{\partial \phi} + \frac{\partial A_z}{\partial z} \tag{3.40}$$

Substituting eqs. (2.28) and (3.20) to (3.22) into eq. (3.39), we obtain the divergence of **A** in spherical coordinates as

$$\nabla \cdot \mathbf{A} = \frac{1}{r^2}\frac{\partial}{\partial r}(r^2 A_r) + \frac{1}{r \sin\theta}\frac{\partial}{\partial \theta}(A_\theta \sin\theta) + \frac{1}{r \sin\theta}\frac{\partial A_\phi}{\partial \phi} \tag{3.41}$$

Note the following properties of the divergence of a vector field:

1. It produces a scalar field (because scalar product is involved).
2. The divergence of a scalar V, div V, makes no sense.
3. $\nabla \cdot (\mathbf{A} + \mathbf{B}) = \nabla \cdot \mathbf{A} + \nabla \cdot \mathbf{B}$

The divergence of a vector field can also be viewed as simply the limit of the field's source strength per unit volume (or source density); it is positive at a *source* point in the field, and negative at a *sink* point, or zero where there is neither sink nor source.

From the definition of the divergence of **A** in eq. (3.32), it is not difficult to expect that

$$\oint_S \mathbf{A} \cdot d\mathbf{S} = \int_v \nabla \cdot \mathbf{A}\, dv \tag{3.42}$$

This is called the *divergence theorem,* otherwise known as the *Gauss-Ostrogradsky theorem*. The theorem states that the total outward flux of a vector field **A** through the *closed* surface S is the same as the volume integral of the divergence of **A**.

To prove the divergence theorem, subdivide volume v into a large number of small cells. If the kth cell has volume Δv_k and is bounded by surface S_k

$$\oint_S \mathbf{A} \cdot d\mathbf{S} = \sum_k \oint_{S_k} \mathbf{A} \cdot d\mathbf{S} = \sum_k \frac{\oint_{S_k} \mathbf{A} \cdot d\mathbf{S}}{\Delta v_k} \Delta v_k \tag{3.43}$$

Since the outward flux to one cell is inward to some neighboring cells, there is cancellation on every interior surface, so the sum of the surface integrals over S_k's is the same as the surface integral over the surface S. Taking the limit of the right-hand side of eq. (3.43) and incorporating eq. (3.32) gives

$$\oint_S \mathbf{A} \cdot d\mathbf{S} = \int_v \nabla \cdot \mathbf{A} \, dv \tag{3.44}$$

which is the divergence theorem. The theorem applies to any volume v bounded by the closed surface S such as that shown in Figure 3.16 provided that **A** and $\nabla \cdot \mathbf{A}$ are continuous in the region. With a little experience, it will soon become apparent that volume integrals are easier to evaluate than surface integrals. For this reason, to determine the flux of **A** through a closed surface we simply find the right-hand side of eq. (3.42) instead of the left-hand side of the equation.

EXAMPLE 3.6

Determine the divergence of these vector fields:

(a) $\mathbf{P} = x^2yz \, \mathbf{a}_x + xz \, \mathbf{a}_z$

(b) $\mathbf{Q} = \rho \sin \phi \, \mathbf{a}_\rho + \rho^2 z \, \mathbf{a}_\phi + z \cos \phi \, \mathbf{a}_z$

(c) $\mathbf{T} = \dfrac{1}{r^2} \cos \theta \, \mathbf{a}_r + r \sin \theta \cos \phi \, \mathbf{a}_\theta + \cos \theta \, \mathbf{a}_\phi$

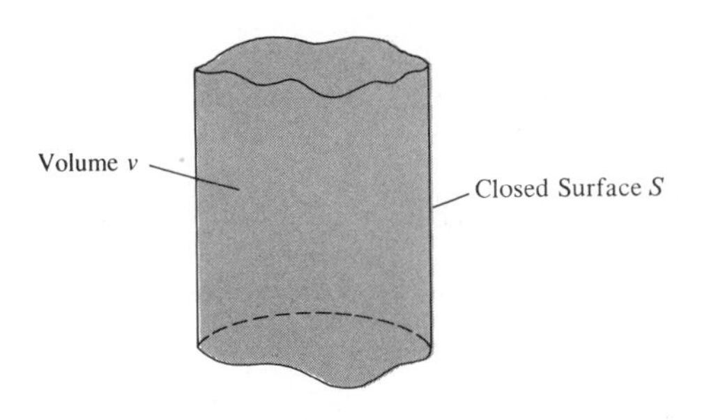

Figure 3.16 Volume v enclosed by surface S.

SOLUTION (a)

$$\nabla \cdot \mathbf{P} = \frac{\partial}{\partial x} P_x + \frac{\partial}{\partial y} P_y + \frac{\partial}{\partial z} P_z$$

$$= \frac{\partial}{\partial x}(x^2yz) + \frac{\partial}{\partial y}(0) + \frac{\partial}{\partial z}(xz)$$

$$= 2xyz + x$$

(b)

$$\nabla \cdot \mathbf{Q} = \frac{1}{\rho}\frac{\partial}{\partial \rho}(\rho Q_\rho) + \frac{1}{\rho}\frac{\partial}{\partial \phi} Q_\phi + \frac{\partial}{\partial z} Q_z$$

$$= \frac{1}{\rho}\frac{\partial}{\partial \rho}(\rho^2 \sin \phi) + \frac{1}{\rho}\frac{\partial}{\partial \phi}(\rho^2 z) + \frac{\partial}{\partial z}(z \cos \phi)$$

$$= 2 \sin \phi + \cos \phi$$

(c)

$$\nabla \cdot \mathbf{T} = \frac{1}{r^2}\frac{\partial}{\partial r}(r^2 T_r) + \frac{1}{r \sin \theta}\frac{\partial}{\partial \theta}(T_\theta \sin \theta) + \frac{1}{r \sin \theta}\frac{\partial}{\partial \phi}(T_\phi)$$

$$= \frac{1}{r^2}\frac{\partial}{\partial r}(\cos \theta) + \frac{1}{r \sin \theta}\frac{\partial}{\partial \theta}(r \sin^2 \theta \cos \phi) + \frac{1}{r \sin \theta}\frac{\partial}{\partial \phi}(\cos \theta)$$

$$= 0 + \frac{1}{r \sin \theta} 2r \sin \theta \cos \theta \cos \phi + 0$$

$$= 2 \cos \theta \cos \phi$$

■

PRACTICE EXERCISE 3.6

Determine the divergence of the following vector fields and evaluate them at the specified points.

(a) $\mathbf{A} = yz\mathbf{a}_x + 4xy\mathbf{a}_y + y\mathbf{a}_z$ at $(1, -2, 3)$

(b) $\mathbf{B} = \rho z \sin \phi \ \mathbf{a}_\rho + 3\rho z^2 \cos \phi \ \mathbf{a}_\phi$ at $(5, \pi/2, 1)$

(c) $\mathbf{C} = 2r \cos \theta \cos \phi \ \mathbf{a}_r + r^{1/2}\mathbf{a}_\phi$ at $(1, \pi/6, \pi/3)$

ANSWER (a) $4x$, 4, (b) $(2 - 3z)z \sin \phi$, -1, (c) $6 \cos \theta \cos \phi$, 2.598.

EXAMPLE 3.7

If $\mathbf{G}(r) = 10e^{-2z}(\rho\mathbf{a}_\rho + \mathbf{a}_z)$, determine the flux of $\mathbf{G}$ out of the entire surface of the cylinder $\rho = 1$, $0 \leq z \leq 1$. Confirm the result using the divergence theorem.

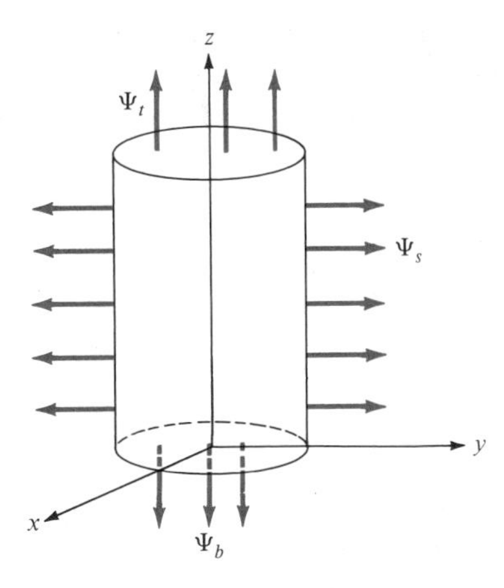

Figure 3.17 For Example 3.7.

SOLUTION If Ψ is the flux of **G** through the given surface, shown in Figure 3.17, then

$$\Psi = \oint \mathbf{G} \cdot d\mathbf{S} = \Psi_t + \Psi_b + \Psi_s$$

where Ψ_t, Ψ_b, and Ψ_s are the fluxes through the top, bottom, and sides (curved surface) of the cylinder as in Figure 3.17.

For Ψ_t, $z = 1$, $d\mathbf{S} = \rho \, d\rho \, d\phi \, \mathbf{a}_z$. Hence,

$$\Psi_t = \int \mathbf{G} \cdot d\mathbf{S} = \int_{\rho=0}^{1} \int_{\phi=0}^{2\pi} 10e^{-2} \rho \, d\rho \, d\phi = 10e^{-2}(2\pi) \left. \frac{\rho^2}{2} \right|_0^1$$

$$= 10\pi e^{-2}$$

For Ψ_b, $z = 0$ and $d\mathbf{S} = \rho \, d\rho \, d\phi(-\mathbf{a}_z)$. Hence,

$$\Psi_b = \int_b \mathbf{G} \cdot d\mathbf{S} = -\int_{\rho=0}^{1} \int_{\phi=0}^{2\pi} 10e^{0} \rho \, d\rho \, d\phi = -10(2\pi) \left. \frac{\rho^2}{2} \right|_0^1$$

$$= -10\pi$$

For Ψ_s, $\rho = 1$, $d\mathbf{S} = \rho \, dz \, d\phi \, \mathbf{a}_\rho$. Hence,

$$\Psi_s = \int_s \mathbf{G} \cdot d\mathbf{S} = \int_{z=0}^{1} \int_{\phi=0}^{2\pi} 10e^{-2z} \rho^2 dz \, d\phi = 10(1)^2(2\pi) \left. \frac{e^{-2z}}{-2} \right|_0^1$$

$$= 10\pi(1 - e^{-2})$$

Thus,

$$\Psi = \Psi_t + \Psi_b + \Psi_s = 10\pi e^{-2} - 10\pi + 10\pi(1 - e^{-2}) = 0$$

Alternatively, since S is a closed surface, we can apply the divergence theorem:

$$\Psi = \oint_S \mathbf{G} \cdot d\mathbf{S} = \int_v (\nabla \cdot \mathbf{G})\, dv$$

But

$$\nabla \cdot \mathbf{G} = \frac{1}{\rho}\frac{\partial}{\partial \rho}(\rho G_\rho) + \frac{1}{\rho}\frac{\partial}{\partial \phi} G_\phi + \frac{\partial}{\partial z} G_z$$

$$= \frac{1}{\rho}\frac{\partial}{\partial \rho}(\rho^2 10 e^{-2z}) - 20e^{-2z} = 0$$

showing that $\mathbf{G}$ has no source. Hence,

$$\Psi = \int_v (\nabla \cdot \mathbf{G})\, dv = 0$$

■

PRACTICE EXERCISE

Determine the flux of $\mathbf{D} = \rho^2 \cos^2\phi\ \mathbf{a}_\rho + z \sin\phi\ \mathbf{a}_\phi$ over the closed surface of the cylinder $0 \le z \le 1$, $\rho = 4$. Verify the divergence theorem for this case.

ANSWER 64π.

3.7 CURL OF A VECTOR AND STOKES'S THEOREM

In Section 3.3, we defined the circulation of a vector field $\mathbf{A}$ around a closed path L as the integral $\oint_L \mathbf{A} \cdot d\mathbf{l}$. We now define the curl of $\mathbf{A}$ as an axial (or rotational) vector whose magnitude is the maximum circulation of $\mathbf{A}$ per unit area as the area tends to zero and whose direction is the normal direction of the area when the area is oriented so as to make the circulation maximum.[2] That is,

$$\text{curl}\ \mathbf{A} = \nabla \times \mathbf{A} = \left(\lim_{\Delta S \to 0} \frac{\oint_L \mathbf{A} \cdot d\mathbf{l}}{\Delta S} \right)_{\max} \mathbf{a}_n \qquad [3.45]$$

where the area ΔS is bounded by the curve L and $\mathbf{a}_n$ is the unit vector normal to the surface ΔS and is determined using the right-hand rule.

To obtain an expression for $\nabla \times \mathbf{A}$ from the definition in eq. (3.45), consider the differential area in the yz-plane as in Figure 3.18. The line integral in eq. (3.45) is obtained as

[2]Because of its rotational nature, some authors use rot $\mathbf{A}$ intead of curl $\mathbf{A}$.

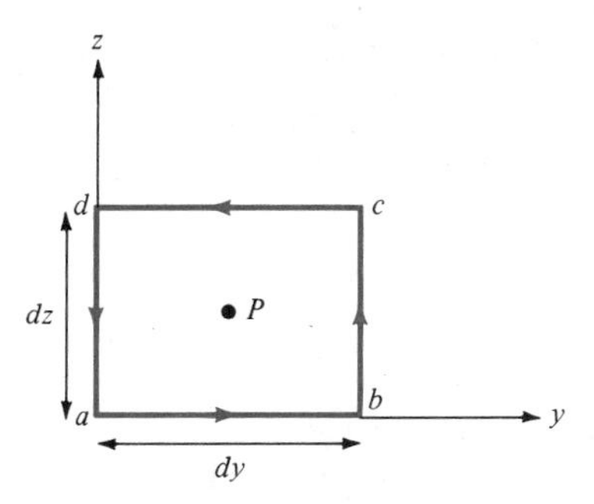

Figure 3.18 Contour used in evaluating the x-component of $\nabla \times \mathbf{A}$ at point $P(x_o, y_o, z_o)$.

$$\oint_L \mathbf{A} \cdot d\mathbf{l} = \left(\int_{ab} + \int_{bc} + \int_{cd} + \int_{da} \right) \mathbf{A} \cdot d\mathbf{l}. \tag{3.46}$$

We expand the field components in a Taylor series expansion about the center point $P(x_o, y_o, z_o)$ as in eq. (3.34) and evaluate eq. (3.46). On side ab, $d\mathbf{l} = dy\, \mathbf{a}_y$ and $z = z_o - dz/2$, so

$$\int_{ab} \mathbf{A} \cdot d\mathbf{l} = dy \left[A_y(x_o, y_o, z_o) - \frac{dz}{2} \frac{\partial A_y}{\partial z} \bigg|_P \right] \tag{3.47}$$

On side bc, $d\mathbf{l} = dz\, \mathbf{a}_z$ and $y = y_o + dy/2$, so

$$\int_{bc} \mathbf{A} \cdot d\mathbf{l} = dz \left[A_z(x_o, y_o, z_o) + \frac{dy}{2} \frac{\partial A_z}{\partial y} \bigg|_P \right] \tag{3.48}$$

On side cd, $d\mathbf{l} = dy\, \mathbf{a}_y$ and $z = z_o + dz/2$, so

$$\int_{cd} \mathbf{A} \cdot d\mathbf{l} = -dy \left[A_y(x_o, y_o, z_o) + \frac{dz}{2} \frac{\partial A_y}{\partial z} \bigg|_P \right] \tag{3.49}$$

On side da, $d\mathbf{l} = dz\, \mathbf{a}_z$ and $y = y_o - dy/2$, so

$$\int_{da} \mathbf{A} \cdot d\mathbf{l} = -dz \left[A_z(x_o, y_o, z_o) - \frac{dy}{2} \frac{\partial A_z}{\partial y} \bigg|_P \right] \tag{3.50}$$

Substituting eqs. (3.47) to (3.50) into eq. (3.46) and noting that $\Delta S = dy\, dz$, we have

$$\lim_{\Delta S \to 0} \oint_L \frac{\mathbf{A} \cdot d\mathbf{l}}{\Delta S} = \frac{\partial A_z}{\partial y} - \frac{\partial A_y}{\partial z}$$

or

$$(\text{curl } \mathbf{A})_x = \frac{\partial A_z}{\partial y} - \frac{\partial A_y}{\partial z} \tag{3.51}$$

The y- and x-components of the curl of **A** can be found in the same way. We obtain

$$(\text{curl } \mathbf{A})_y = \frac{\partial A_x}{\partial z} - \frac{\partial A_z}{\partial x} \tag{3.52a}$$

$$(\text{curl } \mathbf{A})_z = \frac{\partial A_y}{\partial x} - \frac{\partial A_x}{\partial y} \tag{3.52b}$$

The definition of $\nabla \times \mathbf{A}$ in eq. (3.45) is independent of the coordinate system. In Cartesian coordinates the curl of **A** is easily found using

$$\nabla \times \mathbf{A} = \begin{vmatrix} \mathbf{a}_x & \mathbf{a}_y & \mathbf{a}_z \\ \dfrac{\partial}{\partial x} & \dfrac{\partial}{\partial y} & \dfrac{\partial}{\partial z} \\ A_x & A_y & A_z \end{vmatrix} \tag{3.53}$$

or

$$\begin{aligned} \nabla \times \mathbf{A} = \left[\frac{\partial A_z}{\partial y} - \frac{\partial A_y}{\partial z}\right] \mathbf{a}_x + \left[\frac{\partial A_x}{\partial z} - \frac{\partial A_z}{\partial x}\right] \mathbf{a}_y \\ + \left[\frac{\partial A_y}{\partial x} - \frac{\partial A_x}{\partial y}\right] \mathbf{a}_z \end{aligned} \tag{3.54}$$

By transforming eq. (3.54) using point and vector transformation techniques used in Chapter 2, we obtain the curl of **A** in cylindrical coordinates as

$$\nabla \times \mathbf{A} = \frac{1}{\rho} \begin{vmatrix} \mathbf{a}_\rho & \rho\, \mathbf{a}_\phi & \mathbf{a}_z \\ \dfrac{\partial}{\partial \rho} & \dfrac{\partial}{\partial \phi} & \dfrac{\partial}{\partial z} \\ A_\rho & \rho A_\phi & A_z \end{vmatrix}$$

or

$$\begin{aligned} \nabla \times \mathbf{A} = \left[\frac{1}{\rho}\frac{\partial A_z}{\partial \phi} - \frac{\partial A_\phi}{\partial z}\right] \mathbf{a}_\rho + \left[\frac{\partial A_\rho}{\partial z} - \frac{\partial A_z}{\partial \rho}\right] \mathbf{a}_\phi \\ + \frac{1}{\rho}\left[\frac{\partial (\rho A_\phi)}{\partial \rho} - \frac{\partial A_\rho}{\partial \phi}\right] \mathbf{a}_z \end{aligned} \tag{3.55}$$

and in spherical coordinates as

$$\nabla \times \mathbf{A} = \frac{1}{r^2 \sin \theta} \begin{vmatrix} \mathbf{a}_r & r\,\mathbf{a}_\theta & r \sin \theta\, \mathbf{a}_\phi \\ \dfrac{\partial}{\partial r} & \dfrac{\partial}{\partial \theta} & \dfrac{\partial}{\partial \phi} \\ A_r & rA_\theta & r \sin \theta\, A_\phi \end{vmatrix}$$

or

$$\nabla \times \mathbf{A} = \frac{1}{r \sin \theta}\left[\frac{\partial (A_\phi \sin \theta)}{\partial \theta} - \frac{\partial A_\theta}{\partial \phi}\right]\mathbf{a}_r + \frac{1}{r}\left[\frac{1}{\sin \theta}\frac{\partial A_r}{\partial \phi} - \frac{\partial (rA_\phi)}{\partial r}\right]\mathbf{a}_\theta + \frac{1}{r}\left[\frac{\partial (rA_\theta)}{\partial r} - \frac{\partial A_r}{\partial \theta}\right]\mathbf{a}_\phi \qquad [3.56]$$

Note the following properties of the curl:

1. The curl of a vector field is another vector field.
2. The curl of a scalar field V, $\nabla \times V$, makes no sense.
3. $\nabla \times (\mathbf{A} + \mathbf{B}) = \nabla \times \mathbf{A} + \nabla \times \mathbf{B}$
4. $\nabla \times (\mathbf{A} \times \mathbf{B}) = \mathbf{A}(\nabla \cdot \mathbf{B}) - \mathbf{B}(\nabla \cdot \mathbf{A}) + (\mathbf{B} \cdot \nabla\mathbf{A} - (\mathbf{A} \cdot \nabla)\mathbf{B}$
5. The divergence of the curl of a vector field vanishes, that is, $\nabla \cdot (\nabla \times \mathbf{A}) = 0$.
6. The curl of the gradient of a scalar field vanishes, that is, $\nabla \times \nabla V = 0$.

Other properties of the curl are in Appendix A.

The physical significance of the curl of a vector field is evident in eq. (3.45); the curl provides the maximum value of the circulation of the field per unit area (or circulation density) and indicates the direction along which this maximum value occurs. The curl of a vector field $\mathbf{A}$ at a point P may be regarded as a measure of the circulation or how much the field curls around P. For example, Figure 3.19(a) shows

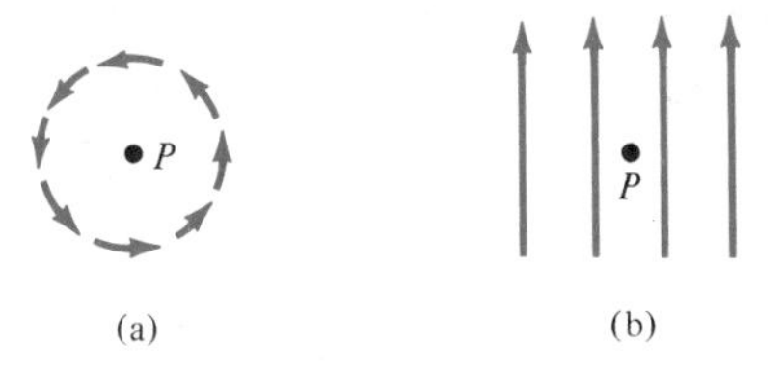

Figure 3.19 Illustration of a curl: **(a)** curl at P points out of the page; **(b)** curl at P is zero.

that the curl of a vector field around P is directed out of the page. Figure 3.19(b) shows a vector field with zero curl.

Also, from the definition of the curl of $\mathbf{A}$ in eq. (3.45), we may expect that

$$\oint_L \mathbf{A} \cdot d\mathbf{l} = \int_S (\nabla \times \mathbf{A}) \cdot d\mathbf{S} \tag{3.57}$$

This is called *Stokes's theorem*. It states that the circulation of $\mathbf{A}$ around a (closed) path L is equal to the surface integral of the curl of $\mathbf{A}$ over the open surface S bounded by L as shown in Figure 3.20 provided that $\mathbf{A}$ and $\nabla \times \mathbf{A}$ are continuous on S.

The proof of Stokes's theorem is similar to that of the divergence theorem. The surface S is subdivided into a large number of cells as in Figure 3.21. If the kth cell has surface area ΔS_k and is bounded by path L_k

$$\oint_L \mathbf{A} \cdot d\mathbf{l} = \sum_k \oint_{L_k} \mathbf{A} \cdot d\mathbf{l} = \sum_k \frac{\oint_{L_k} \mathbf{A} \cdot d\mathbf{l}}{\Delta S_k} \Delta S_k \tag{3.58}$$

As shown in Figure 3.21, there is cancellation on every interior path, so the sum of the line integrals around L_k's is the same as the line integral around the bounding curve L. Therefore, taking the limit of the right-hand side of eq. (3.58) as $\Delta S_k \to 0$ and incorporating eq. (3.45) leads to

$$\oint_L \mathbf{A} \cdot d\mathbf{l} = \int_S (\nabla \times \mathbf{A}) \cdot d\mathbf{S}$$

which is Stokes's theorem.

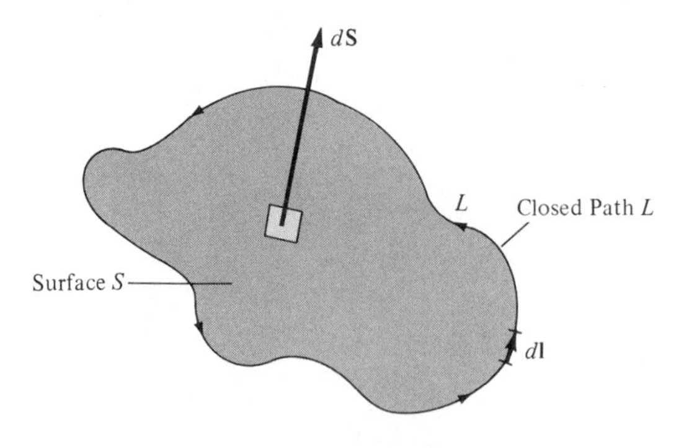

Figure 3.20 Determining the sense of $d\mathbf{l}$ and $d\mathbf{S}$ involved in Stokes's theorem.

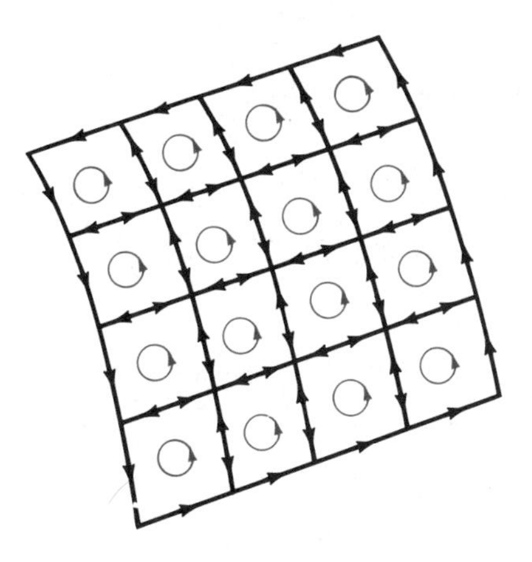

Figure 3.21 Illustration of Stokes's theorem.

The direction of $d\mathbf{l}$ and $d\mathbf{S}$ in eq. (3.57) must be chosen using the right-hand rule or right-handed screw rule. Using the right-hand rule, if we let the fingers point in the direction of $d\mathbf{l}$, the thumb will indicate the direction of $d\mathbf{S}$ (see Fig. 3.20). Note that whereas the divergence theorem relates a surface integral to a volume integral, Stokes's theorem relates a line integral (circulation) to a surface integral.

EXAMPLE 3.8

Determine the curl of the vector fields in Example 3.6.

SOLUTION

(a)

$$\nabla \times \mathbf{P} = \left(\frac{\partial P_z}{\partial y} - \frac{\partial P_y}{\partial z}\right)\mathbf{a}_x + \left(\frac{\partial P_x}{\partial z} - \frac{\partial P_z}{\partial x}\right)\mathbf{a}_y + \left(\frac{\partial P_y}{\partial x} - \frac{\partial P_x}{\partial y}\right)\mathbf{a}_z$$

$$= (0 - 0)\,\mathbf{a}_x + (x^2y - z)\,\mathbf{a}_y + (0 - x^2z)\,\mathbf{a}_z$$

$$= (x^2y - z)\mathbf{a}_y - x^2z\mathbf{a}_z$$

(b)

$$\nabla \times \mathbf{Q} = \left[\frac{1}{\rho}\frac{\partial Q_z}{\partial \phi} - \frac{\partial Q_\phi}{\partial z}\right]\mathbf{a}_\rho + \left[\frac{\partial Q_\rho}{\partial z} - \frac{\partial Q_z}{\partial \rho}\right]\mathbf{a}_\phi + \frac{1}{\rho}\left[\frac{\partial}{\partial \rho}(\rho Q_\phi) - \frac{\partial Q_\rho}{\partial \phi}\right]\mathbf{a}_z$$

$$= \left(\frac{-z}{\rho}\sin\phi - \rho^2\right)\mathbf{a}_\rho + (0 - 0)\mathbf{a}_\phi + \frac{1}{\rho}(3\rho^2 z - \rho\cos\phi)\mathbf{a}_z$$

$$= -\frac{1}{\rho}(z\sin\phi + \rho^3)\mathbf{a}_\rho + (3\rho z - \cos\phi)\mathbf{a}_z$$

(c)
$$\nabla \times \mathbf{T} = \frac{1}{r \sin\theta}\left[\frac{\partial}{\partial\theta}(T_\phi \sin\theta) - \frac{\partial}{\partial\phi}T_\theta\right]\mathbf{a}_r$$

$$+ \frac{1}{r}\left[\frac{1}{\sin\theta}\frac{\partial}{\partial\phi}T_r - \frac{\partial}{\partial r}(rT_\phi)\right]\mathbf{a}_\theta + \frac{1}{r}\left[\frac{\partial}{\partial r}(rT_\theta) - \frac{\partial}{\partial\theta}T_r\right]\mathbf{a}_\phi$$

$$= \frac{1}{r\sin\theta}\left[\frac{\partial}{\partial\theta}(\cos\theta\sin\theta) - \frac{\partial}{\partial\phi}(r\sin\theta\cos\phi)\right]\mathbf{a}_r$$

$$+ \frac{1}{r}\left[\frac{1}{\sin\theta}\frac{\partial}{\partial\phi}\frac{(\cos\theta)}{r^2} - \frac{\partial}{\partial r}(r\cos\theta)\right]\mathbf{a}_\theta$$

$$+ \frac{1}{r}\left[\frac{\partial}{\partial r}(r^2\sin\theta\cos\phi) - \frac{\partial}{\partial\theta}\frac{(\cos\theta)}{r^2}\right]\mathbf{a}_\phi$$

$$= \frac{1}{r\sin\theta}(\cos 2\theta + r\sin\theta\sin\phi)\mathbf{a}_r + \frac{1}{r}(0 - \cos\theta)\mathbf{a}_\theta$$

$$+ \frac{1}{r}\left(2r\sin\theta\cos\phi + \frac{\sin\theta}{r^2}\right)\mathbf{a}_\phi$$

$$= \left(\frac{\cos 2\theta}{r\sin\theta} + \sin\phi\right)\mathbf{a}_r - \frac{\cos\theta}{r}\mathbf{a}_\theta + \left(2\cos\phi + \frac{1}{r^3}\right)\sin\theta\,\mathbf{a}_\phi$$ ■

PRACTICE EXERCISE 3.8

Determine the curl of the vector fields in Practice Exercise 3.6 and evaluate them at the specified points.

ANSWER

(a) $\mathbf{a}_x + y\mathbf{a}_y + (4y - z)\mathbf{a}_z$, $\mathbf{a}_x - 2\mathbf{a}_y - 11\mathbf{a}_z$

(b) $-6\rho z\cos\phi\,\mathbf{a}_\rho + \rho\sin\phi\,\mathbf{a}_\phi + (6z - 1)z\cos\phi\,\mathbf{a}_z$, $5\mathbf{a}_\phi$

(c) $\dfrac{\cot\theta}{r^{1/2}}\mathbf{a}_r - \left(2\cot\theta\sin\phi + \dfrac{3}{2r^{1/2}}\right)\mathbf{a}_\theta + 2\sin\theta\cos\phi\,\mathbf{a}_\phi$,

$1.732\,\mathbf{a}_r - 4.5\,\mathbf{a}_\theta + 0.5\,\mathbf{a}_\phi$.

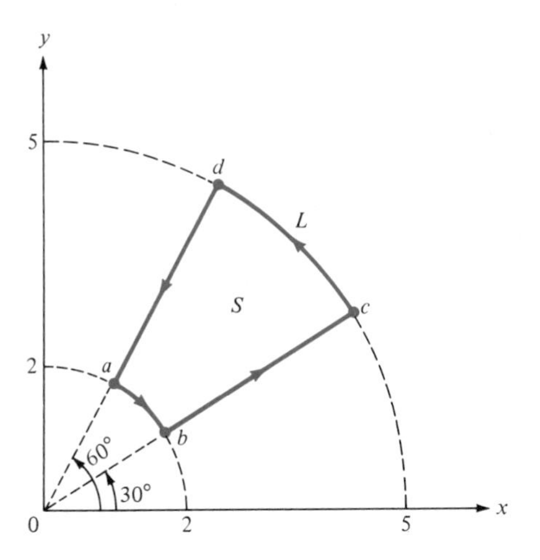

Figure 3.22 For Example 3.9.

EXAMPLE 3.9

If $\mathbf{A} = \rho \cos \phi \, \mathbf{a}_\rho + \sin \phi \, \mathbf{a}_\phi$, evaluate $\oint \mathbf{A} \cdot d\mathbf{l}$ around the path shown in Figure 3.22. Confirm this using Stokes's theorem.

SOLUTION

Let

$$\oint_L \mathbf{A} \cdot d\mathbf{l} = \left[\int_a^b + \int_b^c + \int_c^d + \int_d^a \right] \mathbf{A} \cdot d\mathbf{l}$$

where path L has been divided into segments ab, bc, cd, and da as in Figure 3.22.

Along ab, $\rho = 2$ and $d\mathbf{l} = \rho \, d\phi \, \mathbf{a}_\phi$. Hence,

$$\int_a^b \mathbf{A} \cdot d\mathbf{l} = \int_{\phi = 60°}^{30°} \rho \sin \phi \, d\phi = 2(-\cos \phi) \Big|_{60°}^{30°} = -(\sqrt{3} - 1)$$

Along bc, $\phi = 30°$ and $d\mathbf{l} = d\rho \, \mathbf{a}_\rho$. Hence,

$$\int_b^c \mathbf{A} \cdot d\mathbf{l} = \int_{\rho = 2}^{5} \rho \cos \phi \, d\rho = \cos 30° \frac{\rho^2}{2} \Big|_2^5 = \frac{21\sqrt{3}}{4}$$

Along cd, $\rho = 5$ and $d\mathbf{l} = \rho \, d\phi \, \mathbf{a}_\phi$. Hence,

$$\int_c^d \mathbf{A} \cdot d\mathbf{l} = \int_{\phi = 30°}^{60°} \rho \sin \phi \, d\phi = 5(-\cos \phi) \Big|_{30°}^{60°} = \frac{5}{2}(\sqrt{3} - 1)$$

Along da, $\phi = 60°$ and $d\mathbf{l} = d\rho \, \mathbf{a}_\rho$. Hence,

$$\int_d^a \mathbf{A} \cdot d\mathbf{l} = \int_{\rho=5}^{2} \rho \cos \phi \, d\rho = \cos 60° \frac{\rho^2}{2} \Bigg|_5^2 = -\frac{21}{4}$$

Putting all these together results in

$$\oint_L \mathbf{A} \cdot d\mathbf{l} = -\sqrt{3} + 1 + \frac{21\sqrt{3}}{4} + \frac{5\sqrt{3}}{2} - \frac{5}{2} - \frac{21}{4}$$

$$= \frac{27}{4}(\sqrt{3} - 1) = 4.941$$

Using Stokes's theorem (because L is a closed path)

$$\oint_L \mathbf{A} \cdot d\mathbf{l} = \int_S (\nabla \times \mathbf{A}) \cdot d\mathbf{S}$$

But $d\mathbf{S} = \rho \, d\phi \, d\rho \, \mathbf{a}_z$ and

$$\nabla \times \mathbf{A} = \mathbf{a}_\rho \left[\frac{1}{\rho}\frac{\partial A_z}{\partial \phi} - \frac{\partial A_\phi}{\partial z}\right] + \mathbf{a}_\phi \left[\frac{\partial A_\rho}{\partial z} - \frac{\partial A_z}{\partial \rho}\right] + \mathbf{a}_z \frac{1}{\rho}\left[\frac{\partial}{\partial \rho}(\rho A_\phi) - \frac{\partial A_\rho}{\partial \phi}\right]$$

$$= (0 - 0)\mathbf{a}_\rho + (0 - 0)\mathbf{a}_\phi + \frac{1}{\rho}(1 + \rho) \sin \phi \, \mathbf{a}_z$$

Hence:

$$\int_S (\nabla \times \mathbf{A}) \cdot d\mathbf{S} = \int_{\phi=30°}^{60°} \int_{\rho=2}^{5} \frac{1}{\rho}(1 + \rho) \sin \phi \, \rho \, d\rho \, d\phi$$

$$= \int_{30°}^{60°} \sin \phi \, d\phi \int_2^5 (1 + \rho) d\rho$$

$$= -\cos \phi \Bigg|_{30°}^{60°} \left(\rho + \frac{\rho^2}{2}\right) \Bigg|_2^5$$

$$= \frac{27}{4}(\sqrt{3} - 1) = 4.941$$

■

PRACTICE EXERCISE 3.9

Use Stokes's theorem to confirm your result in Practice Exercise 3.2.

ANSWER 1.

EXAMPLE 3.10

For a vector field **A**, show explicitly that $\nabla \cdot \nabla \times \mathbf{A} = 0$; that is, the divergence of the curl of any vector field is zero.

SOLUTION

This vector identity along with the one in Practice Exercise 3.10 is very useful in EM. For simplicity, assume that **A** is in Cartesian coordinates.

$$\nabla \cdot \nabla \times \mathbf{A} = \left(\frac{\partial}{\partial x}, \frac{\partial}{\partial y}, \frac{\partial}{\partial z}\right) \cdot \begin{vmatrix} \frac{\partial}{\partial x} & \frac{\partial}{\partial y} & \frac{\partial}{\partial z} \\ A_x & A_y & A_z \end{vmatrix}$$

$$= \left(\frac{\partial}{\partial x}, \frac{\partial}{\partial y}, \frac{\partial}{\partial z}\right) \cdot \left[\left(\frac{\partial A_z}{\partial y} - \frac{\partial A_y}{\partial z}\right), -\left(\frac{\partial A_z}{\partial x} - \frac{\partial A_x}{\partial z}\right), \left(\frac{\partial A_y}{\partial x} - \frac{\partial A_x}{\partial y}\right)\right]$$

$$= \frac{\partial}{\partial x}\left(\frac{\partial A_z}{\partial y} - \frac{\partial A_y}{\partial z}\right) - \frac{\partial}{\partial y}\left(\frac{\partial A_z}{\partial x} - \frac{\partial A_x}{\partial z}\right) + \frac{\partial}{\partial z}\left(\frac{\partial A_y}{\partial x} - \frac{\partial A_x}{\partial y}\right)$$

$$= \frac{\partial^2 A_z}{\partial x\, \partial y} - \frac{\partial^2 A_y}{\partial x\, \partial z} - \frac{\partial^2 A_z}{\partial y\, \partial x} + \frac{\partial^2 A_x}{\partial y\, \partial z} + \frac{\partial^2 A_y}{\partial z\, \partial x} - \frac{\partial^2 A_x}{\partial z\, \partial y}$$

$$= 0$$

because $\dfrac{\partial^2 A_z}{\partial x\, \partial y} = \dfrac{\partial^2 A_z}{\partial y\, \partial x}$ and so on. ■

PRACTICE EXERCISE 3.10

For a scalar field V, show that $\nabla \times \nabla V = 0$; that is, the curl of the gradient of any scalar field vanishes.

ANSWER Proof.

3.8 LAPLACIAN OF A SCALAR

For practical reasons, it is expedient to introduce a single operator which is the composite of gradient and divergence operators. This operator is known as the *Laplacian*.

The Laplacian of a scalar field V, written as $\nabla^2 V$, is defined as the divergence of the gradient of V. Thus, in Cartesian coordinates,

$$\text{Laplacian } V = \nabla \cdot \nabla V = \nabla^2 V$$

$$= \left[\frac{\partial}{\partial x}\mathbf{a}_x + \frac{\partial}{\partial y}\mathbf{a}_y + \frac{\partial}{\partial z}\mathbf{a}_z\right] \cdot \left[\frac{\partial V}{\partial x}\mathbf{a}_x + \frac{\partial V}{\partial y}\mathbf{a}_y + \frac{\partial V}{\partial z}\mathbf{a}_z\right] \quad [3.59]$$

that is,

$$\boxed{\nabla^2 V = \frac{\partial^2 V}{\partial x^2} + \frac{\partial^2 V}{\partial y^2} + \frac{\partial^2 V}{\partial z^2}} \quad [3.60]$$

Notice that the Laplacian of a scalar field is another scalar field.

The Laplacian of V in other coordinate systems can be obtained from eq. (3.60) by transformation. In cylindrical coordinates,

$$\boxed{\nabla^2 V = \frac{1}{\rho}\frac{\partial}{\partial \rho}\left(\rho\frac{\partial V}{\partial \rho}\right) + \frac{1}{\rho^2}\frac{\partial^2 V}{\partial \phi^2} + \frac{\partial^2 V}{\partial z^2}} \quad [3.61]$$

and in spherical coordinates,

$$\boxed{\nabla^2 V = \frac{1}{r^2}\frac{\partial}{\partial r}\left(r^2\frac{\partial V}{\partial r}\right) + \frac{1}{r^2 \sin\theta}\frac{\partial}{\partial \theta}\left(\sin\theta\,\frac{\partial V}{\partial \theta}\right) + \frac{1}{r^2 \sin^2\theta}\frac{\partial^2 V}{\partial \phi^2}} \quad [3.62]$$

A scalar field V is said to be *harmonic* in a given region if its Laplacian vanishes in that region. In other words, if

$$\nabla^2 V = 0 \quad [3.63]$$

is satisfied in the region, the solution for V in eq. (3.63) is harmonic (it is of the form of sine or cosine). Equation (3.63) is called *Laplace's equation*. Solving this equation will be our major task in Chapter 6.

We have only considered the Laplacian of a scalar. Since the Laplacian operator ∇^2 is a scalar operator, it is also possible to define the Laplacian of a vector $\mathbf{A}$. In this context, $\nabla^2\mathbf{A}$ should not be viewed as the divergence of the gradient of $\mathbf{A}$, which makes no sense. Rather, $\nabla^2\mathbf{A}$ is defined as the gradient of the divergence of $\mathbf{A}$ minus the curl of the curl of $\mathbf{A}$. That is,

$$\boxed{\nabla^2\mathbf{A} = \nabla(\nabla \cdot \mathbf{A}) - \nabla \times \nabla \times \mathbf{A}} \quad [3.64]$$

This equation can be applied in finding $\nabla^2 \mathbf{A}$ in any coordinate system. In the Cartesian system (and only in that system), eq. (3.64) becomes

$$\nabla^2 \mathbf{A} = \nabla^2 A_x \mathbf{a}_x + \nabla^2 A_y \mathbf{a}_y + \nabla^2 A_z \mathbf{a}_z \tag{3.65}$$

EXAMPLE 3.11

Find the Laplacian of the scalar fields of Example 3.3; that is,

(a) $V = e^{-z} \sin 2x \cosh y$

(b) $U = \rho^2 z \cos 2\phi$

(c) $W = 10r \sin^2\theta \cos \phi$

SOLUTION

The Laplacian in the Cartesian system can be found by taking the first derivative and later the second derivative.

(a)
$$\begin{aligned}
\nabla^2 V &= \frac{\partial^2 V}{\partial x^2} + \frac{\partial^2 V}{\partial y^2} + \frac{\partial^2 V}{\partial z^2} \\
&= \frac{\partial}{\partial x}(2e^{-z}\cos 2x \cosh y) + \frac{\partial}{\partial y}(e^{-z}\cos 2x \sinh y) \\
&\quad + \frac{\partial}{\partial z}(-e^{-z}\sin 2x \cosh y) \\
&= -4e^{-z}\sin 2x \cosh y + e^{-z}\sin 2x \cosh y + e^{-z}\sin 2x \cosh y \\
&= -2e^{-z}\sin 2x \cosh y
\end{aligned}$$

(b)
$$\begin{aligned}
\nabla^2 U &= \frac{1}{\rho}\frac{\partial}{\partial \rho}\left(\rho \frac{\partial U}{\partial \rho}\right) + \frac{1}{\rho^2}\frac{\partial^2 U}{\partial \phi^2} + \frac{\partial^2 U}{\partial z^2} \\
&= \frac{1}{\rho}\frac{\partial}{\partial \rho}(2\rho^2 z \cos 2\phi) - \frac{1}{\rho^2} 4\rho^2 z \cos 2\phi + 0 \\
&= 4z \cos 2\phi - 4z \cos 2\phi \\
&= 0
\end{aligned}$$

(c)
$$\begin{aligned}
\nabla^2 W &= \frac{1}{r^2}\frac{\partial}{\partial r}\left(r^2 \frac{\partial W}{\partial r}\right) + \frac{1}{r^2 \sin\theta}\frac{\partial}{\partial \theta}\left(\sin\theta \frac{\partial W}{\partial \theta}\right) + \frac{1}{r^2\sin^2\theta}\frac{\partial^2 W}{\partial \phi^2} \\
&= \frac{1}{r^2}\frac{\partial}{\partial r}(10r^2\sin^2\theta \cos\phi) + \frac{1}{r^2\sin\theta}\frac{\partial}{\partial \theta}(10r \sin 2\theta \sin\theta \cos\phi) \\
&\quad - \frac{10r \sin^2\theta \cos\phi}{r^2\sin^2\theta}
\end{aligned}$$

$$= \frac{20 \sin^2\theta \cos\phi}{r} + \frac{20r \cos 2\theta \sin\theta \cos\phi}{r^2 \sin\theta}$$

$$+ \frac{10r \sin 2\theta \cos\theta \cos\phi}{r^2 \sin\theta} - \frac{10 \cos\phi}{r}$$

$$= \frac{10 \cos\phi}{r}(2 \sin^2\theta + 2 \cos 2\theta + 2 \cos^2\theta - 1)$$

$$= \frac{10 \cos\phi}{r}(1 + 2 \cos 2\theta)$$

■

PRACTICE EXERCISE 3.11

Determine the Laplacian of the scalar fields of Practice Exercise 3.3 , that is,

(a) $U = x^2y + xyz$

(b) $V = \rho z \sin\phi + z^2\cos^2\phi + \rho^2$

(c) $f = \cos\theta \sin\phi \ln r + r^2\phi$

ANSWER (a) $2y$, (b) $4 + 2\cos^2\phi - \frac{2z^2}{\rho^2}\cos 2\phi$, (c) $\frac{1}{r^2}\cos\theta \sin\phi\,(1 - 2\ln r \operatorname{cosec}^2\theta \ln r) + 6\phi$.

†3.9 CLASSIFICATION OF VECTOR FIELDS

A vector field is uniquely characterized by its divergence and curl. Thus, all vector fields can be classified in terms of their vanishing or nonvanishing divergence or curl as follows:

(a) $\nabla \cdot \mathbf{A} = 0,\ \nabla \times \mathbf{A} = 0$

(b) $\nabla \cdot \mathbf{A} \neq 0,\ \nabla \times \mathbf{A} = 0$

(c) $\nabla \cdot \mathbf{A} = 0,\ \nabla \times \mathbf{A} \neq 0$

(d) $\nabla \cdot \mathbf{A} \neq 0,\ \nabla \times \mathbf{A} \neq 0$

Figure 3.23 illustrates typical fields in these four categories.

For a given vector field **A**, if

$$\nabla \cdot \mathbf{A} = 0 \tag{3.66}$$

then **A** is said to be *solenoidal* or *divergenceless*. Such a field has neither source nor sink of flux. From the divergence theorem,

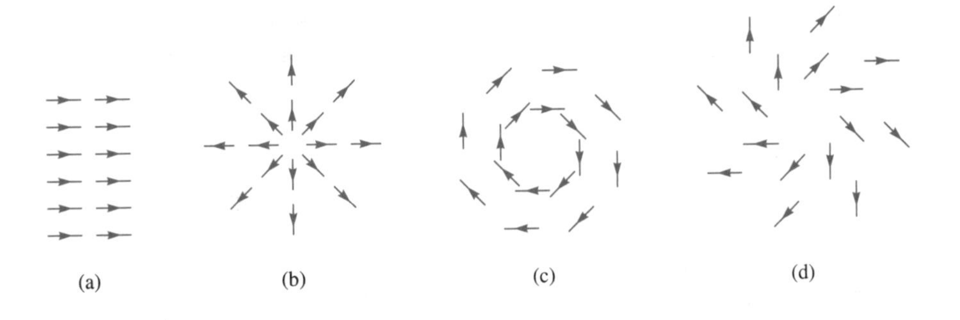

Figure 3.23 Typical fields with vanishing and nonvanishing divergence or curl.
(a) $\mathbf{A} = k\mathbf{a}_x$, $\nabla \cdot \mathbf{A} = 0$, $\nabla \times \mathbf{A} = 0$,
(b) $\mathbf{A} = k\mathbf{r}$, $\nabla \cdot \mathbf{A} = 3k$, $\nabla \times \mathbf{A} = 0$,
(c) $\mathbf{A} = \mathbf{k} \times \mathbf{r}$, $\nabla \cdot \mathbf{A} = 0$, $\nabla \times \mathbf{A} = 2\mathbf{k}$,
(d) $\mathbf{A} = \mathbf{k} \times \mathbf{r} + c\mathbf{r}$, $\nabla \cdot \mathbf{A} = 3c$, $\nabla \times \mathbf{A} = 2\mathbf{k}$.

$$\oint_S \mathbf{A} \cdot d\mathbf{S} = \int_v \nabla \cdot \mathbf{A}\, dv = 0 \tag{3.67}$$

Hence, flux lines of **A** entering any closed surface must also leave it. Examples of solenoidal fields are incompressible fluids, magnetic fields, and conduction current density under steady state conditions. In general, the field of curl **F** (for any **F**) is purely solenoidal because $\nabla \cdot (\nabla \times \mathbf{F}) = 0$, as shown in Example 3.10. Thus, a solenoidal field **A** can always be expressed in terms of another vector **F**; that is,

if $\quad \nabla \cdot \mathbf{A} = 0$

then $\quad \oint_S \mathbf{A} \cdot d\mathbf{S} = 0 \quad$ and $\quad \mathbf{F} = \nabla \times \mathbf{A} \qquad$ [3.68]

A vector field **A** is said to be *irrotational* (or *potential*) if

$$\nabla \times \mathbf{A} = 0 \tag{3.69}$$

That is, a *curl-free* vector is irrotational.[3] From Stokes's theorem

$$\int_S (\nabla \times \mathbf{A}) \cdot d\mathbf{S} = \oint_L \mathbf{A} \cdot d\mathbf{l} = 0 \tag{3.70}$$

Thus in an irrotational field **A**, the circulation of **A** around a closed path is identically zero. This implies that the line integral of **A** is independent of the chosen path. Therefore, an irrotational field is also known as a *conservative field*. Examples of

[3]In fact, curl was once known as *rotation*, and curl **A** is written as rot **A** in some textbooks. This is one reason to use the term *irrotational*.

irrotational fields include the electrostatic field and the gravitational field. In general, the field of gradient V (for any scalar V) is purely irrotational since (see Practice Exercise 3.10)

$$\nabla \times (\nabla V) = 0 \tag{3.71}$$

Thus, an irrotational field **A** can always be expressed in terms of a scalar field V; that is,

$$\text{if} \quad \nabla \times \mathbf{A} = 0$$
$$\text{then} \quad \oint_L \mathbf{A} \cdot d\mathbf{l} = 0 \quad \text{and} \quad \mathbf{A} = -\nabla V \tag{3.72}$$

For this reason, **A** may be called a *potential* field and V the scalar potential of **A**. The negative sign in eq. (3.72) has been inserted for physical reasons that will become evident in Chapter 4.

A vector **A** is uniquely prescribed within a region by its divergence and its curl. If we let

$$\nabla \cdot \mathbf{A} = \rho_v \tag{3.73a}$$

and

$$\nabla \times \mathbf{A} = \boldsymbol{\rho}_S \tag{3.73b}$$

ρ_v can be regarded as the source density of **A** and $\boldsymbol{\rho}_S$ its circulation density. Any vector **A** satisfying eq. (3.73) with both ρ_v and $\boldsymbol{\rho}_S$ vanishing at infinity can be written as the sum of two vectors: one irrotational (zero curl), the other solenoidal (zero divergence). This is called *Helmholtz's theorem*. Thus we may write

$$\mathbf{A} = -\nabla V + \nabla \times \mathbf{B} \tag{3.74}$$

If we let $\mathbf{A}_i = -\nabla V$ and $\mathbf{A}_s = \nabla \times \mathbf{B}$, it is evident from Example 3.10 and Practice Exercise 3.10 that $\nabla \times \mathbf{A}_i = 0$ and $\nabla \times \mathbf{A}_s = 0$, showing that $\mathbf{A}_i$ is irrotational and $\mathbf{A}_s$ is solenoidal. Finally, it is evident from eqs. (3.64) and (3.73) that any vector field has a Laplacian that satisfies

$$\nabla^2 \mathbf{A} = \nabla \rho_v - \nabla \times \boldsymbol{\rho}_S \tag{3.75}$$

EXAMPLE 3.12

Show that the vector field **A** is conservative if **A** possesses one of these two properties:

(a) The line integral of the tangential component of **A** along a path extending from a point P to a point Q is independent of the path.

(b) The line integral of the tangential component of **A** around any closed path is zero.

SOLUTION (a) If **A** is conservative, $\nabla \times \mathbf{A} = 0$, so there exists a potential V such that

$$\mathbf{A} = -\nabla V = -\left[\frac{\partial V}{\partial x}\mathbf{a}_x + \frac{\partial V}{\partial y}\mathbf{a}_y + \frac{\partial V}{\partial z}\mathbf{a}_z\right]$$

Hence,

$$\int_P^Q \mathbf{A} \cdot d\mathbf{l} = -\int_P^Q \left[\frac{\partial V}{\partial x}dx + \frac{\partial V}{\partial y}dy + \frac{\partial V}{\partial z}dz\right]$$

$$= -\int_P^Q \left[\frac{\partial V}{\partial x}\frac{dx}{ds} + \frac{\partial V}{\partial y}\frac{dy}{ds} + \frac{\partial V}{\partial z}\frac{dz}{ds}\right] ds$$

$$= -\int_P^Q \frac{dV}{ds}\, ds = -\int_P^Q dV$$

or

$$\int_P^Q \mathbf{A} \cdot d\mathbf{l} = V(P) - V(Q)$$

showing that the line integral depends only on the end points of the curve. Thus, for a conservative field, $\int_P^Q \mathbf{A} \cdot d\mathbf{l}$ is simply the difference in potential at the end points.

(b) If the path is closed, that is, if P and Q coincide, then

$$\oint \mathbf{A} \cdot d\mathbf{l} = V(P) - V(P) = 0$$

■

PRACTICE EXERCISE 3.12

Show that $\mathbf{B} = (y + z \cos xz)\mathbf{a}_x + x\mathbf{a}_y + x \cos xz\, \mathbf{a}_z$ is conservative, without computing any integrals.

ANSWER Proof.

SUMMARY

1. The differential displacements in the Cartesian, cylindrical, and spherical systems are respectively

$$d\mathbf{l} = dx\, \mathbf{a}_x + dy\, \mathbf{a}_y + dz\, \mathbf{a}_z$$

$$d\mathbf{l} = d\rho\, \mathbf{a}_\rho + \rho\, d\phi\, \mathbf{a}_\phi + dz\, \mathbf{a}_z$$

$$d\mathbf{l} = dr\, \mathbf{a}_r + r\, d\theta\, \mathbf{a}_\theta + r \sin\theta\, d\phi\, \mathbf{a}_\phi$$

Note that $d\mathbf{l}$ is always taken to be in the positive direction; the direction of the displacement is taken care of by the limits of integration.

2. The differential normal areas in the three systems are respectively

$$d\mathbf{S} = dy\, dz\, \mathbf{a}_x$$
$$dx\, dz\, \mathbf{a}_y$$
$$dx\, dy\, \mathbf{a}_z$$
$$d\mathbf{S} = \rho\, d\phi\, dz\, \mathbf{a}_\rho$$
$$d\rho\, dz\, \mathbf{a}_\phi$$
$$\rho\, d\rho\, d\phi\, \mathbf{a}_z$$
$$d\mathbf{S} = r^2 \sin\theta\, d\theta\, d\phi\, \mathbf{a}_r$$
$$r \sin\theta\, dr\, d\phi\, \mathbf{a}_\theta$$
$$r\, dr\, d\theta\, \mathbf{a}_\phi$$

Note that $d\mathbf{S}$ can be in the positive or negative direction depending on the surface under consideration.

3. The differential volumes in the three systems are

$$dv = dx\, dy\, dz$$
$$dv = \rho\, d\rho\, d\phi\, dz$$
$$dv = r^2 \sin\theta\, dr\, d\theta\, d\phi$$

4. The line integral of vector $\mathbf{A}$ along a path L is given by $\int_L \mathbf{A} \cdot d\mathbf{l}$. If the path is closed, the line integral becomes the circulation of $\mathbf{A}$ around L; that is, $\oint_L \mathbf{A} \cdot d\mathbf{l}$.
5. The flux or surface integral of a vector $\mathbf{A}$ across a surface S is defined as $\int_S \mathbf{A} \cdot d\mathbf{S}$. When the surface S is closed, the surface integral becomes the net outward flux of $\mathbf{A}$ across S; that is, $\oint \mathbf{A} \cdot d\mathbf{S}$.
6. The volume integral of a scalar ρ_v over a volume v is defined as $\int_v \rho_v\, dv$.
7. Vector differentiation is performed using the vector differential operator ∇. The gradient of a scalar field V is denoted by ∇V, the divergence of a vector field $\mathbf{A}$ by $\nabla \cdot \mathbf{A}$, the curl of $\mathbf{A}$ by $\nabla \times \mathbf{A}$, and the Laplacian of V by $\nabla^2 V$.
8. The divergence theorem, $\oint_S \mathbf{A} \cdot d\mathbf{S} = \int_v \nabla \cdot \mathbf{A}\, dv$, relates a surface integral over a closed surface to a volume integral.
9. Stokes's theorem, $\oint_L \mathbf{A} \cdot d\mathbf{l} = \int_S (\nabla \times \mathbf{A}) \cdot d\mathbf{S}$, relates a line integral over a closed path to a surface integral.
10. If Laplace's equation, $\nabla^2 V = 0$, is satisfied by a scalar field V in a given region, V is said to be harmonic in that region.
11. A vector field is solenoidal if $\nabla \cdot \mathbf{A} = 0$; it is irrotational or conservative if $\nabla \times \mathbf{A} = 0$.
12. A summary of the vector calculus operations in the three coordinate systems is provided on the inside back cover of the text.
13. The vector identities $\nabla \cdot \nabla \times \mathbf{A} = 0$ and $\nabla \times \nabla V = 0$ are very useful in EM. Other vector identities are in Appendix A.10.

REVIEW QUESTIONS

3.1 Consider the differential volume of Figure 3.24. Match the items in the left column with those in the right column.

(a) $d\mathbf{l}$ from A to B (i) $dy\, dz\, \mathbf{a}_x$

(b) $d\mathbf{l}$ from A to D (ii) $-dx\, dz\, \mathbf{a}_y$

(c) $d\mathbf{l}$ from A to E (iii) $dx\, dy\, \mathbf{a}_z$

(d) $d\mathbf{S}$ for face $ABCD$ (iv) $-dx\, dy\, \mathbf{a}_z$

(e) $d\mathbf{S}$ for face $AEHD$ (v) $dx\, \mathbf{a}_x$

(f) $d\mathbf{S}$ for face $DCGH$ (vi) $dy\, \mathbf{a}_y$

(g) $d\mathbf{S}$ for face $ABFE$ (vii) $dz\, \mathbf{a}_z$

3.2 For the differential volume in Figure 3.25, match the items in the left list with those in the right list.

(a) $d\mathbf{l}$ from E to A (i) $-\rho\, d\phi\, dz\, \mathbf{a}_\rho$

(b) $d\mathbf{l}$ from B to A (ii) $-d\rho\, dz\, \mathbf{a}_\phi$

(c) $d\mathbf{l}$ from D to A (iii) $-\rho\, d\rho\, d\phi\, \mathbf{a}_z$

(d) $d\mathbf{S}$ for face $ABCD$ (iv) $\rho\, d\rho\, d\phi\, \mathbf{a}_z$

(e) $d\mathbf{S}$ for face $AEHD$ (v) $d\rho\, \mathbf{a}_\rho$

(f) $d\mathbf{S}$ for face $ABFE$ (vi) $\rho\, d\phi\, \mathbf{a}_\phi$

(g) $d\mathbf{S}$ for face $DCGH$ (vii) $dz\, \mathbf{a}_z$

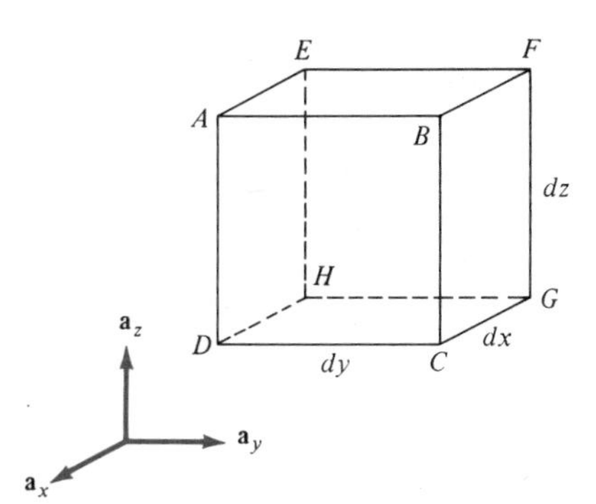

Figure 3.24 For Review Question 3.1.

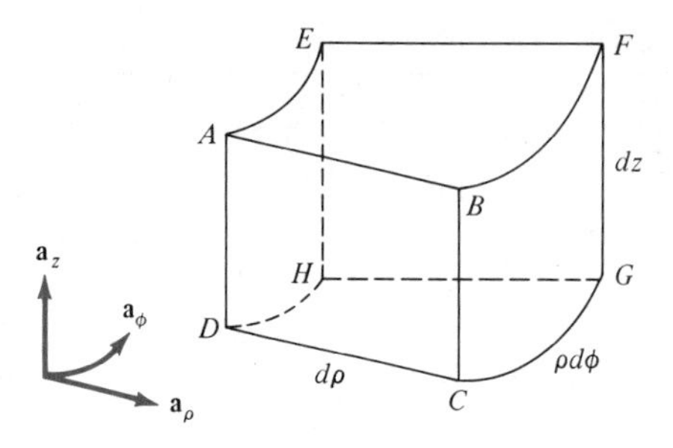

Figure 3.25 For Review Question 3.2.

3.3 A differential volume in spherical coordinates is shown in Figure 3.26. For the volume element, match the items in the left column with those in the right column.

(a) $d\mathbf{l}$ from A to D	(i) $-r^2 \sin\theta \, d\theta \, d\phi \, \mathbf{a}_r$
(b) $d\mathbf{l}$ from E to A	(ii) $-r \sin\theta \, dr \, d\phi \, \mathbf{a}_\theta$
(c) $d\mathbf{l}$ from A to B	(iii) $r \, dr \, d\theta \, \mathbf{a}_\phi$
(d) $d\mathbf{S}$ for face $EFGH$	(iv) $dr \, \mathbf{a}_r$
(e) $d\mathbf{S}$ for face $AEHD$	(v) $r \, d\theta \, \mathbf{a}_\theta$
(f) $d\mathbf{S}$ for face $ABFE$	(vi) $r \sin\theta \, d\phi \, \mathbf{a}_\phi$

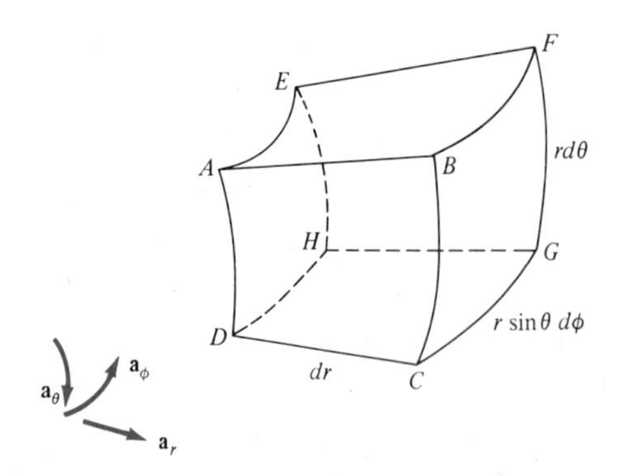

Figure 3.26 For Review Question 3.3 (and also for Practice Exercise 3.1).

3.4 If $\mathbf{r} = x\mathbf{a}_x + y\mathbf{a}_y + z\mathbf{a}_z$, the position vector of point (x, y, z), and $r = |\mathbf{r}|$, which of the following is incorrect?

(a) $\nabla r = \mathbf{r}/r$

(b) $\nabla \cdot \mathbf{r} = 1$

(c) $\nabla^2(\mathbf{r} \cdot \mathbf{r}) = 6$

(d) $\nabla \times \mathbf{r} = 0$

3.5 Assuming that V is a scalar field and $\mathbf{A}$ is a vector field, classify the following fields as (a) vector, (b) scalar, or (c) meaningless.

(i) ∇V

(ii) $\nabla(\nabla^2 V)$

(iii) $\nabla \cdot \mathbf{A}$

(iv) $\nabla(\nabla \times \mathbf{A})$

(v) $\nabla \times \mathbf{A}$

(vi) $\nabla \times (\nabla \cdot V)$

(vii) $\nabla \times (\nabla^2 V)$

3.6 The vector field

$$\mathbf{F} = 2\rho \cos\phi \, \mathbf{a}_\rho - \rho \sin \phi \, \mathbf{a}_\phi$$

does not circulate.

(a) Yes

(b) No

(c) It depends

3.7 Given field $\mathbf{A} = 3x^2yz \, \mathbf{a}_x + x^3z \, \mathbf{a}_y + (x^3y - 2z)\mathbf{a}_z$, it can be said that $\mathbf{A}$ is

(a) Harmonic

(b) Divergenceless

(c) Solenoidal

(d) Rotational

(e) Conservative

3.8 The surface current density $\mathbf{J}$ in a rectangular waveguide is plotted in Figure 3.27. It is evident from the figure that $\mathbf{J}$ diverges at the top wall of the guide whereas it is divergenceless at the side wall.

(a) True (b) False

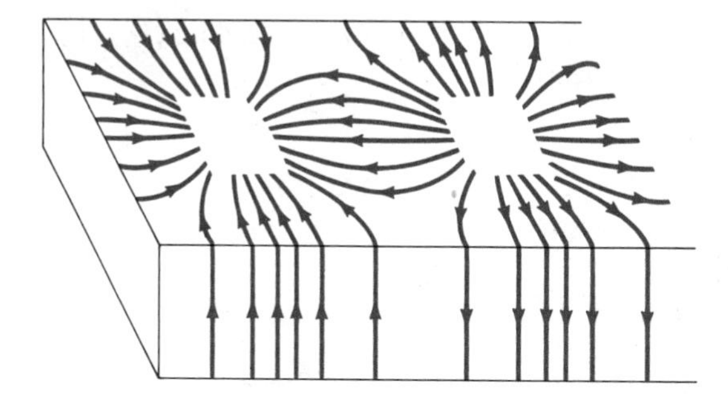

Figure 3.27 For Review Question 3.8.

3.9 Stokes's theorem is applicable only when a closed path exists and the vector field and its derivatives are continuous within the path.

(a) True

(b) False

(c) Not necessarily

3.10 If a vector field **Q** is solenoidal, which of these is true?

(a) $\oint_L \mathbf{Q} \cdot d\mathbf{l} = 0$

(b) $\oint_S \mathbf{Q} \cdot d\mathbf{S} = 0$

(c) $\nabla \times \mathbf{Q} = 0$

(d) $\nabla \times \mathbf{Q} \neq 0$

(e) $\nabla^2 \mathbf{Q} = 0$

Answers: 3.1a-(vi), b-(vii), c-(v), d-(i), e-(ii), f-(iv), g-(iii), 3.2a-(vi), b-(v), c-(vii), d-(ii), e-(i), f-(iv), g-(iii), 3.3a-(v), b-(vi), c-(iv), d-(iii), e-(i), f-(ii), 3.4b, 3.5(i)-a, (ii)-a, (iii)-b, (iv)-a, (v)-a, (vi)-c, (vii)-c, 3.6b, 3.7e, 3.8a, 3.9a, 3.10b.

PROBLEMS

3.1 By using the differential length dl, calculate the length of the curve described by

(a) $\rho = 10$, $30° \leq \phi \leq 90°$, $z =$ constant

(b) $r = 5$, $\pi/4 \leq \theta \leq 3\pi/4$, $\phi =$ constant

(c) The edge of the surface $r = 10$, $\pi/6 \leq \theta \leq \pi/2$, $\pi/4 \leq \phi \leq \pi/2$

3.2 By using the differential surface area dS, determine the area of the following surfaces:

(a) $z = 0,\ 0 \leq \rho \leq 10,\ 0 \leq \phi \leq \pi$

(b) $\rho = 50,\ 0 \leq z \leq 4,\ \pi/4 \leq \phi \leq \pi/2$

(c) $r = 5,\ \theta \leq \pi/2$ (hemispherical)

(d) $\theta = 30°,\ 0 \leq r \leq 10,\ 0 \leq \phi \leq 2\pi$ (conical)

(e) $r = 8.26,\ 30° \leq \theta \leq 60°,\ 0 \leq \phi < 360°$

3.3 Using the differential volume dv, calculate the volume described by

(a) $1 \leq \rho \leq 2,\ -1 \leq z \leq 1,\ \pi/6 \leq \phi \leq 3\pi/4$

(b) $3 \leq r \leq 5,\ \pi/3 \leq \theta \leq \pi/2,\ 5\pi/6 \leq \phi \leq \pi$

(c) $0 \leq x \leq 2,\ 0 \leq y \leq x,\ -4 \leq z \leq 4$

3.4 If $\mathbf{F} = 4y\ \mathbf{a}_x - 3x\ \mathbf{a}_y$, calculate the circulation of $\mathbf{F}$ around

(a) The circle $x^2 + y^2 = 4$

(b) The square of side 2 with center at the origin

3.5 If the integral $\int_A^B \mathbf{F} \cdot d\mathbf{l}$ is regarded as the work done in moving a particle from A to B, find the work done by the force field

$$\mathbf{F} = 2xy\ \mathbf{a}_x + (x^2 - z^2)\mathbf{a}_y - 3xz^2\mathbf{a}_z$$

on a particle that travels from $A(0, 0, 0)$ to $B(2, 1, 3)$ along

(a) The segment $(0, 0, 0) \rightarrow (0, 1, 0) \rightarrow (2, 1, 0) \rightarrow (2, 1, 3)$

(b) The straight line $(0, 0, 0)$ to $(2, 1, 3)$

3.6 If

$$\mathbf{H} = (x - y)\mathbf{a}_x + (x^2 + zy)\mathbf{a}_y + 5yz\ \mathbf{a}_z$$

evaluate $\int H \cdot d\mathbf{l}$ along the contour of Figure 3.28.

3.7 Evaluate the integral

$$\int_L \frac{-y\mathbf{a}_x + x\mathbf{a}_y}{\sqrt{x^2 + y^2}} \cdot d\mathbf{l}$$

where L is a circle of radius a in the xy-plane, with center at the origin, traced out counterclockwise. (*Hint:* Think before you write!)

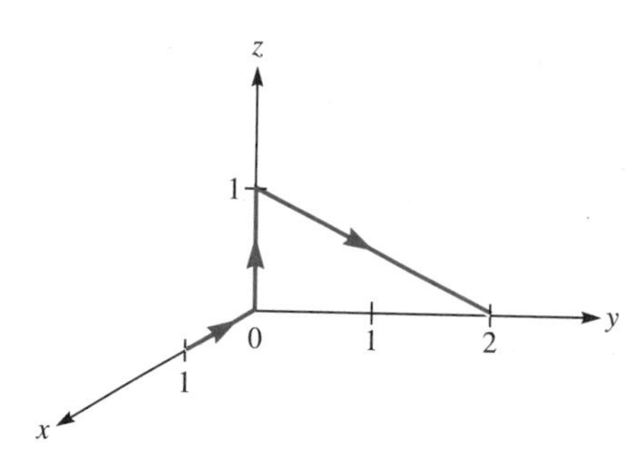

Figure 3.28 For Problem 3.6.

3.8 Given that $\rho_s = x^2 + xy$, calculate $\int_S \rho_s dS$ over the region $y \leq x^2$, $0 < x < 1$.

3.9 Find the flux of the field

$$\mathbf{A} = 4xz\mathbf{a}_x - y^2\mathbf{a}_y + yz\mathbf{a}_z$$

out of the unit cube described by $0 \leq x \leq 1$, $0 \leq y \leq 1$, $0 \leq z \leq 1$.

3.10 Given that $\mathbf{A} = r^2\mathbf{a}_r + \sin\theta\, \mathbf{a}_\phi$, evaluate $\oint \mathbf{A} \cdot d\mathbf{S}$ over a hemisphere of radius 5, $z \geq 0$.

3.11 Find the volume cut from the sphere radius $r = a$ by the cone $\theta = \alpha$. Calculate the volume when $\alpha = \pi/3$ and $\alpha = \pi/2$.

***3.12** Determine the volume of the region bounded by the surface $z = e^{-(x^2 + y^2)}$, the cylinder $x^2 + y^2 = 1$, and the plane $z = 0$.

3.13 If $g(\theta, \phi) = r^2$, evaluate $\int g(\theta, \phi)dv$ over a hemisphere of radius 1, $z \geq 0$ centered at the origin.

***3.14** (a) By substituting eqs. (3.17) and (3.18) into eq. (3.16) and using eq. (2.9), show that in cylindrical coordinates

$$\nabla = \mathbf{a}_\rho \frac{\partial}{\partial \rho} + \mathbf{a}_\phi \frac{1}{\rho}\frac{\partial}{\partial \phi} + \mathbf{a}_z \frac{\partial}{\partial z}$$

(b) By similarly subtituting eqs. (3.20) to (3.22) into eq. (3.16), show that in spherical coordinates

$$\nabla = \mathbf{a}_r \frac{\partial}{\partial r} + \mathbf{a}_\theta \frac{1}{r}\frac{\partial}{\partial \theta} + \mathbf{a}_\phi \frac{1}{r \sin\theta}\frac{\partial}{\partial \phi}$$

3.15 Determine the gradient of the following scalar fields:

(a) $U = 4xz^2 + 3yz$

(b) $V = e^{(2x + 3y)} \cos 5z$

(c) $W = 2\rho(z^2 + 1) \cos \phi$

(d) $T = 5\rho e^{-2z}\sin \phi$

(e) $H = r^2\cos \theta \cos \phi$

(f) $Q = \dfrac{\sin \theta \sin \phi}{r^3}$

3.16 A surface is described by $xy + yz + zx = 5$; find the unit normal vectors at $(-1, 3, 4)$.

3.17 The temperature in an auditorium is given by $T = x^2 + y^2 - z$. A mosquito located at (1, 1, 2) in the auditorium desires to fly in such a direction that it will get warm as soon as possible. In what direction must it fly?

3.18 Find the divergence and curl of the following vectors:

(a) $\mathbf{A} = e^{xy}\mathbf{a}_x + \sin xy\, \mathbf{a}_y + \cos^2 xz\, \mathbf{a}_z$

(b) $\mathbf{B} = \rho z^2\cos \phi\, \mathbf{a}_\rho + z \sin^2\phi\, \mathbf{a}_z$

(c) $\mathbf{C} = r \cos \theta\, \mathbf{a}_r - \dfrac{1}{r} \sin \theta\, \mathbf{a}_\theta + 2r^2\sin \theta\, \mathbf{a}_\phi$

3.19 Find the divergence and the curl of the following vector fields:

(a) $\mathbf{A} = (x + xyz)\mathbf{a}_x + (x^2 + y^2 - xz)\mathbf{a}_y + (xy + x^2)\mathbf{a}_z$

(b) $\mathbf{B} = \dfrac{z}{\rho}\mathbf{a}_\rho + z\rho \sin \phi\, \mathbf{a}_\phi + \cos^2 \phi\, \mathbf{a}_z$

(c) $\mathbf{C} = (r^2 + \sin \theta \cos \phi)\mathbf{a}_r + r \sin \theta \cos \phi\, \mathbf{a}_\theta + \left(\dfrac{1}{r} \tan \theta + \sin \phi\right)\mathbf{a}_\phi$

3.20 If $\mathbf{r} = x\mathbf{a}_x + y\mathbf{a}_y + z\mathbf{a}_z$, $r = |\mathbf{r}|$, n is an integer, $\mathbf{U}$ is any vector, and $\mathbf{A}$ is a constant vector, prove the following:

(a) grad $r = \dfrac{\mathbf{r}}{r}$

(b) grad $\left(\dfrac{1}{r}\right) = \dfrac{-\mathbf{r}}{r^3}$

(c) grad $(\mathbf{A} \cdot \mathbf{r}) = \mathbf{A}$

(d) $(\mathbf{U} \cdot \text{grad})\, \mathbf{r} = \mathbf{U}$

(e) div $\mathbf{r} = 3$

(f) div $r^n\mathbf{r} = (n + 3)r^n$

(g) curl $\mathbf{r} = 0$

(h) Laplacian $\dfrac{1}{r} = 0$

(i) grad $r^n = nr^{n-2}\mathbf{r}$

3.21 If n, $\mathbf{r}$, and $\mathbf{A}$ remain as defined in Problem 3.20, show that

(a) div $[\mathbf{r}\ \Phi(r)] = 3\ \Phi(r) + r\ \Phi'(r)$

(b) div $(\mathbf{r}/r^3) = 0$

(c) Lap $(r^n) = n(n + 1)r^{n-2}$

(d) Lap $(r^n\mathbf{r}) = n(n + 3)r^{n-2}\mathbf{r}$

3.22 The heat flow vector $\mathbf{H} = k\nabla T$, where T is the temperature and k is the thermal conductivity. Show that where

$$T = 50 \sin \frac{\pi x}{2} \cosh \frac{\pi y}{2}$$

then $\nabla \cdot \mathbf{H} = 0$.

3.23 If $\mathbf{r} = x\mathbf{a}_x + y\mathbf{a}_y + z\mathbf{a}_z$ and $\mathbf{T} = 2zy\mathbf{a}_x + xy^2\mathbf{a}_y + x^2yz\mathbf{a}_z$, determine

(a) $(\nabla \cdot \mathbf{r})\mathbf{T}$

(b) $(\mathbf{r} \cdot \nabla)\mathbf{T}$

(c) $\nabla \cdot \mathbf{r}(\mathbf{r} \cdot \mathbf{T})$

(d) $(\mathbf{r} \cdot \nabla)r^2$

3.24 Find the Laplacian of the following scalar fields:

(a) $V_1 = [x^2 + y^2 + z^2]^{-3/2}$

(b) $V_2 = xze^{-y} + x^2 \ln y$

(c) $V_3 = 10\rho^2 \sin 2\phi$

(d) $V_4 = \rho z(\cos \phi + \sin \phi)$

(e) $V_5 = 5r^2 \sin \theta \cos \phi$

(f) $V_6 = \dfrac{\sin\theta}{r^2}$

***3.25** Find the Laplacian of the vector fields in Practice Exercise 3.6. Calculate the Laplacian of the specified points.

3.26 If $V = x^2y^2z^2$ and $\mathbf{A} = x^2y\ \mathbf{a}_x + xz^3\mathbf{a}_y - y^2z^2\mathbf{a}_z$, find: (a) $\nabla^2 V$, (b) $\nabla^2\mathbf{A}$, (c) grad div $\mathbf{A}$, (d) curl curl $\mathbf{A}$.

***3.27** Given that $\mathbf{F} = x^2y\ \mathbf{a}_x - y\ \mathbf{a}_y$, find

(a) $\oint_L \mathbf{F} \cdot d\mathbf{l}$ where L is shown in Figure 3.29.

(b) $\int_S (\nabla \times \mathbf{F}) \cdot d\mathbf{S}$ where S is the area bounded by L.

(c) Is Stokes's theorem satisfied?

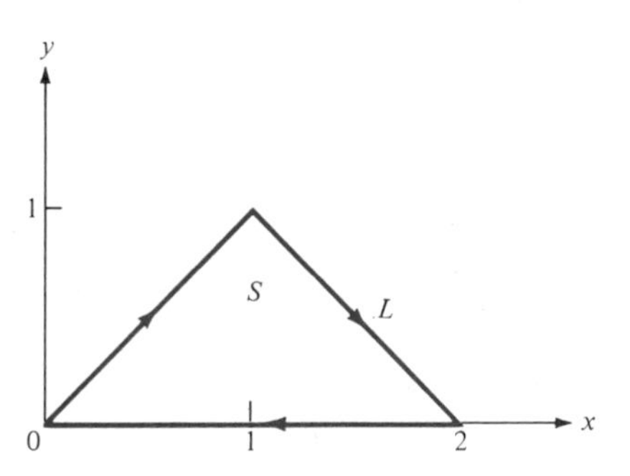

Figure 3.29 For Problem 3.27.

3.28 (a) Show that if $\mathbf{A} = -y\,\mathbf{a}_x + x\,\mathbf{a}_y$, the circulation of **A** around the boundary of a surface has a value that is twice the area of that surface.

(b) Show that the flux of the radius vector $\mathbf{r} = x\,\mathbf{a}_x + y\,\mathbf{a}_y + z\,\mathbf{a}_z = r\,\mathbf{a}_r$ through any closed surface S bounding a volume v is $3v$.

3.29 Let $\mathbf{F} = x^2\mathbf{a}_x + y\,\mathbf{a}_y + z\,\mathbf{a}_z$. Compute

(a) $\oint_S \mathbf{F} \cdot d\mathbf{S}$

(b) $\int_v \nabla \cdot \mathbf{F}\, dv$

where S is the surface of the cubical volume v bounded by planes $x = 0$, $x = 1$, $y = 0$, $y = 1$, $z = 0$, and $z = 1$.

3.30 Given that $\mathbf{E} = \dfrac{1}{r^4}\sin^2\phi\,\mathbf{a}_r$, evaluate

(a) $\oint_S \mathbf{E} \cdot d\mathbf{S}$

(b) $\int_v (\nabla \cdot \mathbf{E})dv$

over the region between the spherical surfaces $r = 2$ and $r = 4$.

3.31 The moment of inertia about the z-axis of a rigid body is proportional to

$$\int_v (x^2 + y^2)\,dx\,dy\,dz.$$

Express this as the flux of some vector field **A** through the surface of the body.

***3.32** Let $\mathbf{A} = \rho\sin\phi\,\mathbf{a}_\rho + \rho^2\mathbf{a}_\phi$. Evaluate $\oint_L \mathbf{A} \cdot d\mathbf{l}$ given that

(a) L is the contour of Figure 3.30(a)

(b) L is the contour of Figure 3.30(b)

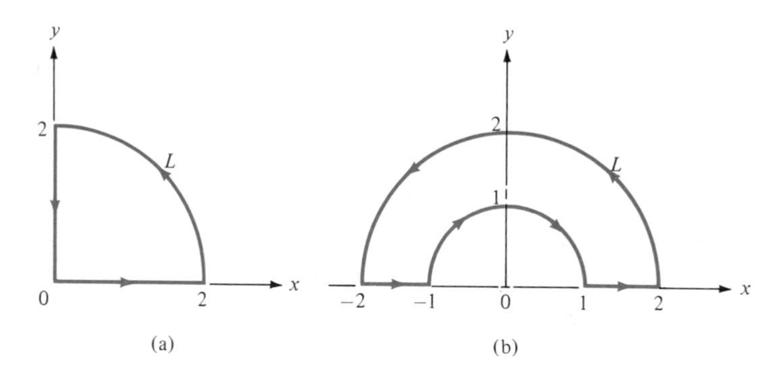

Figure 3.30 For Problem 3.32.

3.33 Calculate the total outward flux of vector

$$\mathbf{F} = \rho^2 \sin \phi \, \mathbf{a}_\rho + z \cos \phi \, \mathbf{a}_\phi + \rho z \mathbf{a}_z$$

through the hollow cylinder defined by $2 \leq \rho \leq 3$, $0 \leq z \leq 5$.

3.34 Find the flux of the curl of field

$$\mathbf{T} = \frac{1}{r^2} \cos \theta \, \mathbf{a}_r + r \sin \theta \cos \phi \, \mathbf{a}_\theta + \cos \theta \, \mathbf{a}_\phi$$

through the hemisphere $r = 4$, $z \leq 0$.

****3.35** A vector field is given by

$$\mathbf{Q} = \frac{\sqrt{x^2 + y^2 + z^2}}{\sqrt{x^2 + y^2}} [(x - y)\mathbf{a}_x + (x + y)\mathbf{a}_y$$

Evaluate the following integrals:

(a) $\int_L \mathbf{Q} \cdot d\mathbf{l}$ where L is the circular edge of the volume in the form of an ice-cream cone shown in Figure 3.31

(b) $\int_{S_1} (\nabla \times \mathbf{Q}) \cdot d\mathbf{S}$ where S_1 is the top surface of the volume

(c) $\int_{S_2} (\nabla \times \mathbf{Q}) \cdot d\mathbf{S}$ where S_2 is the slanting surface of the volume

(d) $\int_{S_1} \mathbf{Q} \cdot d\mathbf{S}$

(e) $\int_{S_2} \mathbf{Q} \cdot d\mathbf{S}$

(f) $\int_v \nabla \cdot \mathbf{Q} \, dv$

How do your results in parts (a) to (f) compare?

....

**Double asterisks indicate problems of highest difficulty.

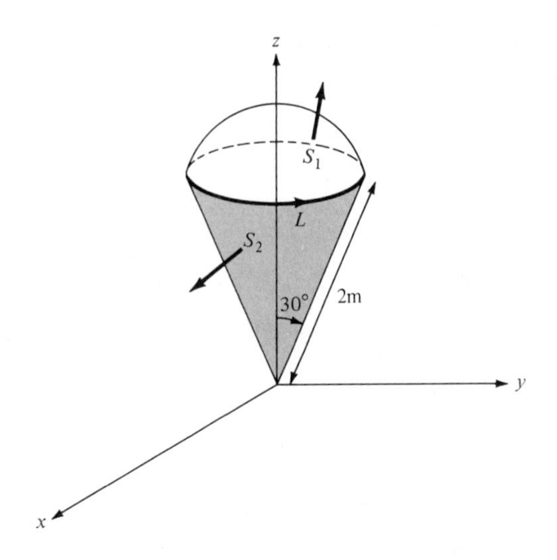

Figure 3.31 Volume in form of ice-cream cone for Problem 3.35.

***3.36** A rigid body spins about a fixed axis through its center with angular velocity $\boldsymbol{\omega}$. If $\mathbf{u}$ is the velocity at any point in the body, show that $\boldsymbol{\omega} = 1/2\ \nabla \times \mathbf{u}$.

3.37 If V is a scalar and $\mathbf{A}$ is a vector.

(a) Prove that $\nabla \cdot (V\mathbf{A}) = V\,\nabla \cdot \mathbf{A} + \mathbf{A} + \nabla V$.

(b) Show in Cartesian coordinates that curl curl $\mathbf{A} = \nabla(\nabla \cdot \mathbf{A}) - \nabla^2\mathbf{A}$.

***3.38** Verify the vector identity

$$\int \nabla \times \mathbf{A}\, dv = -\oint \mathbf{A} \times d\mathbf{S}$$

for $\mathbf{A} = 5x^2y\mathbf{a}_x + 3xy^2\mathbf{a}_y$ and the volume defined by $0 < x < 2$, $-1 < y < 1$, and $-5 < z < 5$.

3.39 Given the vector field

$$\mathbf{R} = (2x^2y + yz)\mathbf{a}_x + (xy^2 - xz^3)\mathbf{a}_y + (cxyz - 2x^2y^2)\mathbf{a}_z$$

determine the value of c for $\mathbf{R}$ to be solenoidal.

3.40 If the vector field

$$\mathbf{T} = (\alpha xy + \beta z^3)\mathbf{a}_x + (3x^2 - \gamma z)\mathbf{a}_y + (3xz^2 - y)a_z$$

is irrotational, determine α, β, and γ. Find $\nabla \cdot \mathbf{T}$ at $(2, -1, 0)$.

***3.41** (a) Show that the following line integrals are independent of the path of integration:

(i) $\int (x - yz)dx + (y^2 - zx)dy + (z^2 - xy)dz$

(ii) $\int (z^2 dx + 2y\, dy + 2xz\, dz)$

(b) Confirm this by evaluating the integrals along

(i) The straight line from (0, 0, 0) to (1, 1, 1)

(ii) The line segments from (0, 0, 0) to (1, 0, 0), from (1, 0, 0) to (1, 1, 0), from (1, 1, 0) to (1, 1, 1)

PART 2

Electrostatics

CHAPTER

4 Electrostatic Fields

The man who does things makes many mistakes, but he never makes the biggest mistake of all—doing nothing.

—BENJAMIN FRANKLIN

4.1 INTRODUCTION

Having mastered some essential mathematical tools needed for this course, we are now prepared to study the basic concepts of EM. We shall begin with those fundamental concepts that are applicable to static (or time-invariant) electric fields in free space (or vacuum). An electrostatic field is produced by a static charge distribution. A typical example of such a field is found in a cathode-ray tube.

Before we commence our study of electrostatics, it might be helpful to examine briefly the importance of such a study. Electrostatics is a fascinating subject that has grown up in diverse areas of application. Electric power transmission, X-ray machines, and lightning protection are associated with strong electric fields and will require a knowledge of electrostatics to understand and design suitable equipment. The devices used in solid-state electronics are based on electrostatics. These include resistors, capacitors, and active devices such as bipolar and field effect transistors, which are based on control of electron motion by electrostatic fields. Almost all computer peripheral devices, with the exception of magnetic memory, are based on electrostatic fields. Touch pads, capacitance keyboards, cathode-ray tubes, liquid crystal displays, and electrostatic printers are typical examples. In medical work, diagnosis is often carried out with the aid of electrostatics, as incorporated in electrocardiograms, electroencephalograms, and other recordings of organs with electrical activity including eyes, ears, and stomachs. In industry, electrostatics is applied in a variety of forms such as paint spraying, electrodeposition, electrochemical machining, and separation of fine particles. Electrostatics is used in agriculture to sort seeds, direct sprays to plants, measure the moisture content of crops, spin cotton, and speed baking of bread and smoking of meat.[1,2]

We begin our study of electrostatics by investigating the two fundamental laws governing electrostatic fields: (1) Coulomb's law, and (2) Gauss's law. Both of these laws are based on experimental studies and they are interdependent. Although Coulomb's law is applicable in finding the electric field due to any charge configuration, it is eaiser to use Gauss's law when charge distribution is symmetrical. Based on Coulomb's law, the concept of electric field intensity will be introduced and applied to cases involving point, line, surface, and volume charges. Special problems that can be solved with much effort using Coulomb's law will be solved with ease by applying Gauss's law. Throughout our discussion in this chapter, we will assume that the electric field is in a vacuum or free space. Electric field in material space will be covered in the next chapter.

4.2 COULOMB'S LAW AND FIELD INTENSITY

Coulomb's law is an experimental law formulated in 1785 by the French colonel, Charles Augustin de Coulomb. It deals with the force a point charge exerts on another point charge. By a *point charge* we mean a charge that is located on a body whose dimensions are much smaller than other relevant dimensions. For example, a collection of electric charges on a pinhead may be regarded as a point charge. Charges are generally measured in coulombs (C). One coulomb is approximately equivalent to 6×10^{18} electrons; it is a very large unit of charge because one electron charge $e = -1.6019 \times 10^{-19}$ C.

Coulomb's law states that the force F between two point charges Q_1 and Q_2 is

1. Along the line joining them
2. Directly proportional to the product Q_1Q_2 of the charges
3. Inversely proportional to the square of the distance R between them.[3]

Expressed mathematically,

$$F = \frac{k\, Q_1 Q_2}{R^2} \tag{4.1}$$

....

[1]For various applications of electrostatics, see J. M. Crowley, *Fundamentals of Applied Electrostatics*. New York: John Wiley & Sons, 1986; A. D. Moore, ed., *Electrostatics and Its Applications*. New York: John Wiley & Sons, 1973; and C. E. Jowett, *Electrostatics in the Electronics Environment*. New York: John Wiley & Sons, 1976.

[2]An interesting story on the magic of electrostatics is found in B. Bolton, *Electromagnetism and Its Applications*. London: Van Nostrand, 1980, p. 2.

[3]Further details of experimental verification of Coulomb's law can be found in W. F. Magie, *A Source Book in Physics*. Cambridge: Harvard Univ. Press, 1963, pp. 408–420.

where k is the proportionality constant. In SI units, charges Q_1 and Q_2 are in coulombs (C), the distance R is in meters (m), and the force F is in newtons (N) so that $k = 1/4\pi\varepsilon_o$. The constant ε_o is known as the *permittivity of free space* (in farads per meter) and has the value

$$\varepsilon_o = 8.854 \times 10^{-12} \simeq \frac{10^{-9}}{36\pi} \text{ F/m}$$
$$\text{or } k = \frac{1}{4\pi\varepsilon_o} \simeq 9 \times 10^9 \text{ m/F} \tag{4.2}$$

Thus eq. (4.1) becomes

$$F = \frac{Q_1Q_2}{4\pi\varepsilon_o R^2} \tag{4.3}$$

If point charges Q_1 and Q_2 are located at points having position vectors $\mathbf{r}_1$ and $\mathbf{r}_2$, then the vector force $\mathbf{F}_{12}$ on Q_2 due to Q_1, shown in Figure 4.1, is given by

$$\mathbf{F}_{12} = \frac{Q_1Q_2}{4\pi\varepsilon_o R^2}\mathbf{a}_{R_{12}} \tag{4.4}$$

where

$$\mathbf{R}_{12} = \mathbf{r}_2 - \mathbf{r}_1 \tag{4.5a}$$

$$R = |\mathbf{R}_{12}| \tag{4.5b}$$

$$\mathbf{a}_{R_{12}} = \frac{\mathbf{R}_{12}}{R} \tag{4.5c}$$

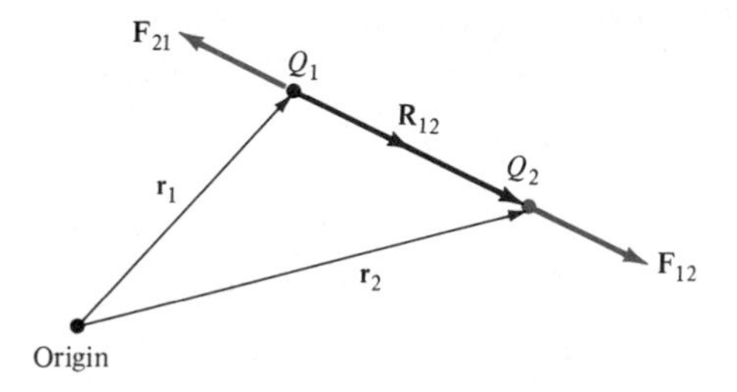

Figure 4.1 Coulomb vector force on point changes Q_1 and Q_2.

By substituting eq. (4.5) into eq. (4.4), we may write eq. (4.4) as

$$\mathbf{F}_{12} = \frac{Q_1 Q_2}{4\pi\varepsilon_o R^3}\,\mathbf{R}_{12} \qquad [4.6a]$$

or

$$\mathbf{F}_{12} = \frac{Q_1 Q_2\,(\mathbf{r}_2 - \mathbf{r}_1)}{4\pi\varepsilon_o|\mathbf{r}_2 - \mathbf{r}_1|^3} \qquad [4.6b]$$

It is worthwhile to note that

1. As shown in Figure 4.1, the force $\mathbf{F}_{21}$ on Q_1 due to Q_2 is given by

$$\mathbf{F}_{21} = |\mathbf{F}_{12}|\mathbf{a}_{R_{21}} = |\mathbf{F}_{12}|\,(-\mathbf{a}_{R_{12}})$$

or

$$\mathbf{F}_{21} = -\mathbf{F}_{12} \qquad [4.7]$$

since

$$\mathbf{a}_{R_{21}} = -\mathbf{a}_{R_{12}}$$

2. Like charges (charges of the same sign) repel each other while unlike charges attract. This is illustrated in Figure 4.2.
3. The distance R between the charged bodies Q_1 and Q_2 must be large compared with the linear dimensions of the bodies; that is, Q_1 and Q_2 must be point charges.
4. Q_1 and Q_2 must be static (at rest).
5. The signs of Q_1 and Q_2 must be taken into account in eq. (4.4).

If we have more than two point charges, we can use the *principle of superposition* to determine the force on a particular charge. The principle states that if there are N charges $Q_1, Q_2, \ldots, Q_N$ located respectively at points with position vectors $\mathbf{r}_1$,

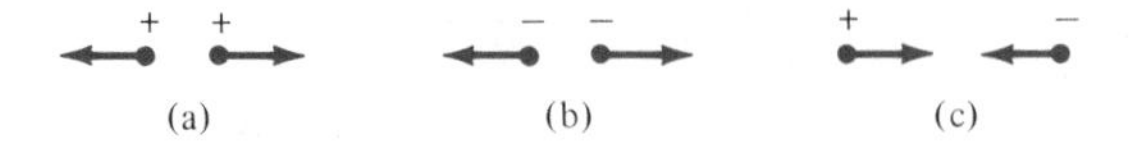

Figure 4.2 (**a**), (**b**) Like charges repel; (**c**) unlike charges attract.

$\mathbf{r}_2, \ldots, \mathbf{r}_N$, the resultant force $\mathbf{F}$ on a charge Q located at point $\mathbf{r}$ is the vector sum of the forces exerted on Q by each of the charges $Q_1, Q_2, \ldots, Q_N$. Hence:

$$\mathbf{F} = \frac{QQ_1(\mathbf{r} - \mathbf{r}_1)}{4\pi\varepsilon_o|\mathbf{r} - \mathbf{r}_1|^3} + \frac{QQ_2(\mathbf{r} - \mathbf{r}_2)}{4\pi\varepsilon_o|\mathbf{r} - \mathbf{r}_2|^3} + \cdots + \frac{QQ_N(\mathbf{r} - \mathbf{r}_n)}{4\pi\varepsilon_o|\mathbf{r} - \mathbf{r}_N|^3}$$

or

$$\boxed{\mathbf{F} = \frac{Q}{4\pi\varepsilon_o}\sum_{k=1}^{N}\frac{Q_k(\mathbf{r} - \mathbf{r}_k)}{|\mathbf{r} - \mathbf{r}_k|^3}} \qquad [4.8]$$

We now define the *electric field intensity* (or *electric field strength*) as the force per unit charge when placed in the electric field. Thus

$$\mathbf{E} = \lim_{Q \to 0} \frac{\mathbf{F}}{Q} \qquad [4.9]$$

or simply

$$\boxed{\mathbf{E} = \frac{\mathbf{F}}{Q}} \qquad [4.10]$$

The electric field intensity $\mathbf{E}$ is obviously in the direction of the force $\mathbf{F}$ and is measured in newtons/coulomb or volts/meter. The electric field intensity at point $\mathbf{r}$ due to a point charge located at $\mathbf{r}'$ is readily obtained from eqs. (4.6) and (4.10) as

$$\boxed{\mathbf{E} = \frac{Q}{4\pi\varepsilon_o R^2}\mathbf{a}_R = \frac{Q\,(\mathbf{r} - \mathbf{r}')}{4\pi\varepsilon_o|\mathbf{r} - \mathbf{r}'|^3}} \qquad [4.11]$$

For N point charges $Q_1, Q_2, \ldots, Q_N$ located at $\mathbf{r}_1, \mathbf{r}_2, \ldots, \mathbf{r}_N$, the electric field intensity at point $\mathbf{r}$ is obtained from eqs. (4.8) and (4.10) as

$$\mathbf{E} = \frac{Q_1(\mathbf{r} - \mathbf{r}_1)}{4\pi\varepsilon_o|\mathbf{r} - \mathbf{r}_1|^3} + \frac{Q_2(\mathbf{r} - \mathbf{r}_2)}{4\pi\varepsilon_o|\mathbf{r} - \mathbf{r}_2|^3} + \cdots + \frac{Q_N(\mathbf{r} - \mathbf{r}_N)}{4\pi\varepsilon_o|\mathbf{r} - \mathbf{r}_N|^3}$$

or

$$\mathbf{E} = \frac{1}{4\pi\varepsilon_o}\sum_{k=1}^{N}\frac{Q_k(\mathbf{r} - \mathbf{r}_k)}{|\mathbf{r} - \mathbf{r}_k|^3} \qquad [4.12]$$

EXAMPLE 4.1

Point charges 1 mC and −2 mC are located at (3, 2, −1) and (−1, −1, 4) respectively. Calculate the electric force on a 10 nC charge located at (0, 3, 1) and the electric field intensity at that point.

SOLUTION

$$\mathbf{F} = \sum_{k=1,2} \frac{QQ_k}{4\pi\varepsilon_o R^2}\,\mathbf{a}_R = \sum_{k=1,2} \frac{QQ_k(\mathbf{r} - \mathbf{r}_k)}{4\pi\varepsilon_o|\mathbf{r} - \mathbf{r}_k|^3}$$

$$= \frac{Q}{4\pi\varepsilon_o}\left\{\frac{10^{-3}[(0, 3, 1) - (3, 2, -1)]}{|(0, 3, 1) - (3, 2, -1)|^3} - \frac{2.10^{-3}[(0, 3, 1) - (-1, -1, 4)]}{|(0, 3, 1) - (-1, -1, 4)|^3}\right\}$$

$$= \frac{10^{-3} \cdot 10 \cdot 10^{-9}}{4\pi \cdot \dfrac{10^{-9}}{36\pi}}\left[\frac{(-3, 1, 2)}{(9 + 1 + 4)^{3/2}} - \frac{2(1, 4, -3)}{(1 + 16 + 9)^{3/2}}\right]$$

$$= 9 \cdot 10^{-2}\left[\frac{(-3, 1, 2)}{14\sqrt{14}} + \frac{(-2, -8, 6)}{26\sqrt{26}}\right]$$

$$\mathbf{F} = -6.507\mathbf{a}_x - 3.817\mathbf{a}_y + 7.506\mathbf{a}_z \text{ mN}$$

At that point,

$$\mathbf{E} = \frac{\mathbf{F}}{Q} = \sum_{k=1,2} \frac{Q_k(\mathbf{r} - \mathbf{r}_k)}{4\pi\varepsilon_o|\mathbf{r} - \mathbf{r}_k|^3}$$

$$= (-6.507, -3.817, 7.506) \cdot \frac{10^{-3}}{10 \cdot 10^{-9}}$$

$$\mathbf{E} = -650.7\mathbf{a}_x - 381.7\mathbf{a}_y + 750.6\mathbf{a}_z \text{ kV/m}$$

■

PRACTICE EXERCISE 4.1

Point charges 5 nC and −2 nC are located at (2, 0, 4) and (−3, 0, 5) respectively.

(a) Determine the force on a 1 nC point charge located at (1, −3, 7).
(b) Find the electric field **E** at (1, −3, 7).

ANSWER (a) $-1.004\mathbf{a}_x - 1.284\mathbf{a}_y + 1.4\mathbf{a}_z$ nN, (b) $-1.004\mathbf{a}_x - 1.284\mathbf{a}_y + 1.4\mathbf{a}_z$ V/m.

EXAMPLE 4.2

Two point charges of equal mass m, charge Q are suspended at a common point by two threads of negligible mass and length ℓ. Show that at equilibrium the inclination angle α of each thread to the vertical is given by

$$Q^2 = 16\pi\, \varepsilon_o mg\ell^2 \sin^2 \alpha \tan \alpha$$

If α is very small, show that

$$\alpha = \sqrt[3]{\frac{Q^2}{16\pi\varepsilon_o mg\ell^2}}$$

SOLUTION Consider the system of charges as shown in Figure 4.3 where F_e is the electric or coulomb force, T is the tension in each thread, and mg is the weight of each charge. At A or B

$$T \sin \alpha = F_e$$

$$T \cos \alpha = mg$$

Hence,

$$\frac{\sin \alpha}{\cos \alpha} = \frac{F_e}{mg} = \frac{1}{mg} \cdot \frac{Q^2}{4\pi\varepsilon_o r^2}$$

But

$$r = 2\ell \sin \alpha$$

Hence,

$$Q^2 \cos \alpha = 16\pi\varepsilon_o mg\ell^2 \sin^3 \alpha$$

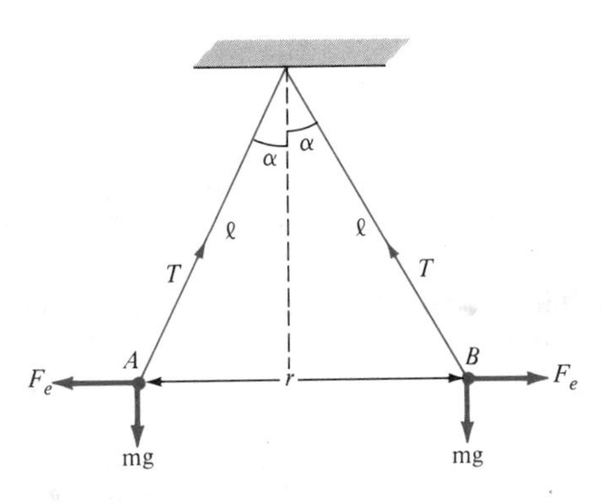

Figure 4.3 Suspended charged particles; for Example 4.2.

or

$$Q^2 = 16\pi\varepsilon_o mg\ell^2 \sin^2\alpha \tan\alpha$$

as required. When α is very small

$$\tan\alpha \simeq \alpha \simeq \sin\alpha$$

and so

$$Q^2 = 16\pi\varepsilon_o mg\ell^2\alpha^3$$

or

$$\alpha = \sqrt[3]{\frac{Q^2}{16\pi\varepsilon_o mg\ell^2}}$$

■

PRACTICE EXERCISE 4.2

Three identical small spheres of mass m are suspended by threads of negligible masses and equal length ℓ from a common point. A charge Q is divided equally between the spheres and they come to equilibrium at the corners of a horizontal equilateral triangle whose sides are d. Show that

$$Q^2 = 12\pi\varepsilon_o mgd^3\left[\ell^2 - \frac{d^2}{3}\right]^{-1/2}$$

where g = acceleration due to gravity.

ANSWER Proof.

EXAMPLE 4.3

A practical application of electrostatics is in electrostatic separation of solids. For example, Florida phosphate ore, consisting of small particles of quartz and phosphate rock, can be separated into its components by applying a uniform electric field as in Figure 4.4. Assuming zero initial velocity and displacement, determine the separation between the particles after falling 80 cm. Take E = 500 kV/m and Q/m = 9 μC/kg for both positively and negatively charged particles.

SOLUTION Ignoring the coulombic force between particles, the electrostatic force is acting horizontally while the gravitational force (weight) is acting vertically on the particles. Thus,

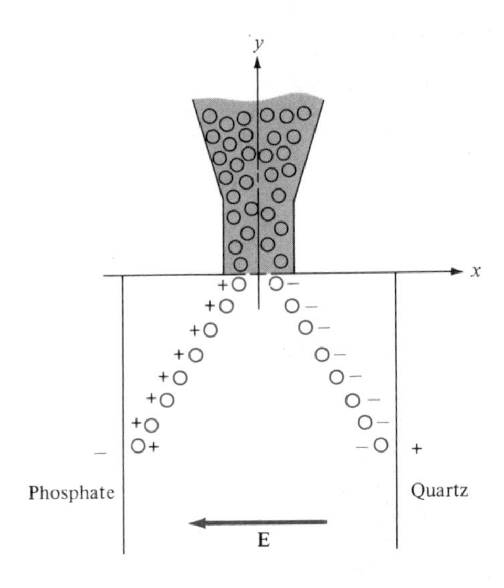

Figure 4.4 Electrostatic separation of solids; for Example 4.3.

$$Q\mathbf{E} = m \frac{d^2x}{dt^2} \mathbf{a}_x$$

or

$$\frac{d^2x}{dt^2} = \frac{Q}{m} E$$

Integrating twice gives

$$x = \frac{Q}{2m} Et^2 + c_1t + c_2 \qquad [4.3.1]$$

where c_1 and c_2 are integration constants. Similarly,

$$-mg = m \frac{d^2y}{dt^2}$$

or

$$\frac{d^2y}{dt^2} = -g$$

Integrating twice, we get

$$y = -1/2gt^2 + c_3t + c_4 \qquad [4.3.2]$$

Since the initial displacement is zero,

$$x(t = 0) = 0 \rightarrow c_2 = 0$$

$$y(t = 0) = 0 \rightarrow c_4 = 0$$

Also, due to zero initial velocity,

$$\left.\frac{dx}{dt}\right|_{t=0} = 0 \rightarrow c_1 = 0$$

$$\left.\frac{dy}{dt}\right|_{t=0} = 0 \rightarrow c_3 = 0$$

Thus

$$x = \frac{QE}{2m} t^2 \qquad y = -\frac{1}{2} gt^2$$

When $y = -80$ cm $= -0.8$ m

$$t^2 = \frac{0.8 \times 2}{9.8} = 0.1633$$

and

$$x = 1/2 \times 9 \times 10^{-6} \times 5 \times 10^5 \times 0.1633 = 0.3673\ m$$

The separation between the particles is $2x = 73.47$ cm. ■

PRACTICE EXERCISE 4.3

An ion rocket emits positive cesium ions from a wedge-shape electrode into the region described by $x > |y|$. The electric field is $\mathbf{E} = -400\ \mathbf{a}_x + 200\ \mathbf{a}_y$ kV/m. The ions have single electronic charges $e = -1.6019 \times 10^{-19}$ C and mass $m = 2.22 \times 10^{-25}$ kg and travel in a vacuum with zero initial velocity. If the emission is confined to -40 cm $< y < 40$ cm, find the largest value of x which can be reached.

ANSWER 0.8 m.

4.3 ELECTRIC FIELDS DUE TO CONTINUOUS CHARGE DISTRIBUTIONS

So far we have only considered forces and electric fields due to point charges, which are essentially charges occupying very small physical space. It is also possible to have continuous charge distribution along a line, on a surface, or in a volume as illustrated in Figure 4.5.

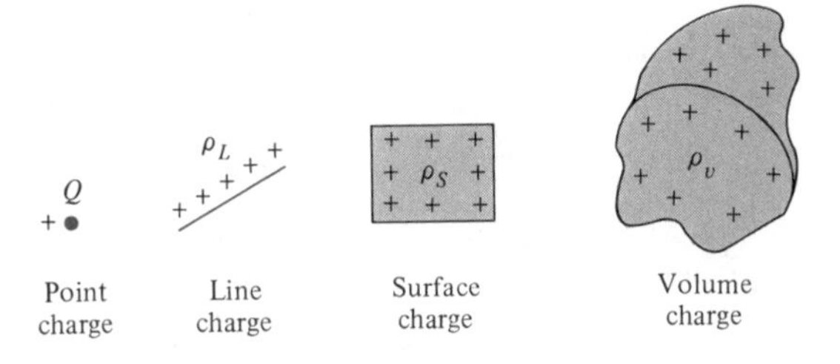

Figure 4.5 Various charge distributions.

It is customary to denote the line charge density, surface charge density, and volume charge density by ρ_L (in C/m), ρ_S (in C/m^2), and ρ_v (in C/m^3), respectively. These must not be confused with ρ (without subscript) used for radial distance in cylindrical coordinates. The electric field intensity due to each of the charge distributions ρ_L, ρ_S, and ρ_v may be regarded as the summation of the field contributed by the numerous point charges making up the charge distribution. Thus by replacing Q in eq. (4.11) with charge element $\rho_L\, dl$, $\rho_S\, dS$, or $\rho_v\, dv$ and integrating, we get

$$\mathbf{E} = \int \frac{\rho_L\, dl}{4\pi\varepsilon_o R^2}\, \mathbf{a}_R \qquad \text{(line charge)} \tag{4.13}$$

$$\mathbf{E} = \int \frac{\rho_S\, dS}{4\pi\varepsilon_o R^2}\, \mathbf{a}_R \qquad \text{(surface charge)} \tag{4.14}$$

$$\mathbf{E} = \int \frac{\rho_v\, dv}{4\pi\varepsilon_o R^2}\, \mathbf{a}_R \qquad \text{(volume charge)} \tag{4.15}$$

It should be noted that R^2 and $\mathbf{a}_R$ vary as the integrals in eqs. (4.13) to (4.15) are evaluated. We shall now apply these formulas to some specific charge distributions.

A. A Line Charge

Consider a line charge with uniform charge density ρ_L extending from A to B along the z-axis as shown in Figure 4.6. The charge element dQ associated with element $dl = dz$ of the line is

$$dQ = \rho_L\, dl = \rho_L\, dz \tag{4.16}$$

and hence the total charge Q is

$$Q = \int_{z_A}^{z_B} \rho_L\, dz \tag{4.17}$$

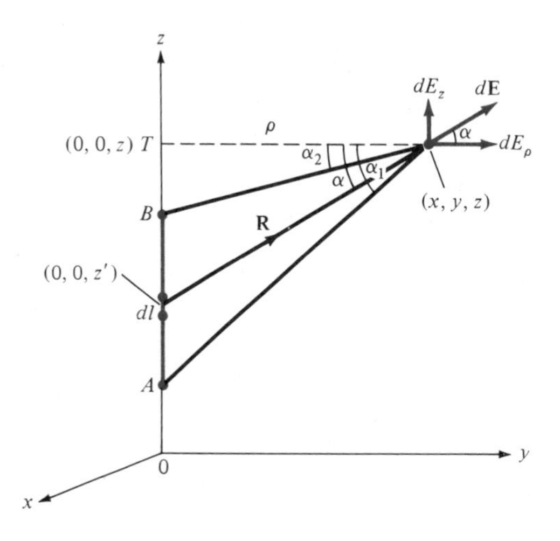

Figure 4.6 Evaluation of the **E** field due to a line charge.

The electric field intensity **E** at an arbitrary point $P(x, y, z)$ can be found using eq. (4.13). It is important that we learn to derive and substitute each term in eqs. (4.13) to (4.14) for a given charge distribution. It is customary to denote the field point[4] by (x, y, z) and the source point by (x', y', z'). Thus from Figure 4.6,

$$dl = dz'$$

$$\mathbf{R} = (x, y, z) - (0, 0, z') = x\mathbf{a}_x + y\mathbf{a}_y + (z - z')\mathbf{a}_z$$

or

$$\mathbf{R} = \rho\mathbf{a}_\rho + (z - z')\,\mathbf{a}_z$$

$$R^2 = |\mathbf{R}|^2 = x^2 + y^2 + (z - z')^2 = \rho^2 + (z - z')^2$$

$$\frac{\mathbf{a}_R}{R^2} = \frac{\mathbf{R}}{|\mathbf{R}|^3} = \frac{\rho\mathbf{a}_\rho + (z - z')\mathbf{a}_z}{[\rho^2 + (z - z')^2]^{3/2}}$$

Substituting all this into eq. (4.13), we get

$$\mathbf{E} = \frac{\rho_L}{4\pi\varepsilon_o}\int \frac{\rho\mathbf{a}_\rho + (z - z')\,\mathbf{a}_z}{[\rho^2 + (z - z')^2]^{3/2}}\,dz' \qquad [4.18]$$

To evaluate this, it is convenient that we define α, α_1, and α_2 as in Figure 4.6.

$$R = [\rho^2 + (z - z')^2]^{1/2} = \rho \sec \alpha$$

$$z' = OT - \rho \tan \alpha, \qquad dz' = -\rho \sec^2\alpha \, d\alpha$$

[4]The field point is the point at which the field is to be evaluated.

Hence, eq. (4.18) becomes

$$\begin{aligned}\mathbf{E} &= \frac{-\rho_L}{4\pi\varepsilon_o}\int_{\alpha_1}^{\alpha_2} \frac{\rho\,\sec^2\alpha\,[\cos\alpha\,\mathbf{a}_\rho + \sin\alpha\,\mathbf{a}_z]\,d\alpha}{\rho^2\sec^2\alpha} \\ &= -\frac{\rho_L}{4\pi\varepsilon_o\rho}\int_{\alpha_1}^{\alpha_2}[\cos\alpha\,\mathbf{a}_\rho + \sin\alpha\,\mathbf{a}_z]\,d\alpha\end{aligned} \qquad [4.19]$$

Thus for a *finite line charge*,

$$\mathbf{E} = \frac{\rho_L}{4\pi\varepsilon_o\rho}[-(\sin\alpha_2 - \sin\alpha_1)\mathbf{a}_\rho + (\cos\alpha_2 - \cos\alpha_1)\mathbf{a}_z] \qquad [4.20]$$

As a special case, for an *infinite line charge*, point B is at $(0, 0, \infty)$ and A at $(0, 0, -\infty)$ so that $\alpha_1 = \pi/2$, $\alpha_2 = -\pi/2$; the z-component vanishes and eq. (4.20) becomes

$$\boxed{\mathbf{E} = \frac{\rho_L}{2\pi\varepsilon_o\rho}\,\mathbf{a}_\rho} \qquad [4.21]$$

Bear in mind that eq. (4.21) is obtained for an infinite line charge along the z-axis so that ρ and $\mathbf{a}_\rho$ have their usual meaning. If the line is not along the z-axis, ρ is the perpendicular distance from the line to the point of interest and $\mathbf{a}_\rho$ is a unit vector along that distance directed from the line charge to the field point.

B. A Surface Charge

Consider an infinite sheet of charge in the xy-plane with uniform charge density ρ_S. The charge associated with an elemental area dS is

$$dQ = \rho_S\,dS$$

and hence the total charge is

$$Q = \int \rho_S\,dS \qquad [4.22]$$

From eq. (4.14), the contribution to the $\mathbf{E}$ field at point $P(0, 0, h)$ by the elemental surface 1 shown in Figure 4.7 is

$$d\mathbf{E} = \frac{dQ}{4\pi\varepsilon_o R^2}\,\mathbf{a}_R \qquad [4.23]$$

From Figure 4.7,

$$\mathbf{R} = \rho\,(-\mathbf{a}_\rho) + h\mathbf{a}_z, \qquad R = |\mathbf{R}| = [\rho^2 + h^2]^{1/2}$$

$$\mathbf{a}_R = \frac{\mathbf{R}}{R}, \qquad dQ = \rho_S\,dS = \rho_S\,\rho\,d\phi\,d\rho$$

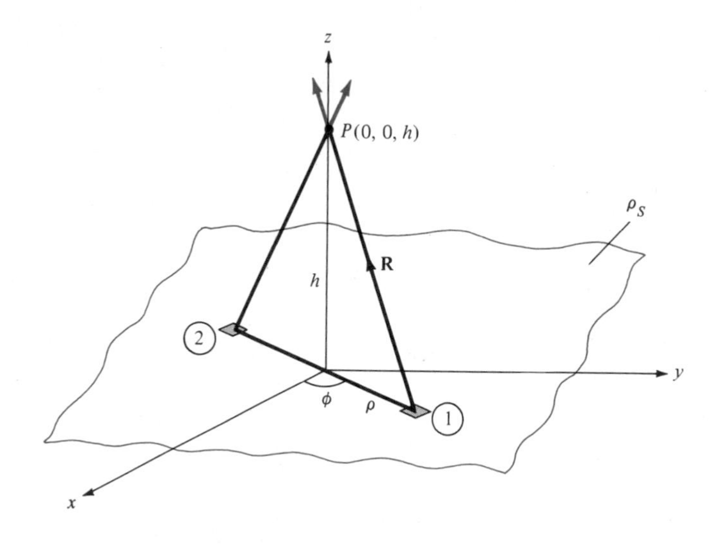

Figure 4.7 Evaluation of the **E** field due to an infinite sheet of charge.

Substitution of these terms into eq. (4.23) gives

$$d\mathbf{E} = \frac{\rho_S\, \rho d\phi\, d\rho\, [-\rho \mathbf{a}_\rho + h\mathbf{a}_z]}{4\pi\varepsilon_o\, [\rho^2 + h^2]^{3/2}} \tag{4.24}$$

Due to the symmetry of the charge distribution, for every element 1, there is a corresponding element 2 whose contribution along $\mathbf{a}_\rho$ cancels that of element 1, as illustrated in Figure 4.7. Thus the contributions to E_ρ add up to zero so that **E** has only z-component. This can also be shown mathematically by replacing $\mathbf{a}_\rho$ with $\cos\phi\, \mathbf{a}_x + \sin\phi\, \mathbf{a}_y$. Integration of $\cos\phi$ or $\sin\phi$ over $0 < \phi < 2\pi$ gives zero. Therefore,

$$\mathbf{E} = \int d\mathbf{E}_z = \frac{\rho_S}{4\pi\varepsilon_o} \int_{\phi=0}^{2\pi} \int_{\rho=0}^{\infty} \frac{h\rho\, d\rho\, d\phi}{[\rho^2 + h^2]^{3/2}}\, \mathbf{a}_z$$

$$= \frac{\rho_S h}{4\pi\varepsilon_o} 2\pi \int_0^\infty [\rho^2 + h^2]^{-3/2} \frac{1}{2}\, d(\rho^2)\, \mathbf{a}_z$$

$$= \frac{\rho_S h}{2\varepsilon_o} \left\{ -[\rho^2 + h^2]^{-1/2} \right\}_0^\infty \mathbf{a}_z$$

$$\mathbf{E} = \frac{\rho_S}{2\varepsilon_o}\, \mathbf{a}_z \tag{4.25}$$

that is, **E** has only z-component if the charge is in the xy-plane. In general, for an *infinite sheet* of charge

$$\boxed{\mathbf{E} = \frac{\rho_S}{2\varepsilon_o}\mathbf{a}_n} \tag{4.26}$$

where $\mathbf{a}_n$ is a unit vector normal to the sheet. From eq. (4.25) or (4.26), we notice that the electric field is normal to the sheet and it is surprisingly independent of the distance between the sheet and the point of observation P. In a parallel plate capacitor, the electric field existing between the two plates having equal and opposite charges is given by

$$\mathbf{E} = \frac{\rho_S}{2\varepsilon_o}\mathbf{a}_n + \frac{-\rho_S}{2\varepsilon_o}(-\mathbf{a}_n) = \frac{\rho_S}{\varepsilon_o}\mathbf{a}_n \tag{4.27}$$

C. A Volume Charge

Let the volume charge distribution with uniform charge density ρ_v be as shown in Figure 4.8. The charge dQ associated with the elemental volume dv is

$$dQ = \rho_v \, dv$$

and hence the total charge in a sphere of radius a is

$$\begin{aligned} Q &= \int \rho_v \, dv = \rho_v \int dv \\ &= \rho_v \frac{4\pi a^3}{3} \end{aligned} \tag{4.28}$$

The electric field $d\mathbf{E}$ at $P(0, 0, z)$ due to the elementary volume charge is

$$d\mathbf{E} = \frac{\rho_v \, dv}{4\pi\varepsilon_o R^2}\mathbf{a}_R$$

where $\mathbf{a}_R = \cos\alpha\,\mathbf{a}_z + \sin\alpha\,\mathbf{a}_\rho$. Due to the symmetry of the charge distribution, the contributions to E_x or E_y add up to zero. We are left with only E_z, given by

$$E_z = \mathbf{E} \cdot \mathbf{a}_z = \int dE \cos\alpha = \frac{\rho_v}{4\pi\varepsilon_o}\int \frac{dv \cos\alpha}{R^2} \tag{4.29}$$

Again, we need to derive expressions for dv, R^2, and $\cos\alpha$.

$$dv = r'^2 \sin\theta' \, dr' \, d\theta' \, d\phi' \tag{4.30}$$

Applying the cosine rule to Figure 4.8, we have

$$R^2 = z^2 + r'^2 - 2zr'\cos\theta'$$

$$r'^2 = z^2 + R^2 - 2zR\cos\alpha$$

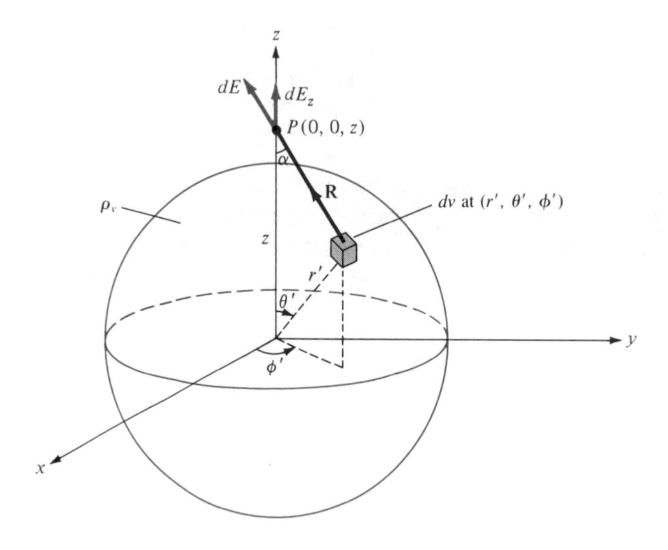

Figure 4.8 Evaluation of the **E** field due to a volume charge distribution.

It is convenient to evaluate the integral in eq. (4.29) in terms of R and r'. Hence we express $\cos\theta'$, $\cos\alpha$, and $\sin\theta'\,d\theta'$ in terms of R and r', that is,

$$\cos\alpha = \frac{z^2 + R^2 - r'^2}{2zR} \qquad [4.31a]$$

$$\cos\theta' = \frac{z^2 + r'^2 - R^2}{2zr'} \qquad [4.31b]$$

Differentiating eq. (4.31b) with respect to θ' keeping z and r' fixed, we obtain

$$\sin\theta'\,d\theta' = \frac{R\,dR}{z\,r'} \qquad [4.32]$$

Substituting eqs. (4.30) to (4.32) into eq. (4.29) yields

$$\begin{aligned}
E_z &= \frac{\rho_v}{4\pi\varepsilon_o}\int_{\phi'=0}^{2\pi} d\phi' \int_{r'=0}^{a}\int_{R=z-r'}^{z+r'} r'^2\,\frac{R\,dR}{zr'}\,dr'\,\frac{z^2+R^2-r'^2}{2zR}\,\frac{1}{R^2} \\
&= \frac{\rho_v 2\pi}{8\pi\varepsilon_o z^2}\int_{r'=0}^{a}\int_{R=z-r'}^{z+r'} r'\left[1+\frac{z^2-r'^2}{R^2}\right]dR\,dr' \\
&= \frac{\rho_v\pi}{4\pi\varepsilon_o z^2}\int_0^a r'\left[R-\frac{(z^2-r'^2)}{R}\right]_{z-r'}^{z+r'} dr' \\
&= \frac{\rho_v\pi}{4\pi\varepsilon_o z^2}\int_0^a 4r'^2\,dr' = \frac{1}{4\pi\varepsilon_o}\,\frac{1}{z^2}\left(\frac{4}{3}\pi a^3\rho_v\right)
\end{aligned}$$

or

$$\mathbf{E} = \frac{Q}{4\pi\varepsilon_o z^2}\,\mathbf{a}_z \tag{4.33}$$

This result is obtained for **E** at $P(0, 0, z)$. Due to the symmetry of the charge distribution, the electric field at $P(r, \theta, \phi)$ is readily obtained from eq. (4.33) as

$$\mathbf{E} = \frac{Q}{4\pi\varepsilon_o r^2}\,\mathbf{a}_r \tag{4.34}$$

which is identical to the electric field at the same point due to a point charge Q located at the origin or the center of the spherical charge distribution. The reason for this will become obvious as we cover Gauss's law in Section 4.5.

EXAMPLE 4.4

A circular ring of radius a carries a uniform charge ρ_L C/m and is placed on the xy-plane with axis the same as the z-axis.

(a) Show that

$$\mathbf{E}(0, 0, h) = \frac{\rho_L a h}{2\varepsilon_o[h^2 + a^2]^{3/2}}\,\mathbf{a}_z$$

(b) What values of h gives the maximum value of **E**?

(c) If the total charge on the ring is Q, find **E** as $a \rightarrow 0$.

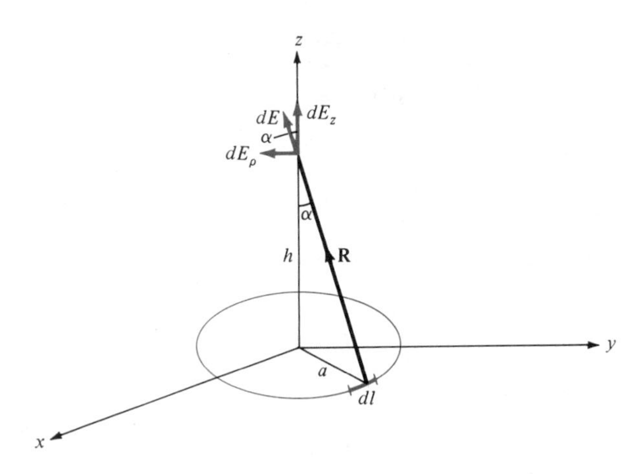

Figure 4.9 Charged ring; for Example 4.4.

SOLUTION (a) Consider the system as shown in Figure 4.9. Again the trick in finding **E** using eq. (4.13) is deriving each term in the equation. In this case,

$$dl = a\, d\phi, \qquad \mathbf{R} = a\,(-\mathbf{a}_\rho) + h\mathbf{a}_z$$

$$R = |\mathbf{R}| = [a^2 + h^2]^{1/2}, \qquad \mathbf{a}_R = \frac{\mathbf{R}}{R}$$

or

$$\frac{\mathbf{a}_R}{R^2} = \frac{\mathbf{R}}{|\mathbf{R}|^3} = \frac{-a\mathbf{a}_\rho + h\mathbf{a}_z}{[a^2 + h^2]^{3/2}}$$

Hence

$$\mathbf{E} = \frac{\rho_L}{4\pi\varepsilon_o} \int_{\phi=0}^{2\pi} \frac{(-a\mathbf{a}_\rho + h\mathbf{a}_z)}{[a^2 + h^2]^{3/2}}\, a\, d\phi$$

By symmetry, the contributions along $\mathbf{a}_\rho$ add up to zero. This is evident from the fact that for every element dl there is a corresponding element diametrically opposite it that gives an equal but opposite dE_ρ so that the two contributions cancel each other. Thus we are left with the z-component. That is,

$$\mathbf{E} = \frac{\rho_L a h \mathbf{a}_z}{4\pi\varepsilon_o[h^2 + a^2]^{3/2}} \int_0^{2\pi} d\phi = \frac{\rho_L a h \mathbf{a}_z}{2\varepsilon_o[h^2 + a^2]^{3/2}}$$

as required.

(b)

$$\frac{d|\mathbf{E}|}{dh} = \frac{\rho_L a}{2\varepsilon_o} \left\{ \frac{[h^2 + a^2]^{3/2}(1) - \frac{3}{2}(h)2h[h^2 + a^2]^{1/2}}{[h^2 + a^2]^3} \right\}$$

For maximum **E**, $\frac{d|\mathbf{E}|}{dh} = 0$, which implies that

$$[h^2 + a^2]^{1/2}\,[h^2 + a^2 - 3h^2] = 0$$

$$a^2 - 2h^2 = 0 \qquad \text{or} \qquad h = \pm \frac{a}{\sqrt{2}}$$

(c) Since the charge is uniformly distributed, the line charge density is

$$\rho_L = \frac{Q}{2\pi a}$$

so that

$$\mathbf{E} = \frac{Qh}{4\pi\varepsilon_o[h^2 + a^2]^{3/2}}\, \mathbf{a}_z$$

As $a \to 0$

$$\mathbf{E} = \frac{Q}{4\pi\varepsilon_o h^2}\,\mathbf{a}_z$$

or in general

$$\mathbf{E} = \frac{Q}{4\pi\varepsilon_o r^2}\,\mathbf{a}_R$$

which is the same as that of a point charge as one would expect. ■

PRACTICE EXERCISE 4.4

A circular disk of radius a is uniformly charged with ρ_S C/m^2. If the disk lies on the $z = 0$ plane with its axis along the z-axis,

(a) Show that at point $(0, 0, h)$

$$\mathbf{E} = \frac{\rho_S}{2\varepsilon_o}\left\{1 - \frac{h}{[h^2 + a^2]^{1/2}}\right\}\mathbf{a}_z$$

(b) From this, derive the $\mathbf{E}$ field due to an infinite sheet of charge on the $z = 0$ plane.

ANSWER (a) Proof, (b) $\dfrac{\rho_S}{2\varepsilon_o}\mathbf{a}_z$

EXAMPLE 4.5

The finite sheet $0 \leq x \leq 1$, $0 \leq y \leq 1$ on the $z = 0$ plane has a charge density $\rho_S = xy(x^2 + y^2 + 25)^{3/2}$ nC/m^2. Find

(a) The total charge on the sheet

(b) The electric field at $(0, 0, 5)$

(c) The force experienced by a -1 mC charge located at $(0, 0, 5)$

SOLUTION (a) $Q = \displaystyle\int \rho_S\, dS = \int_0^1\int_0^1 xy(x^2 + y^2 + 25)^{3/2}\, dx\, dy$ nC

Since $x\,dx = 1/2\; d(x^2)$, we now integrate with respect to x^2 (or change variables: $x^2 = u$ so that $x\,dx = du/2$).

$$Q = \frac{1}{2} \int_0^1 y \int_0^1 (x^2 + y^2 + 25)^{3/2} \, d(x^2) \, dy \text{ nC}$$

$$= \frac{1}{2} \int_0^1 y \frac{2}{5} (x^2 + y^2 + 25)^{5/2} \Big|_0^1 \, dy$$

$$= \frac{1}{5} \int_0^1 \frac{1}{2} [(y^2 + 26)^{5/2} - (y^2 + 25)^{5/2}] \, d(y^2)$$

$$= \frac{1}{10} \cdot \frac{2}{7} \left[(y^2 + 26)^{7/2} - (y^2 + 25)^{7/2}\right] \Big|_0^1$$

$$= \frac{1}{35} [(27)^{7/2} + (25)^{7/2} - 2(26)^{7/2}]$$

$$Q = 33.15 \text{ nC}$$

(b) $\mathbf{E} = \displaystyle\int \frac{\rho_S \, dS \, \mathbf{a}_R}{4\pi\varepsilon_o r^2} = \int \frac{\rho_S \, dS \, (\mathbf{r} - \mathbf{r}')}{4\pi\varepsilon_o |\mathbf{r} - \mathbf{r}'|^3}$

where $\mathbf{r} - \mathbf{r}' = (0, 0, 5) - (x, y, 0) = (-x, -y, 5)$. Hence,

$$\mathbf{E} = \int_0^1 \int_0^1 \frac{10^{-9} xy(x^2 + y^2 + 25)^{3/2}(-x\mathbf{a}_x - y\mathbf{a}_y + 5\mathbf{a}_z) \, dx \, dy}{4\pi \cdot \dfrac{10^{-9}}{36\pi} (x^2 + y^2 + 25)^{3/2}}$$

$$= 9\left[-\int_0^1 x^2 \, dx \int_0^1 y \, dy \, \mathbf{a}_x - \int_0^1 x \, dx \int_0^1 y^2 dy \, \mathbf{a}_y + 5 \int_0^1 x \, dx \int_0^1 y \, dy \, \mathbf{a}_z\right]$$

$$= 9\left(\frac{-1}{6}, \frac{-1}{6}, \frac{5}{4}\right)$$

$$= (-1.5, -1.5, 11.25) \text{ V/m}$$

(c) $\mathbf{F} = q\mathbf{E} = (1.5, 1.5, -11.25)$ mN ■

PRACTICE EXERCISE 4.5

A square plate described by $-2 \leq x \leq 2$, $-2 \leq y \leq 2$, $z = 0$ carries a charge $12 \, |y|$ mC/m^2. Find the total charge on the plate and the electric field intensity at $(0, 0, 10)$.

ANSWER 192 mC, 16.46 $\mathbf{a}_z$ MV/m.

EXAMPLE 4.6

Planes $x = 2$ and $y = -3$ respectively carry charges 10 nC/m^2 and 15 nC/m^2. If the line $x = 0$, $z = 2$ carries charge 10π nC/m, calculate $\mathbf{E}$ at $(1, 1, -1)$ due to the three charge distributions.

SOLUTION Let

$$\mathbf{E} = \mathbf{E}_1 + \mathbf{E}_2 + \mathbf{E}_3$$

where $\mathbf{E}_1$, $\mathbf{E}_2$, and $\mathbf{E}_3$ are respectively the contributions to $\mathbf{E}$ at point $(1, 1, -1)$ due to the infinite sheet 1, infinite sheet 2, and infinite line 3 as shown in Figure 4.10(a). Applying eqs. (4.26) and (4.21) gives

$$\mathbf{E}_1 = \frac{\rho_{S_1}}{2\varepsilon_o}(-\mathbf{a}_x) = -\frac{10 \cdot 10^{-9}}{2 \cdot \dfrac{10^{-9}}{36\pi}}\mathbf{a}_x = -180\pi\mathbf{a}_x$$

$$\mathbf{E}_2 = \frac{\rho_{S_2}}{2\varepsilon_o}\mathbf{a}_y = \frac{15 \cdot 10^{-9}}{2 \cdot \dfrac{10^{-9}}{36\pi}}\mathbf{a}_y = 270\pi\mathbf{a}_y$$

and

$$\mathbf{E}_3 = \frac{\rho_L}{2\pi\varepsilon_o\rho}\mathbf{a}_\rho$$

where $\mathbf{a}_\rho$ (not regular $\mathbf{a}_\rho$ but with a similar meaning) is a unit vector along LP perpendicular to the line charge and ρ is the length LP to be determined from Figure 4.10(b). Figure 4.10(b) results from Figure 4.10(a) if we consider plane $y = 1$ on which $\mathbf{E}_3$ lies. From Figure 4.10(b), the distance vector from L to P is

$$\mathbf{R} = -3\mathbf{a}_z + \mathbf{a}_x$$

$$\rho = |\mathbf{R}| = \sqrt{10}, \qquad \mathbf{a}_\rho = \frac{\mathbf{R}}{|\mathbf{R}|} = \frac{1}{\sqrt{10}}\mathbf{a}_x - \frac{3}{\sqrt{10}}\mathbf{a}_z$$

Hence,

$$\mathbf{E}_3 = \frac{10\pi \cdot 10^{-9}}{2\pi \cdot \dfrac{10^{-9}}{36\pi}} \cdot \frac{1}{10}(\mathbf{a}_x - 3\mathbf{a}_z)$$

$$= 18\pi\,(\mathbf{a}_x - 3\mathbf{a}_z)$$

Thus by adding $\mathbf{E}_1$, $\mathbf{E}_2$, and $\mathbf{E}_3$, we obtain the total field as

$$\mathbf{E} = -162\pi\mathbf{a}_x + 270\pi\mathbf{a}_y - 54\pi\mathbf{a}_z \text{ V/m}$$

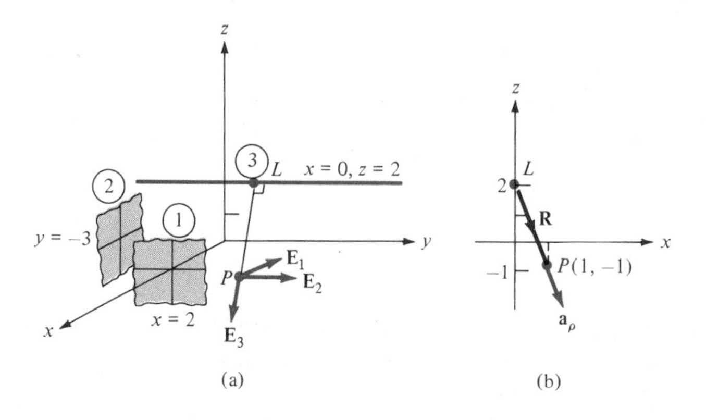

Figure 4.10 For Example 4.6: **(a)** three charge distributions; **(b)** finding ρ and $\mathbf{a}_\rho$ on plane $y = 1$.

Note that to obtain $\mathbf{a}_r$, $\mathbf{a}_\rho$, or $\mathbf{a}_n$ which we always need for finding **F** or **E**, we must go from the charge (at position vector $\mathbf{r}'$) to the field point (at position vector $\mathbf{r}$); hence $\mathbf{a}_r$, $\mathbf{a}_\rho$, or $\mathbf{a}_n$ is a unit vector along $\mathbf{r} - \mathbf{r}'$. Observe this carefully in Figures 4.6 to 4.10. ■

PRACTICE EXERCISE 4.6

In Example 4.6 if the line $x = 0$, $z = 2$ is rotated through 90° about the point (0, 2, 2) so that it becomes $x = 0$, $y = 2$, find **E** at $(1, 1, -1)$.

ANSWER $-282.7\mathbf{a}_x + 564.5\mathbf{a}_y$ V/m.

EXAMPLE 4.7

The three sides of an equiliteral triangle have uniform line charges 2 μC/m, 1 μC/m, and 1 μC/m. Find the electric field at the center of the triangle if each side is 50 cm long.

SOLUTION It may be worthwhile to derive **E** at a point P due to a finite line charge ρ_L C/m as shown in Figure 4.11(a) and then apply this general case to the special case of our problem illustrated in Figure 4.11(b).

Due to symmetry, **E** only has a component along $\mathbf{a}_\rho$; that is,

$$\mathbf{E} = \int_{-\ell/2}^{\ell/2} \frac{\rho_L \, dz}{4\pi\varepsilon_o R^2} \cos\alpha \, \mathbf{a}_\rho = \int_{-\ell/2}^{\ell/2} \frac{\rho_L \rho \, dz}{4\pi\varepsilon_o [\rho^2 + z^2]^{3/2}} \mathbf{a}_\rho$$

For the integration, let $z = \rho \tan\alpha$

$$dz = \rho \sec^2\alpha \, d\alpha, \qquad \rho^2 + z^2 = \rho^2(1 + \tan^2\alpha) = \rho^2 \sec^2\alpha$$

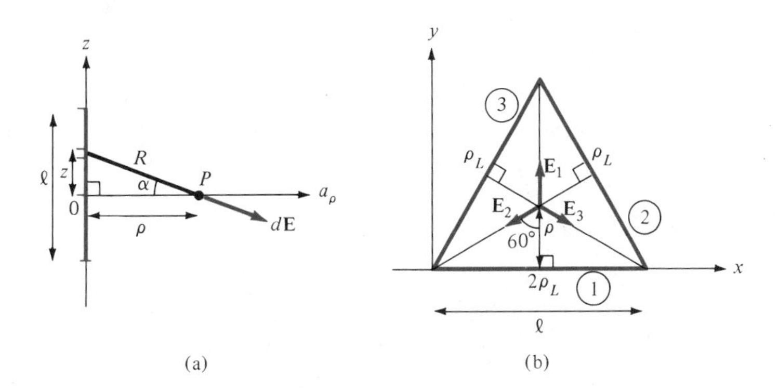

Figure 4.11 For Example 4.7: **(a)** **E** due to a finite line charge (general); **(b)** **E** due to a triangular loop.

$$\mathbf{E} = \frac{\rho_L \mathbf{a}_\rho}{4\pi\varepsilon_o} \int \frac{\rho^2 \sec^2\alpha \, d\alpha}{\rho^3 \sec^3\alpha} = \frac{\rho_L \mathbf{a}_\rho}{4\pi\varepsilon_o \rho} \int \cos\alpha \, d\alpha$$

$$= \frac{\rho_L \mathbf{a}_\rho}{4\pi\varepsilon_o \rho} \sin\alpha$$

Since $z = \rho \tan\alpha$, $\sin\alpha = \dfrac{z}{[\rho^2 + z^2]^{1/2}}$

Hence,

$$\mathbf{E} = \frac{\rho_L}{4\pi\varepsilon_o} \mathbf{a}_\rho \cdot \frac{z}{\rho[\rho^2 + z^2]^{1/2}} \Bigg|_{-\ell/2}^{\ell/2}$$

$$= \frac{\rho_L \ell}{2\pi\varepsilon_o \rho\, [\ell^2 + 4\rho^2]^{1/2}} \mathbf{a}_\rho \qquad \text{(general, for finite line).}$$

If we let $\ell \to \infty$, we obtain eq. (4.21), namely:

$$\mathbf{E} = \frac{\rho_L}{2\pi\varepsilon_o \rho} \mathbf{a}_\rho \qquad \text{(special case, for infinite line).}$$

We now apply the formula for **E** due to the finite line charge to the problem shown in Figure 4.11(b). Let $\rho_L = 1$ μC/m and the total **E** field

$$\mathbf{E} = \mathbf{E}_1 + \mathbf{E}_2 + \mathbf{E}_3$$

where $\mathbf{E}_1$, $\mathbf{E}_2$, and $\mathbf{E}_3$ are the contributions to **E** due to the sides of the triangle labeled 1, 2, and 3 respectively.

$$\mathbf{E}_1 = \frac{2\rho_L \ell}{2\pi\varepsilon_o \rho\, [\ell^2 + 4\rho^2]^{1/2}} \mathbf{a}_y$$

$$\mathbf{E}_2 = \frac{\rho_L \ell}{2\pi\varepsilon_o\rho\,[\ell^2 + 4\rho^2]^{1/2}}(-\mathbf{a}_x\sin 60° - \mathbf{a}_y\cos 60°)$$

$$\mathbf{E}_3 = \frac{\rho_L \ell}{2\pi\varepsilon_o\rho\,[\ell^2 + 4\rho^2]^{1/2}}(\mathbf{a}_x\sin 60° - \mathbf{a}_y\cos 60°)$$

Adding $\mathbf{E}_1$, $\mathbf{E}_2$, and $\mathbf{E}_3$ gives

$$\mathbf{E} = \frac{\rho_L \ell}{2\pi\varepsilon_o\rho\,[\ell^2 + 4\rho^2]^{1/2}}(2 - 2\cos 60°)\mathbf{a}_y$$

$$= \frac{\rho_L \ell}{2\pi\varepsilon_o\rho\,[\ell^2 + 4\rho^2]^{1/2}}\mathbf{a}_y$$

But $\rho_L = 1\ \mu\text{C/m}$, $\rho = (\ell/2)\tan 30° = \ell/(2\sqrt{3})$, $\ell = 0.5$ m. Hence,

$$\mathbf{E} = \frac{3\rho_L}{2\pi\varepsilon_o\ell}\mathbf{a}_y = \frac{3 \cdot 10^{-6}}{2\pi\,\dfrac{10^{-9}}{36\pi}(0.5)}\mathbf{a}_y$$

or

$$\mathbf{E} = 108\mathbf{a}_y\ \text{kV/m}.$$

Note that if the charge densities were equal, $\mathbf{E}$ would be zero at the center of the triangle. ■

PRACTICE EXERCISE 4.7

Two opposite sides of a square loop of length L carry charge Q and $-Q$ uniformly distributed on them while the other two opposite sides are uncharged. Calculate the electric field strength at the center of the square.

ANSWER $\dfrac{\sqrt{2}\,Q}{\pi\varepsilon_o L^2}$

4.4 ELECTRIC FLUX DENSITY

The flux due to the electric field $\mathbf{E}$ can be calculated using the general definition of flux in eq. (3.13). For practical reasons, however, this quantity is not usually considered as the most useful flux in electrostatics. Also, eqs. (4.11) to (4.15) show that the electric field intensity is dependent on the medium in which the charge is placed (free space in this chapter). Suppose a new vector field $\mathbf{D}$ independent of the medium is defined by

$$\boxed{\mathbf{D} = \varepsilon_o \mathbf{E}} \tag{4.35}$$

We define *electric flux* Ψ in terms of **D** using eq. (3.13), namely,

$$\Psi = \int \mathbf{D} \cdot d\mathbf{S} \tag{4.36}$$

In SI units, one line of electric flux emanates from +1 C and terminates on −1 C. Therefore, the electric flux is measured in coulombs. Hence, the vector field **D** is called the *electric flux density* and is measured in coulombs per square meter. For historical reasons, the electric flux density is also called *electric displacement*.

From eq. (4.35), it is apparent that all the formulas derived for **E** from Coulomb's law in Sections 4.2 and 4.3 can be used in calculating **D**, except that we have to multiply those formulas by ε_o. For example, for an infinite sheet of charge, eqs. (4.26) and (4.35) give

$$\mathbf{D} = \frac{\rho_S}{2} \mathbf{a}_n \tag{4.37}$$

and for a volume charge distribution, eqs. (4.15) and (4.35) give

$$\mathbf{D} = \int \frac{\rho_v \, dv}{4\pi R^2} \mathbf{a}_R \tag{4.38}$$

Note from eqs.(4.37) and (4.38) that **D** is a function of charge and position only; it is independent of the medium.

EXAMPLE 4.8

Determine **D** at (4, 0, 3) if there is a point charge -5π mC at (4, 0, 0) and a line charge 3π mC/m along the y-axis.

SOLUTION

Let $\mathbf{D} = \mathbf{D}_Q + \mathbf{D}_L$ where $\mathbf{D}_Q$ and $\mathbf{D}_L$ are flux densities due to the point charge and line charge respectively as shown in Figure 4.12.

$$\mathbf{D}_Q = \varepsilon_o \mathbf{E} = \frac{Q}{4\pi R^2} \mathbf{a}_R = \frac{Q\,(\mathbf{r} - \mathbf{r}')}{4\pi|\mathbf{r} - \mathbf{r}'|^3}$$

where $\mathbf{r} - \mathbf{r}' = (4, 0, 3) - (4, 0, 0) = (0, 0, 3)$. Hence,

$$\mathbf{D}_Q = \frac{-5\pi \cdot 10^{-3}(0, 0, 3)}{4\pi|(0, 0, 3)|^3} = -0.138 \, \mathbf{a}_z \text{ mC/m}^2$$

Also

$$\mathbf{D}_L = \frac{\rho_L}{2\pi\rho} \mathbf{a}_\rho$$

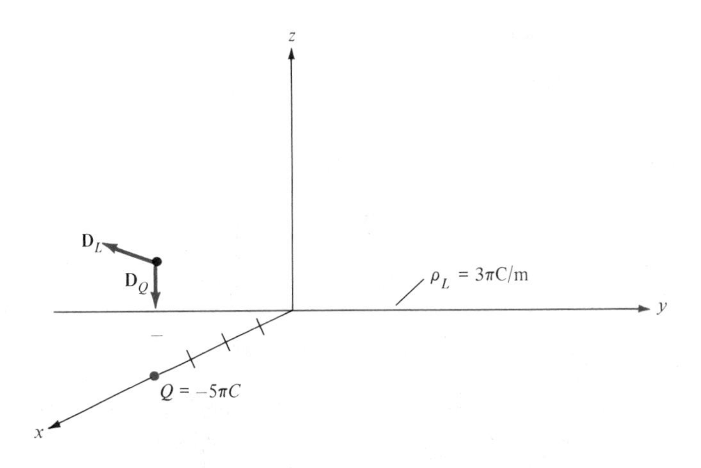

Figure 4.12 Flux density **D** due to a point charge and an infinite line charge.

In this case

$$\mathbf{a}_\rho = \frac{(4, 0, 3) - (0, 0, 0)}{|(4, 0, 3) - (0, 0, 0)|} = \frac{(4, 0, 3)}{5}$$

$$\rho = |(4, 0, 3) - (0, 0, 0)| = 5$$

Hence,

$$\mathbf{D}_L = \frac{3\pi}{2\pi(25)} (4\mathbf{a}_x + 3\mathbf{a}_z) = 0.24\mathbf{a}_x + 0.18\mathbf{a}_z \text{ mC/m}^2$$

Thus

$$\mathbf{D} = \mathbf{D}_Q + \mathbf{D}_L$$
$$= 240\mathbf{a}_x + 42\mathbf{a}_z \ \mu\text{C/m}^2$$

■

PRACTICE EXERCISE 4.8

A point charge of 30 nC is located at the origin while plane $y = 3$ carries charge 10 nC/m^2. Find **D** at (0, 4, 3).

ANSWER $5.076\mathbf{a}_y + 0.0573\mathbf{a}_z$ nC/m^2.

4.5 GAUSS'S LAW—MAXWELL'S EQUATION

Gauss's[5] law constitutes one of the fundamental laws of electromagnetism. Essentially, it states that the total electric flux Ψ through any *closed* surface is equal to the total charge enclosed by that surface. Thus

$$\Psi = Q_{\text{enc}} \tag{4.39}$$

that is,

$$\Psi = \oint d\Psi = \oint_S \mathbf{D} \cdot d\mathbf{S}$$

$$= \text{Total charge enclosed } Q = \int \rho_v \, dv \tag{4.40}$$

or

$$\boxed{Q = \oint_S \mathbf{D} \cdot d\mathbf{S} = \int_v \rho_v \, dv} \tag{4.41}$$

By applying divergence theorem to the middle term in eqs. (4.41)

$$\oint_S \mathbf{D} \cdot d\mathbf{S} = \int_v \nabla \cdot \mathbf{D} \, dv \tag{4.42}$$

Comparing the two volume integrals in eqs. (4.41) and (4.42) results in

$$\boxed{\rho_v = \nabla \cdot \mathbf{D}} \tag{4.43}$$

which is the first of the four *Maxwell's equations* to be derived. Equation (4.43) states that the volume charge density is the same as the divergence of the electric flux density. This should not be surprising to us from the way we defined the divergence of a vector in eq. (3.32) and from the fact that ρ_v at a point is simply the charge per unit volume at that point.

Note that:

1. Equations (4.41) and (4.43) are basically stating Gauss's law in different ways; eq. (4.41) is the integral form, whereas eq. (4.43) is the differential or point form of Gauss's law.

[5] Karl Friedrich Gauss (1777–1855), a German mathematician, developed the divergence theorem of Section 3.6, popularly known by his name. He was the first physicist to measure electric and magnetic quantities in absolute units. For details on Gauss's measurements, see W. F. Magie, *A Source Book in Physics*. Cambridge: Harvard Univ. Press, 1963, pp. 519–524.

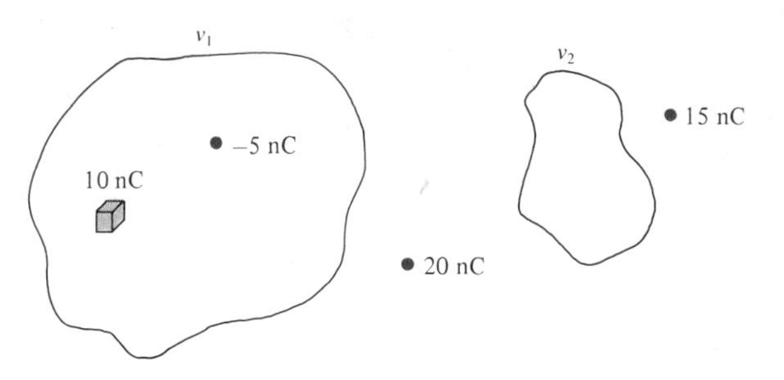

Figure 4.13 Illustration of Gauss's law; flux leaving v_1 is 5 nC and that leaving v_2 is 0 C.

2. Gauss's law is an alternative statement of Coulomb's law; proper application of the divergence theorem to Coulomb's law results in Gauss's law.
3. Gauss's law provides an easy means of finding **E** or **D** for symmetrical charge distributions such as a point charge, an infinite line charge, an infinite cylindrical surface charge, and a spherical distribution of charge. A continuous charge distribution has rectangular symmetry if it depends only on x (or y or z), cylindrical symmetry if it depends only on ρ, or spherical symmetry if it depends only on r (independent of θ and ϕ). It must be stressed that whether the charge distribution is symmetric or not, Gauss's law always holds. For example, consider the charge distribution in Figure 4.13 where v_1 and v_2 are closed surfaces (or volumes). The total flux leaving v_1 is $10 - 5 = 5$ nC because only 10 nC and -5 nC charges are enclosed by v_1. Although charges 20 nC and 15 nC outside v_1 do contribute to the flux crossing v_1, the net flux crossing v_1, according to Gauss's law, is irrespective of those charges outside v_1. Similarly, the total flux leaving v_2 is zero because no charge is enclosed by v_2. Thus we see that Gauss's law, $\Psi = Q_{\text{enclosed}}$, is still obeyed even though the charge distribution is not symmetric. However, we cannot use the law to determine **E** or **D** when the charge distribution is not symmetric; we must resort to Coulomb's law to determine **E** or **D** in that case.

4.6 APPLICATIONS OF GAUSS'S LAW

The procedure for applying Gauss's law to calculate the electric field involves first knowing whether symmetry exists. Once symmetric charge distribution exists, we construct a mathematical closed surface (known as a *Gaussian surface*). The surface is chosen such that **D** is normal or tangential to the Gaussian surface. When **D** is normal to the surface, $\mathbf{D} \cdot d\mathbf{S} = D\, dS$ because **D** is contant on the surface. When **D** is tangential to the surface, $\mathbf{D} \cdot d\mathbf{S} = 0$. Thus we must choose a surface that has some of the symmetry exhibited by the charge distribution. We shall now apply these basic ideas to the following cases.

A. Point Charge

Suppose a point charge Q is located at the origin. To determine **D** at a point P, it is easy to see that choosing a spherical surface containing P will satisfy symmetry conditions. Thus, a spherical surface centered at the origin is the Gaussian surface in this case and is shown in Figure 4.14.

Since **D** is everywhere normal to the Gaussian surface, that is, $\mathbf{D} = D_r \mathbf{a}_r$, applying Gauss's law ($\Psi = Q_{\text{enclosed}}$) gives

$$Q = \oint \mathbf{D} \cdot d\mathbf{S} = D_r \oint dS = D_r\, 4\pi r^2 \qquad [4.44]$$

where $\oint dS = \int_{\phi=0}^{2\pi} \int_{\theta=0}^{\pi} r^2 \sin\theta\, d\theta\, d\phi = 4\pi r^2$ is the surface area of the Gaussian surface. Thus

$$\mathbf{D} = \frac{Q}{4\pi r^2}\, \mathbf{a}_r \qquad [4.45]$$

as expected from eqs. (4.11) and (4.35).

B. Infinite Line Charge

Suppose the infinite line of uniform charge ρ_L C/m lies along the z-axis. To determine **D** at a point P, we choose a cylindrical surface containing P to satisfy symmetry condition as shown in Figure 4.15. **D** is constant on and normal to the cylindrical Gaussian surface; that is, $\mathbf{D} = D_\rho \mathbf{a}_\rho$. If we apply Gauss's law to an arbitrary length ℓ of the line

$$\rho_L \ell = Q = \oint \mathbf{D} \cdot d\mathbf{S} = D_\rho \oint dS = D_\rho\, 2\pi\rho\ell \qquad [4.46]$$

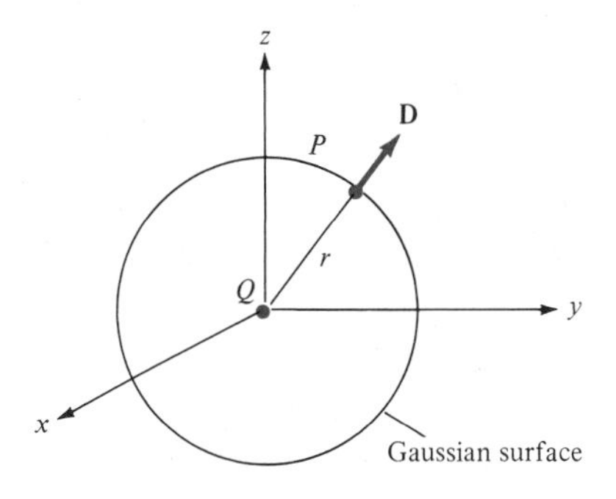

Figure 4.14 Gaussian surface about a point charge.

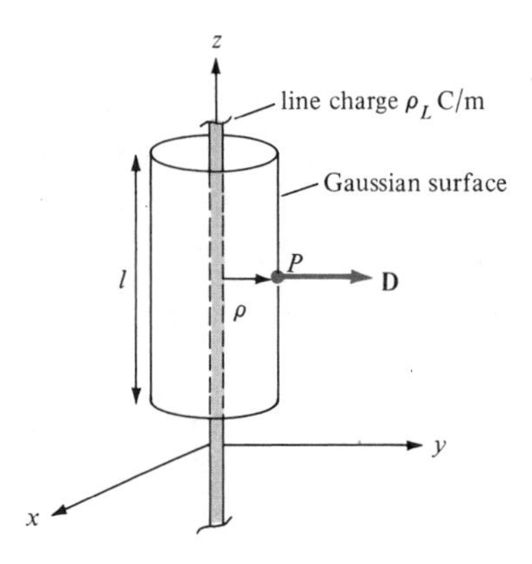

Figure 4.15 Gaussian surface about an infinite line charge.

where $\int dS = 2\pi\rho\ell$ is the surface area of the Gaussian surface. Note that $\int \mathbf{D} \cdot d\mathbf{S}$ evaluated on the top and bottom surfaces of the cylinder is zero since **D** has no z-component; that means that **D** is tangential to those sufaces. Thus

$$\mathbf{D} = \frac{\rho_L}{2\pi\rho}\,\mathbf{a}_\rho \qquad [4.47]$$

as expected from eqs. (4.21) and (4.35).

C. Infinite Sheet of Charge

Consider the infinite sheet of uniform charge ρ_S C/m^2 lying on the $z = 0$ plane. To determine **D** at point P, we choose a rectangular box that is cut symmetrically by the sheet of charge and has two of its faces parallel to the sheet as shown in Figure 4.16. As **D** is normal to the sheet, $\mathbf{D} = D_z\mathbf{a}_z$, and applying Gauss's law gives

$$\rho_S \int dS = Q = \oint \mathbf{D} \cdot d\mathbf{S} = D_z\left[\int_{\text{top}} dS + \int_{\text{bottom}} dS\right] \qquad [4.48]$$

Note that $\mathbf{D} \cdot d\mathbf{S}$ evaluated on the sides of the box is zero because **D** has no components along $\mathbf{a}_x$ and $\mathbf{a}_y$. If the top and bottom area of the box each has area A, eq. (4.48) becomes

$$\rho_S A = D_z\,(A + A) \qquad [4.49]$$

and thus

$$\mathbf{D} = \frac{\rho_S}{2}\,\mathbf{a}_z$$

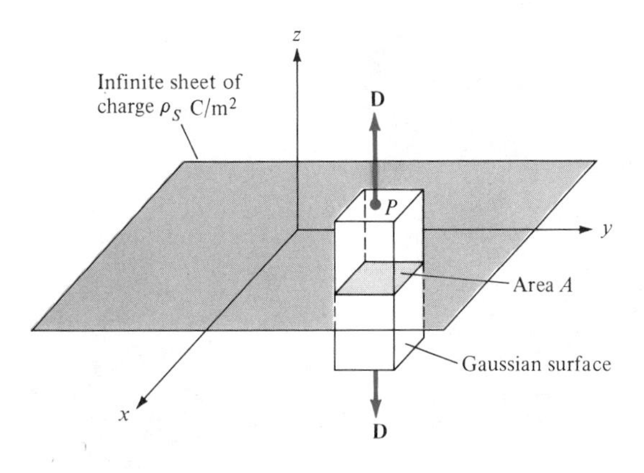

Figure 4.16 Gaussian surface about an infinite line sheet of charge.

or

$$\mathbf{E} = \frac{\mathbf{D}}{\varepsilon_o} = \frac{\rho_S}{2\varepsilon_o}\,\mathbf{a}_z \qquad [4.50]$$

as expected from eq. (4.25).

D. Uniformly Charged Sphere

Consider a sphere of radius a with a uniform charge ρ_v C/m^3. To determine **D** everywhere, we construct Gaussian surfaces for cases $r \leq a$ and $r \geq a$ separately. Since the charge has spherical symmetry, it is obvious that a spherical surface is an appropriate Gaussian surface.

For $r \leq a$, the total charge enclosed by the spherical surface of radius r, as shown in Figure 4.17(a), is

$$\begin{aligned} Q_{\text{enc}} &= \int \rho_v \, dv = \rho_v \int dv = \rho_v \int_{\phi=0}^{2\pi} \int_{\theta=0}^{\pi} \int_{r=0}^{r} r^2 \sin\theta \, dr \, d\theta \, d\phi \qquad [4.51] \\ &= \rho_v \frac{4}{3} \pi r^3 \end{aligned}$$

and

$$\begin{aligned} \Psi &= \oint \mathbf{D} \cdot d\mathbf{S} = D_r \oint dS = D_r \int_{\phi=0}^{2\pi} \int_{\theta=0}^{\pi} r^2 \sin\theta \, d\theta \, d\phi \\ &= D_r \, 4\pi r^2 \qquad [4.52] \end{aligned}$$

Hence, $\Psi = Q_{\text{enc}}$ gives

$$D_r \, 4\pi r^2 = \frac{4\pi r^3}{3} \rho_v$$

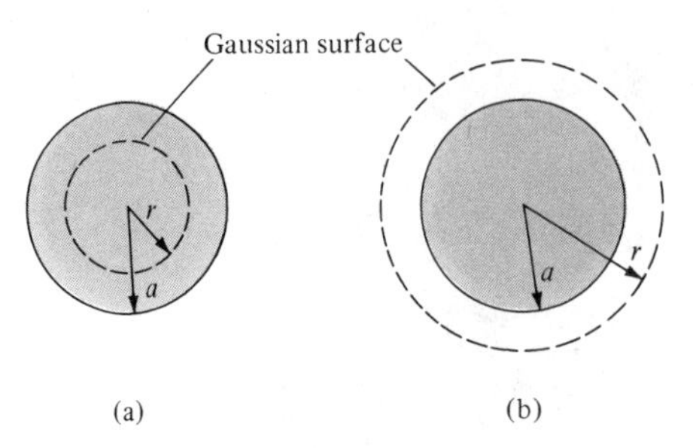

Figure 4.17 Gaussian surface for a uniformly charged sphere when: **(a)** $r \geq a$ and **(b)** $r \leq a$.

or

$$\mathbf{D} = \frac{r}{3}\rho_v\, \mathbf{a}_r \qquad 0 < r \leqslant a \tag{4.53}$$

For $r \geq a$, the Gaussian surface is shown in Figure 4.17(b). The charge enclosed by the surface is the entire charge in this case, that is,

$$\begin{aligned} Q_{\text{enc}} &= \int \rho_v\, dv = \rho_v \int dv = \rho_v \int_{\phi=0}^{2\pi} \int_{\theta=0}^{\pi} \int_{r=0}^{a} r^2 \sin\theta\, dr\, d\theta\, d\phi \\ &= \rho_v \frac{4}{3}\pi a^3 \end{aligned} \tag{4.54}$$

while

$$\Psi = \oint \mathbf{D} \cdot d\mathbf{S} = D_r\, 4\pi r^2 \tag{4.55}$$

just as in eq. (4.52). Hence:

$$D_r\, 4\pi r^2 = \frac{4}{3}\pi a^3 \rho_v$$

or

$$\mathbf{D} = \frac{a^3}{3r^2}\rho_v \mathbf{a}_r \qquad r \geqslant a \tag{4.56}$$

Thus from eqs. (4.53) and (4.56), $\mathbf{D}$ everywhere is given by

$$\mathbf{D} = \begin{bmatrix} \dfrac{r}{3}\rho_v\, \mathbf{a}_r & 0 < r \leqslant a \\ \dfrac{a^3}{3r^2}\rho_v\, \mathbf{a}_r & r \geqslant a \end{bmatrix} \tag{4.57}$$

and $|\mathbf{D}|$ is as sketched in Figure 4.18.

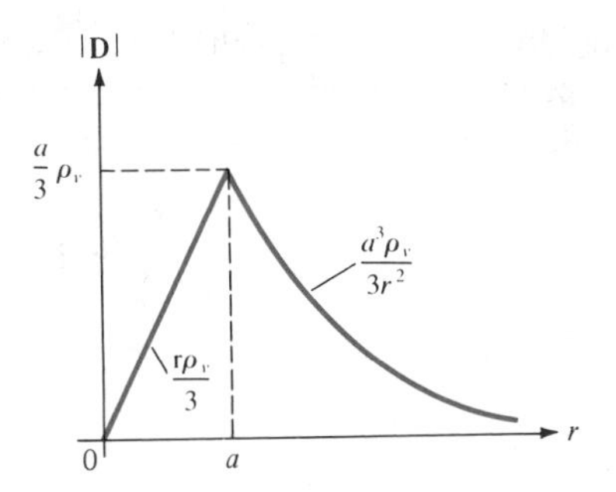

Figure 4.18 Sketch of $|\mathbf{D}|$ against r for a uniformly charged sphere.

Notice from eqs. (4.44), (4.46), (4.48), and (4.52) that the ability to take **D** out of the integral sign is the key to finding **D** using Gauss's law. In other words, **D** must be constant on the Gaussian surface.

EXAMPLE 4.9

Given that $\mathbf{D} = z\rho \cos^2\phi \, \mathbf{a}_z$ C/m^2, calculate the charge density at $(1, \pi/4, 3)$ and the total charge enclosed by the cylinder of radius 1m with $-2 \leq z \leq 2$m.

SOLUTION

$$\rho_v = \nabla \cdot \mathbf{D} = \frac{\partial D_z}{\partial z} = \rho \cos^2\phi$$

At $(1, \pi/4, 3)$, $\rho_v = 1 \cdot \cos^2(\pi/4) = 0.5$ C/m^3. The total charge enclosed by the cylinder can be found in two different ways.

Method 1: This method is based directly on the definition of the total volume charge.

$$Q = \int_v \rho_v \, dv = \int_v \rho \cos^2 \phi \, \rho \, d\phi \, d\rho \, dz$$

$$= \int_{z=-2}^{2} dz \int_{\phi=0}^{2\pi} \cos^2\phi \, d\phi \int_{\rho=0}^{1} \rho^2 \, d\rho = 4(\pi)(1/3)$$

$$= \frac{4\pi}{3} \text{ C}$$

Method 2: Alternatively, we can use Gauss's law.

$$Q = \Psi = \oint \mathbf{D} \cdot d\mathbf{S} = \left[\int_s + \int_t + \int_b\right] \mathbf{D} \cdot d\mathbf{S}$$

$$= \Psi_s + \Psi_t + \Psi_b$$

where Ψ_s, Ψ_t, and Ψ_b are the flux through the sides, the top surface, and the bottom surface of the cylinder respectively (see Figure 3.17). Since **D** does not have component along $\mathbf{a}_\rho$, $\Psi_s = 0$, for Ψ_t, $d\mathbf{S} = \rho\, d\phi\, d\rho\, \mathbf{a}_z$ so

$$\Psi_t = \int_{\rho=0}^{1} \int_{\phi=0}^{2\pi} z\rho \cos^2\phi\, \rho\, d\phi\, d\rho \bigg|_{z=2} = 2 \int_0^1 \rho^2 d\rho \int_0^{2\pi} \cos^2\phi\, d\phi$$

$$= 2\left(\frac{1}{3}\right)\pi = \frac{2\pi}{3}$$

and for Ψ_b, $d\mathbf{S} = -\rho\, d\phi\, d\rho\, \mathbf{a}_z$, so

$$\Psi_b = -\int_{\rho=0}^{1} \int_{\phi=0}^{2\pi} z\rho \cos^2\phi\, \rho\, d\phi\, d\rho \bigg|_{z=-2} = 2 \int_0^1 \rho^2 d\rho \int_0^{2\pi} \cos^2\phi\, d\phi$$

$$= \frac{2\pi}{3}$$

Thus

$$Q = \Psi = 0 + \frac{2\pi}{3} + \frac{2\pi}{3} = \frac{4\pi}{3}\, C$$

as obtained previously. ■

PRACTICE EXERCISE 4.9

If $\mathbf{D} = (2y^2 + z)\mathbf{a}_x + 4xy\mathbf{a}_y + x\mathbf{a}_z$ C/m², find

(a) The volume charge density at $(-1, 0, 3)$

(b) The flux through the cube defined by $0 \le x \le 1$, $0 \le y \le 1$, $0 \le z \le 1$

(c) The total charge enclosed by the cube

ANSWER (a) -4 C/m³, (b) 2 C, (c) 2 C.

EXAMPLE 4.10

A charge distribution with spherical symmetry has density

$$\rho_v = \begin{cases} \dfrac{\rho_o r}{R}, & 0 \le r \le R \\ 0, & r > R \end{cases}$$

Determine **E** everywhere.

SOLUTION Since symmetry exists, we can apply Gauss's law to find **E**.

$$\varepsilon_o \oint \mathbf{E} \cdot d\mathbf{S} = Q_{\text{enc}} = \int \rho_v \, dv$$

(a) For $r < R$

$$\varepsilon_o E_r \, 4\pi r^2 = Q_{\text{enc}} = \int_0^r \int_0^\pi \int_0^{2\pi} \rho_v \, r^2 \sin\theta \, d\phi \, d\theta \, dr$$

$$= \int_0^r 4\pi r^2 \frac{\rho_o r}{R} \, dr = \frac{\rho_o \pi r^4}{R}$$

or

$$\mathbf{E} = \frac{\rho_o r^2}{4\varepsilon_o R} \mathbf{a}_r$$

(b) For $r > R$,

$$\varepsilon_o E_r 4\pi r^2 = Q_{\text{enc}} = \int_0^r \int_0^\pi \int_0^{2\pi} \rho_v r^2 \sin\theta \, d\phi \, d\theta \, dr$$

$$= \int_0^R \frac{\rho_o r}{R} 4\pi r^2 \, dr + \int_R^r 0 \cdot 4\pi r^2 \, dr$$

$$= \pi \rho_o R^3$$

or

$$\mathbf{E} = \frac{\rho_o R^3}{4\varepsilon_o r^2} \mathbf{a}_r$$

■

PRACTICE EXERCISE 4.10

A charge distribution in free space has $\rho_v = 2r$ nC/m^3 for $0 \leq r \leq 10$ m and zero otherwise. Determine **E** at $r = 2$ m and $r = 12$ m.

ANSWER $226\mathbf{a}_r$ V/m, $3.927\mathbf{a}_r$ kV/m.

4.7 ELECTRIC POTENTIAL

From our discussions in the preceding sections, the electric field intensity **E** due to a charge distribution can be obtained from Coulomb's law in general or from Gauss's law when the charge distribution is symmetric. Another way of obtaining **E** is from the electric scalar potential *V* to be defined in this section. In a sense, this way of finding **E** is easier because it is easier to handle scalars than vectors.

Suppose we wish to move a point charge Q from point A to point B in an electric field $\mathbf{E}$ as shown in Figure 4.19. From Coulomb's law, the force on Q is $\mathbf{F} = Q\mathbf{E}$ so that the *work done* in displacing the charge by $d\mathbf{l}$ is

$$dW = -\mathbf{F} \cdot d\mathbf{l} = -Q\mathbf{E} \cdot d\mathbf{l} \tag{4.58}$$

The negative sign indicates that the work is being done by an external agent. Thus the total work done, or the potential energy required, in moving Q from A to B is

$$\boxed{W = -Q \int_A^B \mathbf{E} \cdot d\mathbf{l}} \tag{4.59}$$

Dividing W by Q in eq. (4.59) gives the potential energy per unit charge. This quantity, denoted by V_{AB}, is known as the *potential difference* between points A and B. Thus

$$\boxed{V_{AB} = \frac{W}{Q} = -\int_A^B \mathbf{E} \cdot d\mathbf{l}} \tag{4.60}$$

Note that

1. In determining V_{AB}, A is the initial point while B is the final point.
2. If V_{AB} is negative, there is a loss in potential energy in moving Q from A to B; this implies that the work is being done by the field. However, if V_{AB} is positive, there is a gain in potential energy in the movement; an external agent performs the work.
3. V_{AB} is independent of the path taken (to be shown a little later).
4. V_{AB} is measured in joules per coulomb, commonly referred to as volts (V).

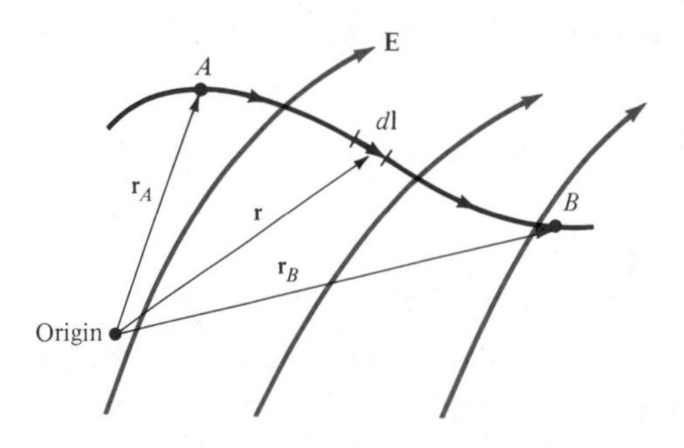

Figure 4.19 Dispacement of point charge Q in an electrostatic field $\mathbf{E}$.

As an example, if the **E** field in Figure 4.19 is due to a point charge Q located at the origin, then

$$\mathbf{E} = \frac{Q}{4\pi\varepsilon_o r^2}\,\mathbf{a}_r \tag{4.61}$$

so eq. (4.60) becomes

$$V_{AB} = -\int_{r_A}^{r_B} \frac{Q}{4\pi\varepsilon_o r^2}\,\mathbf{a}_r \cdot dr\,\mathbf{a}_r \tag{4.62a}$$

$$= \frac{Q}{4\pi\varepsilon_o}\left[\frac{1}{r_B} - \frac{1}{r_A}\right]$$

or

$$V_{AB} = V_B - V_A \tag{4.62b}$$

where V_B and V_A are the *potentials* (or *absolute potentials*) at B and A respectively. Thus the potential difference V_{AB} may be regarded as the potential at B with reference to A. In problems involving point charges, it is customary to choose infinity as reference; that is, we assume the potential at infinity is zero. Thus if $V_A = 0$ as $r_A \to \infty$ in eq. (4.62), the potential at any point ($r_B \to r$) due to a point charge Q located at the origin is

$$\boxed{V = \frac{Q}{4\pi\varepsilon_o r}} \tag{4.63}$$

pt charge located at origin

Note from eq. (4.62a) that because **E** points in the radial direction, any contribution from a displacement in the θ or ϕ direction is wiped out by the dot product $\mathbf{E} \cdot d\mathbf{l} = E \cos\theta\, dl = E\, dr$. Hence the potential difference V_{AB} is independent of the path as asserted earlier.

In general, the potential at any point is defined as the potential difference between that point and a chosen point at which the potential is zero. In other words, by assuming zero potential at infinity, the potential at a distance r from the point charge is the work done per unit charge by an external agent in transferring a test charge from infinity to that point. Thus

$$V = -\int_{\infty}^{r} \mathbf{E} \cdot d\mathbf{l} \tag{4.64}$$

If the point charge Q in eq. (4.63) is not located at the origin but at a point whose position vector is $\mathbf{r}'$, the potential $V(x, y, z)$ or simply $V(\mathbf{r})$ at $\mathbf{r}$ becomes

$$V(\mathbf{r}) = \frac{Q}{4\pi\varepsilon_o|\mathbf{r} - \mathbf{r}'|} \tag{4.65}$$

We have considered the electric potential due to a point charge. The same basic ideas apply to other types of charge distribution because any charge distribution can be regarded as consisting of point charges. The superposition principle, which we applied to electric fields, applies to potentials. For n point charges $Q_1, Q_2, \ldots, Q_n$ located at points with position vectors $\mathbf{r}_1, \mathbf{r}_2, \ldots, \mathbf{r}_n$, the potential at $\mathbf{r}$ is

$$V(\mathbf{r}) = \frac{Q_1}{4\pi\varepsilon_o|\mathbf{r} - \mathbf{r}_1|} + \frac{Q_2}{4\pi\varepsilon_o|\mathbf{r} - \mathbf{r}_2|} + \cdots + \frac{Q_n}{4\pi\varepsilon_o|\mathbf{r} - \mathbf{r}_n|}$$

or

$$V(\mathbf{r}) = \frac{1}{4\pi\varepsilon_o} \sum_{k=1}^{n} \frac{Q_k}{|\mathbf{r} - \mathbf{r}_k|} \quad \text{(point charges)} \tag{4.66}$$

For continuous charge distributions, we replace Q_k in eq. (4.66) with charge element $\rho_L\, dl$, $\rho_S\, dS$, or $\rho_v\, dv$ and the summation becomes an integration, so the potential at $\mathbf{r}$ becomes

$$V(\mathbf{r}) = \frac{1}{4\pi\varepsilon_o} \int_L \frac{\rho_L(\mathbf{r}')dl'}{|\mathbf{r} - \mathbf{r}'|} \quad \text{(line charge)} \tag{4.67}$$

$$V(\mathbf{r}) = \frac{1}{4\pi\varepsilon_o} \int_S \frac{\rho_S(\mathbf{r}')dS'}{|\mathbf{r} - \mathbf{r}'|} \quad \text{(surface charge)} \tag{4.68}$$

$$V(\mathbf{r}) = \frac{1}{4\pi\varepsilon_o} \int_v \frac{\rho_v(\mathbf{r}')dv'}{|\mathbf{r} - \mathbf{r}'|} \quad \text{(volume charge)} \tag{4.69}$$

where the primed coordinates are used customarily to denote source point location and the unprimed coordinates refer to field point (the point at which V is to be determined).

The following points should be noted:

1. We recall that in obtaining eqs. (4.65) to (4.69), the zero potential (reference) point has been chosen arbitrarily to be at infinity. If any other point is chosen as reference, eq. (4.65), for example, becomes

$$V = \frac{Q}{4\pi\varepsilon_o r} + C \tag{4.70}$$

   where C is a constant that is determined at the chosen point of reference. The same idea applies to eqs. (4.66) to (4.69).

2. The potential at a point can be determined in two ways depending on whether the charge distribution or $\mathbf{E}$ is known. If the charge distribution is not known, we use one of eqs. (4.65) to (4.70) depending on the charge distribution. If $\mathbf{E}$ is known, we simply use

$$V = -\int \mathbf{E} \cdot d\mathbf{l} + C. \tag{4.71}$$

The potential difference V_{AB} can be found generally from

$$V_{AB} = V_B - V_A = -\int_A^B \mathbf{E} \cdot d\mathbf{l} = \frac{W}{Q} \tag{4.72}$$

EXAMPLE 4.11

Two point charges $-4\ \mu$C and $5\ \mu$C are located at $(2, -1, 3)$ and $(0, 4, -2)$ respectively. Find the potential at $(1, 0, 1)$ assuming zero potential at infinity.

SOLUTION Let

$$Q_1 = -4\mu\text{C}, \qquad Q_2 = 5\ \mu\text{C}$$

$$V(\mathbf{r}) = \frac{Q_1}{4\pi\varepsilon_o|\mathbf{r} - \mathbf{r}_1|} + \frac{Q_2}{4\pi\varepsilon_o|\mathbf{r} - \mathbf{r}_2|} + C_o$$

If $V(\infty) = 0$, $C_o = 0$,

$$|\mathbf{r} - \mathbf{r}_1| = |(1, 0, 1) - (2, -1, 3)| = |(-1, 1, -2)| = \sqrt{6}$$

$$|\mathbf{r} - \mathbf{r}_2| = |(1, 0, 1) - (0, 4, -2)| = |(1, -4, 3)| = \sqrt{26}$$

Hence

$$V(1, 0, 1) = \frac{10^{-6}}{4\pi \times \dfrac{10^{-9}}{36\pi}} \left[\frac{-4}{\sqrt{6}} + \frac{5}{\sqrt{26}}\right]$$

$$= 9 \times 10^3\,(-1.633 + 0.9806)$$

$$= -5.872 \text{ kV}$$

PRACTICE EXERCISE 4.11

If point charge $3\ \mu$C is located at the origin in addition to the two charges of example 4.11, find the potential at $(-1, 5, 2)$ assuming $V(\infty) = 0$.

ANSWER 10.23 kV.

EXAMPLE 4.12

A point charge 5 nC is located at $(-3, 4, 0)$ while line $y = 1$, $z = 1$ carries uniform charge 2 nC/m.

(a) If $V = 0$ V at $O(0, 0, 0)$, find V at $A(5, 0, 1)$.

(b) If $V = 100$ V at $B(1, 2, 1)$, find V at $C(-2, 5, 3)$.

(c) If $V = -5$ V at O, find V_{BC}.

SOLUTION Let the potential at any point be

$$V = V_Q + V_L$$

where V_Q and V_L are the contributions to V at that point due to the point charge and the line charge respectively. For the point charge,

$$V_Q = -\int \mathbf{E} \cdot d\mathbf{l} = -\int \frac{Q}{4\pi\varepsilon_o r^2} \mathbf{a}_r \cdot dr\, \mathbf{a}_r$$

$$= \frac{Q}{4\pi\varepsilon_o r} + C_1$$

For the infinite line charge,

$$V_L = -\int \mathbf{E} \cdot d\mathbf{l} = -\int \frac{\rho_L}{2\pi\varepsilon_o \rho} \mathbf{a}_\rho \cdot d\rho\, \mathbf{a}_\rho$$

$$= -\frac{\rho_L}{2\pi\varepsilon_o} \ln \rho + C_2$$

Hence,

$$V = -\frac{\rho_L}{2\pi\varepsilon_o} \ln \rho + \frac{Q}{4\pi\varepsilon_o r} + C$$

where $C = C_1 + C_2 =$ constant, ρ is the perpendicular distance from the line $y = 1$, $z = 1$ to the field point, and r is the distance from the point charge to the field point.

(a) If $V = 0$ at $O(0, 0, 0)$, and V at $A(5, 0, 1)$ is to be determined, we must first determine the values of ρ and r at O and A. Finding r is easy; we use eq. (2.31). To find ρ for any point (x, y, z), we utilize the fact that ρ is the perpendicular distance from (x, y, z) to line $y = 1$, $z = 1$, which is parallel to the x-axis. Hence ρ is the distance between (x, y, z) and $(x, 1, 1)$ because the distance vector between the two points is perpendicular to $\mathbf{a}_x$. Thus

$$\rho = |(x, y, z) - (x, 1, 1)| = \sqrt{(y - 1)^2 + (z - 1)^2}$$

Applying this for ρ and eq. (2.31) for r at points O and A, we obtain

$$\rho_O = |(0, 0, 0) - (0, 1, 1)| = \sqrt{2}$$

$$r_O = |(0, 0, 0) - (-3, 4, 0)| = 5$$

$$\rho_A = |(5, 0, 1) - (5, 1, 1)| = 1$$

$$r_A = |(5, 0, 1) - (-3, 4, 0)| = 9$$

Hence,

$$V_O - V_A = -\frac{\rho_L}{2\pi\varepsilon_o} \ln \frac{\rho_O}{\rho_A} + \frac{Q}{4\pi\varepsilon_o}\left[\frac{1}{r_O} - \frac{1}{r_A}\right]$$

$$= \frac{-2 \cdot 10^{-9}}{2\pi \cdot \dfrac{10^{-9}}{36\pi}} \ln \frac{\sqrt{2}}{1} + \frac{5 \cdot 10^{-9}}{4\pi \cdot \dfrac{10^{-9}}{36\pi}}\left[\frac{1}{5} - \frac{1}{9}\right]$$

$$0 - V_A = -36 \ln \sqrt{2} + 45 \left(\frac{1}{5} - \frac{1}{9}\right)$$

or

$$V_A = 36 \ln \sqrt{2} - 4 = 8.477 \text{ V}$$

Notice that we have avoided calculating the constant C by subtracting one potential from another and that it does not matter which one is subtracted from which.

(b) If $V = 100$ at $B(1, 2, 1)$ and V at $C(-2, 5, 3)$ is to be determined, we find

$$\rho_B = |(1, 2, 1) - (1, 1, 1)| = 1$$

$$r_B = |(1, 2, 1) - (-3, 4, 0)| = \sqrt{21}$$

$$\rho_C = |(-2, 5, 3) - (-2, 1, 1)| = \sqrt{20}$$

$$r_C = |(-2, 5, 3) - (-3, 4, 0)| = \sqrt{11}$$

$$V_C - V_B = -\frac{\rho_L}{2\pi\varepsilon_o} \ln \frac{\rho_O}{\rho_B} + \frac{Q}{4\pi\varepsilon_o}\left[\frac{1}{r_C} - \frac{1}{r_B}\right]$$

$$V_C - 100 = -36 \ln \frac{\sqrt{20}}{1} + 45 \cdot \left[\frac{1}{\sqrt{11}} - \frac{1}{\sqrt{21}}\right]$$

$$= -50.175 \text{ V}$$

or

$$V_C = 49.825 \text{ V}$$

(c) To find the potential difference between two points, we do not need a potential reference if a common reference is assumed.

$$V_{BC} = V_C - V_B$$

$$= -\ 50.175 \text{ V}$$

as obtained in part (b). ■

PRACTICE EXERCISE 4.12

A point charge of 5 nC is located at the origin. If $V = 2$ V at $(0, 6, -8)$, find

(a) The potential at $A(-3, 2, 6)$

(b) The potential at $B(1, 5, 7)$

(c) The potential difference V_{AB}

ANSWER (a) 3.929 V, (b) 2.696 V, (c) −1.233 V.

EXAMPLE 4.13

A line of length ℓ carries charge λ C/m. Show that the potential in the median plane can be written as

$$V = \frac{\lambda}{4\pi\varepsilon_o} \ln \frac{1 + \sin\alpha}{1 - \sin\alpha}$$

where $\tan\alpha = \dfrac{\ell}{2r}$.

SOLUTION Consider the line charge as in Figure 4.20. At point P, the potential is given by eq. (4.67); that is,

$$V = \int \frac{\rho_L\, dl}{4\pi\varepsilon_o R}$$

where $dl = dz$, $\rho_L = \lambda$, and $R = \sqrt{r^2 + z^2}$. Hence,

$$V = \int_{-\ell/2}^{\ell/2} \frac{\lambda}{4\pi\varepsilon_o} \frac{dz}{[r^2 + z^2]^{1/2}}$$

To evaluate this integral, we change variables. Let $\tan\theta = z/r$ as in Figure 4.20, that is,

$$z = r\tan\theta, \qquad dz = r\sec^2\theta\, d\theta$$

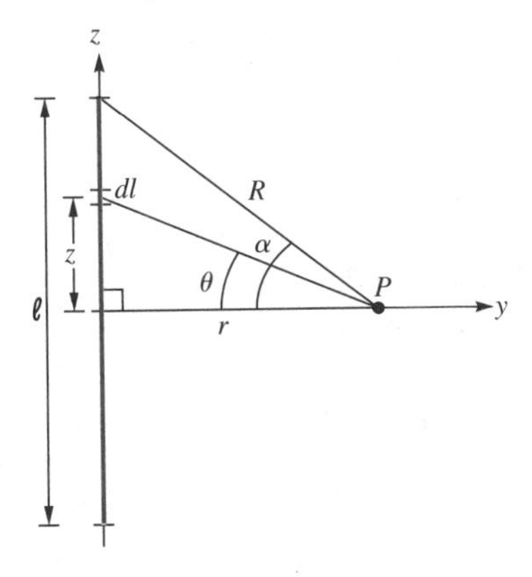

Figure 4.20 For example 4.13.

$$V = \frac{\lambda}{4\pi\varepsilon_o}\int_{-\alpha}^{\alpha}\frac{r\sec^2\theta\, d\theta}{r\sec\theta}$$

$$= \frac{\lambda}{4\pi\varepsilon_o}\int_{-\alpha}^{\alpha}\sec\theta\, d\theta = \frac{\lambda}{4\pi\varepsilon_o}\ln(\sec\theta + \tan\theta)\Big|_{-\alpha}^{\alpha}$$

$$= \frac{\lambda}{4\pi\varepsilon_o}\ln\frac{1+\sin\alpha}{1-\sin\alpha}$$

as required. ■

PRACTICE EXERCISE 4.13

A thin square loop carries a uniform charge of ρ_L. Show that the potential at the center of the loop is

$$V = \frac{2\rho_L}{\pi\varepsilon_o}\ln(1+\sqrt{2})$$

ANSWER Proof.

4.8 RELATIONSHIP BETWEEN E AND V—MAXWELL'S EQUATION

As shown in the previous section, the potential difference between points A and B is independent of the path taken. Hence,

$$V_{BA} = -V_{AB}$$

that is, $V_{BA} + V_{AB} = \oint \mathbf{E} \cdot d\mathbf{l} = 0$

or

$$\boxed{\oint \mathbf{E} \cdot d\mathbf{l} = 0} \tag{4.73}$$

This shows that the line integral of **E** along a closed path as shown in Figure 4.21 must be zero. Physically, this implies that no net work is done in moving a charge along a closed path in an electrostatic field. Applying Stokes's theorem to eq. (4.73) gives

$$\oint \mathbf{E} \cdot d\mathbf{l} = \int (\nabla \times \mathbf{E}) \cdot d\mathbf{S} = 0$$

or

$$\boxed{\nabla \times \mathbf{E} = 0} \tag{4.74}$$

For a more rigorous proof of the equation, see Problem 4.34. Any vector field that satisfies eq. (4.73) or (4.74) is said to be conservative, or irrotational, as discussed in Section 3.8. Thus an electrostatic field is a conservative field. Equation (4.73) or (4.74) is referred to as *Maxwell's equation* (the second Maxwell's equation to be derived) for static electric fields. Equation (4.73) is the integral form, and eq. (4.74) is the differential form; they both depict the conservative nature of an electrostatic field.

From the way we defined potential, $V = -\int \mathbf{E} \cdot d\mathbf{l}$, it follows that

$$dV = -\mathbf{E} \cdot d\mathbf{l} = -E_x\,dx - E_y\,dy - E_z\,dz$$

But

$$dV = \frac{\partial V}{\partial x}dx + \frac{\partial V}{\partial y}dy + \frac{\partial V}{\partial z}dz$$

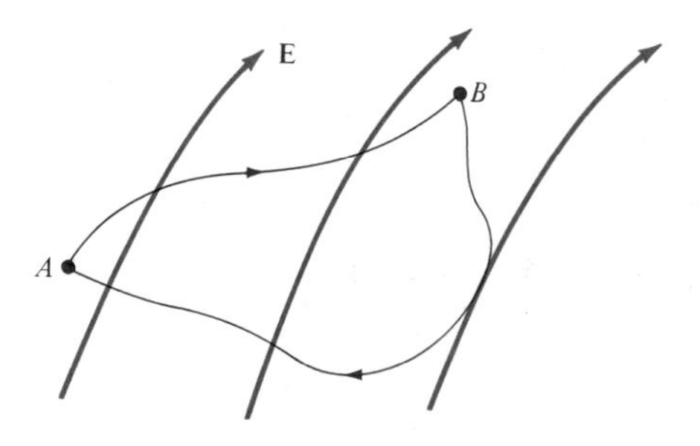

Figure 4.21 Conservative nature of an electrostatic field.

Comparing the two expressions for dV, we obtain

$$E_x = -\frac{\partial V}{\partial x}, \qquad E_y = -\frac{\partial V}{\partial y}, \qquad E_z = -\frac{\partial V}{\partial z} \tag{4.75}$$

Thus:

$$\boxed{\mathbf{E} = -\nabla V} \tag{4.76}$$

that is, the electric field intensity is the gradient of V. The negative sign shows that the direction of **E** is opposite to the direction in which V increases; **E** is directed from higher to lower levels of V. Since the curl of the gradient of a scalar function is always zero ($\nabla \times \nabla V = 0$), eq. (4.74) obviously implies that **E** must be a gradient of some scalar function. Thus eq. (4.76) could have been obtained from eq. (4.74).

Equation (4.76) shows another way to obtain the **E** field apart from using Coulomb's or Gauss's law. That is, if the potential field V is known, the **E** can be found using eq. (4.76). One may wonder how one function V can possibly contain all the information that the three components of **E** carry. The three components of **E** are not independent of one another: They are explicitly interrelated by the condition $\nabla \times \mathbf{E} = 0$. What the potential formulation does is to exploit this feature to maximum advantage, reducing a vector problem to a scalar one.

EXAMPLE 4.14

Given the potential $V = \dfrac{10}{r^2} \sin\theta \cos\phi$,

(a) Find the electric flux density **D** at $(2, \pi/2, 0)$.

(b) Calculate the work done in moving a 10-μC charge from point $A(1, 30°, 120°)$ to $B(4, 90°, 60°)$.

SOLUTION

(a) $\mathbf{D} = \varepsilon_o \mathbf{E}$

But

$$\mathbf{E} = -\nabla V = -\left[\frac{\partial V}{\partial r}\mathbf{a}_r + \frac{1}{r}\frac{\partial V}{\partial \theta}\mathbf{a}_\theta + \frac{1}{r\sin\theta}\frac{\partial V}{\partial \phi}\mathbf{a}_\phi\right]$$

$$= \frac{20}{r^3}\sin\theta\cos\phi\,\mathbf{a}_r - \frac{10}{r^3}\cos\theta\cos\phi\,\mathbf{a}_\theta + \frac{10}{r^3}\sin\phi\,\mathbf{a}_\phi$$

At $(2, \pi/2, 0)$,

$$\mathbf{D} = \varepsilon_o\mathbf{E}\,(r = 2, \theta = \pi/2, \phi = 0) = \varepsilon_o\left(\frac{20}{8}\mathbf{a}_r - 0\mathbf{a}_\theta + 0\mathbf{a}_\phi\right)$$

$$= 2.5\varepsilon_o\mathbf{a}_r \text{ C/m}^2 = 22.1\ \mathbf{a}_r \text{ pC/m}^2$$

(b) The work done can be found in two ways, using either **E** or V.

Method 1:

$$W = -Q \int \mathbf{E} \cdot d\mathbf{l} \quad \text{or} \quad -\frac{W}{Q} = \int \mathbf{E} \cdot d\mathbf{l}$$

and because the electrostatic field is conservative, the path of integration is immaterial. Hence the work done in moving Q from $A(1, 30°, 120°)$ to $B(4, 90°, 60°)$ is the same as that in moving Q from A to A', from A' to B', and from B' to B where

$$\begin{array}{ccc} A(1, 30°, 120°) & & B(4, 90°, 60°) \\ \downarrow\ d\mathbf{l} = dr\,\mathbf{a}_r & d\mathbf{l} = r\,d\theta\,\mathbf{a}_\theta & \uparrow\ d\mathbf{l} = r\sin\theta\,d\phi\,\mathbf{a}_\phi \\ A'(4, 30°, 120°) & \rightarrow & B'(4, 90°, 120°). \end{array}$$

That is, instead of moving Q directly from A and B, it is moved from $A \to A'$, $A' \to B'$, $B' \to B$ so that only one variable is changed at a time. This makes the line integral a lot easier to evaluate. Thus

$$\frac{-W}{Q} = -\frac{1}{Q}(W_{AA'} + W_{A'B'} + W_{B'B})$$

$$= \left(\int_{AA'} + \int_{A'B'} + \int_{B'B}\right) \mathbf{E} \cdot d\mathbf{l}$$

$$= \int_{r=1}^{4} \frac{20 \sin\theta \cos\phi}{r^3}\, dr \,\Bigg|_{\theta = 30°,\ \phi = 120°}$$

$$+ \int_{\theta = 30°}^{90°} \frac{-10\cos\theta\cos\phi}{r^3}\, r\, d\theta \,\Bigg|_{r = 4,\ \phi = 120°}$$

$$+ \int_{\phi = 120°}^{60°} \frac{10 \sin\phi}{r^3}\, r \sin\theta\, d\phi \,\Bigg|_{r = 4,\ \theta = 90°}$$

$$= 20\left(\frac{1}{2}\right)\left(\frac{-1}{2}\right)\left[-\frac{1}{2r^2}\Bigg|_{r=1}^{4}\right]$$

$$- \frac{10}{16}\frac{(-1)}{2} \sin\theta \Bigg|_{30°}^{90°} + \frac{10}{16}(1)\left[-\cos\phi\Bigg|_{120°}^{60°}\right]$$

$$-\frac{W}{Q} = \frac{-75}{32} + \frac{5}{32} - \frac{10}{16}$$

or

$$W = \frac{45}{16} Q = 28.125\ \mu\text{J}$$

Method 2:

Since V is known, this method is a lot easier.

$$W = -Q \int_A^B \mathbf{E} \cdot d\mathbf{l} = QV_{AB}$$

$$= Q(V_B - V_A)$$

$$= 10 \left(\frac{10}{16} \sin 90° \cos 60° - \frac{10}{1} \sin 30° \cos 120° \right) \cdot 10^{-6}$$

$$= 10 \left(\frac{10}{32} - \frac{-5}{2} \right) \cdot 10^{-6}$$

$$= 28.125\ \mu\text{J as obtained before.}$$

■

PRACTICE EXERCISE 4.14

Given that $\mathbf{E} = (3x^2 + y)\,\mathbf{a}_x + x\mathbf{a}_y$ kV/m, find the work done in moving a $-2\ \mu$C charge from (0, 5, 0) to (2, −1, 0) by taking the path

(a) (0, 5, 0) → (2, 5, 0) → (2, −1, 0)

(b) $y = 5 - 3x$

ANSWER (a) 12 mJ, (b) 12 mJ.

4.9 AN ELECTRIC DIPOLE

An *electric dipole* is formed when two point charges of equal magnitude but opposite sign are separated by a small distance. The importance of the field due to a dipole will be evident in the subsequent chapters.

Consider the dipole shown in Figure 4.22. The potential at point $P(r, \theta, \phi)$ is given by

$$V = \frac{Q}{4\pi\varepsilon_o}\left[\frac{1}{r_1} - \frac{1}{r_2}\right] = \frac{Q}{4\pi\varepsilon_o}\left[\frac{r_2 - r_1}{r_1 r_2}\right] \qquad [4.77]$$

where r_1 and r_2 are the distances between P and $+Q$ and P and $-Q$ respectively. If $r \gg d$, $r_2 - r_1 \simeq d \cos\theta$, $r_2 r_1 \simeq r^2$, and eq. (4.77) becomes

$$V = \frac{Q}{4\pi\varepsilon_o}\frac{d\cos\theta}{r^2} \qquad [4.78]$$

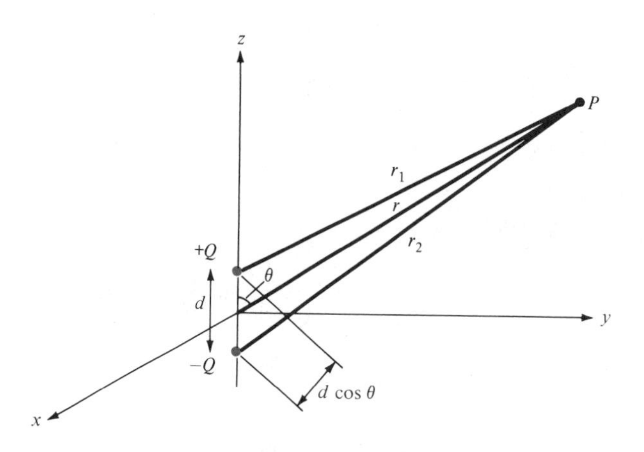

Figure 4.22 An electric dipole.

Since $d \cos\theta = \mathbf{d} \cdot \mathbf{a}_r$, where $\mathbf{d} = d\mathbf{a}_z$, if we define

$$\mathbf{p} = Q\mathbf{d} \tag{4.79}$$

as the *dipole moment*, eq. (4.78) may be written as

$$V = \frac{\mathbf{p} \cdot \mathbf{a}_r}{4\pi\varepsilon_o r^2} \tag{4.80}$$

Note that the dipole moment $\mathbf{p}$ is directed from $-Q$ to $+Q$. If the dipole center is not at the origin but at $\mathbf{r}'$, eq. (4.80) becomes

$$\boxed{V(\mathbf{r}) = \frac{\mathbf{p} \cdot (\mathbf{r} - \mathbf{r}')}{4\pi\varepsilon_o|\mathbf{r} - \mathbf{r}'|^3}} \tag{4.81}$$

The electric field due to the dipole with center at the origin, shown in Figure 4.22, can be obtained readily from eqs. (4.76) and (4.78) as

$$\mathbf{E} = -\nabla V = -\left[\frac{\partial V}{\partial r}\mathbf{a}_r + \frac{1}{r}\frac{\partial V}{\partial \theta}\mathbf{a}_\theta\right]$$

$$= \frac{Qd\cos\theta}{2\pi\varepsilon_o r^3}\mathbf{a}_r + \frac{Qd\sin\theta}{4\pi\varepsilon_o r^3}\mathbf{a}_\theta$$

or

$$\boxed{\mathbf{E} = \frac{p}{4\pi\varepsilon_o r^3}(2\cos\theta\,\mathbf{a}_r + \sin\theta\,\mathbf{a}_\theta)} \tag{4.82}$$

where $p = |\mathbf{p}| = Qd$.

Notice that a point charge is a *monopole* and its electric field varies inversely as r^2 while its potential field varies inversely as r (see eqs. (4.61) and (4.63)). From eqs. (4.80) and (4.82), we notice that the electric field due to a dipole varies inversely as r^3 while its potential varies inversely as r^2. The electric fields due to successive higher-order multipoles (such as a *quadrupole* consisting of two dipoles or an *octupole* consisting of two quadrupoles) vary inversely as $r^4, r^5, r^6, \ldots$ while their corresponding potentials vary inversely as $r^3, r^4, r^5, \ldots$.

EXAMPLE 4.15

Two dipoles with dipole moments $-5\mathbf{a}_z$ nC · m and $9\mathbf{a}_z$ nC · m are located at points $(0, 0, -2)$ and $(0, 0, 3)$ respectively. Find the potential at the origin.

SOLUTION

$$V = \sum_{k=1}^{2} \frac{\mathbf{p}_k \cdot \mathbf{r}_k}{4\pi\varepsilon_o r_k^3}$$

$$= \frac{1}{4\pi\varepsilon_o}\left[\frac{\mathbf{p}_1 \cdot \mathbf{r}_1}{r_1^3} + \frac{\mathbf{p}_2 \cdot \mathbf{r}_2}{r_2^3}\right]$$

where

$$\mathbf{p}_1 = -5\mathbf{a}_z, \qquad \mathbf{r}_1 = (0, 0, 0) - (0, 0, -2) = 2\mathbf{a}_z, \qquad r_1 = |\mathbf{r}_1| = 2$$

$$\mathbf{p}_2 = 9\mathbf{a}_z, \qquad \mathbf{r}_2 = (0, 0, 0) - (0, 0, 3) = -3\mathbf{a}_z, \qquad r_2 = |\mathbf{r}_2| = 3$$

Hence,

$$V = \frac{1}{4\pi \cdot \dfrac{10^{-9}}{36\pi}}\left[\frac{-10}{2^3} - \frac{27}{3^3}\right] \cdot 10^{-9}$$

$$= -20.25 \text{ V}$$

■

PRACTICE EXERCISE 4.15

An electric dipole of 100 $\mathbf{a}_z$ pC · m is located at the origin. Find V and $\mathbf{E}$ at points

(a) $(0, 0, 10)$

(b) $(1, \pi/3, \pi/2)$

ANSWER (a) 9 mV, $1.8\mathbf{a}_r$ mV/m, (b) 0.45 V, $0.9\mathbf{a}_r + 0.7794\mathbf{a}_\theta$ V/m.

†4.10 ELECTRIC FLUX LINES AND EQUIPOTENTIAL SURFACES

The idea of *electric flux* lines (or *electric lines of force* as they are sometimes called) was introduced by Michael Faraday (1791–1867) in his experimental investigation as a way of visualizing the electric field. An electric flux line is an imaginary path or line drawn in such a way that its direction at any point is the direction of the electric field at that point. In other words, they are the lines to which the electric field density **D** is tangential at every point.

The following properties of electric flux lines should be noted:

1. The lines always start at positive charges and terminate at negative charges. Positive charges are thus regarded as *sources* and negative charges as *sinks* of electric flux lines.
2. No two flux lines can intersect except at singular or equilibrium points; that is, those points at which the resultant **D** is zero.
3. The electric flux density **D** is tangential to the flux lines at every point.

To determine the analytic expression for flux lines, we let $\mathbf{D} = D_x\mathbf{a}_x + D_y\mathbf{a}_y$. Since property (3) above must hold, it is evident from Figure 4.23 that at any point (x, y):

$$\frac{\Delta y}{\Delta x} = \frac{D_y}{D_x} \qquad [4.83]$$

In general, for a three-dimensional field, $\mathbf{D} = D_x\mathbf{a}_x + D_y\mathbf{a}_y + D_z\mathbf{a}_z$, so in Cartesian coordinates, the flux lines are given by

$$\boxed{\frac{dx}{D_x} = \frac{dy}{D_y} = \frac{dz}{D_z}} \quad \text{(Cartesian)} \qquad [4.84]$$

Similarly, it can be shown that in cylindrical coordinates, where $\mathbf{D} = (D_\rho, D_\phi, D_z)$, the flux lines are given by

$$\boxed{\frac{d\rho}{D_\rho} = \frac{\rho\, d\phi}{D_\phi} = \frac{dz}{D_z}} \quad \text{(cylindrical)} \qquad [4.85]$$

and in spherical coordinates, $\mathbf{D} = (D_r, D_\theta, D_\phi)$, so the flux lines satisfy

$$\boxed{\frac{dr}{D_r} = \frac{r\, d\theta}{D_\theta} = \frac{r \sin\theta\, d\phi}{D_\phi}} \quad \text{(spherical)} \qquad [4.86]$$

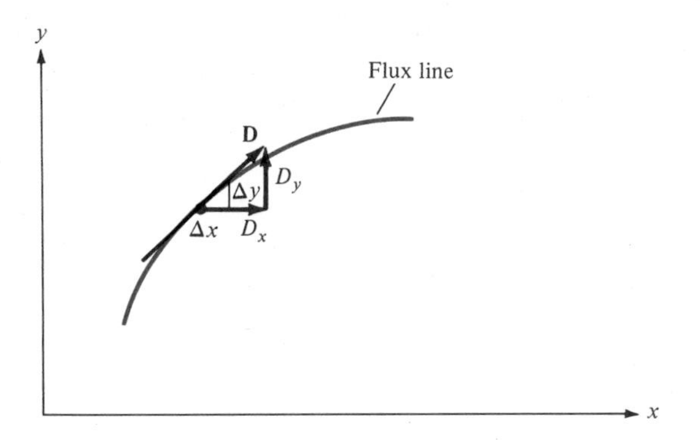

Figure 4.23 The tangent to the electric flux line.

Any surface on which the potential is the same throughout is known as an *equipotential surface*. The intersection of an equipotential surface and a plane results in a path or line known as an *equipotential line*. No work is done in moving a charge from one point to another along an equipotential line or surface ($V_A - V_B = 0$) and hence

$$\int \mathbf{E} \cdot d\mathbf{l} = 0 \tag{4.87}$$

on the line or surface. From eq. (4.87), we may conclude that the lines of force or flux lines (or the direction of **E**) are always normal to equipotential surfaces. Examples of equipotential surfaces for point charge and a dipole are shown in Figure 4.24. Note from these examples that the direction of **E** is everywhere normal to the equipotential lines. We shall see the importance of equipotential surfaces when we discuss conducting bodies in electric fields; it will suffice to say at this point that such bodies are equipotential volumes.

There are two ways to determine equipotential lines depending on whether V or **E** is known. If V is known, the equipotential lines or surfaces are given by

$$V = \text{constant} \tag{4.88}$$

On the other hand, if **E** is given, equipotential lines may be determined in a manner similar to our determination of the flux lines. From calculus, if a line has slope m, a normal line must have slope $-1/m$. Thus, for example, if $\mathbf{D} = (D_x, D_y, D_z)$, we know from eq. (4.83) that the flux lines on a z = constant plane are given by

$$\frac{dy}{dx} = \frac{D_y}{D_x} \tag{4.89}$$

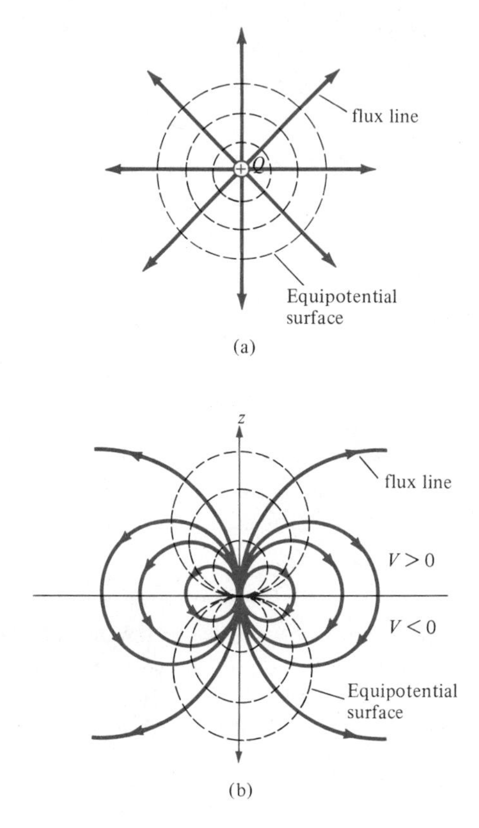

Figure 4.24 Equipotential surfaces for **(a)** a point charge and **(b)** an electric dipole.

Equipotential lines on the same z = constant plane are normal to the flux lines of eq. (4.89) and given by

$$\boxed{\frac{dy}{dx} = -\frac{D_x}{D_y}} \qquad [4.90]$$

The same idea applies to **D** in cylindrical and spherical coordinate systems.

A typical application of field mapping (flux lines and equipotential surfaces) is found in the diagnosis of the human heart. The human heart beats in response to an electric field potential difference across it. The heart can be characterized as a dipole with the field map similar to that of Figure 4.24(b). Such a field map is useful in detecting abnormal heart position.[6] In Section 14.2, we will discuss a numerical technique for field mapping.

....

[6]For more information on this, see R. Plonsey, *Bioelectric Phenomena*. New York: McGraw-Hill, 1969.

EXAMPLE 4.16

For the electric field intensity $\mathbf{E} = y\mathbf{a}_x + x\mathbf{a}_y$,

(a) Determine the general equation of the flux lines and equipotential lines.

(b) Find the lines that pass through (1, 4) and **E** at that point.

SOLUTION

(a) Since $\mathbf{D} = \varepsilon_o \mathbf{E}$, from eq. (4.84),

$$\frac{dy}{dx} = \frac{E_y}{E_x} = \frac{x}{y}$$

Hence,

$$\int y\, dy = \int x\, dx$$

or

$$y^2 = x^2 + C$$

where C is a constant. The flux lines are hyperbolas as shown in Figure 4.25. They are obtained by choosing C to be ± 1, ± 2, and so on. The asymptotes to the hyperbolas as obtained by setting $C = 0$ so that $y = \pm x$.

For the equipotential lines:

$$\frac{dy}{dx} = \frac{-1}{\text{slope of flux lines}}$$

$$= -\frac{E_x}{E_y} = \frac{-y}{x}$$

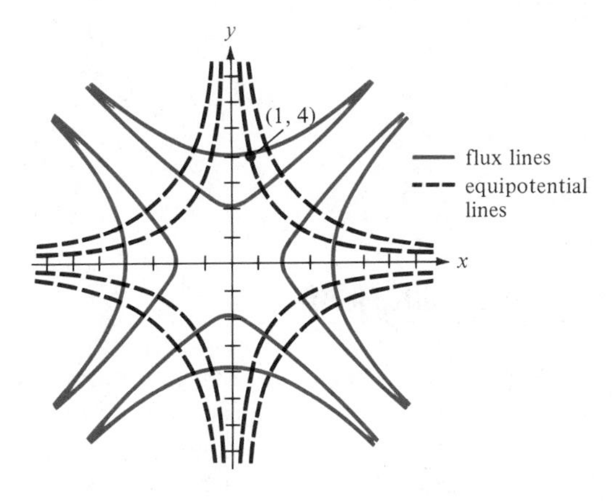

Figure 4.25 Sketch of the equipotential lines of Example 4.16.

Hence

$$-\int \frac{dy}{y} = \int \frac{dx}{x} \rightarrow \ln C_o = \ln y + \ln x$$

or

$$C_o = xy$$

where C_o is a constant. Thus equipotential lines are hyperbolas $C_o = xy$, which are perpendicular to the flux lines $C = y^2 - x^2$ as shown in Figure 4.25.

(b) At (1, 4) $x = 1$, $y = 4$:

$$16 = 1 + C \rightarrow C = 15$$

Hence the flux lines through point (1, 4) are $y^2 = x^2 + 15$. At (1, 4), $\mathbf{E} = 4\mathbf{a}_x + \mathbf{a}_y$. Note from Figure 4.25 that **E** is tangential to the flux lines at that point as expected.

At point (1, 4), $x = 1$ and $y = 4$, so $C_o = 4$. Hence, hyperbola $xy = 4$ is the equipotential line through (1, 4). Notice that in fact $C_o = V = xy$ is the potential that gives rise to **E** except for the negative sign: $\mathbf{E} = -\nabla V = -\nabla(xy) = -(y\mathbf{a}_x + x\mathbf{a}_y)$. ■

PRACTICE EXERCISE 4.16

(a) Obtain the equation of the flux lines for

$$\mathbf{E} = y^2\mathbf{a}_x + 2xy\mathbf{a}_y$$

Determine which of the lines passes through (2, −1, 3).

(b) Find the equipotential surfaces due to

(i) $V = x^2 + y^2$

(ii) $\mathbf{E} = y^2\mathbf{a}_x + 2xy\mathbf{a}_y$

ANSWER (a) $y^2 = 2x^2 + C_o$, $y^2 = 2x^2 - 7$

(b) (i) $x^2 + y^2 = \rho^2 = C_1$, (ii) $xy^2 = C_2$.

4.11 ENERGY DENSITY IN ELECTROSTATIC FIELDS

To determine the energy present in an assembly of charges, we must first determine the amount of work necessary to assemble them. Suppose we wish to position three point charges Q_1, Q_2, and Q_3 in an initially empty space shown shaded in Figure 4.26. No work is required to transfer Q_1 from infinity to P_1 because the space is initially charge free and there is no electric field (from eq. (4.59), $W = 0$). The work done in trans-

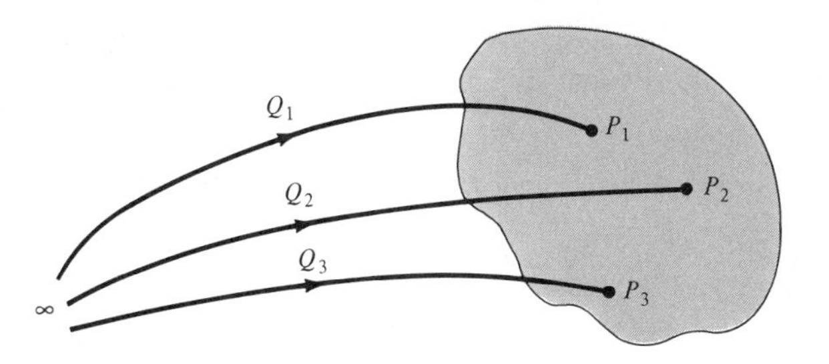

Figure 4.26 Assembling of charges.

ferring Q_2 from infinity to P_2 is equal to the product of Q_2 and the potential V_{21} at P_2 due to Q_1. Similarly, the work done in positioning Q_3 at P_3 is equal to $Q_3(V_{32} + V_{31})$, where V_{32} and V_{31} are the potentials at P_3 due to Q_2 and Q_1 respectively. Hence the total work done in positioning the three charges is

$$\begin{aligned} W_E &= W_1 + W_2 + W_3 \\ &= 0 + Q_2V_{21} + Q_3(V_{31} + V_{32}) \end{aligned} \tag{4.91}$$

If the charges were positioned in reverse order,

$$\begin{aligned} W_E &= W_3 + W_2 + W_1 \\ &= 0 + Q_2V_{23} + Q_1(V_{12} + V_{13}) \end{aligned} \tag{4.92}$$

where V_{23} is the potential at P_2 due to Q_3, V_{12} and V_{13} are respectively the potentials at P_1 due to Q_2 and Q_3. Adding eqs. (4.91) and (4.92) gives

$$\begin{aligned} 2W_E &= Q_1(V_{12} + V_{13}) + Q_2(V_{21} + V_{23}) + Q_3(V_{31} + V_{32}) \\ &= Q_1V_1 + Q_2V_2 + Q_3V_3 \end{aligned}$$

or

$$W_E = \frac{1}{2}(Q_1V_1 + Q_2V_2 + Q_3V_3) \tag{4.93}$$

where V_1, V_2, and V_3 are total potentials at P_1, P_2, and P_3 respectively. In general, if there are n point charges, eq. (4.93) becomes

$$\boxed{W_E = \frac{1}{2}\sum_{k=1}^{n} Q_k V_k} \quad \text{(in joules)} \tag{4.94}$$

If, instead of point charges, the region has a continuous charge distribution, the summation in eq. (4.94) becomes integration; that is,

$$W_E = \frac{1}{2}\int \rho_L V \, dl \qquad \text{(line charge)} \tag{4.95}$$

$$W_E = \frac{1}{2}\int \rho_S V \, dS \qquad \text{(surface charge)} \tag{4.96}$$

or

$$W_E = \frac{1}{2}\int \rho_v V \, dv \qquad \text{(volume charge)} \tag{4.97}$$

Since $\rho_v = \nabla \cdot \mathbf{D}$, eq. (4.97) can be further developed to yield

$$W_E = \frac{1}{2}\int_v (\nabla \cdot \mathbf{D}) \, V \, dv \tag{4.98}$$

But for any vector $\mathbf{A}$ and scalar V, the identity

$$\nabla \cdot V\mathbf{A} = \mathbf{A} \cdot \nabla V + V(\nabla \cdot \mathbf{A})$$

or

$$(\nabla \cdot \mathbf{A})V = \nabla \cdot V\mathbf{A} - \mathbf{A} \cdot \nabla V \tag{4.99}$$

holds. Applying the identity in eqs. (4.99) to (4.98), we get

$$W_E = \frac{1}{2}\int_v (\nabla \cdot V\mathbf{D}) \, dv - \frac{1}{2}\int_v (\mathbf{D} \cdot \nabla V) \, dv \tag{4.100}$$

By applying divergence theorem to the first term on the right-hand side of this equation, we have

$$W_E = \frac{1}{2}\oint_S (V\mathbf{D}) \cdot d\mathbf{S} - \frac{1}{2}\int_v (\mathbf{D} \cdot \nabla V) \, dv \tag{4.101}$$

From Section 4.9, we recall that V varies as $1/r$ and $\mathbf{D}$ as $1/r^2$ for point charges; V varies as $1/r^2$ and $\mathbf{D}$ as $1/r^3$ for dipoles; and so on. Hence, $V\mathbf{D}$ in the first term on the right-hand side of eq. (4.101) must vary at least as $1/r^3$ while dS varies as r^2. Consequently, the first integral in eq. (4.101) must tend to zero as the surface S becomes large. Hence, eq. (4.101) reduces to

$$W_E = -\frac{1}{2}\int_v (\mathbf{D} \cdot \nabla V) \, dv = \frac{1}{2}\int_v (\mathbf{D} \cdot \mathbf{E}) \, dv \tag{4.102}$$

and since $\mathbf{E} = -\nabla V$ and $\mathbf{D} = \varepsilon_o \mathbf{E}$

$$\boxed{W_E = \frac{1}{2}\int \mathbf{D} \cdot \mathbf{E} \, dv = \frac{1}{2}\int \varepsilon_o E^2 \, dv} \tag{4.103}$$

From this, we can define electrostatic energy density w_E (in J/m^3) as

$$w_E = \frac{dW_E}{dv} = \frac{1}{2}\mathbf{D}\cdot\mathbf{E} = \frac{1}{2}\varepsilon_o E^2 = \frac{D^2}{2\varepsilon_o} \quad [4.104]$$

so eq. (4.102) may be written as

$$W_E = \int w_E \, dv \quad [4.105]$$

EXAMPLE 4.17

Three point charges -1 nC, 4 nC, and 3 nC are located at (0, 0, 0), (0, 0, 1), and (1, 0, 0) respectively. Find the energy in the system.

SOLUTION

$$\begin{aligned} W &= W_1 + W_2 + W_3 \\ &= 0 + Q_2V_{21} + Q_3(V_{31} + V_{32}) \\ &= Q_2 \cdot \frac{Q_1}{4\pi\varepsilon_o\,|\,(0,0,1) - (0,0,0)\,|} \\ &\quad + \frac{Q_3}{4\pi\varepsilon_o}\left[\frac{Q_1}{|(1,0,0) - (0,0,0)|} + \frac{Q_2}{|(1,0,0) - (0,0,1)|}\right] \\ &= \frac{1}{4\pi\varepsilon_o}\left(Q_1Q_2 + Q_1Q_3 + \frac{Q_2Q_3}{\sqrt{2}}\right) \\ &= \frac{1}{4\pi \cdot \dfrac{10^{-9}}{36\pi}}\left(-4 - 3 + \frac{12}{\sqrt{2}}\right) \cdot 10^{-18} \\ &= 9\left(\frac{12}{\sqrt{2}} - 7\right) \text{ nJ} = 13.37 \text{ nJ} \end{aligned}$$

Alternatively,

$$\begin{aligned} W &= \frac{1}{2}\sum_{k=1}^{3} Q_kV_k = \frac{1}{2}(Q_1V_1 + Q_2V_2 + Q_3V_3) \\ &= \frac{Q_1}{2}\left[\frac{Q_2}{4\pi\varepsilon_o(1)} + \frac{Q_3}{4\pi\varepsilon_o(1)}\right] + \frac{Q_2}{2}\left[\frac{Q_1}{4\pi\varepsilon_o(1)} + \frac{Q_3}{4\pi\varepsilon_o(\sqrt{2})}\right] \\ &\quad + \frac{Q_3}{2}\left[\frac{Q_1}{4\pi\varepsilon_o(1)} + \frac{Q_2}{4\pi\varepsilon_o(\sqrt{2})}\right] \end{aligned}$$

$$= \frac{1}{4\pi\varepsilon_o}\left(Q_1Q_2 + Q_1Q_3 + \frac{Q_2Q_3}{\sqrt{2}}\right)$$

$$= 9\left(\frac{12}{\sqrt{2}} - 7\right) \text{ nJ} = 13.37 \text{ nJ}$$

as obtained previously.

PRACTICE EXERCISE 4.17

Point charges $Q_1 = 1$ nC, $Q_2 = -2$nC, $Q_3 = 3$ nC, and $Q_4 = -4$ nC are positioned one at a time and in that order at (0, 0, 0), (1, 0, 0), (0, 0, −1), and (0, 0, 1) respectively. Calculate the energy in the system after each charge is positioned.

ANSWER 0, −18 nJ, −29.18 nJ, −68.27 nJ.

EXAMPLE 4.18

A charge distribution with spherical symmetry has density

$$\rho_v = \begin{cases} \rho_o, & 0 \leq r \leq R \\ 0, & r > R \end{cases}$$

Determine V everywhere and the energy stored in region $r < R$.

SOLUTION The **D** field has already been found in Section 4.6D using Gauss's law.

(a) For $r \geqslant R$, $\mathbf{E} = \dfrac{\rho_o R^3}{3\varepsilon_o r^2}\mathbf{a}_r$.

Once **E** is known, V is determined as

$$V = -\int \mathbf{E} \cdot d\mathbf{l} = -\frac{\rho_o R^3}{3\varepsilon_o}\int \frac{1}{r^2}\,dr$$

$$= \frac{\rho_o R^3}{3\varepsilon_o r} + C_1, \qquad r \geqslant R$$

Since $V(r = \infty) = 0$, $C_1 = 0$.

(b) For $r \leqslant R$, $\mathbf{E} = \dfrac{\rho_o r}{3\varepsilon_o}\mathbf{a}_r$.

Hence,

$$V = -\int \mathbf{E} \cdot d\mathbf{l} = -\frac{\rho_o}{3\varepsilon_o} \int r\, dr$$

$$= -\frac{\rho_o r^2}{6\varepsilon_o} + C_2$$

From part (a) $V(r = R) = \dfrac{\rho_o R^2}{3\varepsilon_o}$. Hence,

$$\frac{R^2 \rho_o}{3\varepsilon_o} = \frac{\rho_o R^2}{6\varepsilon_o} + C_2 \rightarrow C_2 = \frac{R^2 \rho_o}{2\varepsilon_o}$$

and

$$V = \frac{\rho_o}{6\varepsilon_o}(3R^2 - r^2)$$

Thus from parts (a) and (b)

$$V = \begin{cases} \dfrac{\rho_o R^3}{3\varepsilon_o r}, & r \geq R \\[2ex] \dfrac{\rho_o}{6\varepsilon_o}(3R^2 - r^2), & r \leq R \end{cases}$$

(c) The energy stored is given by

$$W = \frac{1}{2} \int \mathbf{D} \cdot \mathbf{E}\, dv = \frac{1}{2} \varepsilon_o \int E^2\, dv$$

For $r \leq R$,

$$\mathbf{E} = \frac{\rho_o r}{3\varepsilon_o} \mathbf{a}_r$$

Hence,

$$W = \frac{1}{2} \varepsilon_o \frac{\rho_o^2}{9\varepsilon_o^2} \int_{r=0}^{R} \int_{\theta=0}^{\pi} \int_{\phi=0}^{2\pi} r^2 \cdot r^2 \sin\theta\, d\phi\, d\theta\, dr$$

$$= \frac{\rho_o^2}{18\varepsilon_o} 4\pi \cdot \frac{r^5}{5}\bigg|_0^R = \frac{2\pi\rho_o^2 R^5}{45\varepsilon_o} \text{ J}$$

■

PRACTICE EXERCISE 4.18

If $V = x - y + xy + 2z$ V, find **E** at (1, 2, 3) and the electrostatic energy stored in a cube of side 2 m centered at the origin.

ANSWER $-3\mathbf{a}_x - 2\mathbf{a}_z$ V/m, 0.2358 nJ.

SUMMARY

1. The two fundamental laws for electrostatic fields (Coulomb's and Gauss's) are presented in this chapter. Coulomb's law of force states that

$$\mathbf{F} = \frac{Q_1 Q_2}{4\pi\varepsilon_o R^2} \mathbf{a}_R$$

2. Based on Coulomb's law, we define the electric field intensity **E** as the force per unit charge; that is,

$$\mathbf{E} = \frac{Q}{4\pi\varepsilon_o R^2} \mathbf{a}_R = \frac{Q\,\mathbf{R}}{4\pi\varepsilon R^3} \qquad \text{(point charge only)}$$

For a continuous charge distribution, the total charge is given by

$$Q = \int \rho_L \, dl \qquad \text{for line charge}$$

$$Q = \int \rho_S \, dS \qquad \text{for surface charge}$$

$$Q = \int \rho_v \, dv \qquad \text{for volume charge}$$

The **E** field due to a continuous charge distribution is obtained from the formula for point charge by replacing Q with $dQ = \rho_L \, dl$, $dQ = \rho_S \, dS$ or $dQ = \rho_v \, dv$ and integrating over the line, surface, or volume respectively.

3. For an infinite line charge,

$$\mathbf{E} = \frac{\rho_L}{2\pi\varepsilon_o \rho} \mathbf{a}_\rho$$

and for an infinite sheet of charge,

$$\mathbf{E} = \frac{\rho_S}{2\varepsilon_o} \mathbf{a}_n$$

4. The electric flux density **D** is related to the electric field intensity (in free space) as

$$\mathbf{D} = \varepsilon_o \mathbf{E}$$

The electric flux through a surface S is

$$\Psi = \int_S \mathbf{D} \cdot d\mathbf{S}$$

5. Gauss's law states that the net electric flux penetrating a closed surface is equal to the total charge enclosed, that is, $\Psi = Q_{enc}$. Hence,

$$\Psi = \oint \mathbf{D} \cdot d\mathbf{S} = Q_{enc} = \int \rho_v \, dv$$

or

$$\rho_v = \nabla \cdot \mathbf{D} \qquad \text{(1st Maxwell's equation to be derived)}$$

When charge distribution is symmetric so that a Gaussian surface (where $\mathbf{D} = D_n \mathbf{a}_n$ is contant) can be found, Gauss's law is useful in determining **D**; that is,

$$D_n \oint dS = Q_{\text{enc}} \qquad \text{or} \qquad D_n = \frac{Q_{\text{enc}}}{S}$$

6. The total work done, or the electric potential energy, to move a point charge Q from point A to B in an electric field E is

$$W = -Q \int_A^B \mathbf{E} \cdot d\mathbf{l}$$

7. The potential at $\mathbf{r}$ due to a point charge Q at $\mathbf{r}'$ is

$$V(\mathbf{r}) = \frac{Q}{4\pi\varepsilon_o|\mathbf{r} - \mathbf{r}'|} + C$$

where C is evaluated at a given reference potential point; for example, $C = 0$ if $V(\mathbf{r} \rightarrow \infty) = 0$. To determine the potential due to a continuous charge distribution, we replace Q in the formula for point charge by $dQ = \rho_L\, dl$, $dQ = \rho_S\, dS$, or $dQ = \rho_v\, dv$ and integrate over the line, surface, or volume respectively.

8. If the charge distribution is not known, but the field intensity **E** is given, we find the potential using

$$V = -\int \mathbf{E} \cdot d\mathbf{l} + C$$

9. The potential difference V_{AB}, the potential at B with reference to A, is

$$V_{AB} = -\int_A^B \mathbf{E} \cdot d\mathbf{l} = \frac{W}{Q} = V_B - V_A$$

10. Since an electrostatic field is conservative (the net work done along a closed path in a static **E** field is zero),

$$\oint \mathbf{E} \cdot d\mathbf{l} = 0$$

or

$$\nabla \times \mathbf{E} = 0 \qquad \text{(2nd Maxwell's equation to be derived)}$$

11. Given the potential field, the corresponding electric field is found using

$$\mathbf{E} = -\nabla V$$

12. For an electric dipole centered at $\mathbf{r}'$ with dipole moment $\mathbf{p}$, the potential at $\mathbf{r}$ is given by

$$V(\mathbf{r}) = \frac{\mathbf{p} \cdot (\mathbf{r} - \mathbf{r}')}{4\pi\varepsilon_o|\mathbf{r} - \mathbf{r}'|^3}$$

13. **D** is tangential to the electric flux lines at every point. An equipotential surface (or line) is one on which V = constant. At every point, the equipotential line is orthogonal to the electric flux line.

14. The electrostatic energy due to n point charges is

$$W_E = \frac{1}{2} \sum_{k=1}^{n} Q_k V_k$$

For a continuous volume charge distribution,

$$W_E = \frac{1}{2} \int_v \mathbf{D} \cdot \mathbf{E}\, dv = \frac{1}{2} \int \varepsilon_o |\mathbf{E}|^2\, dv$$

REVIEW QUESTIONS

4.1 Point charges $Q_1 = 1$ nC and $Q_2 = 2$ nC are at a distance apart. Which of the following statements are incorrect?

(a) The force on Q_1 is repulsive.

(b) The force on Q_2 is the same in magnitude as that on Q_1.

(c) As the distance between them decreases, the force on Q_1 increases linearly.

(d) The force on Q_2 is along the line joining them.

(e) A point charge $Q_3 = -3$nC located at the midpoint between Q_1 and Q_2 experiences no net force.

4.2 Plane $z = 10$ m carries charge 20 nC/m^2. The electric field intensity at the origin is:

(a) $-10\, \mathbf{a}_z$ V/m

(b) $-18\pi\, \mathbf{a}_z$ V/m

(c) $-72\pi\, \mathbf{a}_z$ V/m

(d) $-360\pi\, \mathbf{a}_z$ V/m

4.3 Point charges 30 nC, -20 nC, and 10 nC are located at $(-1, 0, 2)$, $(0, 0, 0)$, and $(1, 5, -1)$ respectively. The total flux leaving a cube of side 6 m centered at the origin is:

(a) -20 nC

(b) 10 nC

(c) 20 nC

(d) 30 nC

(e) 60 nC

4.4 The electric flux density on a spherical surface $r = b$ is the same for a point charge Q located at the origin and for charge Q uniformly distributed on surface $r = a(a < b)$.

(a) Yes

(b) No

(c) Not necessarily

4.5 The work done by the force $\mathbf{F} = 4\mathbf{a}_x - 3\mathbf{a}_y + 2\mathbf{a}_z$ N in giving a 1 nC charge a displacement of $10\mathbf{a}_x + 2\mathbf{a}_y - 7\mathbf{a}_z$ m is

(a) 103 nJ

(b) 60 nJ

(c) 64 nJ

(d) 20 nJ

4.6 By saying that the electrostatic field is conservative, we do *not* mean that

(a) It is the gradient of a scalar potential.

(b) Its circulation is identically zero.

(c) Its curl is identically zero.

(d) The work done in a closed path inside the field is zero.

(e) The potential difference between any two points is zero.

4.7 Suppose a uniform electric field exists in the room in which you are working, such that the lines of force are horizontal and at right angles to one wall. As you walk toward the wall from which the lines of force emerge into the room, are you walking toward

(a) Points of higher potential?

(b) Points of lower potential?

(c) Points of the same potential (equipotential line)?

4.8 A charge Q is uniformly distributed throughout a sphere of radius a. Taking the potential at infinity as zero, the potential at $r = b < a$ is

(a) $-\int_{\infty}^{b} \frac{Qr}{4\pi\varepsilon_o a^3}\, dr$

(b) $-\int_{\infty}^{b} \frac{Q}{4\pi\varepsilon_o r^2}\, dr$

(c) $-\int_{\infty}^{a} \frac{Q}{4\pi\varepsilon_o r^2}\, dr - \int_{a}^{b} \frac{Qr}{4\pi\varepsilon_o a^3}\, dr$

(d) $-\int_{\infty}^{a} \frac{Q}{4\pi\varepsilon_o r^2}\, dr$

4.9 A potential field is given by $V = 3x^2y - yz$. Which of the following is not true?

(a) At point $(1, 0, -1)$, V and $\mathbf{E}$ vanish.

(b) $x^2y = 1$ is an equipotential line on the xy-plane.

(c) The equipotential surface $V = -8$ passes through point $P(2, -1, 4)$.

(d) The electric field at P is $12\mathbf{a}_x - 8\mathbf{a}_y - \mathbf{a}_z$ V/m.

(e) A unit normal to the equipotential surface $V = -8$ at P is $-0.83\mathbf{a}_x + 0.55\mathbf{a}_y + 0.07\mathbf{a}_z$.

4.10 An electric potential field is produced by point charges 1 μC and 4 μC located at $(-2, 1, 5)$ and $(1, 3, -1)$ respectively. The energy stored in the field is

(a) 2.57 mJ

(b) 5.14 mJ

(c) 10.28 mJ

(d) None of the above

Answers: 4.1c,e, 4.2d, 4.3b, 4.4a, 4.5d, 4.6e, 4.7a, 4.8c, 4.9a, 4.10b.

PROBLEMS

4.1 (a) State Coulomb's law.

(b) Three point charges $Q_1 = 1$ mC, $Q_2 = 2$ mC, and $Q_3 = -3$ mC are respectively located at $(0, 0, 4)$, $(-2, 6, 1)$, and $(3, -4, -8)$. Calculate the force on Q_1.

4.2 Define electric field intensity. Three identical point charges of 10 nC each are located at the vertices of an equilateral triangle of side 10 cm. Calculate the magnitude of

(a) The force on each charge

(b) The electric field intensity at the center of the triangle

4.3 Q_1 and Q_2 are point charges located at $(0, -4, 3)$ and $(0, 1, 1)$. If Q_1 is 2 nC, find Q_2 such that

(a) The force on a test charge at $(0, -3, 4)$ has no z-component.

(b) The $\mathbf{E}$ at $(0, -3, 4)$ has no y-component.

4.4 Charges $+Q$ and $+3Q$ are separated by a distance 2 m. A third charge is located such that the electrostatic system is in equilibrium. Find the location and the value of the third charge in terms of Q.

4.5 Determine the total charge

(a) On line $0 < x < 5$ m if $\rho_L = 12x^2$ mC/m

(b) On the cylinder $\rho = 3$, $0 < z < 4$ m if $\rho_S = \rho z^2$ nC/m^2

(c) Within the sphere $r = 4$ m if $\rho_v = \dfrac{10}{r \sin \theta}$ C/m^3

4.6 Find the total charge for the configurations labeled A, B, and C in Figure 4.27.

4.7 A ring placed along $y^2 + z^2 = 9$, $x = 0$ carries a uniform charge of 5 nC/m.

(a) Find **E** at $P(4, 0, 0)$.

(b) If two identical point charges Q are placed at $(0, -4, 0)$ and $(0, 4, 0)$ in addition to the ring, find the value of Q such that $\mathbf{E} = 0$ at P.

****4.8** Charge Q is applied to a circular disk of ebonite of radius a by rubbing it while it is rotating. In this way, the surface charge density becomes proportional to the radial distance from the center. Show that the electric field strength on the axis of the disk at an axial distance h from the center is

$$\frac{3Qh}{4\pi\varepsilon_o a^3}\left[\ln\frac{a + \sqrt{a^2 + h^2}}{h} - \frac{1}{\sqrt{a^2 + h^2}}\right]\mathbf{a}_n$$

where $\mathbf{a}_n$ is a unit normal vector to the disk.

4.9 Find **E** at $(0, 0, 4)$ due to a charge of 2 nC distributed uniformly on

(a) The line $0 \leq x \leq 3$

(b) The arc $\rho = 3$, $\pi/4 \leq \phi \leq \pi/2$, $z = 0$

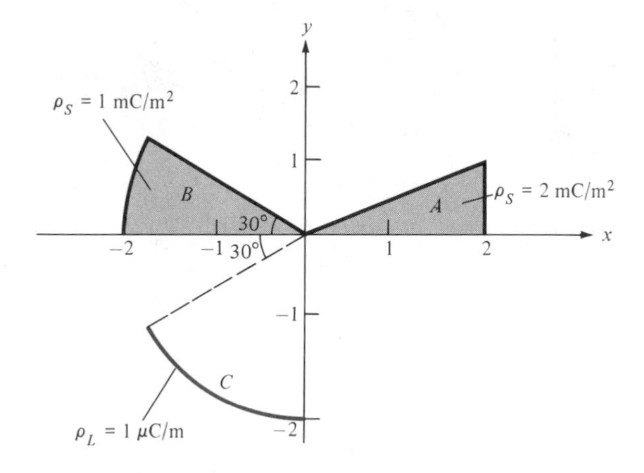

Figure 4.27 Charge distributions for Problem 4.6.

***4.10** (a) Show that the electric field at point $(0, 0, h)$ due to the rectangle described by $-a \leq x \leq a$, $-b \leq y \leq b$, $z = 0$ carrying uniform charge ρ_S C/m^2 is

$$\mathbf{E} = \frac{\rho_S}{\pi \varepsilon_o} \tan^{-1} \left[\frac{ab}{h[a^2 + b^2 + h^2]^{1/2}} \right] \mathbf{a}_z$$

(b) If $a = 2$, $b = 5$, $\rho_S = 10^{-5}$, find the total charge on the plate and the electric field intensity at $(0, 0, 10)$.

4.11 A point charge 100 pC is located at $(4, 1, -3)$ while the x-axis carries charge 2 nC/m. If the plane $z = 3$ also carries charge 5 nC/m^2, find $\mathbf{E}$ at $(1, 1, 1)$.

4.12 The line $y = 1$, $z = -3$ carries charge 30 nC/m while the plane $x = 1$ carries charge 20 nC/m^2. Find $\mathbf{E}$ at the origin.

4.13 Point charges are placed at the corners of a square of size 4 m as shown in Figure 4.28. If $Q = 15$ μC, find $\mathbf{D}$ at $(0, 0, 6)$.

4.14 Define the electric flux density. Point charges $Q_1 = 4$ μC, $Q_2 = -5\mu$C, and $Q_3 = 2$ μC are located at $(0, 0, 1)$, $(-6, 8, 0)$, and $(0, 4, -3)$ respectively. Find $\mathbf{D}$ at the origin.

4.15 A line charge with uniform charge ρ_L C/m lies along the x-axis. The electric flux density at $(-3, 6, 8)$ is 3 nC/m^2.

(a) Find ρ_L.

(b) Determine $\mathbf{D}$ at $(0, 0, 4)$.

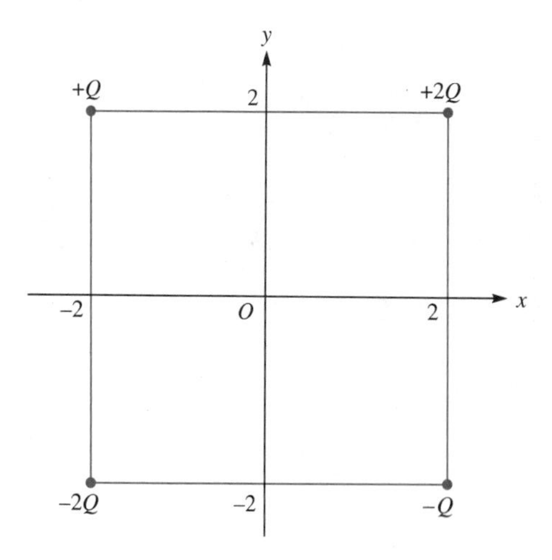

Figure 4.28 For Problem 4.13.

***4.16** State Gauss's law. Deduce Coulomb's law from Gauss's law thereby affirming that Gauss's law is an alternative statement of Coulomb's and that Coulomb's law is implicit in Maxwell's equation $\nabla \cdot \mathbf{D} = \rho_v$.

4.17 Determine the charge density due to each of the following electric flux densities:

(a) $\mathbf{D} = 8xy\mathbf{a}_x + 4x^2\mathbf{a}_y \text{ C/m}^2$

(b) $\mathbf{D} = \rho \sin \phi \, \mathbf{a}_\rho + 2\rho \cos \phi \, \mathbf{a}_\phi + 2z^2\mathbf{a}_z \text{ C/m}^2$

(c) $\mathbf{D} = \dfrac{2 \cos \theta}{r^3}\mathbf{a}_r + \dfrac{\sin \theta}{r^3}\mathbf{a}_\theta \text{ C/m}^2$

4.18 A sphere of radius 10 cm has $\rho_v = \dfrac{r^3}{100} \text{ C/m}^3$. If $\mathbf{D}$ is to vanish for $r > 10$ cm, calculate the value of a point charge that must be placed at the center of the sphere.

4.19 Given that in free space

$$\rho_v = \begin{cases} 10 \text{ nC/m}^3, & 0 < r < 3 \text{ cm} \\ 20 \text{ nC/m}^3, & 3 < r < 5 \text{ cm} \\ 0, & r > 5 \text{ cm} \end{cases}$$

determine $\mathbf{E}$ at

(a) $r = 4$ cm

(b) $r = 6$ cm

4.20 A point charge of 60 mC is placed at the center of a cube. Find the total flux passing through one face of the cube.

4.21 Point charges 5 μC, -3 μC, 2 μC, and 10 μC are located at $(-12, 0, 5)$, $(0, 3, -4)$, $(2, -6, 3)$, and $(3, 0, 0)$ respectively. Calculate the flux through the spherical surfaces at

(a) $r = 1$

(b) $r = 10$

(c) $r = 15$

4.22 If $\mathbf{D} = 2z^2 \sin \dfrac{\phi}{2} \mathbf{a}_\rho + z^2 \cos \dfrac{\phi}{2} \mathbf{a}_\phi + 4z\rho \sin \dfrac{\phi}{2} \mathbf{a}_z \text{ mC/m}^2$, using two different methods, find the total charge enclosed by the object defined $-2 \leq z \leq 1$, $1 \leq \rho \leq 4$, $0 \leq \phi \leq \pi$.

4.23 Given that $\mathrm{D} = 2\rho\, z \cos^2\phi\, \mathbf{a}_\rho - \rho z \sin\phi \cos\phi\, \mathbf{a}_\phi + \rho^2 \cos^2\phi\, \mathbf{a}_z$ C/m^2, calculate the electric flux through the following surfaces of a cylinder:

(a) Surface $\rho = 3,\ 0 \le z \le 5$

(b) Surface $z = 0,\ 0 \le \rho \le 3$

(c) Surface $z = 5,\ 0 \le \rho \le 3$

(d) Calculate the charge enclosed by the whole cylinder $\rho = 3,\ 0 \le z \le 5$.

***4.24** The Thomson model of a hydrogen atom is a sphere of positive charge with an electron (a point charge) at its center. The total positive charge equals the electronic charge e. Prove that when the electron is at a distance r from the center of the sphere of positive charge, it is attracted with a force

$$F = \frac{e^2\, r}{4\pi\varepsilon_o R^3}$$

where R is the radius of the sphere.

4.25 If the electric flux density is $\mathbf{D} = \dfrac{10}{r}\mathbf{a}_r$ nC/m^2, find the total charge within $0 \le r \le 2$ m.

4.26 Find the work done in carrying a 5-C charge from $P(1, 2, -4)$ to $R(3, -5, 6)$ in an electric field

$$\mathbf{E} = \mathbf{a}_x + z^2\mathbf{a}_y + 2yz\mathbf{a}_z \text{ V/m}$$

4.27 Inside $\mathbf{E} = 20\rho z \cos\phi\, \mathbf{a}_\rho - 10\rho z \sin\phi\, \mathbf{a}_\phi + 10\rho^2 \cos\phi\, \mathbf{a}_z$ V/m, find the work done in moving a 1-μC charge

(a) From $A(0, 0, 1)$ to $B(2, 0, 1)$

(b) From $B(2, 0, 1)$ to $C(2, \pi/3, 1)$

(c) From $C(2, \pi/3, 1)$ to $D(2, \pi/3, 3)$

(d) From A to D

4.28 In an electric field $\mathbf{E} = 20r \sin\theta\, \mathbf{a}_r + 10r \cos\theta\, \mathbf{a}_\theta$ V/m, calculate the energy expended in transferring a 10-nC charge

(a) From $A(5, 30°, 0°)$ to $B(5, 90°, 0°)$

(b) From A to $C(10, 30°, 0°)$

(c) From A to $D(5, 30°, 60°)$

(d) From A to $E(10, 90°, 60°)$

4.29 Two point charges $Q_1 = 3$ nC and $Q_2 = -2$ nC are placed at (0, 0, 0) and (0, 0, -1) respectively. Assuming zero potential at infinity, find the potential at

(a) (0, 1, 0)

(b) (1, 1, 1)

4.30 The y-axis carries a uniform charge of 10 nC/m in free space. Given two points $A(-3, 2, 4)$ and $P(6, 1, 0)$, find

(a) V_A if $V_P = 5$ V

(b) V_P if $V_A = 5$ V

4.31 (a) A total charge $Q = 60\ \mu$C is split into two equal charges located at 180° intervals around a circular loop of radius 4 m. Find the potential at the center of the loop.

(b) If Q is split into three equal charges spaced at 120° intervals around the loop, find the potential at the center.

(c) If in the limit $\rho_L = \dfrac{Q}{8\pi}$, find the potential at the center.

***4.32** A circular disk of radius a carries a uniform charge ρ_S C/m^2. Show that the potential at a point on its axis h meters away from its center is

$$V = \frac{\rho_S}{2\varepsilon_o}\left[(h^2 + a^2)^{1/2} - h\right]$$

4.33 Verify that $\mathbf{E} = 10\rho z \cos \mathbf{a}_\rho - 5\rho z \sin \phi\, \mathbf{a}_\phi + 5\rho^2 \cos \phi\, \mathbf{a}_z$ is an electrostatic field by showing that

(a) $\nabla \times \mathbf{E} = 0$

(b) $\oint_L \mathbf{E} \cdot d\mathbf{l} = 0$, where L is the circle $\rho = 1$ located on the $z = 1$ plane.

4.34 Verify eq. (4.74) using eq. (4.12) with the help of the vector identity $\nabla \times V\mathbf{A} = V\,\nabla \times \mathbf{A} + \nabla V \times \mathbf{A}$ and the results of Problems 3.20(g) and (i).

4.35 Find the electric fields due to the following electric potentials:

(a) $V = e^{-x} \sinh y \sin z$

(b) $V = \dfrac{1}{[x^2 + y^2 + z^2]^{3/2}}$

(c) $V = \rho e^{-z} \cos \phi$

(d) $V = \dfrac{\sin \theta \cos \phi}{r^2}$

4.36 Find the potential that gives rise to each of the following electric fields:

(a) $\mathbf{E} = 2xy\mathbf{a}_x + x^2\mathbf{a}_y - \mathbf{a}_z$

(b) $\mathbf{E} = e^{-z}(\sin \phi\, \mathbf{a}_\rho + \cos \phi\, \mathbf{a}_\phi - \rho \sin \phi\, \mathbf{a}_z)$

(c) $\mathbf{E} = \dfrac{\sin \theta}{r}\,\mathbf{a}_r + \dfrac{\ln r}{r} \cos \theta\, \mathbf{a}_\theta$

***4.37** (a) Prove that when a particle of constant mass and charge is accelerated from rest in an electric field, its final velocity is proportional to the square root of the potential difference through which it is accelerated.

(b) Find the magnitude of the proportionality constant if the particle is an electron.

(c) Through what voltage must an electron be accelerated, assuming no change in its mass, to require a velocity one-tenth that of light? (At such velocities, the mass of a body becomes appreciably larger than its "rest mass" and cannot be considered constant.)

***4.38** An electron is projected with an initial velocity $u_o = 10^7$ m/s into the uniform field between the parallel plates of Figure 4.29. It enters the field at the midway between the plates. If the electron just misses the upper plate as it emerges from the field,

(a) Find the electric field intensity.

(b) Calculate its velocity as it emerges from the field. Neglect edge effects.

***4.39** A spherical charge distribution is given by

$$\rho_v = \begin{cases} \rho_o \left(1 - \dfrac{r^2}{a^2}\right), & r \leq a \\ 0, & r > a \end{cases}$$

(a) Find **E** and V for $r \geq a$.

(b) Find **E** and V for $r \leq a$.

(c) Show that the maximum value of **E** is at $r = 0.745a$.

(d) Find where V is maximum and calculate that maximum value.

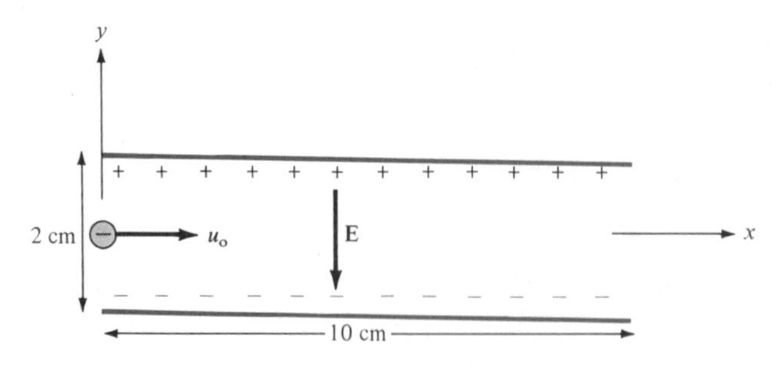

Figure 4.29 For Problem 4.38.

4.40 An electric dipole with $\mathbf{p} = p\mathbf{a}_z$ C · m is placed at $(x, z) = (0, 0)$. If the potential at (0, 1) nm is 9 V, find the potential at (1, 1) nm.

4.41 Point charges Q and $-Q$ are located at $(0, d/2, 0)$ and $(0, -d/2, 0)$. Show that at point (r, θ, ϕ), where $r \gg d$,

$$V = \frac{Qd \sin\theta \sin\phi}{4\pi\varepsilon_o r^2}$$

Find the corresponding $\mathbf{E}$ field.

4.42 (a) What is an electric flux line? Sketch the flux line due to

$$\mathbf{E} = \frac{2\cos\theta}{r^3}\,\mathbf{a}_r + \frac{\sin\theta}{r^3}\,\mathbf{a}_\theta$$

that passes through the point $(4, \pi/4, \pi/2)$. Determine a unit vector tangential to the line at that point.

(b) Point charges -3 nC and 4 nC are located at $(1, 0, 0)$ and $(-1, 0, 0)$. Sketch the equipotential line on the $z = 0$ plane for which $V = 0$.

4.43 In addition to the two point charges of Problem 4.46(b), if a 2-nC charge is brought from infinity to the origin, calculate

(a) The work done

(b) The total energy in the system

4.44 A point charge Q is placed at the origin. Calculate the energy stored in region $r > a$.

4.45 Given that $\mathbf{D} = yz\mathbf{a}_x + xz\mathbf{a}_y + xy\mathbf{a}_z$ nC/m², calculate the total energy stored in the region $0 \leq x, y, z \leq 1$.

4.46 Find the energy stored in the hemispherical region $r \leq 2$m, $0 < \theta < \pi$, where

$$\mathbf{E} = 2r\sin\theta\cos\phi\,\mathbf{a}_r + r\cos\theta\cos\phi\,\mathbf{a}_\theta - r\sin\phi\,\mathbf{a}_\phi \text{ V/m}$$

exists.

4.47 If $V = \rho^2 z\sin\phi$, calculate the energy within the region defined by $1 < \rho < 4$, $-2 < z < 2$, $0 < \phi < \pi/3$.

CHAPTER

5 Electric Fields in Material Space

The average person puts only 25 percent of his energy and ability into his work. The world takes off its hat to those who put in more than 50 percent of their capacity, and stands on its head for those few and far between souls who devote 100 percent.

—ANDREW CARNEGIE

5.1 INTRODUCTION

In the last chapter, we considered electrostatic fields in free space or a space that has no materials in it. Thus what we have developed so far under electrostatics may be regarded as the "vacuum" field theory. By the same token, what we shall develop in this chapter may be regarded as the theory of electric phenomena in material space. As will soon be evident, most of the formulas derived in Chapter 4 are still applicable, though some may require modification.

Just as electric fields can exist in free space, they can exist in material media. Materials are broadly classified in terms of their electrical properties as conductors and nonconductors. Nonconducting materials are usually referred to as *insulators* or *dielectrics*. A brief discussion of the electrical properties of materials in general will be given to provide a basis for understanding the concepts of conduction, electric current, and polarization. Further discussion will be on some properties of dielectric materials such as susceptibility, permittivity, linearity, isotropy, homogeneity, dielectric strength, and relaxation time. The concept of boundary conditions for electric fields existing in two different media will be introduced.

5.2 PROPERTIES OF MATERIALS

In a text of this kind, a discussion on electrical properties of materials may seem out of place. But questions such as why an electron does not leave a conductor surface, why a current-carrying wire remains uncharged, why materials behave differently in an electric field, and why waves travel with less speed in conductors than in dielectrics are easily answered by considering the electrical properties of materials. A thorough discussion on this subject is usually found in texts on physical electronics or electrical engineering. Here, a brief discussion will suffice to help us understand the mechanism by which materials influence an electric field.

In a broad sense, materials may be classified in terms of their *conductivity* σ, in mhos per meter (℧/m) or siemens per meter (S/m), as conductors and nonconductors, or technically as metals and insulators (or dielectrics). The conductivity of a material usually depends on temperature and frequency. A material with *high conductivity* ($\sigma \gg 1$) is referred to as a *metal* whereas one with *low conductivity* ($\sigma \ll 1$) as an *insulator*. A material whose conductivity lies somewhere between those of metals and insulators is called a *semiconductor*. The values of conductivity of some common materials are shown in Table B.1 in Appendix B. From this table, it is clear that materials such as copper and aluminum are metals, silicon and germanium are semiconductors, and glass and rubber are insulators.

The conductivity of metals generally increases with decrease in temperature. At temperatures near absolute zero ($T = 0°\text{K}$), some conductors exhibit infinite conductivity and are called *superconductors*. Lead and aluminum are typical examples of such metals. The conductivity of lead at 4°K is of the order of 10^{20} mhos/m. The interested reader is referred to the literature on superconductivity.[1]

We shall only be concerned with metals and insulators in this text. Microscopically, the major difference between a metal and an insulator lies in the amount of electrons available for conduction of current. Dielectric materials have few electrons available for conduction of current in contrast to metals, which have an abundance of free electrons. Further discussion on the presence of conductors and dielectrics in an electric field will be given in subsequent sections.

5.3 CONVECTION CURRENT

Electric current is generally caused by the motion of electric charges. Convection current, as distinct from conduction current (to be discussed in the next section), does not involve conductors and consequently does not satisfy Ohm's law. It occurs when current flows through an insulating medium such as liquid, rarefied gas, or a vacuum. A beam of electrons in a vacuum tube, for example, is a convection current.

[1]The August 1989 issue of the *Proceedings of IEEE* was devoted to "Applications of Superconductivity."

The current through a given area is defined as the electric charge passing through the area per unit time. That is,

$$I = \frac{dQ}{dt} \tag{5.1}$$

Thus in a current of one ampere, charge is being transferred at a rate of one coulomb per second.

Consider a filament of Figure 5.1. If there is a flow of charge, of density ρ_v, at velocity $\mathbf{u} = u_y\mathbf{a}_y$, from eq. (5.1), the current through the filament is

$$\Delta I = \frac{\Delta Q}{\Delta t} = \rho_v \, \Delta S \, \frac{\Delta \ell}{\Delta t} = \rho_v \, \Delta S \, u_y \tag{5.2}$$

If we define the current density at a given point as the current through a unit normal area at that point, the y-directed current density J_y is given by

$$J_y = \frac{\Delta I}{\Delta S} = \rho_v u_y \tag{5.3}$$

Hence, in general

$$\boxed{\mathbf{J} = \rho_v \mathbf{u}} \tag{5.4}$$

The current I is the *convection current* and J is the *convection current density* in amperes/square meter (A/m^2).

In eq. (5.3), if the current density is not normal to the surface,

$$\Delta I = \mathbf{J} \cdot \Delta \mathbf{S}$$

so that the total current through a prescribed surface S is given by

$$\boxed{I = \int_S \mathbf{J} \cdot d\mathbf{S}} \tag{5.5}$$

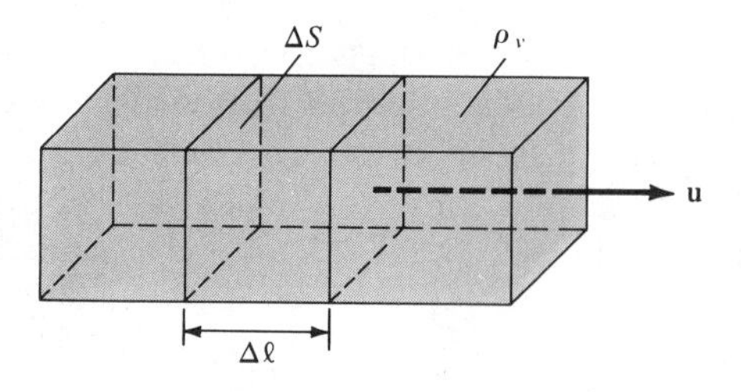

Figure 5.1 Current in a filament.

Compared with the general definition of flux in eq. (3.13), eq. (5.5) shows that the current I through S is merely the flux of the current density $\mathbf{J}$.

5.4 CONDUCTION CURRENT

When an electric field is applied to a conductor, conduction current occurs due to the drift motion of electrons. The conductor atoms remain neutral ($\rho_v = 0$) while the whole conductor itself remains at rest. Though the conductor is at rest, its electrons are not. As the electrons move, they encounter some damping forces called *resistance*. The average drift velocity of the electrons is directly proportional to the applied field. Thus for a conductor, eq. (5.4) becomes

$$\boxed{\mathbf{J} = \sigma\, \mathbf{E}} \qquad [5.6]$$

where σ is the conductivity of the material in mhos per meter or siemens per meter and $\mathbf{J}$ is known as *conduction current density* (in A/m^2) as distinct from convection current density of eq. (5.4). Equation (5.6) is referred to as *Ohm's law*. Based on this law, we will now derive the resistance of a conducting material.

Consider a length ℓ of a conductor of *uniform* cross section S. Suppose a voltage V is applied to the conductor as shown in Figure 5.2. The direction of the electric field $\mathbf{E}$ produced is the same as the direction of the flow of positive charges or current I. This direction is opposite to the direction of the flow of electrons. The electric field applied is uniform and its magnitude is given by

$$E = \frac{V}{\ell} \qquad [5.7]$$

Since the conductor has a uniform cross section,

$$J = \frac{I}{S} \qquad [5.8]$$

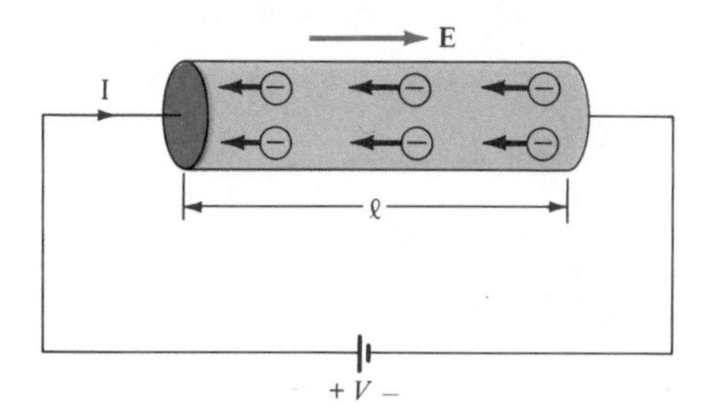

Figure 5.2 A conductor of uniform cross section under an applied $\mathbf{E}$ field.

Substituting eqs. (5.6) and (5.7) into eq. (5.8) gives

$$\frac{I}{S} = \sigma E = \frac{\sigma V}{\ell} \tag{5.9}$$

Hence

$$R = \frac{V}{I} = \frac{\ell}{\sigma S}$$

or

$$\boxed{R = \frac{\rho_c \ell}{S}} \tag{5.10}$$

where $\rho_c = 1/\sigma$ is the *resistivity* of the material. Equation (5.10) is useful in determining the resistance of any conductor of uniform cross section. If the cross section of the conductor is not uniform, eq. (5.10) is not applicable. However, the basic definition of resistance R as the ratio of the potential difference V between the two ends of the conductor to the current I through the conductor still applies. Therefore, applying eqs. (4.60) and (5.5) gives the resistance of a conductor of nonuniform cross-section; that is,

$$\boxed{R = \frac{V}{I} = \frac{\int \mathbf{E} \cdot d\mathbf{l}}{\int \sigma \mathbf{E} \cdot d\mathbf{S}}} \tag{5.11}$$

Note that the negative sign before $V = -\int \mathbf{E} \cdot d\mathbf{l}$ is dropped in eq. (5.11) because $\int \mathbf{E} \cdot d\mathbf{l} < 0$ if $I > 0$. Equation (5.11) will not be utilized until we get to section 6.5

Power P (in watts) is defined as the rate of change of energy W (in joules) or force times velocity. Hence,[2]

$$\int \rho_v \, dv \, \mathbf{E} \cdot \mathbf{u} = \int \mathbf{E} \cdot \rho_v \mathbf{u} \, dv$$

or

$$\boxed{P = \int \mathbf{E} \cdot \mathbf{J} \, dv} \tag{5.12}$$

[2]Note that eq. (5.4) is also applicable to conduction current density in which $\mathbf{J} = \sigma\mathbf{E} = \rho_v \mathbf{u} = \rho_v(-\mu)\mathbf{E}$ where $\mathbf{u} = -\mu\mathbf{E}$ is the drift velocity, μ is the mobility, and ρ_v is the density of conduction electrons.

which is known as *Joule's law*. The power density w_P (in watts/m^3) is given by the integrand in eq. (5.12); that is,

$$w_P = \frac{dP}{dv} = \mathbf{E} \cdot \mathbf{J} = \sigma \, |\mathbf{E}|^2 \qquad [5.13]$$

For a conductor with uniform cross section, $dv = dS\, dl$, so eq. (5.12) becomes

$$P = \int_L E\, dl \int_S J\, dS = VI$$

or

$$P = I^2 R \qquad [5.14]$$

which is the more common form of Joule's law in electric circuit theory.

EXAMPLE 5.1

If $\mathbf{J} = \frac{1}{r^3} (2 \cos \theta \, \mathbf{a}_r + \sin \theta \, \mathbf{a}_\theta)$ A/m^2, calculate the current passing through

(a) A hemispherical shell of radius 20 cm

(b) A spherical shell of radius 10 cm

SOLUTION $I = \int \mathbf{J} \cdot d\mathbf{S}$, where $d\mathbf{S} = r^2 \sin \theta \, d\phi \, d\theta \, \mathbf{a}_r$ in this case.

(a) $$I = \int_{\theta=0}^{\pi/2} \int_{\phi=0}^{2\pi} \frac{1}{r^3} 2 \cos \theta \, r^2 \sin \theta \, d\phi \, d\theta \Bigg|_{r=0.2}$$

$$= \frac{2}{r} 2\pi \int_{\theta=0}^{\pi/2} \sin \theta \, d(\sin \theta) \Bigg|_{r=0.2}$$

$$= \frac{4\pi}{0.2} \frac{\sin^2\theta}{2} \Bigg|_0^{\pi/2} = 10\pi = 31.4 \, A$$

(b) The only difference here is that we have $0 \le \theta \le \pi$ instead of $0 \le \theta \le \pi/2$ and $r = 0.1$. Hence,

$$I = \frac{4\pi}{0.1} \frac{\sin^2\theta}{2} \Bigg|_0^{\pi} = 0$$

Alternatively, for this case

$$I = \oint \mathbf{J} \cdot d\mathbf{S} = \int \nabla \cdot \mathbf{J} \, dv = 0$$

since $\nabla \cdot \mathbf{J} = 0$. ■

PRACTICE EXERCISE 5.1

For the current density $\mathbf{J} = 10z \sin^2\phi \, \mathbf{a}_\rho$ A/m^2, find the current through the cylindrical surface $\rho = 2$, $1 \leq z \leq 5$m.

ANSWER 754 A.

EXAMPLE 5.2

A typical example of convective charge transport is found in the Van de Graaff generator where charge is transported on a moving belt from the base to the dome as shown in Figure 5.3. If a surface charge density 10^{-7} C/m^2 is transported at a velocity of 2 m/s, calculate the charge collected in 5 s. Take the width of the belt as 10 cm.

SOLUTION If ρ_S = surface charge density, u = speed of the belt, and w = width of the belt, the current on the dome is

$$I = \rho_S \, uw$$

The total charge collected in $t = 5$ s is

$$Q = It = \rho_S uwt = 10^{-7} \times 2 \times 0.1 \times 5$$
$$= 100 \text{ nC}$$

■

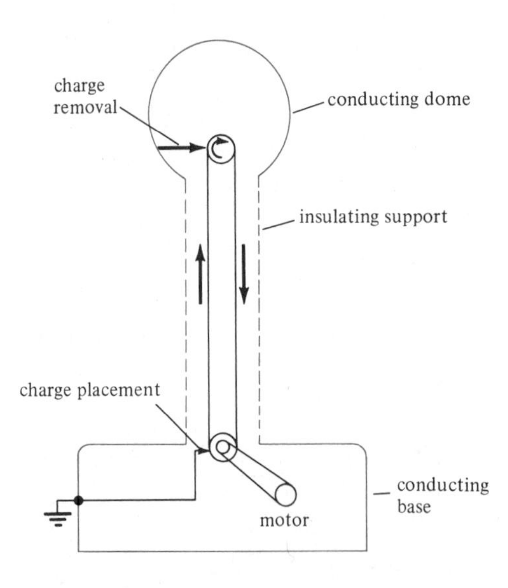

Figure 5.3 Van de Graaff generator; for Example 5.2.

PRACTICE EXERCISE 5.2

In a Van de Graaff generator, $w = 0.1$ m, $u = 10$ m/s, and the leakage paths have resistance $10^{14}\ \Omega$. If the belt carries charge $0.5\ \mu\text{C/m}^2$, find the potential difference between the dome and the base.

ANSWER 50 MV.

EXAMPLE 5.3

A wire of diameter 1 mm and conductivity 5×10^7 mhos/m has 10^{29} free electrons/m^3 when an electric field of 10 mV/m is applied. Determine

(a) The charge density of free electrons

(b) The current density

(c) The current in the wire

(d) The drift velocity of the electrons. Take the electronic charge as $e = -1.6 \times 10^{-19}$ C.

SOLUTION (In this particular problem, convection and conduction currents are the same.)

(a) $\rho_v = ne = (10^{29})(-1.6 \times 10^{-19}) = -1.6 \times 10^{10}\ \text{C/m}^3$

(b) $J = \sigma E = (5 \times 10^7)(10 \times 10^{-3}) = 500\ \text{kA/m}^2$

(c) $I = JS = (5 \times 10^5)\left(\dfrac{\pi d^2}{4}\right) = \dfrac{5\pi}{4} \cdot 10^{-6} \cdot 10^5 = 0.393\ A$

(d) Since $J = \rho_v u$, $u = \dfrac{J}{\rho_v} = \dfrac{5 \times 10^5}{1.6 \times 10^{10}} = 3.125 \times 10^{-5}$ m/s. ■

PRACTICE EXERCISE 5.3

The free charge density in copper is $1.81 \times 10^{10}\ \text{C/m}^3$. For a current density of $8 \times 10^6\ \text{A/m}^2$, find the electric field intensity and the drift velocity.

ANSWER 0.138 V/m, 4.42×10^{-4} m/s.

EXAMPLE 5.4

A lead ($\sigma = 5 \times 10^6$ mhos/m) bar of square cross section has a hole bored along its length of 4 m so that its cross section becomes that of Figure 5.4. Find the resistance between the square ends.

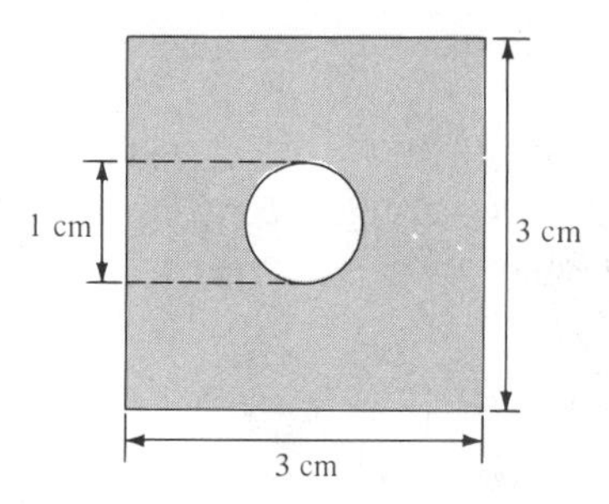

Figure 5.4 Cross section of the lead bar of Example 5.4.

SOLUTION Since the cross section of the bar is uniform, we may apply eq. (5.10); that is,

$$R = \frac{\ell}{\sigma S}$$

where $S = d^2 - \pi r^2 = 3^2 - \pi\left(\frac{1}{2}\right)^2 = 9 - \frac{\pi}{4}\text{ cm}^2$.

Hence,

$$R = \frac{4}{5 \times 10^6(9 - \pi/4) \times 10^{-4}} = 974\ \mu\Omega$$

■

PRACTICE EXERCISE 5.4

If the hole in the lead bar of Example 5.4 is completely filled with copper ($\sigma = 5.8 \times 10^6$ mhos/m), determine the resistance of the composite bar.

ANSWER 876.7 $\mu\Omega$

5.5 POLARIZATION IN DIELECTRICS

In Section 5.2, we noticed that the main difference between a conductor and a dielectric lies in the availability of free electrons in the atomic outermost shells to conduct current. Although the charges in a dielectric are not able to move about freely, they are bound by finite forces and we may certainly expect a displacement when an external force is applied.

To understand the macroscopic effect of an electric field on a dielectric, consider an atom of the dielectric as consisting of a negative charge $-Q$ (electron cloud) and a positive charge $+Q$ (nucleus) as in Figure 5.5(a). A similar picture can be adopted for

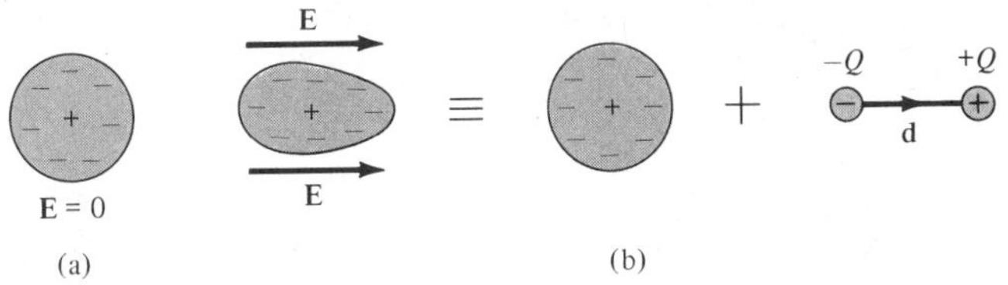

Figure 5.5 Polarization of a nonpolar atom or molecule.

a dielectric molecule; we can treat the nuclei in molecules as point charges and the electronic structure as a single cloud of negative charge. Since we have equal amounts of positive and negative charge, the whole atom or molecule is electrically neutral. When an electric field **E** is applied, the positive charge is displaced from its equilibrium position in the direction of **E** by the force $\mathbf{F}_+ = Q\mathbf{E}$ while the negative charge is displaced in the opposite direction by the force $\mathbf{F}_- = Q\mathbf{E}$. A dipole results from the displacement of the charges and the dielectric is said to be *polarized*. In the polarized state, the electron cloud is distorted by the applied electric field **E**. This distorted charge distribution is equivalent, by the principle of superposition, to the original distribution plus a dipole whose moment is

$$\mathbf{p} = Q\mathbf{d} \tag{5.15}$$

where **d** is the distance vector from $-Q$ to $+Q$ of the dipole as in Figure 5.5(b). If there are N dipoles in a volume Δv of the dielectric, the total dipole moment due to the electric field is

$$Q_1\mathbf{d}_1 + Q_2\mathbf{d}_2 + \cdots + Q_N\mathbf{d}_N = \sum_{k=1}^{N} Q_k\mathbf{d}_k \tag{5.16}$$

As a measure of intensity of the polarization, we define *polarization* **P** (in coulombs/meter square) as the dipole moment per unit volume of the dielectric; that is,

$$\mathbf{P} = \frac{\lim\limits_{\Delta v \to 0} \sum\limits_{k=1}^{N} Q_k\mathbf{d}_k}{\Delta v} \tag{5.17}$$

Thus we conclude that the major effect of the electric field **E** on a dielectric is the creation of dipole moments which align themselves in the direction of **E**. This type of dielectric is said to be *nonpolar*. Examples of such dielectrics are hydrogen, oxygen, nitrogen, and the rare gases. Nonpolar dielectric molecules do not possess dipoles until the application of the electric field as we have noticed. Other types of molecules such as water, sulfur dioxide, and hydrochloric acid have built-in permanent dipoles which are randomly oriented as shown in Figure 5.6(a) and are said to be *polar*. When an

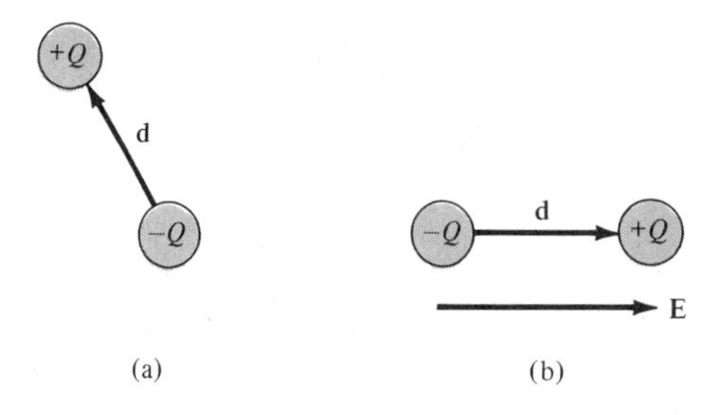

Figure 5.6 Polarization of a polar molecule: **(a)** permanent dipole ($\mathbf{E} = 0$), **(b)** alignment of permanent dipole ($\mathbf{E} \neq 0$).

electric field **E** is applied to a polar molecule, the permanent dipole experiences a torque tending to align its dipole moment parallel with **E** as in Figure 5.6(b).

Let us now calculate the field due to an arbitrary distribution of dipoles based on our definition of **P** in eq. (5.17). Consider the dielectric material shown in Figure 5.7 as consisting of dipoles with dipole moment **P** per unit volume. According to eq. (4.80), the potential dV at an exterior point O due to the dipole moment $\mathbf{P}\ dv'$ is

$$dV = \frac{\mathbf{P} \cdot \mathbf{a}_R\ dv'}{4\pi\varepsilon_o R^2} \tag{5.18}$$

where $R^2 = (x - x')^2 + (y - y')^2 + (z - z')^2$ and R is the distance between the volume element dv' at (x', y', z') and the field point (x, y, z). We can transform eq. (5.18) into a form which facilitates physical interpretation. It is readily shown (see Section 7.7) that the gradient of $1/R$ with respect to the primed coordinates is

$$\nabla' \frac{1}{R} = \frac{\mathbf{a}_R}{R^2}$$

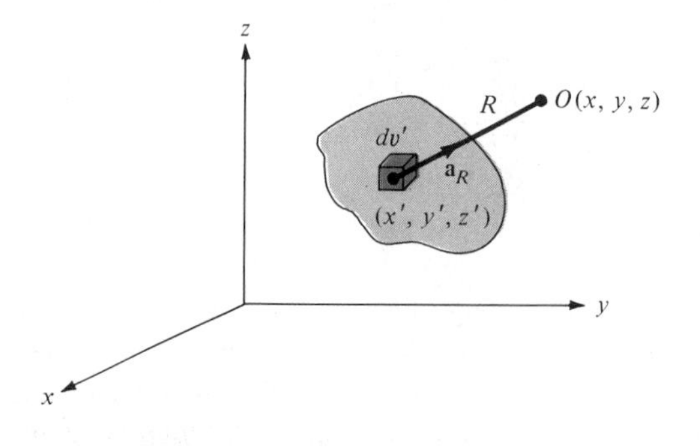

Figure 5.7 A block of dielectric material with dipole moment **p** per unit volume.

Thus,

$$\frac{\mathbf{P} \cdot \mathbf{a}_R}{R^2} = \mathbf{P} \cdot \nabla' \left(\frac{1}{R} \right)$$

Applying the vector identity $\nabla' \cdot f\,\mathbf{A} = f\,\nabla' \cdot \mathbf{A} + \mathbf{A} \cdot \nabla' f$,

$$\frac{\mathbf{P} \cdot \mathbf{a}_R}{R^2} = \nabla' \cdot \frac{\mathbf{P}}{R} - \frac{\nabla' \cdot \mathbf{P}}{R}$$

Substituting this into eq. (5.18) and integrating over the entire volume v' of the dielectric, we obtain

$$V = \int_{v'} \frac{1}{4\pi\varepsilon_o} \left[\nabla' \cdot \frac{\mathbf{P}}{R} - \frac{1}{R} \nabla' \cdot \mathbf{P} \right] dv'$$

Applying divergence theorem to the first term leads finally to

$$V = \int_{S'} \frac{\mathbf{P} \cdot \mathbf{a}_n'}{4\pi\varepsilon_o R} \, dS' + \int_{v'} \frac{-\nabla' \cdot \mathbf{P}}{4\pi\varepsilon_o R} \, dv' \qquad \textbf{[5.19]}$$

where $\mathbf{a}_n'$ is the outward unit normal to surface dS' of the dielectric. Comparing the two terms on the right side of eq. (5.19) with eqs. (4.68) and (4.69) shows that the two terms denote the potential due to surface and volume charge distributions with densities (upon dropping the primes)

$$\rho_{ps} = \mathbf{P} \cdot \mathbf{a}_n \qquad \textbf{[5.20a]}$$

$$\rho_{pv} = -\nabla \cdot \mathbf{P} \qquad \textbf{[5.20b]}$$

In other words, eq. (5.19) reveals that where polarization occurs, an equivalent volume charge density ρ_{pv} is formed throughout the dielectric while an equivalent surface charge density ρ_{ps} is formed over the surface of the dielectric. We refer to ρ_{ps} and ρ_{pv} as *bound* (or *polarization*) *surface* and *volume charge densities*, respectively, as distinct from *free* surface and volume charge densities ρ_S and ρ_v. Bound charges are those which are not free to move within the dielectric material; they are caused by the displacement which occurs on a molecular scale during polarization. Free charges are those which are capable of moving over macroscopic distance as electrons in a conductor; they are the stuff we control. The total positive bound charge on surface S bounding the dielectric is

$$Q_b = \oint \mathbf{P} \cdot d\mathbf{S} = \int \rho_{ps} \, dS \qquad \textbf{[5.21a]}$$

while the charge that remains inside S is

$$-Q_b = \int_v \rho_{pv} \, dv = -\int_v \nabla \cdot \mathbf{P} \, dv \qquad \textbf{[5.21b]}$$

Thus the total charge of the dielectric material remains zero, that is,

$$\text{Total charge} = \oint_S \rho_{ps}\, dS + \int_v \rho_{pv}\, dv = Q_b - Q_b = 0$$

This is expected because the dielectric was electrically neutral before polarization.

If ρ_v is the free charge volume density, the total volume charge density ρ_t is given by

$$\rho_t = \rho_v + \rho_{pv} = \nabla \cdot \varepsilon_o \mathbf{E} \tag{5.22}$$

Hence

$$\begin{aligned} \rho_v &= \nabla \cdot \varepsilon_o \mathbf{E} - \rho_{pv} \\ &= \nabla \cdot (\varepsilon_o \mathbf{E} + \mathbf{P}) \\ &= \nabla \cdot \mathbf{D} \end{aligned} \tag{5.23}$$

where

$$\boxed{\mathbf{D} = \varepsilon_o \mathbf{E} + \mathbf{P}} \tag{5.24}$$

Thus we conclude that the net effect of the dielectric on the electric field **E** is to increase **D** inside it by amount **P**. In other words, due to the application of **E** to the dielectric material, the flux density is greater than it would be in free space. It should be noted that the definition of **D** in eq. (4.35) for free space is a special case of that in eq. (5.24) because **P** = 0 in free space.

We would expect that the polarization **P** would vary directly as the applied electric field **E**. For some dielectrics, this is usually the case and we have

$$\boxed{\mathbf{P} = \chi_e \varepsilon_o \mathbf{E}} \tag{5.25}$$

where χ_e, known as the *electric susceptibility* of the material, is more or less a measure of how susceptible (or sensitive) a given dielectric is to electric fields.

5.6 DIELECTRIC CONSTANT AND STRENGTH

By substituting eq. (5.25) into eq. (5.24), we obtain

$$\mathbf{D} = \varepsilon_o(1 + \chi_e)\, \mathbf{E} = \varepsilon_o \varepsilon_r \mathbf{E} \tag{5.26}$$

or

$$\boxed{\mathbf{D} = \varepsilon \mathbf{E}} \tag{5.27}$$

where

$$\varepsilon = \varepsilon_o \varepsilon_r \tag{5.28}$$

and

$$\varepsilon_r = 1 + \chi_e = \frac{\varepsilon}{\varepsilon_o} \tag{5.29}$$

In eqs. (5.26) to (5.29), ε is called the *permittivity* of the dielectric, ε_o is the permittivity of free space, defined in eq. (4.2) as approximately $10^{-9}/36\pi$ F/m, and ε_r is called the *dielectric constant* or *relative permittivity*. Note that ε_r is the ratio of the permittivity of the dielectric to that of free space, hence the name "relative permittivity." It should also be noticed that ε_r and χ_e are dimensionless whereas ε and ε_o are in farads/meter. The approximate values of the dielectric constants of some common materials are given in Table B.2 in Appendix B. The values given in Table B.2 are for static or low frequency ($<$ 1000 Hz) fields; the values may change at high frequencies. Note from the table that ε_r is always greater or equal to unity. For free space and nondielectric materials (such as metals) $\varepsilon_r = 1$.

The theory of dielectrics we have discussed so far assumes ideal dielectrics. Practically speaking, no dielectric is ideal. When the electric field in a dielectric is sufficiently large, it begins to pull electrons completely out of the molecules, and the dielectric becomes conducting. *Dielectric breakdown* is said to have occurred when a dielectric becomes conducting. Dielectric breakdown occurs in all kinds of dielectric materials (gases, liquids, or solids) and depends on the nature of the material, temperature, humidity, and the amount of time that the field is applied. The minimum value of the electric field at which dielectric breakdown occurs is called the *dielectric strength* of the dielectric material. In other words, the dielectric strength is the maximum electric field which a dielectric can tolerate or withstand without breakdown. Table B.2 also lists the dielectric strength of some common dielectrics. Since our theory of dielectrics does not apply after dielectric breakdown has taken place, we shall always assume ideal dielectric and avoid dielectric breakdown.

†5.7 LINEAR, ISOTROPIC, AND HOMOGENEOUS DIELECTRICS

Although eqs. (5.18) to (5.24) are for dielectric materials in general, eqs. (5.25) to (5.27) are only for linear, isotropic materials. A material is said to be *linear* if **D** varies linearly with **E** and *nonlinear* otherwise. Materials for which ε (or σ) does not vary in the region being considered and is therefore the same at all points (i.e., independent of x, y, z) are said to be *homogeneous*. They are said to be *inhomogeneous* (or nonhomogeneous) when ε is dependent of the space coordinates. The atmosphere is a typical example of an inhomogeneous medium; its permittivity varies with altitude. Materials for which **D** and **E** are in the same direction are said to be *isotropic*. That is, isotropic dielectrics are those which have the same properties in all directions. For *anisotropic* (or nonisotropic) materials, **D**, **E**, and **P** are not parallel; ε or χ_e has nine components which are collectively referred to as a *tensor*. For example, instead of eq. (5.27), we have

$$\begin{bmatrix} D_x \\ D_y \\ D_z \end{bmatrix} = \begin{bmatrix} \varepsilon_{xx} & \varepsilon_{xy} & \varepsilon_{xz} \\ \varepsilon_{yz} & \varepsilon_{yy} & \varepsilon_{yz} \\ \varepsilon_{zx} & \varepsilon_{zy} & \varepsilon_{zz} \end{bmatrix} \begin{bmatrix} E_x \\ E_y \\ E_z \end{bmatrix} \tag{5.30}$$

for anisotropic materials. Crystalline materials and magnetized plasma are anisotropic.

Thus, a dielectric material (in which $\mathbf{D} = \varepsilon\mathbf{E}$ applies) is linear if ε does not change with the applied **E** field, homogeneous if ε does not change from point to point, and isotropic if ε does not change with direction. The same idea applies to a conducting material in which $\mathbf{J} = \sigma\mathbf{E}$ applies. The material is linear if σ does not vary with **E**, homogeneous if σ is the same at all points, and isotropic if σ does not vary with direction.

For most of the time, we will only be concerned with linear, isotropic, and homogeneous media. For such media, all formulas derived in Chapter 4 for free space can be applied by merely replacing ε_o with $\varepsilon_o\varepsilon_r$. Thus Coulomb's law of eq. (4.4), for example, becomes

$$\mathbf{F} = \frac{Q_1 Q_2}{4\pi\varepsilon_o\varepsilon_r R^2}\,\mathbf{a}_R \tag{5.31}$$

and eq. (4.103) becomes

$$W = \frac{1}{2}\int \varepsilon_o\varepsilon_r E^2\, dv \tag{5.32}$$

when applied to a dielectric medium.

EXAMPLE 5.5

A dielectric cube of side L and center at the origin has a radial polarization given by $\mathbf{P} = a\mathbf{r}$, where a is a constant and $\mathbf{r} = x\mathbf{a}_x + y\mathbf{a}_y + z\mathbf{a}_z$. Find all bound charge densities and show explicitly that the total bound charge vanishes.

SOLUTION

For each of the six faces of the cube, there is a surface charge ρ_{ps}. For the face located at $x = L/2$, for example,

$$\rho_{ps} = \mathbf{P} \cdot \mathbf{a}_x \bigg|_{x = L/2} = ax \bigg|_{x = L/2} = aL/2$$

The total bound surface charge is

$$Q_s = \int \rho_{ps}\, dS = 6\int_{-L/2}^{L/2}\int_{-L/2}^{L/2} \rho_{ps}\, dy\, dz = \frac{6aL}{2}L^2$$

$$= 3aL^3$$

The bound volume charge density is given by

$$\rho_{pv} = -\nabla \cdot \mathbf{P} = -(a + a + a) = -3a$$

and the total bound volume charge is

$$Q_v = \int \rho_{pv}\, dv = -3a \int dv = -3aL^3$$

Hence the total charge is

$$Q_t = Q_s + Q_v = 3aL^3 - 3aL^3 = 0$$

■

PRACTICE EXERCISE 5.5

A thin rod of cross section A extends along the x-axis from $x = 0$ to $x = L$. The polarization of the rod is along its length and is given by $P_x = ax^2 + b$. Calculate ρ_{pv} and ρ_{ps} at each end. Show explicitly that the total bound charge vanishes in this case.

ANSWER 0, $-2aL$, $-b$, $aL^2 + b$, proof.

EXAMPLE 5.6

The electric field intensity in polystyrene ($\varepsilon_r = 2.55$) filling the space between the plates of a parallel-plate capacitor is 10 kV/m. The distance between the plates is 1.5 mm. Calculate:

(a) D

(b) P

(c) The surface charge density of free charge on the plates

(d) The surface density of polarization charge

(e) The potential difference between the plates

SOLUTION

(a) $D = \varepsilon_o \varepsilon_r E = \dfrac{10^{-9}}{36\pi} \cdot (2.55) \cdot 10^4 = 225.4 \text{ nC/m}^2$

(b) $P = \chi_e \varepsilon_o E = (1.55) \cdot \dfrac{10^{-9}}{36\pi} \cdot 10^4 = 137 \text{ nC/m}^2$

(c) $\rho_S = \mathbf{D} \cdot \mathbf{a}_n = D_n = 225.4 \text{ nC/m}^2$

(d) $\rho_{ps} = \mathbf{P} \cdot \mathbf{a}_n = P_n = 132 \text{ nC/m}^2$

(e) $V = Ed = 10^4 (1.5 \times 10^{-3}) = 15 \text{ V}$ ■

PRACTICE EXERCISE 5.6

A parallel-plate capacitor with plate separation of 2 mm has a 1-kV voltage applied to its plates. If the space between its plates is filled with polystrene ($\varepsilon_r = 2.55$), find **E**, **P**, and ρ_{ps}.

ANSWER $500\mathbf{a}_x$ kV/m, $6.853\mathbf{a}_x$ μC/m^2, 6.853 μC/m^2.

EXAMPLE 5.7

A dielectric sphere ($\varepsilon_r = 5.7$) of radius 10 cm has a point charge 2 pC placed at its center. Calculate

(a) The surface density of polarization charge on the surface of the sphere

(b) The force exerted by the charge on a −4-pC point charge placed on the sphere

SOLUTION (a) We apply Coulomb's or Gauss's law to obtain

$$\mathbf{E} = \frac{Q}{4\pi\varepsilon_o\varepsilon_r r^2}\,\mathbf{a}_r$$

$$\mathbf{P} = \chi_e\varepsilon_o\mathbf{E} = \frac{\chi_e Q}{4\pi\varepsilon_r r^2}\,\mathbf{a}_r$$

$$\rho_{ps} = \mathbf{P} \cdot \mathbf{a}_r = \frac{(\varepsilon_r - 1)\,Q}{4\pi\varepsilon_r r^2} = \frac{(4.7)\,2 \times 10^{-12}}{4\pi(5.7)\,100 \times 10^{-4}}$$

$$= 13.12 \text{ pC/m}^2$$

(b) Using Coulomb's law, we have

$$\mathbf{F} = \frac{Q_1 Q_2}{4\pi\varepsilon_o\varepsilon_r r^2}\,\mathbf{a}_r = \frac{(-4)(2) \times 10^{-24}}{4\pi \cdot \dfrac{10^{-9}}{36\pi}(5.7)\,100 \times 10^{-4}}\,\mathbf{a}_r$$

$$= -\,1.263\,\mathbf{a}_r \text{ pN}$$ ■

PRACTICE EXERCISE 5.7

In a dielectric material, $E_x = 5$ V/m and $\mathbf{P} = \dfrac{1}{10\pi}(3\mathbf{a}_x - \mathbf{a}_y + 4\mathbf{a}_z)$ nC/m^2.

Calculate

(a) χ_e
(b) $\mathbf{E}$
(c) $\mathbf{D}$

ANSWER (a) 2.16, (b) $5\mathbf{a}_x - 1.67\mathbf{a}_y + 6.67\mathbf{a}_z$ V/m, (c) $139.7\mathbf{a}_x - 46.6\mathbf{a}_y + 186.3\mathbf{a}_z$ pC/m^2.

EXAMPLE 5.8

Find the force with which the plates of a parallel-plate capacitor attract each other. Also determine the pressure on the surface of the plate due to the field.

SOLUTION From eq. (4.26), the electric field intensity on the surface of each plate is

$$\mathbf{E} = \frac{\rho_S}{2\varepsilon}\,\mathbf{a}_n$$

where $\mathbf{a}_n$ is a unit normal to the plate and ρ_S is the surface charge density. The total force on each plate is

$$\mathbf{F} = Q\mathbf{E} = \rho_S S \cdot \frac{\rho_S}{2\varepsilon}\,\mathbf{a}_n = \frac{\rho_S^2\, S}{2\varepsilon_o\varepsilon_r}\,\mathbf{a}_n$$

or

$$F = \frac{\rho_S^2\, S}{2\varepsilon} = \frac{Q^2}{2\varepsilon S}$$

The pressure or force/area is $\dfrac{\rho_S^2}{2\varepsilon_o\varepsilon_r}$. ■

PRACTICE EXERCISE 5.8

Shown in Figure 5.8 is a potential measuring device known as an *electrometer*. It is basically a parallel-plate capacitor with the guarded plate being suspended

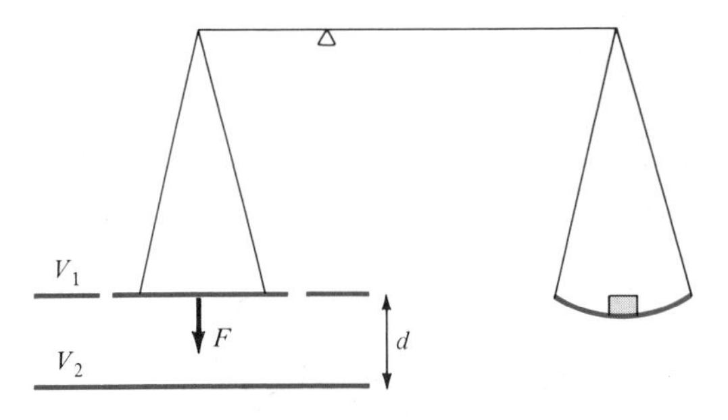

Figure 5.8 An electrometer; for Practice Exercise 5.8.

from a balance arm so that the force F on it is measurable in terms of weight. If S is the area of each plate, show that

$$V_1 - V_2 = \left[\frac{2\,Fd^2}{\varepsilon_o S}\right]^{1/2}$$

ANSWER Proof.

5.8 CONTINUITY EQUATION AND RELAXATION TIME

Due to the principle of charge conservation, the time rate of decrease of charge within a given volume must be equal to the net outward current flow through the closed surface of the volume. Thus current I_{out} coming out of the closed surface is

$$I_{out} = \oint \mathbf{J} \cdot d\mathbf{S} = \frac{-dQ_{in}}{dt} \quad [5.33]$$

where Q_{in} is the total charge enclosed by the closed surface. Invoking divergence theorem

$$\oint_S \mathbf{J} \cdot d\mathbf{S} = \int_v \nabla \cdot \mathbf{J}\, dv \quad [5.34]$$

But

$$\frac{-dQ_{in}}{dt} = -\frac{d}{dt}\int_v \rho_v\, dv = -\int_v \frac{\partial \rho_v}{\partial t}\, dv \quad [5.35]$$

Substituting eqs. (5.34) and (5.35) into eq. (5.33) gives

$$\int_v \nabla \cdot \mathbf{J}\, dv = -\int_v \frac{\partial \rho_v}{\partial t}\, dv \quad [5.36]$$

or

$$\nabla \cdot \mathbf{J} = -\frac{\partial \rho_v}{\partial t} \tag{5.37}$$

which is called the *continuity of current equation*. It must be kept in mind that the continuity equation is derived from the principle of conservation of charge and essentially states that there can be no accumulation of charge at any point. For steady currents, $\partial \rho_v / dt = 0$ and hence $\nabla \cdot \mathbf{J} = 0$ showing that the total charge leaving a volume is the same as the total charge entering it. Kirchhoff's current law follows from this.

Having discussed the continuity equation and the properties σ and ε of materials, it is appropriate to consider the effect of introducing charge at some *interior* point of a given material (conductor or dielectric). We make use of eq. (5.37) in conjunction with Ohm's law

$$\mathbf{J} = \sigma \mathbf{E} \tag{5.38}$$

and Gauss's law

$$\nabla \cdot \mathbf{E} = \frac{\rho_v}{\varepsilon} \tag{5.39}$$

Substituting eqs. (5.38) and (5.39) into eq. (5.37) yields

$$\nabla \cdot \sigma \mathbf{E} = \frac{\sigma \rho_v}{\varepsilon} = -\frac{\partial \rho_v}{\partial t} \tag{5.40}$$

or

$$\frac{\partial \rho_v}{\partial t} + \frac{\sigma}{\varepsilon} \rho_v = 0 \tag{5.41}$$

This is a homogeneous linear ordinary differential equation. By separating variables in eq. (5.41), we get

$$\frac{\partial \rho_v}{\rho_v} = -\frac{\sigma}{\varepsilon} \partial t \tag{5.42}$$

and integrating both sides gives

$$\ln \rho_v = -\frac{\sigma t}{\varepsilon} + \ln \rho_{vo}$$

where $\ln \rho_{vo}$ is a constant of integration. Thus

$$\rho_v = \rho_{vo} e^{-t/T_r} \tag{5.43}$$

where

$$\boxed{T_r = \frac{\varepsilon}{\sigma}} \tag{5.44}$$

In eq. (5.43), ρ_{vo} is the initial charge density (i.e., ρ_v at $t = 0$). The equation shows that as a result of introducing charge at some interior point of the material there is a decay of volume charge density ρ_v. Associated with the decay is charge movement from the interior point at which it was introduced to the surface of the material. The time constant T_r (in seconds) is known as the *relaxation time* or *rearrangement time*. It is the time it takes a charge placed in the interior of a material to drop to $e^{-1} = 36.8$ percent of its initial value. It is short for good conductors and long for good dielectrics. For example, for copper $\sigma = 5.8 \times 10^7$ mhos/m, $\varepsilon_r = 1$, and

$$\begin{aligned} T_r &= \frac{\varepsilon_r \varepsilon_o}{\sigma} = 1 \cdot \frac{10^{-9}}{36\pi} \cdot \frac{1}{5.8 \times 10^7} \\ &= 1.53 \times 10^{-19} \text{ s} \end{aligned} \tag{5.45}$$

showing a rapid decay of charge placed inside copper. This implies that for good conductors, the relaxation time is so short that most of the charge will vanish from any interior point and appear at the surface (as surface charge). On the other hand, for fused quartz, for instance, $\sigma = 10^{-17}$ mhos/m, $\varepsilon_r = 5.0$,

$$\begin{aligned} T_r &= 5 \cdot \frac{10^{-9}}{36\pi} \cdot \frac{1}{10^{-17}} \\ &= 51.2 \text{ days} \end{aligned} \tag{5.46}$$

showing a very large relaxation time. Thus for good dielectrics, one may consider the introduced charge to remain wherever placed.

5.9 BOUNDARY CONDITIONS

So far, we have considered the existence of the electric field in a homogeneous medium. If the field exists in a region consisting of two different media, the conditions that the field must satisfy at the interface separating the media are called *boundary conditions*. These conditions are helpful in determining the field on one side of the boundary if the field on the other side is known. Obviously, the conditions will be dictated by the types of material the media are made of. We shall consider the boundary conditions at an interface separating

- dielectric (ε_{r1}) and dielectric (ε_{r2})
- conductor and dielectric
- conductor and free space

To determine the boundary conditions, we need to use Maxwell's equations for electrostatic fields:

$$\oint \mathbf{E} \cdot d\mathbf{l} = 0 \tag{5.47}$$

and

$$\oint \mathbf{D} \cdot d\mathbf{S} = Q_{\text{enc}} \tag{5.48}$$

Also we need to decompose the electric field intensity **E** into two orthogonal components:

$$\mathbf{E} = \mathbf{E}_t + \mathbf{E}_n \tag{5.49}$$

where $\mathbf{E}_t$ and $\mathbf{E}_n$ are respectively the tangential and normal components of **E** to the interface of interest. A similar decomposition can be done for the electric flux density **D**.

A. Dielectric-Dielectric Boundary Conditions

Consider the **E** field existing in a region consisting of two different dielectrics characterized by $\varepsilon_1 = \varepsilon_o\varepsilon_{r1}$ and $\varepsilon_2 = \varepsilon_o\varepsilon_{r2}$ as shown in Figure 5.9(a). $\mathbf{E}_1$ and $\mathbf{E}_2$ in media 1 and 2 respectively can be decomposed as

$$\mathbf{E}_1 = \mathbf{E}_{1t} + \mathbf{E}_{1n} \tag{5.50a}$$

$$\mathbf{E}_2 = \mathbf{E}_{2t} + \mathbf{E}_{2n} \tag{5.50b}$$

We apply eq. (5.47) to the closed path *abcda* of Figure 5.9(a) assuming that the path is very small with respect to the variation of **E**. We obtain

$$0 = E_{1t}\,\Delta w - E_{1n}\frac{\Delta h}{2} - E_{2n}\frac{\Delta h}{2} - E_{2t}\,\Delta w + E_{2n}\frac{\Delta h}{2} + E_{1n}\frac{\Delta h}{2} \tag{5.51}$$

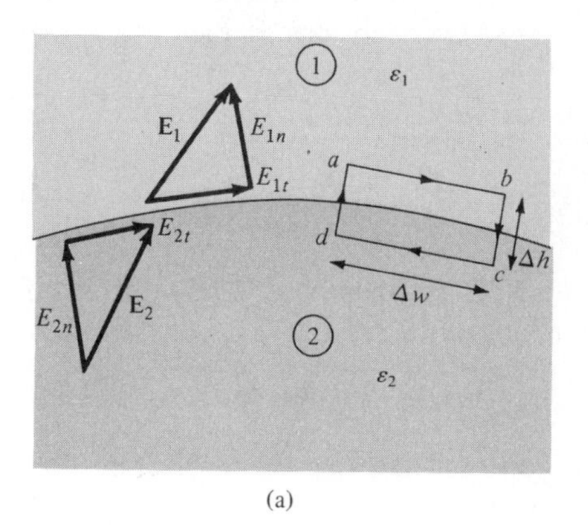

(a)

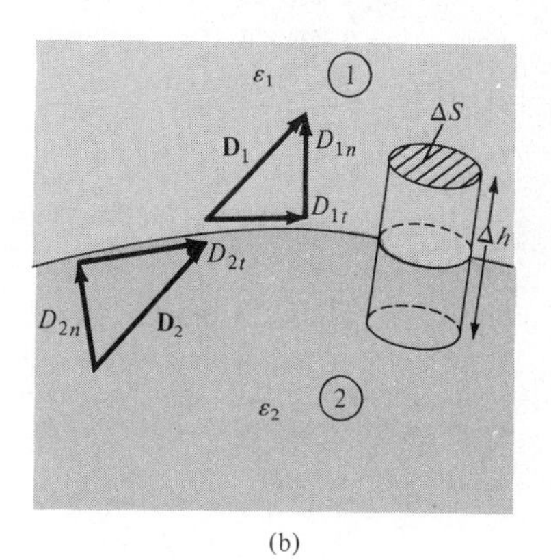

(b)

Figure 5.9 Dielectric-dielectric boundary.

where $E_t = |\mathbf{E}_t|$ and $E_n = |\mathbf{E}_n|$. As $\Delta h \to 0$, eq. (5.51) becomes

$$\boxed{E_{1t} = E_{2t}} \tag{5.52}$$

Thus the tangential components of **E** are the same on the two sides of the boundary. In other words, $\mathbf{E}_t$ undergoes no change on the boundary and it is said to be *continuous* across the boundary. Since $\mathbf{D} = \varepsilon\mathbf{E} = \mathbf{D}_t + \mathbf{D}_n$, eq. (5.52) can be written as

$$\frac{D_{1t}}{\varepsilon_1} = E_{1t} = E_{2t} = \frac{D_{2t}}{\varepsilon_2} \tag{5.53}$$

or

$$\frac{D_{1t}}{\varepsilon_1} = \frac{D_{2t}}{\varepsilon_2} \tag{5.54}$$

that is, D_t undergoes some change across the interface. Hence D_t is said to be *discontinuous* across the interface.

Similarly, we apply eq. (5.48) to the pillbox (Gaussian surface) of Figure 5.9(b). Allowing $\Delta h \to 0$ gives

$$\Delta Q = \rho_s \, \Delta S = D_{1n} \, \Delta S - D_{2n} \, \Delta S \tag{5.55}$$

or

$$\boxed{D_{1n} - D_{2n} = \rho_S} \tag{5.56}$$

where ρ_S is the free charge density placed deliberately at the boundary. It should be borne in mind that eq. (5.56) is based on the assumption that **D** is directed from region 2 to region 1 and eq. (5.56) must be applied accordingly. If no free charges exist at the interface (i.e., charges are not deliberately placed there), $\rho_S = 0$ and eq. (5.56) becomes

$$\boxed{D_{1n} = D_{2n}} \tag{5.57}$$

Thus the normal component of **D** is continuous across the interface; that is, D_n undergoes no change at the boundary. Since $\mathbf{D} = \varepsilon\mathbf{E}$, eq. (5.57) can be written as

$$\varepsilon_1 E_{1n} = \varepsilon_2 E_{2n} \tag{5.58}$$

showing that the normal component of **E** is discontinuous at the boundary. Equations (5.52) and (5.56), or (5.57) are collectively referred to as *boundary conditions;* they must be satisfied by an electric field at the boundary separating two different dielectrics.

As mentioned earlier, the boundary conditions are usually applied in finding the electric field on one side of the boundary given the field on the other side. Besides this, we can use the boundary conditions to determine the "refraction" of the electric field across the interface. Consider $\mathbf{D}_1$ or $\mathbf{E}_1$ and $\mathbf{D}_2$ or $\mathbf{E}_2$ making angles θ_1 and θ_2 with the *normal* to the interface as illustrated in Figure 5.10. Using eq. (5.52), we have

$$E_1 \sin \theta_1 = E_{1t} = E_{2t} = E_2 \sin \theta_2$$

or

$$E_1 \sin \theta_1 = E_2 \sin \theta_2 \tag{5.59}$$

Similarly, by applying eq. (5.57) or (5.58), we get

$$\varepsilon_1 E_1 \cos \theta_1 = D_{1n} = D_{2n} = \varepsilon_2 E_2 \cos \theta_2$$

or

$$\varepsilon_1 E_1 \cos \theta_1 = \varepsilon_2 E_2 \cos \theta_2 \tag{5.60}$$

Dividing eq. (5.59) by eq. (5.60) gives

$$\frac{\tan \theta_1}{\varepsilon_1} = \frac{\tan \theta_2}{\varepsilon_2} \tag{5.61}$$

Since $\varepsilon_1 = \varepsilon_o \varepsilon_{r1}$ and $\varepsilon_2 = \varepsilon_o \varepsilon_{r2}$, eq. (5.61) becomes

$$\boxed{\frac{\tan \theta_1}{\tan \theta_2} = \frac{\varepsilon_{r1}}{\varepsilon_{r2}}} \tag{5.62}$$

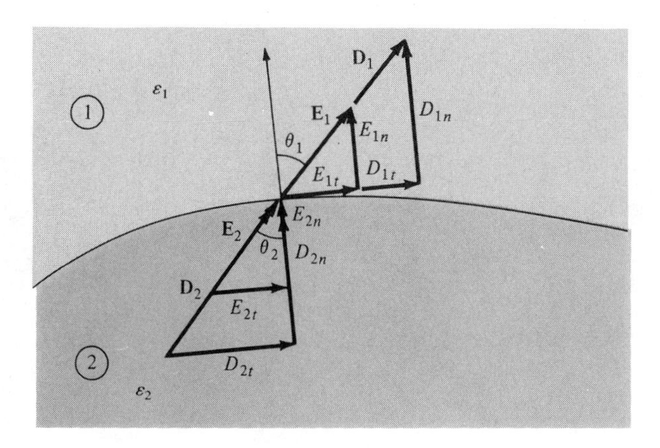

Figure 5.10 Refraction of **D** or **E** at a dielectric-dielectric boundary.

This is the *law of refraction* of the electric field at a boundary free of charge (since $\rho_S = 0$ is assumed at the interface). Thus, in general, an interface between two dielectrics produces bending of the flux lines as a result of unequal polarization charges which accumulate on the sides of the interface.

B. Conductor-Dielectric Boundary Conditions

This is the case shown in Figure 5.11. The conductor is assumed to be perfect (i.e., $\sigma \rightarrow \infty$ or $\rho_c \rightarrow 0$). Although such a conductor is not practically realizable, we may regard conductors such as copper and silver as though they were perfect conductors. From Ohm's law, $\mathbf{J} = \sigma \mathbf{E}$, to maintain a finite current density $\mathbf{J}$ in a perfect conductor requires that the electric field inside the conductor must vanish. In other words, $\mathbf{E} \rightarrow 0$ because $\sigma \rightarrow \infty$ in a conductor. If some charges are introduced in the interior of such a conductor, the charges will move to the conductor surface and redistribute themselves in such a manner that the field inside the conductor vanishes. According to Gauss's law, if $\mathbf{E} = 0$, the charge density ρ_v must be zero. We conclude that a perfect conductor cannot contain an electrostatic field within it.

To determine the boundary conditions for a conductor-dielectric interface, we follow the same procedure used for dielectric-dielectric interface except that we incorporate the fact that $\mathbf{E} = 0$ inside the conductor. Applying eq. (5.47) to the closed path *abcda* of Figure 5.11(a) gives

$$0 = 0 \cdot \Delta w + 0 \cdot \frac{\Delta h}{2} + E_n \cdot \frac{\Delta h}{2} - E_t \cdot \Delta w - E_n \cdot \frac{\Delta h}{2} - 0 \cdot \frac{\Delta h}{2}$$

As $\Delta h \rightarrow 0$,

$$E_t = 0 \qquad [5.63]$$

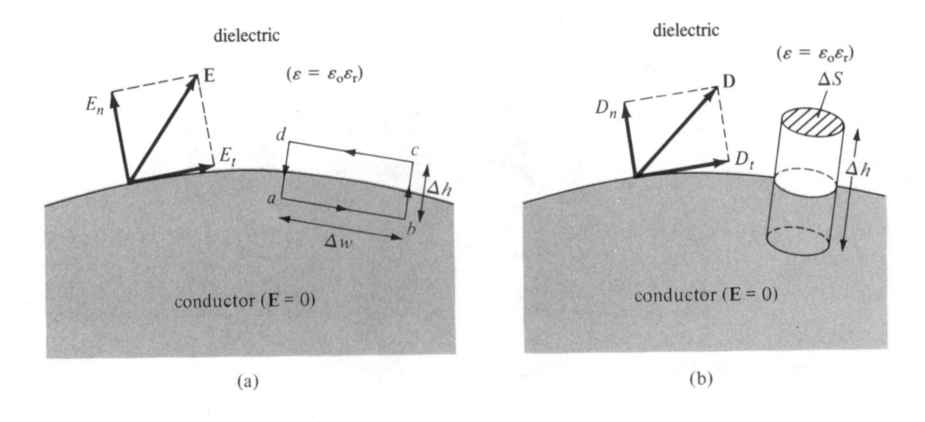

Figure 5.11 Conductor-dielectric boundary.

Similarly, by applying eq. (5.48) to the pillbox of Figure 5.11(b) and letting $\Delta h \rightarrow 0$, we get

$$\Delta Q = D_n \cdot \Delta S - 0 \cdot \Delta S \tag{5.64}$$

because $\mathbf{D} = \varepsilon \mathbf{E} = 0$ inside the conductor. Equation (5.64) may be written as

$$D_n = \frac{\Delta Q}{\Delta S} = \rho_S$$

or

$$D_n = \rho_S \tag{5.65}$$

Thus under static conditions, the following conclusions can be made about a perfect conductor:

1. No electric field may exist *within* a conductor; that is,

$$\boxed{\rho_v = 0, \qquad \mathbf{E} = 0} \tag{5.66}$$

2. Since $\mathbf{E} = -\nabla V = 0$, there can be no potential difference between any two points in the conductor; that is, a conductor is an equipotential body.
3. The electric field $\mathbf{E}$ can only be external to the conductor and *normal* to its surface; that is,

$$\boxed{D_t = \varepsilon_o \varepsilon_r E_t = 0, \qquad D_n = \varepsilon_o \varepsilon_r E_n = \rho_S} \tag{5.67}$$

An important application of the fact that $\mathbf{E} = 0$ inside a conductor is in *electrostatic screening* or *shielding*. If conductor A kept at zero potential surrounds conductor B as shown in Figure 5.12, B is said to be electrically screened by A from other electric systems, such as conductor C, outside A. Similarly, conductor C outside A is screened by A from B. Thus conductor A acts like a screen or shield and the electrical conditions inside and outside the screen are completely independent of each other.

C. Conductor–Free-Space Boundary Conditions

This is a special case of the conductor-dielectric conditions and is illustrated in Figure 5.13. The boundary conditions at the interface between a conductor and free space can be obtained from eq. (5.67) by replacing ε_r by 1 (because free space may be regarded

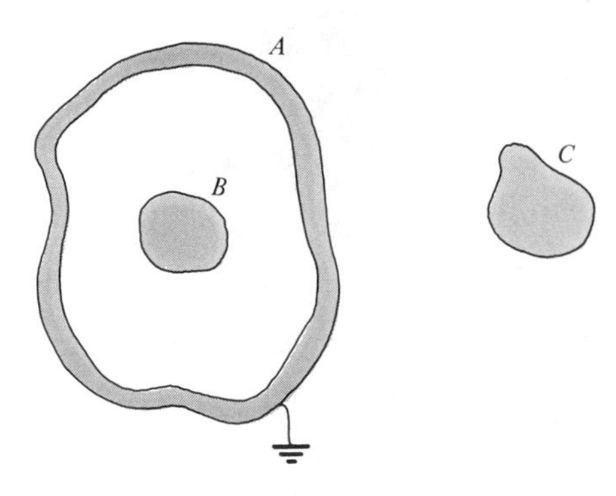

Figure 5.12 Electrostatic screening.

as a special dielectric for which $\varepsilon_r = 1$). We expect the electric field **E** to be external to the conductor and normal to its surface. Thus the boundary conditions are

$$\boxed{D_t = \varepsilon_o E_t = 0, \quad D_n = \varepsilon_o E_n = \rho_S} \tag{5.68}$$

It should be noted again that eq. (5.68) implies that **E** field must approach a conducting surface normally.

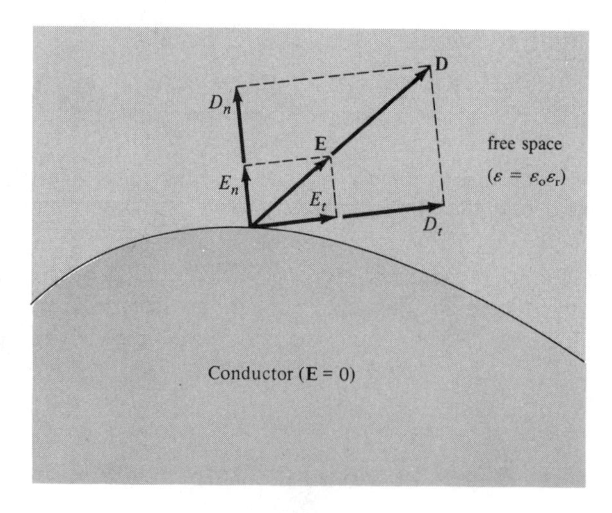

Figure 5.13 Conductor–free-space boundary.

EXAMPLE 5.9

Two extensive homogeneous isotropic dielectrics meet on plane $z = 0$. For $z \geq 0$, $\varepsilon_{r1} = 4$ and for $z \leq 0$, $\varepsilon_{r2} = 3$. A uniform electric field $\mathbf{E}_1 = 5\mathbf{a}_x - 2\mathbf{a}_y + 3\mathbf{a}_z$ kV/m exists for $z \geq 0$. Find

(a) $\mathbf{E}_2$ for $z \leq 0$

(b) The angles E_1 and E_2 make with the interface

(c) The energy densities in J/m^3 in both dielectrics

(d) The energy within a cube of side 2 m centered at (3, 4, −5)

SOLUTION

Let the problem be as illustrated in Figure 5.14.

(a) Since $\mathbf{a}_z$ is normal to the boundary plane, we obtain the normal components as

$$E_{1n} = \mathbf{E}_1 \cdot \mathbf{a}_n = \mathbf{E}_1 \cdot \mathbf{a}_z = 3$$

$$\mathbf{E}_{1n} = 3\mathbf{a}_z$$

$$\mathbf{E}_{2n} = (\mathbf{E}_2 \cdot \mathbf{a}_z)\mathbf{a}_z$$

Also

$$\mathbf{E} = \mathbf{E}_n + \mathbf{E}_t$$

Hence,

$$\mathbf{E}_{1t} = \mathbf{E}_1 - \mathbf{E}_{1n} = 5\mathbf{a}_x - 2\mathbf{a}_y$$

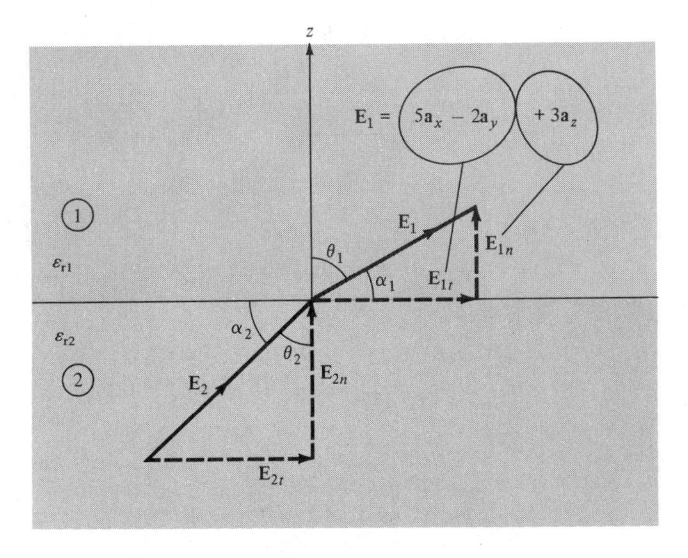

Figure 5.14 For Example 5.9.

Thus

$$\mathbf{E}_{2t} = \mathbf{E}_{1t} = 5\mathbf{a}_x - 2\mathbf{a}_y$$

Similarly,

$$\mathbf{D}_{2n} = \mathbf{D}_{1n} \rightarrow \varepsilon_{r2}\mathbf{E}_{2n} = \varepsilon_{r1}\mathbf{E}_{1n}$$

or

$$\mathbf{E}_{2n} = \frac{\varepsilon_{r1}}{\varepsilon_{r2}}\mathbf{E}_{1n} = \frac{4}{3}(3\mathbf{a}_z) = 4\mathbf{a}_z$$

Thus

$$\begin{aligned} \mathbf{E}_2 &= \mathbf{E}_{2t} + \mathbf{E}_{2n} \\ &= 5\mathbf{a}_x - 2\mathbf{a}_y + 4\mathbf{a}_z \text{ kV/m} \end{aligned}$$

(b) Let α_1 and α_2 be the angles $\mathbf{E}_1$ and $\mathbf{E}_2$ make with the interface while θ_1 and θ_2 are the angles they make with the normal to the interface as shown in Figure 5.14; that is,

$$\begin{aligned} \alpha_1 &= 90 - \theta_1 \\ \alpha_2 &= 90 - \theta_2 \end{aligned}$$

Since $E_{1n} = 3$ and $E_{1t} = \sqrt{25 + 4} = \sqrt{29}$

$$\tan \theta_1 = \frac{E_{1t}}{E_{1n}} = \frac{\sqrt{29}}{3} = 1.795 \rightarrow \theta_1 = 60.9°$$

Hence,

$$\alpha_1 = 29.1°$$

Alternatively,

$$\mathbf{E}_1 \cdot \mathbf{a}_n = |\mathbf{E}_1| \cdot 1 \cdot \cos \theta_1$$

or

$$\cos \theta_1 = \frac{3}{\sqrt{38}} = 0.4867 \rightarrow \theta_1 = 60.9°$$

Similarly,

$$E_{2n} = 4 \qquad E_{2t} = E_{1t} = \sqrt{29}$$

$$\tan \theta_2 = \frac{E_{2t}}{E_{2n}} = \frac{\sqrt{29}}{4} = 1.346 \rightarrow \theta_2 = 53.4°$$

Hence,

$$\alpha_2 = 36.6°$$

Note that $\dfrac{\tan \theta_1}{\tan \theta_2} = \dfrac{\varepsilon_{r1}}{\varepsilon_{r2}}$ is satisfied.

(c) The energy densities are given by

$$w_{E1} = \frac{1}{2}\,\varepsilon_1|\mathbf{E}_1|^2 = \frac{1}{2} \cdot 4 \cdot \frac{10^{-9}}{36\pi} \cdot (25 + 4 + 9) \times 10^6$$
$$= 672\ \mu\text{J/m}^3$$

$$w_{E2} = \frac{1}{2}\,\varepsilon_2|\mathbf{E}_2|^2 = \frac{1}{2} \cdot 3 \cdot \frac{10^{-9}}{36\pi}\,(25 + 4 + 16) \times 10^6$$
$$= 597\ \mu\text{J/m}^3$$

(d) At the center $(3, 4, -5)$ of the cube of side 2 m, $z = -5 < 0$; that is, the cube is in region 2 with $2 \le x \le 4$, $3 \le y \le 5$, $-6 \le z \le -4$. Hence

$$W_E = \int w_{E2}\,dv = \int_{x=2}^{4} \int_{y=3}^{5} \int_{z=-6}^{-4} w_{E2}\,dz\,dy\,dx = w_{E2}(2)(2)(2)$$
$$= 597 \times 8\ \mu\text{J} = 4.776\ \text{mJ}$$

■

PRACTICE EXERCISE 5.9

A homogeneous dielectric ($\varepsilon_r = 2.5$) fills region 1 ($x \le 0$) while region 2 ($x \ge 0$) is free space.

(a) If $\mathbf{D}_1 = 12\mathbf{a}_x - 10\mathbf{a}_y + 4\mathbf{a}_z$ nC/m^2, find $\mathbf{D}_2$ and θ_2.

(b) If $E_2 = 12$ V/m and $\theta_2 = 60°$, find E_1 and θ_1. Take θ_1 and θ_2 as defined in the previous example.

ANSWER (a) $12\mathbf{a}_x - 4\mathbf{a}_y + 1.6\mathbf{a}_z$ nC/m^2, $19.75°$, (b) 10.67 V/m, $77°$

EXAMPLE 5.10

Region $y \le 0$ consists of a perfect conductor while region $y \ge 0$ is a dielectric medium ($\varepsilon_{1r} = 2$) as in Figure 5.15. If there is a surface charge of 2 nC/m^2 on the conductor, determine **E** and **D** at

(a) A$(3, -2, 2)$

(b) B$(-4, 1, 5)$

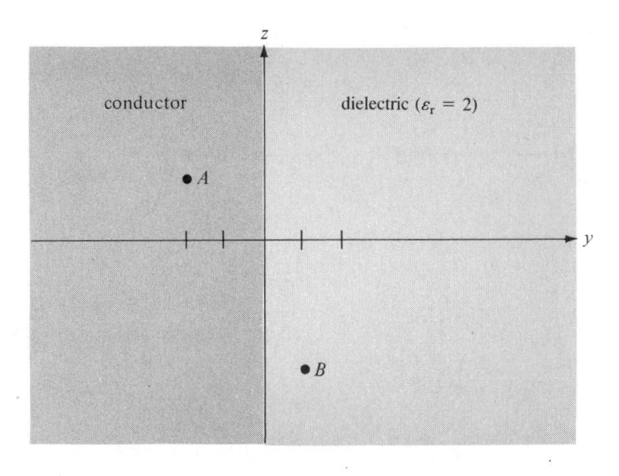

Figure 5.15 For Example 5.10.

SOLUTION (a) Point A (3, −2, 2) is in the conductor since $y = -2 < 0$ at A. Hence,

$$\mathbf{E} = 0 = \mathbf{D}$$

(b) Point B (−4, 1, 5) is in the dielectric medium since $y = 1 > 0$ at B.

$$D_n = \rho_S = 2 \text{ nC/m}^2$$

Hence,

$$\mathbf{D} = 2\mathbf{a}_y \text{ nC/m}^2$$

and

$$\mathbf{E} = \frac{\mathbf{D}}{\varepsilon_o \varepsilon_r} = 2 \times 10^{-9} \times \frac{36\pi}{2} \times 10^9 \, \mathbf{a}_y = 36\pi \mathbf{a}_y$$
$$= 113.1\mathbf{a}_y \text{ V/m}$$

■

PRACTICE EXERCISE 5.10

It is found that $\mathbf{E} = 60\mathbf{a}_x + 20\mathbf{a}_y - 30\mathbf{a}_z$ mV/m at a particular point on the interface between air and a conducting surface. Find $\mathbf{D}$ and ρ_S at that point.

ANSWER $0.531\mathbf{a}_x + 0.177\mathbf{a}_y - 0.265\mathbf{a}_z$ pC/m^2, 0.619 pC/m^2.

SUMMARY

1. Materials can be classified roughly as conductors ($\sigma \gg 1$, $\varepsilon_r = 1$) and dielectrics $\sigma \ll 1$, $\varepsilon_r \geq 1$) in terms of their electrical properties σ and ε_r, where σ is the conductivity and ε_r is the dielectric constant or relative permittivity.
2. Electric current is the flux of electric current density through a surface; that is,

$$I = \int \mathbf{J} \cdot d\mathbf{S}$$

3. The resistance of a conductor of uniform cross section is

$$R = \frac{\ell}{\sigma S}$$

4. The macroscopic effect of polarization on a given volume of a dielectric material is to "paint" its surface with a bound charge $Q_b = \oint_S \rho_{ps}\, dS$ and leave within it an accumulaton of bound charge $Q_b = \int_v \rho_{pv}\, dv$ where $\rho_{ps} = \mathbf{P} \cdot \mathbf{a}_n$ and $\rho_{pv} = -\nabla \cdot \mathbf{P}$.
5. In a dielectric medium, the **D** and **E** fields are related as $\mathbf{D} = \varepsilon\mathbf{E}$, where $\varepsilon = \varepsilon_o\varepsilon_r$ is the permittivity of the medium.
6. The electric susceptibility χ_e ($= \varepsilon_r - 1$) of a dielectric measures the sensitivity of the material to an electric field.
7. A dielectric material is linear if $\mathbf{D} = \varepsilon\mathbf{E}$ holds, that is, if ε is independent of **E**. It is homogeneous if ε is independent of position. It is isotropic if ε is a scalar.
8. The principle of charge conservation, the basis of Kirchhoff's current law, is stated in the continuity equation

$$\nabla \cdot \mathbf{J} + \frac{\partial \rho_v}{\partial t} = 0$$

9. The relaxation time, $T_r = \varepsilon/\sigma$, of a material is the time taken by a charge placed in its interior to decrease by a factor of $e^{-1} \simeq 37$ percent.
10. Boundary conditions must be satisfied by an electric field existing in two different media separated by an interface. For a dielectric-dielectric interface

$$E_{1t} = E_{2t}$$

$$D_{1n} - D_{2n} = \rho_S \qquad \text{or} \qquad D_{1n} = D_{2n} \qquad \text{if} \qquad \rho_S = 0$$

For a dielectric-conductor interface,

$$E_t = 0 \qquad D_n = \varepsilon E_n = \rho_S$$

because $\mathbf{E} = 0$ inside the conductor.

REVIEW QUESTIONS

5.1 Which is *not* an example of convection current?

(a) A moving charged belt

(b) Electronic movement in a vacuum tube

(c) An electron beam in a television tube

(d) Electric current flowing in a copper wire

5.2 When a steady potential difference is applied across the ends of a conducting wire,

(a) All electrons move with a constant velocity.

(b) All electrons move with a constant acceleration.

(c) The random electronic motion will, on the average, be equivalent to a constant velocity of each electron.

(d) The random electronic motion will, on the average, be equivalent to a nonzero constant acceleration of each electron.

5.3 The formula $R = \ell/(\sigma S)$ is for thin wires.

(a) True

(b) False

(c) Not necessarily

5.4 Sea water has $\varepsilon_r = 80$. Its permittivity is

(a) 81

(b) 79

(c) 5.162×10^{-10} F/m

(d) 7.074×10^{-10} F/m

5.5 Both ε_o and χ_e are dimensionless.

(a) True (b) False

5.6 If $\nabla \cdot \mathbf{D} = \varepsilon \nabla \cdot \mathbf{E}$ and $\nabla \cdot \mathbf{J} = \sigma \nabla \cdot \mathbf{E}$ in a given material, the material is said to be

(a) Linear

(b) Homogeneous

(c) Isotropic

(d) Linear and homogeneous

(e) Linear and isotropic

(f) Isotropic and homogeneous

5.7 The relaxation time of mica ($\sigma = 10^{-15}$ mhos/m, $\varepsilon_r = 6$) is

(a) 5×10^{-10} s

(b) 10^{-6} s

(c) 5 hours

(d) 10 hours

(e) 15 hours

5.8 The uniform fields shown in Figure 5.16 are near a dielectric-dielectric boundary but on opposite sides of it. Which configurations are correct? Assume that the boundary is charge-free and that $\varepsilon_2 > \varepsilon_1$.

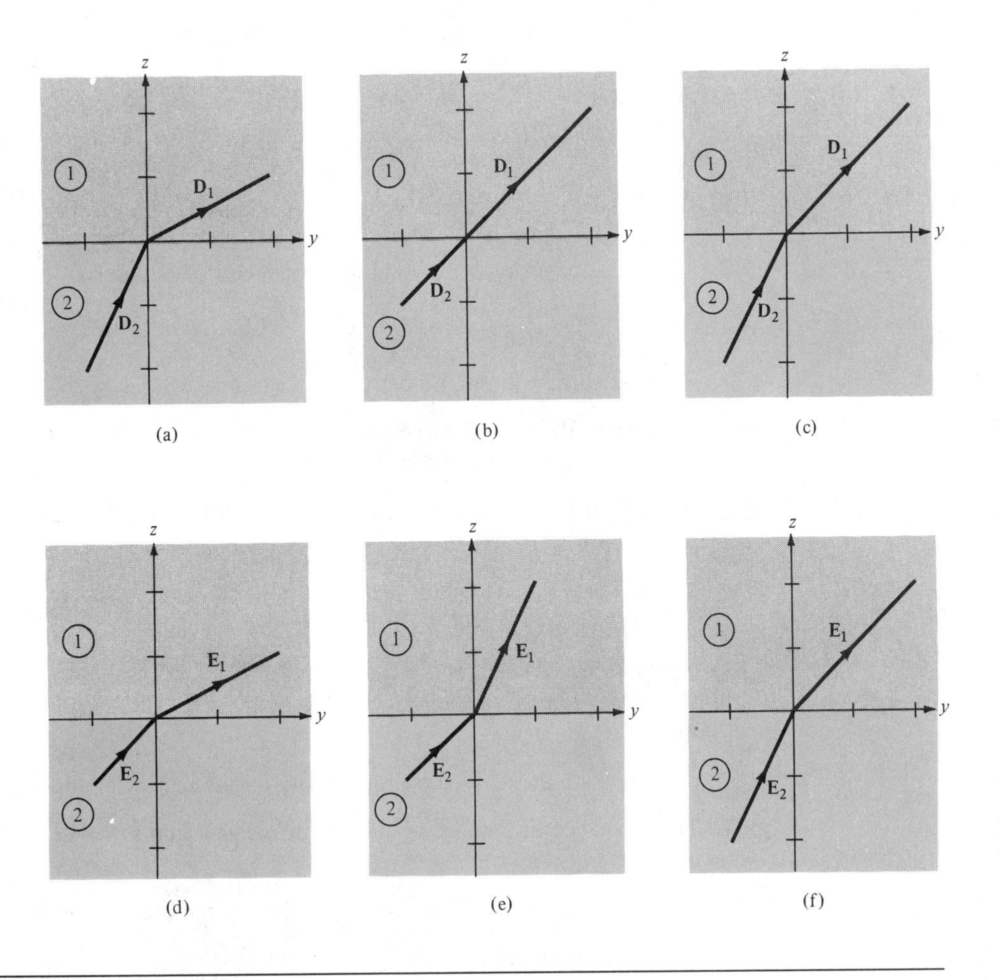

Figure 5.16 For Review Question 5.8.

5.9 Which of the following statements are incorrect?

(a) The conductivities of conductors and insulators vary with temperature and frequency.

(b) A conductor is an equipotential body and $\mathbf{E}$ is always tangential to the conductor.

(c) Nonpolar molecules have no permanent dipoles.

(d) In a linear dielectric, P varies linearly with E.

5.10 The electric conditions (charge and potential) inside and outside an electric screening are completely independent of one another.

(a) True

(b) False

Answers: 5.1d, 5.2c, 5.3c, 5.4d, 5.5b, 5.6d, 5.7e, 5.8e, 5.9b, 5.10a.

PROBLEMS

5.1 Determine the total current in a wire of radius 1.6 mm if $\mathbf{J} = \dfrac{500\mathbf{a}_z}{\rho}$ A/m^2.

5.2 The current density in a cylindrical conductor of radius a is

$$\mathbf{J} = 10e^{-(1 - \rho/a)}\mathbf{a}_z \text{ A/m}^2$$

Find the current through the cross section of the conductor.

5.3 The charge $10^{-4}e^{-3t}$ C is removed from a sphere through a wire. Find the current in the wire at $t = 0$ and $t = 2.5$ s.

5.4 A wire carries a constant current of 2 A. How many coulombs pass its cross section in 6 s? How many electrons?

5.5 The belt of a Van de Graaff generator is 50 cm wide and travels with a speed of 25 m/s.

(a) Ignoring leakage, at what rate in coulombs/second must charge be sprayed on one face of the belt to correspond to a current of 10 μA in the collecting sphere?

(b) Calculate the surface charge density on the belt assuming it to be uniform.

5.6 The resistance of round long wire of diameter 3 mm is 4.04 Ω/km. If a current of 40 A flows through the wire, find

(a) The conductivity of the wire and identify the material of the wire

(b) The electric current density in the wire

5.7 The current density in 1 kg of copper wire is 0.8 A/mm^2. How large is the heat dissipation in the wire? Take the density and resistivity of copper as 8.9 g/cm^3 and $1.75 \times 10^{-8}\ \Omega \cdot$ m respectively.

5.8 A rod whose cross section is an equilateral triangle of side 4 cm is 3 m long. Calculate the resistance between the ends of the rod if it is made of hard rubber ($\sigma = 10^{-15}$ mhos/m).

5.9 A composite conductor 10 m long consists of an inner core of steel of radius 1.5 cm and an outer sheath of copper whose thickness is 0.5 cm.

(a) Determine the resistance of the conductor.

(b) If the total current in the conductor is 60 A, what current flows in each metal?

(c) Find the resistance of a solid copper conductor of the same length and cross-sectional areas as the sheath. Take the resistivities of copper and steel as 1.77×10^{-8} and $11.8 \times 10^{-8}\ \Omega \cdot$ m respectively.

5.10 A hollow cylinder of length 2 m has its cross section as shown in Figure 5.17. If the cylinder is made of carbon ($\sigma = 10^5$ mhos/m), determine the resistance between the ends of the cylinder. Take $a = 3$ cm, $b = 5$ cm.

5.11 At a particular temperature and pressure, a helium gas contains 5×10^{25} atoms/m^3. If a 10-kV/m field applied to the gas causes an average electron cloud shift of 10^{-18}m, find the dielectric constant of helium.

5.12 A dielectric material contains 2×10^{19} polar molecules/m^3, each of dipole moment 1.8×10^{-27} C · m. Assuming that all the dipoles are aligned in the direction of the electric field $\mathbf{E} = 10^5\ \mathbf{a}_x$ V/m, find $\mathbf{P}$ and ε_r.

5.13 In a slab of dielectric material for which $\varepsilon = 2.4\varepsilon_o$ and $V = 300z^2$ V, find: (a) $\mathbf{D}$ and ρ_v, (b) $\mathbf{P}$ and ρ_{pv}.

5.14 (a) Show that $\mathbf{P} = (\varepsilon - \varepsilon_o)\mathbf{E}$ and $\mathbf{D} = \dfrac{\varepsilon_r}{\varepsilon_r - 1}\mathbf{P}$.

(b) Given that $\chi_e = 2.4$ and $D = 300\ \mu$C/m^2, find ε_r, E, and P.

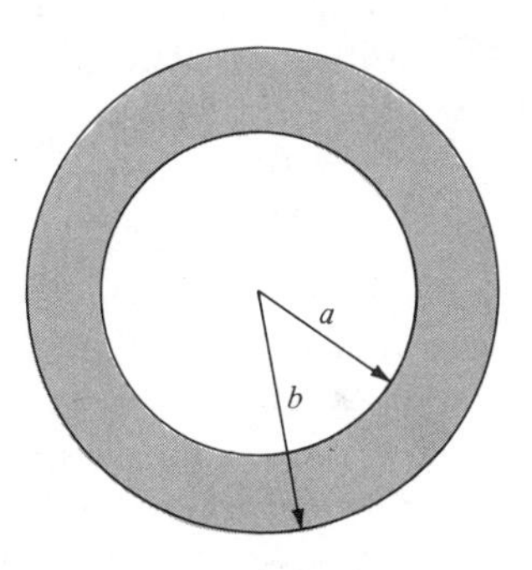

Figure 5.17 For Problems 5.10 and 5.15.

5.15 Consider Figure 5.17 as a spherical dielectric shell so that $\varepsilon = \varepsilon_o \varepsilon_r$ for $a < r < b$ and $\varepsilon = \varepsilon_o$ for $0 < r < a$. If a charge Q is placed at the center of the shell, find

(a) $\mathbf{P}$ for $a < r < b$

(b) ρ_{pv} for $a < r < b$

(c) ρ_{ps} at $r = a$ and $r = b$

5.16 Given that $\mathbf{J} = 10^4(x^2 + y^2)\mathbf{a}_z$ A/m^2, determine

(a) The current density at $(-3, 4, 6)$

(b) The rate of increase in the volume charge density at $(1, -2, 3)$

(c) The current crossing a disk of radius 5 mm placed on the xy-plane and centered at the origin

5.17 Let $\mathbf{J} = \dfrac{e^{-10^3 t}}{\rho^2}\mathbf{a}_\rho$ A/m^2 be the current density in a given region. At $t = 10$ ms, calculate

(a) The amount of current passing through surface $\rho = 2$ m, $0 \leq z \leq 3$ m, $0 \leq \phi < 2\pi$

(b) The charge density ρ_v on the surface

5.18 Calculate the rearrangement time for

(a) Polystyrene ($\sigma = 10^{-16}$ mhos/m, $\varepsilon_r = 2.55$)

(b) Moist earth ($\sigma = 10^{-4}$ mhos/m, $\varepsilon_r = 20$)

(c) Brass ($\sigma = 1.6 \times 10^7$ mhos/m, $\varepsilon_r = 1$)

5.19 The relaxation time of a material with dielectric constant of 6 is 53 s. Calculate the conductivity of the material and identify the material.

5.20 Lightning strikes a dielectric sphere of radius 20 mm for which $\varepsilon_r = 2.5$, $\sigma = 5 \times 10^{-6}$ mhos/m and deposits uniformly a charge of 10 μC. Determine the initial charge density and the charge density 2 μs later.

5.21 Region $z < 0$ contains a perfect dielectric for which $\varepsilon_r = 2.5$, while region $z > 0$ is characterized by $\varepsilon_r = 4$. Let $\mathbf{E}_1 = -30\mathbf{a}_x + 50\mathbf{a}_y + 70\mathbf{a}_z$ V/m and find: (a) $\mathbf{D}_2$, (b) $\mathbf{P}_2$, (c) the angle between $\mathbf{E}_1$ and the normal to the surface.

5.22 Regions 1 ($0 \leq y \leq 1$) and 2 ($1 \leq y \leq 2$) consist of infinite dielectric slabs as shown in Figure 5.18. If the electric field makes an angle 30° with the y-axis in region 1, find the angle it makes with the y-axis in

(a) Region 2

(b) Free space

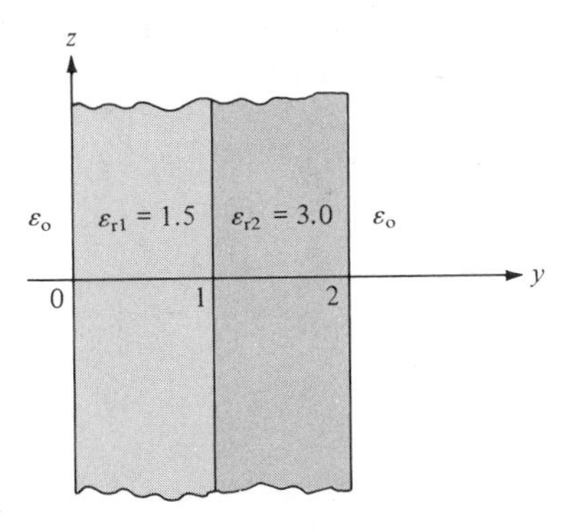

Figure 5.18 For Problem 5.22.

5.23 The cylindrical surface $\rho = 3$ separates two homogeneous dielectric regions 1 ($\rho \leq 3$) and 2 ($\rho \geq 3$) with $\varepsilon_{r1} = 2.5$ and $\varepsilon_{r2} = 10$ respectively. Given that $\mathbf{E}_1 = 2\mathbf{a}_\rho + 5\mathbf{a}_\phi - 4\mathbf{a}_z$ kV/m, find: (a) $\mathbf{P}_1$ and ρ_{pv1}, (b) $\mathbf{E}_2$ and $\mathbf{D}_2$.

5.24 A conducting sphere of radius a is half-embedded in a liquid dielectric medium of permittivity ε_1 as in Figure 5.19. The region above the liquid is a gas of permittivity ε_2. If the total free charge on the sphere is Q, determine the electric field intensity everywhere.

***5.25** Two parallel sheets of glass ($\varepsilon_r = 8.5$) mounted vertically are separated by a uniform air-gap between their inner surface. The sheets, properly sealed, are immersed in oil ($\varepsilon_r = 3.0$) as shown in Figure 5.20. A uniform electric field of strength 2000 V/m in the horizontal direction exists in the oil. Calculate the magnitude and direction of the electric field in the glass and in the enclosed air-gap when (a) the field is normal to the glass surfaces, and (b) the field in the oil makes an angle of 75° with a normal to the glass surfaces. Ignore edge effects.

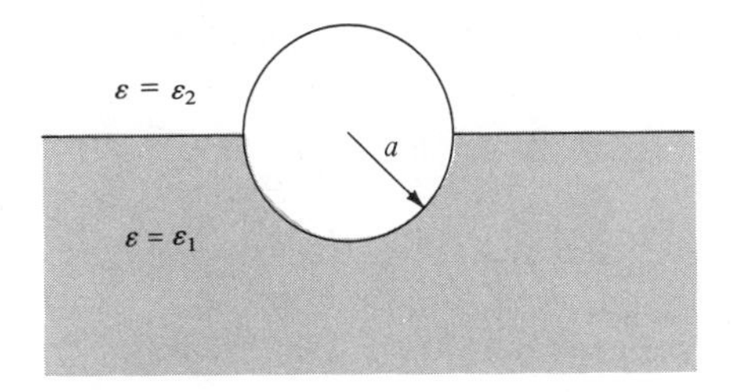

Figure 5.19 For Problem 5.24.

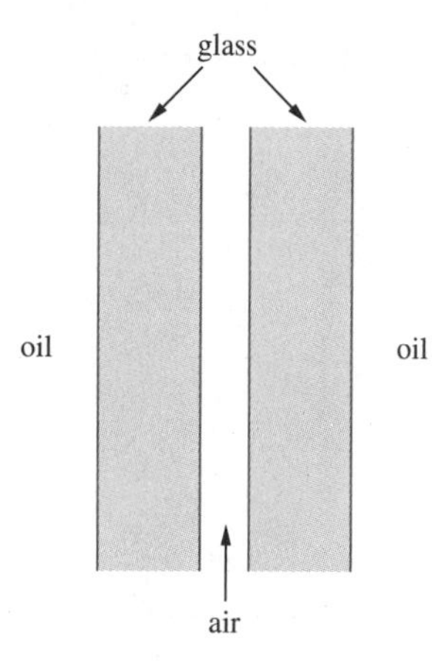

Figure 5.20 For Problem 5.25.

5.26 (a) Given that $\mathbf{E} = 15\mathbf{a}_x - 8\mathbf{a}_z$ V/m at a point on a conductor surface, what is the surface charge density at that point? Assume $\varepsilon = \varepsilon_o$.

(b) Region $y \geq 2$ is occupied by a conductor. If the surface charge on the conductor is -20 nC/m^2, find $\mathbf{D}$ just outside the conductor.

5.27 A silver-coated sphere of radius 5 cm carries a total charge of 12 nC uniformly distributed on its surface in free space. Calculate (a) $|\mathbf{D}|$ on the surface of the sphere, (b) $\mathbf{D}$ external to the sphere, and (c) the total energy stored in the field.

5.28 A solid conducting sphere ($\sigma = 5 \times 10^7$ mhos/m) of radius 40 cm is centered at the origin. If a total charge of 50 nC is on the sphere, under static conditions, determine

(a) $\mathbf{J}$ inside the conductor

(b) ρ_S just on its surface

5.29 Prove that at the interface between two conducting media (not perfect conductors), the boundary conditions for current density are

$$J_{2n} = J_{1n} \qquad \text{and} \qquad \frac{J_{1t}}{\sigma} = \frac{J_{2t}}{\sigma}$$

where n and t denote normal and tangential components respectively.

CHAPTER

Electrostatic Boundary-Value Problems 6

The father of success is work. The mother of success is ambition. The oldest son is common sense. Some of the other boys are: perseverance, honesty, thoroughness, foresight, enthusiasm, cooperation. The oldest daughter is character. Some of the sisters are: cheerfulness, loyalty, courtesy, care, economy, sincerity, harmony. The baby is opportunity. Get acquainted with the "old man" and you will be able to get along pretty well with the rest of the family.

—YEOMAN SHIELD

6.1 INTRODUCTION

The procedure for determining the electric field **E** in the preceding chapters has generally been using either Coulomb's law or Gauss's law when the charge distribution is known, or using $\mathbf{E} = -\nabla V$ when the potential V is known throughout the region. In most practical situations, however, neither the charge distribution nor the potential distribution is known.

In this chapter, we shall consider practical electrostatic problems where only electrostatic conditions (charge and potential) at some boundaries are known and it is desired to find **E** and V throughout the region. Such problems are usually tackled using Poisson's[1] or Laplace's[2] equation or the method of images, and they are usually referred to as *boundary-value* problems. The concepts of resistance and capacitance will be covered. We shall use Laplace's equation in deriving the resistance of an object and the capacitance of a capacitor.

....

[1]After Simeon Denis Poisson (1781–1840), a French mathematical physicist.

[2]After Pierre Simon de Laplace (1749–1829), a French astronomer and mathematician.

6.2 POISSON'S AND LAPLACE'S EQUATIONS

Poisson's and Laplace's equations are easily derived from Gauss's law (for a linear material medium)

$$\nabla \cdot \mathbf{D} = \nabla \cdot \varepsilon \mathbf{E} = \rho_v \tag{6.1}$$

and

$$\mathbf{E} = -\nabla V \tag{6.2}$$

Substituting eq. (6.2) into eq. (6.1) gives

$$\nabla \cdot (-\varepsilon \nabla V) = \rho_v \tag{6.3}$$

for an inhomogeneous medium. For a homogeneous medium, eq. (6.3) becomes

$$\boxed{\nabla^2 V = -\frac{\rho_v}{\varepsilon}} \tag{6.4}$$

homogeneous medium

This is known as *Poisson's equation*. A special case of this equation occurs when $\rho_v = 0$ (i.e., for a charge-free region). Equation (6.4) then becomes

$$\boxed{\nabla^2 V = 0} \tag{6.5}$$

charge free region

which is known as *Laplace's equation*. Note that in taking ε out of the left-hand side of eq. (6.3) to obtain eq. (6.4), we have assumed that ε is constant throughout the region in which V is defined; for an inhomogeneous region, ε is not constant and eq. (6.4) does not follow eq. (6.3). Equation (6.3) is Poisson's equation for an inhomogeneous medium; it becomes Laplace's equation for an inhomogeneous medium when $\rho_v = 0$.

Recall that the Laplacian operator ∇^2 was derived in Section 3.8. Thus Laplace's equation in Cartesian, cylindrical, or spherical coordinates respectively is given by

$$\boxed{\frac{\partial^2 V}{\partial x^2} + \frac{\partial^2 V}{\partial y^2} + \frac{\partial^2 V}{\partial z^2} = 0} \tag{6.6}$$

$$\boxed{\frac{1}{\rho}\frac{\partial}{\partial \rho}\left(\rho \frac{\partial V}{\partial \rho}\right) + \frac{1}{\rho^2}\frac{\partial^2 V}{\partial \phi^2} + \frac{\partial^2 V}{\partial z^2} = 0} \tag{6.7}$$

$$\frac{1}{r^2}\frac{\partial}{\partial r}\left(r^2\frac{\partial V}{\partial r}\right)+\frac{1}{r^2\sin\theta}\frac{\partial}{\partial\theta}\left(\sin\theta\frac{\partial V}{\partial\theta}\right)+\frac{1}{r^2\sin^2\theta}\frac{\partial^2 V}{\partial\phi^2}=0 \quad [6.8]$$

depending on whether the potential is $V(x, y, z)$, $V(\rho, \phi, z)$ or $V(r, \theta, \phi)$. Poisson's equation in those coordinate systems may be obtained by simply replacing zero on the right-hand side of eqs. (6.6), (6.7), and (6.8) with $-\rho_v/\varepsilon$.

Laplace's equation is of primary importance in solving electrostatic problems involving a set of conductors maintained at different potentials. Examples of such problems include capacitors and vacuum tube diodes. Laplace's and Poisson's equations are not only useful in solving electrostatic field problem; they are used in various other field problems. For example, V would be interpreted as magnetic potential in magnetostatics, as temperature in heat conduction, as stress function in fluid flow, and as pressure head in seepage.

†6.3 UNIQUENESS THEOREM

Since there are several methods (analytical, graphical, numerical, experimental, etc.) of solving a given problem, we may wonder whether solving Laplace's equation in different ways gives different solutions. Therefore, before we begin to solve Laplace's equation, we should answer this question: If a solution of Laplace's equation satisfies a given set of boundary conditions, is this the only possible solution? The answer is yes: there is only one solution. We say that the solution is unique. Thus any solution of Laplace's equation which satisfies the same boundary conditions must be the only solution regardless of the method used. This is known as the *uniqueness theorem*. The theorem applies to any solution of Poisson's or Laplace's equation in a given region or closed surface.

The theorem is proved by contradiction. We assume that there are two solutions V_1 and V_2 of Laplace's equation both of which satisfy the prescribed boundary conditions. Thus

$$\nabla^2 V_1 = 0, \qquad \nabla^2 V_2 = 0 \quad [6.9a]$$

$$V_1 = V_2 \qquad \text{on the boundary} \quad [6.9b]$$

We consider their difference

$$V_d = V_2 - V_1 \quad [6.10]$$

which obeys

$$\nabla^2 V_d = \nabla^2 V_2 - \nabla^2 V_1 = 0 \quad [6.11a]$$

$$V_d = 0 \qquad \text{on the boundary} \quad [6.11b]$$

according to eq. (6.9). From the divergence theorem.

$$\int_v \nabla \cdot \mathbf{A}\, dv = \oint_S \mathbf{A} \cdot d\mathbf{S} \tag{6.12}$$

We let $\mathbf{A} = V_d\, \nabla V_d$ and use a vector identity

$$\nabla \cdot \mathbf{A} = \nabla \cdot (V_d \nabla V_d) = V_d \nabla^2 V_d + \nabla V_d \cdot \nabla V_d$$

But $\nabla^2 V_d = 0$ according to eq. (6.11), so

$$\nabla \cdot \mathbf{A} = \nabla V_d \cdot \nabla V_d \tag{6.13}$$

Substituting eq. (6.13) into eq. (6.12) gives

$$\int_v \nabla V_d \cdot \nabla V_d\, dv = \oint_S V_d\, \nabla V_d \cdot d\mathbf{S} \tag{6.14}$$

From eqs. (6.9) and (6.11), it is evident that the right-hand side of eq. (6.14) vanishes. Hence:

$$\int_v |\nabla V_d|^2\, dv = 0$$

Since the integration is always positive,

$$\nabla V_d = 0 \tag{6.15a}$$

or

$$V_d = V_2 - V_1 = \text{constant everywhere in } v \tag{6.15b}$$

But eq. (6.15) must be consistent with eq. (6.9b). Hence, $V_d = 0$ or $V_1 = V_2$ everywhere, showing that V_1 and V_2 cannot be different solutions of the same problem. This is the uniqueness theorem: If a solution to Laplace's equation can be found that satisfies the boundary conditions, then the solution is unique. Similar steps can be taken to show that the theorem applies to Poisson's equation and to prove the theorem for the case where the electric field (potential gradient) is specified on the boundary.

Before we begin to solve boundary-value problems, we should bear in mind the three things that uniquely describe a problem:

1. The appropriate differential equation (Laplace's or Poisson's equation in this chapter)
2. The solution region
3. The prescribed boundary conditions

A problem does not have a unique solution and cannot be solved completely if any of the three items is missing.

6.4 GENERAL PROCEDURE FOR SOLVING POISSON'S OR LAPLACE'S EQUATION

The following general procedure may be taken in solving a given boundary-value problem involving Poisson's or Laplace's equation:

1. Solve Laplace's (if $\rho_v = 0$) or Poisson's (if $\rho_v \neq 0$) equation using either (a) direct integration when V is a function of one variable, or (b) separation of variables if V is a function of more than one variable. The solution at this point is not unique but expressed in terms of unknown integration constants to be determined.
2. Apply the boundary conditions to determine a unique solution for V. Imposing the given boundary conditions makes the solution unique.
3. Having obtained V, find $\mathbf{E}$ using $\mathbf{E} = -\nabla V$ and $\mathbf{D}$ from $\mathbf{D} = \varepsilon\mathbf{E}$.
4. If desired, find the charge Q induced on a conductor using $Q = \int \rho_S \, dS$ where $\rho_S = D_n$ and D_n is the component of $\mathbf{D}$ normal to the conductor. If necessary, the capacitance between two conductors can be found using $C = Q/V$.

Solving Laplace's (or Poisson's) equation, as in step 1, is not always as complicated as it may seem. In some cases, the solution may be obtained by mere inspection of the problem. Also a solution may be checked by going backward and finding out if it satisfies both Laplace's (or Poisson's) equation and the prescribed boundary conditions.

EXAMPLE 6.1

Current-carrying components in high-voltage power equipment must be cooled to carry away the heat caused by ohmic losses. A means of pumping is based on the force transmitted to the cooling fluid by charges in an electric field. The electrohydrodynamic (EHD) pumping is modeled in Figure 6.1. The region between the electrodes contains a uniform charge ρ_o which is generated at the left electrode and collected at the right electrode. Calculate the pressure of the pump if $\rho_o = 25$ mC/m^3 and $V_o = 22$ kV.

SOLUTION

Since $\rho_v \neq 0$, we apply Poisson's equation

$$\nabla^2 V = -\frac{\rho_v}{\varepsilon}$$

The boundary conditions $V(z = 0) = V_o$ and $V(z = d) = 0$ show that V depends only on z (there is no ρ or ϕ dependence). Hence

$$\frac{d^2V}{dz^2} = \frac{-\rho_o}{\varepsilon}$$

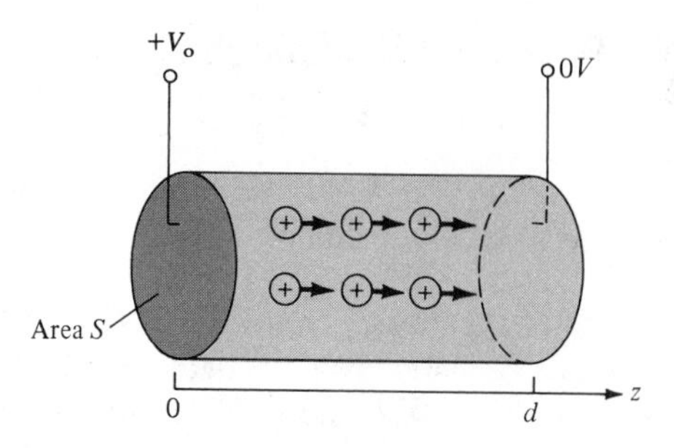

Figure 6.1 An electrohydrodynamic pump; for Example 6.1.

Integrating once gives

$$\frac{dV}{dz} = \frac{-\rho_o z}{\varepsilon} + A$$

Integrating again yields

$$V = -\frac{\rho_o z^2}{2\varepsilon} + Az + B$$

where A and B are integration constants to be determined by applying the boundary conditions. When $z = 0$, $V = V_o$,

$$V_o = -0 + 0 + B \rightarrow B = V_o$$

When $z = d$, $V = 0$,

$$0 = -\frac{\rho_o d^2}{2\varepsilon} + Ad + V_o$$

or

$$A = \frac{\rho_o d}{2\varepsilon} - \frac{V_o}{d}$$

The electric field is given by

$$\mathbf{E} = -\nabla V = -\frac{dV}{dz}\mathbf{a}_z = \left(\frac{\rho_o z}{\varepsilon} - A\right)\mathbf{a}_z$$

$$= \left[\frac{V_o}{d} + \frac{\rho_o}{\varepsilon}\left(z - \frac{d}{2}\right)\right]\mathbf{a}_z$$

The net force is

$$\mathbf{F} = \int \rho_v \mathbf{E}\, dv = \rho_o \int dS \int_{z=0}^{d} \mathbf{E}\, dz$$

$$= \rho_o S \left[\frac{V_o z}{d} + \frac{\rho_o}{2\varepsilon}(z^2 - dz) \right] \Bigg|_0^d \mathbf{a}_z$$

$$\mathbf{F} = \rho_o S V_o \mathbf{a}_z$$

The force per unit area or pressure is

$$p = \frac{F}{S} = \rho_o V_o = 25 \times 10^{-3} \times 22 \times 10^3 = 550 \text{ N/m}^2$$

■

PRACTICE EXERCISE 6.1

In a one-dimensional device, the charge density is given by $\rho_v = \rho_o x/a$. If $\mathbf{E} = 0$ at $x = 0$ and $V = 0$ at $x = a$, find V and $\mathbf{E}$.

ANSWER $\frac{\rho_o}{6\varepsilon a}(a^3 - x^3)$, $\frac{\rho_o x^2}{2a\varepsilon}\mathbf{a}_x$

EXAMPLE 6.2

The xerographic copying machine is an important application of electrostatics. The surface of the photoconductor is initially charged uniformly as in Figure 6.2(a). When light from the document to be copied is focused on the photoconductor, the charges on the lower surface combine with those on the upper surface to neutralize each other. The image is developed by pouring a charged black powder over the surface of the photoconductor. The electric field attracts the charged powder, which is later transferred to paper and melted to form a permanent image. We want to determine the electric field below and above the surface of the photoconductor.

SOLUTION Consider the modeled version of Figure 6.2(a) as in Figure 6.2(b). Since $\rho_v = 0$ in this case, we apply Laplace's equation. Also the potential depends only on x. Thus

$$\nabla^2 V = \frac{d^2V}{dx^2} = 0$$

Integrating twice gives

$$V = Ax + B$$

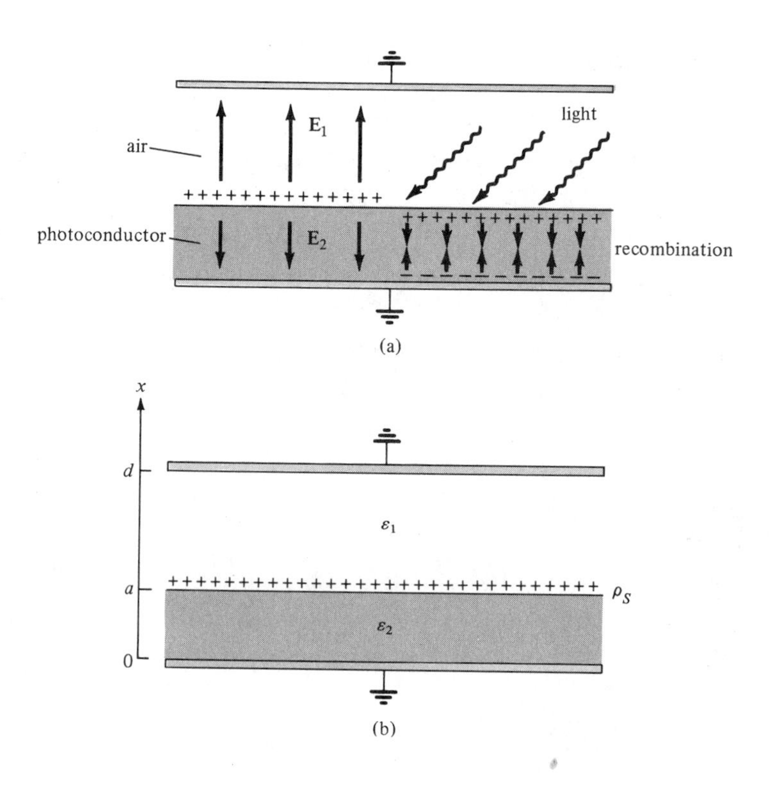

Figure 6.2 For Example 6.2.

Let the potentials above and below be V_1 and V_2 respectively.

$$V_1 = A_1 x + B_1 \qquad x > a \tag{6.2.1a}$$

$$V_2 = A_2 x + B_2 \qquad x < a \tag{6.2.1b}$$

The boundary conditions at the grounded electrodes are

$$V_1(x = d) = 0 \tag{6.2.2a}$$

$$V_2(x = 0) = 0 \tag{6.2.2b}$$

At the surface of the photoconductor,

$$V_1(x = a) = V_2(x = a) \tag{6.2.3a}$$

$$D_{1n} - D_{2n} = \rho_S \Big|_{x = a} \tag{6.2.3b}$$

We use the four conditions in eqs. (6.2.2) and (6.2.3) to determine the four unknown constants A_1, A_2, B_1, and B_2. From eqs. (6.2.1) and (6.2.2),

$$0 = A_1 d + B_1 \rightarrow B_1 = -A_1 d \tag{6.2.4a}$$

$$0 = 0 + B_2 \rightarrow B_2 = 0 \tag{6.2.4b}$$

From eqs. (6.2.1) and (6.2.3a),

$$A_1 a + B_1 = A_2 a \tag{6.2.5}$$

To apply eq. (6.2.3b), recall that $\mathbf{D} = \varepsilon\mathbf{E} = -\varepsilon\nabla V$ so that

$$\rho_S = D_{1n} - D_{2n} = \varepsilon_1 E_{1n} - \varepsilon_2 E_{2n} = -\varepsilon_1 \frac{dV_1}{dx} + \varepsilon_2 \frac{dV_2}{dx}$$

or

$$\rho_S = -\varepsilon_1 A_1 + \varepsilon_2 A_2 \tag{6.2.6}$$

Solving for A_1 and A_2 in eqs. (6.2.4) to (6.2.6), we obtain

$$\mathbf{E}_1 = -A_1\mathbf{a}_x = \frac{\rho_S \mathbf{a}_x}{\varepsilon_1\left[1 + \dfrac{\varepsilon_2}{\varepsilon_1}\dfrac{d}{a} - \dfrac{\varepsilon_2}{\varepsilon_1}\right]}$$

$$\mathbf{E}_2 = -A_2\mathbf{a}_x = \frac{-\rho_S\left(\dfrac{d}{a} - 1\right)\mathbf{a}_x}{\varepsilon_1\left[1 + \dfrac{\varepsilon_2}{\varepsilon_1}\dfrac{d}{a} - \dfrac{\varepsilon_2}{\varepsilon_1}\right]}$$

■

PRACTICE EXERCISE 6.2

For the model of Figure 6.2(b), if $\rho_S = 0$ and the upper electrode is maintained at V_o while the lower electrode is grounded, show that

$$\mathbf{E}_1 = \frac{-V_o\,\mathbf{a}_x}{d - a + \dfrac{\varepsilon_1}{\varepsilon_2}a}, \qquad \mathbf{E}_2 = \frac{-V_o\,\mathbf{a}_x}{a + \dfrac{\varepsilon_2}{\varepsilon_1}d - \dfrac{\varepsilon_2}{\varepsilon_1}a}$$

EXAMPLE 6.3

Semi-infinite conducting planes $\phi = 0$ and $\phi = \pi/6$ are separated by an infinitesimal insulating gap as in Figure 6.3. If $V(\phi = 0) = 0$ and $V(\phi = \pi/6) = 100$ V, calculate V and $\mathbf{E}$ in the region between the planes.

SOLUTION As V depends only on ϕ, Laplace's equation in cylindrical coordinates becomes

$$\nabla^2 V = \frac{1}{\rho^2}\frac{d^2V}{d\phi^2} = 0$$

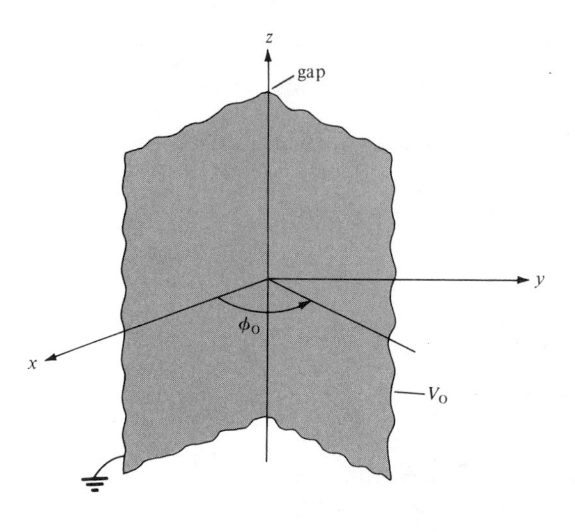

Figure 6.3 Potential $V(\phi)$ due to semi-infinite conducting planes.

Since $\rho = 0$ is excluded due to the insulating gap, we can multiply by ρ^2 to obtain

$$\frac{d^2V}{d\phi^2} = 0$$

which is integrated twice to give

$$V = A\phi + B$$

We apply the boundary conditions to determine constants A and B. When $\phi = 0$, $V = 0$,

$$0 = 0 + B \rightarrow B = 0.$$

When $\phi = \phi_o$, $V = V_o$,

$$V_o = A\phi_o \rightarrow A = \frac{V_o}{\phi_o}$$

Hence:

$$V = \frac{V_o}{\phi_o}\phi$$

and

$$\mathbf{E} = -\nabla V = -\frac{1}{\rho}\frac{dV}{d\phi}\mathbf{a}_\phi = -\frac{V_o}{\rho\phi_o}\mathbf{a}_\phi$$

Substituting $V_o = 100$ and $\phi_o = \pi/6$ gives

$$V = \frac{600}{\pi}\phi \qquad \text{and} \qquad \mathbf{E} = \frac{600}{\pi\rho}\mathbf{a}_\phi$$

Check: $\nabla^2 V = 0$, $V(\phi = 0) = 0$, $V(\phi = \pi/6) = 100$. ■

PRACTICE EXERCISE 6.3

Two conducting plates of size 1×5 m are inclined at 45° to each other with a gap of width 4 mm separating them as shown in Figure 6.4. Determine an approximate value of the charge per plate if the plates are maintained at a potential difference of 50 V. Assume that the medium between them has $\varepsilon_r = 1.5$.

ANSWER 22.2 nC.

EXAMPLE 6.4

Two conducting cones ($\theta = \pi/10$ and $\theta = \pi/6$) of infinite extent are separated by an infinitesimal gap at $r = 0$. If $V(\theta = \pi/10) = 0$ and $V(\theta = \pi/6) = 50$ V, find V and $\mathbf{E}$ between the cones.

SOLUTION Consider the coaxial cone of Figure 6.5, where the gap serves as an insulator between the two conducting cones. V depends only on θ, so Laplace's equation in spherical coordinates becomes

$$\nabla^2 V = \frac{1}{r^2 \sin\theta}\frac{d}{d\theta}\left[\sin\theta\frac{dV}{d\theta}\right] = 0$$

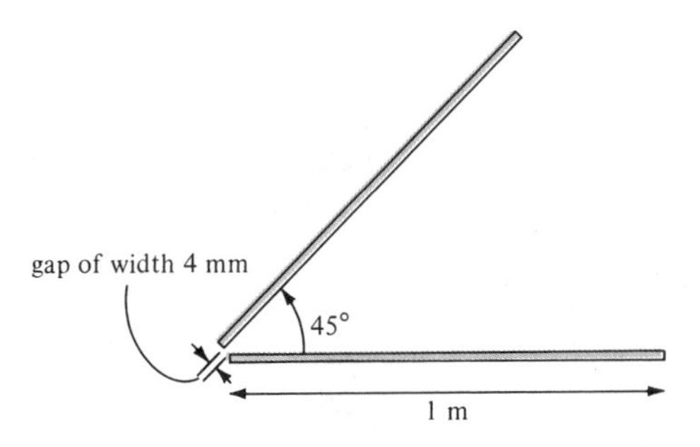

Figure 6.4 For Practice Exercise 6.3.

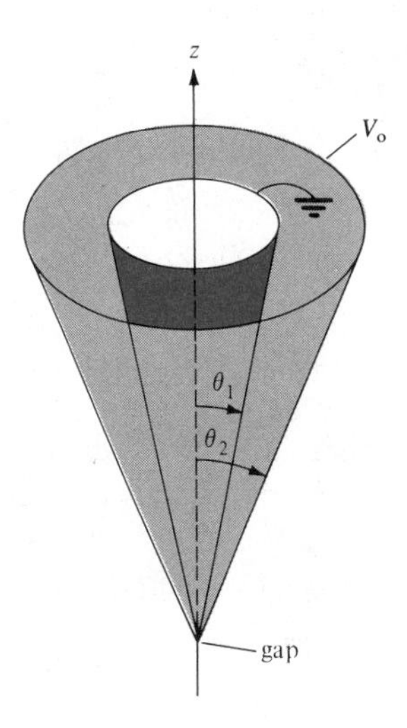

Figure 6.5 Potential $V(\phi)$ due to conducting cones.

Since $r = 0$ and $\theta = 0, \pi$ are excluded, we can multiply by $r^2 \sin\theta$ to get

$$\frac{d}{d\theta}\left[\sin\theta \frac{dV}{d\theta}\right] = 0$$

Integrating once gives

$$\sin\theta \frac{dV}{d\theta} = A$$

or

$$\frac{dV}{d\theta} = \frac{A}{\sin\theta}$$

Integrating this results in

$$\begin{aligned} V = A\int\frac{d\theta}{\sin\theta} &= A\int\frac{d\theta}{2\cos\theta/2\,\sin\theta/2} \\ &= A\int\frac{1/2\,\sec^2\theta/2\,d\theta}{\tan\theta/2} \\ &= A\int\frac{d(\tan\theta/2)}{\tan\theta/2} \\ &= A\ln(\tan\theta/2) + B \end{aligned}$$

We now apply the boundary conditions to determine the integration constants A and B.

$$V(\theta = \theta_1) = 0 \rightarrow 0 = A \ln (\tan \theta_1/2) + B$$

or

$$B = -A \ln (\tan \theta_1/2)$$

Hence

$$V = A \ln \left[\frac{\tan \theta/2}{\tan \theta_1/2}\right]$$

Also

$$V(\theta = \theta_2) = V_o \rightarrow V_o = A \ln \left[\frac{\tan \theta_2/2}{\tan \theta_1/2}\right]$$

or

$$A = \frac{V_o}{\ln \left[\dfrac{\tan \theta_2/2}{\tan \theta_1/2}\right]}$$

Thus

$$V = \frac{V_o \ln \left[\dfrac{\tan \theta/2}{\tan \theta_1/2}\right]}{\ln \left[\dfrac{\tan \theta_2/2}{\tan \theta_1/2}\right]}$$

$$\mathbf{E} = -\nabla V = -\frac{1}{r}\frac{dV}{d\theta}\,\mathbf{a}_\theta = -\frac{A}{r \sin \theta}\,\mathbf{a}_\theta$$

$$= -\frac{V_o}{r \sin \theta \ln \left[\dfrac{\tan \theta_2/2}{\tan \theta_1/2}\right]}\,\mathbf{a}_\theta$$

Taking $\theta_1 = \pi/10$, $\theta_2 = \pi/6$, and $V_o = 50$ gives

$$V = \frac{50 \ln \left[\dfrac{\tan \theta/2}{\tan \pi/20}\right]}{\ln \left[\dfrac{\tan \pi/12}{\tan \pi/20}\right]} = 95.1 \ln \left[\frac{\tan \theta/2}{0.1584}\right] \text{ V}$$

and

$$\mathbf{E} = -\frac{95.1}{r \sin \theta} \mathbf{a}_\theta \text{ V/m}$$

Check: $\nabla^2 V = 0$, $V(\theta = \pi/10) = 0$, $V(\theta = \pi/6) = V_o$. ■

PRACTICE EXERCISE 6.4

A large conducting cone ($\theta = 45°$) is placed on a conducting plane with a tiny gap separating it from the plane as shown in Figure 6.6. If the cone is connected to a 50-V source, find V and $\mathbf{E}$ at $(-3, 4, 2)$.

ANSWER 22.13 V, 11.36 $\mathbf{a}_\theta$ V/m.

EXAMPLE 6.5

(a) Determine the potential function for the region inside the rectangular trough of infinite length whose cross section is shown in Figure 6.7.

(b) For $V_o = 100$ V and $b = 2a$, find the potential at $x = a/2$, $y = 3a/4$.

SOLUTION (a) The potential V in this case depends on x and y. Laplace's equation becomes

$$\nabla^2 V = \frac{\partial^2 V}{\partial x^2} + \frac{\partial^2 V}{\partial y^2} = 0 \tag{6.5.1}$$

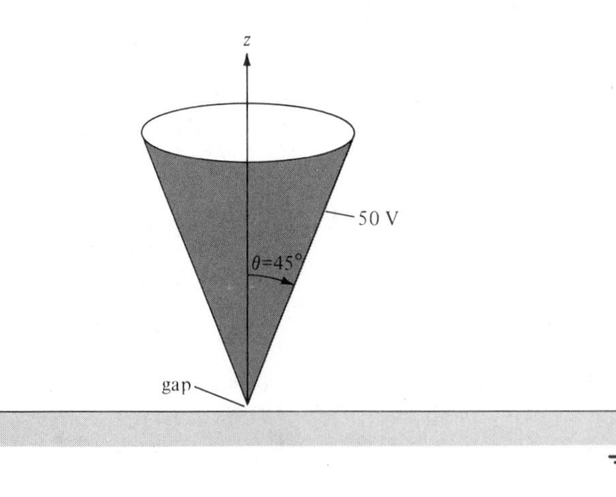

Figure 6.6 For Practice Exercise 6.4.

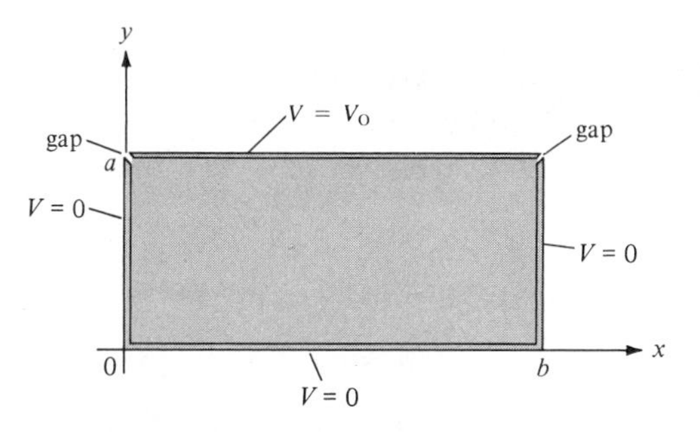

Figure 6.7 Potential $V(x, y)$ due to a conducting rectangular trough.

We have to solve this equation subject to the following boundary conditions:

$$V(x = 0, 0 \leq y \leq a) = 0 \quad [6.5.2a]$$

$$V(x = b, 0 \leq y \leq a) = 0 \quad [6.5.2b]$$

$$V(0 \leq x \leq b, y = 0) = 0 \quad [6.5.2c]$$

$$V(0 \leq x \leq b, y = a) = V_o \quad [6.5.2d]$$

We solve eq. (6.5.1) by the method of *separation of variables;* that is, we seek a product solution of V. Let

$$V(x, y) = X(x)\,Y(y) \quad [6.5.3]$$

when X is a function of x only and Y is a function of y only. Substituting eq. (6.5.3) into eq. (6.5.1) yields

$$X''Y + Y''X = 0$$

Dividing through by XY and separating X from Y gives

$$-\frac{X''}{X} = \frac{Y''}{Y} \quad [6.5.4a]$$

Since the left-hand side of this equation is a function of x only and the right-hand side is a function of y only, for the equality to hold, both sides must be equal to a constant λ; that is

$$-\frac{X''}{X} = \frac{Y''}{Y} = \lambda \quad [6.5.4b]$$

The constant λ is known as the *separation constant*. From eq. (6.5.4b), we obtain

$$X'' + \lambda X = 0 \quad [6.5.5a]$$

and

$$Y'' - \lambda Y = 0 \quad [6.5.5b]$$

Thus the variables have been separated at this point and we refer to eq. (6.5.5) as *separated equations*. We can solve for $X(x)$ and $Y(y)$ separately and then substitute our solutions into eq. (6.5.3). To do this requires that the boundary conditions in eq. (6.5.2) be separated, if possible. We separate them as follows:

$$V(0, y) = X(0)Y(y) = 0 \rightarrow X(0) = 0 \qquad [6.5.6a]$$

$$V(b, y) = X(b)Y(y) = 0 \rightarrow X(b) = 0 \qquad [6.5.6b]$$

$$V(x, 0) = X(x)Y(0) = 0 \rightarrow Y(0) = 0 \qquad [6.5.6c]$$

$$V(x, a) = X(0)Y(a) = V_o \text{ (inseparable)} \qquad [6.5.6d]$$

To solve for $X(x)$ and $Y(y)$ in eq. (6.5.5), we impose the boundary conditions in eq. (6.5.6). We consider possible values of λ that will satisfy both the separated equations in eq. (6.5.5) and the conditions in eq. (6.5.6).

Case 1: If $\lambda = 0$, then eq. (6.5.5a) becomes

$$X'' = 0 \qquad \text{or} \qquad \frac{d^2X}{dx^2} = 0$$

which, upon integrating twice, yields

$$X = Ax + B \qquad [6.5.7]$$

The boundary conditions in eqs. (6.5.6a) and (6.5.6b) imply that

$$X(x = 0) = 0 \rightarrow 0 = 0 + B \quad \text{or} \quad B = 0$$

and

$$X(x = b) = 0 \rightarrow 0 = A \cdot b + 0 \quad \text{or} \quad A = 0$$

because $b \neq 0$. Hence our solution for X in eq. (6.5.7) becomes

$$X(x) = 0$$

which makes $V = 0$ in eq. (6.5.3). Thus we regard $X(x) = 0$ as a trivial solution and we conclude that $\lambda \neq 0$.

Case 2: If $\lambda < 0$, say $\lambda = -\alpha^2$, then eq. (6.5.5a) becomes

$$X'' - \alpha^2 X = 0 \qquad \text{or} \qquad (D^2 - \alpha^2)\, X = 0$$

where $D = \dfrac{d}{dx}$

that is,

$$DX = \pm\, \alpha X \qquad [6.5.8]$$

showing that we have two possible solutions corresponding to the plus and minus signs. For the plus sign, eq. (6.5.8) becomes

$$\frac{dX}{dx} = \alpha X \qquad \text{or} \qquad \frac{dX}{X} = \alpha\, dx$$

Hence

$$\int \frac{dX}{X} = \int \alpha\, dx \qquad \text{or} \qquad \ln X = \alpha x + \ln A_1$$

where $\ln A_1$ is a constant of integration. Thus

$$X = A_1 e^{\alpha x} \tag{6.5.9a}$$

Similarly, for the minus sign, solving eq. (6.5.8) gives

$$X = A_2 e^{-\alpha x} \tag{6.5.9b}$$

The total solution consists of what we have in eqs. (6.5.9a) and (6.5.9b); that is,

$$X(x) = A_1 e^{\alpha x} + A_2 e^{-\alpha x} \tag{6.5.10}$$

Since $\cosh \alpha x = (e^{\alpha x} + e^{-\alpha x})/2$ and $\sinh \alpha x = (e^{\alpha x} - e^{-\alpha x})/2$, or $e^{\alpha x} = \cosh \alpha x + \sinh \alpha x$ and $e^{-\alpha x} = \cosh \alpha x - \sinh \alpha x$, eq. (6.5.10) can be written as

$$X(x) = B_1 \cosh \alpha x + B_2 \sinh \alpha x \tag{6.5.11}$$

where $B_1 = A_1 + A_2$ and $B_2 = A_1 - A_2$. In view of the given boundary conditions, we prefer eq. (6.5.11) to eq. (6.5.10) as the solution. Again, eqs. (6.5.6a) and (6.5.6b) require that

$$X(x = 0) = 0 \rightarrow 0 = B_1 \cdot (1) + B_2 \cdot (0) \qquad \text{or} \qquad B_1 = 0$$

and

$$X(x = b) = 0 \rightarrow 0 = 0 + B_2 \sinh \alpha b$$

Since $\alpha \neq 0$ and $b \neq 0$, $\sinh \alpha b$ cannot be zero. This is due to the fact that $\sinh x = 0$ if and only if $x = 0$ as shown in Figure 6.8. Hence $B_2 = 0$ and

$$X(x) = 0$$

This is also a trivial solution and we conclude that λ cannot be less than zero.

Case 3: If $\lambda > 0$, say $\lambda = \beta^2$, then eq. (6.5.5a) becomes

$$X'' + \beta^2 X = 0$$

that is,

$$(D^2 + \beta^2)X = 0 \qquad \text{or} \qquad DX = \pm j\beta X \tag{6.5.12}$$

where $j = \sqrt{-1}$. From eqs. (6.5.8) and (6.5.12), we notice that the difference between Cases 2 and 3 is replacing α by $j\beta$. By taking the same procedure as in Case 2, we obtain the solution as

$$X(x) = C_0 e^{j\beta x} + C_1 e^{-j\beta x} \tag{6.5.13a}$$

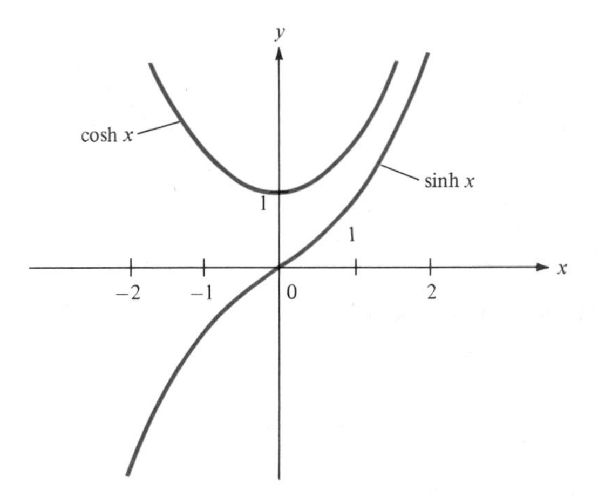

Figure 6.8 Sketch of cosh x and sinh x showing that sinh $x = 0$ if and only if $x = 0$.

Since $e^{j\beta x} = \cos \beta x + j \sin \beta x$ and $e^{-j\beta x} = \cos \beta x - j \sin \beta x$, eq. (6.5.13a) can be written as

$$X(x) = g_o \cos \beta x + g_1 \sin \beta x \qquad [6.5.13b]$$

where $g_o = C_o + C_1$ and $g_1 = C_o - jC_1$.

In view of the given boundary conditions, we prefer to use eq. (6.5.13b). Imposing the conditions in eqs. (6.5.6a) and (6.5.6b) yields

$$X(x = 0) = 0 \rightarrow 0 = g_o \cdot (1) + 0 \qquad \text{or} \qquad g_o = 0$$

and

$$X(x = b) = 0 \rightarrow 0 = 0 + g_1 \sin \beta b$$

Suppose $g_1 \neq 0$ (otherwise we get a trivial solution), then

$$\sin \beta b = 0 = \sin n\pi$$

$$\beta = \frac{n\pi}{b}, \qquad n = 1, 2, 3, 4, \ldots \qquad [6.5.14]$$

Note that, unlike sinh x, which is zero only when $x = 0$, sin x is zero at an infinite number of points as shown in Figure 6.9. It should also be noted that $n \neq 0$ because $\beta \neq 0$; we have already considered the possibility $\beta = 0$ in Case 1 where we ended up with a trivial solution. Also we do not need to consider $n = -1, -2, -3, -4, \ldots$ because $\lambda = \beta^2$ would remain the same for positive and negative values of n. Thus for a given n, eq. (6.5.13b) becomes

$$X_n(x) = g_n \sin \frac{n\pi x}{b} \qquad [6.5.15]$$

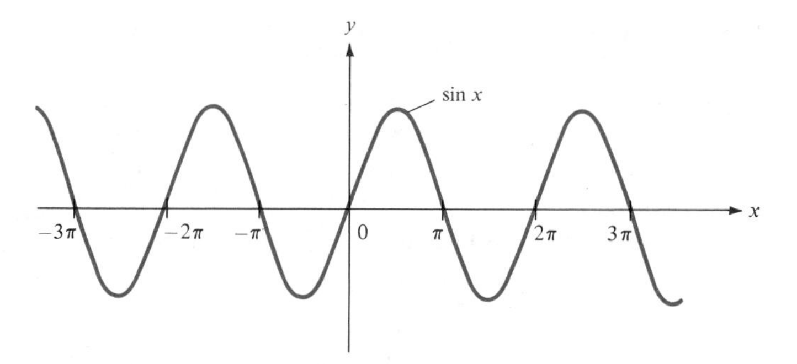

Figure 6.9 Sketch of sin x showing that $\sin x = 0$ at infinite number of points.

Having found $X(x)$ and

$$\lambda = \beta^2 = \frac{n^2\pi^2}{b^2} \tag{6.5.16}$$

we solve eq. (6.5.5b) which is now

$$Y'' - \beta^2 Y = 0$$

The solution to this is similar to eq. (6.5.11) obtained in Case 2 that is,

$$Y(y) = h_o \cosh \beta y + h_1 \sinh \beta y$$

The boundary condition in eq. (6.5.6c) implies that

$$Y(y = 0) = 0 \rightarrow 0 = h_o \cdot (1) + 0 \qquad \text{or} \qquad h_o = 0$$

Hence our solution for $Y(y)$ becomes

$$Y_n(y) = h_n \sinh \frac{n\pi y}{b} \tag{6.5.17}$$

Substituting eqs. (6.5.15) and (6.5.17), which are the solutions to the separated equations in eq. (6.5.5), into the product solution in eq. (6.5.3) gives

$$V_n(x, y) = g_n h_n \sin \frac{n\pi x}{b} \sinh \frac{n\pi y}{b}$$

This shows that there are many possible solutions V_1, V_2, V_3, V_4, and so on, for $n = 1, 2, 3, 4$, and so on.

By the *superposition theorem,* if $V_1, V_2, V_3, \ldots, V_n$ are solutions of Laplace's equation, the linear combination

$$V = c_1V_1 + c_2V_2 + c_3V_3 + \cdots + c_nV_n$$

(where $c_1, c_2, c_3, \ldots, c_n$ are constants) is also a solution of Laplace's equation. Thus the solution to eq. (6.5.1) is

$$V(x, y) = \sum_{n=1}^{\infty} c_n \sin \frac{n\pi x}{b} \sinh \frac{n\pi y}{b} \tag{6.5.18}$$

where $c_n = g_n h_n$ are the coefficients to be determined from the boundary condition in eq. (6.5.6d). Imposing this condition gives

$$V(x, y = a) = V_o = \sum_{n=1}^{\infty} c_n \sin \frac{n\pi x}{b} \sinh \frac{n\pi a}{b} \tag{6.5.19}$$

which is a Fourier series expansion of V_o. Multiplying both sides of eq. (6.5.19) by $\sin m\pi x/b$ and integrating over $0 < x < b$ gives

$$\int_0^b V_o \sin \frac{m\pi x}{b}\, dx = \sum_{n=1}^{\infty} c_n \sinh \frac{n\pi a}{b} \int_0^b \sin \frac{m\pi x}{b} \sin \frac{n\pi x}{b}\, dx \tag{6.5.20}$$

By the orthogonality property of the sine or cosine function (see Appendix A.9),

$$\int_0^{\pi} \sin mx \sin nx\, dx = \begin{cases} 0, & m \neq n \\ \pi/2, & m = n \end{cases}$$

Incorporating this property in eq. (6.5.20) means that all terms on the right-hand side of eq. (6.5.20) will vanish except one term in which $m = n$. Thus eq. (6.5.20) reduces to

$$\int_0^b V_o \sin \frac{n\pi x}{b}\, dx = c_n \sinh \frac{n\pi a}{b} \int_0^b \sin^2 \frac{n\pi x}{b}\, dx$$

$$-V_o \frac{b}{n\pi} \cos \frac{n\pi x}{b} \Bigg|_0^b = c_n \sinh \frac{n\pi a}{b} \frac{1}{2} \int_0^b \left(1 - \cos \frac{2n\pi x}{b}\right) dx$$

$$\frac{V_o b}{n\pi}(1 - \cos n\pi) = c_n \sinh \frac{n\pi a}{b} \cdot \frac{b}{2}$$

or

$$c_n \sinh \frac{n\pi a}{b} = \frac{2V_o}{n\pi}(1 - \cos n\pi)$$

$$= \begin{cases} \dfrac{4V_o}{n\pi}, & n = 1, 3, 5, \ldots \\ 0, & n = 2, 4, 6, \ldots \end{cases}$$

that is,

$$c_n = \begin{cases} \dfrac{4V_o}{n\pi \sinh \dfrac{n\pi a}{b}}, & n = \text{odd} \\ 0, & n = \text{even} \end{cases} \qquad [6.5.21]$$

Substituting this into eq. (6.5.18) gives the complete solution as

$$V(x, y) = \frac{4V_o}{\pi} \sum_{n=1,3,5}^{\infty} \frac{\sin \dfrac{n\pi x}{b} \sinh \dfrac{n\pi y}{b}}{n \sinh \dfrac{n\pi a}{b}} \qquad [6.5.22]$$

Check: $\nabla^2 V = 0$, $V(x = 0, y) = 0 = V(x = b, y) = V(x, y = 0)$, $V(x, y = a) = V_o$. The solution in eq. (6.5.22) should not be a surprise; it can be guessed by mere observation of the potential system in Figure 6.7. From this figure, we notice that along x, V varies from 0 (at $x = 0$) to 0 (at $x = b$) and only a sine function can satisfy this requirement. Similarly, along y, V varies from 0 (at $y = 0$) to V_o (at $y = a$) and only a hyperbolic sine function can satisfy this. Thus we should expect the solution as in eq. (6.5.22).

To determine the potential for each point (x, y) in the trough, we take the first few terms of the convergent infinite series in eq. (6.5.22). Taking four or five terms may be sufficient.

(b) For $x = a/2$ and $y = 3a/4$, where $b = 2a$, we have

$$V\left(\frac{a}{2}, \frac{3a}{4}\right) = \frac{4V_o}{\pi} \sum_{n=1,3,5}^{\infty} \frac{\sin n\pi/4 \sinh 3n\pi/8}{n \sinh n\pi/2}$$

$$= \frac{4V_o}{\pi} \left[\frac{\sin \pi/4 \sinh 3\pi/8}{\sinh \pi/2} + \frac{\sin 3\pi/4 \sinh 9\pi/8}{3 \sinh 3\pi/2} + \frac{\sin 5\pi/4 \sinh 15\pi/4}{5 \sinh 5\pi/4} + \cdots \right]$$

$$= \frac{4V_o}{\pi} (0.4517 + 0.0725 - 0.01985 - 0.00645 + 0.00229 + \cdots)$$

$$= 0.6374V_o$$

It is instructive to consider a special case when $A = b = 1$ m and $V_o = 100$ V. The potentials at some specific points are calculated using eq. (6.5.22) and the result is displayed in Figure 6.10(a). The corresponding flux lines and equipotential lines are shown in Figure 6.10(b). A simple FORTRAN program based on eq. (6.5.22) is displayed in Figure 6.11. This self-explanatory program can be used to calculate $V(x, y)$ at any point within the trough. In Figure 6.11, $V(x = b/4, y = 3a/4)$ is typically calculated and found to be 43.2 volts. ■

PRACTICE EXERCISE 6.5

For the problem in Example 6.5, take $V_o = 100$ V, $b = 2a = 2$ m, find V and $\mathbf{E}$ at

(a) $(x, y) = (a, a/2)$

(b) $(x, y) = (3a/2, a/4)$

ANSWER (a) 44.51 V, $-99.25\,\mathbf{a}_y$ V/m, (b) 16.5 V, $20.6\,\mathbf{a}_x - 70.34\,\mathbf{a}_y$ V/m.

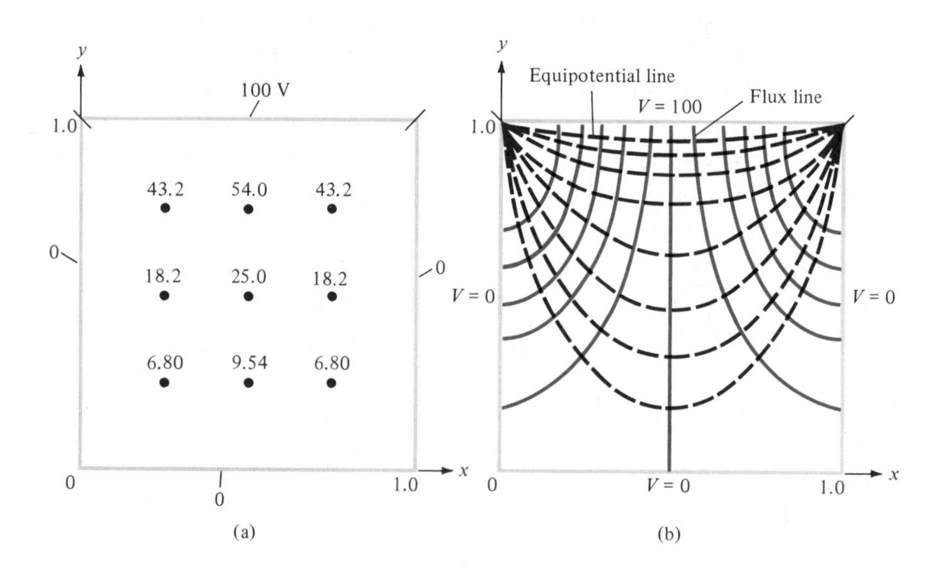

Figure 6.10 For Example 6.5: **(a)** $V(x, y)$ calculated at some points, **(b)** sketch of flux lines and equipotential lines.

```
0001  C     LAPLACE'S SOLUTION
0002  C     - - - - - - - - - - - - - - - - - -
0003  C     THIS PROGRAM SOLVES THE TWO-DIMENSIONAL
0004  C     BOUNDARY-VALUE PROBLEM DESCRIBED IN
0005  C     FIG. 6.7
0006  C     A AND B ARE THE DIMENSIONS OF THE TROUGH
0007  C     X AND Y ARE THE COORDINATES OF THE POINT
            OF INTEREST
0008
0009        VO=100.0
0010        A=1.0
0011        B=A
0012
0013        X=B/4.
0014        Y=3. *A/4.
0015        PIE=3.141592654
0016        C=4.0*VO/PIE
0017        SUM=0.0
0018        DO 20 K =1, 10
0019          N=2*K-1
0020          AN=FLOAT (N)
0021          A1=SIN(AN*PIE*X/B)
0022          A2=SINH(AN*PIE*Y/B)
0023          A3=AN*SINH(AN*PIE*A/B)
0024          TERM=C*A1*A2/A3
0025          SUM=SUM+TERM
0026          PRINT *,N,TERM,SUM
0027          WRITE(6,10) N,TERM,SUM
0028  10      FORMAT (2X, 'N=', I5, 2X, 'TERM=',
              F12.6, 2X, 'SUM=', F12.6,/)
0029  20    CONTINUE
0030        STOP
0031        END
```

Figure 6.11 FORTRAN program for Example 6.5.

EXAMPLE 6.6

In the last example, find the potential distribution if V_o is not constant but

(a) $V_o = 10 \sin 3\pi x/b$, $y = a$, $0 \leq x \leq b$

(b) $V_o = 2 \sin \dfrac{\pi x}{b} + \dfrac{1}{10} \sin \dfrac{5\pi x}{b}$, $y = a$, $0 \leq x \leq b$

SOLUTION (a) In the last example, every step before eq. (6.5.19) remains the same; that is, the solution is of the form

$$V(x, y) = \sum_{n=1}^{\infty} c_n \sin \frac{n\pi x}{b} \sinh \frac{n\pi y}{b} \tag{6.6.1}$$

as per eq. (6.5.18). But instead of eq. (6.5.19), we now have

$$V(y = a) = V_o = 10 \sin \frac{3\pi x}{b} = \sum_{n=1}^{\infty} c_n \sin \frac{n\pi x}{b} \sinh \frac{n\pi a}{b}$$

By equating the coefficients of the sine terms on both sides, we obtain

$$c_n = 0, \qquad n \neq 3$$

For $n = 3$,

$$10 = c_3 \sinh \frac{3\pi a}{b}$$

or

$$c_3 = \frac{10}{\sinh \dfrac{3\pi a}{b}}$$

Thus the solution in eq. (6.6.1) becomes

$$V(x, y) = 10 \sin \frac{3\pi x}{b} \frac{\sinh \dfrac{3\pi y}{b}}{\sinh \dfrac{3\pi a}{b}}.$$

(b) Similarly, instead of eq. (6.5.19), we have

$$V_o = V(y = a)$$

or

$$2 \sin \frac{\pi x}{b} + \frac{1}{10} \sinh \frac{5\pi x}{b} = \sum_{n=1}^{\infty} c_n \sinh \frac{n\pi x}{b} \sinh \frac{n\pi a}{b}$$

Equating the coefficient of the sine terms:

$$c_n = 0, \qquad n \neq 1, 5$$

For $n = 1$,

$$2 = c_1 \sinh \frac{\pi a}{b} \qquad \text{or} \qquad c_1 = \frac{2}{\sinh \dfrac{\pi a}{b}}$$

For $n = 5$,

$$\frac{1}{10} = c_5 \sinh \frac{5\pi a}{b} \qquad \text{or} \qquad c_5 = \frac{1}{10 \sinh \dfrac{5\pi a}{b}}$$

Hence,

$$V(x, y) = \frac{2 \sin \dfrac{\pi x}{b} \sinh \dfrac{\pi y}{b}}{\sinh \dfrac{\pi a}{b}} + \frac{\sin \dfrac{5\pi x}{b} \sinh \dfrac{5\pi y}{b}}{10 \sinh \dfrac{5\pi a}{b}}$$

■

PRACTICE EXERCISE 6.6

In Example 6.5, suppose everything remains the same except that V_o is replaced by $V_o \sin \dfrac{7\pi x}{b}$, $0 \leq x \leq b$, $y = a$. Find $V(x, y)$.

ANSWER

$$\frac{V_o \sin \dfrac{7\pi x}{b} \sinh \dfrac{7\pi y}{b}}{\sinh \dfrac{7\pi a}{b}}$$

EXAMPLE 6.7

Obtain the separated differential equations for potential distribution $V(\rho, \phi, z)$ in a charge-free region.

SOLUTION

This example, like Example 6.5, further illustrates the method of separation of variables. Since the region is free of charge, we need to solve Laplace's equation in cylindrical coordinates; that is,

$$\nabla^2 V = \frac{1}{\rho} \frac{\partial}{\partial \rho}\left(\rho \frac{\partial V}{\partial \rho}\right) + \frac{1}{\rho^2} \frac{\partial^2 V}{\partial \phi^2} + \frac{\partial^2 V}{\partial z^2} = 0 \qquad \text{[6.7.1]}$$

We let

$$V(\rho, \phi, z) = R(\rho)\, \Phi(\phi)\, Z(z) \qquad [6.7.2]$$

where R, Φ, and Z are, respectively, functions of ρ, ϕ, and z. Substituting eq. (6.7.2) into eq. (6.7.1) gives

$$\frac{\Phi Z}{\rho} \frac{d}{d\rho}\left(\frac{\rho\, dR}{d\rho}\right) + \frac{RZ}{\rho^2} \frac{d^2\Phi}{d\phi^2} + R\Phi \frac{d^2Z}{dz^2} = 0 \qquad [6.7.3]$$

We divide through by $R\Phi Z$ to obtain

$$\frac{1}{\rho R} \frac{d}{d\rho}\left(\frac{\rho\, dR}{d\rho}\right) + \frac{1}{\rho^2 \Phi} \frac{d^2\Phi}{d\phi^2} = -\frac{1}{Z} \frac{d^2Z}{dz^2} \qquad [6.7.4]$$

The right-hand side of this equation is solely a function of z whereas the left-hand side does not depend on z. For the two sides to be equal, they must be constant; that is,

$$\frac{1}{\rho R} \frac{d}{d\rho}\left(\frac{\rho\, dR}{d\rho}\right) + \frac{1}{\rho^2 \Phi} \frac{d^2\Phi}{d\phi^2} = -\frac{1}{Z} \frac{d^2Z}{dz^2} = -\lambda^2 \qquad [6.7.5]$$

where $-\lambda^2$ is a separation constant. Equation (6.7.5) can be separated into two parts:

$$\frac{1}{Z} \frac{d^2Z}{dz^2} = \lambda^2 \qquad [6.7.6]$$

or

$$Z'' - \lambda^2 Z = 0 \qquad [6.7.7]$$

and

$$\frac{\rho}{R} \frac{d}{d\rho}\left(\frac{\rho\, dR}{d\rho}\right) + \lambda^2\rho^2 + \frac{1}{\Phi} \frac{d^2\Phi}{d\phi^2} = 0 \qquad [6.7.8]$$

Equation (6.7.8) can be written as

$$\frac{\rho^2}{R} \frac{d^2R}{d\rho^2} + \frac{\rho}{R} \frac{dR}{d\rho} + \lambda^2\rho^2 = -\frac{1}{\Phi} \frac{d^2\Phi}{d\phi^2} = \mu^2 \qquad [6.7.9]$$

where μ^2 is another separation constant. Equation (6.7.9) is separated as

$$\Phi'' + \mu^2\Phi = 0 \qquad [6.7.10]$$

and

$$\rho^2 R'' + \rho R' + (\rho^2\lambda^2 - \mu^2)\, R = 0 \qquad [6.7.11]$$

Equations (6.7.7), (6.7.10), and (6.7.11) are the required separated differential equations. Equation (6.7.7) has a solution similar to the solution obtained in Case 2 of Example 6.5; that is,

$$Z(z) = c_1 \cosh \lambda z + c_2 \sinh \lambda z \qquad [6.7.12]$$

The solution to eq. (6.7.10) is similar to the solution obtained in Case 3 of Example 6.5; that is,

$$\Phi(\phi) = c_3 \cos \mu\phi + c_4 \sin \mu\phi \qquad [6.7.13]$$

Equation (6.7.11) is known as *Bessel's differential equation* and its solution is beyond the scope of this text.[3] ■

PRACTICE EXERCISE 6.7

Repeat Example 6.7 for $V(r, \theta, \phi)$.

ANSWER If $V(r, \theta, \phi) = R(r)\, F(\theta)\, \Phi(\phi)$, $\Phi'' + \lambda^2\Phi = 0$, $R'' + \dfrac{2}{r} R' - \dfrac{\mu^2}{r^2} R = 0$, $F'' + \cot\theta\, F' + (\mu^2 - \lambda^2 \operatorname{cosec}^2 \theta)\, F = 0$.

6.5 RESISTANCE AND CAPACITANCE

In Section 5.4 the concept of resistance was covered and we derived eq. (5.10) for finding the resistance of a conductor of uniform cross section. If the cross section of the conductor is not uniform, eq. (5.10) becomes invalid and the resistance is obtained from eq. (5.11):

$$R = \frac{V}{I} = \frac{\int \mathbf{E} \cdot d\mathbf{l}}{\oint \sigma \mathbf{E} \cdot d\mathbf{S}} \qquad [6.16]$$

The problem of finding the resistance of a conductor of nonuniform cross section can be treated as a boundary-value problem. Using eq. (6.16), the resistance R (or conductance $G = 1/R$) of a given conducting material can be found by following these steps:

1. Choose a suitable coordinate system.
2. Assume V_o as the potential difference between conductor terminals.
3. Solve Laplace's equation $\nabla^2 V$ to obtain V. Then determine $\mathbf{E}$ from $\mathbf{E} = -\nabla V$ and I from $I = \int \sigma \mathbf{E} \cdot d\mathbf{S}$.
4. Finally, obtain R as V_o/I.

[3]For a complete solution of Laplace's equation in cylindrical or spherical coordinates, see, for example, D. T. Paris and F. K. Hurd, *Basic Electromagnetic Theory*. New York: McGraw-Hill, 1969, pp. 150–159.

In essence, we assume V_o, find I, and determine $R = V_o/I$. Alternatively, it is possible to assume current I_o, find the corresponding potential difference V, and determine R from $R = V/I_o$. As will be discussed shortly, the capacitance of a capacitor is obtained using a similar technique.

Generally speaking, in order to have a capacitor we must have two (or more) conductors carrying equal but opposite charges. This implies that all the flux lines leaving one conductor must necessarily terminate at the surface of the other conductor. The conductors are sometimes referred to as the *plates* of the capacitor. The plates may be separated by free space or a dielectric.

Consider the two-conductor capacitor of Figure 6.12. The conductors are maintained at a potential difference V given by

$$V = V_1 - V_2 = -\int_2^1 \mathbf{E} \cdot d\mathbf{l} \qquad [6.17]$$

where $\mathbf{E}$ is the electric field existing between the conductors and conductor 1 is assumed to carry a positive charge. (Note that the $\mathbf{E}$ field is always normal to the conducting surfaces.)

We define the *capacitance* C of the capacitor as the ratio of the magnitude of the charge on one of the plates to the potential difference between them; that is,

$$C = \frac{Q}{V} = \frac{\varepsilon \oint \mathbf{E} \cdot d\mathbf{S}}{\int \mathbf{E} \cdot d\mathbf{l}} \qquad [6.18]$$

The negative sign before $V = -\int \mathbf{E} \cdot d\mathbf{l}$ has been dropped because we are interested in the absolute value of V. The capacitance C is a physical property of the capacitor and

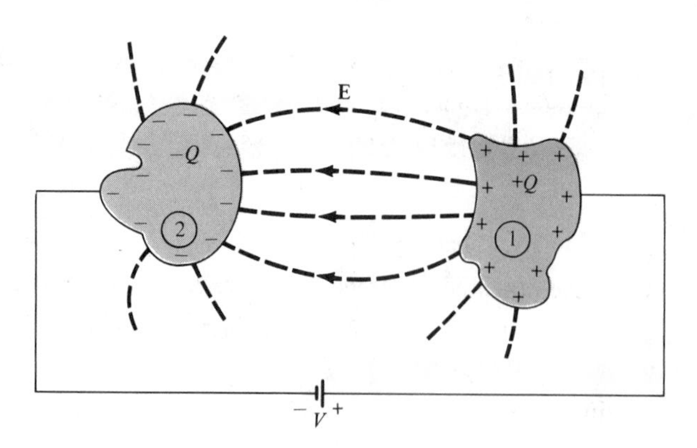

Figure 6.12 A two-conductor capacitor.

is measured in farads (F). Using eq. (6.18), C can be obtained for any given two-conductor capacitance by following either of these methods:

1. Assuming Q and determining V in terms of Q (involving Gauss's law)
2. Assuming V and determining Q in terms of V (involving solving Laplace's equation)

We shall use the former method here, and the latter method will be illustrated in Examples 6.10 and 6.11. The former method involves taking the following steps:

1. Choose a suitable coordinate system.
2. Let the two conducting plates carry charges $+Q$ and $-Q$.
3. Determine $\mathbf{E}$ using Coulomb's or Gauss's law and find V from $V = -\int \mathbf{E} \cdot d\mathbf{l}$. The negative sign may be ignored in this case because we are interested in the absolute value of V.
4. Finally, obtain C from $C = Q/V$.

We will now apply this mathematically attractive procedure to determine the capacitance of some important two-conductor configurations.

A. Parallel-Plate Capacitor

Consider the parallel-plate capacitor of Figure 6.13(a). Suppose that each of the plates has an area S and they are separated by a distance d. We assume that plates 1 and 2 respectively carry charges $+Q$ and $-Q$ uniformly distributed on them so that

$$\boxed{\rho_S = \frac{Q}{S}} \tag{6.19}$$

An ideal parallel-plate capacitor is one in which the plate separation d is very small compared with the dimensions of the plate. Assuming such an ideal case, the fringing field at the edge of the plates, as illustrated in Figure 6.13(b), can be ignored so that the field between them is considered uniform. If the space between the plates is filled with a homogeneous dielectric with permittivity ε and we ignore flux fringing at the edges of the plates, from eq. (4.27), $\mathbf{D} = -\rho_S \mathbf{a}_x$ or

$$\begin{aligned} \mathbf{E} &= \frac{\rho_S}{\varepsilon}(-\mathbf{a}_x) \\ &= -\frac{Q}{\varepsilon S}\mathbf{a}_x \end{aligned} \tag{6.20}$$

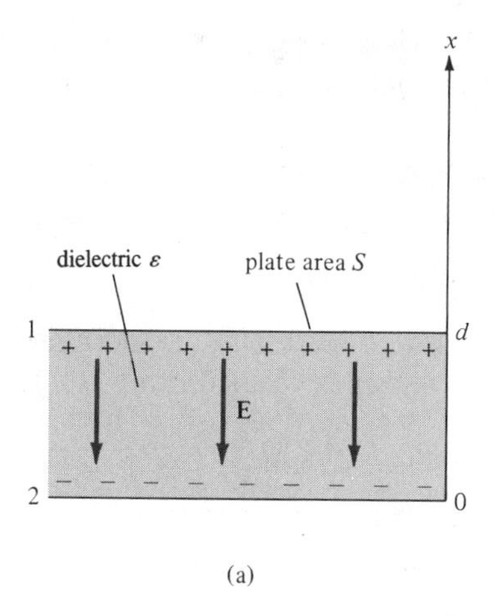

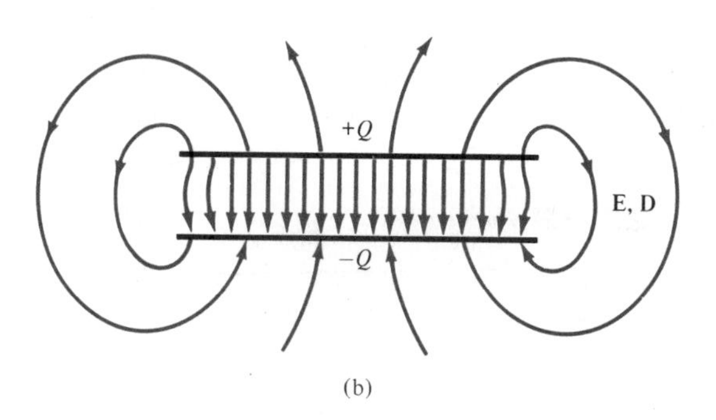

Figure 6.13 **(a)** Parallel-plate capacitor, **(b)** fringing effect due to a parallel-plate capacitor.

Hence

$$V = -\int_2^1 \mathbf{E} \cdot d\mathbf{l} = -\int_0^d \left[-\frac{Q}{\varepsilon S}\,\mathbf{a}_x \right] \cdot dx\,\mathbf{a}_x = \frac{Qd}{\varepsilon S} \tag{6.21}$$

and thus for a parallel-plate capacitor

$$\boxed{C = \frac{Q}{V} = \frac{\varepsilon S}{d}} \tag{6.22}$$

This formula offers a means of measuring the dielectric constant ε_r of a given dielectric. By measuring the capacitance C of a parallel-plate capacitor with the space between the plates filled with the dielectric and the capacitance C_o with air between the plates, we find ε_r from

$$\varepsilon_r = \frac{C}{C_o} \tag{6.23}$$

Using eq. (4.103), it can be shown that the energy stored in a capacitor is given by

$$\boxed{W_E = \frac{1}{2} CV^2 = \frac{1}{2} QV = \frac{Q^2}{2C}} \tag{6.24}$$

To verify this for a parallel-plate capacitor, we substitute eq. (6.20) into eq. (4.103) and obtain

$$\begin{aligned} W_E &= \frac{1}{2} \int_v \varepsilon \frac{Q^2}{\varepsilon^2 S^2} \, dv = \frac{\varepsilon Q^2 Sd}{2\varepsilon^2 S^2} \\ &= \frac{Q^2}{2} \left(\frac{d}{\varepsilon S} \right) = \frac{Q^2}{2C} = \frac{1}{2} QV \end{aligned}$$

as expected.

B. Coaxial Capacitor

This is essentially a coaxial cable or coaxial cylindrical capacitor. Consider length L of two coaxial conductors of inner radius a and outer radius b ($b > a$) as shown in Figure 6.14. Let the space between the conductors be filled with a homogeneous dielectric with permittivity ε. We assume that conductors 1 and 2 respectively carry $+Q$ and $-Q$ uniformly distributed on them. By applying Gauss's law to an arbitrary Gaussian cylindrical surface of radius ρ ($a < \rho < b$), we obtain

$$Q = \varepsilon \oint \mathbf{E} \cdot d\mathbf{S} = \varepsilon E_\rho 2\pi \rho L \tag{6.25}$$

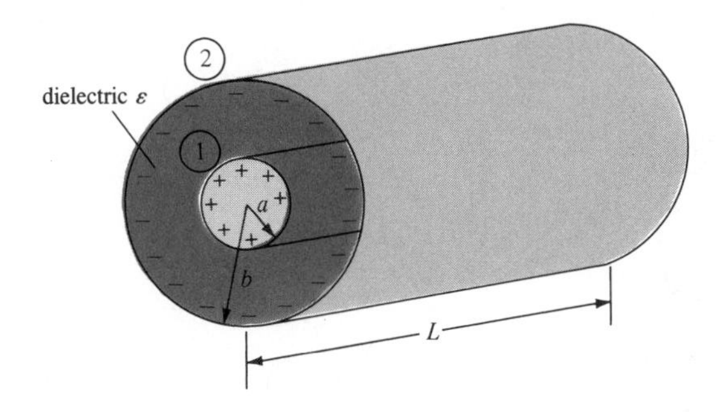

Figure 6.14 Coaxial capacitor.

Hence:

$$\mathbf{E} = \frac{Q}{2\pi\varepsilon\rho L}\,\mathbf{a}_\rho \qquad [6.26]$$

Neglecting flux fringing at the cylinder ends,

$$V = -\int_2^1 \mathbf{E}\cdot d\mathbf{l} = -\int_b^a \left[\frac{Q}{2\pi\varepsilon\rho L}\,\mathbf{a}_\rho\right]\cdot d\rho\,\mathbf{a}_\rho \qquad [6.27a]$$

$$= \frac{Q}{2\pi\varepsilon L}\ln\frac{b}{a} \qquad [6.27b]$$

Thus the capacitance of a coaxial cylinder is given by

$$\boxed{C = \frac{Q}{V} = \frac{2\pi\varepsilon L}{\ln\dfrac{b}{a}}} \qquad [6.28]$$

C. Spherical Capacitor

This is the case of two concentric spherical conductors. Consider the inner sphere of radius a and outer sphere of radius b ($b > a$) separated by a dielectric medium with permittivity ε as shown in Figure 6.15. We assume charges $+Q$ and $-Q$ on the inner and outer spheres respectively. By applying Gauss's law to an arbitrary Gaussian spherical surface of radius r ($a < r < b$),

$$Q = \varepsilon \oint \mathbf{E}\cdot d\mathbf{S} = \varepsilon E_r 4\pi r^2 \qquad [6.29]$$

that is,

$$\mathbf{E} = \frac{Q}{4\pi\varepsilon r^2}\,\mathbf{a}_r \qquad [6.30]$$

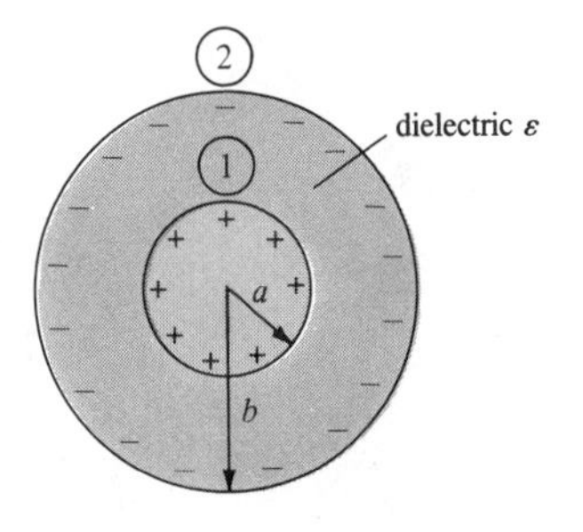

Figure 6.15 Spherical capacitor.

The potential difference between the conductors is

$$V = -\int_2^1 \mathbf{E} \cdot d\mathbf{l} = -\int_b^a \left[\frac{Q}{4\pi\varepsilon r^2}\,\mathbf{a}_r\right] \cdot dr\,\mathbf{a}_r$$

$$= \frac{Q}{4\pi\varepsilon}\left[\frac{1}{a} - \frac{1}{b}\right] \qquad [6.31]$$

Thus the capacitance of the spherical capacitor is

$$C = \frac{Q}{V} = \frac{4\pi\varepsilon}{\dfrac{1}{a} - \dfrac{1}{b}} \qquad [6.32]$$

By letting $b \to \infty$, $C = 4\pi\varepsilon a$, which is the capacitance of a spherical capacitor whose outer plate is infinitely large. Such is the case of a spherical conductor at a large distance from other conducting bodies—the *isolated sphere*. Even an irregularly shaped object of about the same size as the sphere will have nearly the same capacitance. This fact is useful in estimating the stray capacitance of an isolated body or piece of equipment.

Recall from network theory that if two capacitors with capacitance C_1 and C_2 are in series (i.e., they have the same charge on them) as shown in Figure 6.16(a), the total capacitance is

$$\frac{1}{C} = \frac{1}{C_1} + \frac{1}{C_2}$$

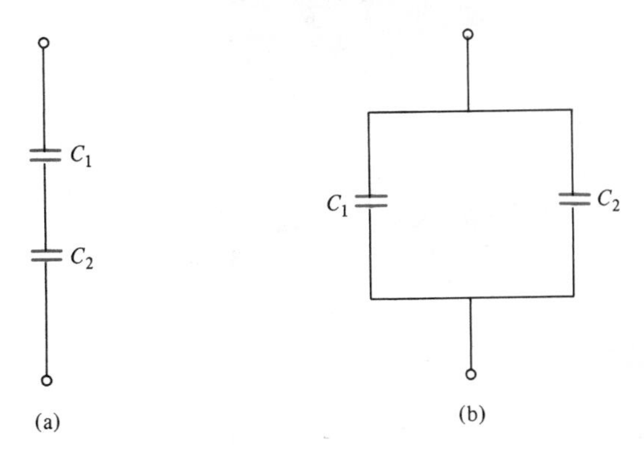

Figure 6.16 Capacitors in **(a)** series, and **(b)** parallel.

or

$$C = \frac{C_1 C_2}{C_1 + C_2} \tag{6.33}$$

If the capacitors are in parallel (i.e., they have the same voltage across their plates) as shown in Figure 6.16(b), the total capacitance is

$$C = C_1 + C_2 \tag{6.34}$$

Let us reconsider the expressions for finding the resistance R and the capacitance C of an electrical system. The expressions were given in eqs. (6.16) and (6.18):

$$R = \frac{V}{I} = \frac{\int \mathbf{E} \cdot d\mathbf{l}}{\oint \sigma \ \mathbf{E} \cdot d\mathbf{S}} \tag{6.16}$$

$$C = \frac{Q}{V} = \frac{\varepsilon \oint \mathbf{E} \cdot d\mathbf{S}}{\int \mathbf{E} \cdot d\mathbf{l}} \tag{6.18}$$

The product of these expressions yields

$$\boxed{RC = \frac{\varepsilon}{\sigma}} \tag{6.35}$$

which is the relaxation time T_r of the medium separating the conductors. It should be remarked that eq. (6.35) is valid only when the medium is homogeneous; this is easily inferred from eqs. (6.16) and (6.18). Assuming homogeneous media, the resistance of various capacitors mentioned earlier can be readily obtained using eq. (6.35). The following examples are provided to illustrate this idea.

For a parallel-plate capacitor,

$$C = \frac{\varepsilon S}{d}, \qquad R = \frac{d}{\sigma S} \tag{6.36}$$

For a cylindrical capacitor,

$$C = \frac{2\pi\varepsilon L}{\ln \frac{b}{a}}, \qquad R = \frac{\ln \frac{b}{a}}{2\pi\sigma L} \tag{6.37}$$

For a spherical capacitor,

$$C = \frac{4\pi\varepsilon}{\frac{1}{a} - \frac{1}{b}}, \qquad R = \frac{\frac{1}{a} - \frac{1}{b}}{4\pi\sigma} \tag{6.38}$$

And finally for an isolated spherical conductor,

$$C = 4\pi\varepsilon a, \qquad R = \frac{1}{4\pi\sigma a} \qquad [6.39]$$

It should be noted that the resistance R in each of eqs. (6.35) to (6.39) is not the resistance of the capacitor plate but the leakage resistance between the plates; therefore, σ in those equations is the conductivity of the dielectric medium separating the plates.

EXAMPLE 6.8

A metal bar of conductivity σ is bent to form a flat 90° sector of inner radius a, outer radius b, and thickness t as shown in Figure 6.17. Show that (a) the resistance of the bar between the vertical curved surfaces at $\rho = a$ and $\rho = b$ is

$$R = \frac{2 \ln \frac{b}{a}}{\sigma \pi t}$$

and (b) the resistance between the two horizontal surfaces at $z = 0$ and $z = t$ is

$$R' = \frac{4t}{\sigma\pi(b^2 - a^2)}$$

SOLUTION

(a) Between the vertical curved ends located at $\rho = a$ and $\rho = b$, the bar has a nonuniform cross section and hence eq. (5.10) does not apply. We have to use eq. (6.16). Let a potential difference V_o be maintained between the curved surfaces at $\rho = a$ and $\rho = b$ so that $V(\rho = a) = 0$ and $V(\rho = b) = V_o$. We solve for V in Laplace's equation $\nabla^2 V = 0$ in cylindrical coordinates. Since $V = V(\rho)$,

$$\nabla^2 V = \frac{1}{\rho}\frac{d}{d\rho}\left(\rho \frac{dV}{d\rho}\right) = 0$$

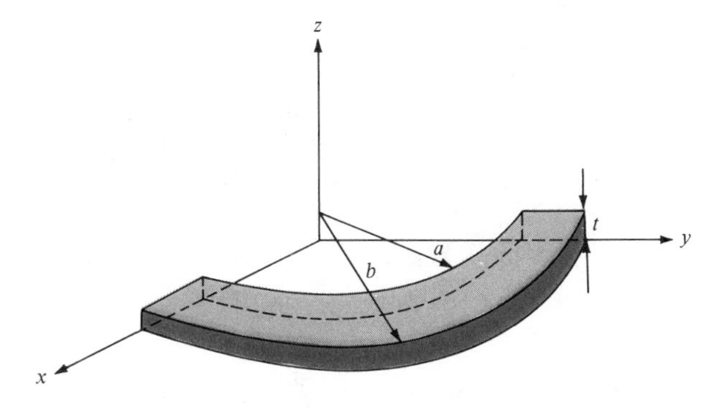

Figure 6.17 Metal bar of Example 6.8.

As $\rho = 0$ is excluded, upon multiplying by ρ and integrating once, this becomes

$$\rho \frac{dV}{d\rho} = A$$

or

$$\frac{dV}{d\rho} = \frac{A}{\rho}$$

Integrating once again yields

$$V = A \ln \rho + B$$

where A and B are constants of integration to be determined from the boundary conditions.

$$V(\rho = a) = 0 \rightarrow 0 = A \ln a + B \quad \text{or} \quad B = -A \ln a$$

$$V(\rho = b) = V_o \rightarrow V_o = A \ln b + B = A \ln \frac{b}{a} \quad \text{or} \quad A = \frac{V_o}{\ln \frac{b}{a}}$$

Hence,

$$V = A \ln \rho - A \ln a = A \ln \frac{\rho}{a} = \frac{V_o}{\ln \frac{b}{a}} \ln \frac{\rho}{a}$$

$$\mathbf{E} = -\nabla V = -\frac{dV}{d\rho}\mathbf{a}_\rho = -\frac{A}{\rho}\mathbf{a}_\rho = -\frac{V_o}{\rho \ln \frac{b}{a}}\mathbf{a}_\rho$$

$$\mathbf{J} = \sigma \mathbf{E}, \quad d\mathbf{S} = -\rho \, d\phi \, dz \, \mathbf{a}_\rho$$

$$I = \int \mathbf{J} \cdot d\mathbf{S} = \int_{\phi=0}^{\pi/2} \int_{z=0}^{t} \frac{V_o \sigma}{\rho \ln \frac{b}{a}} dz \, \rho \, d\phi = \frac{\pi}{2} \frac{t V_o \sigma}{\ln \frac{b}{a}}$$

Thus

$$R = \frac{V_o}{I} = \frac{2 \ln \frac{b}{a}}{\sigma \pi t}$$

as required.

(b) Let V_o be the potential difference between the two horizontal surfaces so that $V(z = 0) = 0$ and $V(z = t) = V_o$. $V = V(z)$, so Laplace's equation $\nabla^2 V = 0$ becomes

$$\frac{d^2V}{dz^2} = 0$$

Integrating twice gives

$$V = Az + B$$

We apply the boundary conditions to determine A and B:

$$V(z = 0) = 0 \rightarrow 0 = 0 + B \qquad \text{or} \qquad B = 0$$

$$V(z = t) = V_o \rightarrow V_o = At \qquad \text{or} \qquad A = \frac{V_o}{t}$$

Hence,

$$V = \frac{V_o}{t} z$$

$$\mathbf{E} = -\nabla V = -\frac{dV}{dz}\mathbf{a}_z = -\frac{V_o}{t}\mathbf{a}_z$$

$$\mathbf{J} = \sigma\mathbf{E} = -\frac{\sigma V_o}{t}\mathbf{a}_z, \quad d\mathbf{S} = -\rho\, d\phi\, d\rho\, \mathbf{a}_z$$

$$I = \int \mathbf{J} \cdot d\mathbf{S} = \int_{\rho = a}^{b} \int_{\phi = 0}^{\pi/2} \frac{V_o\sigma}{t} \rho\, d\phi\, d\rho$$

$$= \frac{V_o\sigma}{t} \cdot \frac{\pi}{2} \frac{\rho^2}{2} \bigg|_a^b = \frac{V_o\, \sigma\, \pi\, (b^2 - a^2)}{4t}$$

Thus

$$R' = \frac{V_o}{I} = \frac{4t}{\sigma\pi(b^2 - a^2)}$$

Alternatively, for this case, the cross section of the bar is uniform between the horizontal surfaces at $z = 0$ and $z = t$ and eq. (5.10) holds. Hence,

$$R' = \frac{\ell}{\sigma S} = \frac{t}{\sigma \frac{\pi}{4}(b^2 - a^2)}$$

$$= \frac{4t}{\sigma\pi(b^2 - a^2)}$$

as required. ■

PRACTICE EXERCISE 6.8

A disc of thickness t has radius b and a central hole of radius a. Taking the conductivity of the disc as σ, find the resistance between

(a) The hole and the rim of the disc
(b) The two flat sides of the disc

ANSWER (a) $\dfrac{\ln b/a}{2\pi t\sigma}$, (b) $\dfrac{t}{\sigma\pi(b^2 - a^2)}$.

EXAMPLE 6.9

A coaxial cable contains an insulating material of conductivity σ. If the radius of the central wire is a and that of the sheath is b, show that the conductance of the cable per unit length is (see eq. (6.37))

$$G = \frac{2\pi\sigma}{\ln b/a}$$

SOLUTION Consider length L of the coaxial cable as shown in Figure 6.14. Let V_o be the potential difference between the inner and outer conductors so that $V(\rho = a) = 0$ and $V(\rho = b) = V_o$. V and $\mathbf{E}$ can be found just as in part (a) of the last example. Hence:

$$\mathbf{J} = \sigma\mathbf{E} = \frac{-\sigma V_o}{\rho \ln b/a}\,\mathbf{a}_\rho, \quad d\mathbf{S} = -\rho d\phi\, dz\, \mathbf{a}_\rho$$

$$I = \int \mathbf{J} \cdot d\mathbf{S} = \int_{\phi=0}^{2\pi} \int_{z=0}^{L} \frac{V_o\sigma}{\rho \ln b/a}\,\rho\, dz\, d\phi$$

$$= \frac{2\pi L\sigma V_o}{\ln b/a}$$

The resistance per unit length is

$$R = \frac{V_o}{I} \cdot \frac{1}{L} = \frac{\ln b/a}{2\pi\sigma}$$

and the conductance per unit length is

$$G = \frac{1}{R} = \frac{2\pi\sigma}{\ln b/a}$$

as required. ■

PRACTICE EXERCISE 6.9

A coaxial cable contains an insulating material of conductivity σ_1 in its upper half and another material of conductivity σ_2 in its lower half (similar to the situation in Figure 6.19b). If the radius of the central wire is a and that of the sheath is b, show that the leakage resistance of length ℓ of the cable is

$$R = \frac{1}{\pi\ell(\sigma_1 + \sigma_2)} \ln \frac{b}{a}$$

ANSWER Proof.

EXAMPLE 6.10

Conducting spherical shells with radii a = 10 cm and b = 30 cm are maintained at a potential difference of 100 V such that $V(r = b) = 0$ and $V(r = a) = 100$ V. Determine V and **E** in the region between the shells. If $\varepsilon_r = 2.5$ in the region, determine the total charge induced on the shells and the capacitance of the capacitor.

SOLUTION Consider the spherical shells shown in Figure 6.18. V depends only on r and hence Laplace's equation becomes

$$\nabla^2 V = \frac{1}{r^2}\frac{d}{dr}\left[r^2 \frac{dV}{dr}\right] = 0$$

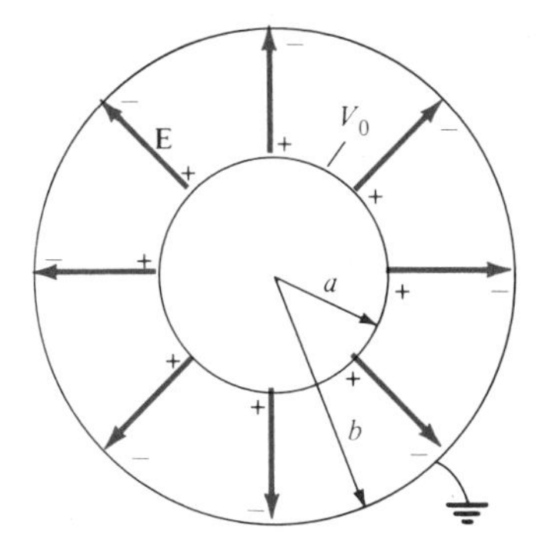

Figure 6.18 Potential $V(r)$ due to conducting spherical shells.

Since $r \neq 0$ in the region of interest, we multiply through by r^2 to obtain

$$\frac{d}{dr}\left[r^2 \frac{dV}{dr}\right] = 0$$

Integrating once gives

$$r^2 \frac{dV}{dr} = A$$

or

$$\frac{dV}{dr} = \frac{A}{r^2}$$

Integrating again gives

$$V = -\frac{A}{r} + B$$

As usual, constants A and B are determined from the boundary conditions.

When $r = b$, $V = 0 \rightarrow 0 = -\frac{A}{b} + B$ or $B = \frac{A}{b}$.

Hence,

$$V = A\left[\frac{1}{b} - \frac{1}{r}\right]$$

Also when $r = a$, $V = V_o \rightarrow V_o = A\left[\frac{1}{b} - \frac{1}{a}\right]$

or

$$A = \frac{V_o}{\frac{1}{b} - \frac{1}{a}}$$

Thus,

$$V = V_o \frac{\left[\frac{1}{r} - \frac{1}{b}\right]}{\frac{1}{a} - \frac{1}{b}}$$

$$\mathbf{E} = -\nabla V = -\frac{dV}{dr}\,\mathbf{a}_r = -\frac{A}{r^2}\,\mathbf{a}_r$$

$$= \frac{V_o}{r^2\left[\dfrac{1}{a} - \dfrac{1}{b}\right]}\,\mathbf{a}_r$$

$$Q = \int \varepsilon\mathbf{E} \cdot d\mathbf{S} = \int_{\theta=0}^{\pi}\int_{\phi=0}^{2\pi} \frac{\varepsilon_o\varepsilon_r V_o}{r^2\left[\dfrac{1}{a} - \dfrac{1}{b}\right]}\, r^2 \sin\theta\, d\phi\, d\theta$$

$$= \frac{4\pi\varepsilon_o\varepsilon_r V_o}{\dfrac{1}{a} - \dfrac{1}{b}}$$

The capacitance is easily determined as

$$C = \frac{Q}{V_o} = \frac{4\pi\varepsilon}{\dfrac{1}{a} - \dfrac{1}{b}}$$

which is the same as we obtained in eq. (6.32); there in Section 6.5, we assumed Q and found the corresponding V_o, but here we assumed V_o and found the corresponding Q to determine C. Substituting $a = 0.1$ m, $b = 0.3$ m, $V_o = 100$ V yields

$$V = 100\,\frac{\left[\dfrac{1}{r} - \dfrac{10}{3}\right]}{10 - 10/3} = 15\left[\frac{1}{r} - \frac{10}{3}\right] V$$

Check: $\nabla^2 V = 0$, $V(r = 0.3 \text{ m}) = 0$, $V(r = 0.1 \text{ m}) = 100$.

$$\mathbf{E} = \frac{100}{r^2\,[10 - 10/3]}\,\mathbf{a}_r = \frac{15}{r^2}\,\mathbf{a}_r \text{ V/m}$$

$$Q = \pm\, 4\pi \cdot \frac{10^{-9}}{36\pi} \cdot \frac{(2.5)\cdot(100)}{10 - 10/3}$$

$$= \pm\, 4.167 \text{ nC}$$

The positive charge is induced on the inner shell; the negative charge is induced on the outer shell. Also

$$C = \frac{|Q|}{V_o} = \frac{4.167 \times 10^{-9}}{100} = 41.67 \text{ pF}$$

■

PRACTICE EXERCISE 6.10

If Figure 6.19 represents the cross sections of two spherical capacitors, determine their capacitances. Let $a = 1$ mm, $b = 3$ mm, $c = 2$ mm, $\varepsilon_{r1} = 2.5$, and $\varepsilon_{r2} = 3.5$.

ANSWER (a) 0.53 pF, (b) 0.5 pF

EXAMPLE 6.11

In Section 6.5, it was mentioned that the capacitance $C = Q/V$ of a capacitor can be found by either assuming Q and finding V or by assuming V and finding Q. The former approach was used in Section 6.5 while we have used the latter method in the last example. Using the latter method, derive eq. (6.22).

SOLUTION Assume that the parallel plates in Figure 6.13 are maintained at a potential difference V_o so that $V(x = 0)$ and $V(x = d) = V_o$. This necessitates solving a one-dimensional boundary-value problem; that is, we solve Laplace's equation

$$\nabla^2 V = \frac{d^2V}{dx^2} = 0$$

Integrating twice gives

$$V = Ax + B$$

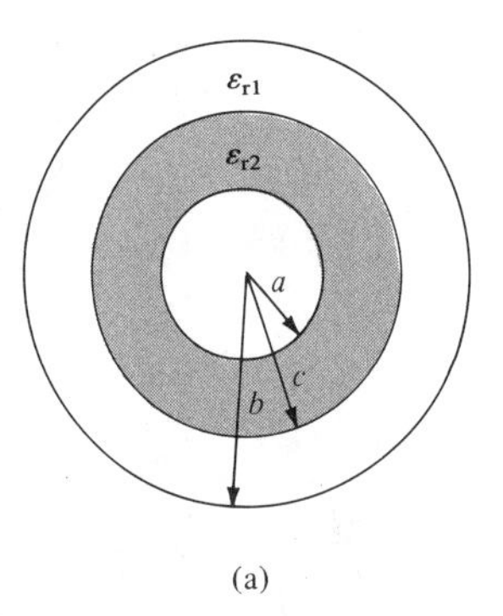

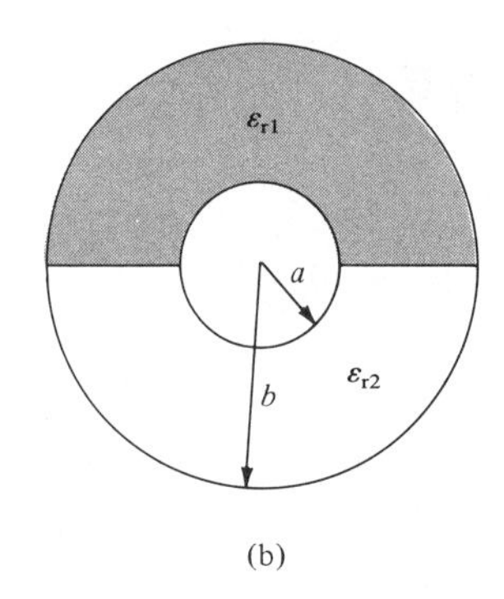

Figure 6.19 For Practice Exercises 6.9, 6.10, and 6.12.

where A and B are integration constants to be determined from the boundary conditions. At $x = 0$, $V = 0 \rightarrow 0 = 0 + B$, or $B = 0$, and at $x = d$, $V = V_o \rightarrow V_o = Ad + 0$ or $A = V_o/d$.

Hence,

$$V = \frac{V_o}{d} x$$

Notice that this solution satisfies Laplace's equation and the boundary conditions.

We have assumed the potential difference between the plates to be V_o. Our goal is to find the charge Q on either plate so that we can eventually find the capacitance $C = Q/V_o$. The charge on either plate is

$$Q = \int \rho_S \, dS.$$

But $\rho_S = \mathbf{D} \cdot \mathbf{a}_n = \varepsilon \mathbf{E} \cdot \mathbf{a}_n$, where

$$\mathbf{E} = -\nabla V = -\frac{dV}{dx} \mathbf{a}_x = -A\mathbf{a}_x = -\frac{V_o}{d} \mathbf{a}_x$$

On the lower plates, $\mathbf{a}_n = \mathbf{a}_x$, so

$$\rho_S = -\frac{\varepsilon V_o}{d} \quad \text{and} \quad Q = -\frac{\varepsilon V_o S}{d}$$

On the upper plates, $\mathbf{a}_n = -\mathbf{a}_x$, so

$$\rho_S = \frac{\varepsilon V_o}{d} \quad \text{and} \quad Q = \frac{\varepsilon V_o S}{d}$$

As expected, Q is equal but opposite on each plate. Thus

$$C = \frac{|Q|}{V_o} = \frac{\varepsilon S}{d}$$

which is in agreement with eq. (6.22). ■

PRACTICE EXERCISE 6.11

Derive the formula for the capacitance $C = Q/V_o$ of a cylindrical capacitor in eq. (6.28) by assuming V_o and finding Q.

EXAMPLE 6.12

Determine the capacitance of each of the capacitors in Figure 6.20. Take $\varepsilon_{r1} = 4$, $\varepsilon_{r2} = 6$, $d = 5$ mm, $S = 30$ cm^2.

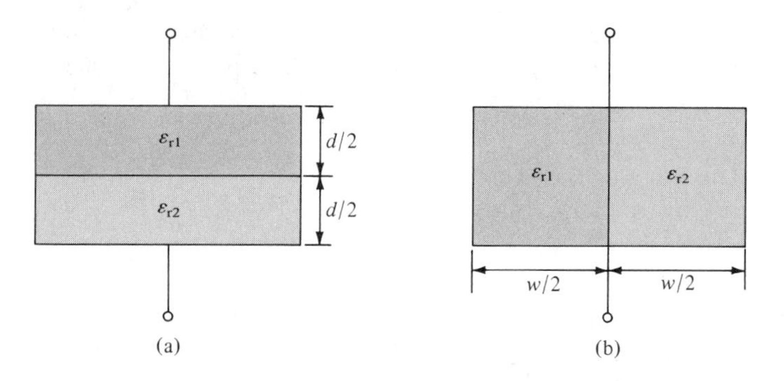

Figure 6.20 For Example 6.12.

SOLUTION (a) Since **D** and **E** are normal to the dielectric interface, the capacitor in Figure 6.20(a) can be treated as consisting of two capacitors C_1 and C_2 in series as in Figure 6.16(a).

$$C_1 = \frac{\varepsilon_o \varepsilon_{r1} S}{d/2} = \frac{2\varepsilon_o \varepsilon_{r1} S}{d}, \qquad C_2 = \frac{2\varepsilon_o \varepsilon_{r2} S}{d}$$

The total capacitor C is given by

$$C = \frac{C_1 C_2}{C_1 + C_2} = \frac{2\varepsilon_o S}{d} \frac{(\varepsilon_{r1}\varepsilon_{r2})}{\varepsilon_{r1} + \varepsilon_{r2}} \tag{6.12.1}$$

$$= 2 \cdot \frac{10^{-9}}{36\pi} \cdot \frac{30 \times 10^{-4}}{5 \times 10^{-3}} \cdot \frac{4 \times 6}{10}$$

$$C = 25.46 \text{ pF}$$

(b) In this case, **D** and **E** are parallel to the dielectric interface. We may treat the capacitor as consisting of two capacitors C_1 and C_2 in parallel (the same voltage across C_1 and C_2) as in Figure 6.16(b).

$$C_1 = \frac{\varepsilon_o \varepsilon_{r1} S/2}{d} = \frac{\varepsilon_o \varepsilon_{r1} S}{2d}, \qquad C_2 = \frac{\varepsilon_1 \varepsilon_{r2} S}{2d}$$

The total capacitance is

$$C = C_1 + C_2 = \frac{\varepsilon_o S}{2d} (\varepsilon_{r1} + \varepsilon_{r2}) \tag{6.12.2}$$

$$= \frac{10^{-9}}{36\pi} \cdot \frac{30 \times 10^{-4}}{2 \cdot (5 \times 10^{-3})} \cdot 10$$

$$C = 26.53 \text{ pF}$$

Notice that when $\varepsilon_{r1} = \varepsilon_{r2} = \varepsilon_r$, eqs. (6.12.1) and (6.12.2) agree with eq. (6.22) as expected. ■

PRACTICE EXERCISE 6.12

Determine the capacitance of 10 m length of the cylindrical capacitors shown in Figure 6.19. Take a = 1 mm, b = 3 mm, c = 2 mm, $\varepsilon_{r1} = 2.5$, and $\varepsilon_{r2} = 3.5$.

ANSWER (a) 1.41 nF, (b) 1.52 nF.

EXAMPLE 6.13

A cylindrical capacitor has radii a = 1 cm and b = 2.5 cm. If the space between the plates is filled with an inhomogeneous dielectric with $\varepsilon_r = (10 + \rho)/\rho$, where ρ is in cm, find the capacitance per meter of the capacitor.

SOLUTION The procedure is the same as that taken in Section 6.5 except that eq. (6.27a) now becomes

$$V = -\int_b^a \frac{Q}{2\pi\varepsilon_o\varepsilon_r\rho L}\,d\rho = -\frac{Q}{2\pi\varepsilon_o L}\int_b^a \frac{d\rho}{\rho\left(\dfrac{10+\rho}{\rho}\right)}$$

$$= \frac{-Q}{2\pi\varepsilon_o L}\int_b^a \frac{d\rho}{10+\rho} = \frac{-Q}{2\pi\varepsilon_o L}\ln(10+\rho)\bigg|_b^a$$

$$= \frac{Q}{2\pi\varepsilon_o L}\ln\frac{10+b}{10+a}$$

Thus the capacitance per meter is (L = 1 m)

$$C = \frac{Q}{V} = \frac{2\pi\varepsilon_o}{\ln\dfrac{10+b}{10+a}} = 2\pi\cdot\frac{10^{-9}}{36\pi}\cdot\frac{1}{\ln\dfrac{12.5}{11.0}}$$

$$C = 434.6 \text{ pF/m}$$

■

PRACTICE EXERCISE 6.13

A spherical capacitor with a = 1.5 cm, b = 4 cm has an inhomogeneous dielectric of $\varepsilon = 10\varepsilon_o/r$. Calculate the capacitance of the capacitor.

ANSWER 1.13 nF.

6.6 METHOD OF IMAGES

The method of images, introduced by Lord Kelvin in 1848, is commonly used to determine V, $\mathbf{E}$, $\mathbf{D}$, and ρ_S due to charges in the presence of conductors. By this method, we avoid solving Poisson's or Laplace's equation but rather utilize the fact that a conducting surface is an equipotential. Although the method does not apply to all electrostatic problems, it can reduce a formidable problem to a simple one.

The image theory states that a given charge configuration above an infinite grounded perfect conducting plane may be replaced by the charge configuration itself, its image, and an equipotential surface in place of the conducting plane. Typical examples of point, line, and volume charge configurations are portrayed in Figure 6.21(a), and their corresponding image configurations are in Figure 6.21(b).

In applying the image method, two conditions must always be satisfied:

1. The image charge(s) must be located in the conducting region.
2. The image charge(s) must be located such that on the conducting surface(s) the potential is zero or constant.

The first condition is necessary to satisfy Poisson's equation, and the second condition ensures that the boundary conditions are satisfied. Let us now apply the image theory to two specific problems.

A. A Point Charge Above a Grounded Conducting Plane

Consider a point charge Q placed at a distance h from a perfect conducting plane of infinite extent as in Figure 6.22(a). The image configuration is in Figure 6.22(b). The electric field at point $P(x, y, z)$ is given by

$$\mathbf{E} = \mathbf{E}_+ + \mathbf{E}_- \tag{6.40}$$

$$= \frac{Q\,\mathbf{r}_1}{4\pi\varepsilon_o r_1^3} + \frac{-Q\,\mathbf{r}_2}{4\pi\varepsilon_o r_2^3} \tag{6.41}$$

The distance vectors $\mathbf{r}_1$ and $\mathbf{r}_2$ are given by

$$\mathbf{r}_1 = (x, y, z) - (0, 0, h) = (x, y, z - h) \tag{6.42}$$

$$\mathbf{r}_2 = (x, y, z) - (0, 0, -h) = (x, y, z + h) \tag{6.43}$$

so eq. (6.41) becomes

$$\mathbf{E} = \frac{Q}{4\pi\varepsilon_o}\left[\frac{x\mathbf{a}_x + y\mathbf{a}_y + (z - h)\mathbf{a}_z}{[x^2 + y^2 + (z - h)^2]^{3/2}} - \frac{x\mathbf{a}_x + y\mathbf{a}_y + (z + h)\mathbf{a}_z}{[x^2 + y^2 + (z + h)^2]^{3/2}}\right] \tag{6.44}$$

It should be noted that when $z = 0$, $\mathbf{E}$ has only the z-component, confirming that $\mathbf{E}$ is normal to the conducting surface.

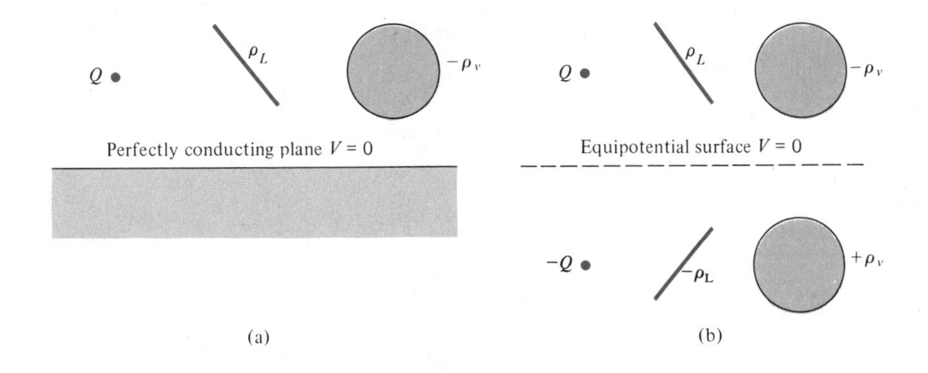

Figure 6.21 Image system: **(a)** charge configurations above a perfectly conducting plane; **(b)** image configuration with the conducting plane replaced by equipotential surface.

The potential at P is easily obtained from eq. (6.41) or (6.44) using $V = -\int \mathbf{E} \cdot d\mathbf{l}$. Thus

$$V = V_+ + V_-$$

$$= \frac{Q}{4\pi\varepsilon_o r_1} + \frac{-Q}{4\pi\varepsilon_o r_2}$$

$$V = \frac{Q}{4\pi\varepsilon_o}\left[\frac{1}{[x^2 + y^2 + (z - h)^2]^{1/2}} - \frac{1}{[x^2 + y^2 + (z + h)^2]^{1/2}}\right] \qquad [6.45]$$

for $z \geq 0$ and $V = 0$ for $z \leq 0$. Note that $V(z = 0) = 0$.

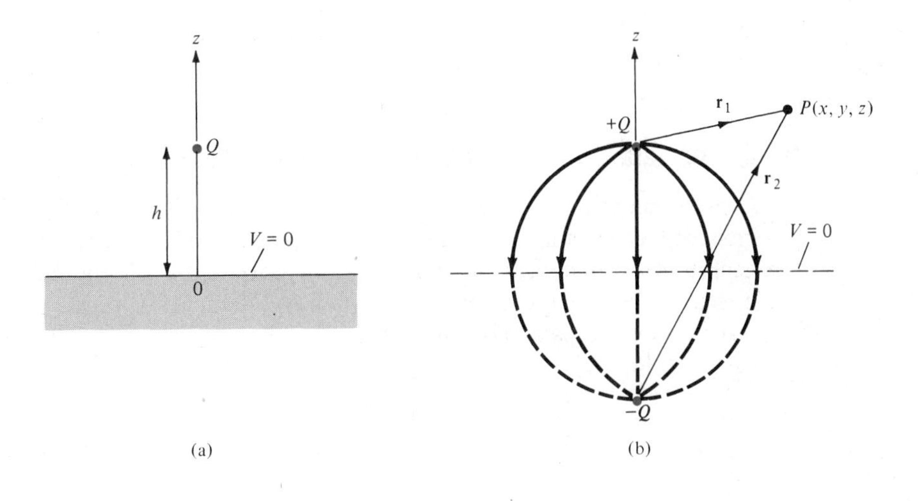

Figure 6.22 **(a)** Point charge and grounded conducting plane, **(b)** image configuration and field lines.

The surface charge density of the induced charge can also be obtained from eq. (6.44) as

$$\rho_S = D_n = \varepsilon_o E_n \Bigg|_{z=0}$$

$$= \frac{-Qh}{2\pi[x^2 + y^2 + h^2]^{3/2}} \tag{6.46}$$

The total induced charge on the conducting plane is

$$Q_i \int \rho_S \, dS = \int_{-\infty}^{\infty} \int_{-\infty}^{\infty} \frac{-Qh \, dx \, dy}{2\pi[x^2 + y^2 + h^2]^{3/2}} \tag{6.47}$$

By changing variables, $\rho^2 = x^2 + y^2$, $dx\,dy = \rho\, d\rho\, d\phi$.

$$Q_i = -\frac{Qh}{2\pi} \int_0^{2\pi} \int_0^{\infty} \frac{\rho \, d\rho \, d\phi}{[\rho^2 + h^2]^{3/2}} \tag{6.48}$$

Integrating over ϕ gives 2π, and letting $\rho\, d\rho = \frac{1}{2} d\,(\rho^2)$, we obtain

$$Q_i = -\frac{Qh}{2\pi} 2\pi \int_0^{\infty} [\rho^2 + h^2]^{-3/2} \frac{1}{2}\, d(\rho^2)$$

$$= \frac{Qh}{[\rho^2 + h^2]^{1/2}} \Bigg|_0^{\infty} \tag{6.49}$$

$$= -Q$$

as expected, because all flux lines terminating on the conductor would have terminated on the image charge if the conductor were absent.

B. A Line Charge Above a Grounded Conducting Plane

Consider an infinite charge with density ρ_L C/m located at a distance h from the grounded conducting plane $z = 0$. The same image system of Figure 6.22(b) applies to the line charge except that Q is replaced by ρ_L. The infinite line charge ρ_L may be assumed to be at $x = 0$, $z = h$ and the image $-\rho_L$ at $x = 0$, $z = -h$ so that the two are parallel to the y-axis. The electric field at point P is given (from eq. 4.21) by

$$\mathbf{E} = \mathbf{E}_+ + \mathbf{E}_- \tag{6.50}$$

$$= \frac{\rho_L}{2\pi\varepsilon_o\rho_1}\mathbf{a}_{\rho 1} + \frac{-\rho_L}{2\pi\varepsilon_o\rho_2}\mathbf{a}_{\rho 2} \tag{6.51}$$

The distance vectors $\boldsymbol{\rho}_1$ and $\boldsymbol{\rho}_2$ are given by

$$\boldsymbol{\rho}_1 = (x, y, z) - (0, y, h) = (x, 0, z - h) \tag{6.52}$$

$$\boldsymbol{\rho}_2 = (x, y, z) - (0, y, -h) = (x, 0, z + h) \tag{6.53}$$

so eq. (6.51) becomes

$$\mathbf{E} = \frac{\rho_L}{2\pi\varepsilon_o}\left[\frac{x\mathbf{a}_x + (z - h)\mathbf{a}_z}{x^2 + (z - h)^2} - \frac{x\mathbf{a}_x + (z + h)\mathbf{a}_z}{x^2 + (z + h)^2}\right] \tag{6.54}$$

Again, notice that when $z = 0$, $\mathbf{E}$ has only the z-component, confirming that $\mathbf{E}$ is normal to the conducting surface.

The potential at P is obtained from eq. (6.51) or (6.54) using $V = -\int \mathbf{E} \cdot d\mathbf{l}$. Thus

$$\begin{aligned} V &= V_+ + V_- \\ &= -\frac{\rho_L}{2\pi\varepsilon_o}\ln \rho_1 - \frac{-\rho_L}{2\pi\varepsilon_o}\ln \rho_2 \\ &= -\frac{\rho_L}{2\pi\varepsilon_o}\ln\frac{\rho_1}{\rho_2} \end{aligned} \tag{6.55}$$

Substituting $\rho_1 = |\boldsymbol{\rho}_1|$ and $\rho_2 = |\boldsymbol{\rho}_2|$ in eqs. (6.52) and (6.53) into eq. (6.55) gives

$$V = -\frac{\rho_L}{2\pi\varepsilon_o}\ln\left[\frac{x^2 + (z - h)^2}{x^2 + (z + h)^2}\right]^{1/2} \tag{6.56}$$

for $z \geq 0$ and $V = 0$ for $z \leq 0$. Note that $V(z = 0) = 0$.

The surface charge induced on the conducting plane is given by

$$\rho_s = D_n = \varepsilon_o E_z \Big|_{z=0} = \frac{-\rho_L h}{\pi(x^2 + h^2)} \tag{6.57}$$

The induced charge per length on the conducting plane is

$$\rho_i = \int \rho_s\, dx = -\frac{\rho_L h}{\pi}\int_{-\infty}^{\infty}\frac{dx}{x^2 + h^2} \tag{6.58}$$

By letting $x = h \tan \alpha$, eq. (6.58) becomes

$$\begin{aligned} \rho_i &= -\frac{\rho_L h}{\pi}\int_{-\pi/2}^{\pi/2}\frac{d\alpha}{h} \\ &= -\rho_L \end{aligned} \tag{6.59}$$

as expected.

EXAMPLE 6.14

A point charge Q is located at point $(a, 0, b)$ between two semi-infinite conducting planes intersecting at right angles as in Figure 6.23. Determine the potential at point $P(x, y, z)$ and the force on Q.

SOLUTION The image configuration is shown in Figure 6.24. Three image charges are necessary to satisfy the conditions in Section 6.6. From Figure 6.24(a), the potential at point $P(x, y, z)$ is the superposition of the potentials at P due to the four point charges; that is,

$$V = \frac{Q}{4\pi\varepsilon_o}\left[\frac{1}{r_1} - \frac{1}{r_2} + \frac{1}{r_3} - \frac{1}{r_4}\right]$$

where

$$r_1 = [(x - a)^2 + y^2 + (z - b)^2]^{1/2}$$

$$r_2 = [(x + a)^2 + y^2 + (z - b)^2]^{1/2}$$

$$r_3 = [(x + a)^2 + y^2 + (z + b)^2]^{1/2}$$

$$r_4 = [(x - a)^2 + y^2 + (z + b)^2]^{1/2}$$

From Figure 6.24(b), the net force on Q

$$\begin{aligned}\mathbf{F} &= \mathbf{F}_1 + \mathbf{F}_2 + \mathbf{F}_3 \\ &= -\frac{Q^2}{4\pi\varepsilon_o(2b)^2}\,\mathbf{a}_z - \frac{Q^2}{4\pi\varepsilon_o(2a)^2}\,\mathbf{a}_x + \frac{Q^2(2a\mathbf{a}_x + 2b\mathbf{a}_z)}{4\pi\varepsilon_o[(2a)^2 + (2b)^2]^{3/2}} \\ &= \frac{Q^2}{16\pi\varepsilon_o}\left\{\left[\frac{a}{[a^2 + b^2]^{3/2}} - \frac{1}{a^2}\right]\mathbf{a}_x + \left[\frac{b}{[a^2 + b^2]^{3/2}} - \frac{1}{b^2}\right]\mathbf{a}_z\right\}\end{aligned}$$

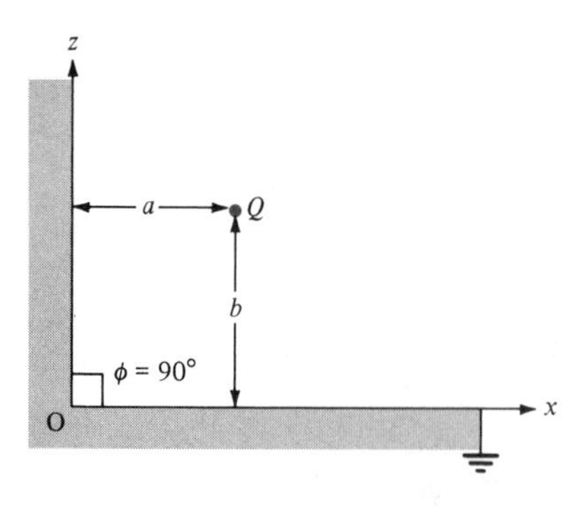

Figure 6.23 Point charge between two semi-infinite conducting planes.

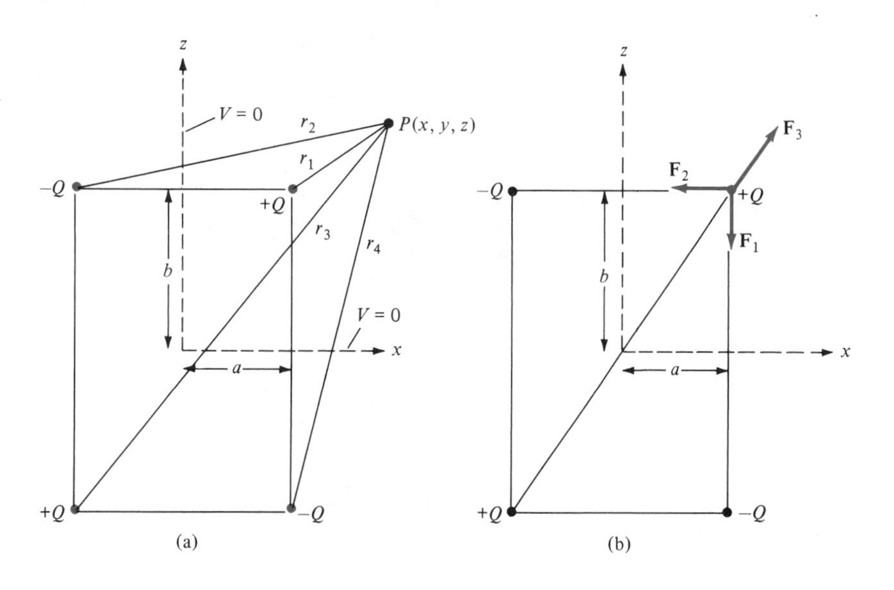

Figure 6.24 Determining **(a)** the potential at P, and **(b)** the force on charge Q.

The electric field due to this system can be determined similarly and the charge induced on the planes can also be found.

In general, when the method of images is used for a system consisting of a point charge between two semi-infinite conducting planes inclined at an angle ϕ (in degrees), the number of images is given by

$$\boxed{N = \left(\frac{360°}{\phi} - 1\right)}$$

because the charge and its images all lie on a circle. For example, when $\phi = 180°$, $N = 1$ as in the case of Figure 6.22; for $\phi = 90°$, $N = 3$ as in the case of Figure 6.23; and for $\phi = 60°$, we expect $N = 5$ as shown in Figure 6.25. ■

PRACTICE EXERCISE 6.14

If the point charge $Q = 10$ nC in Figure 6.25 is 10 cm away from point O and along the line bisecting $\phi = 60°$, find the magnitude of the force on Q due to the charge induced on the conducting walls.

ANSWER 60.53 μN.

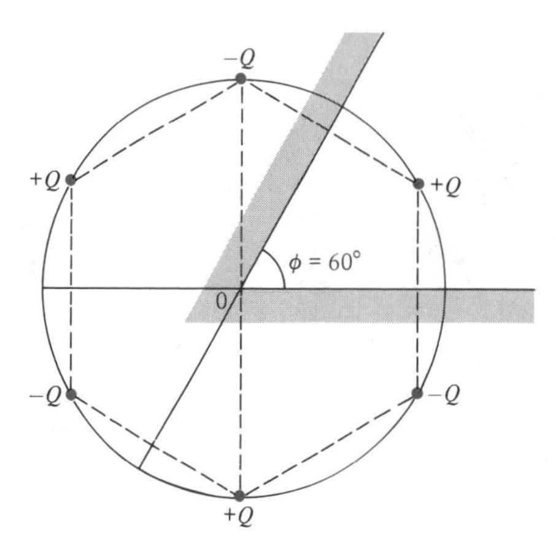

Figure 6.25 Point charge between two semi-infinite conducting walls inclined at $\phi = 60°$ to each.

SUMMARY

1. Boundary-value problems are those in which the potentials at the boundaries of a region are specified and we are to determine the potential field within the region. They are solved using Poisson's equation if $\rho_v \neq 0$ or Laplace's equation if $\rho_v = 0$.
2. In a nonhomogeneous region, Poisson's equation is

$$\nabla \cdot \varepsilon \nabla V = -\rho_v$$

For a homogeneous region, ε is independent of space variables. Poisson's equation becomes

$$\nabla^2 V = -\frac{\rho_v}{\varepsilon}$$

In a charge-free region ($\rho_v = 0$), Poisson's equation becomes Laplace's equation; that is,

$$\nabla^2 V = 0$$

3. We solve the differential equation resulting from Poisson's or Laplace's equation by integrating twice if V depends on one variable or by the method of separation of variables if V is a function of more than one variable. We then apply the prescribed boundary conditions to obtain a unique solution.
4. The uniqueness theorem states that if V satisfies Poisson's or Laplace's equation and the prescribed boundary condition, V is the only possible solution for that given problem. This enables us to find the solution to a given problem via any expedient means because we are assured of one, and only one, solution.
5. The problem of finding the resistance R of an object or the capacitance C of a capacitor may be treated as a boundary-value problem. To determine R, we assume

a potential difference V_o between the ends of the object, solve Laplace's equation, find $I = \int \sigma \mathbf{E} \cdot d\mathbf{S}$, and obtain $R = V_o/I$. Similarly, to determine C, we assume a potential difference of V_o between the plates of the capacitor, solve Laplace's equation, find $Q = \int \varepsilon \mathbf{E} \cdot d\mathbf{S}$, and obtain $C = Q/V_o$.

6. A boundary-value problem involving an infinite conducting plane or wedge may be solved using the method of images. This basically entails replacing the charge configuration by itself, its image, and an equipotential surface in place of the conducting plane. Thus the original problem is replaced by "an image problem," which is solved using techniques covered in Chapters 4 and 5.

REVIEW QUESTIONS

6.1 Equation $\nabla \cdot (-\varepsilon \nabla V) = \rho_v$ may be regarded as Poisson's equation for an inhomogeneous medium.

(a) True (b) False

6.2 In cylindrical coordinates, equation

$$\frac{\partial^2 \psi}{\partial \rho^2} + \frac{1}{\rho}\frac{\partial \psi}{\partial \rho} + \frac{\partial^2 \psi}{\partial z^2} + 10 = 0$$

is called

(a) Maxwell's equation

(b) Laplace's equation

(c) Poisson's equation

(d) Helmholtz's equation

(e) Lorentz's equation

6.3 Two potential functions V_1 and V_2 satisfy Laplace's equation within a closed region and assume the same values on its surface. V_1 must be equal to V_2.

(a) True

(b) False

(c) Not necessarily

6.4 Which of the following potentials does not satisfy Laplace's equation?

(a) $V = 2x + 5$

(b) $V = 10\,xy$

(c) $V = r \cos \phi$

(d) $V = \dfrac{10}{r}$

(e) $V = \rho \cos \phi + 10$

6.5 Which of the following is not true?

(a) $-5\cos 3x$ is a solution to $\phi''(x) + 9\phi(x) = 0$

(b) $10\sin 2x$ is a solution to $\phi''(x) - 4\phi(x) = 0$

(c) $-4\cosh 3y$ is a solution to $R''(y) - 9R(y) = 0$

(d) $\sinh 2y$ is a solution to $R''(y) - 4R(y) = 0$

(e) $\dfrac{g''(x)}{g(x)} = -\dfrac{h''(y)}{h(y)} = f(z) = -1$ where $g(x) = \sin x$ and $h(y) = \sinh y$

6.6 If $V_1 = X_1Y_1$ is a product solution of Laplace's equation, which of these are not solutions of Laplace's equation?

(a) $-10X_1Y_1$

(b) $X_1Y_1 + 2xy$

(c) $X_1Y_1 - x + y$

(d) $X_1 + Y_1$

(e) $(X_1 - 2)(Y_1 + 3)$

6.7 The capacitance of a capacitor filled by a linear dielectric is independent of the charge on the plates and the potential difference between the plates.

(a) True (b) False

6.8 A parallel-plate capacitor connected to a battery stores twice as much charge with a given dielectric as it does with air as dielectric, the susceptibility of the dielectric is

(a) 0

(b) 1

(c) 2

(d) 3

(e) 4

6.9 A potential difference V_o is applied to a mercury column in a cylindrical container. The mercury is now poured into another cylindrical container of half the radius and the same potential difference V_o applied across the ends. As a result of this change of shape, the resistance will be increased

(a) 2 times

(b) 4 times

(c) 8 times

(d) 16 times

6.10 Two conducting plates are inclined at an angle 30° to each other with a point charge between them. The number of image charges is

(a) 12

(b) 11

(c) 6

(d) 5

(e) 3

Answers: 6.1a, 6.2c, 6.3a, 6.4c, 6.5b, 6.6d,e, 6.7a, 6.8b, 6.9d, 6.10b.

PROBLEMS

6.1 Show that the exact solution of the equation

$$\frac{d^2V}{dx^2} = f(x) \qquad 0 < x < L$$

subject to

$$V(x = 0) = V_1 \qquad V(x = L) = V_2$$

is

$$V(x) = \left[V_2 - V_1 - \int_0^L \int_0^\lambda f(\mu)\, d\mu\, d\lambda\right]\frac{x}{L} + V_1 + \int_0^x \int_0^\lambda f(\mu)\, d\mu\, d\lambda$$

6.2 A certain material occupies the space between two conducting slabs located at $y = \pm 2$ cm. When heated, the material emits electrons such that $\rho_v = 50(1 - y^2)\ \mu\text{C/m}^3$. If the slabs are both held at 30 kV, find the potential distribution within the slabs. Take $\varepsilon = 3\varepsilon_o$.

6.3 In cylindrical coordinates, $\rho_v = 10/\rho$ pC/m^3. If $V = 0$ at $\rho = 1$ m and $V = 100$ V at $\rho = 4$ m due to the charge distribution, find

(a) V at $\rho = 3$ m

(b) $\mathbf{E}$ at $\rho = 2$. Take $\varepsilon = \varepsilon_o$.

6.4 Operating a Van de Graaff generator above breakdown so that corona occurs results in charging the surrounding air ($\varepsilon = \varepsilon_o$) with $\rho_v = \rho_o a^4/r^4$, $r \geq a$, where $\rho_o = 1.2$ mC/m^3 is the charge density at the surface of the spherical cap of the generator and $a = 10$ cm is its radius. If the cap is maintained at the voltage 1500 kV, calculate the electric field at the surface. Is this field greater than that obtained if there were no discharge?

6.5 Show in Cartesian coordinates that Poisson's equation in an inhomogeneous medium (see eq. (6.3)) can be written as

$$\varepsilon \nabla^2 V + \nabla V \cdot \nabla \varepsilon = -\rho_v$$

where $\varepsilon = \varepsilon(x, y, z)$.

6.6 Which of the following distributions satisfy Laplace's equation?

(a) $V = V_o \sin\left(\frac{n\pi x}{a}\right) \cos\left(\frac{n\pi y}{a}\right)$

(b) $V = V_o \sin\left(\frac{n\pi x}{a}\right) \cosh\left(\frac{n\pi y}{a}\right)$

(c) $V = V_o \sin\left(\frac{n\pi x}{a}\right) \sin\left(\frac{n\pi y}{a}\right)$

(d) $V = \dfrac{x + y}{(x^2 + y^2 + z^2)^{3/2}}$

Show all work.

6.7 Verify that the following potentials satisfy Laplace's equation:

(a) $V = 15x^2yz - 5y^3z$

(b) $V = \dfrac{\cos\phi}{\rho}$

(c) $V = \dfrac{10 \sin\theta \sin\phi}{r^2}$

6.8 Show that $\mathbf{E} = (E_x, E_y, E_z)$ satisfies Laplace's equation.

6.9 Consider the conducting plates shown in Figure 6.26. If $V(z = 0) = 0$ and $V(z = 2 \text{ mm}) = 50$ V, determine V, $\mathbf{E}$, and $\mathbf{D}$ in the dielectric region ($\varepsilon_r = 1.5$) between the plates and ρ_S on the plates.

6.10 The cylindrical capacitor whose cross section is in Figure 6.27 has inner and outer radii of 5 mm and 15 mm respectively. If $V(\rho = 5 \text{ mm}) = 100$ V and $V(\rho = 15 \text{ mm}) = 0$ V, calculate V, $\mathbf{E}$, and $\mathbf{D}$ at $\rho = 10$ mm and ρ_S on each plate. Take $\varepsilon_r = 2.0$.

6.11 Concentric spherical shells $r = 0.1$ m and $r = 2$ m are maintained at $V = 0$ and $V = 100$ V respectively. Assuming free space between the shells, find V, $\mathbf{E}$, and $\mathbf{D}$.

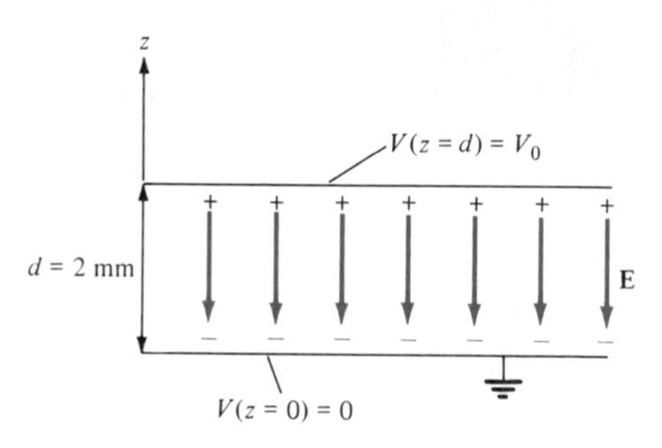

Figure 6.26 For Problem 6.9.

6.12 The region between concentric spherical conducting shells $r = 0.5$ m and $r = 1$ m is charge-free. If $V(r = 0.5) = -50$ V and $V(r = 1) = 50$ V, determine the potential distribution and the electric field strength in the region between the shells.

6.13 Find V and $\mathbf{E}$ at (3, 0, 4) due to the two conducting cones of infinite extent shown in Figure 6.28.

***6.14** The inner and outer electrodes of a diode are coaxial cylinders of radii $a = 0.6$ m and $b = 30$ mm respectively. The inner electrode is maintained at 70 V while the outer electrode is grounded. (a) Assuming that the length of the electrodes $\ell \gg a, b$ and ignoring the effects of space charge, calculate the potential at $\rho = 15$ mm. (b) If an electron is injected radially through a small hole in the inner electrode with velocity 10^7 m/s, find its velocity at $\rho = 15$ mm.

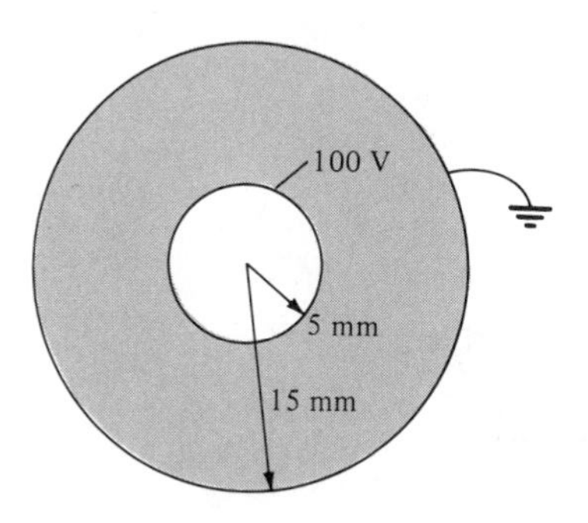

Figure 6.27 Cylindrical capacitor of Problem 6.10.

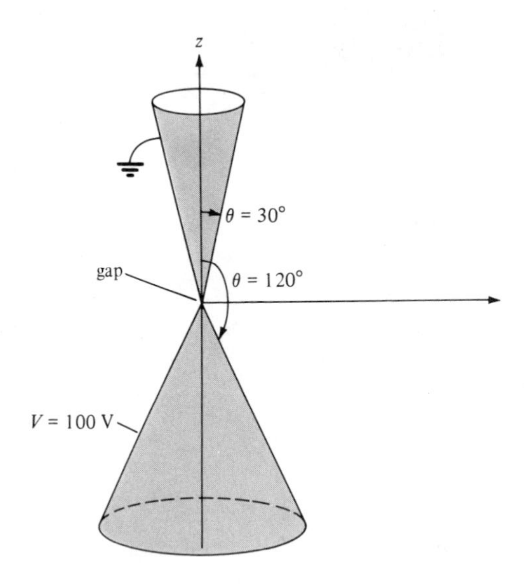

Figure 6.28 Conducting cones of Problem 6.13.

6.15 Another method of finding the capacitance of a capacitor is using energy considerations, that is

$$C = \frac{2W_E}{V_o^2} = \frac{1}{V_o^2}\int \varepsilon|\mathbf{E}|^2\,dv$$

Using this approach, derive eqs. (6.22), (6.28), and (6.32).

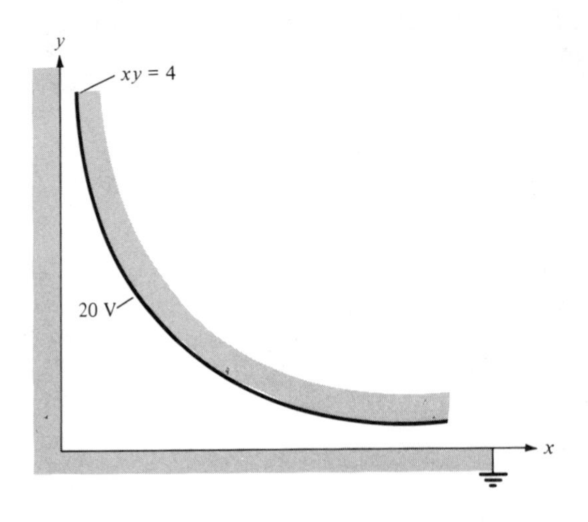

Figure 6.29 For Problem 6.16.

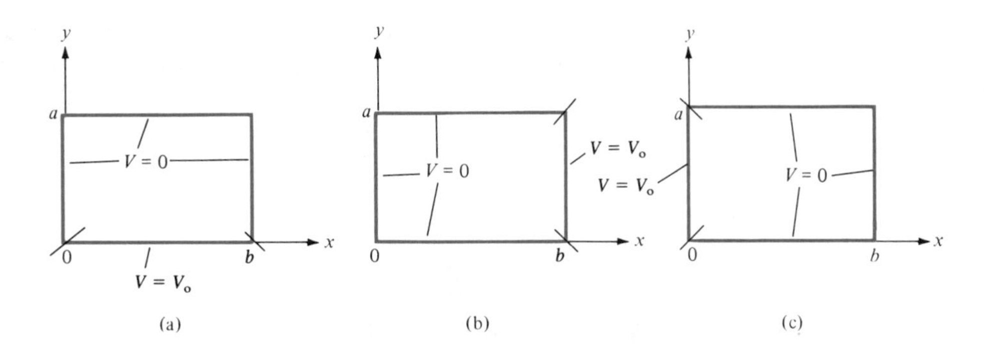

Figure 6.30 For Problem 6.17.

6.16 An electrode with a hyperbolic shape ($xy = 4$) is placed above an earthed right-angle corner as in Figure 6.29. Calculate V and $\mathbf{E}$ at point (1, 2, 0) when the electrode is connected to a 20-V source.

***6.17** Solve Laplace's equation for the two-dimensional electrostatic systems of Figure 6.30 and find the potential $V(x, y)$.

****6.18** Find the potential $V(x, y)$ due to the two-dimensional systems of Figure 6.31.

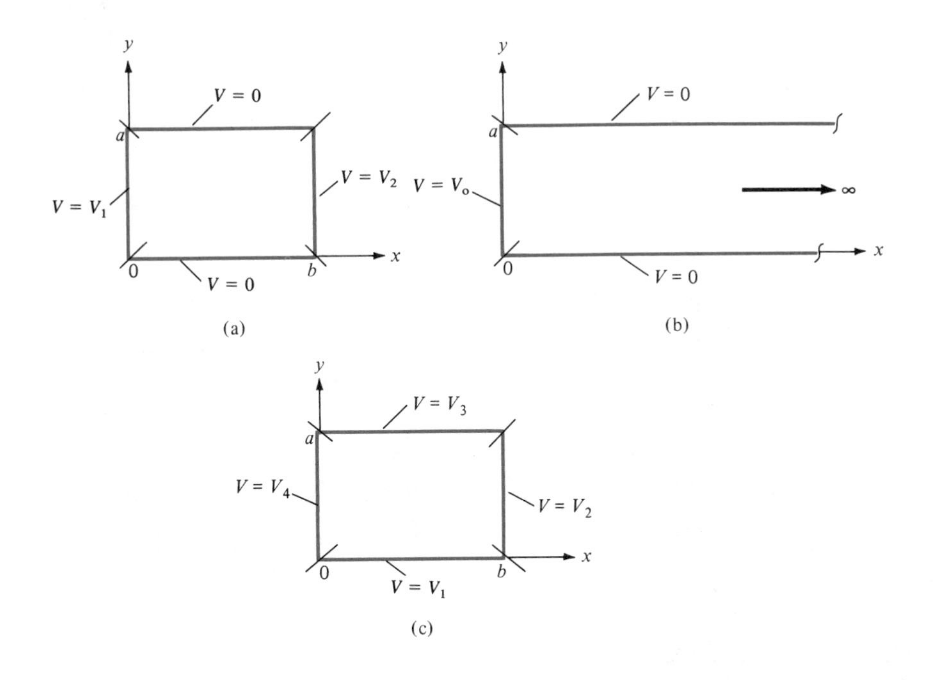

Figure 6.31 For Problem 6.18.

6.19 By letting $V(\rho, \phi) = R(\rho)\Phi(\phi)$ be the solution of Laplace's equation in a region where $\rho \neq 0$, show that the separated differential equations for R and Φ are

$$R'' + \frac{R'}{\rho} - \frac{\lambda}{\rho^2} R = 0$$

and

$$\Phi'' + \lambda\Phi = 0$$

where λ is the separation constant.

6.20 A potential in spherical coordinates is a function of r and θ but not ϕ. Assuming that $V(r, \theta) = R(r)F(\theta)$, obtain the separated differential equations for R and F in a region for which $\rho_v = 0$.

6.21 Show that the resistance of the bar of Figure 6.17 between the vertical ends located at $\phi = 0$ and $\phi = \pi/2$ is

$$R = \frac{\pi}{2\sigma t \ln b/a}$$

***6.22** Show that the resistance of the sector of a spherical shell of conductivity σ, with cross section shown in Figure 6.32 (where $0 \leq \phi < 2\pi$), between its base is

$$R = \frac{1}{2\pi\sigma(1 - \cos\alpha)}\left[\frac{1}{a} - \frac{1}{b}\right]$$

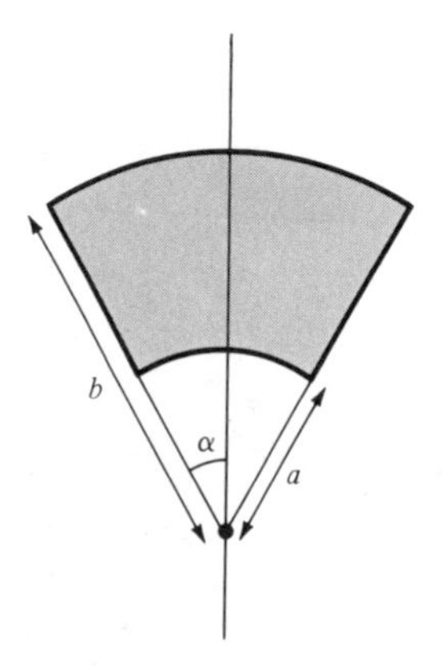

Figure 6.32 For Problem 6.22.

***6.23** A hollow conducting hemisphere of radius a is buried with its flat face lying flush with the earth surface thereby serving as an earthing electrode. If the conductivity of earth is σ, show that the leakage conductance between the electrode and earth is $2\pi a\sigma$.

6.24 In an integrated circuit, a capacitor is formed by growing a silicon dioxide layer ($\varepsilon_r = 4$) of thickness 1 μm over the conducting silicon substrate and covering it with a metal electrode of area S. Determine S if a capacitance of 2 nF is desired.

6.25 The parallel-plate capacitor of Figure 6.33 is quarter-filled with mica ($\varepsilon_r = 6$). Find the capacitance of the capacitor.

***6.26** A cube of side 3 mm and of dielectric constant 4.6 is slid between the plates of a parallel plate capacitor having air as a dielectric. If the capacitor has a face area of 20 cm^2 and a separation distance of 5 mm, calculate the capacitance of the capacitor with the cube in it.

***6.27** An air-filled parallel plate capacitor of length L, width a, and plate separation d has its plates maintained at constant potential difference V_o. If a dielectric slab of dielectric constant ε_r is slid between the plates and is withdrawn until only a length x remains between the plates as in Figure 6.34, show that the force tending to restore the slab to its original position is

$$F = \frac{\varepsilon_0(\varepsilon_r - 1)\, a\, V_o^2}{2d}$$

6.28 A parallel-plate capacitor has plate area 200 cm^2 and plate separation 3 mm. The charge density is 1 μC/m^2 with air as dielectric. Find

(a) The capacitance of the capacitor

(b) The voltage between the plates

(c) The force with which the plates attract each other

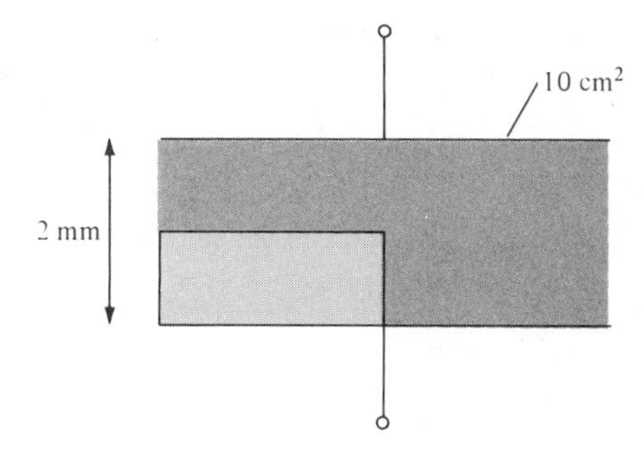

Figure 6.33 For Problem 6.25.

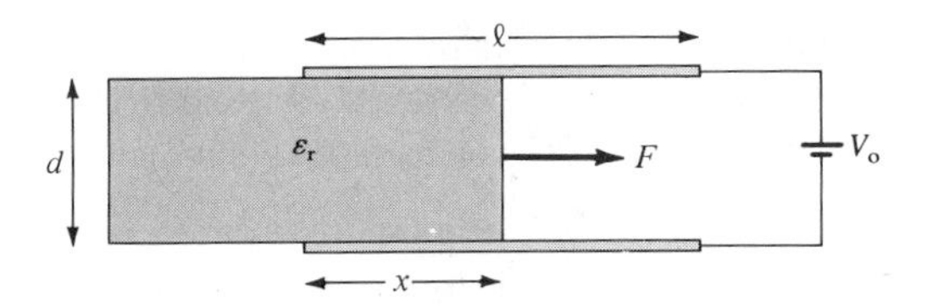

Figure 6.34 For Problem 6.27.

***6.29** A parallel-plate capacitor has its plates at $x = 0, d$ and the space between the plates is filled with an inhomogeneous material with permittivity $\varepsilon = 2d\varepsilon_o/(x + d)$. If the plate at $x = d$ is maintained at V_o while the plate at $x = 0$ is grounded, find

(a) **E**

(b) **P**

(c) ρ_{ps} at $x = 0, d$

(d) The capacitance when $d = 2.5$ mm and each plate has an area of 200 cm^2. Ignore fringing.

6.30 The capacitance of a parallel-plate capacitor with air as a dielectric is C_o. If a slab of thickness Δ ($< d$) and dielectric constant ε_r is slid between the plates, show that the capacitance becomes

$$C = \frac{\varepsilon_r d C_o}{\varepsilon_r d - \chi_e \Delta}$$

6.31 A coaxial cable contains an insulating material of dielectric constant 3.5. The radius of the central wire is 1 mm and that of the sheath is 2 mm. Find the capacitance per kilometer of the cable.

6.32 Determine the capacitance of a conducting sphere of radius 5 cm deeply immersed in sea water ($\varepsilon_r = 80$).

6.33 A conducting sphere of radius 2 cm is surrounded by a concentric conducting sphere of radius 5 cm. If the space between the spheres is filled with sodium chlroide ($\varepsilon_r = 5.9$), calculate the capacitance of the system.

***6.34** In an ink jet printer the drops are charged by surrounding the jet of radius 20 μm with a concentric cylinder of radius 600 μm as in Figure 6.35. Calculate the minimum voltage required to generate a charge 50 fC on the drop if the length of the jet inside the cylinder is 100 μm. Take $\varepsilon = \varepsilon_o$.

6.35 A given length of a cable, the capacitance of which is 10 μF/km and resistance of insulation of 100 MΩ/km, is charged to a voltage of 100 V. How long does it take the voltage to drop to 50 V?

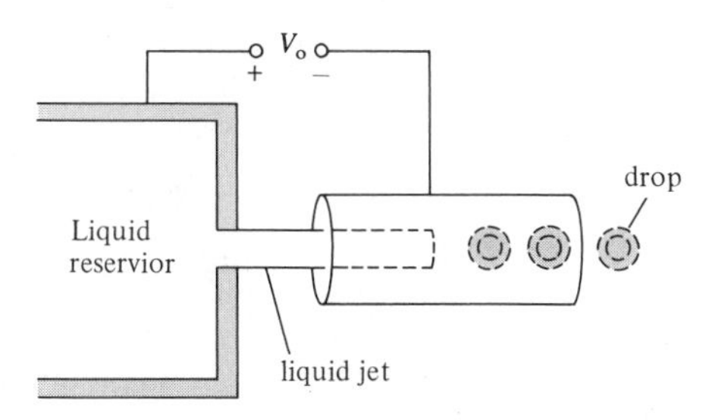

Figure 6.35 Simplified geometry of an ink-jet printer; for Problem 6.34.

6.36 The capacitance per unit length of a two-wire transmission line shown in Figure 6.36 is given by

$$C = \frac{\pi \varepsilon}{\cosh^{-1}\left[\dfrac{d}{2a}\right]}$$

Determine the conductance per unit length.

***6.37** A spherical capacitor has an inner conductor of radius a carrying charge Q and maintained at zero potential. If the outer conductor contracts from a radius b to c under internal forces, prove that the work performed by the electric field as a result of the contraction is

$$W = \frac{Q^2(b - c)}{8\pi\varepsilon bc}$$

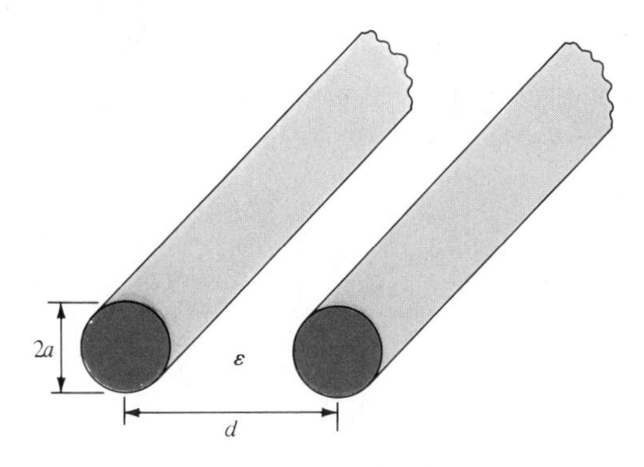

Figure 6.36 For Problem 6.36.

6.38 A spherical capacitor has two concentric spheres of radii a and d. Concentric with these and lying between them is a spherical shell of inner radius b and outer radius c. If the shell is filled with dielectric material of relative permittivity ε_r and $a < b < c < d$, determine the capacitance of the system.

6.39 If the earth is regarded a spherical capacitor, what is its capacitance? Assume the radius of the earth to be approximately 6370 km.

6.40 A point charge of 10 nC is located at point $P(0, 0, 3)$ while the conducting plane $z = 0$ is grounded. Calculate

(a) V and $\mathbf{E}$ at $R(6, 3, 5)$

(b) The force on the charge due to induced charge on the plates

6.41 Two point charges of 3 nC and -4 nC are placed respectively at (0, 0, 1 m) and (0, 0, 2 m) while an infinite conducting plane is at $z = 0$. Determine

(a) The total charge induced on the plane

(b) The magnitude of the force of attraction between the charges and the plane

6.42 Two point charges of 50 nC and -20 nC are located at $(-3, 2, 4)$ and $(1, 0, 5)$ above the conducting ground plane $z = 2$. Calculate (a) the surface charge density at $(7, -2, 2)$, (b) $\mathbf{D}$ at $(3, 4, 8)$, and (c) $\mathbf{D}$ at $(1, 1, 1)$.

****6.43** A point charge of 10 μC is located at (1, 1, 1), and the positive portions of the coordinate planes are occupied by three mutually perpendicular plane conductors maintained at zero potential. Find the force on the charge due to the conductors.

6.44 A line charge of 20 nC/m is located at $y = 1$, $z = 5$ above the conducting plane $z = 0$. Determine E at

(a) (1, 2, 3)

(b) $(2, 4, -8)$

6.45 Infinite planes $x = 3$ and $x = 10$ carry charges -10 nC/m^2 and 5 nC/m^2 respectively. If plane $x = 1$ is grounded, find $\mathbf{E}$ at $(6, 0, -4)$. Take $\varepsilon = \varepsilon_o$.

6.46 A point charge Q is placed between two earthed intersecting conducting planes which are inclined at 45° to each other. Determine the number of image charges and their locations.

PART 3

Magnetostatics

CHAPTER

7

Magnetostatic Fields

Knowledge will forever govern ignorance: And a people who mean to be their own Governors, must arm themselves with the power which only knowledge gives.

—JAMES MADISON

7.1 INTRODUCTION

In Chapters 4 to 6, we limited our discussions to static electric fields characterized by **E** or **D**. We now focus our attention on static magnetic fields, which are characterized by **H** or **B**. There are similarities and dissimilarities between electric and magnetic fields. As **E** and **D** are related according to $\mathbf{D} = \varepsilon\mathbf{E}$ for linear material space, **H** and **B** are related according to $\mathbf{B} = \mu\mathbf{H}$. Table 7.1 further shows the analogy between electric and magnetic field quantities. Some of the magnetic field quantities will be introduced later in this chapter, and others will be presented in the next. The analogy is presented here to show that most of the equations we have derived for the electric fields may be readily used to obtain corresponding equations for magnetic fields if the equivalent analogous quantities are substituted. This way it does not appear as if we are learning new concepts.

A definite link between electric and magnetic fields was established by Oersted[1] in 1820. As we have noticed, an electrostatic field is produced by static or stationary charges. If the charges are moving with constant velocity, a static magnetic (or magnetostatic) field is produced. This movement of charges (or current flow) may be due to magnetization currents as in permanent magnets, electron-beam currents as in vacuum tubes, or conduction currents as in current-carrying wires. In this chapter, we consider magnetic fields in free space due to direct current. Magnetostatic fields in material space are covered in Chapter 8.

Our study of magnetostatics is not a dispensable luxury but an indispensable necessity. The development of the motors, transformers, microphones, compasses,

[1]Hans Christian Oersted (1777–1851), a Danish professor of physics, after thirteen years of frustrating efforts discovered that electricity could produce magnetism.

Table 7.1 **Analogy Between Electric and Magnetic Fields***

Term	Electric	Magnetic
Basic laws	$\mathbf{F} = \dfrac{Q_1 Q_2}{4\pi\varepsilon r^2}\mathbf{a}_r$	$d\mathbf{B} = \dfrac{\mu_o I\, d\mathbf{l} \times \mathbf{a}_R}{4\pi R^2}$
	$\oint \mathbf{D} \cdot d\mathbf{S} = Q_{enc}$	$\oint \mathbf{H} \cdot d\mathbf{l} = I_{enc}$
Force law	$\mathbf{F} = Q\mathbf{E}$	$\mathbf{F} = Q\mathbf{u} \times \mathbf{B}$
Source element	dQ	$Q\mathbf{u} = I\, d\mathbf{l}$
Field intensity	$E = \dfrac{V}{\ell}$ (V/m)	$H = \dfrac{I}{\ell}$ (A/m)
Flux density	$\mathbf{D} = \dfrac{\Psi}{S}$ (C/m^2)	$\mathbf{B} = \dfrac{\Psi}{S}$ (Wb/m^2)
Relationship between fields	$\mathbf{D} = \varepsilon\mathbf{E}$	$\mathbf{B} = \mu\mathbf{H}$
Potentials	$\mathbf{E} = -\nabla V$	$\mathbf{H} = -\nabla V_m \; (\mathbf{J} = 0)$
	$V = \displaystyle\int \frac{\rho_L dl}{4\pi\varepsilon r}$	$\mathbf{A} = \displaystyle\int \frac{\mu I\, d\mathbf{l}}{4\pi R}$
	$\Psi = \int \mathbf{D} \cdot d\mathbf{S}$	$\Psi = \int \mathbf{B} \cdot d\mathbf{S}$
Flux	$\Psi = Q = CV$	$\Psi = LI$
	$I = C\dfrac{dV}{dt}$	$V = L\dfrac{dI}{dt}$
Energy density	$w_E = \dfrac{1}{2}\mathbf{D} \cdot \mathbf{E}$	$w_m = \dfrac{1}{2}\mathbf{B} \cdot \mathbf{H}$
Poisson's equation	$\nabla^2 V = -\dfrac{\rho_v}{\varepsilon}$	$\nabla^2 \mathbf{A} = -\mu\mathbf{J}$

*A similar analogy can be found in R. S. Elliot, "Electromagnetic theory: a simplified representation," *IEEE Trans. Educ.*, vol. E-24, no. 4, Nov. 1981, pp. 294–296.

telephone bell ringers, television focusing controls, advertising displays, magnetically levitated high-speed vehicles, memory stores, magnetic separators, and so on, involve magnetic phenomena and play an important role in our everyday life.[2]

[2]Various applications of magnetism can be found in J. K. Watson, *Applications of Magnetism*. New York: John Wiley & Sons, 1980.

There are two major laws governing magnetostatic fields: (1) Biot-Savart's law,[3] and (2) Ampere's circuit law.[4] Like Coulomb's law, Biot-Savart's law is the general law of magnetostatics. Just as Gauss's law is a special case of Coulomb's law, Ampere's law is a special case of Biot-Savart's law and is easily applied in problems involving symmetrical current distribution. The two laws of magnetostatics are stated and applied first; their derivation is provided later in the chapter.

7.2 BIOT-SAVART'S LAW

The law states that the magnetic field intensity dH produced at a point P, as shown in Figure 7.1, by the differential current element $I\,dl$ is proportional to the product $I\,dl$ and the sine of the angle α between the element and the line joining P to the element and is inversely proportional to the square of the distance R between P and the element; that is,

$$dH \propto \frac{I\,dl \sin\alpha}{R^2} \tag{7.1}$$

or

$$dH = \frac{kI\,dl \sin\alpha}{R^2} \tag{7.2}$$

where k is the constant of proportionality. In SI units, $k = 1/4\pi$, so eq. (7.2) becomes

$$\boxed{dH = \frac{I\,dl \sin\alpha}{4\pi R^2}} \tag{7.3}$$

From the definition of cross product in eq. (1.21), it is easy to notice that eq. (7.3) is better put in vector form as

$$\boxed{d\mathbf{H} = \frac{I\,d\mathbf{l} \times \mathbf{a}_R}{4\pi R^2} = \frac{I\,d\mathbf{l} \times \mathbf{R}}{4\pi R^3}} \tag{7.4}$$

where $R = |\mathbf{R}|$ and $\mathbf{a}_R = \mathbf{R}/R$. Thus the direction of $d\mathbf{H}$ can be determined by the right-hand rule with the right-hand thumb pointing in the direction of the current, the

[3]The experiments and analyses of the effect of a current element were carried out by Ampere and by Jean-Baptiste and Felix Savart, around 1820.

[4]Andre Marie Ampere (1775–1836), a French physicist, developed Oersted's discovery and introduced the concept of current element and the force between current elements.

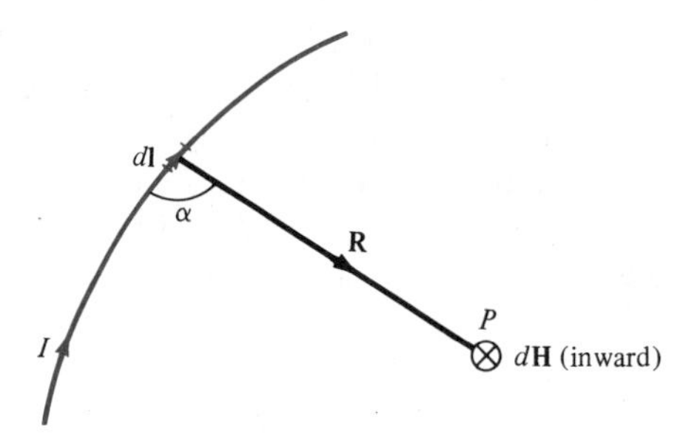

Figure 7.1 Magnetic field $d\mathbf{H}$ at P due to current element $I\,d\mathbf{l}$.

right-hand fingers encircling the wire in the direction of $d\mathbf{H}$ as shown in Figure 7.2(a). Alternatively, we can use the right-handed screw rule to determine the direction of $d\mathbf{H}$: with the screw placed along the wire and pointed in the direction of current flow, the direction of advance of the screw is the direction of $d\mathbf{H}$ as in Figure 7.2(b).

It is customary to represent the direction of the magnetic field intensity $\mathbf{H}$ (or current I) by a small circle with a dot or cross sign depending on whether $\mathbf{H}$ (or I) is out of, or into, the page as illustrated in Figure 7.3.

Just as we can have different charge configurations (see Figure 4.5), we can have different current distributions: line current, surface current, and volume current as shown in Figure 7.4. If we define $\mathbf{K}$ as the surface current density (in amperes/meter) and $\mathbf{J}$ as the volume current density (in amperes/meter square), the source elements are related as

$$I\,d\mathbf{l} \equiv \mathbf{K}\,dS \equiv \mathbf{J}\,dv \qquad [7.5]$$

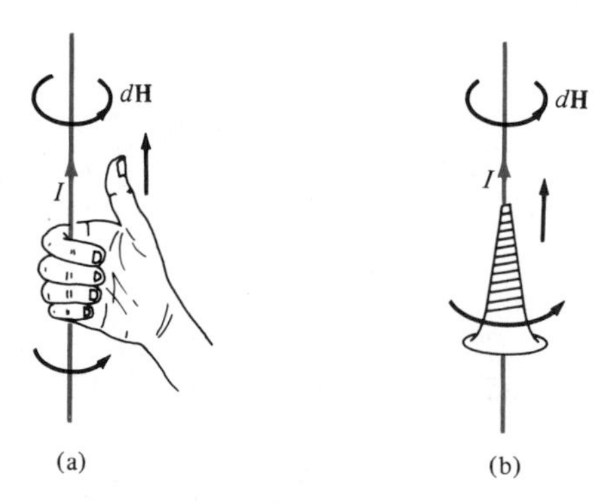

Figure 7.2 Determining the direction of $d\mathbf{H}$ using **(a)** the right-hand rule, or **(b)** the right-handed screw rule.

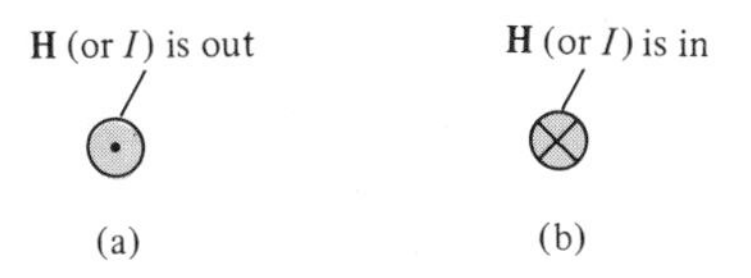

Figure 7.3 Conventional representation of **H** (or I) **(a)** out of the page and **(b)** into the page.

Thus in terms of the distributed current sources, the Biot-Savart law as in eq. (7.4) becomes

$$\mathbf{H} = \int_L \frac{I\, d\mathbf{l} \times \mathbf{a}_R}{4\pi R^2} \quad \text{(line current)} \tag{7.6}$$

$$\mathbf{H} = \int_S \frac{\mathbf{K}\, dS \times \mathbf{a}_R}{4\pi R^2} \quad \text{(surface current)} \tag{7.7}$$

$$\mathbf{H} = \int_v \frac{\mathbf{J}\, dv \times \mathbf{a}_R}{4\pi R^2} \quad \text{(volume current).} \tag{7.8}$$

As an example, let us apply eq. (7.6) to determine the field due to a *straight* current carrying filamentary conductor of finite length AB as in Figure 7.5. We assume that the conductor is along the z-axis with its upper and lower ends respectively subtending angles α_2 and α_1 at P, the point at which **H** is to be determined. Particular note should be taken of this assumption as the formula to be derived will have to be applied accordingly. If we consider the contribution $d\mathbf{H}$ at P due to an element $d\mathbf{l}$ at $(0, 0, z)$,

$$d\mathbf{H} = \frac{I\, d\mathbf{l} \times \mathbf{R}}{4\pi R^3} \tag{7.9}$$

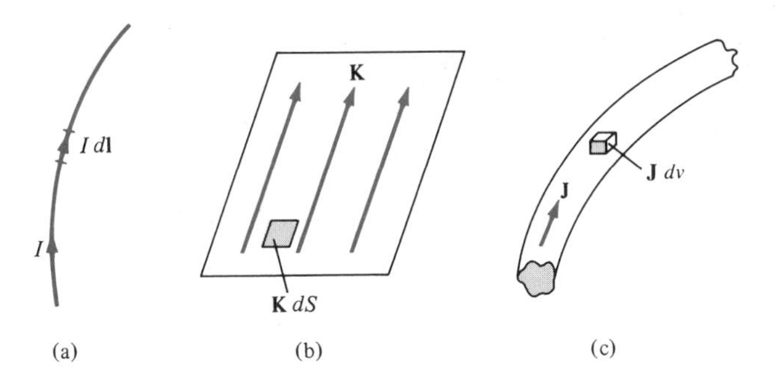

Figure 7.4 Current distributions: **(a)** line current, **(b)** surface current, **(c)** volume current.

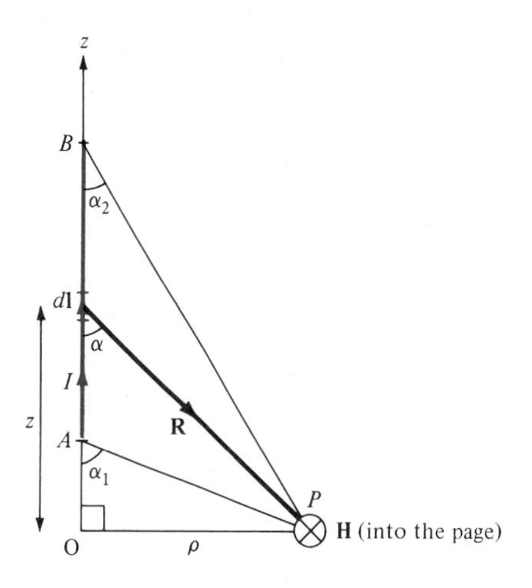

Figure 7.5 Field at point P due to a straight filamentary conductor.

But $d\mathbf{l} = dz\,\mathbf{a}_z$ and $\mathbf{R} = \rho\mathbf{a}_\rho - z\mathbf{a}_z$, so

$$d\mathbf{l} \times \mathbf{R} = \rho\, dz\, \mathbf{a}_\phi \tag{7.10}$$

Hence,

$$\mathbf{H} = \int \frac{I\rho\, dz}{4\pi\,[\rho^2 + z^2]^{3/2}}\,\mathbf{a}_\phi \tag{7.11}$$

Letting $z = \rho \cot \alpha$, $dz = -\rho\, \text{cosec}^2\alpha\, d\alpha$, and eq. (7.11) becomes

$$\begin{aligned}\mathbf{H} &= -\frac{I}{4\pi}\int_{\alpha_1}^{\alpha_2} \frac{\rho^2 \text{cosec}^2\alpha\, d\alpha}{\rho^3 \text{cosec}^3\alpha}\,\mathbf{a}_\phi \\ &= -\frac{I}{4\pi\rho}\mathbf{a}_\phi \int_{\alpha_1}^{\alpha_2} \sin\alpha\, d\alpha\end{aligned}$$

or

$$\boxed{\mathbf{H} = \frac{I}{4\pi\rho}(\cos\alpha_2 - \cos\alpha_1)\,\mathbf{a}_\phi} \tag{7.12}$$

This expression is generally applicable for any straight filamentary conductor of finite length. Notice from eq. (7.12) that $\mathbf{H}$ is always along the unit vector $\mathbf{a}_\phi$ (i.e., along concentric circular paths) irrespective of the length of the wire or the point of interest P. As a special case, when the conductor is *semi-infinite* (with respect to P) so that point A is now at O(0, 0, 0) while B is at (0, 0, ∞); $\alpha_1 = 90°$, $\alpha_2 = 0°$, and eq. (7.12) becomes

$$\mathbf{H} = \frac{I}{4\pi\rho}\mathbf{a}_\phi \tag{7.13}$$

Another special case is when the conductor is *infinite* in length. For this case, point *A* is at (0, 0, −∞) while *B* is at (0, 0, ∞); $\alpha_1 = 180°$, $\alpha_2 = 0°$, so eq. (7.12) reduces to

$$\boxed{\mathbf{H} = \frac{I}{2\pi\rho}\mathbf{a}_\phi} \tag{7.14}$$

To find unit vector $\mathbf{a}_\phi$ in eqs. (7.12) to (7.14) is not always easy. A simple approach is to determine $\mathbf{a}_\phi$ from

$$\boxed{\mathbf{a}_\phi = \mathbf{a}_\ell \times \mathbf{a}_\rho} \tag{7.15}$$

where $\mathbf{a}_\ell$ is a unit vector along the line current and $\mathbf{a}_\rho$ is a unit vector along the perpendicular line from the line current to the field point.

EXAMPLE 7.1

The conducting triangular loop in Figure 7.6(a) carries a current of 10 A. Find **H** at (0, 0, 5) due to side 1 of the loop.

SOLUTION

This example illustrates how eq. (7.12) is applied to any straight, thin, current-carrying conductor. The key point to keep in mind in applying eq. (7.12) is figuring out α_1, α_2, ρ, and $\mathbf{a}_\phi$. To find **H** at (0, 0, 5) due to side 1 of the loop in Figure 7.6(a), consider Figure 7.6(b), where side 1 is treated as a straight conductor. Observe that α_1, α_2, and ρ are assigned in the same manner as in Figure 7.5 on which eq. (7.12) is based.

$$\cos\alpha_1 = \cos 90° = 0, \qquad \cos\alpha_2 = \frac{2}{\sqrt{29}}, \qquad \rho = 5$$

To determine $\mathbf{a}_\phi$ is often the hardest part of applying eq. (7.12). According to eq. (7.15), $\mathbf{a}_\ell = \mathbf{a}_x$ and $\mathbf{a}_\rho = \mathbf{a}_z$, so

$$\mathbf{a}_\phi = \mathbf{a}_x \times \mathbf{a}_z = -\mathbf{a}_y$$

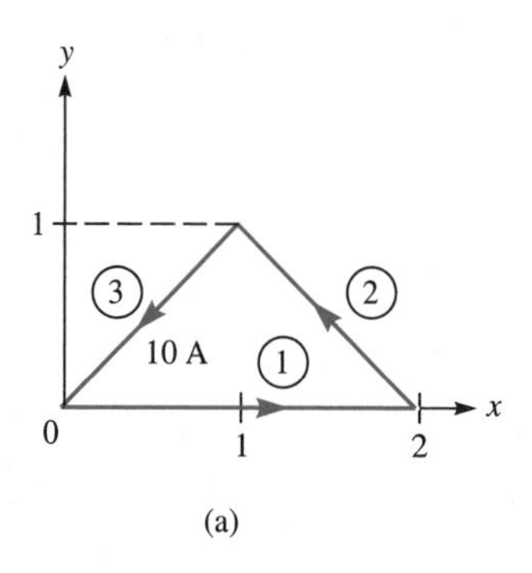

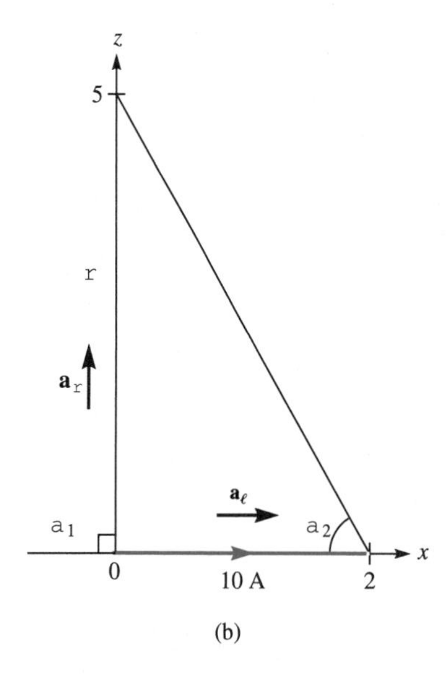

Figure 7.6 For Example 7.1: **(a)** conducting triangular loop, **(b)** side 1 of the loop.

Hence,

$$\mathbf{H}_1 = \frac{I}{4\pi\rho}(\cos\alpha_2 - \cos\alpha_1)\mathbf{a}_\phi = \frac{10}{4\pi(5)}\left(\frac{2}{\sqrt{29}} - 0\right)(-\mathbf{a}_y)$$

$$= -59.1\mathbf{a}_y \text{ mA/m}$$

■

PRACTICE EXERCISE 7.1

Find **H** at (0, 0, 5) due to side 3 of the triangular loop in Figure 7.6(a).

ANSWER $-30.63\mathbf{a}_x + 30.63\mathbf{a}_y$ mA/m.

EXAMPLE 7.2

Find **H** at (−3, 4, 0) due to the current filament shown in Figure 7.7(a).

SOLUTION

Let $\mathbf{H} = \mathbf{H}_x + \mathbf{H}_z$ where $\mathbf{H}_x$ and $\mathbf{H}_z$ are the contributions to the magnetic field intensity at $P(-3, 4, 0)$ due to the portions of the filament along x and z respectively.

$$\mathbf{H}_z = \frac{I}{4\pi\rho}(\cos\alpha_2 - \cos\alpha_1)\,\mathbf{a}_\phi$$

At $P(-3, 4, 0)$, $\rho = (9 + 16)^{1/2} = 5$, $\alpha_1 = 90°$, $\alpha_2 = 0°$, and $\mathbf{a}_\phi$ is obtained as a unit vector along the circular path through P on plane $z = 0$ as in Figure 7.7(b). The direction of $\mathbf{a}_\phi$ is determined using the right-handed screw rule or the right-hand rule. From the geometry in Figure 7.7(b),

$$\mathbf{a}_\phi = \sin\theta\,\mathbf{a}_x + \cos\theta\,\mathbf{a}_y = \frac{4}{5}\mathbf{a}_x + \frac{3}{5}\mathbf{a}_y$$

Alternatively, we can determine $\mathbf{a}_\phi$ from eq. (7.15). At point P, $\mathbf{a}_\ell$ and $\mathbf{a}_\rho$ are as illustrated in Figure 7.7(a) for $\mathbf{H}_z$. Hence,

$$\mathbf{a}_\phi = -\mathbf{a}_z \times \left(-\frac{3}{5}\mathbf{a}_x + \frac{4}{5}\mathbf{a}_y\right) = \frac{4}{5}\mathbf{a}_x + \frac{3}{5}\mathbf{a}_y$$

as obtained before. Thus

$$\mathbf{H}_z = \frac{3}{4\pi(5)}(1 - 0)\frac{(4\mathbf{a}_x + 3\mathbf{a}_y)}{5}$$

$$= 38.2\mathbf{a}_x + 28.65\mathbf{a}_y \text{ mA/m}$$

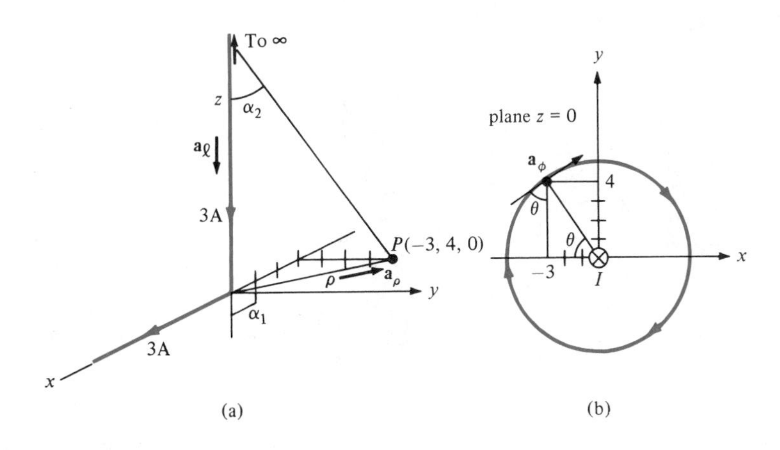

Figure 7.7 For Example 7.2: **(a)** current filament along semi-infinite x and z axes; $\mathbf{a}_\ell$ and $\mathbf{a}_\rho$ for $\mathbf{H}_z$ only; **(b)** determining $\mathbf{a}_\rho$ for $\mathbf{H}_z$.

It should be noted that in this case $\mathbf{a}_\phi$ happens to be the negative of the regular $\mathbf{a}_\phi$ of cylindrical coordinates. $\mathbf{H}_z$ could have also been obtained in cylindrical coordinates as

$$\mathbf{H}_z = \frac{3}{4\pi(5)}(1 - 0)(-\mathbf{a}_\phi)$$

$$= -47.75\mathbf{a}_\phi \text{ mA/m}$$

Similarly, for $\mathbf{H}_x$ at P, $\rho = 4$, $\alpha_2 = 0°$, $\cos\alpha_1 = 3/5$, and $\mathbf{a}_\phi = \mathbf{a}_z$ or $\mathbf{a}_\phi = \mathbf{a}_\ell \times \mathbf{a}_\rho = \mathbf{a}_x \times \mathbf{a}_y = \mathbf{a}_z$. Hence,

$$\mathbf{H}_x = \frac{3}{4\pi(4)}\left(1 - \frac{3}{5}\right)\mathbf{a}_z$$

$$= 23.88\ \mathbf{a}_z \text{ mA/m}$$

Thus

$$\mathbf{H} = \mathbf{H}_x + \mathbf{H}_z = 38.2\mathbf{a}_x + 28.65\mathbf{a}_y + 23.88\mathbf{a}_z \text{ mA/m}$$

or

$$\mathbf{H} = -47.75\mathbf{a}_\phi + 23.88\mathbf{a}_z \text{ mA/m}$$

Notice that although the current filaments appear semi-infinite (they occupy the positive z- and x-axes), it is only the filament along the z-axis that is semi-infinite with respect to point P. Thus $\mathbf{H}_z$ could have been found by using eq. (7.13), but the equation could not have been used to find $\mathbf{H}_x$ because the filament along the x-axis is not semi-infinite with respect to P. ■

PRACTICE EXERCISE 7.2

The positive y-axis (semi-infinite line with respect to the origin) carries a filamentary current of 2 A in the $-\mathbf{a}_y$ direction. Assume it is part of a large circuit. Find $\mathbf{H}$ at

(a) $A(2, 3, 0)$
(b) $B(3, 12, -4)$

ANSWER (a) $145.8\mathbf{a}_z$ mA/m, (b) $48.97\mathbf{a}_x + 36.73\mathbf{a}_z$ mA/m.

EXAMPLE 7.3

A circular loop located on $x^2 + y^2 = 9$, $z = 0$ carries a direct current of 10 A along $\mathbf{a}_\phi$. Determine $\mathbf{H}$ at $(0, 0, 4)$ and $(0, 0, -4)$.

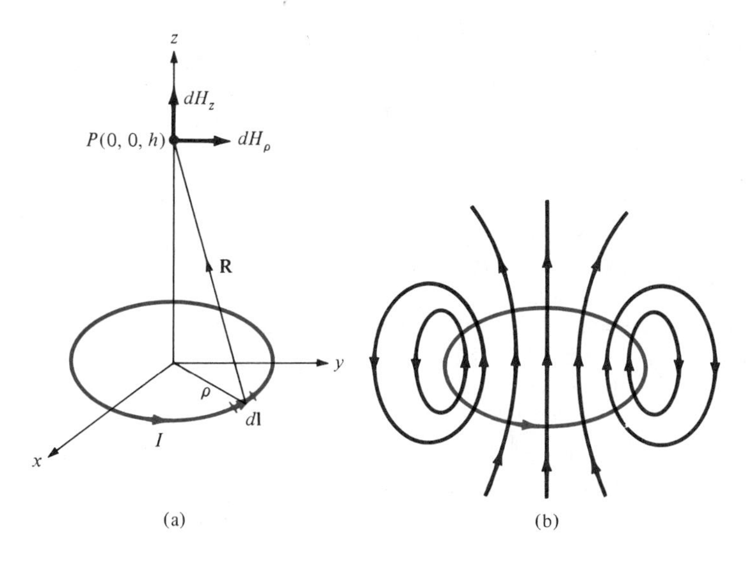

Figure 7.8 For Example 7.3: **(a)** circular current loop, **(b)** flux lines due to the current loop.

SOLUTION Consider the circular loop shown in Figure 7.8(a). The magnetic field intensity $d\mathbf{H}$ at point $P(0, 0, h)$ contributed by current element $I\, d\mathbf{l}$ is given by Biot-Savart's law:

$$d\mathbf{H} = \frac{I\, d\mathbf{l} \times \mathbf{R}}{4\pi R^3}$$

where $d\mathbf{l} = \rho\, d\phi\, \mathbf{a}_\phi$, $\mathbf{R} = (0, 0, h) - (x, y, 0) = -\rho \mathbf{a}_\rho + h\mathbf{a}_z$, and

$$d\mathbf{l} \times \mathbf{R} = \begin{vmatrix} \mathbf{a}_\rho & \mathbf{a}_\phi & \mathbf{a}_z \\ 0 & \rho\, d\phi & 0 \\ -\rho & 0 & h \end{vmatrix} = \rho h\, d\phi\, \mathbf{a}_\rho + \rho^2\, d\phi\, \mathbf{a}_z$$

Hence,

$$d\mathbf{H} = \frac{I}{4\pi[\rho^2 + h^2]^{3/2}} (\rho h\, d\phi\, \mathbf{a}_\rho + \rho^2\, d\phi\, \mathbf{a}_z) = dH_\rho\, \mathbf{a}_\rho + dH_z\, \mathbf{a}_z$$

By symmetry, the contributions along $\mathbf{a}_\rho$ add up to zero because the radial components produced by pairs of current element 180° apart cancel. This may also be shown mathematically by writing $\mathbf{a}_\rho$ in rectangular coordinate systems (i.e., $\mathbf{a}_\rho = \cos\phi\, \mathbf{a}_x + \sin\phi\, \mathbf{a}_y$). Integrating $\cos\phi$ or $\sin\phi$ over $0 \leq \phi \leq 2\pi$ gives zero, thereby showing that $\mathbf{H}_\rho = 0$. Thus

$$\mathbf{H} = \int dH_z\, \mathbf{a}_z = \int_0^{2\pi} \frac{I\rho^2\, d\phi\, \mathbf{a}_z}{4\pi[\rho^2 + h^2]^{3/2}} = \frac{I\rho^2 2\pi \mathbf{a}_z}{4\pi[\rho^2 + h^2]^{3/2}}$$

or

$$\mathbf{H} = \frac{I\rho^2 \mathbf{a}_z}{2[\rho^2 + h^2]^{3/2}}$$

(a) Substituting $I = 10\ A$, $\rho = 3$, $h = 4$ gives

$$\mathbf{H}(0, 0, 4) = \frac{10\ (3)^2\ \mathbf{a}_z}{2[9 + 16]^{3/2}} = 0.36\mathbf{a}_z \text{ A/m}$$

(b) Notice from $d\mathbf{l} \times \mathbf{R}$ above that if h is replaced by $-h$, the z-component of $d\mathbf{H}$ remains the same while the ρ-component still adds up to zero due to the axial symmetry of the loop. Hence

$$\mathbf{H}(0, 0, -4) = \mathbf{H}(0, 0, 4) = 0.36\mathbf{a}_z \text{ A/m}$$

The flux lines due to the circular current loop are sketched in Figure 7.8(b). ■

PRACTICE EXERCISE 7.3

A thin ring of radius 5 cm is placed on plane $z = 1$ cm so that its center is at (0, 0, 1 cm). If the ring carries 50 mA along $\mathbf{a}_\phi$, find $\mathbf{H}$ at

(a) (0, 0, −1 cm)

(b) (0, 0, 10 cm)

ANSWER (a) $400\mathbf{a}_z$ mA/m, (b) $57.3\mathbf{a}_z$ mA/m.

EXAMPLE 7.4

A solenoid of length ℓ and radius a consists of N turns of wire carrying current I. Show that at point P along its axis.

$$\mathbf{H} = \frac{nI}{2}(\cos\theta_2 - \cos\theta_1)\mathbf{a}_z$$

where $n = N/\ell$, θ_1 and θ_2 are the angles subtended at P by the end turns as illustrated in Figure 7.9. Also show that if $\ell \gg a$, at the center of the solenoid,

$$\mathbf{H} = nI\mathbf{a}_z$$

SOLUTION Consider the cross-section of the solenoid as shown in Figure 7.9. Since the solenoid consists of circular loops, we apply the result of Example 7.3. The contribution to the magnetic field H and P by an element of the solenoid of length dz is

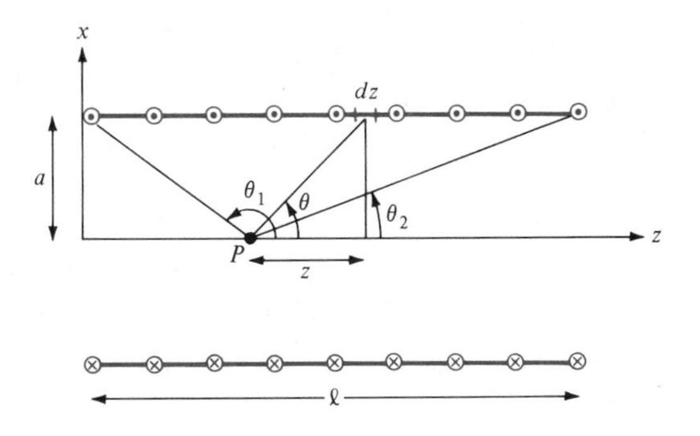

Figure 7.9 For Example 7.4; cross section of a solenoid.

$$dH_z = \frac{I\,dl\,a^2}{2[a^2 + z^2]^{3/2}} = \frac{Ia^2 n\,dz}{2[a^2 + z^2]^{3/2}}$$

where $dl = n\,dz = (N/\ell)\,dz$. From Figure 7.9, $\tan\theta = a/z$; that is,

$$dz = -a \operatorname{cosec}^2\theta\,d\theta = -\frac{[z^2 + a^2]^{3/2}}{a^2}\sin\theta\,d\theta$$

Hence,

$$dH_z = -\frac{nI}{2}\sin\theta\,d\theta$$

or

$$H_z = -\frac{nI}{2}\int_{\theta_1}^{\theta_2}\sin\theta\,d\theta$$

Thus

$$\mathbf{H} = \frac{nI}{2}(\cos\theta_2 - \cos\theta_1)\,\mathbf{a}_z$$

as required. Substituting $n = N/\ell$ gives

$$\mathbf{H} = \frac{NI}{2\ell}(\cos\theta_2 - \cos\theta_1)\,\mathbf{a}_z$$

At the center of the solenoid,

$$\cos\theta_2 = \frac{\ell/2}{[a^2 + \ell^2/4]^{1/2}} = -\cos\theta_1$$

and

$$\mathbf{H} = \frac{In\ell}{2[a^2 + \ell^2/4]^{1/2}}\,\mathbf{a}_z$$

If $\ell \gg a$ or $\theta_2 \simeq 0°$, $\theta_1 \simeq 180°$,

$$\mathbf{H} = nI\mathbf{a}_z = \frac{NI}{\ell}\,\mathbf{a}_z$$

■

PRACTICE EXERCISE 7.4

If the solenoid of Figure 7.9 has 2,000 turns, a length of 75 cm, a radius of 5 cm, and carries a current of 50 mA along $\mathbf{a}_\phi$, find $\mathbf{H}$ at

(a) (0, 0, 0)

(b) (0, 0, 75 cm)

(c) (0, 0, 50 cm)

ANSWER (a) $66.52\mathbf{a}_z$ A/m, (b) $66.52\mathbf{a}_z$ A/m, (c) $131.7\mathbf{a}_z$ A/m.

7.3 AMPERE'S CIRCUIT LAW—MAXWELL'S EQUATION

Ampere's circuit law states that the line integral of the tangential component of $\mathbf{H}$ around a *closed* path is the same as the net current I_{enc} enclosed by the path. In other words, the circulation of $\mathbf{H}$ equals I_{enc}; that is,

$$\boxed{\oint \mathbf{H} \cdot d\mathbf{l} = I_{enc}} \tag{7.16}$$

Ampere's law is similar to Gauss's law and it is easily applied to determine $\mathbf{H}$ when the current distribution is symmetrical. It should be noted that eq. (7.16) always holds whether the current distribution is symmetrical or not but we can only use the equation to determine $\mathbf{H}$ when symmetrical current distribution exists. Ampere's law is a special case of Biot-Savart's law; the former may be derived from the latter.

By applying Stokes's theorem to the left-hand side of eq. (7.16), we obtain

$$I_{enc} = \oint_L \mathbf{H} \cdot d\mathbf{l} = \int_S (\nabla \times \mathbf{H}) \cdot d\mathbf{S} \tag{7.17}$$

But

$$I_{enc} = \int_S \mathbf{J} \cdot d\mathbf{S} \tag{7.18}$$

Comparing the surface integrals in eqs. (7.17) and (7.18) clearly reveals that

$$\boxed{\nabla \times \mathbf{H} = \mathbf{J}} \tag{7.19}$$

This is the third Maxwell's equation to be derived; it is essentially Ampere's law in differential (or point) form whereas eq. (7.16) is the integral form. From eq. (7.19), we should observe that $\nabla \times \mathbf{H} = \mathbf{J} \neq 0$; that is, magnetostatic field is not conservative.

7.4 APPLICATIONS OF AMPERE'S LAW

We now apply Ampere's circuit law to determine **H** for some symmetrical current distributions as we did for Gauss's law. We should consider an infinite line current, an infinite current sheet, and an infinitely long coaxial transmission line.

A. Infinite Line Current

Consider an infinitely long filamentary current I along the z-axis as in Figure 7.10. To determine **H** at an observation point P, we allow a closed path pass through P. This path, on which Ampere's law is to be applied, is known as an *Amperian path* (analogous to the term Gaussian surface). We choose a concentric circle as the

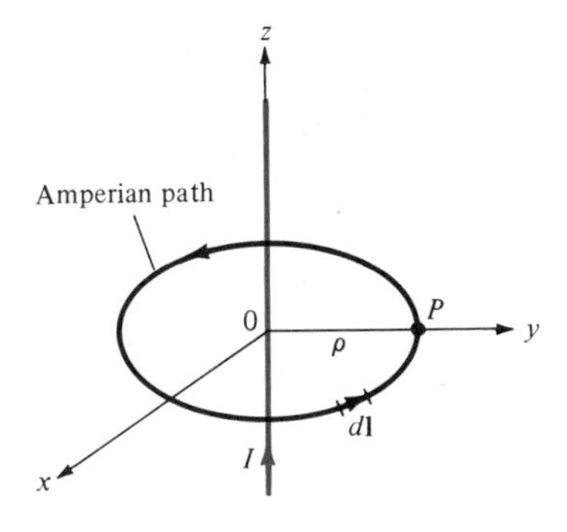

Figure 7.10 Ampere's law applied to an infinite filamentary line current.

Amperian path in view of eq. (7.14), which shows that $\mathbf{H}$ is constant provided ρ is constant. Since this path encloses the whole current I, according to Ampere's law

$$I = \int H_\phi \mathbf{a}_\phi \cdot \rho \, d\phi \, \mathbf{a}_\phi = H_\phi \int \rho \, d\phi = H_\phi \cdot 2\pi\rho$$

or

$$\boxed{\mathbf{H} = \frac{I}{2\pi\rho} \mathbf{a}_\phi} \quad [7.20]$$

as expected from eq. (7.14).

B. Infinite Sheet of Current

Consider an infinite current sheet in the $z = 0$ plane. If the sheet has a uniform current density $\mathbf{K} = K_y \mathbf{a}_y$ A/m as shown in Figure 7.11, applying Ampere's law to the rectangular closed path (Amperian path) gives

$$\oint \mathbf{H} \cdot d\mathbf{l} = I_{\text{enc}} = K_y b \quad [7.21a]$$

To evaluate the integral, we first need to have an idea of what $\mathbf{H}$ is like. To achieve this, we regard the infinite sheet as comprising of filaments; $d\mathbf{H}$ above or below the

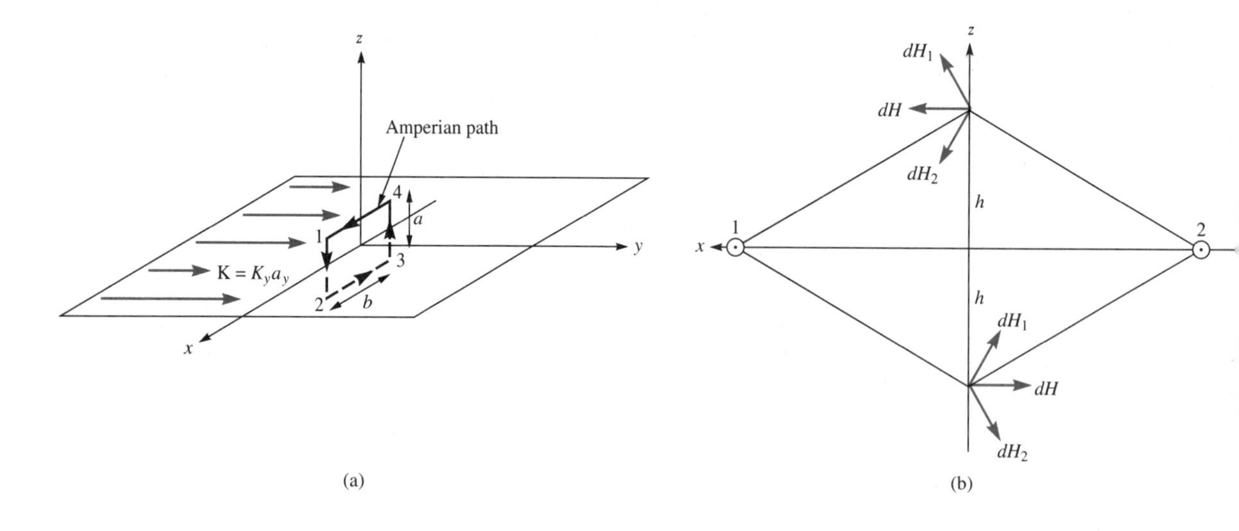

Figure 7.11 Application of Ampere's law to an infinite sheet: **(a)** closed path 1-2-3-4-1, **(b)** symmetrical pair of current filaments with current along $\mathbf{a}_y$.

sheet due to a pair of filamentary currents can be found using eqs. (7.14) and (7.15). As evident in Figure 7.11(b), the resultant $d\mathbf{H}$ has only an x-component. Also, $\mathbf{H}$ on one side of the sheet is the negative of that on the other side. Due to the infinite extent of the sheet, the sheet can be regarded as consisting of such filamentary pairs so that the characteristics of $\mathbf{H}$ for a pair are the same for the infinite current sheets, that is,

$$\mathbf{H} = \begin{cases} H_o \mathbf{a}_x & z > 0 \\ -H_o \mathbf{a}_x & z < 0 \end{cases} \tag{7.21b}$$

where H_o is yet to be determined. Evaluating the line integral of $\mathbf{H}$ in eq. (7.21b) along the closed path in Figure 7.11(a) gives

$$\begin{aligned} \oint \mathbf{H} \cdot d\mathbf{l} &= \left(\int_1^2 + \int_2^3 + \int_3^4 + \int_4^1 \right) \mathbf{H} \cdot d\mathbf{l} \\ &= 0(-a) + (-H_o)(-b) + 0(a) + H_o(b) \\ &= 2H_o b \end{aligned} \tag{7.21c}$$

From eqs. (7.21a and c), we obtain $H_o = \frac{1}{2} K_y$. Substituting H_o in eq. (7.21b) gives

$$\mathbf{H} = \begin{cases} \frac{1}{2} K_y \mathbf{a}_x, & z > 0 \\ -\frac{1}{2} K_y \mathbf{a}_x, & z < 0 \end{cases} \tag{7.22}$$

In general, for an infinite sheet of current density $\mathbf{K}$ A/m,

$$\boxed{\mathbf{H} = \frac{1}{2} \mathbf{K} \times \mathbf{a}_n} \tag{7.23}$$

where $\mathbf{a}_n$ is a unit normal vector directed from the current sheet to the point of interest.

C. Infinitely Long Coaxial Transmission Line

Consider an infinitely long transmission line consisting of two concentric cylinders having their axes along the z-axis. The cross section of the line is shown in Figure 7.12, where the z-axis is out of the page. The inner conductor has radius a and carries current I while the outer conductor has inner radius b and thickness t and carries return current $-I$. We want to determine $\mathbf{H}$ everywhere assuming that current is uniformly distributed in both conductors. Since the current distribution is symmetrical, we apply Ampere's law along the Amperian path for each of the four possible regions: $0 \le \rho \le a$, $a \le \rho \le b$, $b \le \rho \le b + t$, and $\rho \ge b + t$.

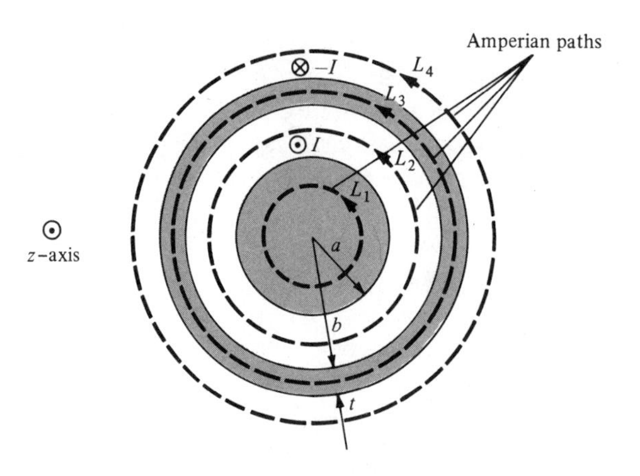

Figure 7.12 Cross section of the transmission line; the positive z direction is out of the page.

For region $0 \leq \rho \leq a$, we apply Ampere's law to path L_1, giving

$$\oint_{L_1} \mathbf{H} \cdot d\mathbf{l} = I_{\text{enc}} = \int \mathbf{J} \cdot d\mathbf{S} \tag{7.24}$$

Since the current is uniformly distributed over the cross section,

$$\mathbf{J} = \frac{I}{\pi a^2}\mathbf{a}_z, \qquad d\mathbf{S} = \rho \, d\phi \, d\rho \, \mathbf{a}_z$$

$$I_{\text{enc}} = \int \mathbf{J} \cdot d\mathbf{S} = \frac{I}{\pi a^2} \iint \rho \, d\phi \, d\rho = \frac{I}{\pi a^2} \pi \rho^2 = \frac{I\rho^2}{a^2}$$

Hence eq. (7.24) becomes

$$H_\phi \int dl = H_\phi \, 2\pi\rho = \frac{I\rho^2}{a^2}$$

or

$$H_\phi = \frac{I\rho}{2\pi a^2} \tag{7.25}$$

For region $a \leq \rho \leq b$, we use path L_2 as the Amperian path,

$$\oint_{L_2} \mathbf{H} \cdot d\mathbf{l} = I_{\text{enc}} = I$$

$$H_\phi 2\pi\rho = I$$

or

$$H_\phi = \frac{I}{2\pi\rho} \tag{7.26}$$

since the whole current I is enclosed by L_2. For region $b \leq \rho \leq b + t$, we use path L_3, getting

$$\oint \mathbf{H} \cdot d\mathbf{l} = H_\phi \cdot 2\pi\rho = I_{enc} \tag{7.27a}$$

where

$$I_{enc} = I + \int \mathbf{J} \cdot d\mathbf{S}$$

and $\mathbf{J}$ in this case is the current density (current per unit area) of the outer conductor and is along $-\mathbf{a}_z$, that is,

$$\mathbf{J} = -\frac{I}{\pi[(b + t)^2 - t^2]}\mathbf{a}_z$$

Thus

$$\begin{aligned} I_{enc} &= I - \frac{I}{\pi[(b + t)^2 - t^2]} \int_{\phi = 0}^{2\pi} \int_{\rho = b}^{\rho} \rho \, d\rho \, d\phi \\ &= I\left[1 - \frac{\rho^2 - b^2}{t^2 + 2bt}\right] \end{aligned}$$

Substituting this in eq. (7.27a), we have

$$H_\phi = \frac{I}{2\pi\rho}\left[1 - \frac{\rho^2 - b^2}{t^2 + 2bt}\right] \tag{7.27b}$$

For region $\rho \geq b + t$, we use path L_4, getting

$$\oint_{L_4} \mathbf{H} \cdot d\mathbf{l} = I - I = 0$$

or

$$H_\phi = 0 \tag{7.28}$$

Putting eqs. (7.25) to (7.28) together gives

$$\mathbf{H} = \begin{cases} \dfrac{I\rho}{2\pi a^2}\mathbf{a}_\phi, & 0 \leq \rho \leq a \\ \dfrac{I}{2\pi\rho}\mathbf{a}_\phi, & a \leq \rho \leq b \\ \dfrac{I}{2\pi\rho}\left[1 - \dfrac{\rho^2 - b^2}{t^2 + 2bt}\right]\mathbf{a}_\phi, & b \leq \rho \leq b + t \\ 0, & \rho \geq b + t \end{cases} \tag{7.29}$$

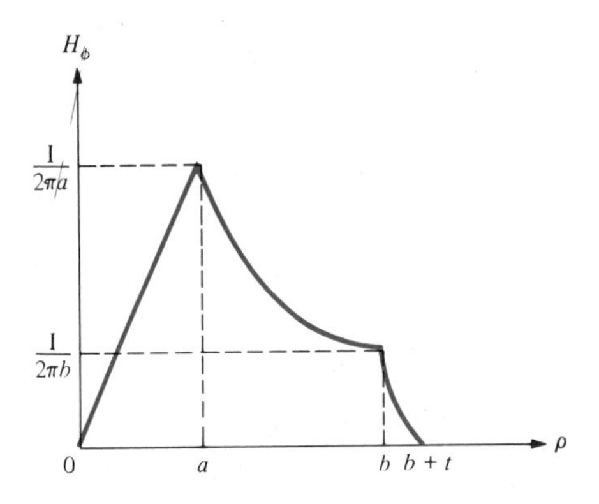

Figure 7.13 Plot of H_ϕ against ρ.

The magnitude of **H** is sketched in Figure 7.13.

Notice from these examples that the ability to take **H** from under the integral sign is the key to using Ampere's law to determine **H**. In other words, Ampere's law can only be used to find **H** due to symmetric current distributions for which it is possible to find a closed path over which **H** is constant in magnitude.

EXAMPLE 7.5

Planes $z = 0$ and $z = 4$ carry current $\mathbf{K} = -10\mathbf{a}_x$ A/m and $\mathbf{K} = 10\mathbf{a}_x$ A/m respectively. Determine **H** at

(a) (1, 1, 1)

(b) (0, −3, 10)

SOLUTION Let the parallel current sheets be as in Figure 7.14. Also let

$$\mathbf{H} = \mathbf{H}_o + \mathbf{H}_4$$

where $\mathbf{H}_o$ and $\mathbf{H}_4$ are the contributions due to the current sheets $z = 0$ and $z = 4$ respectively. We make use of eq. (7.23).

(a) At (1, 1, 1), which is between the plates $(0 < z = 1 < 4)$,

$$\mathbf{H}_o = 1/2\ \mathbf{K} \times \mathbf{a}_n = 1/2\ (-10\mathbf{a}_x) \times \mathbf{a}_z = 5\mathbf{a}_y \text{ A/m}$$

$$\mathbf{H}_4 = 1/2\ \mathbf{K} \times \mathbf{a}_n = 1/2\ (10\mathbf{a}_x) \times (-\mathbf{a}_z) = 5\mathbf{a}_y \text{ A/m}$$

Hence,

$$\mathbf{H} = 10\mathbf{a}_y \text{ A/m}$$

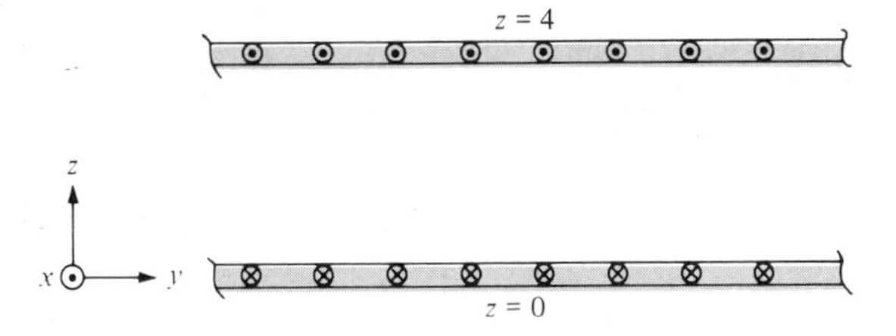

Figure 7.14 For Example 7.5; parallel infinite current sheets.

(b) At (0, −3, 10), which is above the two sheets ($z = 10 > 4 > 0$),

$$\mathbf{H}_o = 1/2\,(-10\mathbf{a}_x) \times \mathbf{a}_z = 5\mathbf{a}_y \text{ A/m}$$

$$\mathbf{H}_4 = 1/2\,(10\mathbf{a}_x) \times \mathbf{a}_z = -5\mathbf{a}_y \text{ A/m}$$

Hence,

$$\mathbf{H} = 0 \text{ A/m}$$

PRACTICE EXERCISE 7.5

Plane $y = 1$ carries current $\mathbf{K} = 50\mathbf{a}_z$ mA/m. Find $\mathbf{H}$ at

(a) (0, 0, 0)

(b) (1, 5, −3)

ANSWER (a) $25\mathbf{a}_x$ mA/m, (b) $-25\mathbf{a}_x$ mA/m.

EXAMPLE 7.6

A toroid whose dimensions are shown in Figure 7.15 has N turns and carries current I. Determine H inside and outside the toroid.

SOLUTION We apply Ampere's circuit law to the Amperian path, which is a circle of radius ρ shown dotted in Figure 7.15. Since N wires cut through this path each carrying current I, the net current enclosed by the Amperian path is NI. Hence,

$$\oint \mathbf{H} \cdot d\mathbf{l} = I_{\text{enc}} \rightarrow H \cdot 2\pi\rho = NI$$

or

$$H = \frac{NI}{2\pi\rho}, \qquad \text{for } \rho_o - a < \rho < \rho_o + a$$

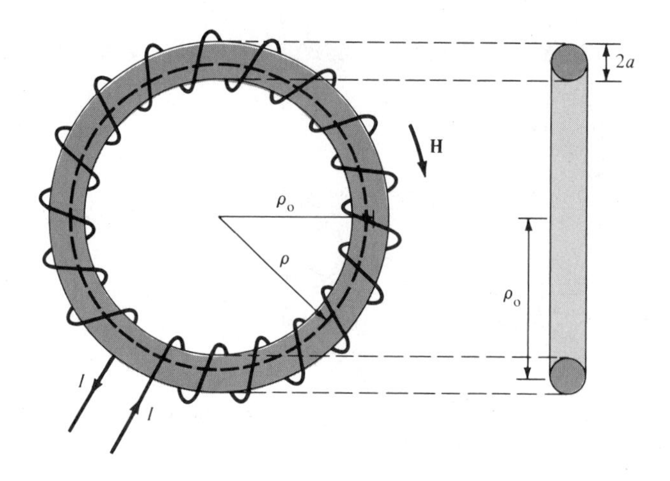

Figure 7.15 For Example 7.6; a toroid with a circular cross section.

where ρ_o is the mean radius of the toroid as shown in Figure 7.15. An approximate value of H is

$$H_{\text{approx}} = \frac{NI}{2\pi\rho_o} = \frac{NI}{\ell}$$

Notice that this is the same as the formula obtained for H for points well inside a very long solenoid ($\ell \gg a$). Thus a straight solenoid may be regarded as a special toroidal coil for which $\rho_o \to \infty$. Outside the toroid, the current enclosed by an Amperian path is $NI - NI = 0$ and hence $H = 0$. ■

PRACTICE EXERCISE 7.6

A toroid of circular cross section whose center is at the origin and axis the same as the z-axis has 1000 turns with $\rho_o = 10$ cm, $a = 1$ cm. If the toroid carries a 100-mA current, find $|\mathbf{H}|$ at

(a) (3 cm, −4 cm, 0)

(b) (6 cm, 9 cm, 0)

ANSWER (a) 0, (b) 147.1 A/m.

7.5 MAGNETIC FLUX DENSITY—MAXWELL'S EQUATION

The magnetic flux density **B** is similar to the electric flux density **D**. As $\mathbf{D} = \varepsilon_o \mathbf{E}$ in free space, the magnetic flux density **B** is related to the magnetic field intensity **H** according to

$$\mathbf{B} = \mu_o \mathbf{H} \tag{7.30}$$

where μ_o is a constant known as the *permeability of free space*. The constant is in henrys/meter (H/m) and has the value of

$$\mu_o = 4\pi \times 10^{-7} \text{ H/m} \tag{7.31}$$

The precise definition of the magnetic field **B**, in terms of the magnetic force, will be given in the next chapter.

The magnetic flux through a surface S is given by

$$\Psi = \int_S \mathbf{B} \cdot d\mathbf{S} \tag{7.32}$$

where the magnetic flux Ψ is in webers (Wb) and the magnetic flux density is in webers/square meter (Wb/m^2) or teslas.

The magnetic flux line is the path to which **B** is tangential at every point in a magnetic field. It is the line along which the needle of a magnetic compass will orient itself if placed in the magnetic field. For example, the magnetic flux lines due to a straight long wire are shown in Figure 7.16. The flux lines are determined using the same principle followed in Section 4.10 for the electric flux lines. The direction of **B** is taken as that indicated as "north" by the needle of the magnetic compass. Notice that each flux line is closed and has no beginning or end. Though Figure 7.16 is for a straight, current-carrying conductor, it is generally true that magnetic flux lines are closed and do not cross each other regardless of the circuit geometry.

In an electrostatic field, the flux passing through a closed surface is the same as the charge enclosed; that is, $\Psi = \oint \mathbf{D} \cdot d\mathbf{S} = Q$. Thus it is possible to have an isolated electric charge as shown in Figure 7.17(a), which also reveals that electric flux lines are not necessarily closed. Unlike electric flux lines, magnetic flux lines always close upon themselves as in Figure 7.17(b). This is due to the fact that *it is not possible to have isolated magnetic poles (or magnetic charges)*. For example, if we desire to have an isolated magnetic pole by dividing a magnetic bar successively into two, we end up

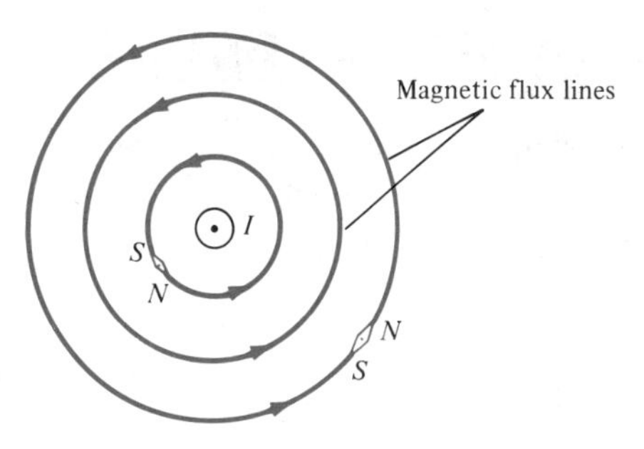

Figure 7.16 Magnetic flux lines due to a straight wire with current coming out of the page.

with pieces each having north and south poles as illustrated in Figure 7.18. We find it impossible to separate the north pole from the south pole. In other words, an isolated magnetic charge does not exist. Thus the total flux through a closed surface in a magnetic field must be zero; that is,

$$\oint \mathbf{B} \cdot d\mathbf{S} = 0 \tag{7.33}$$

This equation is referred to as the *law of conservation of magnetic flux* or *Gauss's law for magnetostatic fields* just as $\oint \mathbf{D} \cdot d\mathbf{S} = Q$ is Gauss's law for electrostatic fields. Although the magnetostatic field is not conservative, magnetic flux is conserved.

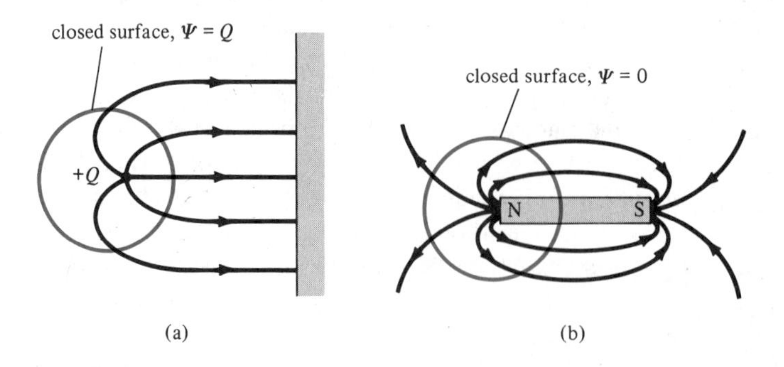

Figure 7.17 Flux leaving a closed surface due to: **(a)** isolated electric charge $\Psi = \oint_S \mathbf{D} \cdot d\mathbf{S} = Q$, **(b)** magnetic charge, $\Psi = \oint_S \mathbf{B} \cdot d\mathbf{S} = 0$.

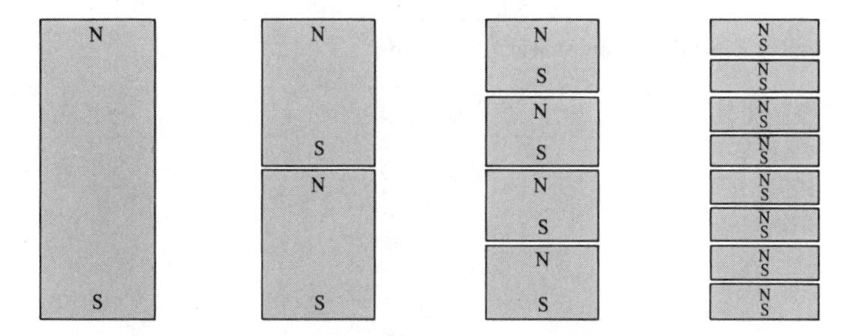

Figure 7.18 Successive division of a bar magnet results in pieces with north and south poles, showing that magnetic poles cannot be isolated.

By applying the divergence theorem to eq. (7.33), we obtain

$$\oint_S \mathbf{B} \cdot d\mathbf{S} = \int_v \nabla \cdot \mathbf{B} \, dv = 0$$

or

$$\boxed{\nabla \cdot \mathbf{B} = 0} \tag{7.34}$$

This equation is the fourth Maxwell's equation to be derived. Equation (7.33) or (7.34) shows that magnetostatic fields have no sources or sinks. Equation (7.34) suggests that magnetic field lines are always continuous. For a mathematical proof of this equation, see Problem 7.28.

7.6 MAXWELL'S EQUATIONS FOR STATIC EM FIELDS

Having derived Maxwell's four equations for static electromagnetic fields, we may take a moment to put them together as in Table 7.2. From the table, we notice that the order in which the equations were derived has been changed for the sake of clarity.

The choice between differential and integral forms of the equations depends on a given problem. It is evident from Table 7.2 that a vector field is defined completely by specifying its curl and divergence. A field can only be electric or magnetic if it satisfies the corresponding Maxwell's equations (see Problem 7.27). It should be noted that Maxwell's equations as in Table 7.2 are only for static EM fields. As will be discussed in Chapter 9, the divergence equations will remain the same for time-varying EM fields but the curl equations will have to be modified.

Table 7.2 Maxwell's Equations for Static EM Fields

Differential (or point) form	Integral form	Remarks
$\nabla \cdot \mathbf{D} = \rho_v$	$\oint_S \mathbf{D} \cdot d\mathbf{S} = \int_v \rho_v \, dv$	Gauss's law
$\nabla \cdot \mathbf{B} = 0$	$\oint_S \mathbf{B} \cdot d\mathbf{S} = 0$	Nonexistence of magnetic monopole
$\nabla \times \mathbf{E} = 0$	$\oint_L \mathbf{E} \cdot d\mathbf{l} = 0$	Conservativeness of electrostatic field
$\nabla \times \mathbf{H} = \mathbf{J}$	$\oint_L \mathbf{H} \cdot d\mathbf{l} = \int_S \mathbf{J} \cdot d\mathbf{S}$	Ampere's law

7.7 MAGNETIC SCALAR AND VECTOR POTENTIALS

We recall that some electrostatic field problems were simplified by relating the electric potential V to the electric field intensity $\mathbf{E}$ ($\mathbf{E} = -\nabla V$). Similarly, we can define a potential associated with magnetostatic field $\mathbf{B}$. In fact, the magnetic potential could be scalar V_m or vector $\mathbf{A}$. To define V_m and $\mathbf{A}$ involves recalling two important identities (see Example 3.9 and Practice Exercise 3.9):

$$\nabla \times (\nabla V) = 0 \quad [7.35a]$$

$$\nabla \cdot (\nabla \times \mathbf{A}) = 0 \quad [7.35b]$$

which must always hold for any scalar field V and vector field $\mathbf{A}$.

Just as $\mathbf{E} = -\nabla V$, we define the *magnetic scalar potential* V_m (in amperes) as related to $\mathbf{H}$ according to

$$\boxed{\mathbf{H} = -\nabla V_m} \qquad \text{if } \mathbf{J} = 0 \quad [7.36]$$

The condition attached to this equation is important and will be explained. Combining eq. (7.36) and eq. (7.19) gives

$$\mathbf{J} = \nabla \times \mathbf{H} = \nabla \times (-\nabla V_m) = 0 \quad [7.37]$$

since V_m must satisfy the condition in eq. (7.35a). Thus the magnetic scalar potential V_m is only defined in a region where $\mathbf{J} = 0$ as in eq. (7.36). We should also note that V_m satisfies Laplace's equation just as V does for electrostatic fields; hence,

$$\nabla^2 V_m = 0, \qquad (\mathbf{J} = 0) \quad [7.38]$$

We know that for a magnetostatic field, $\nabla \cdot \mathbf{B} = 0$ as stated in eq. (7.34). In order to satisfy eqs. (7.34) and (7.35b) simultaneously, we can define the *vector magnetic potential* **A** (in Wb/m) such that

$$\boxed{\mathbf{B} = \nabla \times \mathbf{A}} \tag{7.39}$$

Just as we defined

$$V = \int \frac{dQ}{4\pi\varepsilon_o r} \tag{7.40}$$

we can define

$$\boxed{\mathbf{A} = \int_L \frac{\mu_o I\, d\mathbf{l}}{4\pi R}} \quad \text{for line current} \tag{7.41}$$

$$\boxed{\mathbf{A} = \int_S \frac{\mu_o \mathbf{K}\, dS}{4\pi R}} \quad \text{for surface current} \tag{7.42}$$

$$\boxed{\mathbf{A} = \int_v \frac{\mu_o \mathbf{J}\, dv}{4\pi R}} \quad \text{for volume current} \tag{7.43}$$

Rather than obtaining eqs. (7.41) to (7.43) from eq. (7.40), an alternative approach would be to obtain eqs. (7.41) to 7.43) from eqs. (7.6) to (7.8). For example, we can derive eq. (7.41) from eq. (7.6) in conjunction with eq. (7.39). To do this, we write eq. (7.6) as

$$\mathbf{B} = \frac{\mu_o}{4\pi} \int_L \frac{I\, d\mathbf{l}' \times \mathbf{R}}{R^3} \tag{7.44}$$

where **R** is the distance vector from the line element $d\mathbf{l}'$ at the source point (x', y', z') to the field point (x, y, z) as shown in Figure 7.19 and $R = |\mathbf{R}|$, that is,

$$R = |\mathbf{r} - \mathbf{r}'| = [(x - x')^2 + (y - y')^2 + (z - z')^2]^{1/2} \tag{7.45}$$

Hence,

$$\nabla\left(\frac{1}{R}\right) = -\frac{(x - x')\mathbf{a}_x + (y - y')\mathbf{a}_y + (z - z')\mathbf{a}_z}{[(x - x')^2 + (y - y')^2 + (z - z')^2]^{3/2}} = -\frac{\mathbf{R}}{R^3}$$

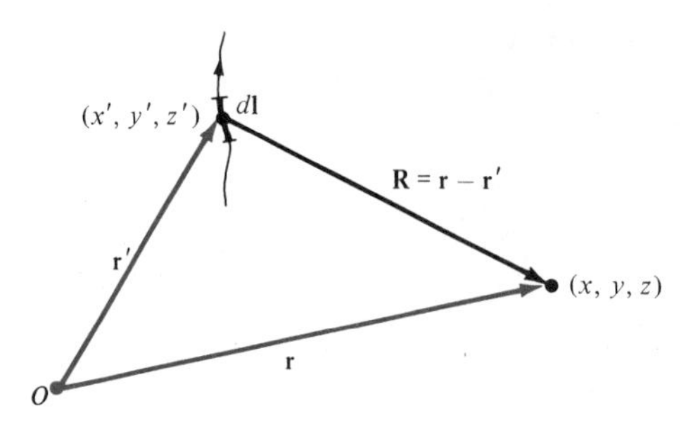

Figure 7.19 Illustration of the source point (x', y', z') and the field point (x, y, z).

or

$$\frac{\mathbf{R}}{R^3} = -\nabla\left(\frac{1}{R}\right) \quad \left(= \frac{\mathbf{a}_R}{R^2}\right) \tag{7.46}$$

Substituting this into eq. (7.44), we obtain

$$\mathbf{B} = -\frac{\mu_o}{4\pi}\int_L I\, d\mathbf{l}' \times \nabla\left(\frac{1}{R}\right) \tag{7.47}$$

We apply the vector identity

$$\nabla \times (f\mathbf{F}) = f\nabla \times \mathbf{F} + (\nabla f) \times \mathbf{F} \tag{7.48}$$

where f is a scalar field and $\mathbf{F}$ is a vector field. Taking $f = 1/R$ and $\mathbf{F} = d\mathbf{l}'$, we have

$$d\mathbf{l}' \times \nabla\left(\frac{1}{R}\right) = \frac{1}{R}\nabla \times d\mathbf{l}' - \nabla \times \left(\frac{d\mathbf{l}'}{R}\right)$$

Since ∇ operates with respect to (x, y, z) while $d\mathbf{l}'$ is a function of (x', y', z'), $\nabla \times d\mathbf{l}' = 0$. Hence,

$$d\mathbf{l}' \times \nabla\left(\frac{1}{R}\right) = -\nabla \times \frac{d\mathbf{l}'}{R} \tag{7.49}$$

With this equation, eq. (7.47) reduces to

$$\mathbf{B} = \nabla \times \int_L \frac{\mu_o I\, d\mathbf{l}'}{4\pi R} \tag{7.50}$$

Comparing eq. (7.50) with eq. (7.39) shows that

$$\mathbf{A} = \int_L \frac{\mu_o I\, d\mathbf{l}'}{4\pi R}$$

verifying eq. (7.41).

By substituting eq. (7.39) into eq. (7.32) and applying Stokes's theorem, we obtain

$$\Psi = \int_S \mathbf{B} \cdot d\mathbf{S} = \int_S (\nabla \times \mathbf{A}) \cdot d\mathbf{S} = \oint_L \mathbf{A} \cdot d\mathbf{l}$$

or

$$\boxed{\Psi = \oint_L \mathbf{A} \cdot d\mathbf{l}} \qquad [7.51]$$

Thus the magnetic flux through a given area can be found using either eq. (7.32) or (7.51). Also, the magnetic field can be determined using either V_m or $\mathbf{A}$; the choice is dictated by the nature of the given problem except that V_m can only be used in a source-free region. The use of the magnetic vector potential provides a powerful, elegant approach to solving EM problems, particularly those relating to antennas. As we shall notice in Chapter 13, it is more convenient to find **B** by first finding **A** in antenna problems.

EXAMPLE 7.7

Given the magnetic vector potential $\mathbf{A} = -\rho^2/4 \ \mathbf{a}_z$ Wb/m, calculate the total magnetic flux crossing the surface $\phi = \pi/2$, $1 \leq \rho \leq 2$ m, $0 \leq z \leq 5$ m.

SOLUTION

We can solve this problem in two different ways: using eq. (7.32) or eq. (7.51).

Method 1:

$$\mathbf{B} = \nabla \times \mathbf{A} = -\frac{\partial A_z}{\partial \rho} \mathbf{a}_\phi = \frac{\rho}{2} \mathbf{a}_\phi, \qquad d\mathbf{S} = d\rho \, dz \, \mathbf{a}_\phi$$

Hence,

$$\Psi = \int \mathbf{B} \cdot d\mathbf{S} = \frac{1}{2} \int_{z=0}^{5} \int_{\rho=1}^{2} \rho \, d\rho \, dz = \frac{1}{4} \rho^2 \Big|_1^2 (5) = \frac{15}{4}$$

$$\Psi = 3.75 \text{ Wb}$$

Method 2:
We use

$$\Psi = \oint_L \mathbf{A} \cdot d\mathbf{l} = \Psi_1 + \Psi_2 + \Psi_3 + \Psi_4$$

where L is the path bounding surface S; Ψ_1, Ψ_2, Ψ_3, and Ψ_4 are respectively the evaluations of $\int \mathbf{A} \cdot d\mathbf{l}$ along the segments of L labeled 1 to 4 in Figure 7.20. Since **A** has only a z-component,

$$\Psi_1 = 0 = \Psi_3$$

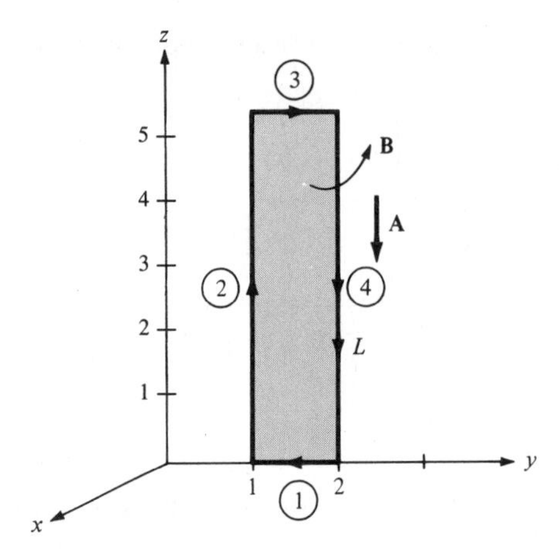

Figure 7.20 For Example 7.7.

That is,

$$\Psi = \Psi_2 + \Psi_4 = -\frac{1}{4}\left[(1)^2 \int_0^5 dz + (2)^2 \int_5^0 dz\right]$$

$$= -\frac{1}{4}(1 - 4)(5) = \frac{15}{4}$$

$$= 3.75 \text{ Wb}$$

as obtained previously. Note that the direction of the path L must agree with that of $d\mathbf{S}$. ■

PRACTICE EXERCISE 7.7

A current distribution gives rise to the vector magnetic potential $\mathbf{A} = x^2y\mathbf{a}_x + y^2x\mathbf{a}_y - 4xyz\mathbf{a}_z$ Wb/m. Calculate

(a) $\mathbf{B}$ at $(-1, 2, 5)$

(b) The flux through the surface defined by $z = 1, 0 \leq x \leq 1, -1 \leq y \leq 4$

ANSWER (a) $20\mathbf{a}_x + 40\mathbf{a}_y + 3\mathbf{a}_z$ Wb/m^2, (b) 20 Wb.

EXAMPLE 7.8

If plane $z = 0$ carries uniform current $\mathbf{K} = K_y\mathbf{a}_y$,

$$\mathbf{H} = \begin{cases} 1/2\ K_y\mathbf{a}_x, & z > 0 \\ -\ 1/2\ K_y\mathbf{a}_x, & z < 0 \end{cases}$$

This was obtained in Section 7.4 using Ampere's law. Obtain this by using the concept of vector magnetic potential.

SOLUTION

Consider the current sheet as in Figure 7.21. From eq. (7.42),

$$d\mathbf{A} = \frac{\mu_o\mathbf{K}\ dS}{4\pi R}$$

In this problem, $\mathbf{K} = K_y\mathbf{a}_y$, $dS = dx'dy'$, and for $z > 0$,

$$R = |\mathbf{R}| = |\ (0, 0, z) - (x', y', 0)\ |$$
$$= [(x')^2 + (y')^2 + z^2]^{1/2} \tag{7.8.1}$$

where the primed coordinates are for the source point while the unprimed coordinates are for the field point. It is necessary (and customary) to distinguish between the two points to avoid confusion (see Figure 7.19). Hence

$$d\mathbf{A} = \frac{\mu_o K_y\ dx'\ dy'\ \mathbf{a}_y}{4\pi[(x')^2 + (y')^2 + z^2]^{1/2}}$$

$$d\mathbf{B} = \nabla \times d\mathbf{A} = -\frac{\partial}{\partial z} dA_y\ \mathbf{a}_x$$

$$= \frac{\mu_o K_y z\ dx'\ dy'\ \mathbf{a}_x}{4\pi[(x')^2 + (y')^2 + z^2]^{3/2}}$$

$$\mathbf{B} = \frac{\mu_o K_y z\mathbf{a}_x}{4\pi} \int_{-\infty}^{\infty} \int_{-\infty}^{\infty} \frac{dx'\ dy'}{[(x')^2 + (y')^2 + z^2]^{3/2}} \tag{7.8.2}$$

In the integrand, we may change coordinates from Cartesian to cylindrical for convenience so that

$$\mathbf{B} = \frac{\mu_o K_y z\mathbf{a}_x}{4\pi} \int_{\rho'=0}^{\infty} \int_{\phi'=0}^{2\pi} \frac{\rho'\ d\phi'\ d\rho'}{[(\rho')^2 + z^2]^{3/2}}$$

$$= \frac{\mu_o K_y z\mathbf{a}_x}{4\pi} 2\pi \int_0^{\infty} [(\rho')^2 + z^2]^{-3/2}\ 1/2\ d[(\rho')^2]$$

$$= \frac{\mu_o K_y z\mathbf{a}_x}{2} \left. \frac{-1}{[(\rho')^2 + z^2)^{1/2}} \right|_{\rho'=0}^{\infty}$$

$$= \frac{\mu_o K_y \mathbf{a}_x}{2}$$

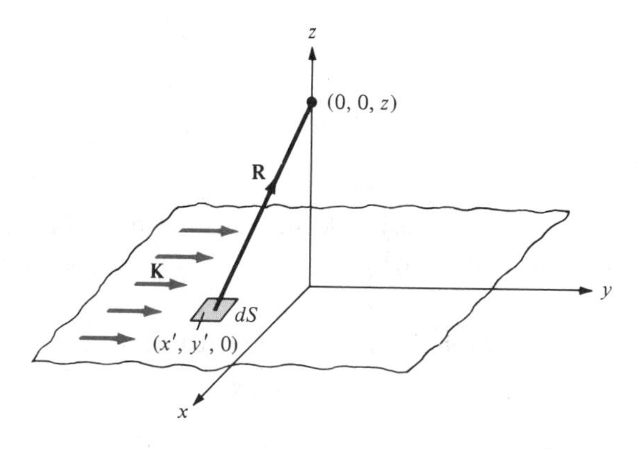

Figure 7.21 For Example 7.8; infinite current sheet.

Hence

$$\mathbf{H} = \frac{\mathbf{B}}{\mu_o} = \frac{K_y}{2}\mathbf{a}_x, \qquad \text{for } z > 0$$

By simply replacing z by $-z$ in eq. (7.8.2) and following the same procedure, we obtain

$$\mathbf{H} = -\frac{K_y}{2}\mathbf{a}_x, \qquad \text{for } z < 0$$

■

PRACTICE EXERCISE 7.8

Repeat Example 7.8 by using Biot-Savart's law to determine **H** at points (0, 0, h) and (0, 0, $-h$).

†7.8 DERIVATION OF BIOT-SAVART'S LAW AND AMPERE'S LAW

Both Biot-Savart's law and Ampere's law may be derived using the concept of magnetic vector potential. The derivation will involve the use of the vector identities in eq. (7.48) and

$$\nabla \times \nabla \times \mathbf{A} = \nabla(\nabla \cdot \mathbf{A}) - \nabla^2\mathbf{A} \tag{7.52}$$

Since Biot-Savart's law as given in eq. (7.4) is basically on line current, we begin our derivation with eqs. (7.39) and (7.41); that is,

$$\mathbf{B} = \nabla \times \oint_L \frac{\mu_o I\, d\mathbf{l}'}{4\pi R} = \frac{\mu_o I}{4\pi}\oint_L \nabla \times \frac{1}{R}\, d\mathbf{l}' \tag{7.53}$$

where R is as defined in eq. (7.45). If the vector identity in eq. (7.48) is applied by letting $\mathbf{F} = d\mathbf{l}$ and $f = 1/R$, eq. (7.53) becomes

$$\mathbf{B} = \frac{\mu_o I}{4\pi} \oint_L \left[\frac{1}{R} \nabla \times d\mathbf{l}' + \left(\nabla \frac{1}{R} \right) \times d\mathbf{l}' \right] \tag{7.54}$$

Since ∇ operates with respect to (x, y, z) and $d\mathbf{l}'$ is a function of (x', y', z'), $\nabla \times d\mathbf{l}' = 0$. Also

$$\frac{1}{R} = [(x - x')^2 + (y - y')^2 + (z - z')^2]^{-1/2} \tag{7.55}$$

$$\nabla \left[\frac{1}{R} \right] = - \frac{(x - x')\mathbf{a}_x + (y - y')\mathbf{a}_y + (z - z')\mathbf{a}_z}{[(x - x')^2 + (y - y')^2 + (z - z')^2]^{3/2}} = - \frac{\mathbf{a}_R}{R^2} \tag{7.56}$$

where $\mathbf{a}_R$ is a unit vector from the source point to the field point. Thus eq. (7.54) (upon dropping the prime in $d\mathbf{l}'$) becomes

$$\mathbf{B} = \frac{\mu_o I}{4\pi} \oint_L \frac{d\mathbf{l} \times \mathbf{a}_R}{R^2} \tag{7.57}$$

which is Biot-Savart's law.

Using the identity in eq. (7.52) with eq. (7.39), we obtain

$$\nabla \times \mathbf{B} = \nabla (\nabla \cdot \mathbf{A}) - \nabla^2 \mathbf{A} \tag{7.58}$$

It can be shown that for a static magnetic field

$$\boxed{\nabla \cdot \mathbf{A} = 0} \tag{7.59}$$

so that upon replacing $\mathbf{B}$ with $\mu_o \mathbf{H}$ and using eq. (7.19), eq. (7.58) becomes

$$\nabla^2 \mathbf{A} = -\mu_o \nabla \times \mathbf{H}$$

or

$$\boxed{\nabla^2 \mathbf{A} = -\mu_o \mathbf{J}} \tag{7.60}$$

which is called the *vector Poisson's equation*. It is similar to Poisson's equation ($\nabla^2 V = -\rho_v/\varepsilon$) in electrostatics. In Cartesian coordinates, eq. (7.60) may be decomposed into three scalar equations:

$$\begin{aligned} \nabla^2 A_x &= -\mu_o J_x \\ \nabla^2 A_y &= -\mu_o J_y \\ \nabla^2 A_z &= -\mu_o J_z \end{aligned} \tag{7.61}$$

which may be regarded as the *scalar Poisson's equations*.

It can also be shown that Ampere's circuit law is consistent with our definition of the magnetic vector potential. From Stokes's theorem and eq. (7.39),

$$\oint_L \mathbf{H} \cdot d\mathbf{l} = \int_S \nabla \times \mathbf{H} \cdot d\mathbf{S}$$

$$= \frac{1}{\mu_o} \int_S \nabla \times (\nabla \times \mathbf{A}) \cdot d\mathbf{S} \tag{7.62}$$

From eqs. (7.52), (7.59), and (7.60),

$$\nabla \times \nabla \times \mathbf{A} = -\nabla^2 \mathbf{A} = \mu_o \mathbf{J}$$

Substituting this into eq. (7.62) yields

$$\oint_L \mathbf{H} \cdot d\mathbf{l} = \int_S \mathbf{J} \cdot d\mathbf{S} = I$$

which is Ampere's circuit law.

SUMMARY

1. The basic laws (Biot-Savart's and Ampere's) that govern magnetostatic fields are discussed. Biot-Savart's law, which is similar to Coulomb's law, states that the magnetic field intensity $d\mathbf{H}$ at $\mathbf{r}$ due to current element $I\, d\mathbf{l}$ at $\mathbf{r}'$ is

$$d\mathbf{H} = \frac{I\, d\mathbf{l} \times \mathbf{R}}{4\pi R^3} \qquad \text{(in A/m)}$$

where $\mathbf{R} = \mathbf{r} - \mathbf{r}'$ and $R = |\mathbf{R}|$. For surface or volume current distribution, we replace $I\, d\mathbf{l}$ with $\mathbf{K}\, dS$ or $\mathbf{J}\, dv$ respectively; that is,

$$I\, d\mathbf{l} = \mathbf{K}\, dS = \mathbf{J}\, dv$$

2. Ampere's circuit law, which is similar to Gauss's law, states that the circulation of $\mathbf{H}$ around a closed path is equal to the current enclosed by the path; that is,

$$\oint \mathbf{H} \cdot d\mathbf{l} = I_{enc} = \int \mathbf{J} \cdot d\mathbf{S}$$

or

$$\nabla \times \mathbf{H} = \mathbf{J} \qquad \text{(3rd Maxwell's equation to be derived).}$$

When current distribution is symmetric so that an Amperian path (on which $\mathbf{H} = H_\phi \mathbf{a}_\phi$ is constant) can be found, Ampere's law is useful in determining $\mathbf{H}$; that is,

$$H_\phi \oint dl = I_{enc} \qquad \text{or} \qquad H_\phi = \frac{I_{enc}}{\ell}$$

3. The magnetic flux through a surface S is given by

$$\Psi = \int_S \mathbf{B} \cdot d\mathbf{S} \qquad \text{(in Wb)}$$

where **B** is the magnetic flux density in Wb/m^2. In free space,

$$\mathbf{B} = \mu_o \mathbf{H}$$

where $\mu_o = 4\pi \times 10^{-7}$ H/m = permeability of free space.

4. Since an isolated or free magnetic monopole does not exist, the net magnetic flux through a closed surface is zero;

$$\Psi = \oint \mathbf{B} \cdot d\mathbf{S} = 0$$

or

$$\nabla \cdot \mathbf{B} = 0 \qquad \text{(4th Maxwell's equation to be derived).}$$

5. At this point, all four Maxwell's equations for static EM fields have been derived, namely:

$$\nabla \cdot \mathbf{D} = \rho_v$$

$$\nabla \cdot \mathbf{B} = 0$$

$$\nabla \times \mathbf{E} = 0$$

$$\nabla \times \mathbf{H} = \mathbf{J}$$

6. The magnetic scalar potential V_m is defined as

$$\mathbf{H} = -\nabla V_m \qquad \text{if } \mathbf{J} = 0$$

and the magnetic vector potential **A** as

$$\mathbf{B} = \nabla \times \mathbf{A}$$

where $\nabla \cdot \mathbf{A} = 0$. With the definition of **A**, the magnetic flux through a surface S can be found from

$$\Psi = \oint_L \mathbf{A} \cdot d\mathbf{l}$$

where L is the closed path defining surface S (see Figure 3.13). Rather than using Biot-Savart's law, the magnetic field due to a current distribution may be found using **A**, a powerful approach that is particularly useful in antenna theory. For a current element $I\, d\mathbf{l}$ at $\mathbf{r}'$, the magnetic vector potential at **r** is

$$\mathbf{A} = \int \frac{\mu_o I\, d\mathbf{l}}{4\pi R}, \qquad R = |\mathbf{r} - \mathbf{r}'|$$

7. Elements of similarity between electric and magnetic fields exist. Some of these are listed in Table 7.1. Corresponding to Poisson's equation $\nabla^2 V = -\rho_v/\varepsilon$, for example, is

$$\nabla^2 \mathbf{A} = -\mu_o \mathbf{J}$$

REVIEW QUESTIONS

7.1 One of the following is not a source of magnetostatic fields:

(a) A dc current in a wire

(b) A permanent magnet

(c) An accelerated charge

(d) An electric field linearly changing with time

(e) A charged disk rotating at uniform speed

7.2 Identify the configuration in Figure 7.22 that is not a correct representation of I and $\mathbf{H}$.

7.3 Consider points A, B, C, D, and E on a circle of radius 2 as shown in Figure 7.23. The items in the right list are the values of $\mathbf{a}_\phi$ at different points on the circle. Match these items with the points in the list on the left.

(a) A

(b) B

(c) C

(d) D

(e) E

(i) $\mathbf{a}_x$

(ii) $-\mathbf{a}_x$

(iii) $\mathbf{a}_y$

(iv) $-\mathbf{a}_y$

(v) $\dfrac{\mathbf{a}_x + \mathbf{a}_y}{\sqrt{2}}$

(vi) $\dfrac{-\mathbf{a}_x - \mathbf{a}_y}{\sqrt{2}}$

(vii) $\dfrac{-\mathbf{a}_x + \mathbf{a}_y}{\sqrt{2}}$

(viii) $\dfrac{\mathbf{a}_x - \mathbf{a}_y}{\sqrt{2}}$

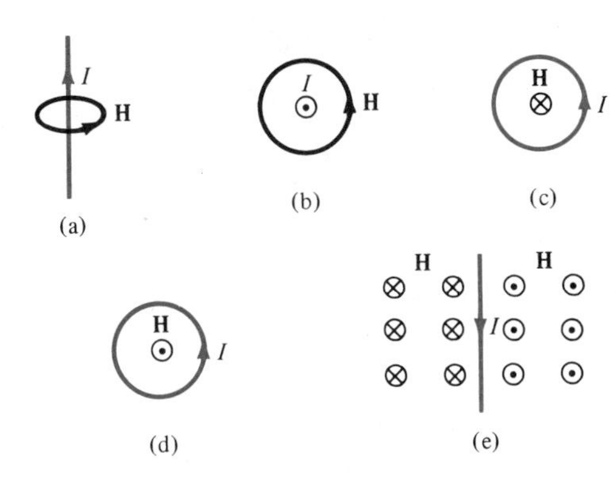

Figure 7.22 For Review Question 7.2.

7.4 The z-axis carries filamentary current of 10π A along $\mathbf{a}_z$. Which of these is incorrect?

(a) $\mathbf{H} = -\mathbf{a}_x$ A/m at (0, 5, 0)

(b) $\mathbf{H} = \mathbf{a}_\phi$ A/m at $(5, \pi/4, 0)$

(c) $\mathbf{H} = -0.8\mathbf{a}_x - 0.6\mathbf{a}_y$ at $(-3, 4, 0)$

(d) $\mathbf{H} = -\mathbf{a}_\phi$ at $(5, 3\pi/2, 0)$

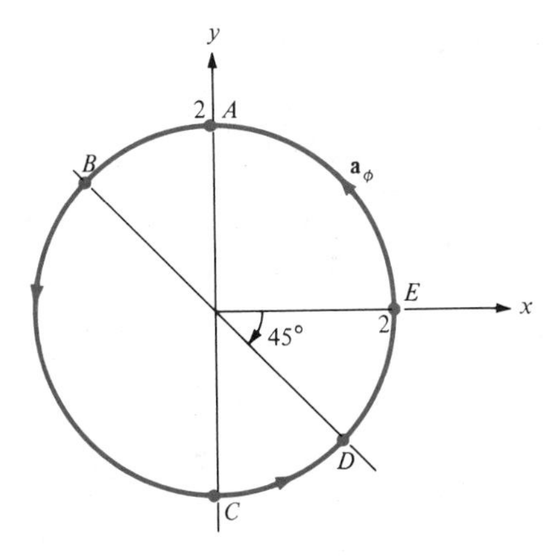

Figure 7.23 For Review Question 7.3.

7.5 Plane $y = 0$ carries a uniform current of $30\mathbf{a}_z$ mA/m. At (1, 10, −2), the magnetic field intensity is

(a) $-15\mathbf{a}_x$ mA/m

(b) $15\mathbf{a}_x$ mA/m

(c) $477.5\mathbf{a}_y$ μA/m

(d) $18.85\mathbf{a}_y$ nA/m

(e) None of the above

7.6 For the currents and closed paths of Figure 7.24, calculate the value of $\oint_L \mathbf{H} \cdot d\mathbf{l}$.

7.7 Which of these statements is not characteristic of a static magnetic field?

(a) It is solenoidal.

(b) It is conservative.

(c) It has no sinks or sources.

(d) Magnetic flux lines are always closed.

(e) The total number of flux lines entering a given region is equal to the total number of flux lines leaving the region.

7.8 Two identical coaxial circular coils carry the same current I but in opposite directions. The magnitude of the magnetic field $\mathbf{B}$ at a point on the axis midway between the coils is

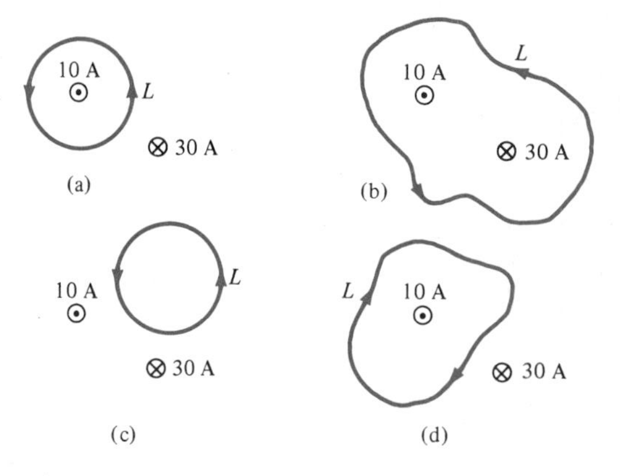

Figure 7.24 For Review Question 7.6.

(a) Zero

(b) The same as that produced by one coil

(c) Twice that produced by one coil

(d) Half that produced by one coil.

7.9 One of these equations is not Maxwell's equation for a static electromagnetic field in a linear homogeneous medium.

(a) $\nabla \cdot \mathbf{B} = 0$

(b) $\nabla \times \mathbf{D} = 0$

(c) $\oint \mathbf{B} \cdot d\mathbf{l} = \mu_o I$

(d) $\oint \mathbf{D} \cdot d\mathbf{S} = Q$

(e) $\nabla^2 \mathbf{A} = \mu_o \mathbf{J}$

7.10 Two bar magnets with their north poles having strength $Q_{m1} = 20$ A · m and $Q_{m2} = 10$ A · m (magnetic charges) are placed inside a volume as shown in Figure 7.25. The magnetic flux leaving the volume is

(a) 200 Wb

(b) 30 Wb

(c) 10 Wb

(d) 0 Wb

(e) −10 Wb

Answers: 7.1c, 7.2c, 7.3 (a)-(ii), (b)-(vi), (c)-(i), (d)-(v), (e)-(iii), 7.4d, 7.5a, 7.6 (a) 10 A, (b) −20 A, (c) 0, (d) −10 A, 7.7b, 7.8a, 7.9e, 7.10d.

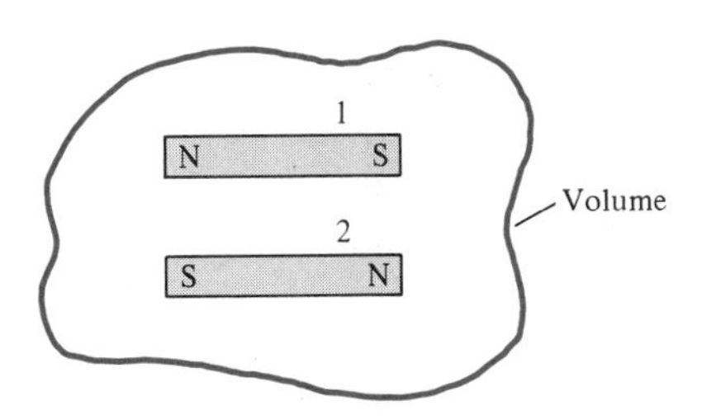

Figure 7.25 For Review Question 7.10.

PROBLEMS

7.1 Calculate **H** at (3m, −6m, 2m) due to a current element of length 2 mm located at the origin in free space that carries current 16 mA in the $+y$ direction.

7.2 Line $x = 0$, $y = 0$, $0 \leq z \leq 10$ m carries current 2 A along $\mathbf{a}_z$. Calculate **H** at points

(a) (5, 0, 0)

(b) (5, 5, 0)

(c) (5, 15, 0)

(d) (5, −15, 0)

7.3 (a) State Biot-Savart's law.

(b) The z- and x-axis respectively carry filamentary currents of 20 A along $\mathbf{a}_z$ and 30 A along $\mathbf{a}_x$. Calculate **H** at (6, 8, −6).

***7.4** (a) Find **H** at (0, 0, 5) due to side 2 of the triangular loop in Figure 7.6(a).

(b) Find **H** at (0, 0, 5) due to the entire loop.

7.5 An infinitely long conductor is bent into an L shape as shown in Figure 7.26. If a direct current of 5 A flows in the current, find the magnetic field intensity at (a) (2, 2, 0), (b) (0, −2, 0), and (c) (0, 0, 2).

7.6 Find **H** at the center C of an equilateral triangular loop of side 4 m carrying 5 A of current as in Figure 7.27.

7.7 A square conducting loop of side $2a$ lies in the $z = 0$ plane and carries a current I in the counterclockwise direction. Show that at the center of the loop

$$\mathbf{H} = \frac{\sqrt{2}I}{\pi a}\mathbf{a}_z$$

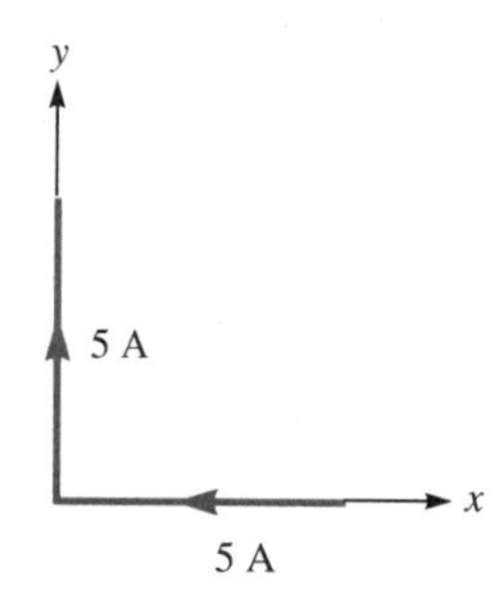

Figure 7.26 Current filament for Problem 7.5.

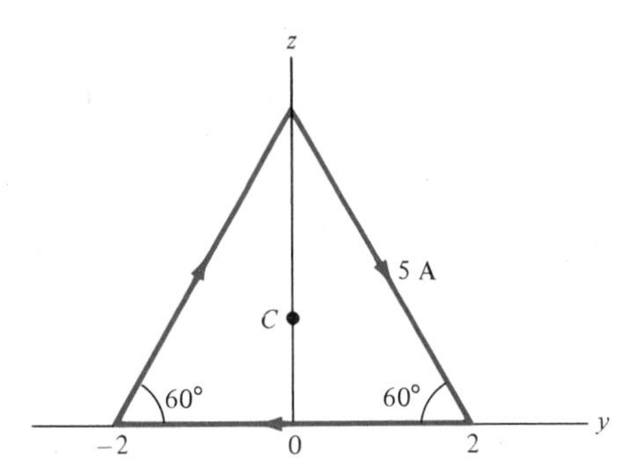

Figure 7.27 Equilateral triangular loop for Problem 7.6.

***7.8** A circuit of thin current-carrying conductor is shown in Figure 7.28. Calculate the magnetic flux density at points $P(0, 2, 0)$ and $Q(2, 0, 0)$.

7.9 A rectangular loop carrying 10 A of current is placed on $z = 0$ plane as shown in Figure 7.29. Evaluate **H** at

(a) (2, 2, 0)

(b) (4, 2, 0)

(c) (4, 8, 0)

(d) (0, 0, 2)

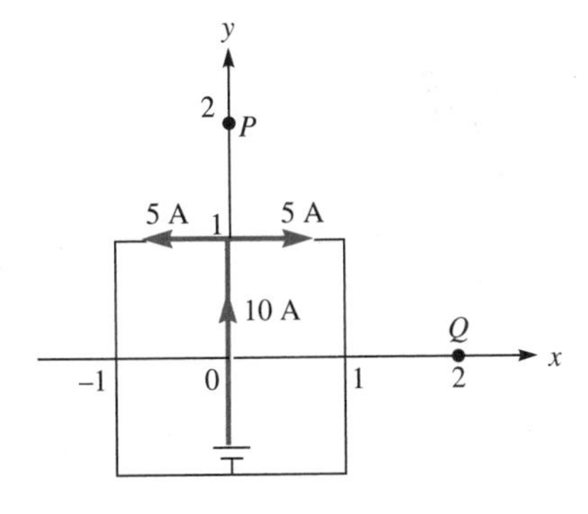

Figure 7.28 For Problem 7.8.

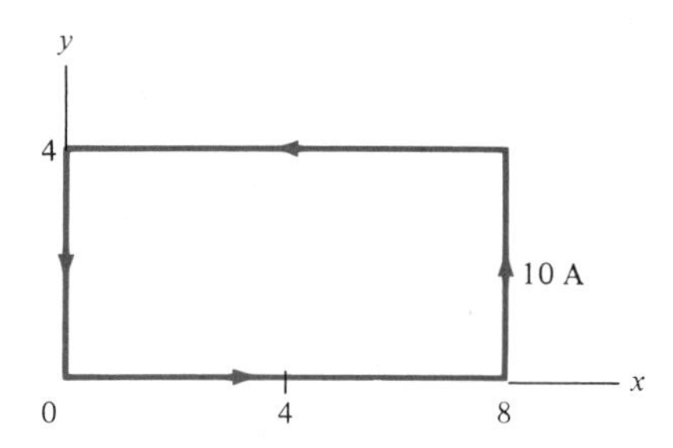

Figure 7.29 Rectangular loop of Problem 7.9.

***7.10** (a) A filamentary loop carrying current I is bent to assume the shape of a regular polygon of n sides. Show that at the center of the polygon

$$H = \frac{nI}{2\pi r} \sin \frac{\pi}{n}$$

where r is the radius of the circle circumscribed by the polygon.

(b) Apply this to cases when $n = 3$ and $n = 4$ and see if your results agree with those for the triangular loop of Problem 7.6 and the square loop of Problem 7.7 respectively.

(c) As n becomes large, show that the result of part (a) becomes that of the circular loop of Example 7.3.

/ **7.11** For the filamentary loop shown in Figure 7.30, find the magnetic field strength at O.

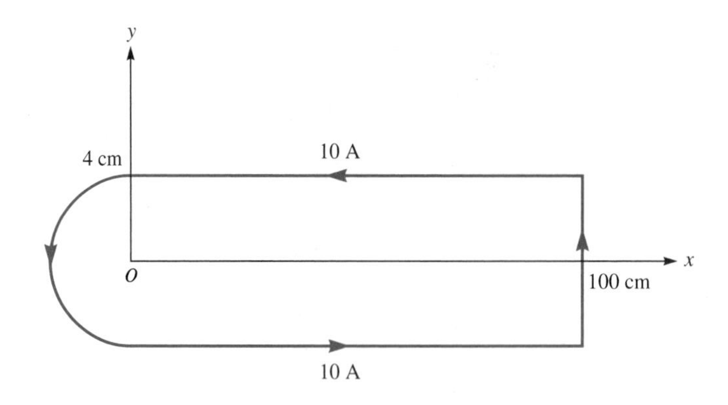

Figure 7.30 Filamentary loop of Problem 7.11; not drawn to scale.

****7.12** A wire in the form of a parabola carries current 3 A. Calculate the magnitude of the magnetic field intensity at its focus if the distance from the focus to the apex (or vertex) is 20 cm.

7.13 Two identical current loops have their centers at (0, 0, 0) and (0, 0, 4) and their axes the same as the z-axis (so that the "Helmholtz coil" is formed). If each loop has radius 2 m and carries current 5 A in $\mathbf{a}_\phi$, calculate $\mathbf{H}$ at

(a) (0, 0, 0)

(b) (0, 0, 2)

7.14 Line $y = 1$, $z = 4$ carries filamentary current 50π mA along $\mathbf{a}_x$ while the $z = 0$ plane carries 20 mA/m along $\mathbf{a}_x$. Find $\mathbf{H}$ at (3, 4, 5).

7.15 (a) State Ampere's circuit law.

(b) An infinitely long wire of radius a is placed along the z-axis and carries current I along $\mathbf{a}_z$. By applying Ampere's circuit law, find $\mathbf{H}$ everywhere. Sketch $|\mathbf{H}|$ as a function of ρ.

7.16 A long circular conductor of radius a carries a current with $\mathbf{J} = J_o\rho\mathbf{a}_z$ A/m², $\rho < a$ when placed along the z-axis. Find $\mathbf{H}$ inside and outside the conductor.

7.17 (a) An infinitely long solid conductor of radius a is placed along the z-axis. If the conductor carries current I in the $+z$ direction, show that

$$\mathbf{H} = \frac{I\rho}{2\pi a^2}\,\mathbf{a}_\phi$$

within the conductor. Find the corresponding current density.

(b) If $I = 3$ A and $a = 2$ cm in part (a), find $\mathbf{H}$ at (0, 1 cm, 0) and (0, 4 cm, 0).

7.18 If $\mathbf{H} = y\mathbf{a}_x - x\mathbf{a}_y$ A/m on plane $z = 0$, (a) determine the current density and (b) verify Ampere's law by taking the circulation of $\mathbf{H}$ around the edge of the rectangle $z = 0$, $0 < x < 3$, $-1 < y < 4$.

7.19 In a conducting medium,

$$\mathbf{H} = y^2z\mathbf{a}_x + 2(x + 1)yz\mathbf{a}_y - (x + 1)z^2\mathbf{a}_z \text{ A/m}$$

Find

(a) $\mathbf{J}$ at (1, 0, −3)

(b) The current passing through $y = 1$, $0 \le x \le 1$, $0 \le z \le 1$

7.20 An infinitely long filamentary wire carries a current of 2 A in the $+z$ direction. Calculate

(a) $\mathbf{B}$ at (−3, 4, 7)

(b) The flux through the square loop described by $2 \le \rho \le 6$, $0 \le z \le 4$, $\phi = 90°$

7.21 The magnetic field $\mathbf{H} = (10^6/\rho) \cos \phi \, \mathbf{a}_\rho$ A/m exists in free space. Find the flux through the surface described by $\rho = 1$ m, $0 \leq \phi \leq \pi/2$, $0 \leq z \leq 2$ m.

7.22 The electric motor shown in Figure 7.31 has field

$$\mathbf{H} = \frac{10^6}{\rho} \sin 2\phi \, \mathbf{a}_\rho \text{ A/m}$$

Calculate the flux per pole passing through the air gap if the axial length of the pole is 20 cm.

7.23 Consider the two-wire transmission line whose cross section is illustrated in Figure 7.32. Each wire is of radius 2 cm and the wires are separated 10 cm. The wire centered at (0, 0) carries current 5 A while the other centered at (10 cm, 0) carries the return current. Find **H** at

(a) (5 cm, 0)

(b) (10 cm, 5 cm)

7.24 A stroke of lightning may be regarded as a filamentary line current. If a lightning stroke with current 50 kA occurs 100 m away from your house, calculate the magnetic flux density at your house due to the lightning stroke.

****7.25** (a) Show that the magnetic flux through the cross section of the toroid of Example 7.6 is

$$\Psi = \mu_o NI[\rho_o - (\rho_o^2 - a^2)^{1/2}]$$

and that when $\rho_o \gg a$,

$$\Psi = \frac{\mu_o NIa^2}{2\rho_o}$$

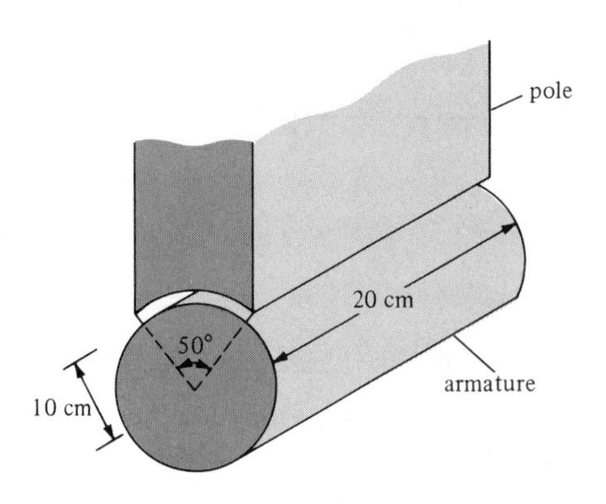

Figure 7.31 Electric motor pole of Problem 7.22.

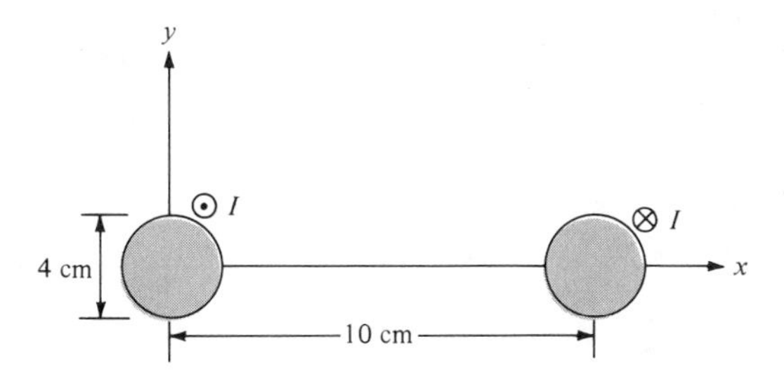

Figure 7.32 Two-wire line of Problem 7.23.

(b) If the cross section of the toroid is a square of side a, show that the magnetic flux becomes

$$\Psi = \frac{\mu_o}{2\pi} NIa \ln \left[\frac{2\rho_o + a}{2\rho_o - a}\right]$$

***7.26** A brass ring with triangular cross section encircles a very long straight wire concentrically as in Figure 7.33. If the wire carries a current I, show that the total number of magnetic flux lines in the ring is

$$\Psi = \frac{\mu_o I h}{2\pi b}\left[b - a \ln \frac{a + b}{b}\right]$$

Calculate Ψ if a = 30 cm, b = 10 cm, h = 5 cm, and I = 10 A.

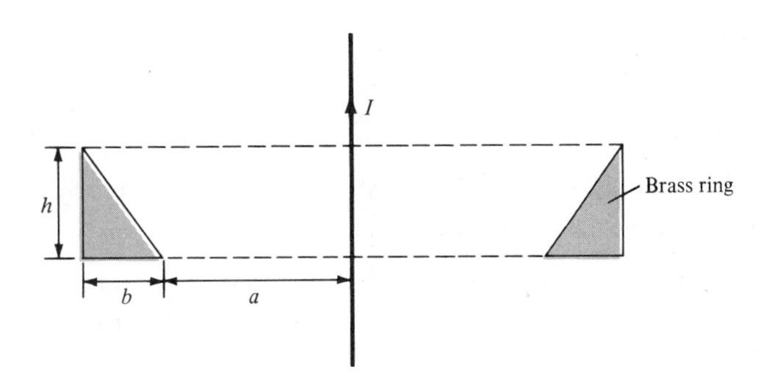

Figure 7.33 Cross section of a brass ring enclosing a long straight wire; for Problem 7.26.

7.27 The following vector fields are arbitrary. Which of them can possibly represent an electrostatic or magnetostatic field in free space?

(a) $\mathbf{A} = x\mathbf{a}_x - y\mathbf{a}_y$

(b) $\mathbf{B} = \dfrac{-y\mathbf{a}_x + x\mathbf{a}_y}{x^2 + y^2}$

(c) $\mathbf{C} = e^{-y}(\cos x\, \mathbf{a}_x - \sin x\, \mathbf{a}_y$

(d) $\mathbf{D} = 5e^{-2z}(\rho\mathbf{a}_\rho + \mathbf{a}_z)$

(e) $\mathbf{E} = \dfrac{1}{r}(2\cos\theta\, \mathbf{a}_r + \sin\theta\, \mathbf{a}_\theta)$

(*Hint:* Maxwell's equations must be satisfied.)

***7.28** Show mathematically that $\nabla \cdot \mathbf{B} = 0$.

(*Hint:* Use eq. (7.44).)

7.29 Given the magnetic vector potential

$$\mathbf{A} = e^{-z}\cos y\, \mathbf{a}_x + (1 + \sin x)\, \mathbf{a}_z \text{ Wb/m},$$

find

(a) The magnetic flux density

(b) The magnetic flux through a square loop described by $0 \leq x, y \leq \pi$, $z = 0$

7.30 The magnetic vector potential of a current distribution in free space is given by

$$\mathbf{A} = 15e^{-\rho}\sin\phi\, \mathbf{a}_z \text{ Wb/m}$$

Find $\mathbf{H}$ at $(3, \pi/4, -10)$. Calculate the flux through $\rho = 5$, $0 \leq \phi \leq \pi/2$, $0 \leq z \leq 10$.

7.31 Show that the vector magnetic potential at distant point $P(x, y, z)$ due to a finite line current flowing through $-\ell \leq z \leq \ell$ along $\mathbf{a}_z$ is

$$\mathbf{A} = \frac{\mu_o I \ell}{2\pi[x^2 + y^2 + z^2]^{1/2}}\, \mathbf{a}_z$$

Find $\mathbf{B}$ from $\mathbf{A}$.

***7.32** Obtain the vector magnetic potential in the region surrounding an infinitely long, straight, filamentary current I along $\mathbf{a}_z$.

7.33 An infinitely long conductor of radius a is placed such that its axis is along the z-axis. The vector magnetic potential, due to a direct current I_o flowing along $\mathbf{a}_z$ in the conductor, is given by

$$\mathbf{A} = \frac{I_o}{4\pi a^2}\, \mu_o(x^2 + y^2)\, \mathbf{a}_z \text{ Wb/m}$$

Find the corresponding $\mathbf{H}$. Confirm your result using Ampere's law.

7.34 Find the current density **J** due to

$$\mathbf{A} = \frac{10}{\rho^2}\mathbf{a}_z \text{ Wb/m}$$

in free space.

7.35 Derive eq. (7.12) using the concept of vector magnetic potential.

7.36 Prove that the magnetic scalar potential at $(0, 0, z)$ due to a circular loop of radius a shown in Figure 7.8(a) is

$$V_m = \frac{I}{2}\left[1 - \frac{z}{[z^2 + a^2]^{1/2}}\right]$$

***7.37** A coaxial transmission line is constructed such that the radius of the inner conductor is a and the outer conductor has radii $3a$ and $4a$. Find the vector magnetic potential within the outer conductor. Assume $A_z = 0$ for $\rho = 3a$.

7.38 An infinitely long filamentary wire along the z-axis carries current 20 A along $\mathbf{a}_z$.

(a) If $V_m = 0$ at $(2, 0°, 5)$, find V_m at $(6, \pi/4, 0)$.

(b) If $V_m = 1$ A at $(0, 7, 10)$, calculate V_m at $(-3, 4, 0)$.

7.39 Plane $z = -2$ carries a current of $50\mathbf{a}_y$ A/m. If $V_m = 0$ at the origin, find V_m at

(a) $(-2, 0, 5)$

(b) $(10, 3, 1)$

7.40 Prove in cylindrical coordinates that

(a) $\nabla \times (\nabla V) = 0$

(b) $\nabla \cdot (\nabla \times \mathbf{A}) = 0$

***7.41** Show in Cartesian coordinates that

(a) $\nabla \times (\Psi \mathbf{F}) = \Psi \nabla \times \mathbf{F} + (\nabla \Psi) \times \mathbf{F}$

(b) $\nabla \times (\nabla \times \mathbf{A}) = \nabla(\nabla \cdot \mathbf{A}) - \nabla^2 \mathbf{A}$

7.42 If $\mathbf{R} = \mathbf{r} - \mathbf{r}'$ and $R = |\mathbf{R}|$, show that

$$\nabla \frac{1}{\mathbf{R}} = -\nabla' \frac{1}{\mathbf{R}} = -\frac{\mathbf{R}}{R^3}$$

where ∇ and ∇' are del operators with respect to (x, y, z) and (x', y', z') respectively.

CHAPTER

Magnetic Forces, Materials, and Devices

8

In essentials, unity; in non-essentials, liberty; in all things, charity.

—JOHN WESLEY

8.1 INTRODUCTION

Having considered the basic laws and techniques commonly used in calculating magnetic field **B** due to current-carrying elements, we are prepared to study the force a magnetic field exerts on charged particles, current elements, and loops. Such a study is important to problems on electrical devices such as ammeters, voltmeters, galvanometers, cyclotrons, plasmas, motors, and magnetohydrodynamic generators. The precise definition of the magnetic field, deliberately sidestepped in the previous chapter, will be given here. The concepts of magnetic moments and dipole will also be considered.

Furthermore, we will consider magnetic fields in material media, as opposed to the magnetic fields in vacuum or free space examined in the previous chapter. The results of the preceding chapter need only some modification to account for the presence of materials in a magnetic field. Further discussions will cover inductors, inductances, magnetic energy, and magnetic circuits.

8.2 FORCES DUE TO MAGNETIC FIELDS

There are at least three ways in which force due to magnetic fields can be experienced. The force can be (a) due to a moving charged particle in a **B** field, (b) on a current element in an external **B** field, or (c) between two current elements.

A. Force on a Charged Particle

According to our discussion in Chapter 4, the electric force $\mathbf{F}_e$ on a stationary or moving electric charge Q in an electric field is given by Coulomb's experimental law and is related to the electric field intensity **E** as

$$\mathbf{F}_e = Q\mathbf{E} \tag{8.1}$$

This shows that if Q is positive, $\mathbf{F}_e$ and $\mathbf{E}$ have the same direction.

A magnetic field can only exert force on a moving charge. From experiments, it is found that the magnetic force $\mathbf{F}_m$ experienced by a charge Q moving with a velocity $\mathbf{u}$ in a magnetic field $\mathbf{B}$ is

$$\mathbf{F}_m = Q\mathbf{u} \times \mathbf{B} \tag{8.2}$$

This clearly shows that $\mathbf{F}_m$ is perpendicular to both $\mathbf{u}$ and $\mathbf{B}$.

From eqs. (8.1) and (8.2), a comparison between the electric force $\mathbf{F}_e$ and the magnetic force $\mathbf{F}_m$ can be made. $\mathbf{F}_e$ is independent of the velocity of the charge and can perform work on the charge and change its kinetic energy. Unlike $\mathbf{F}_e$, $\mathbf{F}_m$ depends on the charge velocity and is normal to it. $\mathbf{F}_m$ cannot perform work because it is at right angles to the direction of motion of the charge ($\mathbf{F}_m \cdot d\mathbf{l} = 0$); it does not cause an increase in kinetic energy of the charge. The magnitude of $\mathbf{F}_m$ is generally small compared to $\mathbf{F}_e$ except at high velocities.

For a moving charge Q in the presence of both electric and magnetic fields, the total force on the charge is given by

$$\mathbf{F} = \mathbf{F}_e + \mathbf{F}_m$$

or

$$\boxed{\mathbf{F} = Q(\mathbf{E} + \mathbf{u} \times \mathbf{B})} \tag{8.3}$$

This is known as the *Lorentz force equation.*[1] It relates mechanical force to electrical force. If the mass of the charged particle moving in $\mathbf{E}$ and $\mathbf{B}$ fields is m, by Newton's second law of motion

$$\mathbf{F} = m\frac{d\mathbf{u}}{dt} = Q\,(\mathbf{E} + \mathbf{u} \times \mathbf{B}) \tag{8.4}$$

The solution to this equation is important in determining the motion of charged particles in $\mathbf{E}$ and $\mathbf{B}$ fields. We should bear in mind that in such fields, energy transfer can only be by means of the electric field. A summary on the force exerted on a charged particle is given in Table 8.1.

Since eq. (8.2) is closely parallel to eq. (8.1), which defines the electric field, some authors and instructors prefer to begin their discussions on magnetostatics from eq. (8.2) just as discussions on electrostatics usually begin with Coulomb's force law.

....

[1]After Hendrik Lorentz (1853–1928), who first applied the equation to electric field motion.

Table 8.1 **Force on a Charged Particle**

State of Particle	E Field	B Field	Combined E and B Fields
Stationary	$Q\mathbf{E}$	—	$Q\mathbf{E}$
Moving	$Q\mathbf{E}$	$Q\mathbf{u} \times \mathbf{B}$	$Q(\mathbf{E} + \mathbf{u} \times \mathbf{B})$

B. Force on a Current Element

To determine the force on a current element $I\,d\mathbf{l}$ of a current-carrying conductor due to the magnetic field **B**, we modify eq. (8.2) using the fact that for convection current (see eq. (5.4)):

$$\mathbf{J} = \rho_v \mathbf{u} \tag{8.5}$$

From eq. (7.5), we recall the relationship between current elements:

$$I\,d\mathbf{l} = \mathbf{K}\,dS = \mathbf{J}\,dv \tag{8.6}$$

Combining eqs. (8.5) and (8.6) yields

$$I\,d\mathbf{l} = \rho_v \mathbf{u}\,dv = dQ\,\mathbf{u}$$

Alternatively, $I\,d\mathbf{l} = \dfrac{dQ}{dt}\,d\mathbf{l} = dQ\,\dfrac{d\mathbf{l}}{dt} = dQ\,\mathbf{u}$

Hence,

$$\boxed{I\,d\mathbf{l} = dQ\,\mathbf{u}} \tag{8.7}$$

This shows that a charge dQ moving with velocity **u** (thereby producing convection current element $dQ\,\mathbf{u}$) is equivalent to a conduction current element $I\,d\mathbf{l}$. Thus the force on a current element $I\,d\mathbf{l}$ in a magnetic field **B** is found from eq. (8.2) by merely replacing $Q\mathbf{u}$ by $I\,d\mathbf{l}$; that is,

$$\boxed{d\mathbf{F} = I\,d\mathbf{l} \times \mathbf{B}} \tag{8.8}$$

If the current I is through a closed path L or circuit, the force on the circuit is given by

$$\mathbf{F} = \oint_L I\,d\mathbf{l} \times \mathbf{B} \tag{8.9}$$

In using eq. (8.8) or (8.9), we should keep in mind that the magnetic field produced by the current element $I\,d\mathbf{l}$ does not exert force on the element itself just as a point charge does not exert force on itself. The **B** field that exerts force on $I\,d\mathbf{l}$ must be due to another element. In other words, the **B** field in eq. (8.8) or (8.9) is external to the current element $I\,d\mathbf{l}$. If instead of the line current element $I\,d\mathbf{l}$, we have surface current elements $\mathbf{K}\,dS$ or a volume current element $\mathbf{J}\,dv$, we simply make use of eq. (8.6) so that eq. (8.8) becomes

$$d\mathbf{F} = \mathbf{K}\,dS \times \mathbf{B} \quad or \quad d\mathbf{F} = \mathbf{J}\,dv \times \mathbf{B} \tag{8.8a}$$

while eq. (8.9) becomes

$$\mathbf{F} = \int_S \mathbf{K}\,dS \times \mathbf{B} \quad or \quad \mathbf{F} = \int_v \mathbf{J}\,dv \times \mathbf{B} \tag{8.9a}$$

C. Force Between Two Current Elements

Let us now consider the force between two elements $I_1\,d\mathbf{l}_1$ and $I_2\,d\mathbf{l}_2$. According to Biot-Savart's law, both current elements produce magnetic fields. So we may find the force $d(d\mathbf{F}_1)$ on element $I_1\,d\mathbf{l}_1$ due to the field $d\mathbf{B}_2$ produced by element $I_2\,d\mathbf{l}_2$ as shown in Figure 8.1. From eq. (8.8),

$$d(d\mathbf{F}_1) = I_1\,d\mathbf{l}_1 \times d\mathbf{B}_2 \tag{8.10}$$

But from Biot-Savart's law,

$$d\mathbf{B}_2 = \frac{\mu_o I_2\,d\mathbf{l}_2 \times \mathbf{a}_{R_{21}}}{4\pi R_{21}^2} \tag{8.11}$$

Hence,

$$d(d\mathbf{F}_1) = \frac{\mu_o I_1\,d\mathbf{l}_1 \times (I_2\,d\mathbf{l}_2 \times \mathbf{a}_{R_{21}})}{4\pi R_{21}^2} \tag{8.12}$$

This equation is essentially the law of force between two current elements and is analogous to Coulomb's law, which expresses the force between two stationary

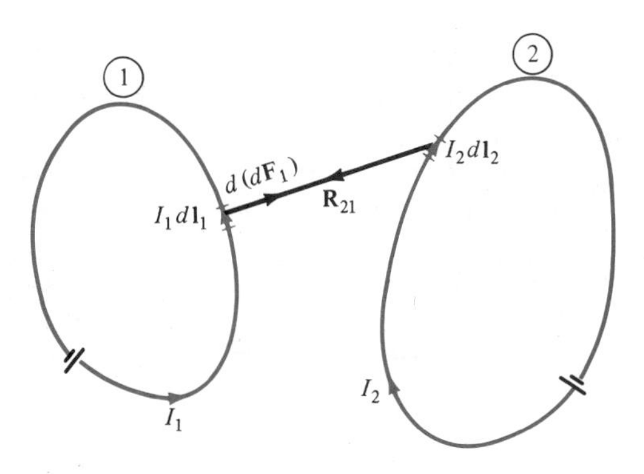

Figure 8.1 Force between two current loops.

charges. From eq. (8.12), we obtain the total force $\mathbf{F}_1$ on current loop 1 due to current loop 2 shown in Figure 8.1 as

$$\mathbf{F}_1 = \frac{\mu_o I_1 I_2}{4\pi} \oint_{L_1} \oint_{L_2} \frac{d\mathbf{l}_1 \times (d\mathbf{l}_2 \times \mathbf{a}_{R_{21}})}{R_{21}^2} \tag{8.13}$$

Although this equation appears complicated, we should remember that it is based on eq. (8.10). It is eq. (8.8) or (8.10) that is of fundamental importance.

The force $\mathbf{F}_2$ on loop 2 due to the magnetic field $\mathbf{B}_1$ from loop 1 is obtained from eq. (8.13) by interchanging subscripts 1 and 2. It can be shown that $\mathbf{F}_2 = -\mathbf{F}_1$; thus $\mathbf{F}_1$ and $\mathbf{F}_2$ obey Newton's third law that action and reaction must be equal and opposite. It is worthwhile to mention that eq. (8.13) was experimentally established by Oersted and Ampere; Biot and Savart (Ampere's colleagues) actually based their law on it.

From eq. (8.8), the magnetic field $\mathbf{B}$ is defined as the force per unit current element. Alternatively, $\mathbf{B}$ may be defined from eq. (8.2) as the vector which satisfies $\mathbf{F}_m/q = \mathbf{u} \times \mathbf{B}$ just as we defined electric field $\mathbf{E}$ as the force per unit charge, $\mathbf{F}_e/q$. Both of these definitions of $\mathbf{B}$ show that $\mathbf{B}$ describes the force properties of a magnetic field.

EXAMPLE 8.1

A charged particle of mass 2 kg and charge 3 C starts at point $(1, -2, 0)$ with velocity $4\mathbf{a}_x + 3\mathbf{a}_z$ m/s in an electric field $12\mathbf{a}_x + 10\mathbf{a}_y$ V/m. At time $t = 1$ s, determine

(a) The acceleration of the particle

(b) Its velocity

(c) Its kinetic energy

(d) Its position

SOLUTION

(a) This is an initial-value problem because initial values are given. According to Newton's second law of motion,

$$\mathbf{F} = m\mathbf{a} = Q\mathbf{E}$$

where $\mathbf{a}$ is the acceleration of the particle. Hence,

$$\mathbf{a} = \frac{Q\mathbf{E}}{m} = \frac{3}{2}(12\mathbf{a}_x + 10\mathbf{a}_y) = 18\mathbf{a}_x + 15\mathbf{a}_y \text{ m/s}^2$$

$$\mathbf{a} = \frac{d\mathbf{u}}{dt} = \frac{d}{dt}(u_x, u_y, u_z) = 18\mathbf{a}_x + 15\mathbf{a}_y$$

(b) Equating components gives

$$\frac{du_x}{dt} = 18 \rightarrow u_x = 18t + A \tag{8.1.1}$$

$$\frac{du_y}{dt} = 15 \rightarrow u_y = 15t + B \tag{8.1.2}$$

$$\frac{du_z}{dt} = 0 \rightarrow u_z = C \tag{8.1.3}$$

where A, B, and C are integration constants. But at $t = 0$, $\mathbf{u} = 4\mathbf{a}_x + 3\mathbf{a}_z$. Hence,

$$u_x(t = 0) = 4 \rightarrow 4 = 0 + A \quad \text{or} \quad A = 4$$

$$u_y(t = 0) = 0 \rightarrow 0 = 0 + B \quad \text{or} \quad B = 0$$

$$u_z(t = 0) = 3 \rightarrow 3 = C$$

Substituting the values of A, B, and C into eqs. (8.1.1) to (8.1.3) gives

$$\mathbf{u}(t) = (u_x, u_y, u_z) = (18t + 4, 15t, 3)$$

Hence

$$\mathbf{u}(t = 1 \text{ s}) = 22\mathbf{a}_x + 15\mathbf{a}_y + 3\mathbf{a}_z \text{ m/s}$$

(c) Kinetic energy (K.E.) $= \frac{1}{2}m\,|\mathbf{u}|^2 = \frac{1}{2}(2)(22^2 + 15^2 + 3^2)$

$$= 718 \text{ J}$$

(d) $\mathbf{u} = \frac{d\mathbf{l}}{dt} = \frac{d}{dt}(x, y, z) = (18t + 4, 15t, 3)$

Equating components yields

$$\frac{dx}{dt} = u_x = 18t + 4 \rightarrow x = 9t^2 + 4t + A_1 \tag{8.1.4}$$

$$\frac{dy}{dt} = u_y = 15t \rightarrow y = 7.5t^2 + B_1 \tag{8.1.5}$$

$$\frac{dz}{dt} = u_z = 3 \rightarrow z = 3t + C_1 \tag{8.1.6}$$

At $t = 0$, $(x, y, z) = (1, -2, 0)$; hence,

$$x(t = 0) = 1 \rightarrow 1 = 0 + A_1 \quad \text{or} \quad A_1 = 1$$

$$y(t = 0) = -2 \rightarrow -2 = 0 + B_1 \quad \text{or} \quad B_1 = -2$$

$$z(t = 0) = 0 \rightarrow 0 = 0 + C_1 \quad \text{or} \quad C_1 = 0.$$

Substituting the values of A_1, B_1, and C_1 into eqs. (8.1.4) to (8.1.6), we obtain

$$(x, y, z) = (9t^2 + 4t + 1, 7.5t^2 - 2, 3t) \tag{8.1.7}$$

Hence, at $t = 1$, $(x, y, z) = (14, 5.5, 3)$.
By eliminating t in eq. (8.1.7), the motion of the particle may be described in terms of x, y, and z. ■

PRACTICE EXERCISE 8.1

A charged particle of mass 1 kg and charge 2 C starts at the origin with zero initial velocity in a region where $\mathbf{E} = 3\mathbf{a}_z$ V/m. Find

(a) The force on the particle
(b) The time it takes to reach point $P(0, 0, 12 \text{ m})$
(c) Its velocity and acceleration at P
(d) Its K.E. at P.

ANSWER (a) $6\mathbf{a}_z$ N, (b) $2s$, (c) $12\mathbf{a}_z$ m/s, $6\mathbf{a}_z \text{m/s}^2$, (d) 72 J.

EXAMPLE 8.2

A charged particle of mass 2 kg and 1 C starts at the origin with velocity $3\mathbf{a}_y$ m/s and travels in a region of uniform magnetic field $\mathbf{B} = 10\mathbf{a}_z$ Wb/m². At $t = 4$ s, calculate

(a) The velocity and acceleration of the particle
(b) The magnetic force on it
(c) Its K.E. and location
(d) Find the particle's trajectory by eliminating t.
(e) Show that its K.E. remains constant.

SOLUTION (a) $\mathbf{F} = m \dfrac{d\mathbf{u}}{dt} = Q\mathbf{u} \times \mathbf{B}$

$$\mathbf{a} = \frac{d\mathbf{u}}{dt} = \frac{Q}{m}\mathbf{u} \times \mathbf{B}$$

Hence

$$\frac{d}{dt}(u_x\mathbf{a}_x + u_y\mathbf{a}_y + u_z\mathbf{a}_z) = \frac{1}{2}\begin{vmatrix} \mathbf{a}_x & \mathbf{a}_y & \mathbf{a}_z \\ u_x & u_y & u_z \\ 0 & 0 & 10 \end{vmatrix} = 5(u_y\mathbf{a}_x - u_x\mathbf{a}_y)$$

By equating components, we get

$$\frac{du_x}{dt} = 5u_y \tag{8.2.1}$$

$$\frac{du_y}{dt} = -5u_x \tag{8.2.2}$$

$$\frac{du_z}{dt} = 0 \rightarrow u_z = C_o \tag{8.2.3}$$

We can eliminate u_x or u_y in eqs. (8.2.1) and (8.2.2) by taking second derivatives of one equation and making use of the other. Thus

$$\frac{d^2u_x}{dt^2} = 5\frac{du_y}{dt} = -25u_x$$

or

$$\frac{d^2u_x}{dt^2} + 25u_x = 0$$

which is a linear differential equation with solution (see Case 3 of Example 6.5)

$$u_x = C_1 \cos 5t + C_2 \sin 5t \tag{8.2.4}$$

From eqs. (8.2.1) and (8.2.4),

$$5u_y = \frac{du_x}{dt} = -5C_1 \sin 5t + 5C_2 \cos 5t \tag{8.2.5}$$

or

$$u_y = -C_1 \sin 5t + C_2 \cos 5t$$

We now determine constants C_o, C_1, and C_2 using the initial conditions. At $t = 0$, $\mathbf{u} = 3\mathbf{a}_y$. Hence,

$$u_x = 0 \rightarrow 0 = C_1 \cdot 1 + C_2 \cdot 0 \rightarrow C_1 = 0$$

$$u_y = 3 \rightarrow 3 = -C_1 \cdot 0 + C_2 \cdot 1 \rightarrow C_2 = 3$$

$$u_z = 0 \rightarrow 0 = C_o$$

Substituting the values of C_o, C_1, and C_2 into eqs. (8.2.3) to (8.2.5) gives

$$\mathbf{u} = (u_x, u_y, u_z) = (3 \sin 5t, 3 \cos 5t, 0) \tag{8.2.6}$$

Hence,

$$\mathbf{u}(t = 4) = (3 \sin 20, 3 \cos 20, 0)$$
$$= 2.739\mathbf{a}_x + 1.224\mathbf{a}_y \text{ m/s}$$

$$\mathbf{a} = \frac{d\mathbf{u}}{dt} = (15 \cos 5t, -15 \sin 5t, 0)$$

and

$$\mathbf{a}(t = 4) = 6.101\mathbf{a}_x - 13.703\mathbf{a}_y \text{ m/s}^2$$

(b)

$$\mathbf{F} = m\mathbf{a} = 12.2\mathbf{a}_x - 27.4\mathbf{a}_y \text{ N}$$

or

$$\mathbf{F} = Q\mathbf{u} \times \mathbf{B} = (1)(2.739\mathbf{a}_x + 1.224\mathbf{a}_y) \times 10\mathbf{a}_z$$
$$= 12.2\mathbf{a}_x - 27.4\mathbf{a}_y \text{ N}$$

(c) K.E. $= 1/2m\,|\mathbf{u}|^2 = 1/2(2)\,(2.739^2 + 1.224^2) = 9\,J$

$$u_x = \frac{dx}{dt} = 3 \sin 5t \rightarrow x = -\frac{3}{5} \cos 5t + b_1 \qquad [8.2.7]$$

$$u_y = \frac{dy}{dt} = 3 \cos 5t \rightarrow y = \frac{3}{5} \sin 5t + b_2 \qquad [8.2.8]$$

$$u_z = \frac{dz}{dt} = 0 \rightarrow z = b_3 \qquad [8.2.9]$$

where b_1, b_2, and b_3 are integration constants. At $t = 0$, $(x, y, z) = (0, 0, 0)$ and hence,

$$x(t = 0) = 0 \rightarrow 0 = -\frac{3}{5} \cdot 1 + b_1 \rightarrow b_1 = 0.6$$

$$y(t = 0) = 0 \rightarrow 0 = \frac{3}{5} \cdot 0 + b_2 \rightarrow b_2 = 0$$

$$z(t = 0) = 0 \rightarrow 0 = b_3$$

Substituting the values of b_1, b_2, and b_3 into eqs. (8.2.7) to (8.2.9), we obtain

$$(x, y, z) = (0.6 - 0.6 \cos 5t,\ 0.6 \sin 5t,\ 0). \qquad [8.2.10]$$

At $t = 4$ s,

$$(x, y, z) = (0.3552,\ 0.5478,\ 0).$$

(d) From eq. (8.2.10), we eliminate t by noting that

$$(x - 0.6)^2 + y^2 = (0.6)^2 (\cos^2 5t + \sin^2 5t) \qquad z = 0$$

or

$$(x - 0.6)^2 + y^2 = (0.6)^2 \qquad z = 0$$

which is a circle on plane $z = 0$, centered at $(0.6, 0, 0)$ and of radius 0.6 m. Thus the particle gyrates in an orbit about a magnetic field line.

(e)

$$\text{K.E.} = \frac{1}{2} m\,|\mathbf{u}|^2 = \frac{1}{2} (2)\,(9 \cos^2 5t + 9 \sin^2 5t) = 9\text{J}$$

which is the same as the K.E. at $t = 0$ and $t = 4$ s. Thus the uniform magnetic field has no effect on the K.E. of the particle.

Note that the angular velocity $\omega = QB/m$ and the radius of the orbit $r = \mu_o/\omega$, where u_o is the initial speed. An interesting application of the idea in this example is found in a common method of focusing a beam of electrons. The method employs a uniform magnetic field directed parallel to the desired beam as shown in Figure 8.2. Each electron emerging from the electron gun follows a helical path and is back on the axis at the same focal point with other electrons. If the screen of a cathode ray tube were at this point, a single spot would appear on the screen. ■

PRACTICE EXERCISE 8.2

A proton of mass m is projected into a uniform field $\mathbf{B} = B_o\mathbf{a}_z$ with an initial velocity $\alpha\mathbf{a}_x + \beta\mathbf{a}_z$. (a) Find the differential equations that the position vector $\mathbf{r} = x\mathbf{a}_x + y\mathbf{a}_y + z\mathbf{a}_z$ must satisfy. (b) Show that a solution to these equations is

$$x = \frac{\alpha}{\omega} \sin \omega t, \qquad y = \frac{\alpha}{\omega} \cos \omega t, \qquad z = \beta t$$

where $\omega = eB_o/m$ and e is the charge on the proton. (c) Show that this solution describes a circular helix in space.

ANSWER (a) $\frac{dx}{dt} = \alpha \cos \omega t$, $\frac{dy}{dt} = -\alpha \sin \omega t$, $\frac{dz}{dt} = \beta$, (b) and (c) Proof.

EXAMPLE 8.3

A charged particle moves with a uniform velocity $4\mathbf{a}_x$ m/s in a region where $\mathbf{E} = 20\ \mathbf{a}_y$ V/m and $\mathbf{B} = B_o\mathbf{a}_z$ Wb/m^2. Determine B_o such that the velocity of the particle remains constant.

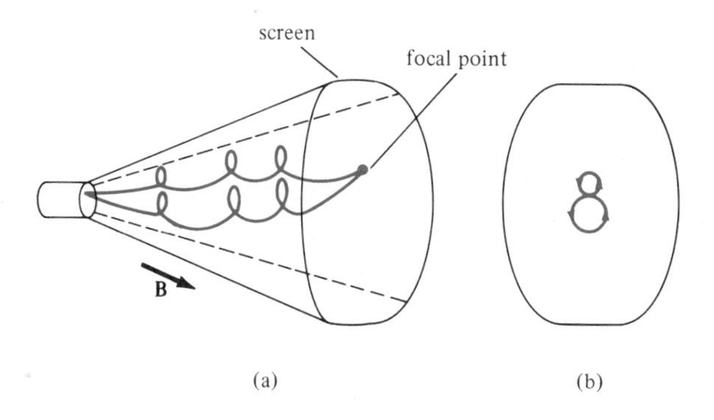

Figure 8.2 Magnetic focusing of a beam of electrons: **(a)** helical paths of electrons, **(b)** end view of paths.

SOLUTION If the particle moves with a constant velocity, it implies that its acceleration is zero. In other words, the particle experiences no net force. Hence,

$$0 = \mathbf{F} = m\mathbf{a} = Q\,(\mathbf{E} + \mathbf{u} \times \mathbf{B})$$

$$0 = Q\,(20\mathbf{a}_y + 4\mathbf{a}_x \times B_o\mathbf{a}_z)$$

or

$$-20\mathbf{a}_y = -4B_o\mathbf{a}_y$$

Thus $B_o = 5$.

This example illustrates an important principle employed in a velocity filter shown in Figure 8.3. In this application, **E**, **B**, and **u** are mutually perpendicular so that $Q\mathbf{u} \times \mathbf{B}$ is directed opposite to $Q\mathbf{E}$, regardless of the sign of the charge. When the magnitudes of the two vectors are equal,

$$QuB = QE$$

or

$$u = \frac{E}{B}$$

This is the required (critical) speed to balance out the two parts of the Lorentz force. Particles with this speed are undeflected by the fields; they are "filtered" through the aperture. Particles with other speeds are deflected down or up, depending on whether their speeds are greater or less than this critical speed. ■

PRACTICE EXERCISE 8.3

Uniform **E** and **B** fields are oriented at right angles to each other. An electron moves with a speed of 8×10^6 m/s at right angles to both fields and passes undeflected through the field.

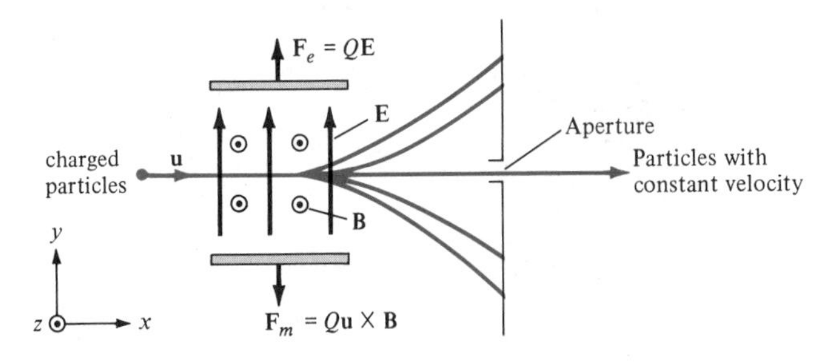

Figure 8.3 A velocity filter for charged particles.

(a) If the magnitude of **B** is 0.5 m Wb/m^2, find the value of **E**.

(b) Will this filter work for positive and negative charges and any value of mass?

ANSWER (a) 4 kV/m, (b) Yes.

EXAMPLE 8.4

A rectangular loop carrying current I_2 is placed parallel to an infinitely long filamentary wire carrying current I_1 as shown in Figure 8.4(a). Show that the force experienced by the loop is given by

$$\mathbf{F} = -\frac{\mu_o I_1 I_2 b}{2\pi}\left[\frac{1}{\rho_o} - \frac{1}{\rho_o + a}\right]\mathbf{a}_\rho \quad \text{N}$$

SOLUTION Let the force on the loop be

$$\mathbf{F}_\ell = \mathbf{F}_1 + \mathbf{F}_2 + \mathbf{F}_3 + \mathbf{F}_4 = I_2 \oint d\mathbf{l}_2 \times \mathbf{B}_1$$

where $\mathbf{F}_1$, $\mathbf{F}_2$, $\mathbf{F}_3$, and $\mathbf{F}_4$ are respectively the forces exerted on sides of the loop labeled 1, 2, 3, and 4 in Figure 8.4(b). Due to the infinitely long wire

$$\mathbf{B}_1 = \frac{\mu_o I_1}{2\pi\rho_o}\,\mathbf{a}_\phi$$

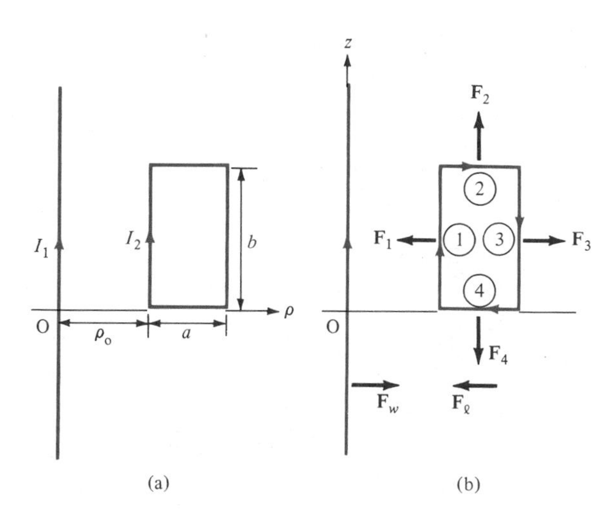

Figure 8.4 For Example 8.4: **(a)** rectangular loop inside the field produced by an infinitely long wire, **(b)** forces acting on the loop and wire.

Hence,

$$\mathbf{F}_1 = I_2 \int d\mathbf{l}_2 \times \mathbf{B}_1 = I_2 \int_{z=0}^{b} dz\, \mathbf{a}_z \times \frac{\mu_o I_1}{2\pi\rho_o}\, \mathbf{a}_\phi$$

$$= -\frac{\mu_o I_1 I_2 b}{2\pi\rho_o}\, \mathbf{a}_\rho \qquad \text{(attractive)}$$

$\mathbf{F}_1$ is attractive because it is directed toward the long wire; that is, $\mathbf{F}_1$ is along $-\mathbf{a}_\rho$ due to the fact that loop side 1 and the long wire carry currents along the same direction. Similarly,

$$\mathbf{F}_3 = I_2 \int d\mathbf{l}_2 \times \mathbf{B}_1 = I_2 \int_{z=b}^{0} dz\, \mathbf{a}_z \times \frac{\mu_o I_1}{2\pi(\rho_o + a)}\, \mathbf{a}_\phi$$

$$= \frac{\mu_o I_1 I_2 b}{2\pi(\rho_o + a)}\, \mathbf{a}_\rho \qquad \text{(repulsive)}$$

$$\mathbf{F}_2 = I_2 \int_{\rho=\rho_o}^{\rho_o + a} d\rho\, \mathbf{a}_\rho \times \frac{\mu_o I_1 \mathbf{a}_\phi}{2\pi\rho}$$

$$= \frac{\mu_o I_1 I_2}{2\pi} \ln \frac{\rho_o + a}{\rho_o}\, \mathbf{a}_z \qquad \text{(parallel)}$$

$$\mathbf{F}_4 = I_2 \int_{\rho=\rho_o + a}^{\rho_o} d\rho\, \mathbf{a}_\rho \times \frac{\mu_o I_1 \mathbf{a}_\phi}{2\pi\rho}$$

$$= -\frac{\mu_o I_1 I_2}{2\pi} \ln \frac{\rho_o + a}{\rho_o}\, \mathbf{a}_z \qquad \text{(parallel)}$$

The total force $\mathbf{F}_\ell$ on the loop is the sum of $\mathbf{F}_1$, $\mathbf{F}_2$, $\mathbf{F}_3$ and $\mathbf{F}_4$; that is,

$$\mathbf{F}_\ell = \frac{\mu_o I_1 I_2 b}{2\pi}\left[\frac{1}{\rho_o} - \frac{1}{\rho_o + a}\right](-\mathbf{a}_\rho)$$

which is an attractive force trying to draw the loop toward the wire. The force $\mathbf{F}_w$ on the wire, by Newton's third law, is $-\mathbf{F}_\ell$; see Figure 8.4(b). ■

PRACTICE EXERCISE 8.4

In Example 8.4, find the force experienced by the infinitely long wire if $I_1 = 10$ A, $I_2 = 5$ A, $\rho_o = 20$ cm, $a = 10$ cm, $b = 30$ cm.

ANSWER $5\mathbf{a}_\rho$ μN.

8.3 MAGNETIC TORQUE AND MOMENT

Now that we have considered the force on a current loop in a magnetic field, we can determine the torque on it. The concept of a current loop experiencing a torque in a magnetic field is of paramount importance in understanding the behavior of orbiting charged particles, d.c. motors, and generators. If the loop is placed parallel to a magnetic field, it experiences a force that tends to rotate it. We define the *torque* **T** (or mechanical moment of force) on the loop as the vector product of the force **F** and the moment arm **r**; that is,

$$\mathbf{T} = \mathbf{r} \times \mathbf{F} \tag{8.14}$$

and its units are Newton-meters (N · m).

Let us apply this to a rectangular loop of length ℓ and width w placed in a uniform magnetic field **B** as shown in Figure 8.5(a). From this figure, we notice that $d\mathbf{l}$ is parallel to **B** along sides 12 and 34 of the loop and no force is exerted on those sides. Thus

$$\begin{aligned} \mathbf{F} &= I \int_2^3 d\mathbf{l} \times \mathbf{B} + I \int_4^1 d\mathbf{l} \times \mathbf{B} \\ &= I \int_0^\ell dz\, \mathbf{a}_z \times \mathbf{B} + I \int_\ell^0 dz\, \mathbf{a}_z \times \mathbf{B} \end{aligned}$$

or

$$\mathbf{F} = \mathbf{F}_o - \mathbf{F}_o = 0 \tag{8.15}$$

where $|\mathbf{F}_o| = IB\ell$ because **B** is uniform. Thus, no force is exerted on the loop as a whole. However, $\mathbf{F}_o$ and $-\mathbf{F}_o$ act at different points on the loop, thereby creating a

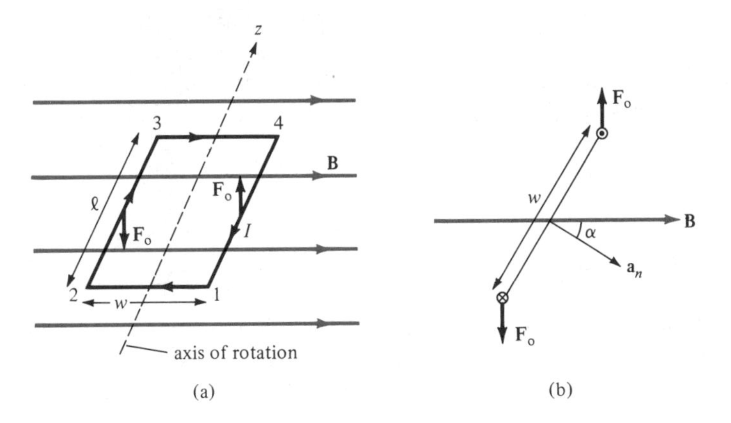

Figure 8.5 Rectangular planar loop in a uniform magnetic field.

couple. If the normal to the plane of the loop makes an angle α with **B**, as shown in the cross-sectional view of Figure 8.5(b), the torque on the loop is

$$|\mathbf{T}| = |\mathbf{F}_o|\, w \sin \alpha$$

or

$$T = BI\ell w \sin \alpha \qquad [8.16]$$

But $\ell w = S$, the area of the loop. Hence,

$$T = BIS \sin \alpha \qquad [8.17]$$

We define the quantity

$$\boxed{\mathbf{m} = IS\mathbf{a}_n} \qquad [8.18]$$

as the *magnetic dipole moment* (in $A \cdot m^2$) of the loop. In eq. (8.18), $\mathbf{a}_n$ is a unit normal vector to the plane of the loop and its direction is determined by the right-hand rule: fingers in the direction of current and thumb along $\mathbf{a}_n$. The magnetic dipole moment is essentially the product of current and area of the loop; its direction is normal to the loop. Introducing eq. (8.18) in eq. (8.17), we obtain

$$\boxed{\mathbf{T} = \mathbf{m} \times \mathbf{B}} \qquad [8.19]$$

This expression is generally applicable in determining the torque on a planar loop of any arbitrary shape although it was obtained using a rectangular loop. The only limitation is that the magnetic field must be uniform. It should be noted that the torque is in the direction of the axis of rotation (the z-axis in the case of Figure 8.5a). It is directed such as to reduce α so that **m** and **B** are in the same direction. In an equilibrium position (when **m** and **B** are in the same direction), the loop is perpendicular to the magnetic field and the torque will be zero as well as the sum of the forces on the loop.

8.4 A MAGNETIC DIPOLE

A bar magnet or a small filamentary current loop is usually referred to as a *magnetic dipole*. The reason for this and what we mean by ''small'' will soon be evident. Let us determine the magnetic field **B** at an observation point $P(r, \theta, \phi)$ due to a circular loop carrying current I as in Figure 8.6. The magnetic vector potential at P is

$$\mathbf{A} = \frac{\mu_o I}{4\pi} \oint \frac{d\mathbf{l}}{r} \qquad [8.20]$$

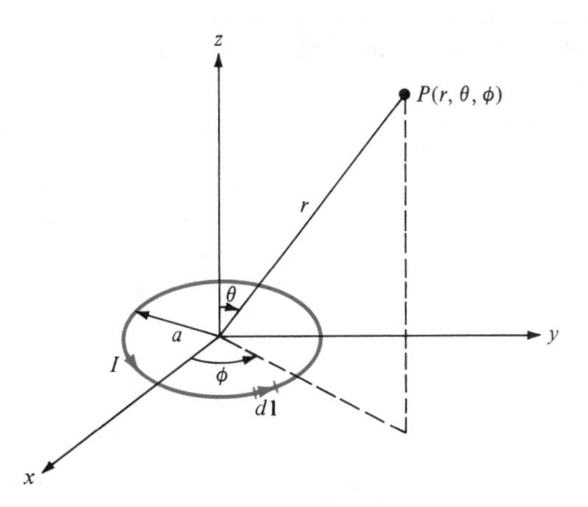

Figure 8.6 Magnetic field at P due to a current loop.

It can be shown that at far-field ($r \gg a$, so that the loop appears small at the observation point), $\mathbf{A}$ has only ϕ-component and it is given by

$$\mathbf{A} = \frac{\mu_o I \pi a^2 \sin\theta \, \mathbf{a}_\phi}{4\pi r^2} \qquad [8.21a]$$

or

$$\mathbf{A} = \frac{\mu_o \, \mathbf{m} \times \mathbf{a}_r}{4\pi r^2} \qquad [8.21b]$$

where $\mathbf{m} = I\pi a^2 \mathbf{a}_z$, the magnetic moment of the loop, and $\mathbf{a}_z \times \mathbf{a}_r = \sin\theta \, \mathbf{a}_\phi$. We determine the magnetic flux density $\mathbf{B}$ from $\mathbf{B} = \nabla \times \mathbf{A}$ as

$$\mathbf{B} = \frac{\mu_o m}{4\pi r^3} (2\cos\theta \, \mathbf{a}_r + \sin\theta \, \mathbf{a}_\theta) \qquad [8.22]$$

It is interesting to compare eqs. (8.21) and (8.22) with similar expressions in eqs. (4.80) and (4.82) for electrical potential V and electric field intensity $\mathbf{E}$ due to an electric dipole. This comparison is done in Table 8.2, in which we notice the striking similarities between $\mathbf{B}$ at far field due to a small current loop and $\mathbf{E}$ at far field due to an electric dipole. It is therefore reasonable to regard a small current loop as a magnetic dipole. The $\mathbf{B}$ lines due to a magnetic dipole are similar to the $\mathbf{E}$ lines due to an electric dipole. Figure 8.7(a) illustrates the $\mathbf{B}$ lines around the magnetic dipole $\mathbf{m} = I\mathbf{S}$.

Table 8.2 **Comparison Between Electric and Magnetic Monopoles and Dipoles**

Electric	Magnetic
$V = \dfrac{Q}{4\pi\varepsilon_o r}$ $\mathbf{E} = \dfrac{Q\mathbf{a}_r}{4\pi\varepsilon_o r^2}$ 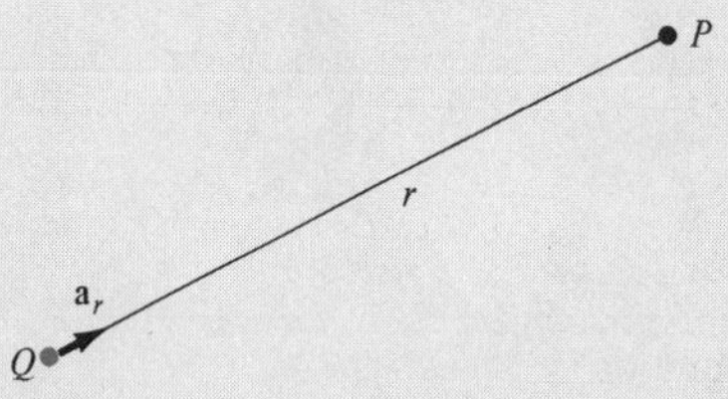 P, r, $\mathbf{a}_r$, Q Monopole (point charge)	Does not exist Q_m Monopole (point charge)
$V = \dfrac{Q\cos\theta}{4\pi\varepsilon_o r^2}$ $\mathbf{E} = \dfrac{Qd}{4\pi\varepsilon_o r^3}(2\cos\theta\,\mathbf{a}_r + \sin\theta\,\mathbf{a}_\theta)$ 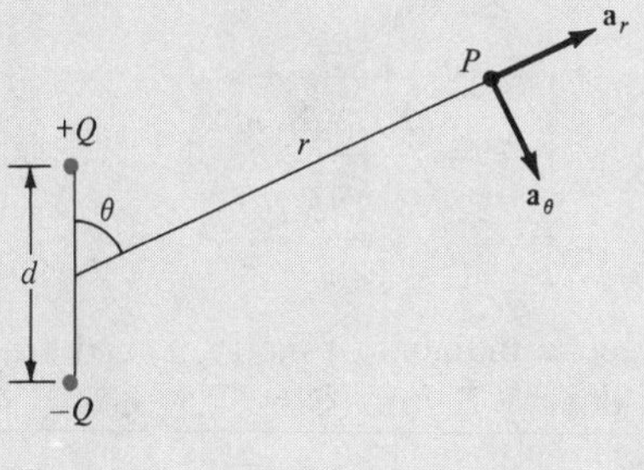 $\mathbf{a}_r$, P, $+Q$, r, θ, $\mathbf{a}_\theta$, d, $-Q$ Dipole (two point charge)	$\mathbf{A} = \dfrac{\mu_o m \sin\theta\,\mathbf{a}_\phi}{4\pi r^2}$ $\mathbf{B} = \dfrac{\mu_o m}{4\pi r^3}(2\cos\theta\,\mathbf{a}_r + \sin\theta\,\mathbf{a}_\theta)$ 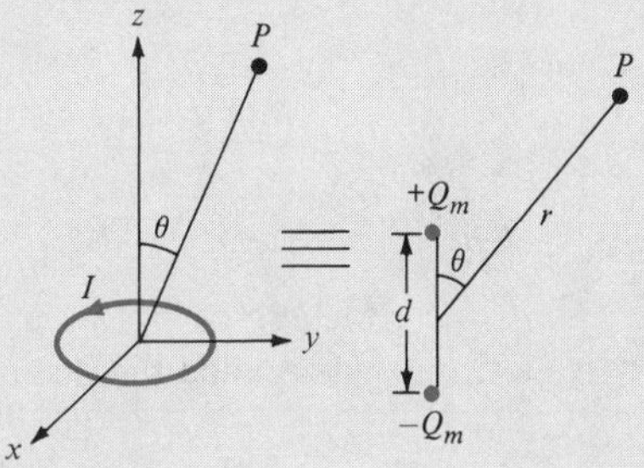 z, P, θ, I, y, x, $\equiv$, P, $+Q_m$, r, θ, d, $-Q_m$ Dipole (small current loop or bar magnet)

A short permanent magnetic bar, shown in Figure 8.7(b), may also be regarded as a magnetic dipole. Observe that the **B** lines due to the bar are similar to those due to a small current loop in Figure 8.7(a).

Consider the bar magnet of Figure 8.8. If Q_m is an isolated magnetic charge (pole strength) and ℓ is the length of the bar, the bar has a dipole moment $Q_m\ell$. (Notice that Q_m does exist; however, it does not exist without an associated $-Q_m$. See Table 8.2.) When the bar is in a uniform magnetic field **B**, it experiences a torque

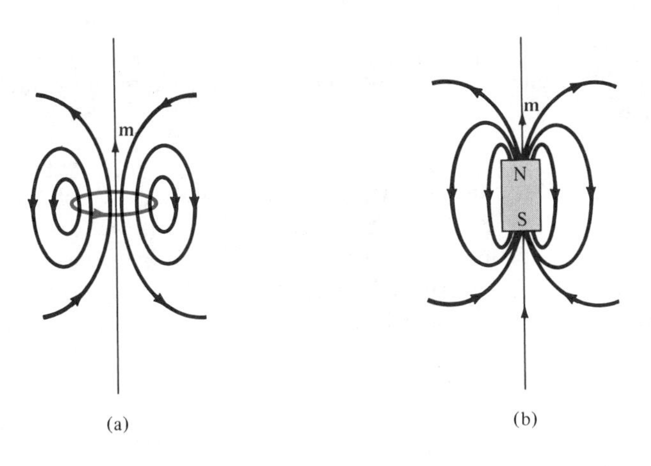

Figure 8.7 The **B** lines due to magnetic dipoles: **(a)** a small current loop with $\mathbf{m} = I\mathbf{S}$, **(b)** a bar magnet with $\mathbf{m} = Q_m\boldsymbol{\ell}$.

$$\mathbf{T} = \mathbf{m} \times \mathbf{B} = Q_m\boldsymbol{\ell} \times \mathbf{B} \quad [8.23]$$

where $\boldsymbol{\ell}$ points in the direction south-to-north. The torque tends to align the bar with the external magnetic field. The force acting on the magnetic charge is given by

$$\mathbf{F} = Q_m\mathbf{B} \quad [8.24]$$

Since both a small current loop and a bar magnet produce magnetic dipoles, they are equivalent if they produce the same torque in a given **B** field; that is, when

$$T = Q_m\ell B = ISB \quad [8.25]$$

Hence,

$$Q_m\ell = IS \quad [8.26]$$

showing that they must have the same dipole moment.

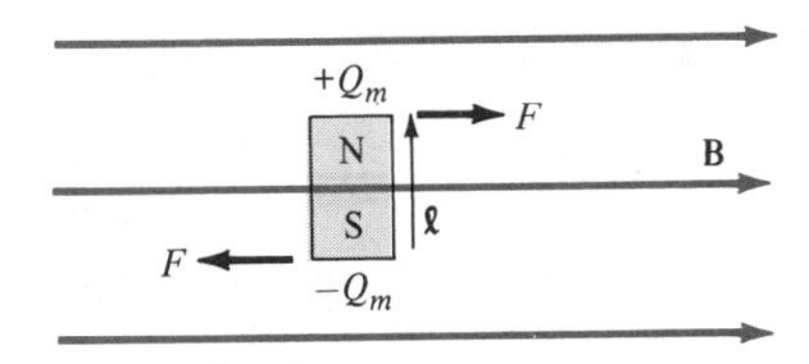

Figure 8.8 A bar magnet in an external magnetic field.

EXAMPLE 8.5

Determine the magnetic moment of an electric circuit formed by the triangular loop of Figure 8.9.

SOLUTION From Problem 1.13(c), the equation of a plane is given by $Ax + By + Cz + D = 0$ where $D = -(A^2 + B^2 + C^2)$. Since points (2, 0, 0), (0, 2, 0), and (0, 0, 2) lie on the plane, these points must satisfy the equation of the plane, and the constants A, B, C, and D can be determined. Doing this gives $x + y + z = 2$ as the plane on which the loop lies. Thus we can use

$$\mathbf{m} = IS\mathbf{a}_n$$

where

$$S = \text{loop area} = \frac{1}{2} \times \text{base} \times \text{height} = \frac{1}{2}(2\sqrt{2})(2\sqrt{2})\sin 60°$$
$$= 4 \sin 60°$$

If we define the plane surface by a function

$$f(x,y,z) = x + y + z - 2 = 0,$$

$$\mathbf{a}_n = \pm \frac{\nabla f}{|\nabla f|} = \pm \frac{(\mathbf{a}_x + \mathbf{a}_y + \mathbf{a}_z)}{\sqrt{3}}$$

We choose the plus sign in view of the direction of the current in the loop (using the right-hand rule, **m** is directed as in Fig. 8.9). Hence

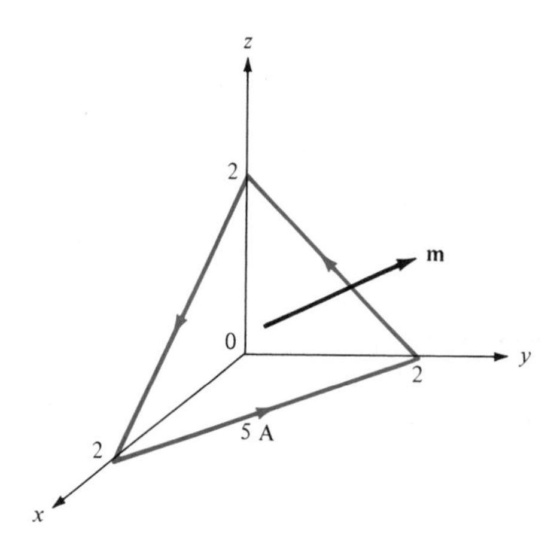

Figure 8.9 Triangular loop of Example 8.5.

$$\mathbf{m} = 5\,(4 \sin 60°)\,\frac{(\mathbf{a}_x + \mathbf{a}_y + \mathbf{a}_z)}{\sqrt{3}}$$

$$= 10(\mathbf{a}_x + \mathbf{a}_y + \mathbf{a}_z)\ \text{A} \cdot \text{m}^2$$

■

PRACTICE EXERCISE 8.5

A rectangular coil of area 10 cm^2 carrying current of 50 A lies on plane $2x + 6y - 3z = 7$ such that the magnetic moment of the coil is directed away from the origin. Calculate its magnetic moment.

SOLUTION $(1.429\mathbf{a}_x + 4.286\mathbf{a}_y - 2.143\mathbf{a}_z) \times 10^{-2}\ \text{A} \cdot \text{m}^2$

EXAMPLE 8.6

A small current loop L_1 with magnetic moment $5\mathbf{a}_z$ $\text{A} \cdot \text{m}^2$ is located at the origin while another small loop current L_2 with magnetic moment $3\mathbf{a}_y$ $\text{A} \cdot \text{m}^2$ is located at $(4, -3, 10)$. Determine the torque on L_2.

SOLUTION The torque $\mathbf{T}_2$ on the loop L_2 is due to the field $\mathbf{B}_1$ produced by loop L_1. Hence,

$$\mathbf{T}_2 = \mathbf{m}_2 \times \mathbf{B}_1$$

Since $\mathbf{m}_1$ for loop L_1 is along $\mathbf{a}_z$, we find $\mathbf{B}_1$ using eq. (8.22):

$$\mathbf{B}_1 = \frac{\mu_o m_1}{4\pi r^3}\,(2 \cos \theta\ \mathbf{a}_r + \sin \theta\ \mathbf{a}_\theta)$$

Using eq. (2.23), we transform $\mathbf{m}_2$ from Cartesian to spherical coordinates:

$$\mathbf{m}_2 = 3\mathbf{a}_y = 3\,(\sin \theta \sin \phi\ \mathbf{a}_r + \cos \theta \sin \phi\ \mathbf{a}_\theta + \cos \phi\ \mathbf{a}_\phi)$$

At $(4, -3, 10)$,

$$r = \sqrt{4^2 + (-3)^2 + 10^2} = 5\sqrt{5}$$

$$\tan \theta = \frac{\rho}{z} = \frac{5}{10} = \frac{1}{2} \rightarrow \sin \theta = \frac{1}{\sqrt{5}}, \qquad \cos \theta = \frac{2}{\sqrt{5}}$$

$$\tan \phi = \frac{y}{x} = \frac{-3}{4} \rightarrow \sin \phi = \frac{-3}{5}, \qquad \cos \phi = \frac{4}{5}$$

Hence,

$$\mathbf{B}_1 = \frac{4\pi \times 10^{-7} \times 5}{4\pi\ 625\ \sqrt{5}} \left(\frac{4}{\sqrt{5}} \mathbf{a}_r + \frac{1}{\sqrt{5}} \mathbf{a}_\theta \right)$$

$$= \frac{10^{-7}}{625} (4\mathbf{a}_r + \mathbf{a}_\theta)$$

$$\mathbf{m}_2 = 3 \left[-\frac{3\mathbf{a}_r}{5\sqrt{5}} - \frac{6\mathbf{a}_\theta}{5\sqrt{5}} + \frac{4\mathbf{a}_\phi}{5} \right]$$

and

$$\mathbf{T} = \frac{10^{-7}\ (3)}{625\ (5\sqrt{5})} (-3\mathbf{a}_r - 6\mathbf{a}_\theta + 4\sqrt{5}\mathbf{a}_\phi) \times (4\mathbf{a}_r + \mathbf{a}_\phi)$$

$$= 4.293 \times 10^{-11}\ (-6\mathbf{a}_r + 38.78\mathbf{a}_\theta + 24\mathbf{a}_\phi)$$

$$= -0.258\mathbf{a}_r + 1.665\mathbf{a}_\theta + 1.03\mathbf{a}_\phi \text{ nN} \cdot \text{m}$$

■

PRACTICE EXERCISE 8.6

If the coil of Practice Exercise 8.5 is surrounded by a uniform field $0.6\mathbf{a}_x + 0.4\mathbf{a}_y + 0.5\mathbf{a}_z$ Wb/m^2,

(a) Find the torque on the coil.

(b) Show that the torque on the coil is maximum if placed on plane $2x - 8y + 4z = \sqrt{84}$. Calculate the value of the maximum torque.

ANSWER (a) $0.03\mathbf{a}_x - 0.02\mathbf{a}_y - 0.02\mathbf{a}_z$ N · m, (b) 0.04387 N · m.

8.5 MAGNETIZATION IN MATERIALS

Our discussion here will parallel that on polarization of materials in an electric field. We shall assume that our atomic model is that of an electron orbiting about a positive nucleus.

We know that a given material is composed of atoms. Each atom may be regarded as consisting of electrons orbiting about a central positive nucleus, the electrons also rotate (or spin) about their own axes. Thus an internal magnetic field is produced by electrons orbiting around the nucleus as in Figure 8.10(a) or electrons spinning as in

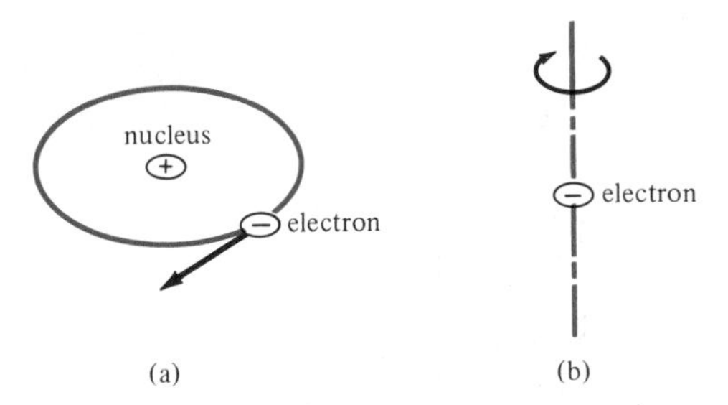

Figure 8.10 (**a**) Electron orbiting around the nucleus; (**b**) electron spin.

Figure 8.10(b). Both of these electronic motions produce internal magnetic fields $\mathbf{B}_i$ that are similar to the magnetic field produced by a current loop of Figure 8.11. The equivalent current loop has a magnetic moment of $\mathbf{m} = I_b S \mathbf{a}_n$, where S is the area of the loop and I_b is the bound current (bound to the atom).

Without an external **B** field applied to the material, the sum of **m**s is zero due to random orientation as in Figure 8.12(a). When an external **B** field is applied, the magnetic moments of the electrons more or less align themselves with **B** so that the net magnetic moment is not zero, as illustrated in Figure 8.12(b). If there are N atoms in a given volume Δv and the kth atom has a magnetic moment $\mathbf{m}_k$, we define the *magnetization* **M** (in amperes/meter) as the magnetic dipole moment per unit volume; that is,

$$\mathbf{M} = \lim_{\Delta v \to 0} \frac{\sum_{k=1}^{N} \mathbf{m}_k}{\Delta v} \qquad [8.27]$$

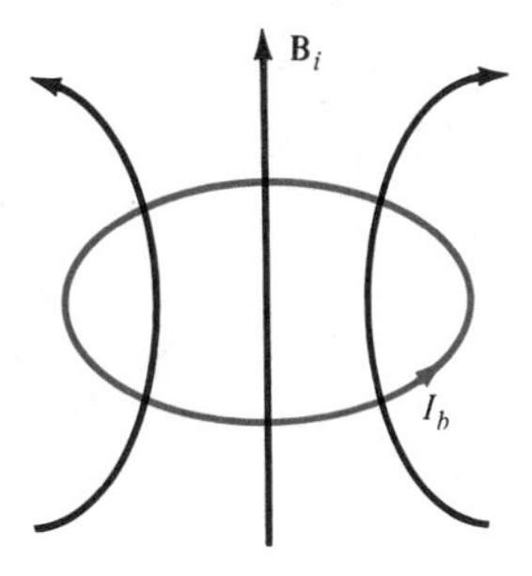

Figure 8.11 Circular current loop equivalent to electronic motion of Figure 8.10.

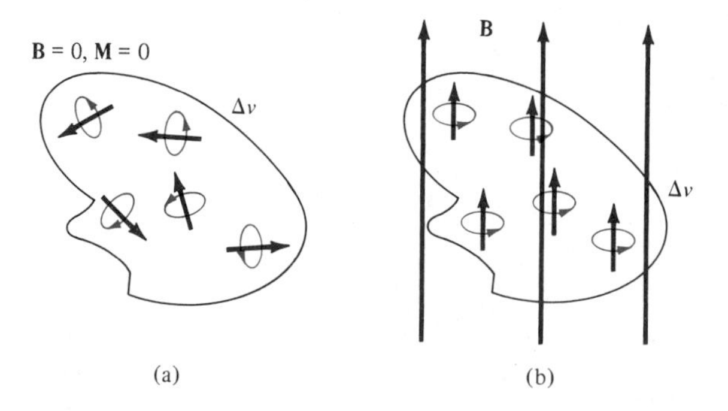

Figure 8.12 Magnetic dipole moment in a volume Δv: **(a)** before **B** is applied, **(b)** after **B** is applied.

A medium for which **M** is not zero everywhere is said to be magnetized. For a differential volume dv', the magnetic moment $d\mathbf{m} = \mathbf{M}\, dv'$. From eq. (8.21b), the vector magnetic potential due to $d\mathbf{m}$ is

$$d\mathbf{A} = \frac{\mu_o \mathbf{M} \times \mathbf{a}_R}{4\pi R^2}\, dv' = \frac{\mu_o \mathbf{M} \times \mathbf{R}}{4\pi R^3}\, dv'$$

According to eq. (7.46),

$$\frac{\mathbf{R}}{R^3} = \nabla' \frac{1}{R}$$

Hence,

$$\mathbf{A} = \frac{\mu_o}{4\pi} \int \mathbf{M} \times \nabla' \frac{1}{R}\, dv' \tag{8.28}$$

Using eq. (7.48) gives

$$\mathbf{M} \times \nabla' \frac{1}{R} = \frac{1}{R} \nabla' \times \mathbf{M} - \nabla' \times \frac{\mathbf{M}}{R}$$

Substituting this into eq. (8.28) yields

$$\mathbf{A} = \frac{\mu_o}{4\pi} \int_{v'} \frac{\nabla' \times \mathbf{M}}{R}\, dv' - \frac{\mu_o}{4\pi} \int_{v'} \nabla' \times \frac{\mathbf{M}}{R}\, dv'$$

Applying the vector identity

$$\int_{v'} \nabla' \times \mathbf{F}\, dv' = -\oint_{S'} \mathbf{F} \times d\mathbf{S}$$

to the second integral, we obtain

$$\mathbf{A} = \frac{\mu_o}{4\pi} \int_{v'} \frac{\nabla' \times \mathbf{M}}{R} \, dv' + \frac{\mu_o}{4\pi} \oint_{S'} \frac{\mathbf{M} \times \mathbf{a}_n}{R} \, dS'$$

$$= \frac{\mu_o}{4\pi} \int_{v'} \frac{\mathbf{J}_b \, dv'}{R} + \frac{\mu_o}{4\pi} \oint_{S'} \frac{\mathbf{K}_b \, dS'}{R} \tag{8.29}$$

Comparing eq. (8.29) with eqs. (7.42) and (7.43) (upon dropping the primes) gives

$$\boxed{\mathbf{J}_b = \nabla \times \mathbf{M}} \tag{8.30}$$

and

$$\boxed{\mathbf{K}_b = \mathbf{M} \times \mathbf{a}_n} \tag{8.31}$$

where $\mathbf{J}_b$ is the *bound volume current density* or *magnetization volume current density* (in amperes per meter square), $\mathbf{K}_b$ is the *bound surface current density* (in amperes per meter), and $\mathbf{a}_n$ is a unit vector normal to the surface. Equation (8.29) shows that the potential of a magnetic body is due to a volume current density $\mathbf{J}_b$ throughout the body and a surface current $\mathbf{K}_b$ on the surface of the body. The vector $\mathbf{M}$ is analogous to the polarization $\mathbf{P}$ in dielectrics and is sometimes called the *magnetic polarization density* of the medium. In another sense, $\mathbf{M}$ is analogous to $\mathbf{H}$ and they both have the same units. In this respect, as $\mathbf{J} = \nabla \times \mathbf{H}$, so is $\mathbf{J}_b = \nabla \times \mathbf{M}$. Also, $\mathbf{J}_b$ and $\mathbf{K}_b$ for a magnetized body are similar to ρ_{pv} and ρ_{ps} for a polarized body. As is evident in eqs. (8.29) to (8.31), $\mathbf{J}_b$ and $\mathbf{K}_b$ can be derived from $\mathbf{M}$; therefore, $\mathbf{J}_b$ and $\mathbf{K}_b$ are not commonly used.

In free space, $\mathbf{M} = 0$ and we have

$$\nabla \times \mathbf{H} = \mathbf{J}_f \quad \text{or} \quad \nabla \times \left(\frac{\mathbf{B}}{\mu_o}\right) = \mathbf{J}_f \tag{8.32}$$

where $\mathbf{J}_f$ is the free current volume density. In a material medium $\mathbf{M} \neq 0$, and as a result, $\mathbf{B}$ changes so that

$$\nabla \times \left(\frac{\mathbf{B}}{\mu_o}\right) = \mathbf{J}_f + \mathbf{J}_b = \mathbf{J}$$

$$= \nabla \times \mathbf{H} + \nabla \times \mathbf{M}$$

or

$$\boxed{\mathbf{B} = \mu_o(\mathbf{H} + \mathbf{M})} \tag{8.33}$$

The relationship in eq. (8.33) holds for all materials whether they are linear or not. The concepts of linearity, isotropy, and homogeneity introduced in Section 5.7 for dielectric media equally apply here for magnetic media. For linear materials, **M** (in A/m) depends linearly on **H** such that

$$\boxed{\mathbf{M} = \chi_m \mathbf{H}} \tag{8.34}$$

where χ_m is a dimensionless quantity (ratio of M to H) called *magnetic susceptibility* of the medium. It is more or less a measure of how susceptible (or sensitive) the material is to a magnetic field. Substituting eq. (8.34) into eq. (8.33) yields

$$\mathbf{B} = \mu_o(1 + \chi_m)\mathbf{H} = \mu\mathbf{H} \tag{8.35}$$

or

$$\boxed{\mathbf{B} = \mu_o \mu_r \mathbf{H}} \tag{8.36}$$

where

$$\boxed{\mu_r = 1 + \chi_m = \frac{\mu}{\mu_o}} \tag{8.37}$$

The quantity $\mu = \mu_o \mu_r$ is called the *permeability* of the material and is measured in the henrys/meter; the henry is the unit of inductance and will be defined a little later. The dimensionless quantity μ_r is the ratio of the permeability of a given material to that of free space and is known as the *relative permeability* of the material.

It should be borne in mind that the relationships in eqs. (8.34) to (8.37) only hold for linear and isotropic materials. If the materials are anisotropic (e.g., crystals), eq. (8.33) still holds but eqs. (8.34) to (8.37) do not apply. In this case, μ has nine terms (similar to ε in eq. 5.30) and consequently, the fields **B**, **H**, and **M** are no longer parallel.

†8.6 CLASSIFICATION OF MAGNETIC MATERIALS

In general, we may use the magnetic susceptibility χ_m or the relative permeability μ_r to classify materials in terms of their magnetic property or behavior. A material is said to be *nonmagnetic* if $\chi_m = 0$ (or $\mu_r = 1$); it is magnetic otherwise. Free space and materials with $\chi_m = 0$ (or $\mu_r \approx 1$) are regarded as nonmagnetic.

Roughly speaking, magnetic materials may be grouped into three major classes: diamagnetic, paramagnetic, and ferromagnetic. This rough classification is depicted in

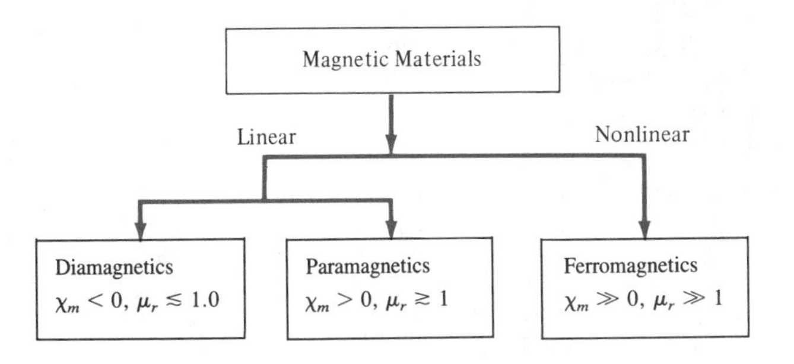

Figure 8.13 Classification of magnetic materials.

Figure 8.13. A material is said to be *diamagnetic* if it has $\mu_r \lesssim 1$ (i.e., very small negative χ_m). It is *paramagnetic* if $\mu_r \gtrsim 1$ (i.e., very small positive χ_m). If $\mu_r \gg 1$ (i.e., very large positive χ_m), the material is *ferromagnetic*. Table B.3 in Appendix B presents the values μ_r for some materials. From the table, it is apparent that for most practical purposes we may assume that $\mu_r \simeq 1$ for diamagnetic and paramagnetic materials. Thus, we may regard diamagnetic and paramagnetic materials as linear and nonmagnetic. Ferromagnetic materials are always nonlinear and magnetic except when their temperatures are above curie temperature (to be explained later). The reason for this will become evident as we have a closer examination of each of these three types of magnetic materials.

Diamagnetism occurs in materials where the magnetic fields due to electronic motions of orbiting and spinning completely cancel each other. Thus, the permanent (or intrinsic) magnetic moment of each atom is zero and the materials are weakly affected by a magnetic field. For most diamagnetic materials (e.g., bismuth, lead, copper, silicon, diamond, sodium chloride), χ_m is of the order of -10^{-5}. In certain types of materials called *superconductors* at temperatures near absolute zero, "perfect diamagnetism" occurs: $\chi_m = -1$ or $\mu_r = 0$ and $B = 0$. Thus superconductors cannot contain magnetic fields.[2] Except for superconductors, diamagnetic materials are seldom used in practice. Although the diamagnetic effect is overshadowed by other stronger effects in some materials, all materials exhibit diamagnetism.

Materials whose atoms have nonzero permanent magnetic moment may be paramagnetic or ferromagnetic. *Paramagnetism* occurs in materials where the magnetic fields produced by orbital and spinning electrons do not cancel completely. Unlike diamagnetism, paramagnetism is temperature-dependent. For most paramagnetic materials (e.g., air, platinum, tungsten, potassium), χ_m is of the order $+10^{-5}$ to $+10^{-3}$ and is temperature-dependent. Such materials find application in masers.

....

[2]An excellent treatment of superconductors is found in M. A. Plonus, *Applied Electromagnetics*. New York: McGraw-Hill, 1978, pp. 375–388. Also, the August 1989 issue of the *Proceedings of IEEE* is devoted to Superconductivity.

Ferromagnetism occurs in materials whose atoms have relatively large permanent magnetic moment. They are called ferromagnetic materials because the best known member is iron. Other members are cobalt, nickel, and their alloys. Ferromagnetic materials are very useful in practice. As distinct from diamagnetic and paramagnetic materials, ferromagnetic materials have the following properties:

1. They are capable of being magnetized very strongly by a magnetic field.
2. They retain a considerable amount of their magnetization when removed from the field.
3. They lose their ferromagnetic properties and become linear paramagnetic materials when the temperature is raised above a certain temperature known as *curie temperature*. Thus if a permanent magnet is heated above its curie temperature (770° C for iron), it loses its magnetization completely.
4. They are nonlinear; that is, the constitutive relation $\mathbf{B} = \mu_o\mu_r\mathbf{H}$ does not hold for ferromagnetic materials because μ_r depends on $\mathbf{B}$ and cannot be represented by a single value.

Thus, the values of μ_r cited in Table B.3 for ferromagnetics are only typical. For example, for nickel $\mu_r = 50$ under some conditions and 600 under other conditions.

As mentioned in section 5.9 for conductors, ferromagnetic materials, such as iron and steel, are used for screening (or shielding) to protect sensitive electrical devices from disturbances from strong magnetic fields. A typical example of an iron shield is shown in Figure 8.14(a) where the compass is protected. Without the iron shield, the compass gives an erroneous reading due to the effect of the external magnetic field as in Figure 8.14(b). For perfect screening, it is required that the shield have infinite permeability.

Even though $\mathbf{B} = \mu_o(\mathbf{H} + \mathbf{M})$ holds for all materials including ferromagnetics, the relationship between $\mathbf{B}$ and $\mathbf{H}$ depends on previous magnetization of a ferromagnetic material—its "magnetic history." Instead of having a linear relationship between $\mathbf{B}$ and $\mathbf{H}$ (i.e., $\mathbf{B} = \mu\mathbf{H}$), it is only possible to represent the relationship by a *magnetization curve* or *B-H* curve.

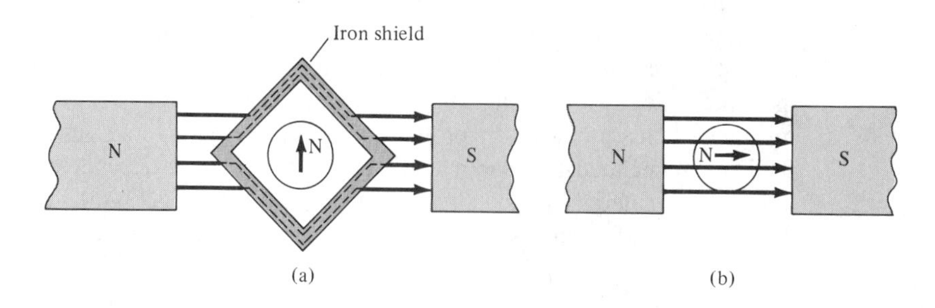

Figure 8.14 Magnetic screening: **(a)** iron shield protecting a small compass, **(b)** compass gives erroneous reading without the shield.

A typical *B-H* curve is shown in Figure 8.15. First of all, note the nonlinear relationship between B and H. Second, at any point on the curve, μ is given by the ratio B/H and not by dB/dH, the slope of the curve.

If we assume that the ferromagnetic material whose *B-H* curve in Figure 8.15 is initially unmagnetized, as H increases (due to increase in current) from O to maximum applied field intensity H_{max}, curve OP is produced. This curve is referred to as the *virgin* or *initial magnetization curve*. After reaching saturation at P, if H is decreased, B does not follow the initial curve but lags behind H. This phenomenon of B lagging behind H is called *hysteresis* (which means "to lag" in Greek).

If H is reduced to zero, B is not reduced to zero but to B_r, which is referred to as the *permanent flux density*. The value of B_r depends on H_{max}, the maximum applied field intensity. The existence of B_r is the cause of having permanent magnets. If H increases negatively (by reversing the direction of current), B becomes zero when H becomes H_c, which is known as the *coercive field intensity*. Materials for which H_c is small are said to be magnetically hard. The value of H_c also depends on H_{max}.

Further increase in H in the negative direction to reach Q and a reverse in its direction to reach P gives a closed curve called a *hysteresis loop*. The shape of hysteresis loops varies from one material to another. Some ferrites, for example, have an almost rectangular hysteresis loop and are used in digital computers as magnetic information storage devices. The area of a hysteresis loop gives the energy loss (hysteresis loss) per unit volume during one cycle of the periodic magnetization of the ferromagnetic material. This energy loss is in the form of heat. It is therefore desirable that materials used in electric generators, motors, and transformers should have tall but narrow hysteresis loops so that hysteresis losses are minimal.

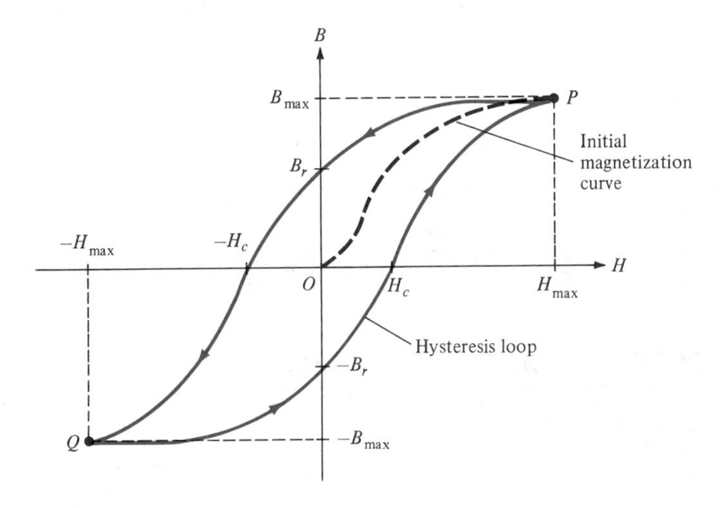

Figure 8.15 Typical magnetization (*B-H*) curve.

EXAMPLE 8.7

Region $0 \leq z \leq 2$ m is occupied by an infinite slab of permeable material ($\mu_r = 2.5$). If $\mathbf{B} = 10y\mathbf{a}_x - 5x\mathbf{a}_y$ mWb/m^2 within the slab, determine: (a) $\mathbf{J}$, (b) $\mathbf{J}_b$, (c) $\mathbf{M}$, (d) $\mathbf{K}_b$ on $z = 0$.

SOLUTION

(a) By definition,

$$\mathbf{J} = \nabla \times \mathbf{H} = \nabla \times \frac{\mathbf{B}}{\mu_o \mu_r} = \frac{1}{4\pi \times 10^{-7}(2.5)} \left(\frac{\partial B_y}{\partial x} - \frac{\partial B_x}{\partial y} \right) \mathbf{a}_z$$

$$= \frac{10^6}{\pi}(-5 - 10)10^{-3}\mathbf{a}_z = -4.775\mathbf{a}_z \text{ kA/m}^2$$

(b) $\mathbf{J}_b = \chi_m \mathbf{J} = (\mu_r - 1)\mathbf{J} = 1.5(-4.775\mathbf{a}_z) \cdot 10^3$

$$= -7.163\mathbf{a}_z \text{ kA/m}^2$$

(c) $\mathbf{M} = \chi_m \mathbf{H} = \chi_m \dfrac{\mathbf{B}}{\mu_o \mu_r} = \dfrac{1.5(10y\mathbf{a}_x - 5x\mathbf{a}_y) \cdot 10^{-3}}{4\pi \times 10^{-7}(2.5)}$

$$= 4.775y\mathbf{a}_x - 2.387x\mathbf{a}_y \text{ kA/m}$$

(d) $\mathbf{K}_b = \mathbf{M} \times \mathbf{a}_n$. Since $z = 0$ is the lower side of the slab occupying $0 \leq z \leq 2$, $\mathbf{a}_n = -\mathbf{a}_z$. Hence,

$$\mathbf{K}_b = (4.775y\mathbf{a}_x - 2.387x\mathbf{a}_y) \times (-\mathbf{a}_z)$$

$$= 2.387x\mathbf{a}_x + 4.775y\mathbf{a}_y \text{ kA/m}$$

■

PRACTICE EXERCISE 8.7

In a certain region ($\mu = 4.6\mu_o$),

$$\mathbf{B} = 10e^{-y}\mathbf{a}_z \text{ mWb/m}^2$$

find: (a) χ_m, (b) $\mathbf{H}$, (c) $\mathbf{M}$.

ANSWER (a) 3.6, (b) $1730e^{-y}\mathbf{a}_z$ A/m, (c) $6228e^{-y}\mathbf{a}_z$ A/m.

8.7 MAGNETIC BOUNDARY CONDITIONS

We define magnetic boundary conditions as the conditions that $\mathbf{H}$ (or $\mathbf{B}$) field must satisfy at the boundary between two different media. Our derivations here are similar to those in Section 5.9. We make use of Gauss's law for magnetic fields

$$\oint \mathbf{B} \cdot d\mathbf{S} = 0 \qquad [8.38]$$

and Ampere's circuit law

$$\oint \mathbf{H} \cdot d\mathbf{l} = I \tag{8.39}$$

Consider the boundary between two magnetic media 1 and 2, characterized respectively by μ_1 and μ_2 as in Figure 8.16. Applying eq. (8.38) to the pillbox (Gaussian surface) of Figure 8.16(a) and allowing $\Delta h \to 0$, we obtain

$$B_{1n}\,\Delta S - B_{2n}\,\Delta S = 0. \tag{8.40}$$

Thus

$$\boxed{\mathbf{B}_{1n} = \mathbf{B}_{2n}} \qquad \text{or} \qquad \mu_1 \mathbf{H}_{1n} = \mu_2 \mathbf{H}_{2n} \tag{8.41}$$

since $\mathbf{B} = \mu\mathbf{H}$. Equation (8.41) shows that the normal component of $\mathbf{B}$ is continuous at the boundary. It also shows that the normal component of $\mathbf{H}$ is discontinuous at the boundary; $\mathbf{H}$ undergoes some change at the interface.

Similarly, we apply eq. (8.39) to the closed path *abcda* of Figure 8.16(b) where surface current K on the boundary is assumed normal to the path. We obtain

$$K \cdot \Delta w = H_{1t} \cdot \Delta w + H_{1n} \cdot \frac{\Delta h}{2} + H_{2n} \cdot \frac{\Delta h}{2}$$

$$- H_{2t} \cdot \Delta w - H_{2n} \cdot \frac{\Delta h}{2} - H_{1n} \cdot \frac{\Delta h}{2} \tag{8.42}$$

As $\Delta h \to 0$, eq. (8.42) leads to

$$H_{1t} - H_{2t} = K \tag{8.43}$$

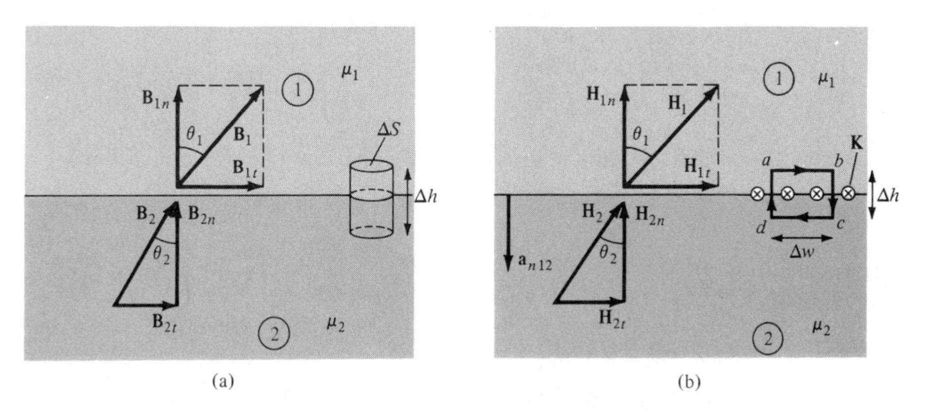

Figure 8.16 Boundary conditions between two magnetic media: **(a)** for $\mathbf{B}$, **(b)** for $\mathbf{H}$.

This shows that the tangential component of H is also discontinuous. Equation (8.43) may be written in terms of B as

$$\frac{B_{1t}}{\mu_1} - \frac{B_{2t}}{\mu_2} = K \tag{8.44}$$

In the general case, eq. (8.43) becomes

$$\boxed{(\mathbf{H}_1 - \mathbf{H}_2) \times \mathbf{a}_{n12} = \mathbf{K}} \tag{8.45}$$

where $\mathbf{a}_{n12}$ is a unit vector normal to the interface and is directed from medium 1 to medium 2. If the boundary is free of current or the media are not conductors (for K is free current density), $K = 0$ and eq. (8.43) becomes

$$\boxed{\mathbf{H}_{1t} = \mathbf{H}_{2t}} \quad \text{or} \quad \frac{\mathbf{B}_{1t}}{\mu_1} = \frac{\mathbf{B}_{2t}}{\mu_2} \tag{8.46}$$

Thus the tangential component of **H** is continuous while that of **B** is discontinuous at the boundary.

If the fields make an angle θ with the normal to the interface, eq. (8.41) results in

$$B_1 \cos \theta_1 = B_{1n} = B_{2n} = B_2 \cos \theta_2 \tag{8.47}$$

while eq. (8.46) produces

$$\frac{B_1}{\mu_1} \sin \theta_1 = H_{1t} = H_{2t} = \frac{B_2}{\mu_2} \sin \theta_2 \tag{8.48}$$

Dividing eq. (8.48) by eq. (8.47) gives

$$\boxed{\frac{\tan \theta_1}{\tan \theta_2} = \frac{\mu_1}{\mu_2}} \tag{8.49}$$

which is (similar to eq. (5.62)), the law of refraction for magnetic flux lines at a boundary with no surface current.

EXAMPLE 8.8

Given that $\mathbf{H}_1 = -2\mathbf{a}_x + 6\mathbf{a}_y + 4\mathbf{a}_z$ A/m in region $y - x - 2 \leq 0$ where $\mu_1 = 5\mu_o$, calculate

(a) $\mathbf{M}_1$ and $\mathbf{B}_1$

(b) $\mathbf{H}_2$ and $\mathbf{B}_2$ in region $y - x - 2 \geq 0$ where $\mu_2 = 2\mu_o$

SOLUTION Since $y - x - 2 = 0$ is a plane, $y - x \leq 2$ or $y \leq x + 2$ is region 1 in Figure 8.17. A point in this region may be used to confirm this. For example, the origin (0, 0) is in this region since $0 - 0 - 2 < 0$. If we let the surface of the plane be described by $f(x, y) = y - x - 2$, a unit vector normal to the plane is given by

$$\mathbf{a}_n = \frac{\nabla f}{|\nabla f|} = \frac{\mathbf{a}_y - \mathbf{a}_x}{\sqrt{2}}$$

(a)

$$\mathbf{M}_1 = \chi_{m1}\mathbf{H}_1 = (\mu_{r1} - 1)\,\mathbf{H}_1 = (5 - 1)(-2, 6, 4)$$
$$= -8\mathbf{a}_x + 24\mathbf{a}_y + 16\mathbf{a}_z \text{ A/m}$$
$$\mathbf{B}_1 = \mu_1\mathbf{H}_1 = \mu_o\mu_{r1}\mathbf{H}_1 = 4\pi \times 10^{-7}(5)(-2, 6, 4)$$
$$= -12.57\mathbf{a}_x + 37.7\mathbf{a}_y + 25.13\mathbf{a}_z \ \mu\text{Wb/m}^2$$

(b)
$$\mathbf{H}_{1n} = (\mathbf{H}_1 \cdot \mathbf{a}_n)\mathbf{a}_n = \left[(-2, 6, 4) \cdot \frac{(-1, 1, 0)}{\sqrt{2}}\right]\frac{(-1, 1, 0)}{\sqrt{2}}$$
$$= -4\mathbf{a}_x + 4\mathbf{a}_y$$

But

$$\mathbf{H}_1 = \mathbf{H}_{1n} + \mathbf{H}_{1t}$$

Hence,

$$\mathbf{H}_{1t} = \mathbf{H}_1 - \mathbf{H}_{1n} = (-2, 6, 4) - (-4, 4, 0)$$
$$= 2\mathbf{a}_x + 2\mathbf{a}_y + 4\mathbf{a}_z$$

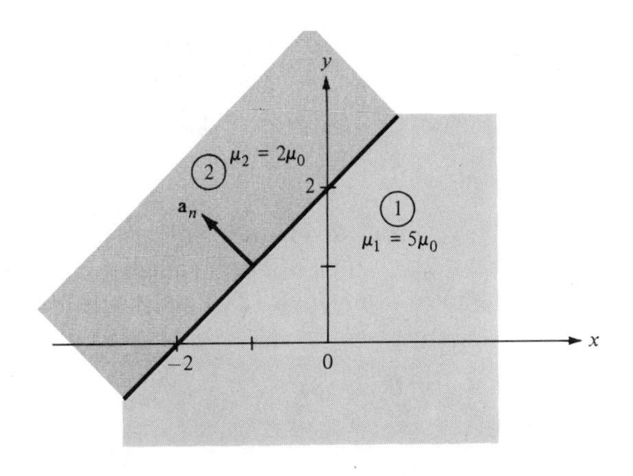

Figure 8.17 For Example 8.8.

Using the boundary conditions, we have

$$\mathbf{H}_{2t} = \mathbf{H}_{1t} = 2\mathbf{a}_x + 2\mathbf{a}_y + 4\mathbf{a}_z$$

$$\mathbf{B}_{2n} = \mathbf{B}_{1n} \rightarrow \mu_2\mathbf{H}_{2n} = \mu_1\mathbf{H}_{1n}$$

or

$$\mathbf{H}_{2n} = \frac{\mu_1}{\mu_2}\mathbf{H}_{1n} = \frac{5}{2}(-4\mathbf{a}_x + 4\mathbf{a}_y) = -10\mathbf{a}_x + 10\mathbf{a}_y$$

Thus

$$\mathbf{H}_2 = \mathbf{H}_{2n} + \mathbf{H}_{2t} = -8\mathbf{a}_x + 12\mathbf{a}_y + 4\mathbf{a}_z \text{ A/m}$$

and

$$\mathbf{B}_2 = \mu_2\mathbf{H}_2 = \mu_o\mu_{r2}\mathbf{H}_2 = (4\pi \times 10^{-7})(2)(-8, 12, 4)$$

$$= -20.11\mathbf{a}_x + 30.16\mathbf{a}_y + 10.05\mathbf{a}_z \ \mu\text{Wb/m}^2$$

■

PRACTICE EXERCISE 8.8

Region 1, described by $3x + 4y \geq 10$, is free space whereas region 2, described by $3x + 4y \leq 10$, is a magnetic material for which $\mu \simeq 10\mu_0$. Assuming that the boundary between the material and free space is current-free, find $\mathbf{B}_2$ if $\mathbf{B}_1 = 0.1\mathbf{a}_x + 0.4\mathbf{a}_y + 0.2\mathbf{a}_z$ Wb/m^2

ANSWER $-1.052\mathbf{a}_x + 1.264\mathbf{a}_y + 2\mathbf{a}_z$ Wb/m^2

EXAMPLE 8.9

The xy-plane serves as the interface between two different media. Medium 1 ($z < 0$) is filled with a material whose $\mu_r = 6$, and medium 2 ($z > 0$) is filled with a material whose $\mu_r = 4$. If the interface carries current $(1/\mu_0)\ \mathbf{a}_y$ mA/m, and $\mathbf{B}_2 = 5\mathbf{a}_x + 8\mathbf{a}_z$ mWb/m^2, find $\mathbf{H}_1$ and $\mathbf{B}_1$.

SOLUTION In the previous example $\mathbf{K} = 0$, so eq. (8.46) was appropriate. In this example, however, $\mathbf{K} \neq 0$, and we must resort to eq. (8.45) in addition to eq. (8.41). Consider the problem as illustrated in Figure 8.18. Let $\mathbf{B}_1 = (B_x, B_y, B_z)$ in mWb/m^2.

$$\mathbf{B}_{1n} = \mathbf{B}_{2n} = 8\mathbf{a}_z \rightarrow B_z = 8 \quad [8.8.1]$$

But

$$\mathbf{H}_2 = \frac{\mathbf{B}_2}{\mu_2} = \frac{1}{4\mu_o}(5\mathbf{a}_x + 8\mathbf{a}_z) \text{ mA/m} \quad [8.8.2]$$

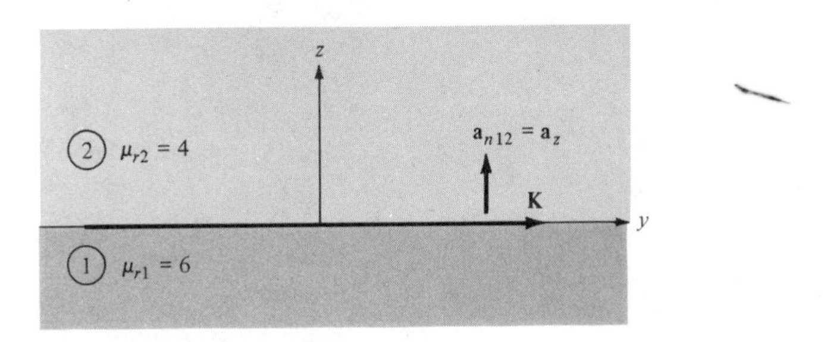

Figure 8.18 For Example 8.9.

and

$$\mathbf{H}_1 = \frac{\mathbf{B}_1}{\mu_1} = \frac{1}{6\mu_o}(B_x\mathbf{a}_x + B_y\mathbf{a}_y + B_z\mathbf{a}_z) \text{ mA/m} \qquad [8.8.3]$$

Having found the normal components, we can find the tangential components using

$$(\mathbf{H}_1 - \mathbf{H}_2) \times \mathbf{a}_{n12} = \mathbf{K}$$

or

$$\mathbf{H}_1 \times \mathbf{a}_{n12} = \mathbf{H}_2 \times \mathbf{a}_{n12} + \mathbf{K} \qquad [8.8.4]$$

Substituting eqs. (8.8.2) and (8.8.3) into eq. (8.8.4) gives

$$\frac{1}{6\mu_o}(B_x\mathbf{a}_x + B_y\mathbf{a}_y + B_z\mathbf{a}_z) \times \mathbf{a}_z = \frac{1}{4\mu_o}(5\mathbf{a}_x + 8\mathbf{a}_z) \times \mathbf{a}_z + \frac{1}{\mu_o}\mathbf{a}_y$$

Equating components yields

$$B_y = 0, \qquad \frac{-B_x}{6} = \frac{-5}{4} + 1 \qquad \text{or} \qquad B_x = \frac{6}{4} = 1.5 \qquad [8.8.5]$$

From eqs. (8.8.1) and (8.8.5),

$$\mathbf{B}_1 = 1.5\mathbf{a}_x + 8\mathbf{a}_z \text{ mWb/m}^2$$

$$\mathbf{H}_1 = \frac{\mathbf{B}_1}{\mu_1} = \frac{1}{\mu_o}(0.25\mathbf{a}_x + 1.33\mathbf{a}_z) \text{ mA/m}$$

and

$$\mathbf{H}_2 = \frac{1}{\mu_o}(1.25\mathbf{a}_x + 2\mathbf{a}_z) \text{ mA/m}$$

Note that H_{1x} is $(1/\mu_o)$ mA/m less than H_{2x} due to the current sheet and also that $B_{1n} = B_{2n}$. ■

PRACTICE EXERCISE 8.9

A unit normal vector from region 2 ($\mu = 2\mu_o$) to region 1 ($\mu = \mu_o$) is $\mathbf{a}_{n21} = (6\mathbf{a}_x + 2\mathbf{a}_y - 3\mathbf{a}_z)/7$. If $\mathbf{H}_1 = 10\mathbf{a}_x + \mathbf{a}_y + 12\mathbf{a}_z$ A/m and $\mathbf{H}_2 = H_{2x}\mathbf{a}_x - 5\mathbf{a}_y + 4\mathbf{a}_z$ A/m, determine

(a) $\mathbf{H}_{2x}$

(b) The surface current density **K** on the interface

(c) The angles $\mathbf{B}_1$ and $\mathbf{B}_2$ make with the normal to the interface.

ANSWER (a) 5.833, (b) $4.86\mathbf{a}_x - 8.64\mathbf{a}_y + 3.95\mathbf{a}_z$ A/m, (c) 76.27°, 77.62°.

8.8 INDUCTORS AND INDUCTANCES

A circuit (or closed conducting path) carrying current I produces a magnetic field **B** which causes a flux $\Psi = \int \mathbf{B} \cdot d\mathbf{S}$ to pass through each turn of the circuit as shown in Figure 8.19. If the circuit has N identical turns, we define the flux linkage λ as

$$\lambda = N\Psi \tag{8.50}$$

Also, if the medium surrounding the circuit is linear, the *flux linkage* λ is proportional to the current I producing it; that is,

$$\lambda \propto I$$

$$\text{or} \quad \lambda = LI \tag{8.51}$$

where L is a constant of proportionality called the *inductance* of the circuit. The inductance L is a property of the physical arrangement of the circuit. A circuit or part of a circuit that has inductance is called an *inductor*. From eqs. (8.50) and (8.51), we may define inductance L of an inductor as the ratio of the magnetic flux linkage λ to the current I through the inductor; that is,

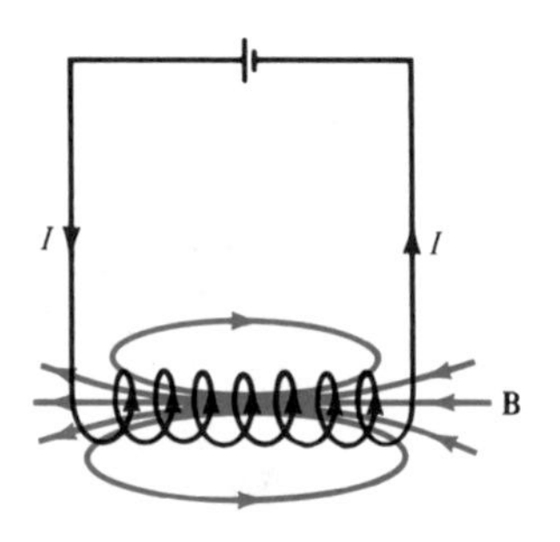

Figure 8.19 Magnetic field **B** produced by a circuit.

$$\boxed{L = \frac{\lambda}{I} = \frac{N\Psi}{I}} \quad [8.52]$$

The unit of inductance is the henry (H) which is the same as webers/ampere. Since henry is a fairly large unit, inductances are usually expressed in millihenrys (mH).

The inductance defined by eq. (8.52) is commonly referred to as *self-inductance* since the linkages are produced by the inductor itself. Like capacitance, we may regard inductance as a measure of how much magnetic energy is stored in an inductor. The magnetic energy (in joules) stored in an inductor is expressed in circuit theory as:

$$W_m = \frac{1}{2}LI^2 \quad [8.53]$$

or

$$\boxed{L = \frac{2W_m}{I^2}} \quad [8.54]$$

Thus the self-inductance of a circuit may be defined or calculated from energy considerations.

If instead of having a single circuit we have two circuits carrying current I_1 and I_2 as shown in Figure 8.20, a magnetic interaction exists between the circuits. Four component fluxes Ψ_{11}, Ψ_{12}, Ψ_{21}, and Ψ_{22} are produced. The flux Ψ_{12}, for example, is the flux passing through circuit 1 due to current I_2 in circuit 2. If $\mathbf{B}_2$ in the field due to I_2 and S_1 is the area of circuit 1, then

$$\Psi_{12} = \int_{S_1} \mathbf{B}_2 \cdot d\mathbf{S} \quad [8.55]$$

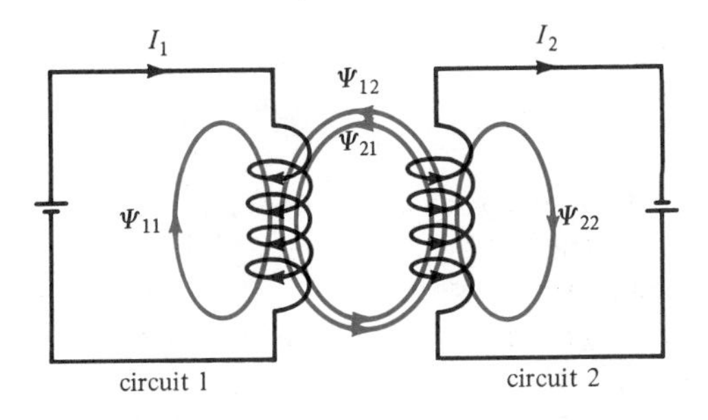

Figure 8.20 Magnetic interaction between two circuits.

We define the *mutual inductance* M_{12} as the ratio of the flux linkage $\lambda_{12} = N_1\Psi_{12}$ on circuit 1 due to current I_2, that is,

$$M_{12} = \frac{\lambda_{12}}{I_2} = \frac{N_1\Psi_{12}}{I_2} \tag{8.56}$$

Similarly, the mutual inductance M_{21} is defined as the flux linkages of circuit 2 per unit current I_1; that is,

$$M_{21} = \frac{\lambda_{21}}{I_1} = \frac{N_2\Psi_{21}}{I_1} \tag{8.57a}$$

It can be shown by using energy concepts that if the medium surrounding the circuits is linear (i.e., in the absence of ferromagnetic material),

$$M_{12} = M_{21} \tag{8.57b}$$

The mutual inductance M_{12} or M_{21} is expressed in henrys and should not be confused with the magnetization vector $\mathbf{M}$ expressed in amperes/meter.

We define the self-inductance of circuits 1 and 2 respectively as

$$L_1 = \frac{\lambda_{11}}{I_1} = \frac{N_1\Psi_1}{I_1} \tag{8.58}$$

and

$$L_2 = \frac{\lambda_{22}}{I_2} = \frac{N_2\Psi_2}{I_2} \tag{8.59}$$

where $\Psi_1 = \Psi_{11} + \Psi_{12}$ and $\Psi_2 = \Psi_{21} + \Psi_{22}$. The total energy in the magnetic field is the sum of the energies due to L_1, L_2, and M_{12} (or M_{21}); that is,

$$\begin{aligned} W_m &= W_1 + W_2 + W_{12} \\ &= \frac{1}{2} L_1 I_1^2 + \frac{1}{2} L_2 I_2^2 \pm M_{12} I_1 I_2 \end{aligned} \tag{8.60}$$

The positive sign is taken if currents I_1 and I_2 flow such that the magnetic fields of the two circuits strengthen each other. If the currents flow such that their magnetic fields oppose each other, the negative sign is taken.

As mentioned earlier, an inductor is a conductor arranged in an appropriate shape to store magnetic energy. Typical examples of inductors are toroids, solenoids, coaxial transmission lines, and parallel-wire transmission lines. The inductance of each of these inductors can be determined by following a procedure similar to that taken in determining the capacitance of a capacitor. For a given inductor, we find the self-inductance L by taking these steps:

1. Choose a suitable coordinate system.
2. Let the inductor carry current I.
3. Determine $\mathbf{B}$ from Biot-Savart's law (or from Ampere's law if symmetry exists) and calculate Ψ from $\Psi = \int \mathbf{B} \cdot d\mathbf{S}$.
4. Finally find L from $L = \dfrac{\lambda}{I} = \dfrac{N\Psi}{I}$.

The mutual inductance between two circuits may be calculated by taking a similar procedure.

In an inductor such as a coaxial or a parallel-wire transmission line, the inductance produced by the flux internal to the conductor is called the *internal inductance* L_{in} while that produced by the flux external to it is called *external inductance* L_{ext}. The total inductance L is

$$L = L_{in} + L_{ext} \tag{8.61}$$

Just as it was shown that for capacitors

$$RC = \frac{\varepsilon}{\sigma} \tag{6.35}$$

it can be shown that

$$\boxed{L_{ext}C = \mu\varepsilon} \tag{8.62}$$

Thus L_{ext} may be calculated using eq. (8.62) if C is known.

A collection of formulas for some fundamental circuit elements is presented in Table 8.3. All formulas can be derived by taking the steps outlined above.[3]

8.9 MAGNETIC ENERGY

Just as the potential energy in an electrostatic field was derived as

$$W_E = \frac{1}{2}\int \mathbf{D} \cdot \mathbf{E}\, dv = \frac{1}{2}\int \varepsilon E^2\, dv \tag{4.103}$$

[3]Additional formulas can be found in standard electrical handbooks or in H. Knoepfel, *Pulsed High Magnetic Fields*. Amsterdam: North-Holland, 1970, pp. 312–324.

Table 8.3 A Collection of Formulas for Inductance of Common Elements

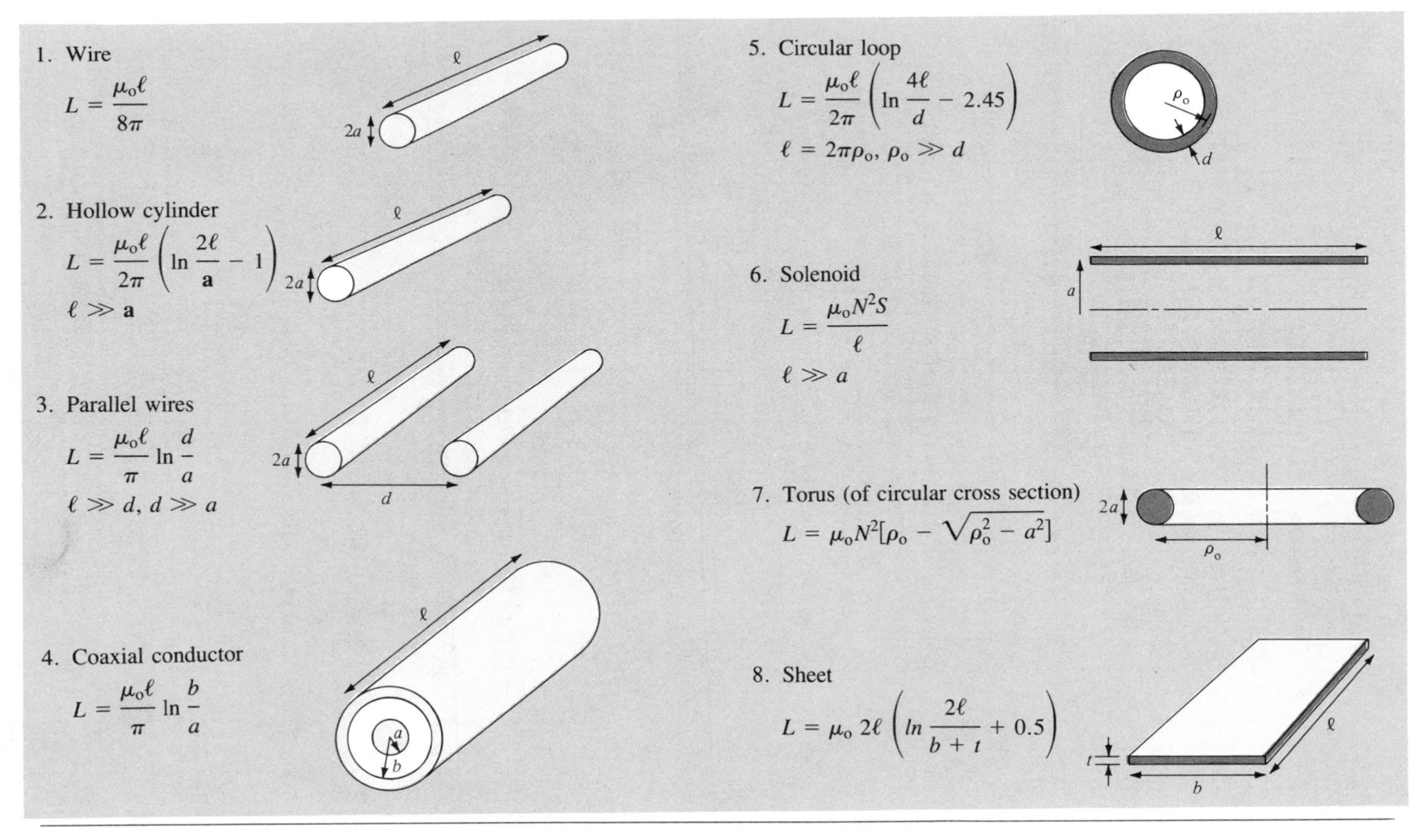

1. Wire

$$L = \frac{\mu_o \ell}{8\pi}$$

2. Hollow cylinder

$$L = \frac{\mu_o \ell}{2\pi}\left(\ln \frac{2\ell}{\mathbf{a}} - 1\right)$$

$\ell \gg \mathbf{a}$

3. Parallel wires

$$L = \frac{\mu_o \ell}{\pi} \ln \frac{d}{a}$$

$\ell \gg d,\ d \gg a$

4. Coaxial conductor

$$L = \frac{\mu_o \ell}{\pi} \ln \frac{b}{a}$$

5. Circular loop

$$L = \frac{\mu_o \ell}{2\pi}\left(\ln \frac{4\ell}{d} - 2.45\right)$$

$\ell = 2\pi \rho_o,\ \rho_o \gg d$

6. Solenoid

$$L = \frac{\mu_o N^2 S}{\ell}$$

$\ell \gg a$

7. Torus (of circular cross section)

$$L = \mu_o N^2 [\rho_o - \sqrt{\rho_o^2 - a^2}]$$

8. Sheet

$$L = \mu_o\, 2\ell \left(ln \frac{2\ell}{b + t} + 0.5\right)$$

we would like to derive a similar expression for the energy in a magnetostatic field. A simple approach is using the magnetic energy in the field of an inductor. From eq. (8.53),

$$W_m = \frac{1}{2} LI^2 \tag{8.53}$$

The energy is stored in the magnetic field **B** of the inductor. We would like to express eq. (8.53) in terms of **B** or **H**.

Consider a differential volume in a magnetic field as shown in Figure 8.21. Let the volume be covered with conducting sheets at the top and bottom surfaces with current ΔI. We assume that the whole region is filled with such differential volumes. From eq. (8.52), each volume has an inductance

$$\Delta L = \frac{\Delta \Psi}{\Delta I} = \frac{\mu H \, \Delta x \, \Delta z}{\Delta I} \tag{8.63}$$

where $\Delta I = H \, \Delta y$. Substituting eq. (8.63) into eq. (8.53), we have

$$\Delta W_m = \frac{1}{2} \Delta L \, \Delta I^2 = \frac{1}{2} \mu H^2 \, \Delta x \, \Delta y \, \Delta z \tag{8.64}$$

or

$$\Delta W_m = \frac{1}{2} \mu H^2 \, \Delta v$$

The magnetostatic energy density w_m (in J/m^3) is defined as

$$w_m = \lim_{\Delta v \to 0} \frac{\Delta W_m}{\Delta v} = \frac{1}{2} \mu H^2$$

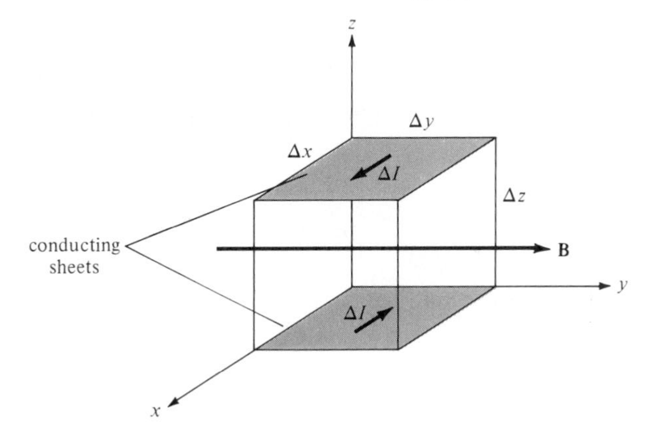

Figure 8.21 A differential volume in a magnetic field.

Hence,

$$w_m = \frac{1}{2}\,\mu H^2 = \frac{1}{2}\,B \cdot H = \frac{B^2}{2\mu} \qquad [8.65]$$

Thus the energy in a magnetostatic field in a linear medium is

$$W_m = \int w_m \, dv$$

or

$$\boxed{W_m = \frac{1}{2}\int \mathbf{B} \cdot \mathbf{H} \, dv = \frac{1}{2}\int \mu H^2 \, dv} \qquad [8.66]$$

which is similar to eq. (4.103) for an electrostatic field.

EXAMPLE 8.10

Calculate the self-inductance per unit length of an infinitely long solenoid.

SOLUTION We recall from Example 7.4 that for an infinitely long solenoid, the magnetic flux inside the solenoid per unit length is

$$B = \mu H = \mu I n$$

where $n = N/\ell$ = number of turns per unit length. If S is the cross-sectional area of the solenoid, the total flux through the cross section is

$$\Psi = BS = \mu I n S$$

Since this flux is only for a unit length of the solenoid, the linkage per unit length is

$$\lambda' = \frac{\lambda}{\ell} = n\Psi = \mu n^2 I S$$

and thus the inductance per unit length is

$$L' = \frac{L}{\ell} = \frac{\lambda'}{I} = \mu n^2 S$$

$$\boxed{L' = \mu n^2 S} \quad \text{H/m}$$

■

PRACTICE EXERCISE 8.10

A very long solenoid with 2 × 2 cm cross section has an iron core ($\mu_r = 1000$) and 4000 turns/meter. If it carries a current of 500 mA, find

(a) Its self-inductance per meter

(b) The energy per meter stored in its field

ANSWER (a) 8.042 H/m, (b) 1.005 J/m.

EXAMPLE 8.11

Determine the self-inductance of a coaxial cable of inner radius a and outer radius b.

SOLUTION The self-inductance of the inductor can be found in two different ways: by taking the four steps given in Section 8.8 or by using eqs. (8.54) and (8.66).

Method 1: Consider the cross section of the cable as shown in Figure 8.22. We recall from eq. (7.29) that by applying Ampere's circuit law, we obtained for region 1 ($0 \leq \rho \leq a$),

$$\mathbf{B}_1 = \frac{\mu I \rho}{2\pi a^2}\, \mathbf{a}_\phi$$

and for region 2 ($a \leq \rho \leq b$),

$$\mathbf{B}_2 = \frac{\mu I}{2\pi \rho}\, \mathbf{a}_\phi$$

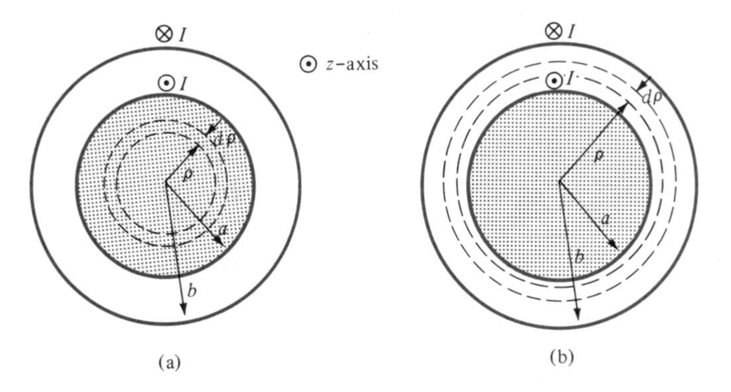

Figure 8.22 Cross section of the coaxial cable: **(a)** for region 1, $0 < \rho < a$, **(b)** for region 2, $a < \rho < b$; for Example 8.11.

We first find the internal inductance L_{in} by considering the flux linkages due to the inner conductor. From Figure 8.22(a), the flux leaving a differential shell of thickness $d\rho$ is

$$d\Psi_1 = B_1\, d\rho\, dz = \frac{\mu I \rho}{2\pi a^2}\, d\rho\, dz$$

The flux linkage is $d\Psi_1$ multiplied by the ratio of the area within the path enclosing the flux to the total area, that is,

$$d\lambda_1 = d\Psi_1 \cdot \frac{I_{enc}}{I} = d\Psi_1 \cdot \frac{\pi \rho^2}{\pi a^2}$$

because I is uniformly distributed over the cross section for d.c. excitation. Thus, the total flux linkages within the differential flux element are

$$d\lambda_1 = \frac{\mu I \rho\, d\rho\, dz}{2\pi a^2} \cdot \frac{\rho^2}{a^2}$$

For length ℓ of the cable,

$$\lambda_1 = \int_{\rho=0}^{a} \int_{z=0}^{\ell} \frac{\mu I \rho^3\, d\rho\, dz}{2\pi a^4} = \frac{\mu I \ell}{8\pi}$$

$$L_{in} = \frac{\lambda_1}{I} = \frac{\mu \ell}{8\pi} \qquad [8.11.1]$$

The inductance per unit length, given by

$$\boxed{L'_{in} = \frac{L_{in}}{\ell} = \frac{\mu}{8\pi}} \qquad \text{H/m} \qquad [8.11.2]$$

is independent of the radius of the conductor or wire. Thus eqs. (8.11.1) and (8.11.2) are also applicable to finding the inductance of any infinitely long straight conductor of finite radius.

We now determine the external inductance L_{ext} by considering the flux linkages between the inner and the outer conductor as in Figure 8.22(b). For a differential shell of thickness $d\rho$,

$$d\Psi_2 = B_2\, d\rho\, dz = \frac{\mu I}{2\pi \rho}\, d\rho\, dz$$

In this case, the total current I is enclosed within the path enclosing the flux. Hence,

$$\lambda_2 = \Psi_2 = \int_{\rho=a}^{b} \int_{z=0}^{\ell} \frac{\mu I\, d\rho\, dz}{2\pi \rho} = \frac{\mu I \ell}{2\pi} \ln \frac{b}{a}$$

$$L_{ext} = \frac{\lambda_2}{I} = \frac{\mu \ell}{2\pi} \ln \frac{b}{a}$$

Thus

$$L = L_{\text{in}} + L_{\text{ext}} = \frac{\mu \ell}{2\pi}\left[\frac{1}{4} + \ln \frac{b}{a}\right]$$

or the inductance per length is

$$\boxed{L' = \frac{L}{\ell} = \frac{\mu}{2\pi}\left[\frac{1}{4} + \ln \frac{b}{a}\right] \qquad \text{H/m}}$$

Method 2: It is easier to use eqs. (8.54) and (8.66) to determine L, that is,

$$W_m = \frac{1}{2} LI^2 \qquad \text{or} \qquad L = \frac{2W_m}{I^2}$$

where

$$W_m = \frac{1}{2}\int \mathbf{B} \cdot \mathbf{H} \, dv = \int \frac{B^2}{2\mu} \, dv$$

Hence

$$L_{\text{in}} = \frac{2}{I^2}\int \frac{B_1^2}{2\mu} \, dv = \frac{1}{I^2 \mu}\iiint \frac{\mu^2 I^2 \rho^2}{4\pi^2 a^4} \rho \, d\rho \, d\phi \, dz$$

$$= \frac{\mu}{4\pi^2 a^4}\int_0^\ell dz \int_0^{2\pi} d\phi \int_0^a \rho^3 \, d\rho = \frac{\mu \ell}{8\pi}$$

$$L_{\text{ext}} = \frac{2}{I^2}\int \frac{B_2^2}{2\mu} \, dv = \frac{1}{I^2 \mu}\iiint \frac{\mu^2 I^2}{4\pi^2 \rho^2} \rho \, d\rho \, d\phi \, dz$$

$$= \frac{\mu}{4\pi^2}\int_0^\ell dz \int_0^{2\pi} d\phi \int_a^b \frac{d\rho}{\rho} = \frac{\mu \ell}{2\pi} \ln \frac{b}{a}$$

and

$$L = L_{\text{in}} + L_{\text{ext}} = \frac{\mu \ell}{2\pi}\left[\frac{1}{4} + \ln \frac{b}{a}\right]$$

as obtained previously. ■

PRACTICE EXERCISE 8.11

Calculate the self-inductance of the coaxial cable of Example 8.11 if the inner conductor is made of an inhomogeneous material having $\mu = 2\mu_o/(1 + \rho)$.

ANSWER

$$\frac{\mu_o \ell}{8\pi} + \frac{\mu_o \ell}{\pi}\left[\ln \frac{b}{a} - \ln \frac{(1 + b)}{(1 + a)}\right]$$

EXAMPLE 8.12

Determine the inductance per unit length of a two-wire transmission line with separation distance d. Each wire has radius a as shown in Figure 6.37.

SOLUTION

We use the two methods of the last example.

Method 1: We determine L_{in} just as we did in the last example. Thus for region $0 \leq \rho \leq a$, we obtain

$$\lambda_1 = \frac{\mu I \ell}{8\pi}$$

as in the last example. For region $a \leq \rho \leq d - a$, the flux linkages between the wires are

$$\lambda_2 = \Psi_2 = \int_{\rho = a}^{d - a} \int_{z = 0}^{\ell} \frac{\mu I}{2\pi\rho} \, d\rho \, dz = \frac{\mu I \ell}{2\pi} \ln \frac{d - a}{a}$$

The flux linkages produced by wire 1 are

$$\lambda_1 + \lambda_2 = \frac{\mu I \ell}{8\pi} + \frac{\mu I \ell}{2\pi} \ln \frac{d - a}{a}$$

By symmetry, the same amount of flux produced by current $-I$ in wire 2. Hence the total linkages are

$$\lambda = 2(\lambda_1 + \lambda_2) = \frac{\mu I \ell}{\pi} \left[\frac{1}{4} + \ln \frac{d - a}{a} \right] = LI$$

If $d \gg a$, the self-inductance per unit length is

$$\boxed{L' = \frac{L}{\ell} = \frac{\mu}{\pi} \left[\frac{1}{4} + \ln \frac{d}{a} \right]} \qquad \text{H/m}$$

Method 2: From the last example,

$$L_{in} = \frac{\mu \ell}{8\pi}$$

Now

$$\begin{aligned} L_{ext} &= \frac{2}{I^2} \int \frac{B^2 \, dv}{2\mu} = \frac{1}{I^2 \mu} \iiint \frac{\mu^2 I^2}{4\pi^2 \rho^2} \rho \, d\rho \, d\phi \, dz \\ &= \frac{\mu}{4\pi^2} \int_0^{\ell} dz \int_0^{2\pi} d\phi \int_a^{d - a} \frac{d\rho}{\rho} \\ &= \frac{\mu \ell}{2\pi} \ln \frac{d - a}{a} \end{aligned}$$

Since the two wires are symmetrical,

$$L = 2\,(L_{in} + L_{ext})$$
$$= \frac{\mu\ell}{\pi}\left[\frac{1}{4} + \ln\frac{d-a}{a}\right]\ \text{H}$$

as obtained previously. ■

PRACTICE EXERCISE 8.12

Two # 10 copper wires (2.588 mm in diameter) are placed parallel in air with a separation distance d between them. If the inductance of each wire is 1.2 μH/m, calculate

(a) L_{in} and L_{ext} per meter for each wire
(b) The separation distance d

ANSWER (a) 0.05, 1.15 μH/m, (b) 40.79 cm.

EXAMPLE 8.13

Two coaxial circular wires of radii a and b ($b > a$) are separated by distance h ($h \gg a, b$) as shown in Figure 8.23. Find the mutual inductance between the wires.

SOLUTION Let current I_1 flow in wire 1. At an arbitrary point P on wire 2, the magnetic vector potential due to wire 1 is given by eq. (8.21a), namely

$$\mathbf{A}_1 = \frac{\mu I_1 a^2 \sin\theta}{4r^2}\,\mathbf{a}_\phi = \frac{\mu I_1 a^2 b\,\mathbf{a}_\phi}{4[h^2 + b^2]^{3/2}}$$

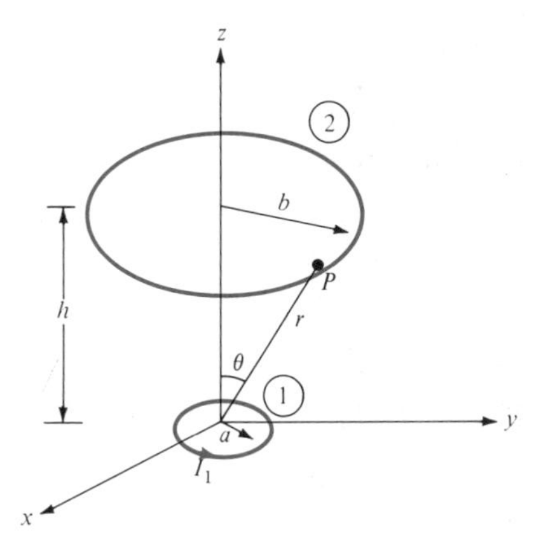

Figure 8.23 Two coaxial circular wires; for Example 8.13.

If $h \gg b$

$$\mathbf{A}_1 \simeq \frac{\mu I_1 a^2 b}{4h^3}\,\mathbf{a}_\phi$$

Hence,

$$\Psi_{12} = \oint \mathbf{A}_1 \cdot d\mathbf{l}_2 = \frac{\mu I_1 a^2 b}{4h^3}\,2\pi b = \frac{\mu \pi I_1 a^2 b^2}{2h^3}$$

and

$$M_{12} = \frac{\Psi_{12}}{I_1} = \frac{\mu \pi a^2 b^2}{2h^3}$$

■

PRACTICE EXERCISE 8.13

Find the mutual inductance of two coplanar concentric circular loops of radii 2 m and 3 m.

ANSWER 2.632 μH.

†8.10 MAGNETIC CIRCUITS

The concept of magnetic circuits is based on solving some magnetic field problems using circuit approach. Magnetic devices such as toroids, transformers, motors, generators, and relays may be considered as magnetic circuits. The analysis of such circuits is made simple if an analogy between magnetic circuits and electric circuits is exploited. Once this is done, we can directly apply concepts in electric circuits to solve their analogous magnetic circuits.

The analogy between magnetic and electric circuits is summarized in Table 8.4 and portrayed in Figure 8.24. The reader is advised to pause and study Table 8.4 and Figure 8.24. First, we notice from the table that two terms are new. We define the *magnetomotive force* (mmf) $\mathcal{F}$ (in ampere-turns) as

$$\boxed{\mathcal{F} = NI = \oint \mathbf{H} \cdot d\mathbf{l}} \qquad [8.67]$$

The source of mmf in magnetic circuits is usually a coil carrying current as in Figure 8.24. We also define *reluctance* $\mathcal{R}$ (in ampere-turns/weber) as

$$\boxed{\mathcal{R} = \frac{\ell}{\mu S}} \qquad [8.68]$$

Table 8.4 Analogy between Electric and Magnetic Circuits

Electric	Magnetic
Conductivity σ	Permeability μ
Field intensity E	Field intensity H
current $I = \int \mathbf{J} \cdot d\mathbf{S}$	Magnetic flux $\Psi = \int \mathbf{B} \cdot d\mathbf{S}$
Current density $J = \dfrac{I}{S} = \sigma E$	Flux density $B = \dfrac{\Psi}{S} = \mu H$
Electromotive force (emf) V	Magnetomotive force (mmf) $\mathcal{F}$
Resistance R	Reluctance $\mathcal{R}$
Conductance $G = \dfrac{1}{R}$	Permeance $\mathcal{P} = \dfrac{1}{\mathcal{R}}$
Ohm's law $R = \dfrac{V}{I} = \dfrac{\ell}{\sigma S}$	Ohm's law $\mathcal{R} = \dfrac{\mathcal{F}}{\Psi} = \dfrac{\ell}{\mu S}$
or $V = E\ell = IR$	or $\mathcal{F} = H\ell = \Psi\mathcal{R} = NI$
Kirchhoff's laws:	Kirchhoff's laws:
$\Sigma I = 0$	$\Sigma \Psi = 0$
$\Sigma V - \Sigma RI = 0$	$\Sigma \mathcal{F} - \Sigma \mathcal{R}\Psi = 0$

where ℓ and S are respectively the mean length and the cross-sectional area of the magnetic core. The reciprocal of reluctance is *permeance* $\mathcal{P}$. The basic relationship for circuit elements is Ohm's law ($V = IR$):

$$\boxed{\mathcal{F} = \Psi\mathcal{R}} \qquad [8.69]$$

Based on this, Kirchhoff's current and voltage laws can be applied to nodes and loops of a given magnetic circuit just as in an electric circuit. The rules of adding voltages

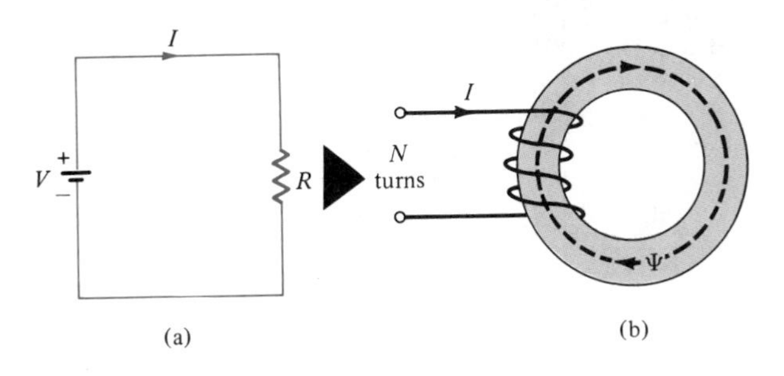

Figure 8.24 Analogy between **(a)** an electric circuit, and **(b)** a magnetic circuit.

and for combining series and parallel resistances also hold for mmfs and reluctances. Thus for n magnetic circuit elements in series

$$\Psi_1 = \Psi_2 = \Psi_3 = \cdots = \Psi_n \tag{8.70}$$

and

$$\mathcal{F} = \mathcal{F}_1 + \mathcal{F}_2 + \cdots + \mathcal{F}_n \tag{8.71}$$

For n magnetic circuit elements in parallel,

$$\Psi = \Psi_1 + \Psi_2 + \Psi_3 + \cdots + \Psi_n \tag{8.72}$$

and

$$\mathcal{F}_1 = \mathcal{F}_2 = \mathcal{F}_3 = \cdots = \mathcal{F}_n \tag{8.73}$$

Some differences between electric and magnetic circuits should be pointed out. Unlike an electric circuit where current I flows, magnetic flux does not flow. Also, conductivity σ is independent of current density J in an electric circuit whereas permeability μ varies with flux density B in a magnetic circuit. This is because ferromagnetic (nonlinear) materials are normally used in most practical magnetic devices. These differences notwithstanding, the magnetic circuit concept serves as an approximate analysis of practical magnetic devices.

†8.11 FORCE ON MAGNETIC MATERIALS

It is of practical interest to determine the force that a magnetic field exerts on a piece of magnetic material in the field. This is useful in electromechanical systems such as electromagnets, relays, rotating machines, and magnetic levitation. Consider, for example, an electromagnet made of iron of constant relative permeability as shown in Figure 8.25. The coil has N turns and carries a current I. If we ignore fringing, the

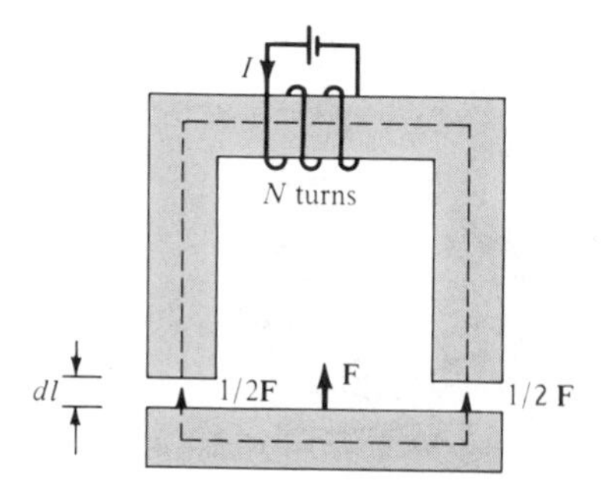

Figure 8.25 An electromagnet.

magnetic field in the air gap is the same as that in iron ($B_{1n} = B_{2n}$). To find the force between the two pieces of iron, we calculate the change in the total energy that would result were the two pieces of the magnetic circuit separated by a differential displacement $d\mathbf{l}$. The work required to effect the displacement is equal to the change in stored energy in the air gap (assuming constant current), that is

$$-F\,dl = dW_m = 2\left[\frac{1}{2}\frac{B^2}{\mu_o} S\,dl\right] \tag{8.74}$$

where S is the cross-sectional area of the gap, the factor 2 accounts for the two air gaps, and the negative sign indicates that the force acts to reduce the air gap (or that the force is attractive). Thus

$$F = -2\left(\frac{B^2 S}{2\mu_o}\right) \tag{8.75}$$

Note that the force is exerted on the lower piece and not on the current-carrying upper piece giving rise to the field. The tractive force across a *single* gap can be obtained from eq. (8.75) as

$$\boxed{F = -\frac{B^2 S}{2\mu_o}} \tag{8.76}$$

Notice the similarity between eq. (8.76) and that derived in Example 5.8 for electrostatic case. Equation (8.76) can be used to calculate the forces in many types of devices including relays, rotating machines, and magnetic levitation. The tractive pressure (in N/m^2) in a magnetized surface is

$$p = \frac{F}{S} = \frac{B^2}{2\mu_o} = \frac{1}{2}BH \tag{8.77}$$

which is the same as the energy density w_m in the air gap.

EXAMPLE 8.14

The toroidal core of Figure 8.26(a) has $\rho_o = 10$ cm and a circular cross section with $a = 1$ cm. If the core is made of steel ($\mu = 1000\,\mu_o$) and has a coil with 200 turns, calculate the amount of current that will produce a flux of 0.5 mWb in the core.

SOLUTION

This problem can be solved in two different ways: using the magnetic field approach (direct), or using the electric circuit analog (indirect).

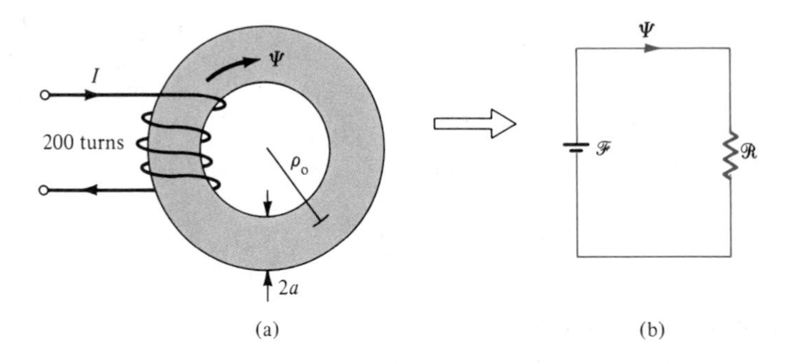

Figure 8.26 (a) Toroidal core of Example 8.14; (b) its equivalent electric circuit analog.

Method 1: Since ρ_o is large compared with a, from Example 7.6,

$$B = \frac{\mu NI}{\ell} = \frac{\mu_o \mu_r NI}{2\pi\rho_o}$$

Hence,

$$\Psi = BS = \frac{\mu_o \mu_r NI \, \pi a^2}{2\pi\rho_o}$$

or

$$I = \frac{2\rho_o \Psi}{\mu_o \mu_r N a^2} = \frac{2\,(10 \times 10^{-2})(0.5 \times 10^{-3})}{4\pi \times 10^{-7}(1000)(200)(1 \times 10^{-4})}$$

$$= \frac{100}{8\pi} \qquad = 3.979 \text{ A}$$

Method 2: The toroidal core in Figure 8.26(a) is analogous to the electric circuit of Figure 8.26(b). From the circuit and Table 8.4,

$$\mathcal{F} = NI = \Psi \mathcal{R} = \Psi \frac{\ell}{\mu S} = \Psi \frac{2\pi\rho_o}{\mu_o \mu_r \pi a^2}$$

or

$$I = \frac{2\rho_o \Psi}{\mu_o \mu_r N a^2} = 3.979 \; A$$

as obtained previously. ■

PRACTICE EXERCISE 8.14

A conductor of radius a is bent into a circular loop of mean radius ρ_o (see Fig. 8.26a). If $\rho_o = 10$ cm and $2a = 1$ cm, calculate the internal inductance of the loop.

ANSWER 31.42 nH.

EXAMPLE 8.15

In the magnetic circuit of Figure 8.27, calculate the current in the coil that will produce a magnetic flux density of 1.5 Wb/m^2 in the air gap assuming that $\mu = 50\mu_o$ and that all branches have the same cross-sectional area of 10 cm^2.

SOLUTION The magnetic circuit of Figure 8.27 is analogous to the electric circuit of Figure 8.28. In Figure 8.27, $\mathcal{R}_1$, $\mathcal{R}_2$, $\mathcal{R}_3$, and $\mathcal{R}_a$ are the reluctances in paths 143, 123, 35 and 16, and 56 (air gap) respectively. Thus

$$\mathcal{R}_1 = \mathcal{R}_2 = \frac{\ell}{\mu_o \mu_r S} = \frac{30 \times 10^{-2}}{(4\pi \times 10^{-7})(50)(10 \times 10^{-4})}$$

$$= \frac{3 \times 10^8}{20\pi}$$

$$\mathcal{R}_3 = \frac{9 \times 10^{-2}}{(4\pi \times 10^{-7})(50)(10 \times 10^{-4})} = \frac{0.9 \times 10^8}{20\pi}$$

$$\mathcal{R}_a = \frac{1 \times 10^{-2}}{(4\pi \times 10^{-7})(1)(10 \times 10^{-4})} = \frac{5 \times 10^8}{20\pi}$$

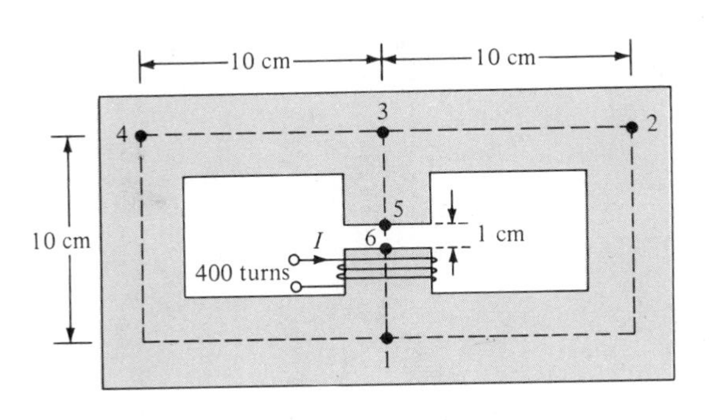

Figure 8.27 Magnetic circuit of Example 8.15.

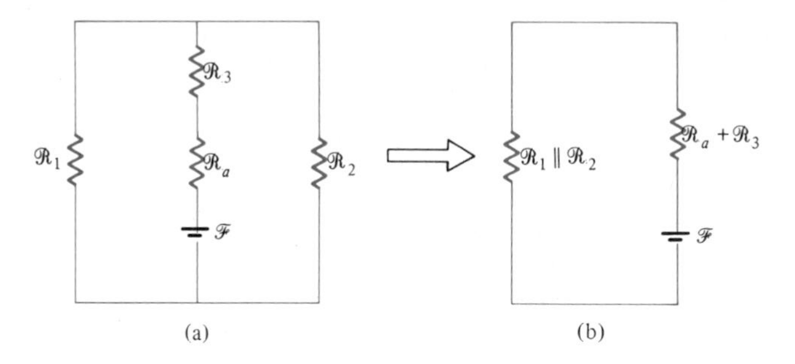

Figure 8.28 Electric circuit analog of the magnetic circuit in Figure 8.27.

We combine $\mathcal{R}_1$ and $\mathcal{R}_2$ as resistors in parallel. Hence,

$$\mathcal{R}_1 \| \mathcal{R}_2 = \frac{\mathcal{R}_1 \mathcal{R}_2}{\mathcal{R}_1 + \mathcal{R}_2} = \frac{\mathcal{R}_1}{2} = \frac{1.5 \times 10^8}{20\pi}$$

The mmf for the air gap is

$$\mathcal{F}_a = \frac{\mathcal{R}_a}{\mathcal{R}_a + \mathcal{R}_3 + \mathcal{R}_1 \| \mathcal{R}_2} \cdot \mathcal{F} = \frac{5}{5 + 0.9 + 1.5} NI$$

But

$$\mathcal{F}_a = H_a \ell_a = \frac{B_a}{\mu_o} \ell_a$$

Hence,

$$\frac{B_a \ell_a}{\mu_o} = \frac{5\, NI}{7.4}$$

or

$$I = \frac{B_a \ell_a}{\mu_o N} \cdot 1.48 = \frac{1.5 \times 10^{-2}(1.48)}{4\pi \times 10^{-7}\,(400)}$$
$$= 44.16\ A$$

■

PRACTICE EXERCISE 8.15

The toroid of Figure 8.26(a) has a coil of 1000 turns wound on its core. If $\rho_o =$ 10 cm and $a = 1$ cm, what current is required to establish a magnetic flux of 0.5 mWb

(a) If the core is nonmagnetic

(b) If the core has $\mu_r = 500$

ANSWER (a) 795.8 A, (b) 1.592 A.

EXAMPLE 8.16

A U-shaped electromagnet shown in Figure 8.29 is designed to lift a 400-kg mass (which includes the mass of the keeper). The iron yoke ($\mu_r = 3000$) has a cross section of 40 cm^2 and mean length of 50 cm, and the air gaps are each 0.1 mm long. Neglecting the reluctance of the keeper, calculate the number of turns in the coil when the excitation current is 1 A.

SOLUTION The tractive force across the two air gaps must balance the weight. Hence

$$F = 2\frac{(B_a^2 S)}{2\mu_o} = mg$$

or

$$B_a^2 = \frac{mg\mu_o}{S} = \frac{400 \times 9.8 \times 4\pi \times 10^{-7}}{40 \times 10^{-4}}$$

$$B_a = 1.11 \text{ Wb/m}^2$$

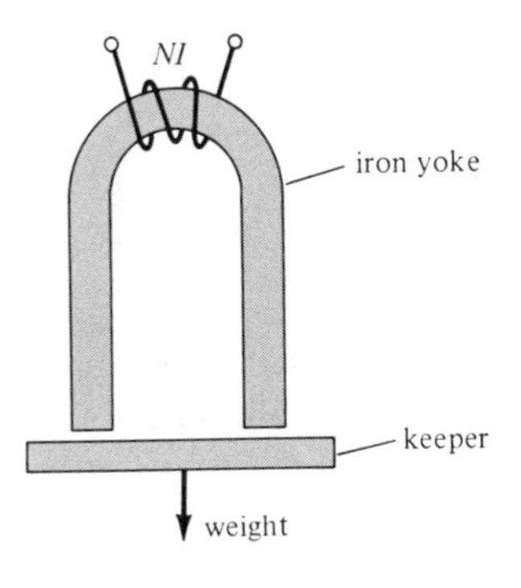

Figure 8.29 U-shaped electromagnet; for Example 8.16.

But

$$\mathcal{F} = NI = \Psi(\mathcal{R}_a + \mathcal{R}_i)$$

$$\mathcal{R}_a = \frac{\ell_a}{\mu S} = \frac{2 \times 0.1 \times 10^{-3}}{4\pi \times 10^{-7} \times 40 \times 10^{-4}} = \frac{6 \times 10^6}{48\pi}$$

$$\mathcal{R}_i = \frac{\ell_i}{\mu_o \mu_r S} = \frac{50 \times 10^{-2}}{4\pi \times 10^{-7} \times 3000 \times 40 \times 10^{-4}} = \frac{5 \times 10^6}{48\pi}$$

$$\mathcal{F}_a = \frac{\mathcal{R}_a}{\mathcal{R}_a + \mathcal{R}_i} \mathcal{F} = \frac{6}{6 + 5} NI = \frac{6}{11} NI$$

Since

$$\mathcal{F}_a = H_a \ell_a = \frac{B_a \ell_a}{\mu_o}$$

$$N = \frac{11}{6} \frac{B_a \ell_a}{\mu_o I} = \frac{11 \times 1.11 \times 0.1 \times 10^{-3}}{6 \times 4\pi \times 10^{-7} \times 1}$$

$$N = 162$$

■

PRACTICE EXERCISE 8.16

Find the force across the air gap of the magnetic circuit of Example 8.15.

ANSWER 895.2 N.

SUMMARY

1. The Lorentz force equation,

$$\mathbf{F} = Q\,(\mathbf{E} + \mathbf{u} \times \mathbf{B}) = m \frac{d\mathbf{u}}{dt}$$

relates the force acting on a particle with charge Q in the presence of EM fields. It expresses the fundamental law relating EM to mechanics.

2. Based on the Lorentz force law, the force experienced by a current element $I d\mathbf{l}$ in a magnetic field $\mathbf{B}$ is

$$d\mathbf{F} = I\, d\mathbf{l} \times \mathbf{B}$$

From this, the magnetic field $\mathbf{B}$ is defined as the force per unit current element.

3. The torque on a current loop with magnetic moment **m** in a uniform magnetic field **B** is

$$\mathbf{T} = \mathbf{m} \times \mathbf{B} = IS\mathbf{a}_n \times \mathbf{B}$$

4. A magnetic dipole is a bar magnet or a small filamental current loop, it is so called due to the fact that its **B** field lines are similar to the **E** field lines of an electric dipole.
5. When a material is subjected to a magnetic field, it becomes magnetized. The magnetization **M** is the magnetic dipole moment per unit volume of the material. For linear material,

$$\mathbf{M} = \chi_m \mathbf{H}$$

where χ_m is the magnetic susceptibility of the material.
6. In terms of their magnetic properties, materials are either linear (diamagnetic or paramagnetic) or nonlinear (ferromagnetic). For linear materials,

$$\mathbf{B} = \mu\mathbf{H} = \mu_o\mu_r\mathbf{H} = \mu_o(1 + \chi_m)\,\mathbf{H} = \mu_o(\mathbf{H} + \mathbf{M})$$

where μ = permeability and $\mu_r = \mu/\mu_o$ = relative permeability of the material. For nonlinear material, $B = \mu(H)\,H$, that is, μ does not have a fixed value; the relationship between B and H is usually represented by a magnetization curve.
7. The boundary conditions that **H** or **B** must satisfy at the interface between two different media are

$$\mathbf{B}_{1n} = \mathbf{B}_{2n}$$

$$(\mathbf{H}_1 - \mathbf{H}_2) \times \mathbf{a}_{n12} = \mathbf{K} \qquad \text{or} \qquad \mathbf{H}_{1t} = \mathbf{H}_{2t} \quad \text{if } \mathbf{K} = 0$$

where $\mathbf{a}_{n12}$ is a unit vector directed from medium 1 to medium 2.
8. Energy in a magnetostatic field is given by

$$W_m = \frac{1}{2}\int \mathbf{B} \cdot \mathbf{H}\, dv$$

For an inductor carrying current I

$$W_m = \frac{1}{2}LI^2$$

Thus the inductance L can be found using

$$L = \frac{\int \mathbf{B} \cdot \mathbf{H}\, dv}{I^2}$$

9. The inductance L of an inductor can also be determined from its basic definition: the ratio of the magnetic flux linkage to the current through the inductor, that is,

$$L = \frac{\lambda}{I} = \frac{N\Psi}{I}$$

Thus by assuming current I, we determine **B** and $\Psi = \int \mathbf{B} \cdot d\mathbf{S}$, and finally find $L = N\Psi/I$.

10. A magnetic circuit can be analyzed in the same way as an electric circuit. We simply keep in mind the similarity between

$$\mathcal{F} = NI = \oint \mathbf{H} \cdot d\mathbf{l} = \Psi \mathcal{R} \qquad \text{and} \qquad V = IR$$

that is,

$$\mathcal{F} \leftrightarrow V, \ \Psi \leftrightarrow I, \ \mathcal{R} \leftrightarrow R$$

Thus we can apply Ohms and Kirchhoff's laws to magnetic circuits just as we apply them to electric circuits.

11. The magnetic pressure (or force per unit surface area) on a piece of magnetic material is

$$P = \frac{F}{S} = \frac{1}{2} BH = \frac{B^2}{2\mu_o}$$

where B is the magnetic field at the surface of the material.

REVIEW QUESTIONS

8.1 Which of the following statements are not true about electric force $\mathbf{F}_e$ and magnetic force $\mathbf{F}_m$ on a charged particle?

(a) $\mathbf{E}$ and $\mathbf{F}_e$ are parallel to each other whereas $\mathbf{B}$ and $\mathbf{F}_m$ are perpendicular to each other.

(b) Both $\mathbf{F}_e$ and $\mathbf{F}_m$ depend on the velocity of the charged particle.

(c) Both $\mathbf{F}_e$ and $\mathbf{F}_m$ can perform work.

(d) Both $\mathbf{F}_e$ and $\mathbf{F}_m$ are produced when a charged particle moves at a constant velocity.

(e) $\mathbf{F}_m$ is generally small in magnitude compared to $\mathbf{F}_e$.

(f) $\mathbf{F}_e$ is an accelerating force whereas $\mathbf{F}_m$ is a purely deflecting force.

8.2 Two thin parallel wires carry currents along the same direction. The force experienced by one due to the other is

(a) Parallel to the lines

(b) Perpendicular to the lines and attractive

(c) Perpendicular to the lines and repulsive

(d) Zero

8.3 The force on differential length $d\mathbf{l}$ at point P in the conducting circular loop in Figure 8.30 is

(a) Outward along OP

(b) Inward along OP

(c) In the direction of the magnetic field

(d) Tangential to the loop at P

8.4 The resultant force on the circular loop in Figure 8.30 has the magnitude of

(a) $2\pi\rho_o IB$

(b) $\pi\rho_o^2 IB$

(c) $2\rho_o IB$

(d) Zero

8.5 What is the unit of magnetic charge?

(a) Ampere-meter square

(b) Coulomb

(c) Ampere

(d) Ampere-meter

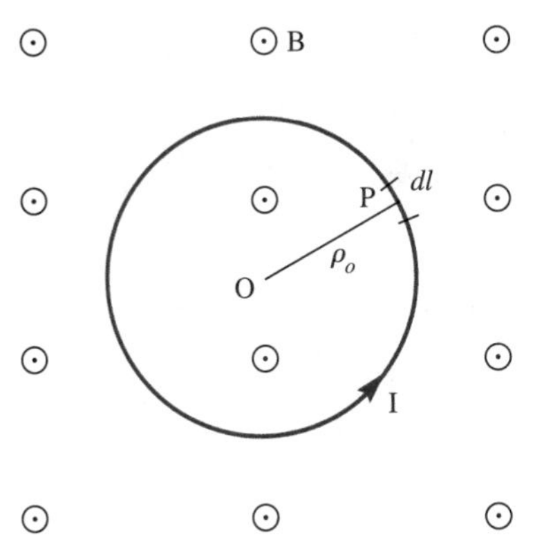

Figure 8.30 For Review Questions 8.3 and 8.4.

8.6 Which of these materials requires the least value of magnetic field strength to magnetize it?

(a) Nickel

(b) Silver

(c) Tungsten

(d) Sodium chloride

8.7 Identify the statement that is not true of ferromagnetic materials.

(a) They have a large χ_m.

(b) They have a fixed value of μ_r.

(c) Energy loss is proportional to the area of the hysteresis loop.

(d) They lose their nonlinearity property above curie temperature.

8.8 Which of these formulas is wrong?

(a) $B_{1n} = B_{2n}$

(b) $B_2 = \sqrt{B_{2n}^2 + B_{2t}^2}$

(c) $H_1 = H_{1n} + H_{1t}$

(d) $\mathbf{a}_{n21} \times (\mathbf{H}_1 - \mathbf{H}_2) = \mathbf{K}$, where $\mathbf{a}_{n21}$ is a unit vector normal to the interface and directed from region 2 to region 1.

8.9 Each of the following pairs consists of an electric circuit term and the corresponding magnetic circuit term. Which pairs are not corresponding?

(a) V and $\mathcal{F}$

(b) G and $\mathcal{P}$

(c) ε and μ

(d) IR and $H\mathcal{R}$

(e) $\Sigma\, I = 0$ and $\Sigma\, \Psi = 0$

8.10 A multilayer coil of 2000 turns of fine wire is 20 mm long and has a thickness 5 mm of winding. If the coil carries a current of 5 mA, the mmf generated is

(a) 10 A-t

(b) 500 A-t

(c) 2000 A-t

(d) None of the above

Answers: 8.1 b,c, 8.2b, 8.3a, 8.4d, 8.5d, 8.6a, 8.7b, 8.8c, 8.9c,d, 8.10a.

PROBLEMS

8.1 A point charge of 10 C moves with a uniform velocity of $2\mathbf{a}_x - 4\mathbf{a}_z$ m/s in an EM field having $\mathbf{E} = \mathbf{a}_x - 3\mathbf{a}_y + 8\mathbf{a}_z$ V/m and $\mathbf{B} = 0.3\mathbf{a}_x + 0.1\mathbf{a}_y$ Wb/m^2. Find

(a) $\mathbf{F}_e$

(b) $\mathbf{F}_m$

(c) The total force on the charge

8.2 A magnetostatic field never delivers energy to a charged particle moving in that field. Explain.

8.3 A charged particle has mass 2 kg and charge 3 C. It starts at point $(1, -2, 0)$ with velocity $4\mathbf{a}_x + 3\mathbf{a}_z$ m/s in an electric field $12\mathbf{a}_x + 10\mathbf{a}_y$ V/m. At time $t = 1$ s, determine

(a) The acceleration of the particle

(b) Its velocity

(c) Its K.E.

(d) Its location

***8.4** A particle with mass 1 kg and charge 2 C starts from rest at point $(2, 3, -4)$ in a region where $\mathbf{E} = -4\mathbf{a}_y$ V/m and $\mathbf{B} = 5\mathbf{a}_x$ Wb/m^2. Calculate

(a) The location of the particle at $t = 1$ s

(b) Its velocity and K.E. at that location

8.5 A -2-mC charge starts at point $(0, 1, 2)$ with a velocity of $5\mathbf{a}_x$ m/s in a magnetic field $\mathbf{B} = 6\mathbf{a}_y$ Wb/m^2. Determine the position and velocity of the particle after 10 s assuming that the mass of the charge is 1 gram. Describe the motion of the charge.

***8.6** By injecting an electron beam normally to the plane edge of a uniform field $B_o\mathbf{a}_z$, electrons can be dispersed according to their velocity as in Figure 8.31.

(a) Show that the electrons would be ejected out of the field in paths parallel to the input beam as shown.

(b) Derive an expression for the exit distance d above entry point.

8.7 (a) Show that two infinitely long, parallel, straight filamentary conductors with a separation distance ρ and currents I_1 and I_2 in opposite directions as in Figure 8.32 will experience a repulsive force per unit length of

$$\mathbf{f} = \frac{\mu_o I_1 I_2}{2\pi\rho}\,\mathbf{a}_\rho \text{ N/m}$$

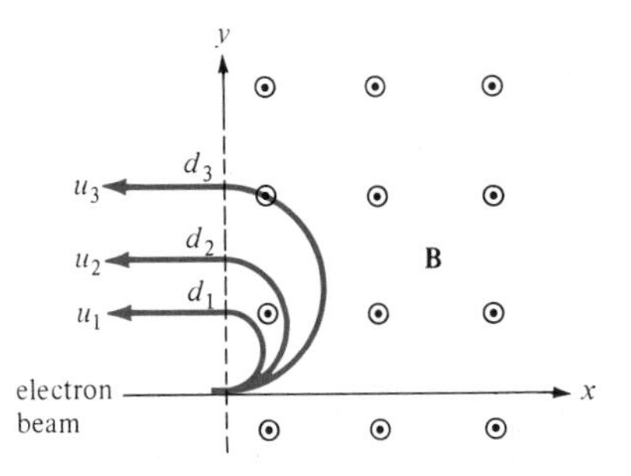

Figure 8.31 For Problem 8.6.

(b) Two infinitely long parallel straight conductors carry currents 5 A and 10 A respectively. If the conductors are separated by 2 m, what is the force per meter on each conductor assuming that the currents flow

(i) In the same direction

(ii) In opposite directions

***8.8** Three infinite lines L_1, L_2, and L_3 defined by $x = 0, y = 0; x = 0, y = 4; x = 3, y = 4$ respectively carry filamentary currents -100 A, 200 A, and 300 A along $\mathbf{a}_z$. Find the force per unit length on

(a) L_2 due to L_1

(b) L_1 due to L_2

(c) L_3 due to L_1

(d) L_3 due to L_1 and L_2. State whether each force is repulsive or attractive.

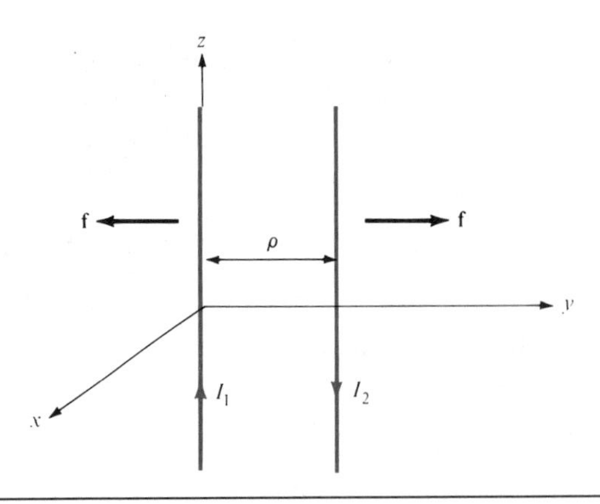

Figure 8.32 For Problem 8.7.

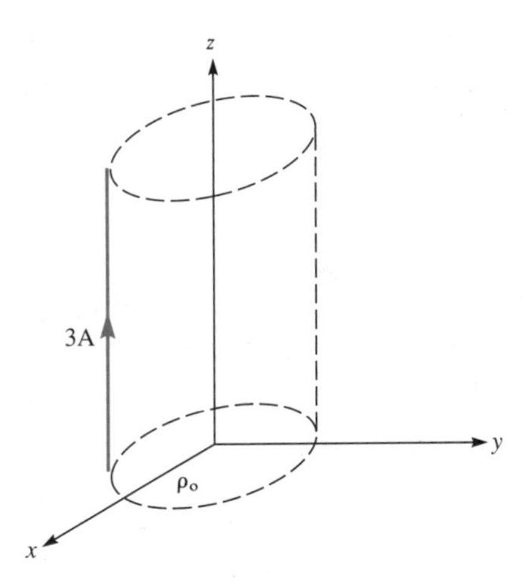

Figure 8.33 For Problem 8.9.

8.9 A conductor 2 m long carrying 3A is placed parallel to the z-axis at distance $\rho_o = 10$ cm as shown in Figure 8.33. If the field in the region is $\cos(\phi/3)\, \mathbf{a}_\rho$ Wb/m^2, how much work is required to rotate the conductor one revolution about the z-axis?

***8.10** A conducting triangular loop carrying a current of 2 A is located close to an infinitely long, straight conductor with a current of 5 A, as shown in Figure 8.34. Calculate (a) the force on side 1 of the triangular loop and (b) the total force on the loop.

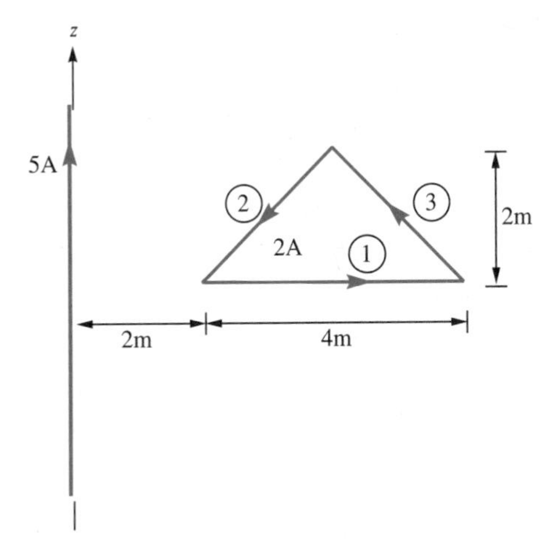

Figure 8.34 For Problem 8.10.

***8.11** A three-phase transmission line consists of three conductors that are supported at points A, B, and C to form an equilateral triangle as shown in Figure 8.35. At one instant, conductors A and B both carry a current of 75 A while conductor C carries a return current of 150 A. Find the force per meter on conductor C at that instant.

***8.12** An infinitely long tube of inner radius a and outer radius b is made of a conducting magnetic material. The tube carries a total current I and is placed along the z-axis. If it is exposed to a constant magnetic field $B_o\mathbf{a}_\rho$, determine the force per unit length acting on the tube.

***8.13** An infinitely long conductor is buried but insulated from an iron mass ($\mu = 2000\mu_o$) as shown in Figure 8.36. Using image theory, estimate the magnetic flux density at point P.

***8.14** Show that eq. (8.22) can be written as

$$\mathbf{B} = \frac{\mu_o}{4\pi}\left[\frac{3(\mathbf{m}\cdot\mathbf{r})\mathbf{r}}{r^5} - \frac{\mathbf{m}}{r^3}\right]$$

8.15 A current-carrying coil may be used to determine the strength of a magnetic field by measuring the torque exerted on the coil in the field. If a coil of 800 turns with area 0.6 cm^2 carrying a current of 10 mA experiences a maximum torque of 50 μN · m, find $|\mathbf{B}|$.

8.16 A galvanometer has a rectangular coil of side 10 by 30 mm pivoted about the center of the shorter side. It is mounted in radial magnetic field so that a constant magnetic field of 0.4 Wb/m^2 always acts across the plane of the coil. If the coil has 1000 turns and carries current 2 mA, find the torque exerted on it.

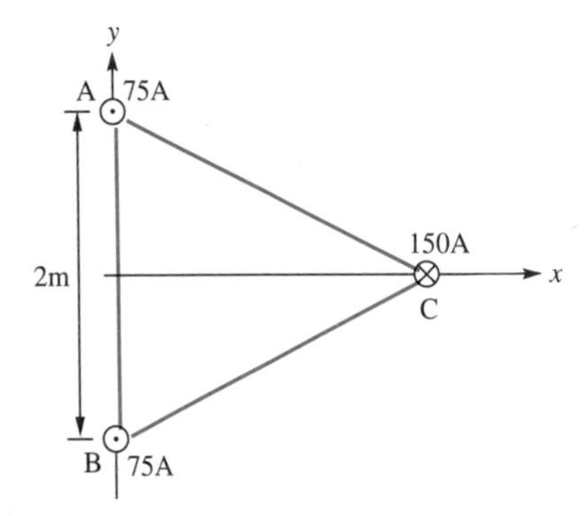

Figure 8.35 For Problem 8.11.

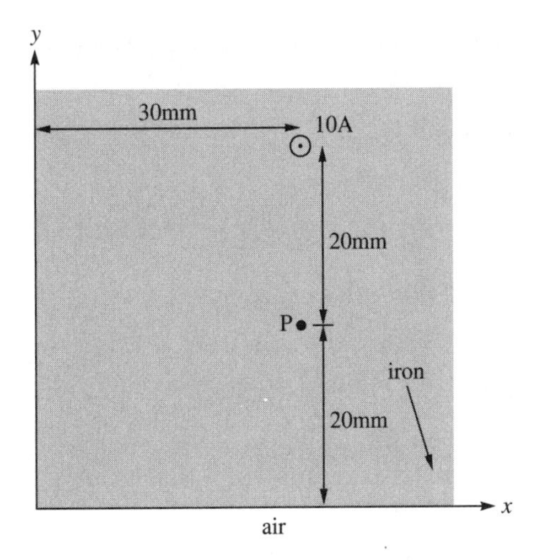

Figure 8.36 For Problem 8.13.

8.17 A small circular loop of radius 10 cm is centered at the origin and placed on the $z = 0$ plane. If the loop carries a current of 1 A along $\mathbf{a}_\phi$, calculate

(a) The magnetic moment of the loop

(b) The magnetic field intensity at (2, 2, 2)

(c) The magnetic flux density at (−6, 8, 10)

8.18 Determine the torque on a small loop with magnetic moment $4\mathbf{a}_z$ A · m^2 located at (2, 2, 2) when placed in the magnetic field produced by the loop of the previous problem.

8.19 A small magnet placed at the origin produces $\mathbf{B} = -0.5\mathbf{a}_z$ mWb/m^2 at (10, 0, 0). Find **B** at

(a) (0, 3, 0)

(b) (3, 4, 0)

(c) (1, 1, −1)

***8.20** If the magnetic moment of loop L_1 in Example 8.6 is changed to $5\mathbf{a}_x$ A · m^2 while other things remain the same, find the torque on loop L_2.
(*Hint:* Use eq. (8.22) or the result of Problem 8.14 to determine $\mathbf{B}_1$.)

8.21 If $\mathbf{B} = 2\mathbf{a}_x - 5\mathbf{a}_y + 4\mathbf{a}_z$ mWb/m^2 in a homogeneous, isotropic material for which $\mu_r = 4$, calculate

(a) The magnetic susceptibility χ_m

(b) The magnetic field intensity **H**

(c) The magnetization **M**

(d) The magnetic energy density w_m

8.22 In a material ($\mu_r = 5$), $\mathbf{H} = z\mathbf{a}_y$ A/m, determine

(a) The magnetization

(b) The free current volume density

(c) The bound current volume density

8.23 An infinitely long wire of radius 2 mm is placed along the z-axis. If the wire is made of a material with $\mu = 5\mu_o$ and carries a current of 4 A along $\mathbf{a}_z$, find

(a) **M**, $\mathbf{J}_b$, and **J** at $\rho = 1$ mm and $\rho = 3$ mm

(b) $\mathbf{K}_b$ on the surface of the wire

***8.24** (a) For the boundary between two magnetic media such as is shown in Figure 8.16, show that the boundary conditions on the magnetization vector are

$$\frac{M_{1t}}{\chi_{m1}} - \frac{M_{2t}}{\chi_{m2}} = K \qquad \text{and} \qquad \frac{\mu_1}{\chi_{m1}} m_{1n} = \frac{\mu_2}{\chi_{m2}} M_{2n}$$

(b) If the boundary is not current-free, show that instead of eq. (8.49), we obtain

$$\frac{\tan \theta_1}{\tan \theta_2} = \frac{\mu_1}{\mu_2}\left[1 + \frac{K\mu_2}{B_2 \sin \theta_2}\right]$$

8.25 Let $\mu_1 = \mu_o$ for region 1 ($\rho \geq 10$ m) and $\mu_2 = 3\mu_o$ for region 2 ($\rho \leq 10$ m). If $\mathbf{H}_2 = 4\mathbf{a}_\rho = 6\mathbf{a}_\phi + 7\mathbf{a}_z$ kA/m, find

(a) $\mathbf{H}_1$

(b) $\mathbf{B}_2$

(c) The angles $\mathbf{H}_1$ and $\mathbf{H}_2$ make with the normal to the surface $\rho = 10$ m

8.26 Region 1 ($y \leq 0$) consists of a magnetic material for which $\mu_{r1} = 2$ while region 2 ($y \geq 0$) is free space. If $\mathbf{B}_1 = 40\mathbf{a}_x + 50\mathbf{a}_y - 30\mathbf{a}_z$ mWb/m^2, calculate

(a) $\mathbf{B}_2$

(b) $\dfrac{H_1}{H_2}$

(c) $\dfrac{\tan \theta_1}{\tan \theta_2}$

8.27 The interface $4x - 5z = 0$ between two magnetic media carries current $35\mathbf{a}_y$ A/m. If $\mathbf{H}_1 = 25\mathbf{a}_x - 30\mathbf{a}_y + 45\mathbf{a}_z$ A/m in region $4x - 5z \leq 0$ where $\mu_{r1} = 5$, calculate $\mathbf{H}_2$ in region $4x - 5z \geq 0$ where $\mu_{r2} = 10$.

8.28 The plane $z = 0$ separates air ($z \geq 0$, $\mu = \mu_o$) from iron ($z \leq 0$, $\mu = 200\mu_o$). Given that

$$\mathbf{H} = 10\mathbf{a}_x + 15\mathbf{a}_y - 3\mathbf{a}_z \text{ A/m}$$

in air, find $\mathbf{B}$ in iron and the angle it makes with the interface.

8.29 Region $0 \leq z \leq 2$ m is filled with an infinite slab of magnetic material ($\mu = 2.5\mu_o$). If the surfaces of the slab at $z = 0$ and $z = 2$ respectively carry surface currents $30\mathbf{a}_x$ A/m and $-40\mathbf{a}_x$ A/m as in Figure 8.37, calculate $\mathbf{H}$ and $\mathbf{B}$ for

(a) $z < 0$

(b) $0 < z < 2$

(c) $z > 2$

8.30 The magnetization curve for an iron alloy is approximately given by $B = \frac{1}{3}H + H^2\mu$ Wb/m^2. Find: (a) μ_r when $H = 210$ A/m, (b) the energy stored per unit volume in the alloy as H increases from 0 to 210 A/m.

8.31 A solenoid has a circular cross section of diameter 10 cm. If it has 1000 turns and is 40 cm long, calculate the approximate value of its self-inductance.

***8.32** (a) If the cross-section of the toroid of Figure 7.15 is a square of side a, show that the self-inductance of the toroid is

$$L = \frac{\mu_o N^2 a}{2\pi} \ln \left[\frac{2\rho_o + a}{2\rho_o - a}\right]$$

(b) If the toroid has a circular cross-section as in Figure 7.15, show that

$$L = \frac{\mu_o N^2 a^2}{2\rho_o}$$

where $\rho_o \gg a$. (*Hint*: See Problem 7.25.)

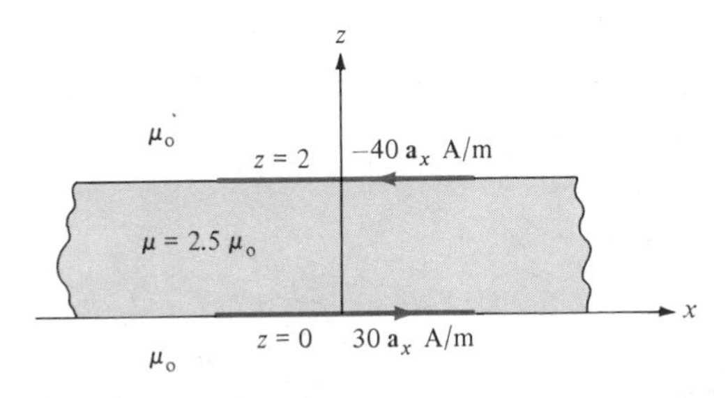

Figure 8.37 For Problem 8.29.

8.33 For the wire in Problem 8.23, find the self-inductance per unit length of the wire and the energy per unit length stored in its field.

8.34 When two parallel identical wires are separated by 3 m, the inductance per unit length is 2.5 μH/m. Calculate the diameter of each wire.

8.35 The core of a toroid is 12 cm^2 and is made of material with $\mu_r = 200$. If the mean radius of the toroid is 50 cm, calculate the number of turns needed to obtain an inductance of 2.5 H.

8.36 Show that the mutual inductance between the rectangular loop and the infinite line current of Figure 8.4 is

$$M_{12} = \frac{\mu b}{2\pi} \ln \left[\frac{a + \rho_o}{\rho_o}\right]$$

Calculate M_{12} when $a = b = \rho_o = 1$ m.

8.37 An infinitely long wire is placed along the axis of the toroidal core of Figure 7.15. Show that the mutual inductance between the core and the wire is

$$M_{12} = \mu_o N \left[\rho_o - \sqrt{\rho_o^2 - a^2}\right]$$

***8.38** Prove that the mutual inductance between the closed wound coaxial solenoids of length ℓ_1 and ℓ_2 ($\ell_1 \gg \ell_2$), turns N_1 and N_2, and radii r_1 and r_2 with $r_1 \simeq r_2$ is

$$M_{12} = \frac{\mu N_1 N_2}{\ell_1} \pi r_1^2$$

8.39 An iron ring has a mean circumference of 60 cm and a cross-sectional area of 12 cm^2. If a magnetic field strength of 700 A/m is needed to produce a magnetic flux of 0.8 mWb, find: (a) the reluctance of the ring, (b) the number of ampere turns required.

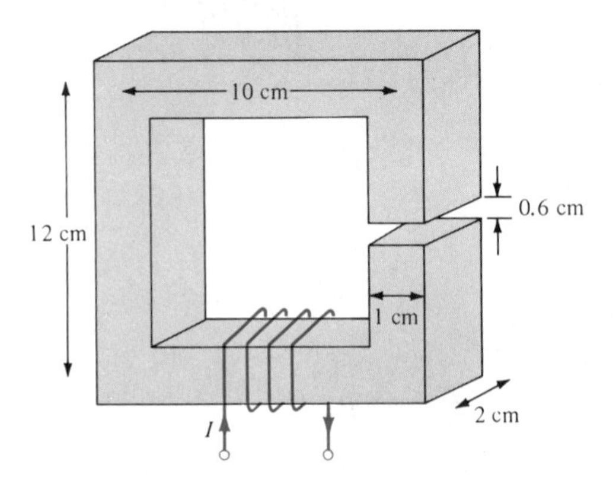

Figure 8.38 For Problem 8.41.

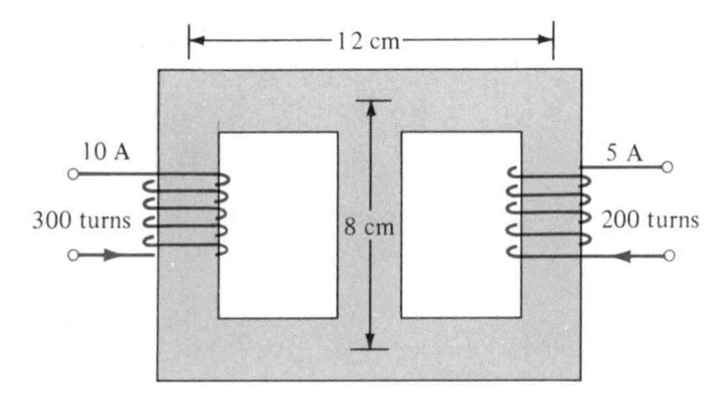

Figure 8.39 For Problem 8.42.

8.40 Refer to Figure 8.27. If the current in the coil is 0.5 A, find the mmf and the magnetic field intensity in the air gap. Assume that $\mu = 500\mu_o$ and that all branches have the same cross-sectional area of 10 cm^2.

8.41 The magnetic circuit of Figure 8.38 has current 10 A in the coil of 2000 turns. Assume that all branches have the same cross section of 2 cm^2 and that the material of the core is iron with $\mu_r = 1500$. Calculate R, $\mathcal{F}$, and Ψ for

(a) The core

(b) The air gap

8.42 A symmetric core of steel with $\mu = 1000\mu_o$ has a uniform cross section of 4 cm^2 except in the central leg with cross section 6 cm^2 as shown in Figure 8.39. The left and right legs have coils with current 10 A and 5 A respectively. Determine the flux density in each leg.

8.43 An electromagnetic relay is modeled as shown in Figure 8.40. What force is on the armature (moving part) of the relay if the flux in the air gap is 2 mWb? The area of the gap is 0.3 cm^2, and its length 1.5 mm.

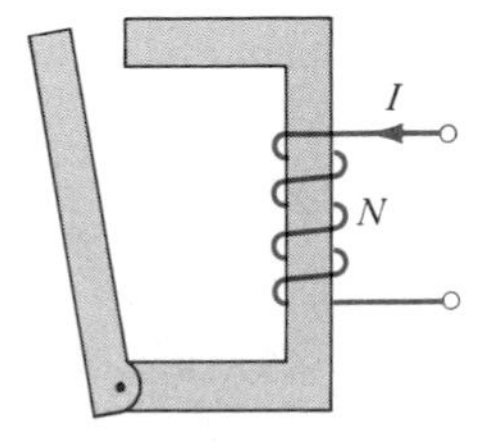

Figure 8.40 For Problem 8.43.

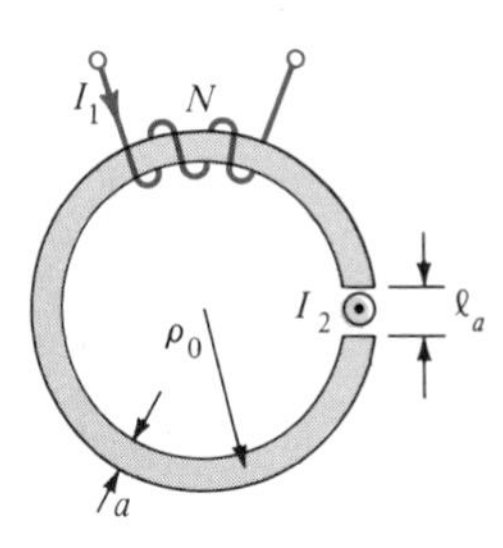

Figure 8.41 For Problem 8.44.

8.44 A toroid with air gap, shown in Figure 8.41, has a square cross section. A long conductor carrying current I_2 is inserted in the air gap. If $I_1 = 200$ mA, $N = 750$, $\rho_o = 10$ cm, $a = 5$ mm, and $\ell_a = 1$ mm, calculate

(a) The force across the gap when $I_2 = 0$ and the relative permeability of the toroid is 300

(b) The force on the conductor when $I_2 = 2$ mA and the permeability of the toroid is infinite.. Neglect fringing in the gap in both cases.

8.45 A section of an electromagnet with a plate below it carrying a load is shown in Figure 8.42. The electromagnet has a contact area of 200 cm^2 per pole with the middle pole having a winding of 1000 turns with $I = 3$ A. Calculate the maximum mass which can be lifted. Assume that the reluctance of the electromagnet and the plate is negligible.

8.46 Figure 8.43 shows the cross section of an electromechanical system in which the plunger moves freely between two nonmagnetic sleeves. Assuming that all legs have the same cross-sectional area S, show that

$$\mathbf{F} = -\frac{2\,N^2I^2\mu_o S}{(a + 2x)}\,\mathbf{a}_x$$

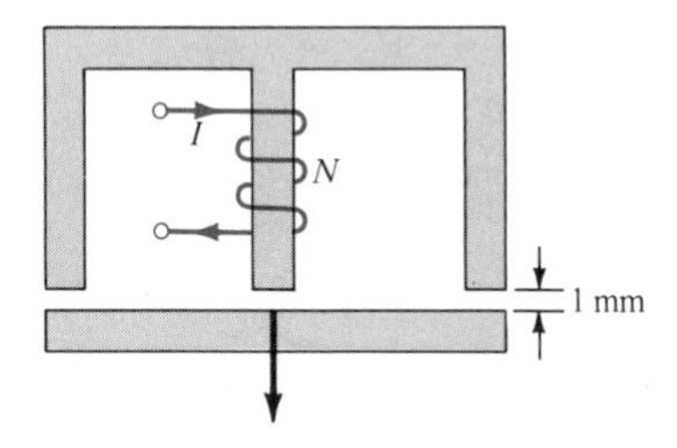

Figure 8.42 For Problem 8.45.

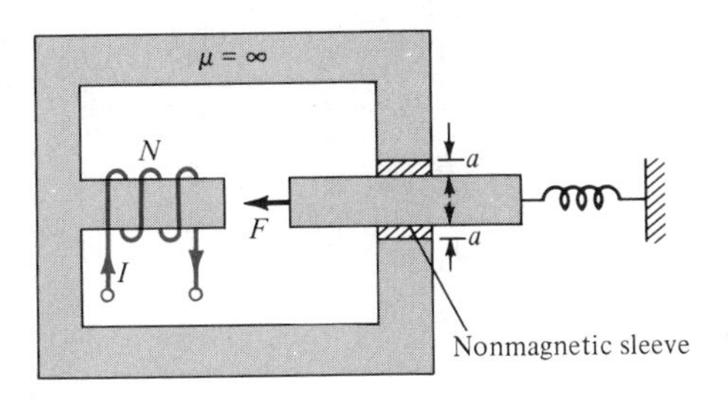

Figure 8.43 For Problem 8.46.

PART 4

Waves and Applications

CHAPTER

Maxwell's Equations

9

Small minds discuss persons. Average minds discuss events. Great minds discuss ideas.

—THIS WEEK

9.1 INTRODUCTION

In Part II (Chapters 4 to 6) of this text, we mainly concentrated our efforts on electrostatic fields denoted by $\mathbf{E}(x, y, z)$; Part III (Chapters 7 and 8) was devoted to magnetostatic fields represented by $\mathbf{H}(x, y, z)$. We have therefore restricted our discussions to static, or time-invariant, EM fields. Henceforth, we shall examine situations where electric and magnetic fields are dynamic, or time-varying. It should be mentioned first that in static EM fields, electric and magnetic fields are independent of each other whereas in dynamic EM fields, the two fields are interdependent. In other words, a time-varying electric field necessarily involves a corresponding time-varying magnetic field. Second, time-varying EM fields, represented by $\mathbf{E}(x, y, z, t)$ and $\mathbf{H}(x, y, z, t)$, are of more practical value than static EM fields. However, familiarity with static fields provides a good background for understanding dynamic fields. Third, recall that electrostatic fields are usually produced by static electric charges whereas magnetostatic fields are due to motion of electric charges with uniform velocity (direct current) or static magnetic charges (magnetic poles); time-varying fields or waves are usually due to accelerated charges or time-varying currents such as shown in Figure 9.1.

Any pulsating current will produce radiation (time-varying fields). It is worth noting that pulsating current of the type shown in Figure 9.1(b) is the cause of radiated emission in digital logic boards. In summary:

stationary charges $\rightarrow$ electrostatic fields
steady currents $\rightarrow$ magnetostatic fields
time-varying currents $\rightarrow$ electromagnetic fields (or waves)

Our aim in this chapter is to lay a firm foundation for our subsequent studies. This will involve introducing two major concepts: (1) electromotive force based on

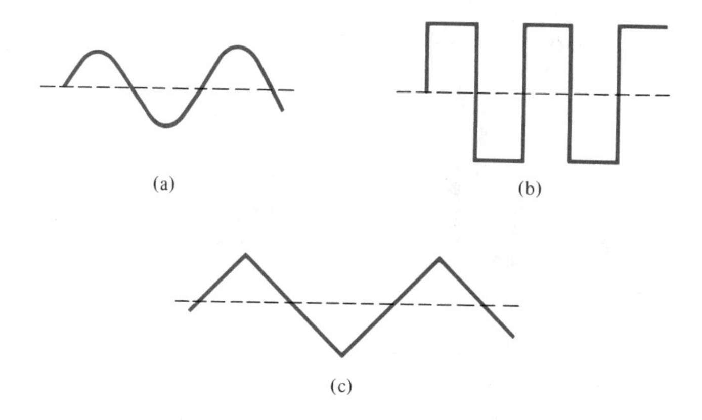

Figure 9.1 Various types of time-varying current: **(a)** sinusoidal, **(b)** rectangular, **(c)** triangular.

Faraday's experiments, and (2) displacement current, which resulted from Maxwell's hypothesis. As a result of these concepts, Maxwell's equations as presented in Section 7.6 and the boundary conditions for static EM fields will be modified to account for the time-variation of the fields. It should be stressed that Maxwell's equations summarize the laws of electromagnetism and shall be the basis of our discussions in the remaining part of the text. For this reason, Section 9.5 should be regarded as the heart of this text.

9.2 FARADAY'S LAW

After Oersted's experimental discovery (upon which Biot-Savart and Ampere based their laws) that a steady current produces a magnetic field, it seemed logical to find out if magnetism would produce electricity. In 1831, about 11 years after Oersted's discovery, Michael Faraday in London and Joseph Henry in New York discovered that a time-varying magnetic field would produce an electric current.[1]

According to Faraday's experiments, a static magnetic field produces no current flow, but a time-varying field produces an induced voltage (called *electromotive force* or simply emf) in a closed circuit, which causes a flow of current. Faraday discovered that the induced emf, V_{emf} (in volts), in any closed circuit is equal to the time rate of change of the magnetic flux linkage by the circuit. This is called *Faraday's law*, and it can be expressed as

$$V_{emf} = -\frac{d\lambda}{dt} = -N\frac{d\Psi}{dt} \qquad [9.1]$$

....

[1]For details on the experiments of Michael Faraday (1791–1867) and Joseph Henry (1797–1878), see W. F. Magie, *A Source Book in Physics*. Cambridge, MA: Harvard Univ. Press, 1963, pp. 472–519.

where N is the number of turns in the circuit and Ψ is the flux through each turn. The negative sign shows that the induced voltage acts in such a way as to oppose the flux producing it. This is known as *Lenz's law,*[2] and it emphasizes the fact that the direction of current flow in the circuit is such that the induced magnetic field produced by the induced current will oppose the original magnetic field.

Recall that we described an electric field as one in which electric charges experience force. The electric fields considered so far are caused by electric charges; in such fields, the flux lines begin and end on the charges. However, there are other kinds of electric fields not directly caused by electric charges. These are emf-produced fields. Sources of emf include electric generators, batteries, thermocouples, fuel cells, and photovoltaic cells, which all convert nonelectrical energy into electrical energy.

Consider the electric circuit of Figure 9.2, where the battery is a source of emf. The electrochemical action of the battery results in an emf-produced field $\mathbf{E}_f$. Due to the accumulation of charge at the battery terminals, an electrostatic field $\mathbf{E}_e$ $(= -\nabla V)$ also exists. The total electric field at any point is

$$\mathbf{E} = \mathbf{E}_f + \mathbf{E}_e \tag{9.2}$$

Note that $\mathbf{E}_f$ is zero outside the battery, $\mathbf{E}_f$ and $\mathbf{E}_e$ have opposite directions in the battery, and the direction of $\mathbf{E}_e$ inside the battery is opposite to that outside it. If we integrate eq. (9.2) over the closed circuit,

$$\oint_L \mathbf{E} \cdot d\mathbf{l} = \oint_L \mathbf{E}_f \cdot d\mathbf{l} + 0 = \int_N^P \mathbf{E}_f \cdot d\mathbf{l} \qquad \text{(through battery)} \tag{9.3a}$$

where $\oint \mathbf{E}_e \cdot d\mathbf{l} = 0$ because $\mathbf{E}_e$ is conservative. The emf of the battery is the line integral of the emf-produced field; that is,

$$V_{\text{emf}} = \int_N^P \mathbf{E}_f \cdot d\mathbf{l} = -\int_N^P \mathbf{E}_e \cdot d\mathbf{l} = IR \tag{9.3b}$$

....

[2]After Heinrich Friedrich Emil Lenz (1804–1865), a Russian professor of physics.

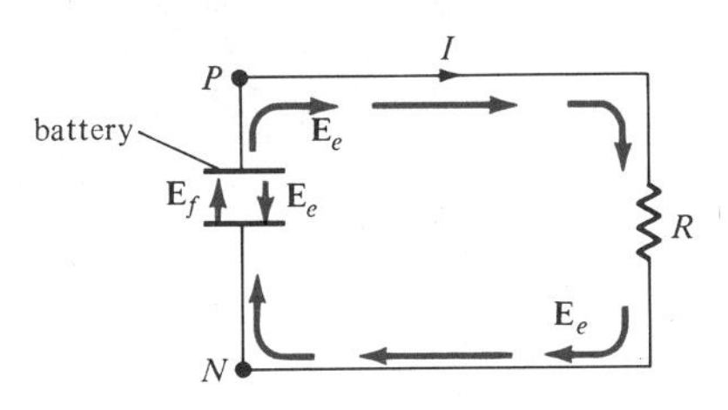

Figure 9.2 A circuit showing emf-producing field $\mathbf{E}_f$ and electrostatic field $\mathbf{E}_e$.

since $\mathbf{E}_f$ and $\mathbf{E}_e$ are equal but opposite within the battery (see Figure 9.2). It may also be regarded as the potential difference $(V_P - V_N)$ between the battery's open-circuit terminals. It is important to note that:

1. An electrostatic field $\mathbf{E}_e$ cannot maintain a steady current in a closed circuit since $\oint_L \mathbf{E}_e \cdot d\mathbf{l} = 0 = IR$.
2. An emf-produced field $\mathbf{E}_f$ is nonconservative.
3. Except in electrostatics, voltage and potential difference are usually not equivalent.

9.3 TRANSFORMER AND MOTIONAL EMFs

Having considered the connection between emf and electric field, we may examine how Faraday's law links electric and magnetic fields. For a circuit with a single turn $(N = 1)$, eq. (9.1) becomes

$$\boxed{V_{\text{emf}} = -\frac{d\Psi}{dt}} \tag{9.4}$$

In terms of $\mathbf{E}$ and $\mathbf{B}$, eq. (9.4) can be written as

$$V_{\text{emf}} = \oint_L \mathbf{E} \cdot d\mathbf{l} = -\frac{d}{dt}\int_S \mathbf{B} \cdot d\mathbf{S} \tag{9.5}$$

where Ψ has been replaced by $\int_S \mathbf{B} \cdot d\mathbf{S}$ and S is the surface area of the circuit bounded by the closed path L. It is evident from eq. (9.5) that in a time-varying situation, both electric and magnetic fields are present and are interrelated. Note that $d\mathbf{l}$ and $d\mathbf{S}$ in eq. (9.5) are in accordance with the right-hand rule as well as Stokes's theorem. This should be observed in Figure 9.3. The variation of flux with time as in eq. (9.1) or eq. (9.5) may be caused in three ways:

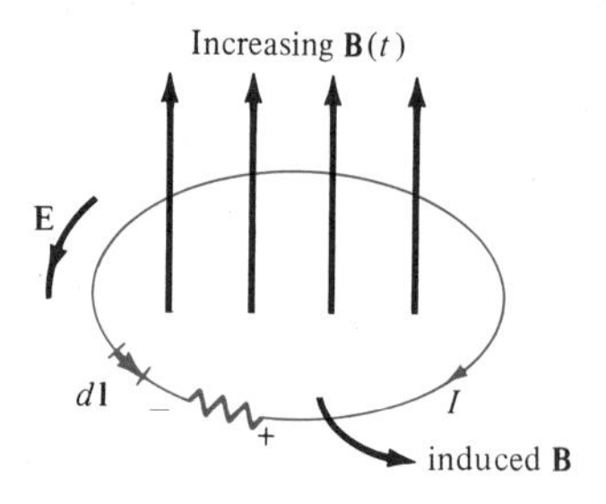

Figure 9.3 Induced emf due to a stationary loop in a time-varying $\mathbf{B}$ field.

1. by having a stationary loop in a time-varying **B** field
2. by having a time-varying loop area in a static **B** field
3. by having a time-varying loop area in a time-varying **B** field.

Each of these will be considered separately.

A. Stationary Loop in Time-Varying B Field (Transformer emf)

This is the case portrayed in Figure 9.3 where a stationary conducting loop is in a time-varying magnetic **B** field. Equation (9.5) becomes

$$V_{\text{emf}} = \oint_L \mathbf{E} \cdot d\mathbf{l} = -\int_S \frac{\partial \mathbf{B}}{\partial t} \cdot d\mathbf{S} \qquad [9.6]$$

This emf induced by the time-varying current (producing the time-varying **B** field) in a stationary loop is often referred to as *transformer emf* in power analysis since it is due to transformer action. By applying Stokes's theorem to the middle term in eq. (9.6), we obtain

$$\int_S (\nabla \times \mathbf{E}) \cdot d\mathbf{S} = -\int_S \frac{\partial \mathbf{B}}{\partial t} \cdot d\mathbf{S} \qquad [9.7]$$

For the two integrals to be equal, their integrands must be equal; that is,

$$\nabla \times \mathbf{E} = -\frac{\partial \mathbf{B}}{\partial t} \qquad [9.8]$$

This is one of the Maxwell's equations for time-varying fields. It shows that the time-varying **E** field is not conservative ($\nabla \times \mathbf{E} \neq 0$). This does not imply that the principles of energy conservation are violated. The work done in taking a charge about a closed path in a time-varying electric field, for example, is due to the energy from the time-varying magnetic field. Observe that Figure 9.3 obeys Lenz's law; the induced current I flows such as to produce a magnetic field that opposes $\mathbf{B}(t)$.

B. Moving Loop in Static B Field (Motional emf)

When a conducting loop is moving in a static **B** field, an emf is induced in the loop. We recall from eq. (8.2) that the force on a charge moving with uniform velocity **u** in a magnetic field **B** is

$$\mathbf{F}_m = Q\mathbf{u} \times \mathbf{B} \qquad [8.2]$$

We define the *motional electric field* $\mathbf{E}_m$ as

$$\mathbf{E}_m = \frac{\mathbf{F}_m}{Q} = \mathbf{u} \times \mathbf{B} \tag{9.9}$$

If we consider a conducting loop, moving with uniform velocity $\mathbf{u}$ as consisting of a large number of free electrons, the emf induced in the loop is

$$\boxed{V_{\text{emf}} = \oint_L \mathbf{E}_m \cdot d\mathbf{l} = \oint_L (\mathbf{u} \times \mathbf{B}) \cdot d\mathbf{l}} \tag{9.10}$$

This type of emf is called *motional emf* or *flux-cutting emf* because it is due to motional action. It is the kind of emf found in electrical machines such as motors, generators, and alternators. Figure 9.4 illustrates a two-pole dc machine with one armature coil and a two-bar commutator. Although the analysis of the dc machine is beyond the scope of this text, it is evident that voltage is generated as the coil rotates within the magnetic field. Another example of motional emf is illustrated in Figure 9.5, where a rod is moving between a pair of rails. In this example, $\mathbf{B}$ and $\mathbf{u}$ are perpendicular, so eq. (9.9) in conjunction with eq. (8.2) becomes

$$\mathbf{F}_m = I\boldsymbol{\ell} \times \mathbf{B} \tag{9.11}$$

or

$$F_m = I\ell B \tag{9.12}$$

and eq. (9.10) becomes

$$V_{\text{emf}} = uB\ell \tag{9.13}$$

By applying Stokes's theorem to eq. (9.10),

$$\int_S (\nabla \times \mathbf{E}_m) \cdot d\mathbf{S} = \int_S \nabla \times (\mathbf{u} \times \mathbf{B}) \cdot d\mathbf{S}$$

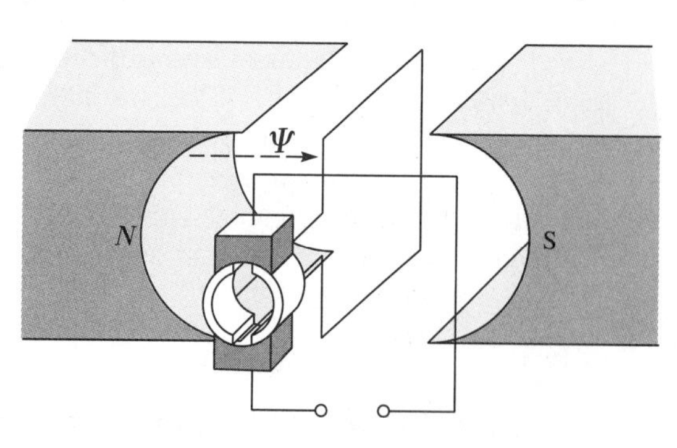

Figure 9.4 A direct-current machine.

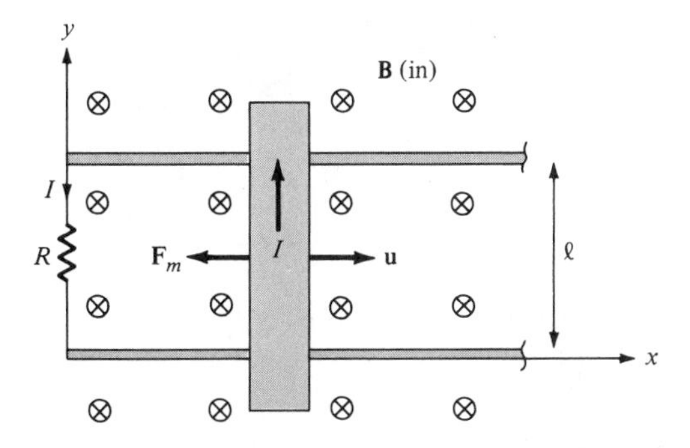

Figure 9.5 Induced emf due to a moving loop in a static **B** field.

or

$$\nabla \times \mathbf{E}_m = \nabla \times (\mathbf{u} \times \mathbf{B}) \tag{9.14}$$

Notice that unlike eq. (9.6), there is no need for a negative sign in eq. (9.10) because Lenz's law is already accounted for.

To apply eq. (9.10) is not always easy; some care must be exercised. The following points should be noted:

1. The integral in eq. (9.10) is zero along the portion of the loop where $\mathbf{u} = 0$. Thus $d\mathbf{l}$ is taken along the rod (the portion of the loop that is cutting the field) where $\mathbf{u}$ has nonzero value.
2. The direction of the induced current is the same as that of $\mathbf{E}_m$ or $\mathbf{u} \times \mathbf{B}$. The limits of the integral in eq. (9.10) are selected in the opposite direction to the induced current thereby satisfying Lenz's law. In eq. (9.13), for example, the integration over L is along $-\mathbf{a}_y$ whereas induced current flows in the rod along $\mathbf{a}_y$.

C. Moving Loop in Time-Varying Field

This is the general case in which a moving conducting loop is in a time-varying magnetic field. Both transformer emf and motional emf are present. Combining eqs. (9.6) and (9.10) gives the total emf as

$$V_{\text{emf}} = \oint_L \mathbf{E} \cdot d\mathbf{l} = -\int_S \frac{\partial \mathbf{B}}{\partial t} \cdot d\mathbf{S} + \oint_L (\mathbf{u} \times \mathbf{B}) \cdot d\mathbf{l} \tag{9.15}$$

or from eqs. (9.8) and (9.14),

$$\nabla \times \mathbf{E} = -\frac{\partial \mathbf{B}}{\partial t} + \nabla \times (\mathbf{u} \times \mathbf{B}) \quad [9.16]$$

Note that eq. (9.15) is equivalent to eq. (9.4), so V_{emf} can be found using either eq. (9.15) or (9.4). In fact, eq. (9.4) can always be applied in place of eqs. (9.6), (9.10), and (9.15).

EXAMPLE 9.1

A conducting bar can slide freely over two conducting rails as shown in Figure 9.6. Calculate the induced voltage in the bar

(a) If the bar is stationed at $y = 8$ cm and $\mathbf{B} = 4 \cos 10^6 t \, \mathbf{a}_z$ mWb/m^2

(b) If the bar slides at a velocity $\mathbf{u} = 20\mathbf{a}_y$ m/s and $\mathbf{B} = 4\mathbf{a}_z$ mWb/m^2

(c) If the bar slides at a velocity $\mathbf{u} = 20\mathbf{a}_y$ m/s and $\mathbf{B} = 4 \cos (10^6 t - y) \, \mathbf{a}_z$ mWb/m^2

SOLUTION (a) In this case, we have transformer emf given by

$$V_{emf} = -\int \frac{\partial \mathbf{B}}{\partial t} \cdot d\mathbf{S} = \int_{y=0}^{0.08} \int_{x=0}^{0.06} 4(10^{-3})(10^6) \sin 10^6 t \, dx \, dy$$

$$= 4(10^3)(0.08)(0.06) \sin 10^6 t$$

$$= 19.2 \sin 10^6 t \text{ V}$$

+ bec B is decreasing, so it should oppose

The polarity of the induced voltage (according to Lenz's law) is such that point P on the bar is at lower potential than Q when $\mathbf{B}$ is increasing.

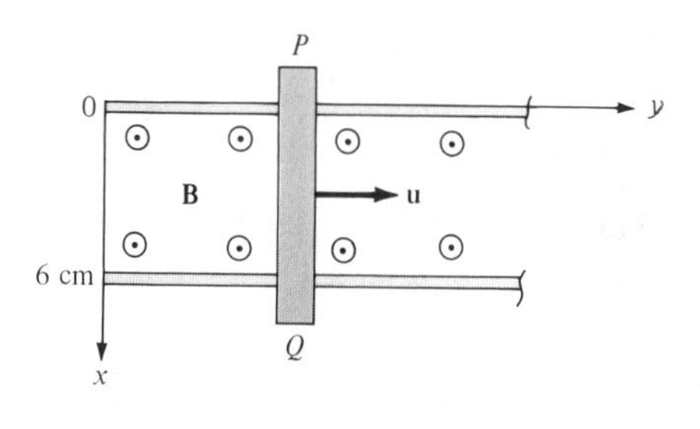

Figure 9.6 For Example 9.1.

(b) This is the case of motional emf:

$$V_{\text{emf}} = \int (\mathbf{u} \times \mathbf{B}) \cdot d\mathbf{l} = \int_{x=\ell}^{0} (u\mathbf{a}_y \times B\mathbf{a}_z) \cdot dx\mathbf{a}_x$$

$$= -uB\ell = -20(4.10^{-3})(0.06)$$

$$= -4.8 \text{ mV}$$

negative bec B is +

(c) Both transformer emf and motional emf are present in this case. This problem can be solved in two ways.

Method 1: Using eq. (9.15)

$$V_{\text{emf}} = -\int \frac{\partial \mathbf{B}}{\partial t} \cdot d\mathbf{S} + \int (\mathbf{u} \times \mathbf{B}) \cdot d\mathbf{l} \qquad [9.1.1]$$

$$= \int_{x=0}^{0.06} \int_0^y 4.10^{-3}(10^6) \sin(10^6 t - y')dy'\, dx$$

$$+ \int_{0.06}^{0} [20\mathbf{a}_y \times 4.10^{-3} \cos(10^6 t - y)\mathbf{a}_z] \cdot dx\, \mathbf{a}_x$$

$$= 240 \cos(10^6 t - y')\Big|_0^y - 80(10^{-3})(0.06) \cos(10^6 t - y)$$

$$= 240 \cos(10^6 t - y) - 240 \cos 10^6 t - 4.8(10^{-3}) \cos(10^6 t - y)$$

$$\simeq 240 \cos(10^6 t - y) - 240 \cos 10^6 t \qquad [9.1.2]$$

because the motional emf is negligible compared with the transformer emf. Using trigonometric identity

$$\cos A - \cos B = -2 \sin \frac{A+B}{2} \sin \frac{A-B}{2}$$

$$V_{\text{emf}} = 480 \sin\left(10^6 t - \frac{y}{2}\right) \sin \frac{y}{2} \text{ V} \qquad [9.1.3]$$

Method 2: Alternatively we can apply eq. (9.4), namely,

$$V_{\text{emf}} = -\frac{\partial \Psi}{\partial t} \qquad [9.1.4]$$

where

$$\Psi = \int \mathbf{B} \cdot d\mathbf{S}$$

$$= \int_{y=0}^{y} \int_{x=0}^{0.06} 4 \cos(10^6 t - y)\, dx\, dy$$

$$= -4(0.06) \sin(10^6 t - y)\Big|_{y=0}^{y}$$

$$= -0.24 \sin(10^6 t - y) + 0.24 \sin 10^6 t \text{ mWb}$$

But

$$\frac{dy}{dt} = u \rightarrow y = ut = 20t$$

Hence,

$$\Psi = -0.24 \sin(10^6 t - 20t) + 0.24 \sin 10^6 t \text{ mWb}$$

$$V_{\text{emf}} = -\frac{\partial \Psi}{\partial t} = 0.24(10^6 - 20) \cos(10^6 t - 20t) - 0.24(10^6) \cos 10^6 t \text{ mV}$$

$$\simeq 240 \cos(10^6 t - y) - 240 \cos 10^6 t \text{ V} \quad [9.1.5]$$

which is the same result in (9.1.2). Notice that in eq. (9.1.1), the dependence of y on time is taken care of in $\int (\mathbf{u} \times \mathbf{B}) \cdot d\mathbf{l}$, and we should not be bothered by it in $\partial \mathbf{B}/\partial t$. Why? Because the loop is assumed stationary when computing the transformer emf. This is a subtle point one must keep in mind in applying eq. (9.1.1). For the same reason, the second method is always easier. ■

PRACTICE EXERCISE 9.1

Consider the loop of Figure 9.5. If $\mathbf{B} = 0.5\mathbf{a}_z$ Wb/m^2, $R = 20\ \Omega$, $\ell = 10$ cm, and the rod is moving with a constant velocity of $8\mathbf{a}_x$ m/s, find

(a) The induced emf in the rod

(b) The current through the resistor

(c) The motional force on the rod

(d) The power dissipated by the resistor.

ANSWER (a) 0.4 V, (b) 20mA, (c) $-\mathbf{a}_x$ mN, (d) 8mW.

EXAMPLE 9.2

The loop shown in Figure 9.7 is inside a uniform magnetic field $\mathbf{B} = 50\ \mathbf{a}_x$ mWb/m^2. If side DC of the loop cuts the flux lines at the frequency of 50 Hz and the loop lies in the yz-plane at time $t = 0$, find

(a) The induced emf at $t = 1$ ms

(b) The induced current at $t = 3$ ms

SOLUTION

(a) Since the **B** field is time-invariant, the induced emf is motional, that is,

$$V_{\text{emf}} = \int (\mathbf{u} \times \mathbf{B}) \cdot d\mathbf{l}$$

where

$$d\mathbf{l} = d\mathbf{l}_{DC} = dz\ \mathbf{a}_z, \qquad \mathbf{u} = \frac{d\mathbf{l}'}{dt} = \frac{\rho\, d\phi}{dt}\mathbf{a}_\phi = \rho\omega\mathbf{a}_\phi$$

$$\rho = AD = 4 \text{ cm}, \qquad \omega = 2\pi f = 100\pi$$

As **u** and $d\mathbf{l}$ are in cylindrical coordinates, we transform **B** into cylindrical coordinates using eq. (2.9):

$$\mathbf{B} = B_o\mathbf{a}_x = B_o\,(\cos\phi\ \mathbf{a}_\rho - \sin\phi\ \mathbf{a}_\phi)$$

where $B_o = 0.05$. Hence,

$$\mathbf{u} \times \mathbf{B} = \begin{vmatrix} \mathbf{a}_\rho & \mathbf{a}_\phi & \mathbf{a}_z \\ 0 & \rho\omega & 0 \\ B_o\cos\phi & -B_o\sin\phi & 0 \end{vmatrix} = -\rho\omega B_o\cos\phi\ \mathbf{a}_z$$

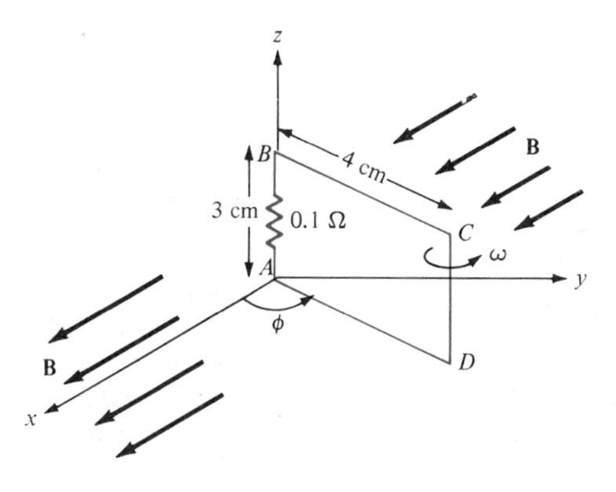

Figure 9.7 For Example 9.2; polarity is for increasing emf.

and

$$(\mathbf{u} \times \mathbf{B}) \cdot d\mathbf{l} = -\rho\omega B_o \cos\phi\, dz = -\,0.04\,(100\pi)(0.05)\cos\phi\, dz$$

$$= -\,0.2\pi \cos\phi\, dz$$

$$V_{emf} = \int_{z=0}^{0.03} -\,0.2\pi \cos\phi\, dz = -\,6\pi\cos\phi \text{ mV}$$

To determine ϕ, recall that

$$\omega = \frac{d\phi}{dt} \rightarrow \phi = \omega t + C_o$$

where C_o is an integration constant. At $t = 0$, $\phi = \pi/2$ because the loop is in the yz-plane at that time, $C_o = \pi/2$. Hence,

$$\phi = \omega t + \frac{\pi}{2}$$

and

$$V_{emf} = -6\pi \cos\left(\omega t + \frac{\pi}{2}\right) = 6\pi \sin(100\pi t) \text{ mV}$$

$$\text{At } t = 1 \text{ ms},\ V_{emf} = 6\pi\sin(0.1\pi) = 5.825 \text{ mV}$$

(b) The current induced is

$$i = \frac{V_{emf}}{R} = 60\pi \sin(100\pi t) \text{ mA}$$

At $t = 3$ ms,

$$i = 60\pi \sin(0.3\pi) \text{ mA} = 0.1525\text{A}$$

PRACTICE EXERCISE 9.2

Rework Example 9.2 with everything the same except that the **B** field is changed to:

(a) $\mathbf{B} = 50\mathbf{a}_y$ mWb/m^2—that is, the magnetic field is oriented along the y-direction

(b) $\mathbf{B} = 0.02t\,\mathbf{a}_x$ Wb/m^2—that is, the magnetic field is time-varying.

ANSWER (a) -17.93 mV, -0.1108 A, (b) 20.5 μV, $-$ 41.92 mA.

EXAMPLE 9.3

The magnetic circuit of Figure 9.8 has a uniform cross section of 10^{-3}m^2. If the circuit is energized by a current $i_1(t) = 3 \sin 100\pi t$ A in the coil of $N_1 = 200$ turns, find the emf induced in the coil of $N_2 = 100$ turns. Assume that $\mu = 500\mu_o$.

SOLUTION The flux in the circuit is

$$\Psi = \frac{\mathcal{F}}{\mathcal{R}} = \frac{N_1 i_1}{\ell/\mu S} = \frac{N_1 i_1 \mu S}{2\pi\rho_o}$$

According to Faraday's law, the emf induced in the second coil is

$$V_2 = -N_2 \frac{d\Psi}{dt} = -\frac{N_1 N_2 \mu S}{2\pi\rho_o} \frac{di_1}{dt}$$

$$= -\frac{100 \cdot (200) \cdot (500) \cdot (4\pi \times 10^{-7}) \cdot (10^{-3}) \cdot 300\pi \cos 100\pi t}{2\pi \cdot (10 \times 10^{-2})}$$

$$= -6\pi \cos 100\pi t \text{ V}$$

■

PRACTICE EXERCISE 9.3

A magnetic core of uniform cross-section 4 cm^2 is connected to a 120-V, 60-Hz generator as shown in Figure 9.9. Calculate the induced emf V_2 in the secondary coil.

ANSWER 72 V.

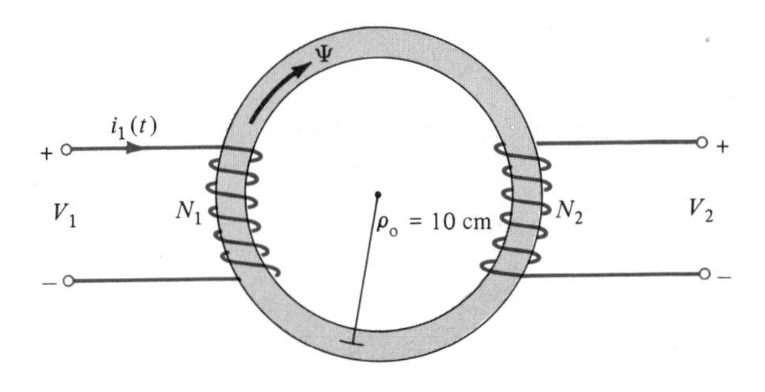

Figure 9.8 Magnetic circuit of Example 9.3.

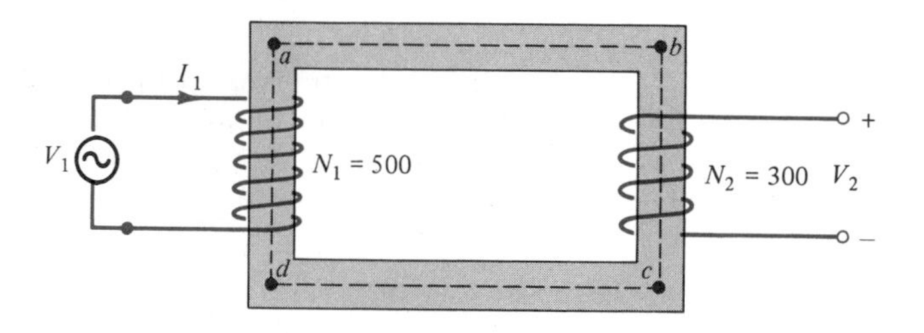

Figure 9.9 For Practice Exercise 9.3.

9.4 DISPLACEMENT CURRENT

In the previous section, we have essentially reconsidered Maxwell's curl equation for electrostatic fields and modified it for time-varying situations to satisfy Faraday's law. We shall now reconsider Maxwell's curl equation for magnetic fields (Ampere's circuit law) for time-varying conditions.

For static EM fields, we recall that

$$\nabla \times \mathbf{H} = \mathbf{J} \tag{9.17}$$

But the divergence of the curl of any vector field is identically zero (see Example 3.10). Hence,

$$\nabla \cdot (\nabla \times \mathbf{H}) = 0 = \nabla \cdot \mathbf{J} \tag{9.18}$$

The continuity of current in eq. (5.37), however, requires that

$$\nabla \cdot \mathbf{J} = -\frac{\partial \rho_v}{\partial t} \neq 0 \tag{9.19}$$

Thus eqs. (9.18) and (9.19) are obviously incompatible for time-varying conditions. We must modify eq. (9.17) to agree with eq. (9.19). To do this, we add a term to eq. (9.17) so that it becomes

$$\nabla \times \mathbf{H} = \mathbf{J} + \mathbf{J}_d \tag{9.20}$$

where $\mathbf{J}_d$ is to be determined and defined. Again, the divergence of the curl of any vector is zero. Hence:

$$\nabla \cdot (\nabla \times \mathbf{H}) = 0 = \nabla \cdot \mathbf{J} + \nabla \cdot \mathbf{J}_d \tag{9.21}$$

In order for eq. (9.21) to agree with eq. (9.19),

$$\nabla \cdot \mathbf{J}_d = -\nabla \cdot \mathbf{J} = \frac{\partial \rho_v}{\partial t} = \frac{\partial}{\partial t}(\nabla \cdot \mathbf{D}) = \nabla \cdot \frac{\partial \mathbf{D}}{\partial t} \tag{9.22a}$$

or

$$\boxed{\mathbf{J}_d = \frac{\partial \mathbf{D}}{dt}} \tag{9.22b}$$

Substituting eq. (9.22b) into eq. (9.20) results in

$$\boxed{\nabla \times \mathbf{H} = \mathbf{J} + \frac{\partial \mathbf{D}}{\partial t}} \tag{9.23}$$

This is Maxwell's equation (based on Ampere's circuit law) for time-varying field. The term $\mathbf{J}_d = \partial \mathbf{D}/\partial t$ is known as *displacement current density* and $\mathbf{J}$ is the conduction current density ($\mathbf{J} = \sigma \mathbf{E}$). The insertion of $\mathbf{J}_d$ into eq. (9.17) was one of the major contributions of Maxwell. Without the term $\mathbf{J}_d$, electromagnetic wave propagation (radio or TV waves, for example) would be impossible. At low frequencies, $\mathbf{J}_d$ is usually neglected compared with $\mathbf{J}$. However, at radio frequencies, the two terms are comparable. At the time of Maxwell, high-frequency sources were not available and eq. (9.23) could not be verified experimentally. It was years later that Hertz succeeded in generating and detecting radio waves thereby verifying eq. (9.23). This is one of the rare situations where mathematical argument paved the way to experimental investigation.

Based on the displacement current density, we define the *displacement current* as

$$I_d = \int \mathbf{J}_d \cdot d\mathbf{S} = \int \frac{\partial \mathbf{D}}{\partial t} \cdot d\mathbf{S} \tag{9.24}$$

We must bear in mind that displacement current is a result of time-varying electric field. A typical example of such current is the current through a capacitor when an alternating voltage source is applied to its plates. This example, shown in Figure 9.10, serves to illustrate the need for the displacement current. Applying an unmodified form of Ampere's circuit law to a closed path L shown in Figure 9.10 (a) gives

$$\oint_L \mathbf{H} \cdot d\mathbf{l} = \int_{S_1} \mathbf{J} \cdot d\mathbf{S} = I_{\text{enc}} = I \tag{9.25}$$

where I is the current through the conductor and S_1 is the flat surface bounded by L. If we use the balloon-shaped surface S_2 which passes between the capacitor plates, as in Figure 9.10(b),

$$\oint_L \mathbf{H} \cdot d\mathbf{l} = \int_{S_2} \mathbf{J} \cdot d\mathbf{S} = I_{\text{enc}} = 0 \tag{9.26}$$

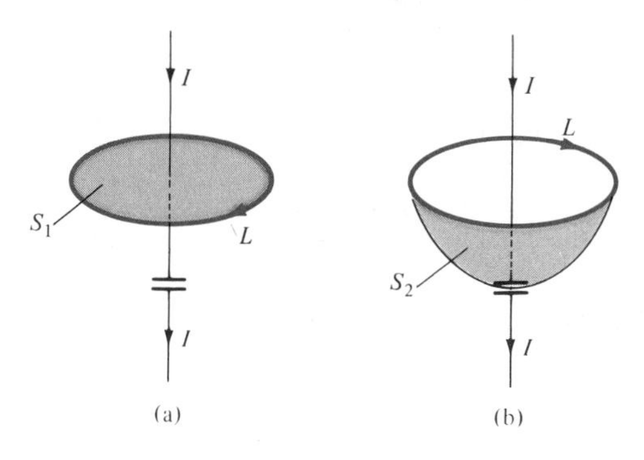

Figure 9.10 Two surfaces of integration showing the need for $\mathbf{J}_d$ in Ampere's circuit law.

because no conduction current ($\mathbf{J} = 0$) flows through S_2. This is contradictory in view of the fact that the same closed path L is used. To resolve the conflict, we need to include the displacement current in Ampere's circuit law. The total current density is $\mathbf{J} + \mathbf{J}_d$. In eq. (9.25), $\mathbf{J}_d = 0$ so that the equation remains valid. In eq. (9.26), $\mathbf{J} = 0$ so that

$$\oint_L \mathbf{H} \cdot d\mathbf{l} = \int_{S_2} \mathbf{J}_d \cdot d\mathbf{S} = \frac{d}{dt}\int_{S_2} \mathbf{D} \cdot d\mathbf{S} = \frac{dQ}{dt} = I \tag{9.27}$$

So we obtain the same current for either surface though it is conduction current in S_1 and displacement current in S_2.

EXAMPLE 9.4

A parallel plate capacitor with plate area of 5 cm^2 and plate separation of 3 mm has a voltage $50 \sin 10^3 t$ V applied to its plates. Calculate the displacement current assuming $\varepsilon = 2\varepsilon_o$.

SOLUTION

$$D = \varepsilon E = \varepsilon \frac{V}{d}$$

$$J_d = \frac{\partial D}{\partial t} = \frac{\varepsilon}{d}\frac{dV}{dt}$$

Hence,

$$I_d = J_d \cdot S = \frac{\varepsilon S}{d}\frac{dV}{dt} = C\frac{dV}{dt}$$

which is the same as the conduction current, given by

$$I_c = \frac{dQ}{dt} = S\frac{d\rho_s}{dt} = S\frac{dD}{dt} = \varepsilon S\frac{dE}{dt} = \frac{\varepsilon S}{d}\frac{dV}{dt} = C\frac{dV}{dt}$$

$$I_d = 2 \cdot \frac{10^{-9}}{36\pi} \cdot \frac{5 \times 10^{-4}}{3 \times 10^{-3}} \cdot 10^3 \times 50 \cos 10^3 t$$

$$= 147.4 \cos 10^3 t \text{ nA}$$

■

PRACTICE EXERCISE 9.4

In free space, $\mathbf{E} = 20 \cos(\omega t - 50x)\, \mathbf{a}_y$ V/m. Calculate

(a) $\mathbf{J}_d$

(b) $\mathbf{H}$

(c) ω

ANSWER (a) $-20\omega\varepsilon_o \sin(\omega t - 50x)\, \mathbf{a}_y$ A/m^2, (b) $0.4\omega\varepsilon_o \cos(\omega t - 50x)\, \mathbf{a}_z$ A/m, (c) 1.5×10^{10} rad/s.

9.5 MAXWELL'S EQUATIONS IN FINAL FORMS

James Clerk Maxwell is regarded as the founder of electromagnetic theory in its present form. Maxwell's celebrated work led to the discovery of electromagnetic waves.[3] Through his theoretical efforts over about five years (when he was between 35 and 40), Maxwell published the first unified theory of electricity and magnetism. The theory comprised all previously known results, both experimental and theoretical, on electricity and magnetism. It further introduced displacement current and predicted the existence of electromagnetic waves. Maxwell's equations were not fully accepted by many scientists until they were later confirmed by Heinrich Rudolf Hertz (1857–1894), a German physics professor. Hertz was successful in generating and detecting radio waves.

The laws of electromagnetism that Maxwell put together in the form of four equations were presented in Table 7.2 in Section 7.6 for static conditions. The more generalized forms of these equations are those for time-varying conditions shown in Table 9.1. We notice from the table that the divergence equations remain the same while the curl equations have been modified. The integral form of Maxwell's equation depicts the underlying physical laws, whereas the differential form is used more frequently in solving problems. For a field to be "qualified" as an electromagnetic

[3]The work of James Clerk Maxwell (1831–1879), a Scottish physicist, can be found in his book, *A Treatise on Electricity and Magnetism*. New York: Dover, vols. 1 and 2, 1954.

Table 9.1 Generalized Forms of Maxwell's Equations

Differential form	Integral form	Remarks
$\nabla \cdot \mathbf{D} = \rho_v$	$\oint_S \mathbf{D} \cdot d\mathbf{S} = \int_v \rho_v \, dv$	Gauss's law
$\nabla \cdot \mathbf{B} = 0$	$\oint_S \mathbf{B} \cdot d\mathbf{S} = 0$	Nonexistence of isolated magnetic charge*
$\nabla \times \mathbf{E} = -\dfrac{\partial \mathbf{B}}{\partial t}$	$\oint_L \mathbf{E} \cdot d\mathbf{l} = -\dfrac{\partial}{\partial t}\int_S \mathbf{B} \cdot d\mathbf{S}$	Faraday's law
$\nabla \times \mathbf{H} = \mathbf{J} + \dfrac{\partial \mathbf{D}}{\partial t}$	$\oint_L \mathbf{H} \cdot d\mathbf{l} = \int_S \left(\mathbf{J} + \dfrac{\partial \mathbf{D}}{\partial t}\right) \cdot d\mathbf{S}$	Ampere's circuital law

*This is also referred to as Gauss's law for magnetic fields.

field, it must satisfy all four Maxwell's equations. The importance of Maxwell's equations cannot be overemphasized because they summarize all known laws of electromagnetism. We shall often refer to them in the remaining part of this text.

Since this section is meant to be a compendium of our discussion in this text, it is worthwhile to mention other equations that go hand in hand with Maxwell's equations. The Lorentz force equation

$$\mathbf{F} = Q(\mathbf{E} + \mathbf{u} \times \mathbf{B}) \tag{9.28}$$

is associated with Maxwell's equations. Also the equation of continuity

$$\nabla \cdot \mathbf{J} = -\frac{\partial \rho_v}{\partial t} \tag{9.29}$$

is implicit in Maxwell's equations. The concepts of linearity, isotropy, and homogeneity of a material medium still apply for time-varying fields; in a linear, homogeneous, and isotropic medium characterized by σ, ε, and μ, the constitutive relations

$$\mathbf{D} = \varepsilon \mathbf{E} = \varepsilon_o \mathbf{E} + \mathbf{P} \tag{9.30a}$$

$$\mathbf{B} = \mu \mathbf{H} = \mu_o(\mathbf{H} + \mathbf{M}) \tag{9.30b}$$

$$\mathbf{J} = \sigma \mathbf{E} + \rho_v \mathbf{u} \tag{9.30c}$$

hold for time-varying fields. Consequently, the boundary conditions

$$E_{1t} = E_{2t} \quad \text{or} \quad (\mathbf{E}_1 - \mathbf{E}_2) \times \mathbf{a}_{n12} = 0 \tag{9.31a}$$

$$H_{1t} - H_{2t} = K \quad \text{or} \quad (\mathbf{H}_1 - \mathbf{H}_2) \times \mathbf{a}_{n12} = \mathbf{K} \tag{9.31b}$$

$$D_{1n} - D_{2n} = \rho_s \quad \text{or} \quad (\mathbf{D}_1 - \mathbf{D}_2) \cdot \mathbf{a}_{n12} = \rho_s \tag{9.31c}$$

$$B_{1n} - B_{2n} = 0 \quad \text{or} \quad (\mathbf{B}_2 - \mathbf{B}_1) \cdot \mathbf{a}_{n12} = 0 \tag{9.31d}$$

remain valid for time-varying fields. However, for a perfect conductor ($\sigma \simeq \infty$) in a time-varying field,

$$\mathbf{E} = 0, \qquad \mathbf{H} = 0, \qquad \mathbf{J} = 0 \tag{9.32}$$

and hence,

$$\mathbf{B}_n = 0, \qquad \mathbf{E}_t = 0 \tag{9.33}$$

For a perfect dielectric ($\sigma \simeq 0$), eq. (9.31) holds except that $\mathbf{K} = 0$. Though eqs. (9.28) to (9.33) are not Maxwell's equations, they are associated with them.

To complete this summary section, we present a structure linking the various potentials and vector fields of the electric and magnetic fields in Figure 9.11. This electromagnetic flow diagram helps with the visualization of the basic relationships between field quantities. It also shows that it is usually possible to find alternative formulations, for a given problem, in a relatively simple manner. It should be noted that in Figures 9.10(b) and (c), we introduce ρ^m as the free magnetic density (similar to ρ_v), which is, of course, zero, $\mathbf{A}_e$ as the magnetic current density (analogous to $\mathbf{J}$). Using terms from stress analysis, the principal relationships are typified as:

(a) compatibility equations

$$\nabla \cdot \mathbf{B} = \rho^m = 0 \tag{9.34}$$

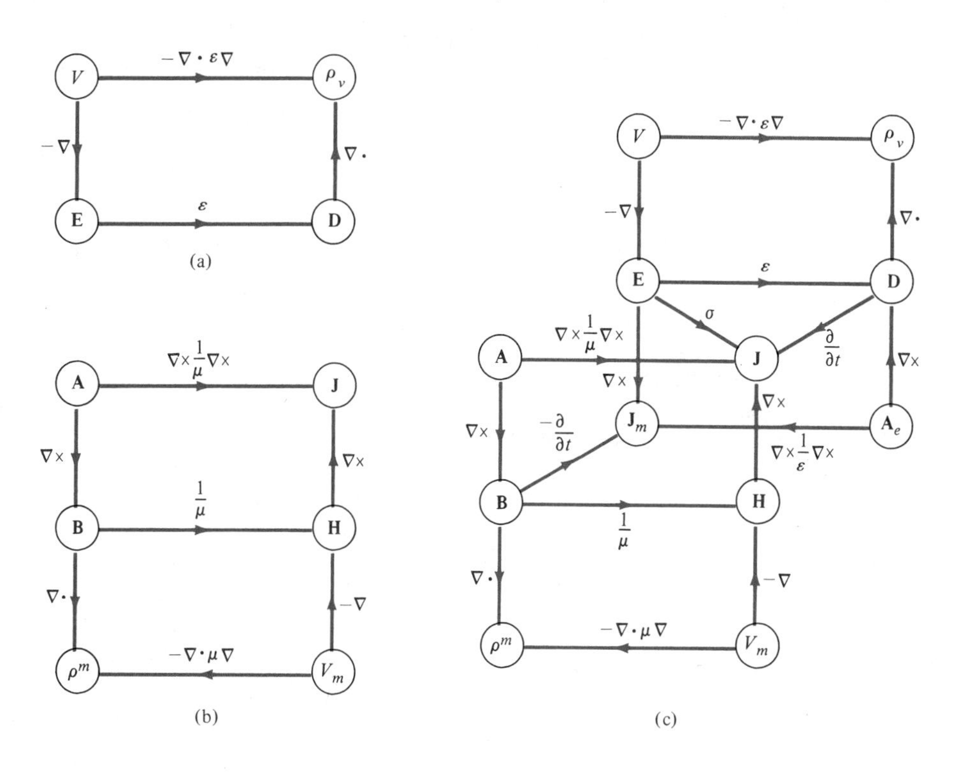

Figure 9.11 Electromagnetic flow diagram showing the relationship between the potentials and vector fields: **(a)** electrostatic system, **(b)** magnetostatic system, **(c)** electromagnetic system. [Adapted with permission from IEE Publishing Dept.]

and

$$\nabla \times \mathbf{E} = -\frac{\partial \mathbf{B}}{\partial t} = \mathbf{J}_m \quad [9.35]$$

(b) constitutive equations

$$\mathbf{B} = \mu \mathbf{H} \quad [9.36]$$

and

$$\mathbf{D} = \varepsilon \mathbf{E} \quad [9.37]$$

(c) equilibrium equations

$$\nabla \cdot \mathbf{D} = \rho_v \quad [9.38]$$

and

$$\nabla \times \mathbf{H} = J + \frac{\partial \mathbf{D}}{\partial t} \quad [9.39]$$

†9.6 TIME-VARYING POTENTIALS

For static EM fields, we obtained the electric scalar potential as

$$V = \int_v \frac{\rho_v \, dv}{4\pi\varepsilon R} \quad [9.40]$$

and the magnetic vector potential as

$$\mathbf{A} = \int_v \frac{\mu J \, dv}{4\pi R} \quad [9.41]$$

We would like to examine what happens to these potentials when the fields are time-varying. Recall that **A** was defined from the fact that $\nabla \cdot \mathbf{B} = 0$, which still holds for time-varying fields. Hence the relation

$$\boxed{\mathbf{B} = \nabla \times \mathbf{A}} \quad [9.42]$$

holds for time-varying situations. Combining Faraday's law in eq. (9.8) with eq. (9.42) gives

$$\nabla \times \mathbf{E} = -\frac{\partial}{\partial t}(\nabla \times \mathbf{A}) \quad [9.43a]$$

or

$$\nabla \times \left(\mathbf{E} + \frac{\partial \mathbf{A}}{\partial t}\right) = 0 \quad [9.43b]$$

Since the curl of the gradient of a scalar field is identically zero (see Practice Exercise 3.10), the solution to eq. (9.43b) is

$$\mathbf{E} + \frac{\partial \mathbf{A}}{\partial t} = -\nabla V \tag{9.44}$$

or

$$\boxed{\mathbf{E} = -\nabla V - \frac{\partial \mathbf{A}}{\partial t}} \tag{9.45}$$

From eqs. (9.42) and (9.45), we can determine the vector fields **B** and **E** provided that the potentials **A** and V are known. However, we still need to find some expressions for **A** and V similar to those in eqs. (9.40) and (9.41) that are suitable for time-varying fields.

From Table 9.1 or eq. (9.38) we know that $\nabla \cdot \mathbf{D} = \rho_v$ is valid for time-varying conditions. By taking the divergence of eq. (9.45) and making use of eqs. (9.37) and (9.38), we obtain

$$\nabla \cdot \mathbf{E} = \frac{\rho_v}{\varepsilon} = -\nabla^2 V - \frac{\partial}{\partial t} (\nabla \cdot \mathbf{A})$$

or

$$\nabla^2 V + \frac{\partial}{\partial t} (\nabla \cdot \mathbf{A}) = -\frac{\rho_v}{\varepsilon} \tag{9.46}$$

Taking the curl of eq. (9.42) and incorporating eqs. (9.23) and (9.45) results in

$$\begin{aligned} \nabla \times \nabla \times \mathbf{A} &= \mu \mathbf{J} + \varepsilon\mu \frac{\partial}{\partial t}\left(-\nabla V - \frac{\partial \mathbf{A}}{\partial t}\right) \\ &= \mu \mathbf{J} - \mu\varepsilon \nabla \left(\frac{\partial V}{\partial t}\right) - \mu\varepsilon \frac{\partial^2 \mathbf{A}}{\partial t^2} \end{aligned} \tag{9.47}$$

where $\mathbf{D} = \varepsilon \mathbf{E}$ and $\mathbf{B} = \mu \mathbf{H}$ have been assumed. By applying the vector identity

$$\nabla \times \nabla \times \mathbf{A} = \nabla (\nabla \cdot \mathbf{A}) - \nabla^2 \mathbf{A} \tag{9.48}$$

to eq. (9.47),

$$\nabla^2 \mathbf{A} - \nabla(\nabla \cdot \mathbf{A}) = -\mu \mathbf{J} + \mu\varepsilon \nabla \left(\frac{\partial V}{\partial t}\right) + \mu\varepsilon \frac{\partial^2 \mathbf{A}}{\partial t^2} \tag{9.49}$$

A vector field is uniquely defined when its curl and divergence are specified. The curl of **A** has been specified by eq. (9.42); for reasons that will be obvious shortly, we may choose the divergence of **A** as

$$\boxed{\nabla \cdot \mathbf{A} = -\mu\varepsilon \frac{\partial V}{\partial t}} \tag{9.50}$$

This choice relates **A** with V and it is called the *Lorentz condition for potentials*. We had this in mind when we chose $\nabla \cdot \mathbf{A} = 0$ for magnetostatic fields in eq. (7.59). By imposing the Lorentz condition of eq. (9.50), eqs. (9.46) and (9.49) respectively become

$$\boxed{\nabla^2 V - \mu\varepsilon \frac{\partial^2 V}{\partial t^2} = -\frac{\rho_v}{\varepsilon}} \tag{9.51}$$

and

$$\boxed{\nabla^2 \mathbf{A} - \mu\varepsilon \frac{\partial^2 \mathbf{A}}{\partial t^2} = -\mu \mathbf{J}} \tag{9.52}$$

which are *wave equations* to be discussed in the next chapter. The reason for choosing the Lorentz condition becomes obvious as we examine eqs. (9.51) and (9.52). It uncouples eqs. (9.46) and (9.49) and also produces a symmetry between eqs. (9.51) and (9.52). It can be shown that the Lorentz condition can be obtained from the continuity equation (see Problem 9.25); therefore, our choice of eq. (9.50) is not arbitrary. Notice that eqs. (6.4) and (7.60) are special static cases of eqs. (9.51) and (9.52) respectively. In other words, potentials V and **A** satisfy Poisson's equations for time-varying conditions. Just as eqs. (9.40) and (9.41) are the solutions, or the integral forms of, eqs. (6.4) and (7.60), it can be shown that the solutions[4] to eqs. (9.51) and (9.52) are

$$V = \int_v \frac{[\rho_v]\, dv}{4\pi\varepsilon R} \tag{9.53}$$

and

$$\mathbf{A} = \int_v \frac{\mu[\mathbf{J}]\, dv}{4\pi R} \tag{9.54}$$

The term $[\rho_v]$ (or $[\mathbf{J}]$) means that the time t in $\rho_v(x, y, z, t)$ (or $\mathbf{J}(x, y, z, t)$) is replaced by the *retarded time* t' given by

$$t' = t - \frac{R}{u} \tag{9.55}$$

where $R = |\mathbf{r} - \mathbf{r}'|$ is the distance between the source point $\mathbf{r}'$ and the observation point $\mathbf{r}$ and

$$u = \frac{1}{\sqrt{\mu\varepsilon}} \tag{9.56}$$

....

[4]For example, see D. K. Cheng, *Field and Wave Electromagnetics*. Reading, MA: Addison-Wesley, 1983, pp. 291–292.

is the velocity of wave propagation. In free space, $u = c \simeq 3 \times 10^8$ m/s is the speed of light in a vacuum. Potentials V and $\mathbf{A}$ in eqs. (9.53) and (9.54) are respectively called the *retarded electric scalar potential* and the *retarded magnetic vector potential*. Given ρ_v and $\mathbf{J}$, V and $\mathbf{A}$ can be determined using eqs. (9.53) and (9.54); from V and $\mathbf{A}$, $\mathbf{E}$ and $\mathbf{B}$ can determined using eqs. (9.45) and (9.42), respectively.

9.7 TIME-HARMONIC FIELDS

So far, our time-dependence of EM fields has been arbitrary. To be specific, we shall assume that the fields are *time-harmonic*. By time-harmonic quantities we mean quantities which vary periodically or sinusoidally with time. Not only is sinusoidal analysis of practical value, it can be extended to most waveforms by Fourier and Laplace transform techniques. Sinusoids are easily expressed in phasors, which are more convenient to work with. Before applying phasors to EM fields, it is worthwhile to have a brief review of the concept of phasor.

A *phasor* z is a complex number that can be written as

$$z = x + jy = r\underline{/\phi} \tag{9.57}$$

or

$$z = r\, e^{j\phi} = r\,(\cos\phi + j\sin\phi) \tag{9.58}$$

where $j = \sqrt{-1}$, x is the real part of z, y is the imaginary part of z, r is the magnitude of z, given by

$$r = |z| = \sqrt{x^2 + y^2} \tag{9.59}$$

and ϕ is the phase of z, given by

$$\phi = \tan^{-1}\frac{y}{x} \tag{9.60}$$

Here x, y, z, r, and ϕ should not be mistaken as the coordinate variables although they look similar (different letters could have been used but it is hard to find better ones). The phasor z can be represented in *rectangular form* as $z = x + jy$ or in *polar form* as $z = r\underline{/\phi} = r\, e^{j\phi}$. The two forms of representing z are related in eqs. (9.57) to (9.60) and illustrated in Figure 9.12. Addition and subtraction of phasors are better performed in rectangular form; multiplication and division are better done in polar form.

Given complex numbers

$$z = x + jy = r\underline{/\phi}, \qquad z_1 = x_1 + jy_1 = r_1\underline{/\phi_1}, \qquad \text{and} \qquad z_2 = x_2 + jy_2 = r_2\underline{/\phi_2}$$

the following basic properties should be noted.

Addition:

$$z_1 + z_2 = (x_1 + x_2) + j(y_1 + y_2) \tag{9.61a}$$

Subtraction:

$$z_1 - z_2 = (x_1 - x_2) + j(y_1 - y_2) \tag{9.61b}$$

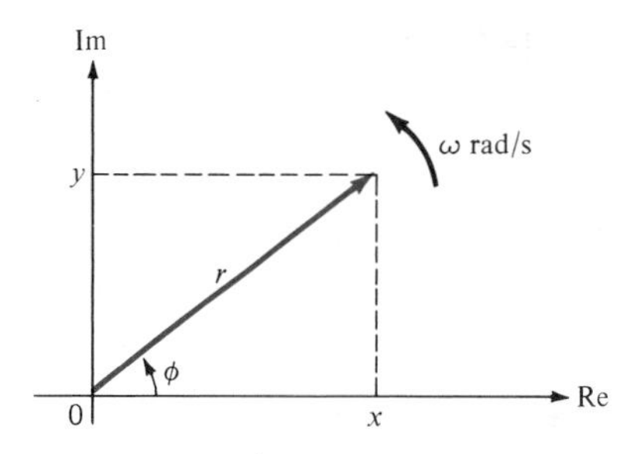

Figure 9.12 Representation of a phasor $z = x + jy = r\underline{/\phi}$.

Multiplication:

$$z_1 z_2 = r_1 r_2 \underline{/\phi_1 + \phi_2} \quad [9.61c]$$

Division:

$$\frac{z_1}{z_2} = \frac{r_1}{r_2} \underline{/\phi_1 - \phi_2} \quad [9.61d]$$

Square Root:

$$\sqrt{z} = \sqrt{r} \underline{/\phi/2} \quad [9.61e]$$

Complex Conjugate:

$$z^* = x - jy = r\underline{/-\phi} = re^{-j\phi} \quad [9.61f]$$

Other properties of complex numbers can be found in Appendix A.2.

To introduce the time element, we let

$$\phi = \omega t + \theta \quad [9.62]$$

where θ may be a function of time or space coordinates or a constant. The real (Re) and imaginary (Im) parts of

$$re^{j\phi} = re^{j\theta}e^{j\omega t} \quad [9.63]$$

are respectively given by

$$\text{Re}\,(re^{j\phi}) = r\cos\,(\omega t + \theta) \quad [9.64a]$$

and

$$\text{Im}\,(re^{j\phi}) = r\sin\,(\omega t + \theta) \quad [9.64b]$$

Thus a sinusoidal current $I(t) = I_o \cos(\omega t + \theta)$, for example, equals the real part of $I_o e^{j\theta} e^{j\omega t}$. The current $I'(t) = I_o \sin(\omega t + \theta)$, which is the imaginary part of $I_o e^{j\theta} e^{j\omega t}$, can also be represented as the real part of $I_o e^{j\theta} e^{j\omega t} e^{-j90°}$ because $\sin \alpha = \cos(\alpha - 90°)$. However, in performing our mathematical operations, we must be consistent in our use of either the real part or the imaginary part of a quantity but not both at the same time.

The complex term $I_o e^{j\theta}$, which results from dropping the time factor $e^{j\omega t}$ in $I(t)$, is called the *phasor* current, denoted by I_s; that is,

$$I_s = I_o e^{j\theta} = I_0 \angle \theta \tag{9.65}$$

where the subscript s denotes the phasor form of $I(t)$. Thus $I(t) = I_o \cos(\omega t + \theta)$, the *instantaneous form,* can be expressed as

$$I(t) = \text{Re}\,(I_s e^{j\omega t}) \tag{9.66}$$

In general, a phasor could be a scalar or vector. If a vector $\mathbf{A}(x, y, z, t)$ is a time-harmonic field, the *phasor form* of $\mathbf{A}$ is $\mathbf{A}_s(x, y, z)$; the two quantities are related as

$$\boxed{\mathbf{A} = \text{Re}\,(\mathbf{A}_s e^{j\omega t})} \tag{9.67}$$

For example, if $\mathbf{A} = A_o \cos(\omega t - \beta x)\, \mathbf{a}_y$, we can write $\mathbf{A}$ as

$$\mathbf{A} = \text{Re}\,(A_o e^{-j\beta x}\, \mathbf{a}_y e^{j\omega t}) \tag{9.68}$$

Comparing this with eq. (9.67) indicates that the phasor form of $\mathbf{A}$ is

$$\mathbf{A}_s = A_o e^{-j\beta x} \mathbf{a}_y \tag{9.69}$$

Notice from eq. (9.67) that

$$\begin{aligned} \frac{\partial \mathbf{A}}{\partial t} &= \frac{\partial}{\partial t} \text{Re}\,(\mathbf{A}_s e^{j\omega t}) \\ &= \text{Re}\,(j\omega \mathbf{A}_s e^{j\omega t}) \end{aligned} \tag{9.70}$$

showing that taking the time derivative of the instantaneous quantity is equivalent to multiplying its phasor form by $j\omega$. That is,

$$\frac{\partial \mathbf{A}}{\partial t} \rightarrow j\omega \mathbf{A}_s \tag{9.71}$$

Similarly,

$$\int \mathbf{A}\, \partial t \rightarrow \frac{\mathbf{A}_s}{j\omega} \tag{9.72}$$

Note that the real part is chosen in eq. (9.67) as in circuit analysis; the imaginary part could equally have been chosen. Also notice the basic difference between the intantaneous form $\mathbf{A}(x, y, z, t)$ and its phasor form $\mathbf{A}_s(x, y, z)$; the former is

Table 9.2 **Time-Harmonic Maxwell's Equations Assuming Time Factor $e^{j\omega t}$**

Point form	Integral form
$\nabla \cdot \mathbf{D}_s = \rho_{vs}$	$\oint \mathbf{D}_s \cdot d\mathbf{S} = \int \rho_{vs}\, dv$
$\nabla \cdot \mathbf{B}_s = 0$	$\oint \mathbf{B}_s \cdot d\mathbf{S} = 0$
$\nabla \times \mathbf{E}_s = -j\omega \mathbf{B}_s$	$\oint \mathbf{E}_s \cdot d\mathbf{l} = -j\omega \int \mathbf{B}_s \cdot d\mathbf{S}$
$\nabla \times \mathbf{H}_s = \mathbf{J}_s + j\omega \mathbf{D}_s$	$\oint \mathbf{H}_s \cdot d\mathbf{l} = \int (\mathbf{J}_s + j\omega \mathbf{D}_s) \cdot d\mathbf{S}$

time-dependent and real whereas the latter is time-invariant and generally complex. It is easier to work with $\mathbf{A}_s$ and obtain $\mathbf{A}$ from $\mathbf{A}_s$ whenever necessary using eq. (9.67).

We shall now apply the phasor concept to time-varying EM fields. The fields quantities $\mathbf{E}(x, y, z, t)$, $\mathbf{D}(x, y, z, t)$, $\mathbf{H}(x, y, z, t)$, $\mathbf{B}(x, y, z, t)$, $\mathbf{J}(x, y, z, t)$, and $\rho_v(x, y, z, t)$ and their derivatives can be expressed in phasor form using eqs. (9.67) and (9.71). In phasor form, Maxwell's equations for time-harmonic EM fields in a linear, isotropic, and homogeneous medium are presented in Table 9.2. From Table 9.2, note that the time factor $e^{j\omega t}$ disappears because it is associated with every term and therefore factors out, resulting in time-independent equations. Herein lies the justification for using phasors; the time factor can be suppressed in our analysis of time-harmonic fields and inserted when necessary. Also note that in Table 9.2, the time factor $e^{j\omega t}$ has been assumed. It is equally possible to have assumed the time factor $e^{-j\omega t}$, in which case we would need to replace every j in Table 9.2 with $-j$.

EXAMPLE 9.5

Evaluate the complex numbers

(a) $$z_1 = \frac{j(3 - j4)^*}{(-1 + j6)(2 + j)^2}$$

(b) $$z_2 = \left[\frac{1 + j}{4 - j8}\right]^{1/2}$$

SOLUTION

(a) This can be solved in two ways: working with z in rectangular form or polar form.

Method 1 (working in rectangular form):
Let

$$z_1 = \frac{z_3 z_4}{z_5 z_6}$$

where

$z_3 = j$
$z_4 = (3 - j4)^* =$ the complex conjugate of $(3 - j4)$
$\quad = 3 + j4$

(To find the complex conjugate of a complex number, simply replace every j with $-j$).

$$z_5 = -1 + j6$$

and

$$z_6 = (2 + j)^2 = 4 - 1 + j4 = 3 + j4$$

Hence,

$$z_3 z_4 = j(3 + j4) = -4 + j3$$

$$z_5 z_6 = (-1 + j6)(3 + j4) = -3 - j4 + j18 - 24$$
$$= -27 + j14$$

and

$$z_1 = \frac{-4 + j3}{-27 + j14}$$

Multiplying and dividing z_1 by $-27 - j14$ (rationalization), we have

$$z_1 = \frac{(-4 + j3)(-27 - j14)}{(-27 + j14)(-27 - j14)} = \frac{150 - j25}{27^2 + 14^2}$$
$$= 0.1622 - j0.027 = 0.1644 \underline{/-9.46^\circ}$$

Method 2 (working in polar form):

$$z_3 = j = 1\underline{/90^\circ}$$

$$z_4 = (3 - j4)^* = (5 \underline{/-53.13^\circ})^* = 5 \underline{/53.13^\circ}$$

$$z_5 = (-1 + j6) = \sqrt{37} \underline{/99.46^\circ}$$

$$z_6 = (2 + j)^2 = (\sqrt{5} \underline{/26.56^\circ})^2 = 5 \underline{/53.13^\circ}$$

Hence,

$$z_1 = \frac{(1 \underline{/90^\circ})(5 \underline{/53.13^\circ})}{(\sqrt{37} \underline{/99.46^\circ})(5 \underline{/53.13^\circ})}$$

$$= \frac{1}{\sqrt{37}} \underline{/90^\circ - 99.46^\circ} = 0.1644 \underline{/-9.46^\circ}$$

$$= 0.1622 - j0.027$$

as obtained before.

(b) Let

$$z_2 = \left[\frac{z_7}{z_8}\right]^{1/2}$$

where

$$z_7 = 1 + j = \sqrt{2}\ \underline{/45^\circ}$$

and

$$z_8 = 4 - j8 = 4\sqrt{5}\ \underline{/-63.4^\circ}$$

Hence

$$\frac{z_7}{z_8} = \frac{\sqrt{2}\ \underline{/45^\circ}}{4\sqrt{5}\ \underline{/-63.4^\circ}} = \frac{\sqrt{2}}{4\sqrt{5}}\ \underline{/45^\circ - -63.4^\circ}$$

$$= 0.1581\ \underline{/108.4^\circ}$$

and

$$z_2 = \sqrt{0.1581}\ \underline{/108.4^\circ}\ \underline{/2}$$

$$= 0.3976\ \underline{/54.2^\circ}$$

■

PRACTICE EXERCISE 9.5

Evaluate these complex numbers:

(a) $j^3 \left[\frac{1 + j}{2 - j}\right]^2$

(b) $6\ \underline{/30^\circ} + j5 - 3 + e^{j45^\circ}$

ANSWER (a) $0.24 + j0.32$, (b) $2.903 + j8.707$.

EXAMPLE 9.6

Given that $\mathbf{A} = 10 \cos (10^8 t - 10x + 60^\circ)\mathbf{a}_z$ and $\mathbf{B}_s = (20/j)\ \mathbf{a}_x + 10\ e^{j2\pi x/3}\ \mathbf{a}_y$, express $\mathbf{A}$ in phasor form and $\mathbf{B}_s$ in instantaneous form.

SOLUTION

$$\mathbf{A} = \text{Re}\left[10\ e^{j(\omega t - 10x + 60^\circ)}\mathbf{a}_z\right]$$

where $\omega = 10^8$. Hence

$$\mathbf{A} = \text{Re}\left[10 e^{j(60^\circ - 10x)}\mathbf{a}_z\ e^{j\omega t}\right] = \text{Re}\left(\mathbf{A}_s e^{j\omega t}\right)$$

or

$$\mathbf{A}_s = 10\, e^{j(60° - 10x)}\mathbf{a}_z$$

If

$$\mathbf{B}_s = \frac{20}{j}\,\mathbf{a}_x + 10e^{j2\pi x/3}\mathbf{a}_y = -j20\mathbf{a}_x + 10\, e^{j2\pi x/3}\mathbf{a}_y$$

$$= 20\, e^{-j\pi/2}\mathbf{a}_x + 10\, e^{j2\pi x/3}\,\mathbf{a}_y$$

$$\mathbf{B} = \text{Re}\,(\mathbf{B}_s e^{j\omega t})$$

$$= \text{Re}\left[20e^{j(\omega t - \pi/2)}\mathbf{a}_x + 10e^{j(\omega t + 2\pi x/3)}\mathbf{a}_y\right]$$

$$= 20\cos(\omega t - \pi/2)\mathbf{a}_x + 10\cos\left(\omega t + \frac{2\pi x}{3}\right)\mathbf{a}_y$$

$$= 20\sin\omega t\,\mathbf{a}_x + 10\cos\left(\omega t + \frac{2\pi x}{3}\right)\mathbf{a}_y$$

■

PRACTICE EXERCISE 9.6

If $\mathbf{P} = 2\sin(10t + x - \pi/4)\,\mathbf{a}_y$ and $\mathbf{Q}_s = e^{jx}(\mathbf{a}_x - \mathbf{a}_z)\sin\pi y$, determine the phasor form of $\mathbf{P}$ and the instantaneous form of $\mathbf{Q}_s$.

ANSWER $2e^{j(x - 3\pi/4)}\mathbf{a}_y$, $\sin\pi\, y\cos(\omega t + x)\,(\mathbf{a}_x - \mathbf{a}_z)$.

EXAMPLE 9.7

The electric field and magnetic field in free space are given by

$$\mathbf{E} = \frac{50}{\rho}\cos(10^6 t + \beta z)\,\mathbf{a}_\phi \text{ V/m}$$

$$\mathbf{H} = \frac{H_o}{\rho}\cos(10^6 t + \beta z)\,\mathbf{a}_\rho \text{ A/m}$$

Express these in phasor form and determine the constants H_o and β such that the fields satisfy Maxwell's equations.

SOLUTION The instantaneous forms of $\mathbf{E}$ and $\mathbf{H}$ are written as

$$\mathbf{E} = \text{Re}\,(\mathbf{E}_s e^{j\omega t}), \qquad \mathbf{H} = \text{Re}\,(\mathbf{H}_s e^{j\omega t}) \tag{9.7.1}$$

where $\omega = 10^6$ and phasors $\mathbf{E}_s$ and $\mathbf{H}_s$ are given by

$$\mathbf{E}_s = \frac{50}{\rho}\, e^{j\beta z}\mathbf{a}_\phi, \qquad \mathbf{H}_s = \frac{H_o}{\rho}\, e^{j\beta z}\mathbf{a}_\rho \tag{9.7.2}$$

For free space, $\rho_v = 0$, $\sigma = 0$, $\varepsilon = \varepsilon_o$, and $\mu = \mu_o$ so Maxwell's equations become

$$\nabla \cdot \mathbf{D} = \varepsilon_o \nabla \cdot \mathbf{E} = 0 \rightarrow \nabla \cdot \mathbf{E}_s = 0 \quad [9.7.3]$$

$$\nabla \cdot \mathbf{B} = \mu_o \nabla \cdot \mathbf{H} = 0 \rightarrow \nabla \cdot \mathbf{H}_s = 0 \quad [9.7.4]$$

$$\nabla \times \mathbf{H} = \sigma \mathbf{E} + \varepsilon_o \frac{\partial \mathbf{E}}{\partial t} \rightarrow \nabla \times \mathbf{H}_s = j\omega\varepsilon_o \mathbf{E}_s \quad [9.7.5]$$

$$\nabla \times \mathbf{E} = -\mu_o \frac{\partial \mathbf{H}}{\partial t} \rightarrow \nabla \times \mathbf{E}_s = -j\omega\mu_o \mathbf{H}_s \quad [9.7.6]$$

Substituting eq. (9.7.2) into eqs. (9.7.3) and (9.7.4), it is readily verified that two Maxwell's equations are satisfied; that is,

$$\nabla \cdot \mathbf{E}_s = \frac{1}{\rho} \frac{\partial}{\partial \phi} (E_{\phi s}) = 0$$

$$\nabla \cdot \mathbf{H}_s = \frac{1}{\rho} \frac{\partial}{\partial \rho} (\rho H_{\rho s}) = 0$$

Now

$$\nabla \times \mathbf{H}_s = \nabla \times \left(\frac{H_o}{\rho} e^{j\beta z} \mathbf{a}_\rho \right) = \frac{jH_o\beta}{\rho} e^{j\beta z} \mathbf{a}_\phi \quad [9.7.7]$$

Substituting eqs. (9.7.2) and (9.7.7) into eq. (9.7.5), we have

$$\frac{jH_o\beta}{\rho} e^{j\beta z} \mathbf{a}_\phi = j\omega\varepsilon_o \frac{50}{\rho} e^{j\beta z} \mathbf{a}_\phi$$

or

$$H_o\beta = 50\, \omega\varepsilon_o \quad [9.7.8]$$

Similarly, substituting eq. (9.7.2) into (9.7.6) gives

$$-j\beta \frac{50}{\rho} e^{j\beta z} \mathbf{a}_\rho = -j\omega\mu_o \frac{H_o}{\rho} e^{j\beta z} \mathbf{a}_\rho$$

or

$$\frac{H_o}{\beta} = \frac{50}{\omega\mu_o} \quad [9.7.9]$$

Multiplying eq. (9.7.8) with eq. (9.7.9) yields

$$H_o^2 = (50)^2 \frac{\varepsilon_o}{\mu_o}$$

or

$$H_o = \pm 50 \sqrt{\varepsilon_o/\mu_o} = \pm \frac{50}{120\pi} = \pm 0.1326$$

Dividing eq. (9.7.8) by eq. (9.7.9), we get

$$\beta^2 = \omega^2 \mu_o \varepsilon_o$$

or

$$\beta = \pm \omega \sqrt{\mu_o \varepsilon_o} = \pm \frac{\omega}{c} = \pm \frac{10^6}{3 \times 10^8}$$
$$= \pm 3.33 \times 10^{-3}.$$

In view of eq. (9.7.8), $H_o = 0.1326$, $\beta = 3.33 \times 10^{-3}$ or $H_o = -0.1326$, $\beta = -3.33 \times 10^{-3}$; only these will satisfy Maxwell's four equations. ■

PRACTICE EXERCISE 9.7

In air, $\mathbf{E} = \frac{\sin \theta}{r} \cos (6 \times 10^7 t - \beta r) \, \mathbf{a}_\phi$ V/m.

Find β and $\mathbf{H}$.

ANSWER 0.2 rad/m, $-\frac{1}{12\pi r^2} \cos \theta \sin (6 \times 10^7 t - 0.2r) \, \mathbf{a}_r - \frac{1}{120\pi r} \sin \theta \times \cos (6 \times 10^7 t - 0.2r) \, \mathbf{a}_\theta$ A/m.

EXAMPLE 9.8

In a medium characterized by $\sigma = 0$, $\mu = \mu_o$, ε_o, and

$$\mathbf{E} = 20 \sin (10^8 t - \beta z) \, \mathbf{a}_y \text{ V/m}$$

calculate β and $\mathbf{H}$.

SOLUTION This problem can be solved directly in time domain or using phasors. As in the previous example, we find β and $\mathbf{H}$ by making $\mathbf{E}$ and $\mathbf{H}$ satisfy Maxwell's four equations.

Method 1 (time domain): Let us solve this problem the harder way—in time domain. It is evident that Gauss's law for electric fields is satisfied; that is,

$$\nabla \cdot \mathbf{E} = \frac{\partial E_y}{\partial y} = 0$$

From Faraday's law,

$$\nabla \times \mathbf{E} = -\mu \frac{\partial \mathbf{H}}{\partial t} \quad \rightarrow \quad \mathbf{H} = -\frac{1}{\mu}\int (\nabla \times \mathbf{E})\, dt$$

But

$$\nabla \times \mathbf{E} = \begin{vmatrix} \frac{\partial}{\partial x} & \frac{\partial}{\partial y} & \frac{\partial}{\partial z} \\ 0 & E_y & 0 \end{vmatrix} = -\frac{\partial E_y}{\partial z}\mathbf{a}_x + \frac{\partial E_y}{\partial x}\mathbf{a}_z$$

$$= 20\beta \cos(10^8 t - \beta z)\, \mathbf{a}_x + 0$$

Hence,

$$\mathbf{H} = -\frac{20\beta}{\mu}\int \cos(10^8 t - \beta z)\, dt\, \mathbf{a}_x$$

$$= -\frac{20\beta}{\mu 10^8} \sin(10^8 t - \beta z)\, \mathbf{a}_x \qquad [9.8.1]$$

It is readily verified that

$$\nabla \cdot \mathbf{H} = \frac{\partial H_x}{\partial x} = 0$$

showing that Gauss's law for magnetic fields is satisfied. Lastly, from Ampere's law

$$\nabla \times \mathbf{H} = \sigma \mathbf{E} + \varepsilon \frac{\partial \mathbf{E}}{\partial t} \quad \rightarrow \quad \mathbf{E} = \frac{1}{\varepsilon}\int (\nabla \times \mathbf{H})\, dt \qquad [9.8.2]$$

because $\sigma = 0$.
But

$$\nabla \times \mathbf{H} = \begin{vmatrix} \frac{\partial}{\partial x} & \frac{\partial}{\partial y} & \frac{\partial}{\partial z} \\ H_x & 0 & 0 \end{vmatrix} = \frac{\partial H_x}{\partial z}\mathbf{a}_y - \frac{\partial H_x}{\partial y}\mathbf{a}_z$$

$$= \frac{20\beta^2}{\mu 10^8} \cos(10^8 t - \beta z)\, \mathbf{a}_y + 0$$

where **H** in eq. (9.8.1) has been substituted. Thus eq. (9.8.2) becomes

$$\mathbf{E} = \frac{20\beta^2}{\mu\varepsilon 10^8}\int \cos(10^8 t - \beta z)\, dt\, \mathbf{a}_y$$

$$= \frac{20\beta^2}{\mu\varepsilon 10^{16}} \sin(10^8 t - \beta z)\, \mathbf{a}_y$$

Comparing this with the given **E**, we have

$$\frac{20\beta^2}{\mu\varepsilon 10^{16}} = 20$$

or

$$\beta = \pm\, 10^8 \sqrt{\mu\varepsilon} = \pm\, 10^8 \sqrt{\mu_o \cdot 4\varepsilon_o} = \pm\frac{10^8(2)}{c} = \pm\frac{10^8(2)}{3 \times 10^8}$$

$$= \pm\frac{2}{3}$$

From eq. (9.8.1),

$$\mathbf{H} = \pm\frac{20\ (2/3)}{4\pi \cdot 10^{-7}(10^8)} \sin\left(10^8 t \pm \frac{2z}{3}\right) \mathbf{a}_x$$

or

$$\mathbf{H} = \pm\frac{1}{3\pi} \sin\left(10^8 t \pm \frac{2z}{3}\right) \mathbf{a}_x \text{ A/m.}$$

Method 2 (using phasors):

$$\mathbf{E} = \text{Im}\,(E_s e^{j\omega t}) \quad \rightarrow \quad \mathbf{E}_s = 20e^{-j\beta z}\,\mathbf{a}_y \tag{9.8.3}$$

where $\omega = 10^8$.
Again

$$\nabla \cdot \mathbf{E}_s = \frac{\partial E_{ys}}{\partial y} = 0$$

$$\nabla \times \mathbf{E}_s = -j\omega\mu\mathbf{H}_s \quad \rightarrow \quad \mathbf{H}_s = \frac{\nabla \times \mathbf{E}_s}{-j\omega\mu}$$

or

$$\mathbf{H}_s = \frac{1}{-j\omega\mu}\left[-\frac{\partial E_{ys}}{\partial z}\mathbf{a}_x\right] = -\frac{20\beta}{\omega\mu}\, e^{-j\beta z}\,\mathbf{a}_x \tag{9.8.4}$$

Notice that $\nabla \cdot \mathbf{H}_s = 0$ is satisfied.

$$\nabla \times \mathbf{H}_s = j\omega\varepsilon\mathbf{E}_s \quad \rightarrow \quad \mathbf{E}_s = \frac{\nabla \times \mathbf{H}_s}{j\omega\varepsilon} \tag{9.8.5}$$

Substituting $\mathbf{H}_s$ in eq. (9.8.4) into eq. (9.8.5) gives

$$\mathbf{E}_s = \frac{1}{j\omega\varepsilon}\frac{\partial H_{xs}}{\partial z}\,\mathbf{a}_y = \frac{20\beta^2 e^{-j\beta z}}{\omega^2\mu\varepsilon}\,\mathbf{a}_y$$

Comparing this with the given $\mathbf{E}_s$ in eq. (9.8.3), we have

$$20 = \frac{20\beta^2}{\omega^2\mu\varepsilon}$$

or

$$\beta = \pm\,\omega\sqrt{\mu\varepsilon} = \pm\,\frac{2}{3}$$

as obtained before. From eq. (9.8.4),

$$\mathbf{H}_s = \pm\,\frac{20(2/3)\,e^{\pm j\beta z}}{10^8(4\pi \times 10^{-7})}\,\mathbf{a}_x = \pm\,\frac{1}{3\pi}\,e^{\pm j\beta z}\,\mathbf{a}_x$$

$$\mathbf{H} = \text{Im}\,(\mathbf{H}_s e^{j\omega t})$$

$$= \pm\,\frac{1}{3\pi}\,\sin\,(10^8 t \pm \beta z)\,\mathbf{a}_x \text{ A/m}$$

as obtained before. It should be noticed that working with phasors provides a considerable simplification compared with working directly in time domain. Also, notice that we have used

$$\mathbf{A} = \text{Im}\,(\mathbf{A}_s e^{j\omega t})$$

because the given $\mathbf{E}$ is in sine form and not cosine. We could have used

$$\mathbf{A} = \text{Re}\,(\mathbf{A}_s e^{j\omega t})$$

in which case sine is expressed in terms of cosine and eq. (9.8.3) would be

$$\mathbf{E} = 20\cos\,(10^8 t - \beta z - 90°)\,\mathbf{a}_y = \text{Re}\,(\mathbf{E}_s e^{j\omega t})$$

or

$$\mathbf{E}_s = 20e^{-j\beta z - j90°}\mathbf{a}_y = -j20e^{-j\beta z}\mathbf{a}_y$$

and we follow the same procedure. ■

PRACTICE EXERCISE 9.8

A medium is characterized by $\sigma = 0$, $\mu = 2\mu_o$ and $\varepsilon = 5\varepsilon_o$. If $\mathbf{H} = 2\cos\,(\omega t - 3y)\,\mathbf{a}_z$ A/m, calculate ω and $\mathbf{E}$.

ANSWER 2.846×10^8 rad/s, $-476.8\cos\,(2.846 \times 10^8 t - 3y)\,\mathbf{a}_x$ V/m.

SUMMARY

1. In this chapter, we have introduced two fundamental concepts: electromotive force (emf), based on Faraday's experiments, and displacement current, which resulted from Maxwell's hypothesis. These concepts call for modifications in Maxwell's curl equations obtained for static EM fields to accommodate the time-dependence of the fields.
2. Faraday's law states that the induced emf is given by ($N = 1$)

$$V_{\text{emf}} = -\frac{\partial \Psi}{\partial t}$$

For transformer emf, $V_{\text{emf}} = -\int \frac{\partial \mathbf{B}}{\partial t} \cdot d\mathbf{S}$

and for motional emf, $V_{\text{emf}} = \int (\mathbf{u} \times \mathbf{B}) \cdot d\mathbf{l}$.

3. The displacement current

$$I_d = \int \mathbf{J}_d \cdot d\mathbf{S}$$

where $\mathbf{J}_d = \frac{\partial \mathbf{D}}{\partial t}$ (displacement current density), is a modification to Ampere's circuital law. This modification attributed to Maxwell predicted electromagnetic waves several years before it was verified experimentally by Hertz.

4. In differential form, Maxwell's equations for dynamic fields are:

$$\nabla \cdot \mathbf{D} = \rho_v$$

$$\nabla \cdot \mathbf{B} = 0$$

$$\nabla \times \mathbf{E} = -\frac{\partial \mathbf{B}}{\partial t}$$

$$\nabla \times \mathbf{H} = \mathbf{J} + \frac{\partial \mathbf{D}}{\partial t}$$

Each differential equation has its integral counterpart (see Tables 9.1 and 9.2) which can be derived from the differential form using Stokes's or divergence theorem. Any EM field must satisfy the four Maxwell's equations simultaneously.

5. Time-varying electric scalar potential $V(x, y, z, t)$ and magnetic vector potential $\mathbf{A}(x, y, z, t)$ are shown to satisfy wave equations if Lorentz's condition is assumed.
6. Time-harmonic fields are those that vary sinusoidally with time. They are easily expressed in phasors which are more convenient to work with. Using the cosine reference, the instantaneous vector quantity $\mathbf{A}(x, y, z, t)$ is related to its phasor form $\mathbf{A}_s(x, y, z)$ according to

$$\mathbf{A}(x, y, z, t) = \text{Re}\left[\mathbf{A}_s(x, y, z)\, e^{j\omega t}\right]$$

REVIEW QUESTIONS

9.1 The flux through each turn of a 100-turn coil is $(t^3 - 2t)$ mWb where t is in seconds. The induced emf at $t = 2$ s is

(a) 1 V

(b) −1 V

(c) 4 mV

(d) 0.4 V

(e) −0.4 V

9.2 Assuming that each loop is stationary and the time-varying magnetic field **B** induces current I, which of the configurations in Figure 9.13 are incorrect?

9.3 Two conducting coils 1 and 2 (identical except that 2 is split) are placed in a uniform magnetic field which decreases at a constant rate as in Figure 9.14. If the plane of the coils are perpendicular to the field lines, which of the following statements its true?

(a) An emf is induced in both coils.

(b) An emf is induced in split coil 2.

(c) Equal joule heating occurs in both coils.

(d) Joule heating does not occur in either coil.

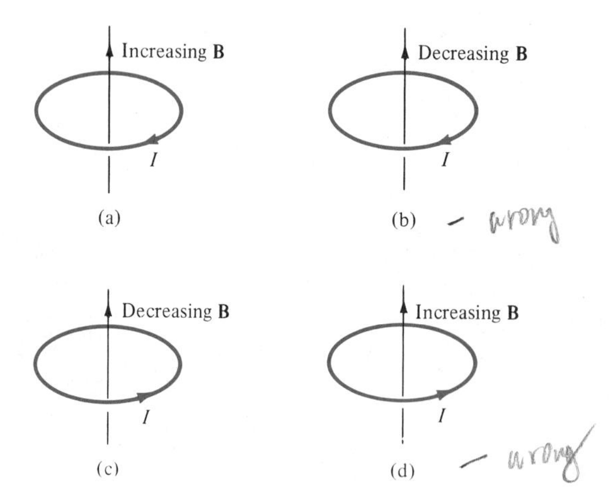

Figure 9.13 For Review Question 9.2.

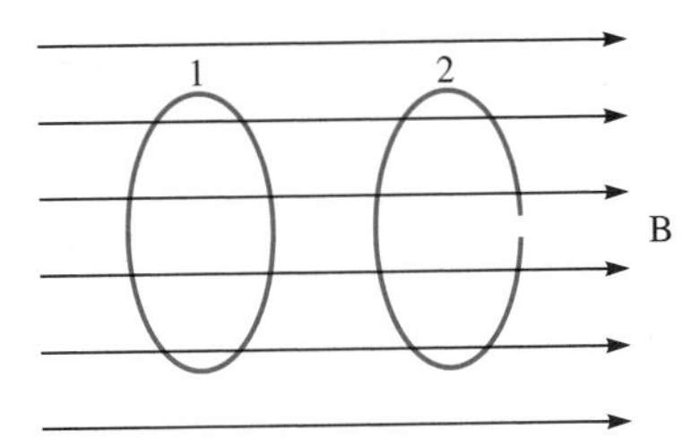

Figure 9.14 For Review Question 9.3.

9.4 A loop is rotating about the y-axis in a magnetic field $\mathbf{B} = B_o \sin \omega t \, \mathbf{a}_x$ Wb/m^2. The voltage induced in the loop is due to

(a) Motional emf

(b) Transformer emf

(c) A combination of motional and transformer emf

(d) None of the above

9.5 A rectangular loop is placed in the time-varying magnetic field $\mathbf{B} = 0.2 \cos 150\pi t \mathbf{a}_z$ Wb/m^2 as shown in Figure 9.15. V_1 is not equal to V_2.

(a) True (b) False

9.6 The concept of displacement current was a major contribution attributed to

(a) Faraday

(b) Lenz

(c) Maxwell

(d) Lorentz

(e) Your professor

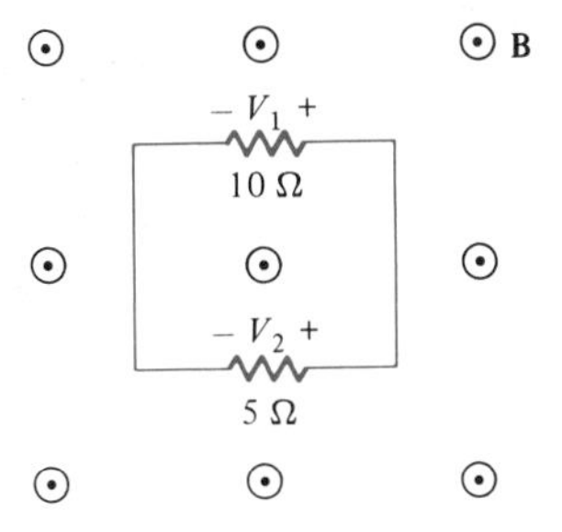

Figure 9.15 For Review Question 9.5 and Problem 9.8.

9.7 Identify which of the following expressions are not Maxwell's equations for time-varying fields:

/ (a) $\nabla \cdot \mathbf{J} + \dfrac{\partial \rho_v}{\partial t} = 0$

/ (b) $\nabla \cdot \mathbf{D} = \rho_v$

(c) $\nabla \cdot \mathbf{E} = -\dfrac{\partial \mathbf{B}}{\partial t}$

/ (d) $\oint \mathbf{H} \cdot d\mathbf{l} = \int \left(\sigma \mathbf{E} + \varepsilon \dfrac{\partial \mathbf{E}}{\partial t}\right) \cdot d\mathbf{S}$

(e) $\oint \mathbf{B} \cdot d\mathbf{S} = 0$

9.8 An EM field is said to be nonexistent or not Maxwellian if it fails to satisfy Maxwell's equations and the wave equations derived from them. Which of the following fields in free space are not Maxwellian?

(a) $\mathbf{H} = \cos x \cos 10^6 t\, \mathbf{a}_x$

(b) $\mathbf{E} = 100 \cos \omega t\, \mathbf{a}_x$

(c) $\mathbf{D} = e^{-10y} \sin (10^5 - 10y)\, \mathbf{a}_z$

(d) $\mathbf{B} = 0.4 \sin 10^4 t\, \mathbf{a}_z$

(e) $\mathbf{H} = 10 \cos \left(10^5 t - \dfrac{z}{10}\right) \mathbf{a}_x$

(f) $\mathbf{E} = \dfrac{\sin \theta}{r} \cos (\omega t - r\omega \sqrt{\mu_o \varepsilon_o})\, \mathbf{a}_\theta$

(g) $\mathbf{B} = (1 - \rho^2) \sin \omega t\, \mathbf{a}_z$

9.9 Which of the following statements is not true of a phasor?

(a) It may be a scalar or a vector.

(b) It is a time-dependent quantity.

(c) A phasor V_s may be represented as $V_o \underline{/\theta}$ or $V_o e^{j\theta}$ where $V_o = |V_s|$.

(d) It is a complex quantity.

9.10 If $\mathbf{E}_s = 10\, e^{j4x}\, \mathbf{a}_y$, which of these is not a correct representation of $\mathbf{E}$?

(a) $\text{Re}\,(\mathbf{E}_s e^{j\omega t})$

(b $\text{Re}\,(\mathbf{E}_s e^{-j\omega t})$

(c) $\text{Im}\,(\mathbf{E}_s e^{j\omega t})$

(d) $10 \cos\,(\omega t + j4x)\, \mathbf{a}_y$

(e) $10 \sin\,(\omega t + 4x)\, \mathbf{a}_y$

Answers: 9.1b, 9.2b, d, 9.3a, 9.4c, 9.5a, 9.6c, 9.7a, b, d, g, 9.8b, 9.9a,c, 9.10d.

PROBLEMS

9.1 A conducting circular loop of radius 20 cm lies in the $z = 0$ plane in a magnetic field $\mathbf{B} = 10 \cos 377t\ \mathbf{a}z$ mWb/m^2. Calculate the induced voltage in the loop.

9.2 A 30-cm by 40-cm rectangular loop rotates at 130 rad/s in a magnetic field 0.06 Wb/m^2 normal to the axis of rotation. If the loop has 50 turns, determine the induced voltage in the loop.

9.3 Find the current I in the circuit of Figure 9.16 if the circuit is located in a time-varying magnetic field:

(a) $\mathbf{B} = 0.05t\mathbf{a}_z$ Wb/m^2

(b) $\mathbf{B} = 5 \sin\,(120\pi t - 2y)\, \mathbf{a}_z$ mWb/m^2

***9.4** A square loop of side a recedes with a uniform velocity $u_o\mathbf{a}_y$ from an infinitely long filament carrying current I along $\mathbf{a}_z$ as shown in Figure 9.17. Assuming that $\rho = \rho_o$ at time $t = 0$, show that the emf induced in the loop at $t > 0$ is

$$V_{\text{emf}} = \frac{\mu_o a^2 u_o I}{2\pi\rho(\rho + a)}$$

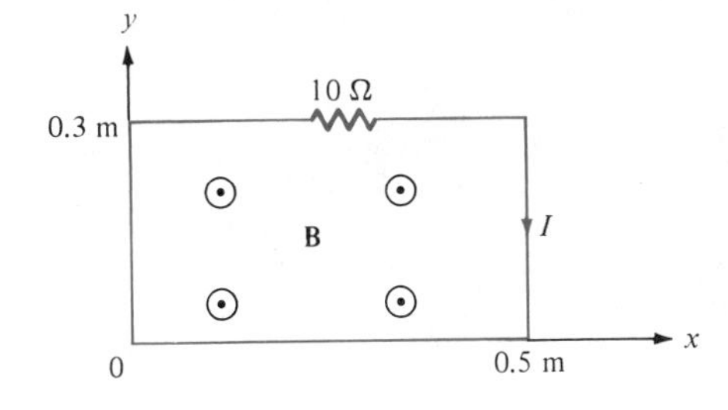

Figure 9.16 For Problem 9.3.

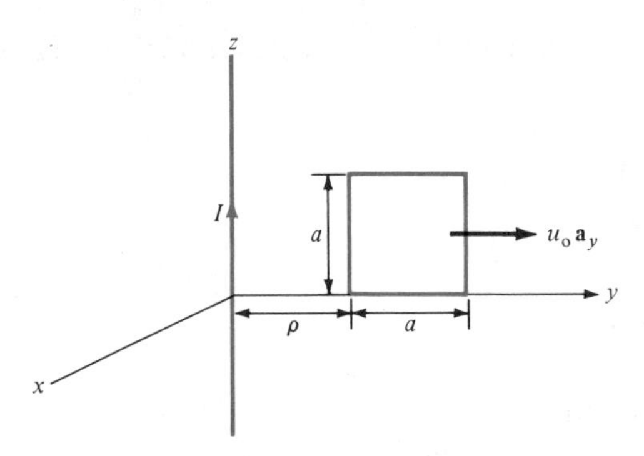

Figure 9.17 For Problem 9.4.

***9.5** A conducting rod moves with a constant velocity of $3\mathbf{a}_z$ m/s parallel to a long straight wire carrying current 15 A as in Figure 9.18. Calculate the emf induced in the rod and state which end is at higher potential.

***9.6** A conducting bar is connected via flexible leads to a pair of rails in a magnetic field $\mathbf{B} = 6 \cos 10t\, \mathbf{a}_x$ mWb/m^2 as in Figure 9.19. If the z-axis is the equilibrium position of the bar and its velocity is $2 \cos 10t\, \mathbf{a}_y$ m/s, find the voltage induced in it.

9.7 A train travels at 80 km/hr between a pair of rails that are 2.5 m apart. Assuming that the vertical component of the earth magnetic field along the rails is 70 μWb/m^2 and that other fields are negligibly small, calculate the voltage developed between the rails.

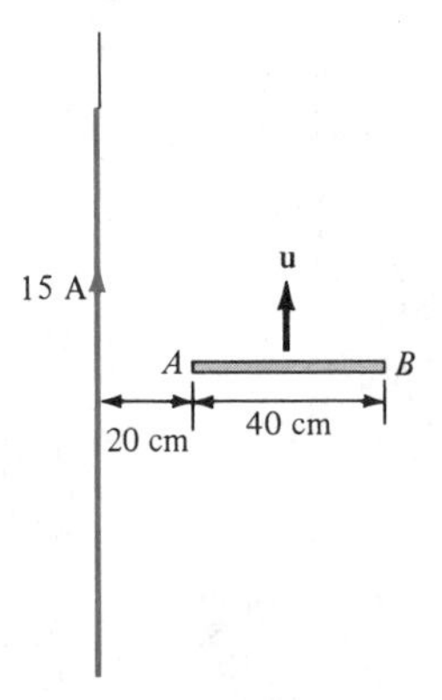

Figure 9.18 For Problem 9.5.

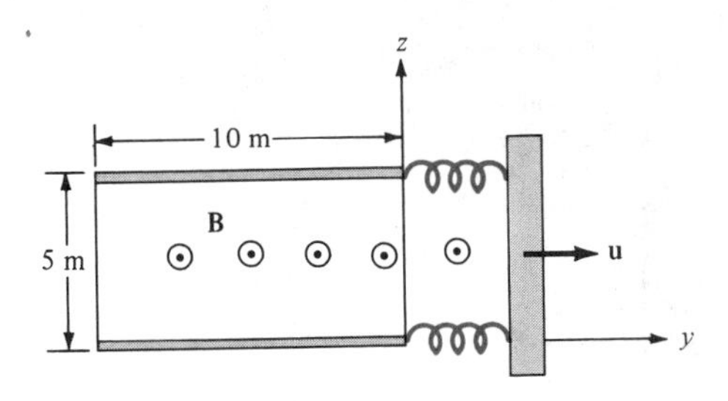

Figure 9.19 For Problem 9.6.

***9.8** If the area of the loop in Figure 9.15 is 10 cm^2, calculate V_1 and V_2.

9.9 As portrayed in Figure 9.20, a bar magnet is thrust towards the center of a coil of 10 turns and resistance 15 Ω. If the magnetic flux through the coil changes from 0.45 Wb to 0.64 Wb in 0.02 s, what is the magnitude and direction (as viewed from the side near the magnet) of the induced current?

9.10 For the transformer of Practice Exercise 9.3, if the current I_1 varies sinusoidally as 10 cos $120\pi t$, $\ell_{ab} = 12$ cm, $\ell_{bc} = 8$ cm, and a 6-mm air-gap is provided in branch dc of the core, determine the induced emf V_2. Take $\mu = 1000\ \mu_o$.

9.11 The cross section of a homopolar generator disk is shown in Figure 9.21. The disk has inner radius $\rho_1 = 2$ cm and outer radius $\rho_2 = 10$ cm and rotates in a uniform magnetic field 15 mWb/m^2 at a speed of 60 rad/s. Calculate the induced voltage.

9.12 A 50-V voltage generator at 20 MHz is connected to the plates of an air dielectric parallel plate capacitor with plate area 2.8 cm^2 and separation distance 0.2 mm. Find the maximum value of displacement current density and displacement current.

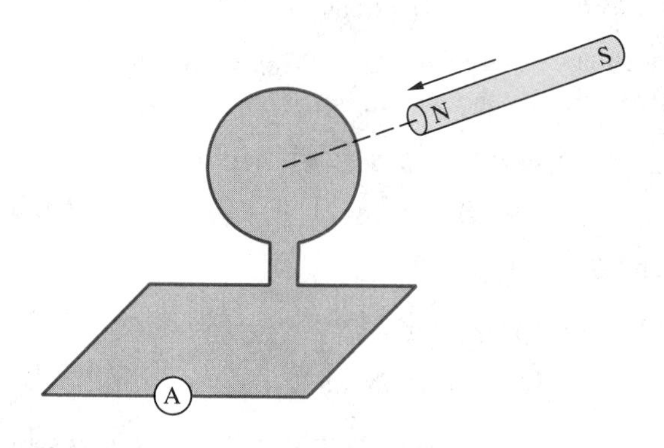

Figure 9.20 For Problem 9.9.

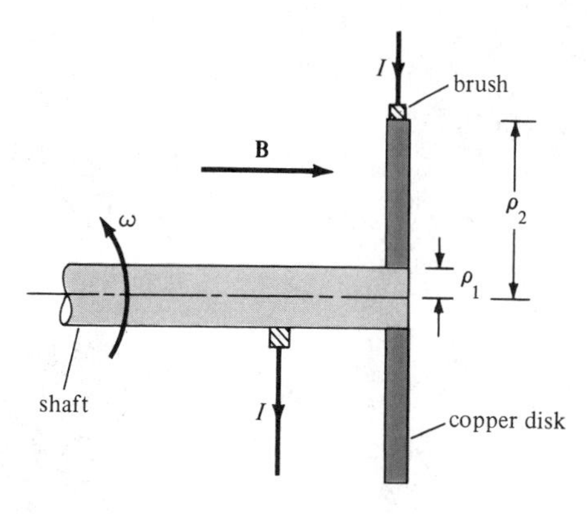

Figure 9.21 For Problem 9.11.

9.13 Assuming that limestone is characterized by $\mu = \mu_o$, $\varepsilon = 5\varepsilon_o$, $\sigma = 2 \times 10^{-4}$ mhos/m, calculate J and J_d at a point in the medium where $E = 20 \cos 10^7 t$ μV/m.

9.14 A conductor with cross-sectional area of 10 cm^2 carries a conduction current 0.2 sin $10^9 t$ mA. Given that $\sigma = 2.5 \times 10^6$ mhos/m and $\varepsilon_r = 6$, calculate the magnitude of the displacement current density.

9.15 If lake water has constitutive parameters $\mu = \mu_o$, $\varepsilon = 80\varepsilon_o$, and $\sigma = 4 \times 10^{-3}$ mhos/m, determine the frequency at which displacement current density is equal in magnitude to the conduction current density.

9.16 (a) Write Maxwell's equations for a linear, homogeneous medium in terms of $\mathbf{E}_s$ and $\mathbf{H}_s$ only assuming the time factor $e^{-j\omega t}$.

(b) In Cartesian coordinates, write the point form of Maxwell's equations in Table 9.2 as eight scalar equations.

9.17 Show that in a source-free region ($\mathbf{J} = 0$, $\rho_v = 0$), Maxwell's equations can be reduced to two. Identify the two all-embracing equations.

9.18 In a material for which $\varepsilon = 2\varepsilon_o$, $\mu = \mu_o$, and $\sigma = 0$,

$$\mathbf{H} = 2 \cos (10^6 t + \beta x)\, \mathbf{a}_z$$

find $\mathbf{J}_d$, $\mathbf{D}$, and β.

9.19 Show that the time-varying field

$$\mathbf{E} = 30 \sin 2x \sin (kz - \omega t)\, \mathbf{a}_y \text{ V/m}$$

where $k^2 = \mu_o\varepsilon_o\omega^2 - 4$, is a genuine EM field—that is, it satisfies Maxwell's equations in free space. Find the corresponding $\mathbf{H}$ and $\mathbf{J}_d$.

9.20 In a certain region with $\sigma = 0$, $\mu = \mu_o$, and $\varepsilon = 6.25\varepsilon_o$, the magnetic field of an EM wave is

$$\mathbf{H} = 0.6 \cos \beta x \cos 10^8 t \; \mathbf{a}_z \text{ A/m}$$

Find β and the corresponding $\mathbf{E}$ using Maxwell's equations.

****9.21** Given the total electromagnetic energy

$$W = \frac{1}{2} \int (\mathbf{E} \cdot \mathbf{D} + \mathbf{H} \cdot \mathbf{B}) dv$$

show from Maxwell's equations that

$$\frac{\partial W}{\partial t} = - \oint_S (\mathbf{E} \times \mathbf{H}) \cdot d\mathbf{S} - \int_v \mathbf{E} \cdot \mathbf{J} \, dv$$

9.22 Given that in free space

$$\mathbf{H} = \frac{10}{\rho} \sin \left(10^8 t - \frac{z}{3}\right) \mathbf{a}_\phi \text{ A/m}$$

determine $\mathbf{J}_d$ and $\mathbf{E}$.

***9.23** The electric field in air is given by $\mathbf{E} = \rho t e^{-\rho - t} \mathbf{a}_\phi$ V/m; find $\mathbf{B}$ and $\mathbf{J}$.

****9.24** In free space ($\rho_v = 0$, $\mathbf{J} = 0$). Show that

$$\mathbf{A} = \frac{\mu_o}{4\pi r} (\cos \theta \; \mathbf{a}_r - \sin \theta \; \mathbf{a}_\theta) e^{j\omega(t - r/c)}$$

satisfies the wave equation in eq. (9.52). Find the corresponding V. Take c as the speed of light in free space.

****9.25** Show that Lorentz condition in eq. (9.50) is merely a restatement of the continuity equation. (*Hint:* $\nabla^2(\nabla \cdot \mathbf{F}) = \nabla \cdot \nabla^2 \mathbf{F}$.)

9.26 Evaluate the following complex numbers and leave your answers in rectangular form:

(a) $\dfrac{3 - j2}{4 - j} - \dfrac{2}{1 + j} + \dfrac{(2 + j)^{1/2}}{(-3 + 5j)^*}$

(b) $-10 \angle -30^\circ + (20 \angle 120^\circ)^*$

(c) $10 \ln(3 - j4)$

9.27 Evalute and express the following complex numbers in polar form:

(a) $\dfrac{2 + j3}{j(-10 + j2)}$

(b) $\dfrac{8 \underline{/30^\circ} - 5 \underline{/60^\circ}}{1 + j}$

(c) $\dfrac{(1 + j3)(4 - j2)}{(1 + j3) + (4 - j2)}$

(d) $\left[\dfrac{(15 - j7)(3 + j2)^*}{(4 + j6)^*(3 \underline{/70^\circ})}\right]^*$

(e) $[(-10 \underline{/30^\circ})(4e^{j\pi/4})]^{1/2}$

9.28 Express the following time-harmonic fields in phasor form:

(a) $\mathbf{A} = 5\sin(\omega t - 2z)\,\mathbf{a}_x$

(b) $\mathbf{B} = 15e^{-2y}\cos(\omega t - y)\,\mathbf{a}_z$

(c) $\mathbf{C} = 5\cos\omega t\,\mathbf{a}_y - 8\sin(\omega t - x)\,\mathbf{a}_z$

(d) $\mathbf{D} = \dfrac{10}{\rho}\cos(10^6 t - 3z)\,\mathbf{a}_\phi$

9.29 Write the instantaneous form of the following phasors:

(a) $\mathbf{A}_s = 5je^{-j20^\circ}\mathbf{a}_z - (3 + j4)x\mathbf{a}_y$

(b) $\mathbf{B}_s = 10e^{-jkz}\mathbf{a}_z + j5e^{jkz + \pi/4}\mathbf{a}_y$

(c) $C_s = \dfrac{2}{j}e^{-j3}\sin x + e^{3x - j4x}$

9.30 By assuming a time factor $e^{j\omega t}$, express each of eqs. (9.42), (9.45), (9.50), (9.51), and (9.52) in phasor form. Repeat this for a time factor $e^{-j\omega t}$.

9.31 The electric field in free space is given by

$$\mathbf{E}_s = 20\sin(k_x x)\sin(k_y y)\,\mathbf{a}_z$$

where $k_x^2 + k_y^2 = \omega^2\varepsilon_o\mu_o$, find **E** and **B**.

CHAPTER

Electromagnetic Wave Propagation

10

An engineer is an unordinary person who can do for one dollar what any ordinary person can do for two dollars.

— ANONYMOUS

10.1 INTRODUCTION

Our first application of Maxwell's equations will be in relation to electromagnetic wave propagation. The existence of EM waves, predicted by Maxwell's equations, was first investigated by Heinrich Hertz. After several calculations and experiments Hertz succeeded in generating and detecting radio waves, which are sometimes called Hertzian waves in his honor.

In general, waves are means of transporting energy or information. Typical examples of EM waves include radio waves, TV signals, radar beams, and light rays. All forms of EM energy share three fundamental characteristics: they all travel at high velocity; in traveling, they assume the properties of waves; and they radiate outward from a source, without benefit of any discernible physical vehicles. The problem of radiation will be addressed in Chapter 13.

In this chapter, our major goal is to solve Maxwell's equations and derive EM wave motion in the following media:

1. Free space ($\sigma = 0$, $\varepsilon = \varepsilon_o$, $\mu = \mu_o$)
2. Lossless dielectrics ($\sigma = 0$, $\varepsilon = \varepsilon_r\varepsilon_o$, $\mu = \mu_r\mu_o$, or $\sigma \ll \omega\varepsilon$)
3. Lossy dielectrics($\sigma \neq 0$, $\varepsilon = \varepsilon_r\varepsilon_o$, $\mu = \mu_r\mu_o$)
4. Good conductors ($\sigma \simeq \infty$, $\varepsilon = \varepsilon_o$, $\mu = \mu_r\mu_o$, or $\sigma \gg \omega\varepsilon$)

where ω is the angular frequency of the wave. Case 3, for lossy dielectrics, is the most general case and will be considered first. Once this general case is solved, we simply derive other cases (1, 2, and 4) from it as special cases by changing the values of σ, ε, and μ. However, before we consider wave motion in those different media, it is appropriate that we study the characteristics of waves in general. This is important for proper understanding of EM waves. The reader who is conversant with the concept of waves may skip Section 10.2. Power considerations, reflection, and transmission between two different media will be discussed later in the chapter.

†10.2 WAVES IN GENERAL

A clear understanding of EM wave propagation depends on a grasp of what waves are in general. A wave is a function of both space and time. Wave motion occurs when a disturbance at point A, at time t_o, is related to what happens at point B, at time $t > t_o$. A wave equation, as exemplified by eqs. (9.51) and (9.52), is a partial differential equation of the second order. In one dimension, a scalar wave equation takes the form of

$$\frac{\partial^2 E}{\partial t^2} - u^2 \frac{\partial^2 E}{\partial z^2} = 0 \tag{10.1}$$

where u is the *wave velocity*. Equation (10.1) is a special case of eq. (9.51) in which the medium is source-free ($\rho_v = 0$, $\mathbf{J} = 0$). It can be solved by following procedure, similar to that in Example 6.5. Its solutions are of the form

$$E^- = f(z - ut) \tag{10.2a}$$

$$E^+ = g(z + ut) \tag{10.2b}$$

or

$$E = f(z - ut) + g(z + ut) \tag{10.2c}$$

where f and g denote any function of $z - ut$ and $z + ut$ respectively. Examples of such functions include $z \pm ut$, $\sin k(z \pm ut)$, $\cos k(z \pm ut)$, and $e^{jk(z \pm ut)}$, where k is a constant. It can easily be shown that these functions all satisfy eq. (10.1).

If we particularly assume harmonic (or sinusoidal) time dependence $e^{j\omega t}$, eq. (10.1) becomes

$$\frac{d^2 E_s}{dz^2} + \beta^2 E_s = 0 \tag{10.3}$$

where $\beta = \omega/u$ and E_s is the phasor form of E. The solution to eq. (10.3) is similar to Case 3 of Example 6.5 (see eq. (6.5.12)). With the time factor inserted, the possible solutions to eq. (10.3) are

$$E^+ = Ae^{j(\omega t - \beta z)} \tag{10.4a}$$

$$E^- = Be^{j(\omega t + \beta z)} \tag{10.4b}$$

and

$$E = Ae^{j(\omega t - \beta z)} + Be^{j(\omega t + \beta z)} \tag{10.4c}$$

where A and B are real constants.

For the moment, let us consider the solution in eq. (10.4a). Taking the imaginary part of this equation, we have

$$E = A \sin (\omega t - \beta z) \tag{10.5}$$

This is a sine wave chosen for simplicity; a cosine wave would have resulted had we taken the real part of eq. (10.4a). Note the following characteristics of the wave in eq. (10.5):

1. It is time-harmonic because we assumed time dependence $e^{j\omega t}$ to arrive at eq. (10.5).
2. A is called the *amplitude* of the wave and has the same units as E.
3. $(\omega t - \beta z)$ is the *phase* (in radians) of the wave; it depends on time t and space variable z.
4. ω is the *angular frequency* (in radians/second); β is the *phase constant* or *wave number* (in radians/meter).

Due to the variation of E with both time t and space variable z, we may plot E as a function of t by keeping z constant and vice versa. The plots of $E(z, t = \text{constant})$ and $E(t, z = \text{constant})$ are shown in Figure 10.1(a) and (b) respectively. From Figure 10.1(a), we observe that the wave takes distance λ to repeat itself and hence λ is called the *wavelength* (in meters). From Figure 10.1(b), the wave takes time T to repeat itself; consequently T is known as the *period* (in seconds). Since it takes time T for the wave to travel distance λ at the speed u, we expect

$$\lambda = uT \tag{10.6a}$$

But $T = 1/f$, where f is the *frequency* (the number of cycles per second) of the wave in Hertz (Hz). Hence,

$$\boxed{u = f\lambda} \tag{10.6b}$$

Because of this fixed relationship between wavelength and frequency, one can identify the position of a radio station within its band by either the frequency or the wavelength. Usually the frequency is preferred. Also, because

$$\omega = 2\pi f \tag{10.7a}$$

$$\beta = \frac{\omega}{u} \tag{10.7b}$$

and

$$T = \frac{1}{f} = \frac{2\pi}{\omega} \tag{10.7c}$$

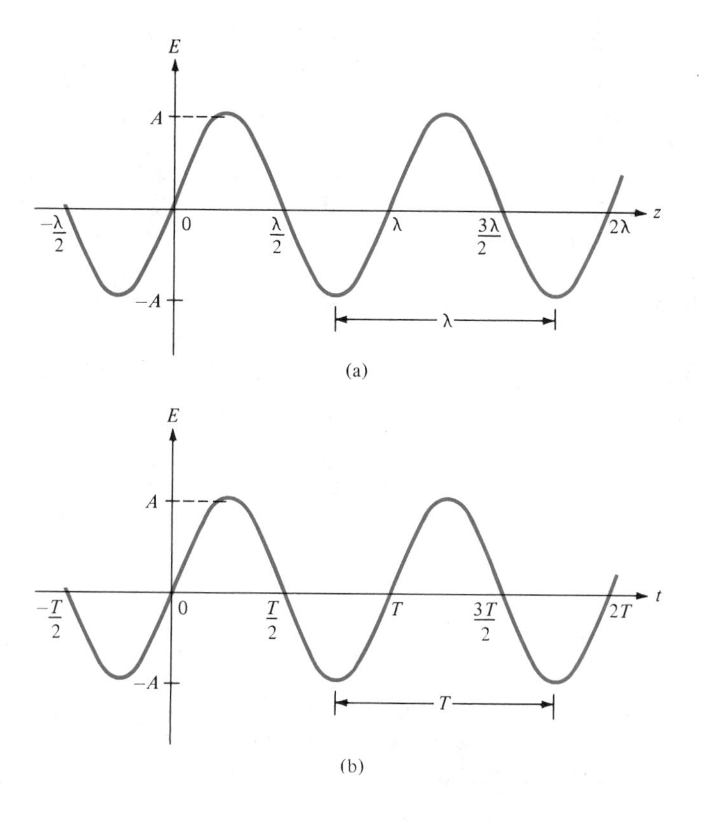

Figure 10.1 Plot of $E(z, t) = A \sin(\omega t - \beta z)$: **(a)** with constant t, **(b)** with constant z.

we expect from eqs. (10.6) and (10.7) that

$$\boxed{\beta = \frac{2\pi}{\lambda}} \qquad [10.8]$$

Equation (10.8) shows that for every wavelength of distance traveled, a wave undergoes a phase change of 2π radians.

We will now show that the wave represented by eq. (10.5) is traveling with a velocity u in the $+z$ direction. To do this, we consider a fixed point P on the wave. We sketch eq. (10.5) at times $t = 0$, $T/4$, and $T/2$ as in Figure 10.2. From the figure, it is evident that as the wave advances with time, point P moves along $+z$ direction. Point P is a point of constant phase, therefore

$$\omega t - \beta z = \text{constant}$$

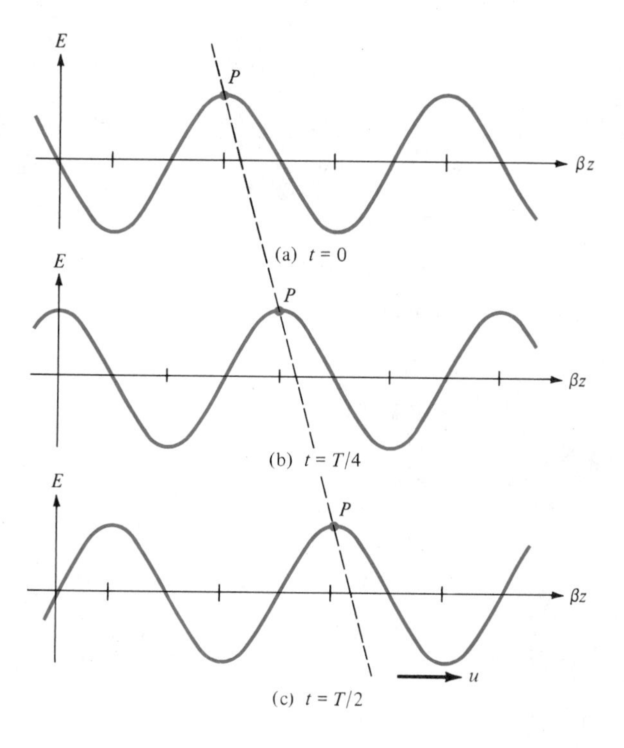

Figure 10.2 Plot of $E(z, t) = A \sin(\omega t - \beta z)$ at time **(a)** $t = 0$, **(b)** $t = T/4$, **(c)** $t = T/2$; P moves along $+z$ direction with velocity u.

or

$$\frac{dz}{dt} = \frac{\omega}{\beta} = u \tag{10.9}$$

which is the same as eq. (10.7b). Equation (10.9) shows that the wave travels with velocity u in the $+z$ direction. Similarly, it can be shown that the wave $B \sin(\omega t + \beta z)$ in eq. (10.4b) is traveling with velocity u in the $-z$ direction.

In summary, we note the following:

1. A wave is a function of both time and space.
2. Though time $t = 0$ is arbitrarily selected as a reference for the wave, a wave is without beginning or end.
3. A negative sign in $(\omega t \pm \beta z)$ is associated with a wave propagating in the $+z$ direction (forward traveling or positive-going wave) whereas a positive sign indicates that a wave is traveling in the $-z$ direction (backward traveling or negative-going wave).

Table 10.1 **Electromagnetic Spectrum**

EM phenomena	Examples of uses	Approximate frequency range
Cosmic rays	Physics, astronomy	10^{14} GHz and above
Gamma rays	Cancer therapy	10^{10}–10^{13} GHz
X-rays	X-ray examination	10^{8}–10^{9} GHz
Ultraviolet radiation	Sterilization	10^{6}–10^{8} GHz
Visible light	Human vision	10^{5}–10^{6} GHz
Infrared radiation	Photography	10^{3}–10^{4} GHz
Microwave waves	Radar, microwave relays, satellite communication	3–300 GHz
Radio waves	UHF television	470–806 MHz
	VHF television, FM radio	54–216 MHz
	Short-wave radio	3–26 MHz
	AM radio	535–1605 kHz

4. Since $\sin(-\psi) = -\sin\psi = \sin(\psi \pm \pi)$, whereas $\cos(-\psi) = \cos\psi$,

$$\sin(\psi \pm \pi/2) = \pm\cos\psi \tag{10.10a}$$

$$\sin(\psi \pm \pi) = -\sin\psi \tag{10.10b}$$

$$\cos(\psi \pm \pi/2) = \mp\sin\psi \tag{10.10c}$$

$$\cos(\psi \pm \pi) = -\cos\psi \tag{10.10d}$$

where $\psi = \omega t \pm \beta z$. With eq. (10.10), any time-harmonic wave can be represented in the form of sine or cosine.

A large number of frequencies visualized in numerical order constitute a *spectrum*. Table 10.1 shows at what frequencies various types of energy in the EM spectrum occur. Frequencies usable for radio communication occur near the lower end of the EM spectrum. As frequency increases, the manifestation of EM energy becomes dangerous to human beings.[1] Microwave ovens, for example, can pose a hazard if not properly shielded. The practical difficulties of using EM energy for communication purposes also increase as frequency increases, until finally it can no longer be used. As communication methods improve, the limit to usable frequency has been pushed

[1]See March 1987 special issue of *IEEE Engineering in Medicine and Biology Magazine* on "Effects of EM Radiation."

higher. Today communication satellites use frequencies near 14 GHz. This is still far below light frequencies, but in the enclosed environment of fiber-optics, light itself can be used for radio communication.[2]

EXAMPLE 10.1

The electric field in free space is given by

$$\mathbf{E} = 50 \cos (10^8 t + \beta x)\, \mathbf{a}_y \text{ V/m}$$

(a) Find the direction of wave propagation.

(b) Calculate β and the time it takes to travel a distance of $\lambda/2$.

(c) Sketch the wave at $t = 0$, $T/4$, and $T/2$.

SOLUTION

(a) From the positive sign in $(\omega t + \beta x)$, we infer that the wave is propagating along $-\mathbf{a}_x$. This will be confirmed in part (c) of this example.

(b) In free space, $u = c$.

$$\beta = \frac{\omega}{c} = \frac{10^8}{3 \times 10^8} = \frac{1}{3}$$

or

$$\beta = 0.3333 \text{ rad/m}$$

If T is the period of the wave, it takes T seconds to travel a distance λ at speed c. Hence to travel a distance of $\lambda/2$ will take

$$t_1 = \frac{T}{2} = \frac{1}{2}\frac{2\pi}{\omega} = \frac{\pi}{10^8} = 31.42\text{ns}$$

Alternatively, because the wave is traveling at the speed of light c,

$$\frac{\lambda}{2} = ct_1 \qquad \text{or} \qquad t_1 = \frac{\lambda}{2c}$$

But

$$\lambda = \frac{2\pi}{\beta} = 6\pi$$

Hence,

$$t_1 = \frac{6\pi}{2(3 \times 10^8)} = 31.42\text{ns}$$

as obtained before.

[2]See October 1980 issue of *IEEE Proceedings* on "Optical-Fiber Communications."

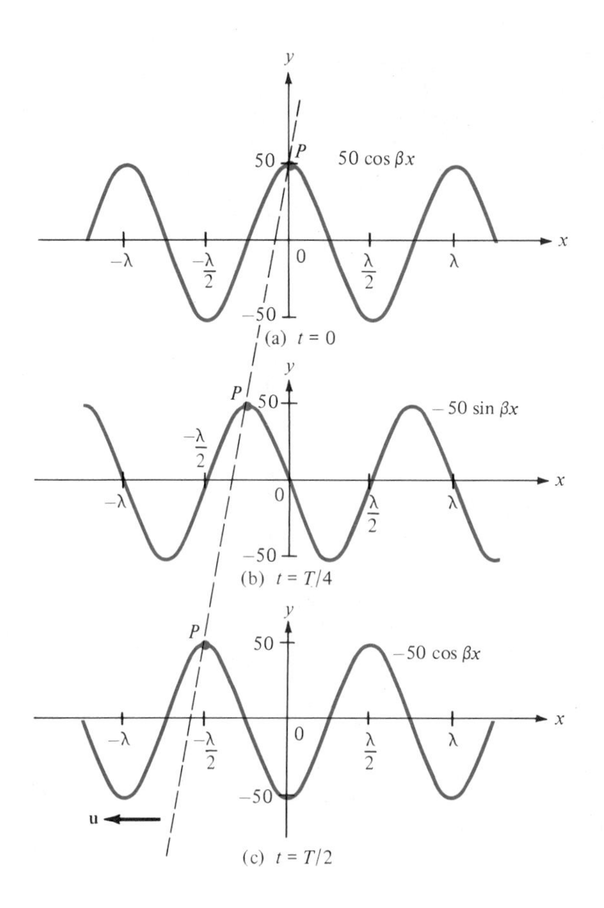

Figure 10.3 For Example 10.1; wave travels along $-\mathbf{a}_x$.

(c) At $t = 0, \quad E_y = 50 \cos \beta x$

At $t = T/4, \; E_y = 50 \cos \left(\omega \cdot \dfrac{2\pi}{4\omega} + \beta x \right) = 50 \cos (\beta x + \pi/2)$

$$= -\, 50 \sin \beta x$$

At $t = T/2, \; E_y = 50 \cos \left(\omega \cdot \dfrac{2\pi}{2\omega} + \beta x \right) = 50 \cos(\beta x + \pi)$

$$= -\, 50 \cos \beta x$$

E_y at $t = 0, T/4, T/2$ is plotted against x as shown in Figure 10.3. Notice that a point P (arbitrarily selected) on the wave moves along $-\mathbf{a}_x$ as t increases with time. This shows that the wave travels along $-\mathbf{a}_x$. ■

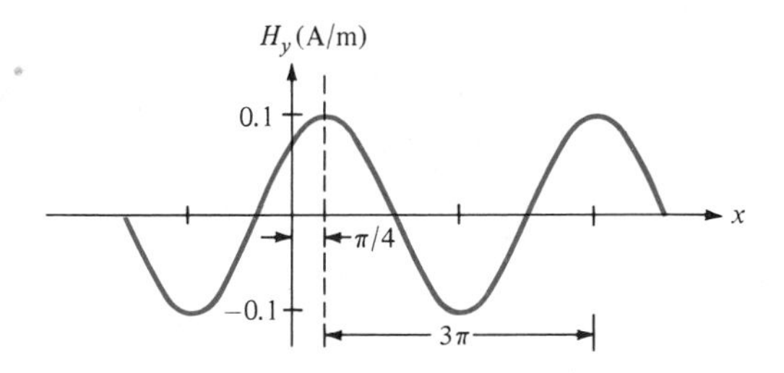

Figure 10.4 For Practice Exercise 10.1(c).

PRACTICE EXERCISE 10.1

In free space, $\mathbf{H} = 0.1 \cos(2 \times 10^8 t - kx)\, \mathbf{a}_y$ A/m. Calculate

(a) k, λ, and T

(b) The time t_1 it takes the wave to travel $\lambda/8$

(c) Sketch the wave at time t_1.

ANSWER (a) 0.667 rad/m, 9.425 m, 31.42 ns, (b) 3.927 ns, (c) see Figure 10.4.

10.3 WAVE PROPAGATION IN LOSSY DIELECTRICS

As mentioned in Section 10.1, wave propagation in lossy dielectrics is a general case from which wave propagation in other types of media can be derived as special cases. Therefore, this section is foundational to the next three sections.

A lossy dielectric is a medium in which an EM wave loses power as it propagates due to poor conduction. In other words, a lossy dielectric is a partially conducting medium (imperfect dielectric or imperfect conductor) with $\sigma \neq 0$, as distinct from a lossless dielectric (perfect or good dielectric) in which $\sigma = 0$.

Consider a linear, isotropic, homogeneous, lossy dielectric medium that is charge-free ($\rho_v = 0$). Assuming and suppressing the time factor $e^{j\omega t}$, Maxwell's equations (see Table 9.2) become

$$\nabla \cdot \mathbf{E}_s = 0 \qquad [10.11]$$

$$\nabla \cdot \mathbf{H}_s = 0 \qquad [10.12]$$

$$\nabla \times \mathbf{E}_s = -j\omega\mu\mathbf{H}_s \qquad [10.13]$$

$$\nabla \times \mathbf{H}_s = (\sigma + j\omega\varepsilon)\mathbf{E}_s \qquad [10.14]$$

Taking the curl of both sides of eq. (10.13) gives

$$\nabla \times \nabla \times \mathbf{E}_s = -j\omega\mu\, \nabla \times \mathbf{H}_s \qquad [10.15]$$

Applying the vector identity

$$\nabla \times \nabla \times \mathbf{A} = \nabla (\nabla \cdot \mathbf{A}) - \nabla^2 \mathbf{A} \tag{10.16}$$

to the left-hand side of eq. (10.15) and invoking eqs. (10.11) and (10.14), we obtain

$$\nabla (\cancel{\nabla \cdot \mathbf{E}_s}) - \nabla^2 \mathbf{E}_s = -j\omega\mu(\sigma + j\omega\varepsilon)\mathbf{E}_s$$

$$0$$

or

$$\boxed{\nabla^2 \mathbf{E}_s - \gamma^2 \mathbf{E}_s = 0} \tag{10.17}$$

where

$$\gamma^2 = j\omega\mu(\sigma + j\omega\varepsilon) \tag{10.18}$$

and γ is called the *propagation constant* (in per meter) of the medium. By a similar procedure, it can be shown that for the **H** field,

$$\nabla^2 \mathbf{H}_s - \gamma^2 \mathbf{H}_s = 0 \tag{10.19}$$

Equations (10.17) and (10.19) are known as homogeneous vector *Helmholtz's equations* or simply vector *wave equations*. In Cartesian coordinates, eq. (10.17), for example, is equivalent to three scalar wave equations, one for each component of **E** along $\mathbf{a}_x$, $\mathbf{a}_y$, and $\mathbf{a}_z$.

Since γ in eqs. (10.17) to (10.19) is a complex quantity, we may let

$$\boxed{\gamma = \alpha + j\beta} \tag{10.20}$$

We obtain α and β from eqs. (10.18) and (10.20) by noting that

$$-\operatorname{Re} \gamma^2 = \beta^2 - \alpha^2 = \omega^2\mu\varepsilon \tag{10.21}$$

and

$$|\gamma^2| = \beta^2 + \alpha^2 = \omega\mu \sqrt{\sigma^2 + \omega^2\varepsilon^2} \tag{10.22}$$

From eqs. (10.21) and (10.22), we obtain

$$\alpha = \omega \sqrt{\frac{\mu\varepsilon}{2}\left[\sqrt{1 + \left[\frac{\sigma}{\omega\varepsilon}\right]^2} - 1\right]} \tag{10.23}$$

$$\beta = \omega \sqrt{\frac{\mu\varepsilon}{2}\left[\sqrt{1 + \left[\frac{\sigma}{\omega\varepsilon}\right]^2} + 1\right]} \tag{10.24}$$

Without loss of generality, if we assume that the wave propagates along $+\mathbf{a}_z$ and that $\mathbf{E}_s$ has only an x-component, then

$$\mathbf{E}_s = E_{xs}(z)\mathbf{a}_x \tag{10.25}$$

Substituting this into eq. (10.17) yields

$$(\nabla^2 - \gamma^2)E_{xs}(z) \tag{10.26}$$

Hence

$$\underbrace{\frac{\partial^2 E_{xs}(z)}{\partial x^2}}_{0} + \underbrace{\frac{\partial^2 E_{xs}(z)}{\partial y^2}}_{0} + \frac{\partial^2 E_{xs}(z)}{\partial z^2} - \gamma^2 E_{xs}(z) = 0$$

or

$$\left[\frac{d^2}{dz^2} - \gamma^2\right] E_{xs}(z) = 0 \tag{10.27}$$

This is a scalar wave equation, a linear homogeneous differential equation, with solution (see Case 2 in Example 6.5)

$$E_{xs}(z) = E_o e^{-\gamma z} + E_o' e^{\gamma z} \tag{10.28}$$

where E_o and E_o' are constants. The fact that the field must be finite at infinity requires that $E_o' = 0$. Alternatively, because $e^{\gamma z}$ denotes a wave traveling along $-\mathbf{a}_z$ whereas we assume wave propagation along $\mathbf{a}_z$, $E_o' = 0$. Whichever way we look at it, $E_o' = 0$. Inserting the time factor $e^{j\omega t}$ into eq. (10.28) and using eq. (10.20), we obtain

$$\mathbf{E}(z, t) = \text{Re}\,(E_{xs}(z)e^{j\omega t}\mathbf{a}_x) = \text{Re}\,(E_o e^{-\alpha z} e^{j(\omega t - \beta z)}\mathbf{a}_x)$$

or

$$\boxed{\mathbf{E}(z, t) = E_o e^{-\alpha z} \cos(\omega t - \beta z)\, \mathbf{a}_x} \tag{10.29}$$

A sketch of $|\mathbf{E}|$ at times $t = 0$ and $t = \Delta t$ is portrayed in Figure 10.5, where it is evident that $\mathbf{E}$ has only an x-component and it is traveling along the $+z$ direction. Having obtained $\mathbf{E}(z, t)$, we obtain $\mathbf{H}(z, t)$ either by taking similar steps to solve eq. (10.19) or by using eq. (10.29) in conjunction with Maxwell's equations as we did in Example 9.8. We will eventually arrive at

$$\mathbf{H}(z, t) = \text{Re}\,(H_o e^{-\alpha z} e^{j(\omega t - \beta z)}\, \mathbf{a}_y) \tag{10.30}$$

where

$$H_o = \frac{E_o}{\eta} \tag{10.31}$$

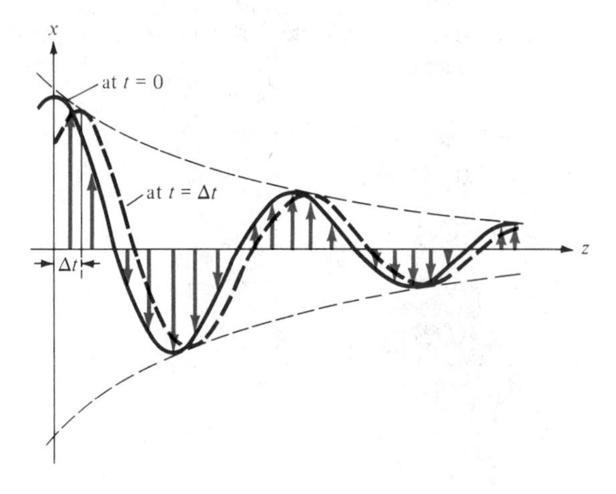

Figure 10.5 *E*-field with *x*-component traveling along $+z$ direction at times $t = 0$ and $t = \Delta t$; arrows indicate instantaneous values of *E*.

and η is a complex quantity known as the *intrinsic impedance* (in ohms) of the medium. It can be shown by following the steps taken in Example 9.8 that

$$\eta = \sqrt{\frac{j\omega\mu}{\sigma + j\omega\varepsilon}} = |\eta| \underline{/\theta_\eta} = |\eta| e^{j\theta_\eta} \tag{10.32}$$

with

$$|\eta| = \frac{\sqrt{\mu/\varepsilon}}{\left[1 + \left(\frac{\sigma}{\omega\varepsilon}\right)^2\right]^{1/4}}, \qquad \tan 2\theta_\eta = \frac{\sigma}{\omega\varepsilon} \tag{10.33}$$

where $0 \leq \theta_\eta \leq 45°$. Substituting eqs. (10.31) and (10.32) into eq. (10.30) gives

$$\mathbf{H} = \text{Re}\left[\frac{E_o}{|\eta| e^{j\theta_\eta}} e^{-\alpha z} e^{j(\omega t - \beta z)} \, \mathbf{a}_y\right]$$

or

$$\mathbf{H} = \frac{E_o}{|\eta|} e^{-\alpha z} \cos(\omega t - \beta z - \theta_\eta) \, \mathbf{a}_y \tag{10.34}$$

Notice from eqs. (10.29) and (10.34) that as the wave propagates along $\mathbf{a}_z$, it decreases or attenuates in amplitude by a factor $e^{-\alpha z}$, and hence α is known as the *attenuation constant* or *attenuation factor* of the medium. It is a measure of the spatial rate of decay of the wave in the medium, measured in nepers per meter (Np/m) or in decibels per meter (dB/m). An attenuation of 1 neper denotes a reduction to e^{-1} of the original value whereas an increase of 1 neper indicates an increase by a factor of e. Hence, for voltages

$$1 \text{ Np} = 20 \log_{10} e = 8.686 \; dB \qquad [10.35]$$

From eq. (10.23), we notice that if $\sigma = 0$, as is the case for a lossless medium and free space, $\alpha = 0$ and the wave is not attenuated as it propagates. The quantity β is a measure of the phase shift per length and is called the *phase constant* or *wave number*. In terms of β, the wave velocity u and wavelength λ are respectively given by (see eqs. (10.7b) and (10.8))

$$u = \frac{\omega}{\beta}, \qquad \lambda = \frac{2\pi}{\beta} \qquad [10.36]$$

We also notice from eqs. (10.29) and (10.34) that **E** and **H** are out of phase by θ_η at any instant of time due to the complex intrinsic impedance of the medium. Thus at any time, **E** leads **H** (or **H** lags **E**) by θ_η. Finally, we notice that the ratio of the magnitude of the conduction current density **J** to that of the displacement current density $\mathbf{J}_d$ in a lossy medium is

$$\frac{|\mathbf{J}_s|}{|\mathbf{J}_{ds}|} = \frac{|\sigma \mathbf{E}_s|}{|j\omega\varepsilon \mathbf{E}_s|} = \frac{\sigma}{\omega\varepsilon} = \tan\theta$$

or

$$\boxed{\tan\theta = \frac{\sigma}{\omega\varepsilon}} \qquad [10.37]$$

where $\tan\theta$ is known as the *loss tangent* and θ is the *loss angle* of the medium as illustrated in Figure 10.6. Although a line of demarcation between good conductors and lossy dielectrics is not easy to make, $\tan\theta$ or θ may be used to determine how lossy a medium is. A medium is said to be a good (lossless or perfect) dielectric if $\tan\theta$ is very small ($\sigma \ll \omega\varepsilon$) or a good conductor if $\tan\theta$ is very large ($\sigma \gg \omega\varepsilon$). From the viewpoint of wave propagation, the characteristic behavior of a medium depends not only on its constitutive parameters σ, ε, and μ but also on the frequency of operation. A medium that is regarded as a good conductor at low frequencies may be a good dielectric at high frequencies. Note from eqs. (10.33) and (10.37) that

$$\theta = 2\theta_\eta \qquad [10.38]$$

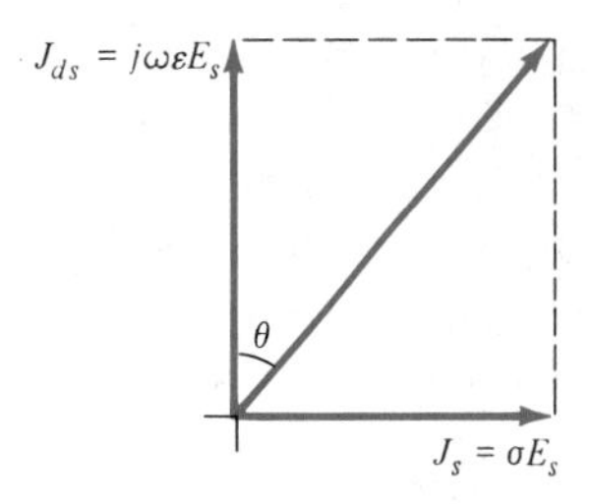

Figure 10.6 Loss angle of a lossy medium.

From eq. (10.14)

$$\nabla \times \mathbf{H}_s = (\sigma + j\omega\varepsilon)\mathbf{E}_s = j\omega\varepsilon \left[1 - \frac{j\sigma}{\omega\varepsilon}\right] \mathbf{E}_s$$
$$= j\omega\varepsilon_c \mathbf{E}_s \qquad [10.39]$$

where

$$\boxed{\varepsilon_c = \varepsilon \left[1 - j\frac{\sigma}{\omega\varepsilon}\right]} \qquad [10.40a]$$

or

$$\varepsilon_c = \varepsilon' - j\varepsilon'' \qquad [10.40b]$$

and $\varepsilon' = \varepsilon$, $\varepsilon'' = \sigma/\omega$; ε_c is called the *complex permittivity* of the medium. We observe that the ratio of ε'' to ε' is the loss tangent of the medium; that is,

$$\tan\theta = \frac{\varepsilon''}{\varepsilon'} = \frac{\sigma}{\omega\varepsilon} \qquad [10.41]$$

In subsequent sections, we will consider wave propagation in other types of media which may be regarded as special cases of what we have considered here. Thus we will simply deduce the governing formulas from those obtained for the general case treated in this section. The student is advised not just to memorize the formulas but to observe how they are easily obtained from the formulas for the general case.

10.4 PLANE WAVES IN LOSSLESS DIELECTRICS

In a lossless dielectric, $\sigma \ll \omega\varepsilon$. It is a special case of that in Section 10.3 except that

$$\boxed{\sigma \simeq 0, \qquad \varepsilon = \varepsilon_0\varepsilon_r, \qquad \mu = \mu_0\mu_r} \qquad [10.42]$$

Substituting these into eqs. (10.23) and (10.24) gives

$$\alpha = 0, \qquad \beta = \omega\sqrt{\mu\varepsilon} \tag{10.43a}$$

$$u = \frac{\omega}{\beta} = \frac{1}{\sqrt{\mu\varepsilon}}, \qquad \lambda = \frac{2\pi}{\beta} \tag{10.43b}$$

Also

$$\eta = \sqrt{\frac{\mu}{\varepsilon}} \underline{/0^\circ} \tag{10.44}$$

and thus **E** and **H** are in time phase with each other.

10.5 PLANE WAVES IN FREE SPACE

This is a special case of what we considered in Section 10.3. In this case,

$$\boxed{\sigma = 0, \qquad \varepsilon = \varepsilon_o, \qquad \mu = \mu_o} \tag{10.45}$$

This may also be regarded as a special case of section 10.4. Thus we simply replace ε by ε_o and μ by μ_o in eq. (10.43) or we substitute eq. (10.45) directly into eqs. (10.23) and (10.24). Either way, we obtain

$$\alpha = 0, \qquad \beta = \omega\sqrt{\mu_o\varepsilon_o} = \frac{\omega}{c} \tag{10.46a}$$

$$u = \frac{1}{\sqrt{\mu_o\varepsilon_o}} = c, \qquad \lambda = \frac{2\pi}{\beta} \tag{10.46b}$$

where $c \simeq 3 \times 10^8$ m/s, the speed of light in a vacuum. The fact that EM wave travels in free space at the speed of light is significant. It shows that light is the manifestation of an EM wave. In other words, light is characteristically electromagnetic.

By substituting the constitutive parameters in eq. (10.45) into eq. (10.33), $\theta_\eta = 0$ and $\eta = \eta_o$, where η_o is called the *intrinsic impedance of free space* and is given by

$$\boxed{\eta_o = \sqrt{\frac{\mu_o}{\varepsilon_o}} = 120\pi \simeq 377\ \Omega} \tag{10.47}$$

$$\mathbf{E} = E_o \cos(\omega t - \beta z)\, \mathbf{a}_x \tag{10.48a}$$

then

$$\mathbf{H} = H_o \cos(\omega t - \beta z)\,\mathbf{a}_y = \frac{E_o}{\eta_o}\cos(\omega t - \beta z)\,\mathbf{a}_y \tag{10.48b}$$

The plots of **E** and **H** are shown in Figure 10.7(a). In general, if $\mathbf{a}_E$, $\mathbf{a}_H$, and $\mathbf{a}_k$ are unit vectors along the **E** field, the **H** field, and the direction of wave propagation; it can be shown that (see Problem 10.13)

$$\mathbf{a}_k \times \mathbf{a}_E = \mathbf{a}_H$$

or

$$\mathbf{a}_k \times \mathbf{a}_H = -\mathbf{a}_E$$

or

$$\boxed{\mathbf{a}_E \times \mathbf{a}_H = \mathbf{a}_k} \tag{10.49}$$

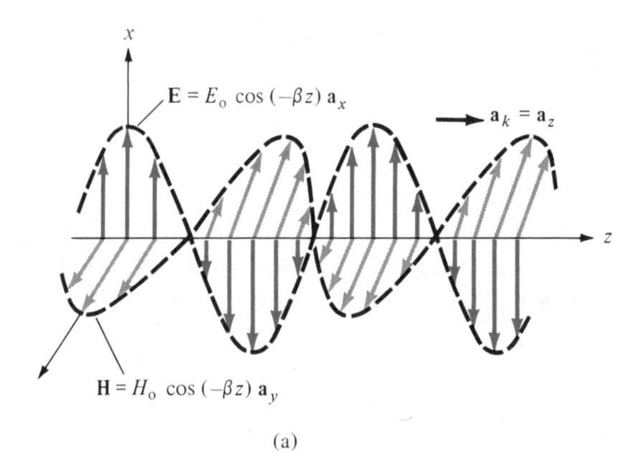

(a)

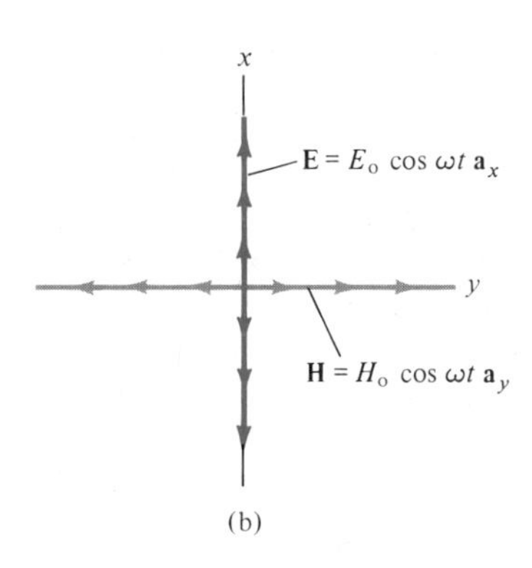

(b)

Figure 10.7 **(a)** Plot of **E** and **H** as functions of z at $t = 0$; **(b)** plot of **E** and **H** at $z = 0$. The arrows indicate instantaneous values.

Both **E** and **H** fields (or EM waves) are everywhere normal to the direction of wave propagation, $\mathbf{a}_k$. That means, the fields lie in a plane that is transverse or orthogonal to the direction of wave propagation. They form an EM wave that has no electric or magnetic field components along the direction of propagation; such a wave is called a *transverse electromagnetic* (TEM) wave. Each of **E** and **H** is called a *uniform plane wave* because **E** (or **H**) has the same magnitude throughout any transverse plane, defined by z = constant. The direction in which the electric field points is the *polarization* of a TEM wave.[3] The wave in eq. (10.29), for example, is polarized in the x-direction. This should be observed in Figure 10.7(b), where an illustration of uniform plane waves is given. A uniform plane wave cannot exist physically because it stretches to infinity and would represent an infinite energy. However, such waves are characteristically simple but fundamentally important. They serve as approximations to practical waves, such as from a radio antenna, at distances sufficiently far from radiating sources. Although our discussion after eq. (10.48) deals with free space, it also applies for any other isotropic medium.

10.6 PLANE WAVES IN GOOD CONDUCTORS

This is another special case of that considered in Section 10.3. A perfect, or good conductor, is one in which $\sigma \gg \omega\varepsilon$ so that $\sigma/\omega\varepsilon \to \infty$; that is,

$$\boxed{\sigma \simeq \infty, \qquad \varepsilon = \varepsilon_o, \qquad \mu = \mu_o\mu_r} \tag{10.50}$$

Hence, eqs. (10.23) and (10.24) become

$$\alpha = \beta = \sqrt{\frac{\omega\mu\sigma}{2}} = \sqrt{\pi f \mu\sigma} \tag{10.51a}$$

$$u = \frac{\omega}{\beta} = \sqrt{\frac{2\omega}{\mu\sigma}}, \qquad \lambda = \frac{2\pi}{\beta} \tag{10.51b}$$

Also,

$$\eta = \sqrt{\frac{\omega\mu}{\sigma}} \underline{/45^\circ} \tag{10.52}$$

and thus **E** leads **H** by 45°. If

$$\mathbf{E} = E_o e^{-\alpha z} \cos(\omega t - \beta z)\, \mathbf{a}_x \tag{10.53a}$$

....

[3]Some texts define polarization differently.

then

$$\mathbf{H} = \frac{E_o}{\sqrt{\dfrac{\omega\mu}{\sigma}}} e^{-\alpha z} \cos(\omega t - \beta z - 45°)\, \mathbf{a}_y \qquad [10.53b]$$

Therefore, as **E** (or **H**) wave travels in a conducting medium, its amplitude is attenuated by the factor $e^{-\alpha z}$. The distance δ, shown in Figure 10.8, through which the wave amplitude decreases by a factor e^{-1} (about 37%) is called *skin depth* or *penetration depth* of the medium; that is,

$$E_o e^{-\alpha\delta} = E_o e^{-1}$$

or

$$\boxed{\delta = \frac{1}{\alpha}} \qquad [10.54a]$$

The skin depth is a measure of the depth to which an EM wave can penetrate the medium. Equation (10.54a) is generally valid for any material medium. For good conductors, eqs. (10.51a) and (10.54a) give

$$\delta = \frac{1}{\sqrt{\pi f \mu \sigma}} \qquad [10.54b]$$

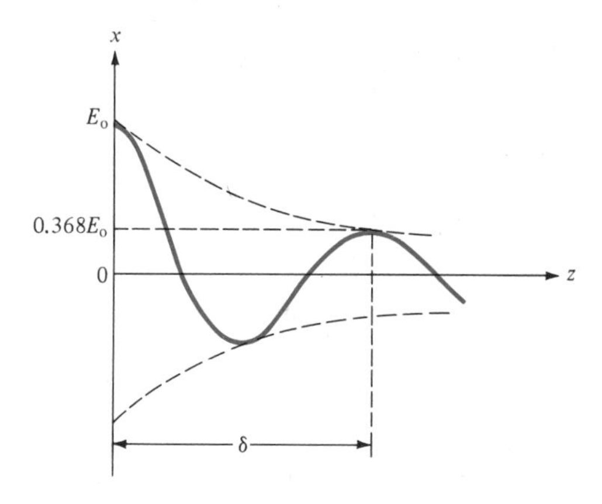

Figure 10.8 Illustration of skin depth.

The illustration in Figure 10.8 for a good conductor is exaggerated. However, for a partially conducting medium, the skin depth can be considerably large. Note from eqs. (10.51a), (10.52), and (10.54b) that for a good conductor,

$$\eta = \frac{1}{\sigma\delta}\sqrt{2}\, e^{j\pi/4} = \frac{1 + j}{\sigma\delta} \qquad [10.55]$$

Also for a good conductors, eq. (10.53a) can be written as

$$\mathbf{E} = E_o e^{-z/\delta} \cos\left(\omega t - \frac{z}{\delta}\right) \mathbf{a}_x$$

showing that δ measures the exponential damping of the wave as it travels through the conductor. The skin depth in copper at various frequencies is shown in Table 10.2. From the table, we notice that the skin depth decreases with increase in frequency. Thus, **E** and **H** can hardly propagate through good conductors.

The phenomenon whereby field intensity in a conductor rapidly decreases is known as *skin effect*. The fields and associated currents are confined to a very thin layer (the skin) of the conductor surface. For a wire of radius a, for example, it is a good approximation at high frequencies to assume that all of the current flows in the circular ring of thickness δ as shown in Figure 10.9. Skin effect appears in different guises in such problems as attenuation in waveguides, effective or ac resistance of transmission lines, and electromagnetic shielding. It is used to advantage in many applications. For example, because the skin depth in silver is very small, the difference in performance between a pure silver component and a silver-plated brass component is negligible, so silver plating is often used to reduce material cost of waveguide components. For the same reason, hollow tubular conductors are used instead of solid conductors in outdoor television antennas. Effective electromagnetic shielding of electrical devices can be provided by conductive enclosures a few skin depths in thickness.

The skin depth is useful in calculating the *ac resistance* due to skin effect. The resistance in eq. (5.10) is called the *dc resistance*, that is,

$$R_{dc} = \frac{\ell}{\sigma S} \qquad [5.10]$$

Table 10.2 **Skin Depth in Copper***

Frequency (Hz)	10	60	100	500	10^4	10^8	10^{10}
Skin depth (mm)	20.8	8.6	6.6	2.99	0.66	6.6×10^{-3}	6.6×10^{-4}

*For copper, $\sigma = 5.8 \times 10^7$ mhos/m, $\mu = \mu_o$, $\delta = 66.1/\sqrt{f}$ (in mm).

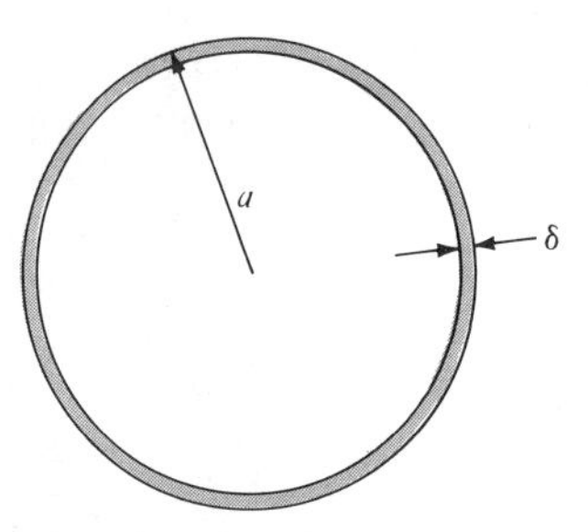

Figure 10.9 Skin depth at high frequencies, $\delta \ll a$.

We define the *surface or skin resistance* R_s (in Ω/m^2) as the real part of the η for a good conductor. Thus from eq. (10.55)

$$R_s = \frac{1}{\sigma\delta} = \sqrt{\frac{\pi f \mu}{\sigma}} \tag{10.56}$$

This is the resistance of a unit width and unit length of the conductor. It is equivalent to the dc resistance for a unit length of the conductor having cross-sectional area $1 \times \delta$. Thus for a given width w and length ℓ, the ac resistance is calculated using the familiar dc resistance relation of eq. (5.10) and assuming a uniform current flow in the conductor of thickness δ, that is,

$$R_{\text{ac}} = \frac{\ell}{\sigma\delta w} = \frac{R_s \ell}{w} \tag{10.57}$$

where $S \simeq \delta w$. For a conductor wire of radius a (see Figure 10.9), $w = 2\pi a$, so

$$\frac{R_{\text{ac}}}{R_{\text{dc}}} = \frac{\dfrac{\ell}{\sigma 2\pi a \delta}}{\dfrac{\ell}{\sigma \pi a^2}} = \frac{a}{2\delta}$$

Since $\delta \ll a$ at high frequencies, this shows than R_{ac} is far greater than R_{dc}. In general, the ratio of the ac to the dc resistance starts at 1.0 for dc and very low frequencies and increases as the frequency increases. Also, although the bulk of the current is nonuniformly distributed over a thickness of 5δ of the conductor, the power loss is the same as though it were uniformly distributed over a thickness of δ and zero elsewhere. This is one more reason why δ is referred to as the skin depth.

EXAMPLE 10.2

A lossy dielectric has an intrinsic impedance of $200 \underline{/30^\circ}\ \Omega$ at a particular frequency. If, at that frequency, the plane wave propagating through the dielectric has the magnetic field component

$$\mathbf{H} = 10\, e^{-\alpha x} \cos\left(\omega t - \frac{1}{2} x\right) \mathbf{a}_y \text{ A/m}$$

find $\mathbf{E}$ and α. Determine the skin depth and wave polarization.

SOLUTION The given wave travels along $\mathbf{a}_x$ so that $\mathbf{a}_k = \mathbf{a}_x$; $\mathbf{a}_H = \mathbf{a}_y$, so

$$-\mathbf{a}_E = \mathbf{a}_k \times \mathbf{a}_H = \mathbf{a}_x \times \mathbf{a}_y = \mathbf{a}_z$$

or

$$\mathbf{a}_E = -\ \mathbf{a}_z$$

Also $H_o = 10$, so

$$\frac{E_o}{H_o} = \eta = 200 \underline{/30^\circ} = 200\ e^{j\pi/6} \rightarrow E_o = 2000 e^{j\pi/6}$$

Except for the amplitude and phase difference, $\mathbf{E}$ and $\mathbf{H}$ always have the same form. Hence

$$\mathbf{E} = \text{Re}\ (2000 e^{j\pi/6} e^{-\gamma x} e^{j\omega t} \mathbf{a}_E)$$

or

$$\mathbf{E} = -2e^{-\alpha x} \cos\left(\omega t - \frac{x}{2} + \frac{\pi}{6}\right) \mathbf{a}_z \text{ kV/m}$$

Knowing that $\beta = 1/2$, we need to determine α. Since

$$\alpha = \omega \sqrt{\frac{\mu\varepsilon}{2}\left[\sqrt{1 + \left[\frac{\sigma}{\omega\varepsilon}\right]^2} = 1\right]}$$

and

$$\beta = \omega \sqrt{\frac{\mu\varepsilon}{2}\left[\sqrt{1 + \left[\frac{\sigma}{\omega\varepsilon}\right]^2} + 1\right]}$$

$$\frac{\alpha}{\beta} = \left[\frac{\sqrt{1 + \left[\frac{\sigma}{\omega\varepsilon}\right]^2} - 1}{\sqrt{1 + \left[\frac{\sigma}{\omega\varepsilon}\right]^2} + 1}\right]^{1/2}$$

But $\dfrac{\sigma}{\omega\varepsilon} = \tan 2\theta_\eta = \tan 60° = \sqrt{3}$. Hence,

$$\frac{\alpha}{\beta} = \left[\frac{2-1}{2+1}\right]^{1/2} = \frac{1}{\sqrt{3}}$$

or

$$\alpha = \frac{\beta}{\sqrt{3}} = \frac{1}{2\sqrt{3}} = 0.2887 \text{ Np/m}$$

and

$$\delta = \frac{1}{\alpha} = 2\sqrt{3} = 3.464 \text{ m}$$

The wave has an E_z component; hence it is polarized along the z-direction. ■

PRACTICE EXERCISE 10.2

A plane wave propagating through a medium with $\varepsilon_r = 8$, $\mu_r = 2$ has $\mathbf{E} = 0.5\, e^{-z/3} \sin(10^8 t - \beta z)\, \mathbf{a}_x$ V/m. Determine

(a) β

(b) The loss tangent

(c) Wave impedance

(d) Wave velocity

ANSWER (a) 1.374 rad/m, (b) 0.5154, (c) $177.72 \underline{/13.63°}\ \Omega$, (d) 7.278×10^7 m/s.

EXAMPLE 10.3

In a lossless medium for which $\eta = 60\pi$, $\mu_r = 1$, and $\mathbf{H} = -0.1 \cos(\omega t - z)\, \mathbf{a}_x + 0.5 \sin(\omega t - z)\mathbf{a}_y$ A/m, calculate ε_r, ω, and $\mathbf{E}$.

SOLUTION In this case, $\sigma = 0$, $\alpha = 0$, and $\beta = 1$, so

$$\eta = \sqrt{\mu/\varepsilon} = \sqrt{\frac{\mu_o}{\varepsilon_o}}\sqrt{\frac{\mu_r}{\varepsilon_r}} = \frac{120\pi}{\sqrt{\varepsilon_r}}$$

or

$$\sqrt{\varepsilon_r} = \frac{120\pi}{\eta} = \frac{120\pi}{60\pi} = 2 \quad \rightarrow \quad \varepsilon_r = 4$$

$$\beta = \omega\sqrt{\mu\varepsilon} = \omega\sqrt{\mu_o\varepsilon_o}\sqrt{\mu_r\varepsilon_r} = \frac{\omega}{c}\sqrt{4} = \frac{2\omega}{c}$$

or

$$\omega = \frac{\beta c}{2} = \frac{1\,(3 \times 10^8)}{2} = 1.5 \times 10^8 \text{ rad/s}$$

From the given **H** field, **E** can be calculated in two ways: using the techniques (based on Maxwell's equations) developed in this chapter or directly using Maxwell's equations as in the last chapter.

Method 1: To use the techniques developed in this chapter, we let

$$\mathbf{E} = \mathbf{H}_1 + \mathbf{H}_2$$

where $\mathbf{H}_1 = -0.1 \cos(\omega t - z)\,\mathbf{a}_x$ and $\mathbf{H}_2 = 0.5 \sin(\omega t - z)\,\mathbf{a}_y$ and the corresponding electric field

$$\mathbf{E} = \mathbf{E}_1 + \mathbf{E}_2$$

where $\mathbf{E}_1 = E_{1o} \cos(\omega t - z)\,\mathbf{a}_{E_1}$ and $\mathbf{E}_2 = E_{2o} \sin(\omega t - z)\,\mathbf{a}_{E_2}$. Notice that although **H** has components along $\mathbf{a}_x$ and $\mathbf{a}_y$, it has no component along the direction of propagation; it is therefore a TEM wave.

For $\mathbf{E}_1$:

$$\mathbf{a}_{E_1} = -(\mathbf{a}_k \times \mathbf{a}_{H_1}) = -(\mathbf{a}_z \times -\mathbf{a}_x) = \mathbf{a}_y$$

$$E_{1o} = \eta\, H_{1o} = 60\pi\,(0.1) = 6\pi$$

Hence

$$\mathbf{E}_1 = 6\pi \cos(\omega t - z)\,\mathbf{a}_y$$

For $\mathbf{E}_2$:

$$\mathbf{a}_{E_2} = -\,(\mathbf{a}_k \times \mathbf{a}_{H_2}) = -\,(\mathbf{a}_z \times \mathbf{a}_y) = \mathbf{a}_x$$

$$E_{2o} = \eta\, H_{2o} = 60\pi\,(0.5) = 30\pi$$

Hence

$$\mathbf{E}_2 = 30\pi \sin(\omega t - z)\,\mathbf{a}_x$$

Adding $\mathbf{E}_1$ and $\mathbf{E}_2$ gives **E**; that is,

$$\mathbf{E} = 94.25 \sin(1.5 \times 10^8 t - z)\,\mathbf{a}_x + 18.85 \cos(1.5 \times 10^8 t - z)\,\mathbf{a}_y \text{ V/m}$$

Method 2: We may apply Maxwell's equations directly.

$$\nabla \times \mathbf{H} = \underbrace{\sigma \mathbf{E}}_{0} + \varepsilon \frac{\partial \mathbf{E}}{\partial t} \quad \rightarrow \quad \mathbf{E} = \frac{1}{\varepsilon} \int \nabla \times H\, dt$$

because $\sigma = 0$. But

$$\nabla \times \mathbf{H} = \begin{vmatrix} \dfrac{\partial}{\partial x} & \dfrac{\partial}{\partial y} & \dfrac{\partial}{\partial z} \\ H_x(z) & H_y(z) & 0 \end{vmatrix} = -\frac{\partial H_y}{\partial z}\,\mathbf{a}_x + \frac{\partial H_x}{\partial z}\,\mathbf{a}_y$$

$$= H_{2o} \cos(\omega t - z)\,\mathbf{a}_x + H_{1o} \sin(\omega t - z)\mathbf{a}_y$$

where $H_{1o} = -0.1$ and $H_{2o} = 0.5$. Hence

$$\mathbf{E} = \frac{1}{\varepsilon}\int \nabla \times \mathbf{H}\, dt = \frac{H_{2o}}{\varepsilon\omega}\sin(\omega t - z)\,\mathbf{a}_x - \frac{H_{1o}}{\varepsilon\omega}\cos(\omega t - z)\,\mathbf{a}_y$$

$$= 94.25 \sin(\omega t - z)\,\mathbf{a}_x + 18.85\cos(\omega t - z)\,\mathbf{a}_y \text{ V/m}$$

as expected. ■

PRACTICE EXERCISE 10.3

A plane wave in a nonmagnetic medium has $\mathbf{E} = 50 \sin(10^8 t + 2z)\,\mathbf{a}_y$ V/m. Find

(a) The direction of wave propagation

(b) λ, f, and ε_r

(c) $\mathbf{H}$

ANSWER (a) along $-z$ direction, (b) 3.142 m, 15.92 MHz, 36, (c) $0.7958 \sin(10^8 t + 2z)\,\mathbf{a}_x$ A/m.

EXAMPLE 10.4

A uniform plane wave propagating in a medium has

$$\mathbf{E} = 2e^{-\alpha z}\sin(10^8 t - \beta z)\,\mathbf{a}_y \text{ V/m}.$$

If the medium is characterized by $\varepsilon_r = 1$, $\mu_r = 20$, and $\sigma = 3$ mhos/m, find α, β, and $\mathbf{H}$.

SOLUTION We need to determine the loss tangent to be able to tell whether the medium is a lossy dielectric or a good conductor.

$$\frac{\sigma}{\omega\varepsilon} = \frac{3}{10^8 \times 1 \times \dfrac{10^{-9}}{36\pi}} = 3393 \gg 1$$

showing that the medium may be regarded as a good conductor at the frequency of operation. Hence,

$$\alpha = \beta = \sqrt{\frac{\mu\omega\sigma}{2}} = \left[\frac{4\pi \times 10^{-7} \times 20(10^8)(3)}{2}\right]^{1/2}$$

$$= 61.4$$

$$\alpha = 61.4 \text{ Np/m}, \qquad \beta = 61.4 \text{ rad/m}$$

Also,

$$|\eta| = \sqrt{\frac{\mu\omega}{\sigma}} = \left[\frac{4\pi \times 10^{-7} \times 20(10^8)}{3}\right]^{1/2}$$
$$= \sqrt{\frac{800\pi}{3}}$$

$$\tan 2\theta_\eta = \frac{\sigma}{\omega\varepsilon} = 3393 \quad \rightarrow \quad \theta_\eta = 45° = \pi/4$$

Hence

$$\mathbf{H} = H_o e^{-\alpha z} \sin\left(\omega t - \beta z - \frac{\pi}{4}\right) \mathbf{a}_H$$

where

$$\mathbf{a}_H = \mathbf{a}_k \times \mathbf{a}_E = \mathbf{a}_z \times \mathbf{a}_y = -\mathbf{a}_x$$

and

$$H_o = \frac{E_o}{|\eta|} = 2\sqrt{\frac{3}{800\pi}} = 69.1 \times 10^{-3}$$

Thus

$$\mathbf{H} = -69.1\, e^{-61.4z} \sin\left(10^8 t - 61.42z - \frac{\pi}{4}\right) \mathbf{a}_x \text{ mA/m}$$

■

PRACTICE EXERCISE 10.4

A plane wave traveling in the $+\,y$-direction in a lossy medium ($\varepsilon_r = 4$, $\mu_r = 1$, $\sigma = 10^{-2}$ mhos/m) has $\mathbf{E} = 30 \cos(10^9\pi\, t + \pi/4)\, \mathbf{a}_z$ V/m at $y = 0$. Find

(a) $\mathbf{E}$ at $y = 1$ m, $t = 2$ ns

(b) The distance traveled by the wave to have a phase shift of 10°

(c) The distance traveled by the wave to have its amplitude reduced by 40%

(d) $\mathbf{H}$ at $y = 2$ m, $t = 2$ ns

ANSWER (a) $2.787\mathbf{a}_z$ V/m, (b) 8.325 mm, (c) 542 mm, (d) $-4.71\mathbf{a}_x$ mA/m.

EXAMPLE 10.5

A plane wave $\mathbf{E} = E_o \cos(\omega t - \beta z)\, \mathbf{a}_x$ is incident on a good conductor at $z = 0$. Find the current density in the conductor.

SOLUTION Since the current density $\mathbf{J} = \sigma\mathbf{E}$, we expect $\mathbf{J}$ to satisfy the wave equation in eq. (10.17), that is,

$$\nabla^2\mathbf{J}_s - \gamma^2\mathbf{J}_s = 0$$

Also the incident $\mathbf{E}$ has only an x-component and varies with z. Hence $\mathbf{J} = J_x(z, t)\,\mathbf{a}_x$ and

$$\frac{d^2}{dz^2}J_{sx} - \gamma^2 J_{sx} = 0$$

which is an ordinary differential equation with solution (see Case 2 of Example 6.5)

$$J_{sx} = Ae^{-\gamma z} + Be^{+\gamma z}$$

The constant B must be zero because J_{sx} is finite as $z \rightarrow \infty$. But in a good conductor, $\sigma \gg \omega\varepsilon$ so that $\alpha = \beta = 1/\delta$. Hence

$$\gamma = \alpha + j\beta = a(1 + j) = \frac{(1 + j)}{\delta}$$

and

$$J_{sx} = Ae^{-z(1 + j)/\delta}$$

or

$$J_{sx} = J_{sx}(0)\; e^{-z(1 + j)/\delta}$$

where J_{sx} (0) is the current density on the conductor surface. ■

PRACTICE EXERCISE 10.5

Due to the current density of Example 10.5, find the magnitude of the total current through a strip of the conductor of infinite depth along z and width w along y.

ANSWER $\dfrac{J_{sx}(0)w\delta}{\sqrt{2}}$

EXAMPLE 10.6

For the copper coaxial cable of Figure 7.12, let $a = 2$mm, $b = 6$ mm, and $t = 1$ mm. Calculate the resistance of 2 m length of the cable at dc and at 100 MHz.

SOLUTION Let

$$R = R_o + R_i$$

where R_o and R_i are the resistances of the inner and outer conductors.

At dc,

$$R_i = \frac{\ell}{\sigma S} = \frac{\ell}{\sigma \pi a^2} = \frac{2}{5.8 \times 10^7 \pi [2 \times 10^{-3}]^2} = 2.744 \text{ m}\Omega$$

$$R_o = \frac{\ell}{\sigma S} = \frac{\ell}{\sigma \pi [[b + t]^2 - b^2]} = \frac{\ell}{\sigma \pi [t^2 + 2bt]}$$

$$= \frac{2}{5.8 \times 10^7 \pi \, [1 + 12] \times 10^{-6}}$$

$$= 0.8429 \text{ m}\Omega$$

Hence $R_{dc} = 2.744 + 0.8429 = 3.587$ mΩ

At $f = 100$ MHz,

$$R_i = \frac{R_s \ell}{w} = \frac{\ell}{\sigma \delta 2\pi a} = \frac{\ell}{2\pi a} \sqrt{\frac{\pi f \mu}{\sigma}}$$

$$= \frac{2}{2\pi \times 2 \times 10^{-3}} \sqrt{\frac{\pi \times 10^8 \times 4\pi \times 10^{-7}}{5.8 \times 10^7}}$$

$$= 0.41 \ \Omega.$$

Since $\delta = 6.6 \ \mu\text{m} \ll t = 1$ mm, $w = 2\pi b$ for the outer conductor. Hence,

$$R_o = \frac{R_s \ell}{w} = \frac{\ell}{2\pi b} \sqrt{\frac{\pi f \mu}{\sigma}}$$

$$= \frac{2}{2\pi \times 6 \times 10^{-3}} \sqrt{\frac{\pi \times 10^8 \times 4\pi \times 10^{-7}}{5.8 \times 10^7}}$$

$$= 0.1384 \ \Omega$$

Hence,

$$R_{ac} = 0.41 + 0.1384 = 0.5484 \ \Omega$$

which is about 150 times greater than R_{dc}. Thus, for the same effective current i, the ohmic loss (i^2R) of the cable at 100 MHz is far greater than dc power loss by a factor of 150. ■

PRACTICE EXERCISE 10.6

For an aluminum wire having a diameter 2.6 mm, calculate the ratio of ac to dc resistance at

(a) 10 MHz

(b) 2 GHz

ANSWER (a) 24.16, (b) 341.7.

10.7 POWER AND THE POYNTING VECTOR

As mentioned before, energy can be transported from one point (where a transmitter is located) to another point (with a receiver) by means of EM waves. The rate of such energy transportation can be obtained from Maxwell's equations:

$$\nabla \times \mathbf{E} = -\mu \frac{\partial \mathbf{H}}{\partial t} \tag{10.58a}$$

$$\nabla \times \mathbf{H} = \sigma \mathbf{E} + \varepsilon \frac{\partial \mathbf{E}}{\partial t} \tag{10.58b}$$

Dotting both sides of eq. (10.58b) with **E** gives

$$\mathbf{E} \cdot (\nabla \times \mathbf{H}) = \sigma E^2 + \mathbf{E} \cdot \varepsilon \frac{\partial \mathbf{E}}{\varepsilon t} \tag{10.59}$$

But for any vector fields **A** and **B** (see Appendix A.10)

$$\nabla \cdot (\mathbf{A} \times \mathbf{B}) = \mathbf{B} \cdot (\nabla \times \mathbf{A}) - \mathbf{A} \cdot (\nabla \times \mathbf{B}).$$

Applying this vector identity to eq. (10.59) (letting $\mathbf{A} = \mathbf{H}$ and $\mathbf{B} = \mathbf{E}$) gives

$$\mathbf{H} \cdot (\nabla \times \mathbf{E}) + \nabla \cdot (\mathbf{H} \times \mathbf{E}) = \sigma E^2 + \mathbf{E} \cdot \varepsilon \frac{\partial \mathbf{E}}{\partial t} \tag{10.60}$$

From eq. (10.58a),

$$\mathbf{H} \cdot (\nabla \times \mathbf{E}) = \mathbf{H} \cdot \left(-\mu \frac{\partial \mathbf{H}}{\partial t}\right) = -\frac{\mu}{2} \frac{\partial}{\partial t} (\mathbf{H} \cdot \mathbf{H}) \tag{10.61}$$

and thus eq. (10.60) becomes

$$-\frac{\mu}{2} \frac{\partial H^2}{\partial t} - \nabla \cdot (\mathbf{E} \times \mathbf{H}) = \sigma E^2 + \frac{1}{2} \varepsilon \frac{\partial E^2}{\partial t}$$

Rearranging terms and taking the volume integral of both sides,

$$\int_v \nabla \cdot (\mathbf{E} \times \mathbf{H}) \, dv = -\frac{\partial}{\partial t} \int_v \left[\frac{1}{2} \varepsilon E^2 + \frac{1}{2} \mu H^2\right] dv - \int_v \sigma E^2 \, dv \tag{10.62}$$

Applying the divergence theorem to the left-hand side gives

$$\oint_S (\mathbf{E} \times \mathbf{H}) \cdot d\mathbf{S} = -\frac{\partial}{\partial t} \int_v \left[\frac{1}{2} \varepsilon E^2 + \frac{1}{2} \mu H^2\right] dv - \int_v \sigma E^2 \, dv \tag{10.63}$$

$$\underset{\text{leaving the volume}}{\text{Total power}} = \underset{\text{and magnetic fields}}{\underset{\text{energy stored in electric}}{\text{Rate of decrease in}}} - \underset{\text{dissipated}}{\text{Ohmic power}} \tag{10.64}$$

Equation (10.63) is referred to as *Poynting's theorem*.[4] The various terms in the equation are identified using energy-conservation arguments for EM fields. The first term on the right-hand side of eq. (10.63) is interpreted as the rate of decrease in energy stored in the electric and magnetic fields. The second term is the power dissipated due to the fact that the medium is conducting ($\sigma \neq 0$). The quantity $\mathbf{E} \times \mathbf{H}$ on the left-hand side of eq. (10.63) is known as the *Poynting vector* $\mathcal{P}$ in watts per square meter (W/m^2); that is,

$$\boxed{\mathcal{P} = \mathbf{E} \times \mathbf{H}} \qquad [10.65]$$

It represents the instantaneous power density vector associated with the EM field at a given point. The integration of the Poynting vector over any closed surface gives the net power flowing out of that surface. Thus, Poynting's theorem can be stated as follows: The net power flowing out of a given volume v is equal to the time rate of decrease in the energy stored within v minus the conduction losses. The theorem is illustrated in Figure 10.10.

It should be noted that $\mathcal{P}$ is normal to both $\mathbf{E}$ and $\mathbf{H}$ and is therefore along the direction of wave propagation $\mathbf{a}_k$ for uniform plane waves. Thus

$$\mathbf{a}_k = \mathbf{a}_E \times \mathbf{a}_H \qquad [10.49]$$

The fact that $\mathcal{P}$ points along $\mathbf{a}_k$ causes $\mathcal{P}$ to be regarded derisively as a "pointing" vector.

Again, if we assume that

$$\mathbf{E}(z, t) = E_o e^{-\alpha z} \cos (\omega t - \beta z)\, \mathbf{a}_x$$

then

$$\mathbf{H}(z, t) = \frac{E_o}{|\eta|} e^{-\alpha z} \cos (\omega t - \beta z - \theta_\eta)\, \mathbf{a}_y$$

and

$$\begin{aligned} \mathcal{P}(z, t) &= \frac{E_o^2}{|\eta|} e^{-2\alpha z} \cos (\omega t - \beta z) \cos (\omega t - \beta z - \theta_\eta)\, \mathbf{a}_z \\ &= \frac{E_o^2}{2|\eta|} e^{-2\alpha z} \left[\cos \theta_\eta + \cos (2\omega t - 2\beta z - \theta_\eta)\right] \mathbf{a}_z \end{aligned} \qquad [10.66]$$

since $\cos A \cos B = \frac{1}{2}\left[\cos (A - B) + \cos (A + B)\right]$. To determine the time-average Poynting vector $\mathcal{P}_{ave}(z)$ (in W/m^2), which is of more practical value than the

....

[4]After J. H. Poynting, "On the transfer of energy in the electromagnetic field," *Phil. Trans.*, vol. 174, 1883, p. 343.

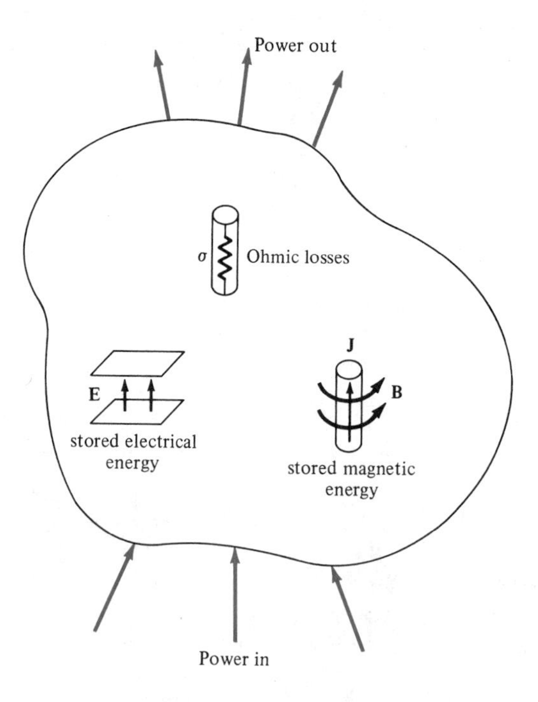

Figure 10.10 Illustration of power balance for EM fields.

instantaneous Poynting vector $\mathcal{P}(z, t)$, we integrate eq. (10.66) over the period $T = 2\pi/\omega$; that is,

$$\mathcal{P}_{\text{ave}}(z) = \frac{1}{T}\int_0^T \mathcal{P}(z, t)\, dt \qquad [10.67]$$

It can be shown (see Prob. 10.26) that this is equivalent to

$$\boxed{\mathcal{P}_{\text{ave}}(z) = \frac{1}{2}\,\text{Re}\,(\mathbf{E}_s \times \mathbf{H}_s^*)} \qquad [10.68]$$

By substituting eq. (10.66) into eq. (10.67), we obtain

$$\boxed{\mathcal{P}_{\text{ave}}(z) = \frac{E_o^2}{2|\eta|} e^{-2\alpha z} \cos\theta_\eta\, \mathbf{a}_z} \qquad [10.69]$$

The total time-average power crossing a given surface S is given by

$$P_{ave} = \int_S \mathcal{P}_{ave} \cdot d\mathbf{S} \qquad [10.70]$$

EXAMPLE 10.7

In a nonmagnetic medium

$$\mathbf{E} = 4 \sin (2\pi \times 10^7 t - 0.8x)\, \mathbf{a}_z \text{ V/m}$$

Find

(a) ε_r, η

(b) The time-average power carried by the wave

(c) The total power crossing 100 cm^2 of plane $2x + y = 5$

SOLUTION

(a) Since $\alpha = 0$ and $\beta \neq \omega/c$, the medium is not free space but a lossless medium.

$$\beta = 0.8, \quad \omega = 2\pi \times 10^7, \quad \mu = \mu_o \text{ (nonmagnetic)}, \quad \varepsilon = \varepsilon_o \varepsilon_r$$

Hence,

$$\beta = \omega\sqrt{\mu\varepsilon} = \omega\sqrt{\mu_o\varepsilon_o\varepsilon_r} = \frac{\omega}{c}\sqrt{\varepsilon_r}$$

or

$$\sqrt{\varepsilon_r} = \frac{\beta c}{\omega} = \frac{0.8\,(3 \times 10^8)}{2\pi \times 10^7} = \frac{12}{\pi}$$

$$\varepsilon_r = 14.59$$

$$\eta = \sqrt{\frac{\mu}{\varepsilon}} = \sqrt{\frac{\mu_o}{\varepsilon_o\varepsilon_r}} = \frac{120\pi}{\sqrt{\varepsilon_r}} = 120\pi \cdot \frac{\pi}{12} = 10\pi^2$$

$$= 98.7\ \Omega$$

(b) $$\mathcal{P} = \mathbf{E} \times \mathbf{H} = \frac{E_o^2}{\eta} \sin^2(\omega t - \beta x)\, \mathbf{a}_x$$

$$\mathcal{P}_{ave} = \frac{1}{T}\int_0^T \mathcal{P}\, dt = \frac{E_o^2}{2\eta}\mathbf{a}_x = \frac{16}{2 \times 10\pi^2}\mathbf{a}_x$$

$$= 81\, \mathbf{a}_x \text{ mW/m}^2$$

(c) On plane $2x + y = 5$ (see Example 3.5 or 8.5),

$$\mathbf{a}_n = \frac{2\mathbf{a}_x + \mathbf{a}_y}{\sqrt{5}}$$

Hence the total power is

$$P_{\text{ave}} = \int \mathcal{P}_{\text{ave}} \cdot d\mathbf{S} = \mathcal{P}_{\text{ave}} \cdot S\,\mathbf{a}_n$$

$$= (81 \times 10^{-3}\mathbf{a}_x) \cdot (100 \times 10^{-4}) \left[\frac{2\mathbf{a}_x + \mathbf{a}_y}{\sqrt{5}}\right]$$

$$= \frac{162 \times 10^{-5}}{\sqrt{5}} = 724.5\ \mu\text{W}$$

■

PRACTICE EXERCISE 10.7

In free space, $\mathbf{H} = 0.2 \cos(\omega t - \beta x)\,\mathbf{a}_z$ A/m. Find the total power passing through:

(a) A square plate of side 10 cm on plane $x + z = 1$

(b) A circular disc of radius 5 cm on plane $x = 1$.

ANSWER (a) 0, (b) 59.22 mW.

10.8 REFLECTION OF A PLANE WAVE AT NORMAL INCIDENCE

So far, we have considered uniform plane waves traveling in unbounded, homogeneous media. When a plane wave from one medium meets a different medium, it is partly reflected and partly transmitted. The proportion of the incident wave that is reflected or transmitted depends on the constitutive parameters (ε, μ, σ) of the two media involved. Here we will assume that the incident wave plane is normal to the boundary between the media; oblique incidence of plane waves will be covered in the next section after we understand the simpler case of normal incidence.

Suppose that a plane wave propagating along the $+z$-direction is incident normally on the boundary $z = 0$ between medium 1 ($z < 0$) characterized by σ_1, ε_1, μ_1 and medium 2 ($z > 0$) characterized by σ_2, ε_2, μ_2 as shown in Figure 10.11. In the figure, subscripts i, r, and t denote incident, reflected, and transmitted waves respectively. The incident, reflected, and transmitted waves shown in Figure 10.11 are obtained as follows:

Incident Wave: ($\mathbf{E}_i$, $\mathbf{H}_i$) is traveling along $+\mathbf{a}_z$ in medium 1. If we suppress the time factor $e^{j\omega t}$ and assume that

$$\mathbf{E}_{is}(z) = E_{io}e^{-\gamma_1 z}\,\mathbf{a}_x \qquad [10.71]$$

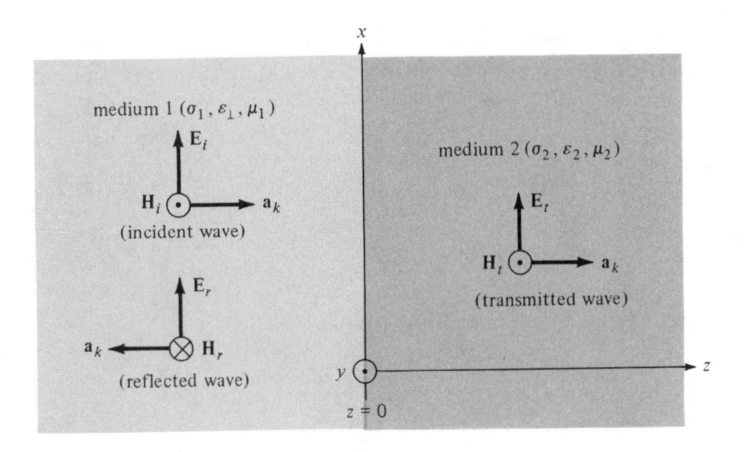

Figure 10.11 A plane wave incident normally on an interface between two different media.

then

$$\mathbf{H}_{is}(z) = H_{io}e^{-\gamma_1 z}\,\mathbf{a}_y = \frac{E_{io}}{\eta_1}\,e^{-\gamma_1 z}\,\mathbf{a}_y \tag{10.72}$$

Reflected Wave: ($\mathbf{E}_r$, $\mathbf{H}_r$) is traveling along $-\mathbf{a}_z$ in medium 1. If

$$\mathbf{E}_{rs}(z) = E_{ro}e^{\gamma_1 z}\,\mathbf{a}_x \tag{10.73}$$

then

$$\mathbf{H}_{rs}(z) = H_{ro}\,e^{\gamma_1 z}(-\mathbf{a}_y) = -\frac{E_{ro}}{\eta_1}\,e^{\gamma_1 z}\,\mathbf{a}_y \tag{10.74}$$

where $\mathbf{E}_{rs}$ has been assumed to be along $\mathbf{a}_x$; we will consistently assume that for normal incident $\mathbf{E}_i$, $\mathbf{E}_r$, and $\mathbf{E}_t$ have the same polarization.

Transmitted Wave: ($\mathbf{E}_t$, $\mathbf{H}_t$) is traveling along $+\mathbf{a}_z$ in medium 2. If

$$\mathbf{E}_{ts}(z) = E_{to}\,e^{-\gamma_2 z}\,\mathbf{a}_x \tag{10.75}$$

then

$$\mathbf{H}_{ts}(z) = H_{to}\,e^{-\gamma_2 z}\,\mathbf{a}_y = \frac{E_{to}}{\eta_2}\,e^{-\gamma_2 z}\,\mathbf{a}_y \tag{10.76}$$

In eqs. (10.71) to (10.76), E_{io}, E_{ro}, and E_{to} are respectively the magnitudes of the incident, reflected, and transmitted electric fields at $z = 0$.

Notice from Figure 10.11 that the total field in medium 1 comprises both the incident and reflected fields, whereas medium 2 has only the transmitted field, that is,

$$\mathbf{E}_1 = \mathbf{E}_i + \mathbf{E}_r, \qquad \mathbf{H}_1 = \mathbf{H}_i + \mathbf{H}_r$$

$$\mathbf{E}_2 = \mathbf{E}_t, \qquad \mathbf{H}_2 = \mathbf{H}_t$$

At the interface $z = 0$, the boundary conditions require that the tangential components of **E** and **H** fields must be continuous. Since the waves are transverse, **E** and **H** fields are entirely tangential to the interface. Hence at $z = 0$, $\mathbf{E}_{1\tan} = \mathbf{E}_{2\tan}$ and $\mathbf{H}_{1\tan} = \mathbf{H}_{2\tan}$ imply that

$$\mathbf{E}_i(0) + \mathbf{E}_r(0) = \mathbf{E}_t(0) \quad \rightarrow \quad E_{io} + E_{ro} = E_{to} \tag{10.77}$$

$$\mathbf{H}_i(0) + \mathbf{H}_r(0) = \mathbf{H}_t(0) \quad \rightarrow \quad \frac{1}{\eta_1}(E_{io} - E_{ro}) = \frac{E_{to}}{\eta_2} \tag{10.78}$$

From eqs. (10.77) and (10.78), we obtain

$$E_{ro} = \frac{\eta_2 - \eta_1}{\eta_2 + \eta_1} E_{io} \tag{10.79}$$

and

$$E_{to} = \frac{2\eta_2}{\eta_2 + \eta_1} E_{io} \tag{10.80}$$

We now define the *reflection coefficient* Γ and the *transmission coefficient* τ from eqs. (10.79) and (10.80) as

$$\boxed{\Gamma = \frac{E_{ro}}{E_{io}} = \frac{\eta_2 - \eta_1}{\eta_2 + \eta_1}} \tag{10.81}$$

and

$$\boxed{\tau = \frac{E_{to}}{E_{io}} = \frac{2\eta_2}{\eta_2 + \eta_1}} \tag{10.82}$$

Note that

1. $1 + \Gamma = \tau$ [10.83]
2. Both Γ and τ are dimensionless and may be complex.
3. $0 \leq |\Gamma| \leq 1$

The case considered above is the general case. Let us now consider a special case when medium 1 is a perfect dielectric (lossless, $\sigma_1 = 0$) and medium 2 is a perfect conductor ($\sigma_2 \simeq \infty$). For this case, $\eta_2 = 0$; hence, $\Gamma = -1$, and $\tau = 0$, showing that the wave is totally reflected. This should be expected because fields in a perfect conductor must vanish, so there can be no transmitted wave ($\mathbf{E}_2 = 0$). The totally reflected wave combines with the incident wave to form a *standing wave*. A standing wave ''stands'' and does not travel; it consists of two traveling waves ($\mathbf{E}_i$ and $\mathbf{E}_r$) of equal amplitudes but in opposite directions. Combining eqs. (10.71) and (10.73) gives the standing wave in medium 1 as

$$\mathbf{E}_{1s} = \mathbf{E}_{is} + \mathbf{E}_{rs} = (E_{io}e^{-\gamma_1 z} + E_{ro}e^{\gamma_1 z})\, \mathbf{a}_x \tag{10.84}$$

But

$$\Gamma = \frac{E_{ro}}{E_{io}} = -1,\ \sigma_1 = 0,\ \alpha_1 = 0,\ \gamma_1 = j\beta_1$$

Hence,

$$\mathbf{E}_{1s} = -E_{io}(e^{j\beta_1 z} - e^{-j\beta_1 z})\, \mathbf{a}_x$$

or

$$\mathbf{E}_{1s} = -2jE_{io} \sin \beta_1 z\, \mathbf{a}_x \tag{10.85}$$

Thus

$$\mathbf{E}_1 = \text{Re}\,(\mathbf{E}_{1s}e^{j\omega t})$$

or

$$\boxed{\mathbf{E}_1 = 2E_{io} \sin \beta_1 z \sin \omega t\, \mathbf{a}_x} \tag{10.86}$$

By taking similar steps, it can be shown that the magnetic field component of the wave is

$$\boxed{\mathbf{H}_1 = \frac{2E_{io}}{\eta_1} \cos \beta_1 z \cos \omega t\, \mathbf{a}_y} \tag{10.87}$$

A sketch of the standing wave in eq. (10.86) is presented in Figure 10.12 for $t = 0$, $T/8$, $T/4$, $3T/8$, $T/2$, and so on, where $T = 2\pi/\omega$. From the figure, we notice that the wave does not travel but oscillates.

When media 1 and 2 are both lossless we have another special case ($\sigma_1 = 0 = \sigma_2$). In this case, η_1 and η_2 are real and so are Γ and τ. Let us consider the following cases:

Case A: If $\eta_2 > \eta_1$, $\Gamma > 0$. Again there is a standing wave in medium 1 but there is also a transmitted wave in medium 2. However, the incident and reflected waves have amplitudes that are not equal in magnitude. It can be shown that the maximum values of $|\mathbf{E}_1|$ occur at

$$-\beta_1 z_{\max} = n\pi$$

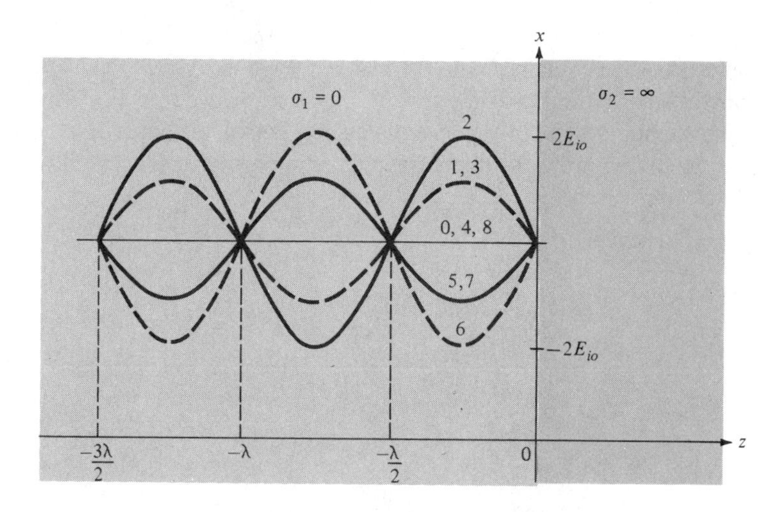

Figure 10.12 Standing waves $E = 2E_{io}\ sin\ \beta_1 z \sin \omega t\ \mathbf{a}_x$; curves 0, 1, 2, 3, 4, · · · are respectively at times $t = 0, T/8, T/4, 3T/8, T/2, \cdots; \lambda = 2\pi/\beta_1$.

or

$$z_{\max} = -\frac{n\pi}{\beta_1} = -\frac{n\lambda_1}{2}, \qquad n = 0, 1, 2, \ldots \tag{10.88}$$

and the minimum values of $|\mathbf{E}_1|$ occur at

$$-\beta_1 z_{\min} = (2n + 1)\frac{\pi}{2}$$

or

$$z_{\min} = -\frac{(2n + 1)\pi}{2\beta_1} = -\frac{(2n + 1)}{4}\lambda_1, \qquad n = 0, 1, 2, \ldots \tag{10.89}$$

Case B: If $\eta_2 < \eta_1$, $\Gamma < 0$. For this case, the locations of $|\mathbf{E}_1|$ maximum are given by eq. (10.89) whereas those of $|\mathbf{E}_1|$ minimum are given by eq. (10.88). All these are illustrated in Figure 10.13. Note that

1. $|\mathbf{H}_1|$ minimum occurs whenever there is $|\mathbf{E}_1|$ maximum and vice versa.
2. The transmitted wave (not shown in Figure 10.13) in medium 2 is a purely traveling wave and consequently there are no maxima or minima in this region.

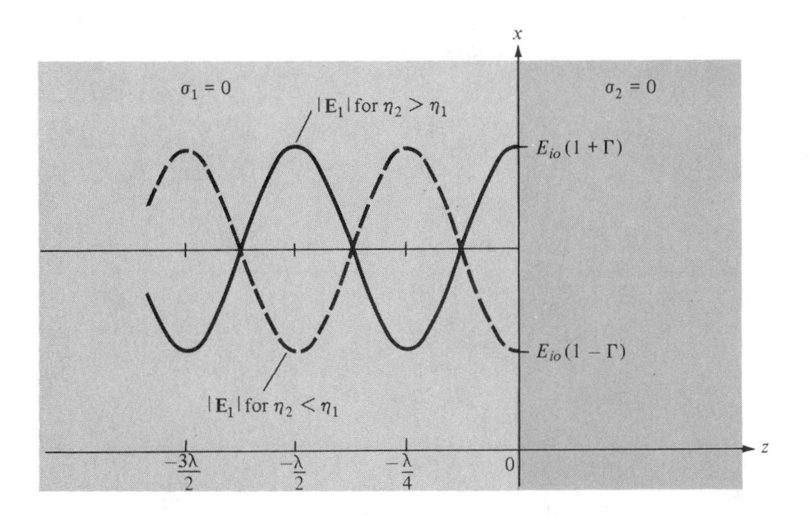

Figure 10.13 Standing waves due to reflection at an interface between two lossless media; $\lambda = 2\pi/\beta_1$.

The ratio of $|\mathbf{E}_1|_{max}$ to $|\mathbf{E}_1|_{min}$ (or $|\mathbf{H}_1|_{max}$ to $|\mathbf{H}_1|_{min}$) is called the *standing-wave ratio s;* that is,

$$s = \frac{|\mathbf{E}_1|_{max}}{|\mathbf{E}_1|_{min}} = \frac{|\mathbf{H}_1|_{max}}{|\mathbf{H}_1|_{min}} = \frac{1 + |\Gamma|}{1 - |\Gamma|} \quad [10.90]$$

or

$$|\Gamma| = \frac{s - 1}{s + 1} \quad [10.91]$$

Since $|\Gamma| \leq 1$, it follows that $1 \leq s \leq \infty$. The standing-wave ratio is dimensionless and it is customarily expressed in decibels (dB) as

$$s \text{ in dB} = 20 \log_{10} s \quad [10.92]$$

EXAMPLE 10.8

In free space ($z \leq 0$), a plane wave with

$$\mathbf{H} = 10 \cos (10^8 t - \beta z) \, \mathbf{a}_x \text{ mA/m}$$

is incident normally on a lossless medium ($\varepsilon = 2\varepsilon_o$, $\mu = 8\mu_o$) in region $z \geq 0$. Determine the reflected wave $\mathbf{H}_r$, $\mathbf{E}_r$ and the transmitted wave $\mathbf{H}_t$, $\mathbf{E}_t$.

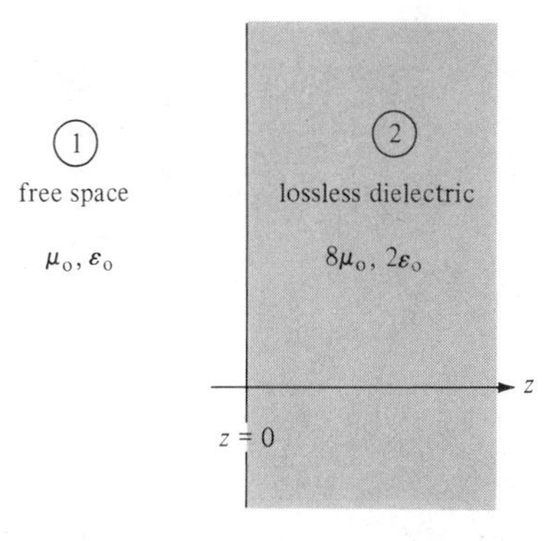

Figure 10.14 For Example 10.8.

SOLUTION This problem can be solved in two different ways.

Method 1: Consider the problem as illustrated in Figure 10.14. For free space,

$$\beta_1 = \frac{\omega}{c} = \frac{10^8}{3 \times 10^8} = \frac{1}{3}$$

$$\eta_1 = \eta_o = 120\pi$$

For the lossless dielectric medium,

$$\beta_2 = \omega\sqrt{\mu\varepsilon} = \omega\sqrt{\mu_o\varepsilon_o}\,\sqrt{\mu_r\varepsilon_r} = \frac{\omega}{c} \cdot (4) = 4\beta_1 = \frac{4}{3}$$

$$\eta_2 = \sqrt{\frac{\mu}{\varepsilon}} = \sqrt{\frac{\mu_o}{\varepsilon_o}}\,\sqrt{\frac{\mu_r}{\varepsilon_r}} = 2\,\eta_o$$

Given that $\mathbf{H}_i = 10 \cos (10^8 t - \beta_1 z)\, \mathbf{a}_x$, we expect that

$$\mathbf{E}_i = E_{io} \cos (10^8 t - \beta_1 z)\, \mathbf{a}_{E_i}$$

where

$$\mathbf{a}_{E_i} = \mathbf{a}_{H_i} \times \mathbf{a}_{k_i} = \mathbf{a}_x \times \mathbf{a}_z = -\mathbf{a}_y$$

and

$$E_{io} = \eta_1 H_{io} = 10\,\eta_o$$

Hence,

$$\mathbf{E}_i = -10\eta_o \cos (10^8 t - \beta_1 z)\, \mathbf{a}_y \text{ mV/m}$$

Now

$$\frac{E_{ro}}{E_{io}} = \Gamma = \frac{\eta_2 - \eta_1}{\eta_2 + \eta_1} = \frac{2\eta_o - \eta_o}{2\eta_o + \eta_o} = \frac{1}{3}$$

$$E_{ro} = \frac{1}{3} E_{io}$$

Thus

$$\mathbf{E}_r = -\frac{10}{3} \eta_o \cos\left(10^8 t + \frac{1}{3} z\right) \mathbf{a}_y \text{ mV/m}$$

from which we easily obtain $\mathbf{H}_r$ as

$$\mathbf{H}_r = -\frac{10}{3} \cos\left(10^8 t + \frac{1}{3} z\right) \mathbf{a}_x \text{ mA/m}$$

Similarly,

$$\frac{E_{to}}{E_{io}} = \tau = 1 + \Gamma = \frac{4}{3} \qquad \text{or} \qquad E_{to} = \frac{4}{3} E_{io}$$

Thus

$$\mathbf{E}_t = E_{to} \cos(10^8 t - \beta_2 z)\, \mathbf{a}_{E_t}$$

where $\mathbf{a}_{E_t} = \mathbf{a}_{E_i} = -\mathbf{a}_y$. Hence,

$$\mathbf{E}_t = -\frac{40}{3} \eta_o \cos\left(10^8 t - \frac{4}{3} z\right) \mathbf{a}_y \text{ mV/m}$$

from which we obtain

$$\mathbf{H}_t = \frac{20}{3} \cos\left(10^8 t - \frac{4}{3} z\right) \mathbf{a}_x \text{ mA/m}$$

Method 2: Alternatively, we can obtain $\mathbf{H}_r$ and $\mathbf{H}_t$ directly from $\mathbf{H}_i$ using (see Problem 10.28)

$$\frac{H_{ro}}{H_{io}} = -\Gamma \qquad \text{and} \qquad \frac{H_{to}}{H_{io}} = \tau \frac{\eta_1}{\eta_2}$$

Thus

$$H_{ro} = -\frac{1}{3} H_{io} = -\frac{10}{3}$$

$$H_{to} = \frac{4}{3} \frac{\eta_o}{2\eta_o} \cdot H_{io} = \frac{2}{3} H_{io} = \frac{20}{3}$$

and

$$\mathbf{H}_r = -\frac{10}{3} \cos (10^8 t + \beta_1 z)\, \mathbf{a}_x \text{ mA/m}$$

$$\mathbf{H}_t = \frac{20}{3} \cos (10^8 t - \beta_2 z)\, \mathbf{a}_x \text{ mA/m}$$

as previously obtained.

Notice that the boundary conditions at $z = 0$, namely,

$$\mathbf{E}_i(0) + \mathbf{E}_r(0) = \mathbf{E}_t(0) = -\frac{40}{3} \eta_o \cos (10^8 t)\, \mathbf{a}_y$$

and

$$\mathbf{H}_i(0) + \mathbf{H}_r(0) = \mathbf{H}_t(0) = \frac{20}{3} \cos (10^8 t)\, \mathbf{a}_x$$

are satisfied. These boundary conditions can always be used to cross-check **E** and **H**. ■

PRACTICE EXERCISE 10.8

A 5 GHz uniform plane wave $\mathbf{E}_{is} = 10\, e^{-j\beta z}\, \mathbf{a}_x$ V/m in free space is incident normally on a large plane, lossless dielectric slab ($z > 0$) having $\varepsilon = 4\varepsilon_o$, $\mu = \mu_o$. Find the reflected wave $\mathbf{E}_{rs}$ and the transmitted wave $\mathbf{E}_{ts}$

ANSWER $-3.333 \exp(j\beta_1 z)\, \mathbf{a}_x$ V/m, $6.667 \exp(-j\beta_2 z)\, \mathbf{a}_x$ V/m where $\beta_2 = 2\beta_1 = 200\pi/3$.

EXAMPLE 10.9

Given a uniform plane wave in air as

$$\mathbf{E}_i = 40 \cos (\omega t - \beta z)\, \mathbf{a}_x + 30 \sin (\omega t - \beta z)\, \mathbf{a}_y \text{ V/m}$$

(a) Find $\mathbf{H}_i$.

(b) If the wave encounters a perfectly conducting plate normal to the z-axis at $z = 0$, find the reflected wave $\mathbf{E}_r$ and $\mathbf{H}_r$.

(c) What are the total **E** and **H** fields for $z \leq 0$?

(d) Calculate the time-average Poynting vectors for $z \leq 0$ and $z \geq 0$.

SOLUTION (a) This is similar to the problem in Example 10.3. We may treat the wave as consisting of two waves $\mathbf{E}_{i1}$ and $\mathbf{E}_{i2}$, where

$$\mathbf{E}_{i1} = 40 \cos (\omega t - \beta z)\, \mathbf{a}_x, \qquad \mathbf{E}_{i2} = 30 \sin (\omega t - \beta z)\, \mathbf{a}_y$$

At atmospheric pressure, air has $\varepsilon_r = 1.0006 \simeq 1$. Thus air may be regarded as free space. Let $\mathbf{H}_i = \mathbf{H}_{i1} + \mathbf{H}_{i2}$.

$$\mathbf{H}_{i1} = H_{i1o} \cos(\omega t - \beta z)\, \mathbf{a}_{H_1}$$

where

$$H_{i1o} = \frac{E_{i1o}}{\eta_o} = \frac{40}{120\pi} = \frac{1}{3\pi}$$

$$\mathbf{a}_{H_1} = \mathbf{a}_k \times \mathbf{a}_E = \mathbf{a}_z \times \mathbf{a}_x = \mathbf{a}_y$$

Hence

$$\mathbf{H}_{i1} = \frac{1}{3\pi} \cos(\omega t - \beta z)\, \mathbf{a}_y$$

Similarly,

$$\mathbf{H}_{i2} = H_{i2o} \sin(\omega t - \beta z)\, \mathbf{a}_{H_2}$$

where

$$H_{i2o} = \frac{E_{i2o}}{\eta_o} = \frac{30}{120\pi} = \frac{1}{4\pi}$$

$$\mathbf{a}_{H_2} = \mathbf{a}_k \times \mathbf{a}_E = \mathbf{a}_z \times \mathbf{a}_y = -\mathbf{a}_x$$

Hence

$$\mathbf{H}_{i2} = -\frac{1}{4\pi} \sin(\omega t - \beta z)\, \mathbf{a}_x$$

and

$$\begin{aligned}\mathbf{H}_i &= \mathbf{H}_{i1} + \mathbf{H}_{i2} \\ &= -\frac{1}{4\pi} \sin(\omega t - \beta z)\, \mathbf{a}_x + \frac{1}{3\pi} \cos(\omega t - \beta z)\, \mathbf{a}_y \text{ mA/m}\end{aligned}$$

This problem can also be solved using Method 2 of Example 10.3.

(b) Since medium 2 is perfectly conducting,

$$\frac{\sigma_2}{\omega \varepsilon_2} \gg 1 \quad \rightarrow \quad \eta_2 \ll \eta_1$$

that is,

$$\Gamma \simeq -1, \qquad \tau = 0$$

showing that the incident **E** and **H** fields are totally reflected.

$$E_{ro} = \Gamma\, E_{io} = -E_{io}$$

Hence,

$$\mathbf{E}_r = -40 \cos (\omega t + \beta z)\, \mathbf{a}_x - 30 \sin (\omega t + \beta z)\, \mathbf{a}_y \text{ V/m}$$

$\mathbf{H}_r$ can be found from $\mathbf{E}_r$ just as we did in part (a) of this example or by using Method 2 of the last example starting with $\mathbf{H}_i$. Whichever approach is taken, we obtain

$$\mathbf{H}_r = \frac{1}{3\pi} \cos (\omega t + \beta z)\, \mathbf{a}_y - \frac{1}{4\pi} \sin (\omega t + \beta z)\mathbf{a}_x \text{ A/m}$$

(c) The total fields in air

$$\mathbf{E}_1 = \mathbf{E}_i + \mathbf{E}_r \qquad \text{and} \qquad \mathbf{H}_1 = \mathbf{H}_i + \mathbf{H}_r$$

can be shown to be standing wave (see Problem 10.39). The total fields in the conductor are

$$\mathbf{E}_2 = \mathbf{E}_t = 0, \qquad \mathbf{H}_2 = \mathbf{H}_t = 0.$$

(d) For $z \leq 0$,

$$\mathcal{P}_{1\text{ave}} = \frac{|\mathbf{E}_{1s}|^2}{2\eta_1}\, \mathbf{a}_k = \frac{1}{2\eta_o} [E_{io}^2 \mathbf{a}_z - E_{ro}^2 \mathbf{a}_z]$$

$$= \frac{1}{240\pi} [(40^2 + 30^2)\mathbf{a}_z - (40^2 + 30^2)\mathbf{a}_z]$$

$$= 0$$

For $z \geq 0$,

$$\mathcal{P}_{2\text{ave}} = \frac{|\mathbf{E}_{2s}|^2}{2\eta_2}\, \mathbf{a}_k = \frac{E_{to}^2}{2\eta_2}\, \mathbf{a}_z = 0$$

because the whole incident power is reflected. ■

PRACTICE EXERCISE 10.9

The plane wave $\mathbf{E} = 50 \sin (\omega t - 5x)\, \mathbf{a}_y$ V/m in a lossless medium ($\mu = 4\mu_o$, $\varepsilon = \varepsilon_o$) encounters a lossy medium ($\mu = \mu_o$, $\varepsilon = 4\varepsilon_o$, $\sigma = 0.1$ mhos/m) normal to the x-axis at $x = 0$. Find

(a) Γ, τ, and s

(b) $\mathbf{E}_r$ and $\mathbf{H}_r$

(c) $\mathbf{E}_t$ and $\mathbf{H}_t$

(d) the time-average Poynting vectors in both regions

ANSWER (a) $0.8186 \angle 171.1°$, $0.2295 \angle 33.56°$, 10.025, (b) $40.93 \sin(\omega t + 5x + 171.9°)$ $\mathbf{a}_y$ V/m, $-54.3 \sin(\omega t + 5x + 171.9°$ $\mathbf{a}_z$ mA/m, (c) $11.47\, e^{-6.021x}$ $\sin(\omega t - 7.826x + 33.56°)$ $\mathbf{a}_y$ V/m, $120.2\, e^{-6.021x} \sin(\omega t - 7.826x - 4.01°)$ $\mathbf{a}_z$ mA/m, (d) $0.5469\, \mathbf{a}_x$ W/m^2, $0.5469 \exp(-12.04x)\, \mathbf{a}_x$ W/m^2.

†10.9 REFLECTION OF A PLANE WAVE AT OBLIQUE INCIDENCE

We now consider a more general situation than that in Section 10.8. To simplify the analysis, we will assume that we are dealing with lossless media. (We may extend our analysis to that of lossy media by merely replacing ε by ε_c.) It can be shown (see Problems 10.13 and 10.14) that a uniform plane wave takes the general form of

$$\begin{aligned} \mathbf{E}(\mathbf{r}, t) &= \mathbf{E}_o \cos(\mathbf{k} \cdot \mathbf{r} - \omega t) \\ &= \text{Re}\left[E_o e^{j(\mathbf{k} \cdot \mathbf{r} - \omega t)}\right] \end{aligned} \quad [10.93]$$

where $\mathbf{r} = x\mathbf{a}_x + y\mathbf{a}_y + z\mathbf{a}_z$ is the radius or position vector and $\mathbf{k} = k_x\mathbf{a}_x + k_y\mathbf{a}_y + k_z\mathbf{a}_z$ is the *wave number vector* or the *propagation vector;* $\mathbf{k}$ is always in the direction of wave propagation. The magnitude of $\mathbf{k}$ is related to ω according to the dispersion relation

$$k^2 = k_x^2 + k_y^2 + k_z^2 = \omega^2 \mu \varepsilon \quad [10.94]$$

Thus, for lossless media, k is essentially the same as β in the previous sections. With the general form of $\mathbf{E}$ as in eq. (10.93), Maxwell's equations reduce to

$$\mathbf{k} \times \mathbf{E} = \omega \mu \mathbf{H} \quad [10.95a]$$

$$\mathbf{k} \times \mathbf{H} = -\omega \varepsilon \mathbf{E} \quad [10.95b]$$

$$\mathbf{k} \cdot \mathbf{H} = 0 \quad [10.95c]$$

$$\mathbf{k} \cdot \mathbf{E} = 0 \quad [10.95d]$$

showing that (i) $\mathbf{E}$, $\mathbf{H}$, and $\mathbf{k}$ are mutually orthogonal, and (ii) $\mathbf{E}$ and $\mathbf{H}$ lie on the plane

$$\mathbf{k} \cdot \mathbf{r} = k_x x + k_y y + k_z z = \text{constant}$$

From eq. (10.95a), the $\mathbf{H}$ field corresponding to the $\mathbf{E}$ field in eq. (10.93) is

$$\mathbf{H} = \frac{1}{\omega \mu} \mathbf{k} \times \mathbf{E} = \frac{\mathbf{a}_k \times \mathbf{E}}{\eta} \quad [10.96]$$

Having expressed $\mathbf{E}$ and $\mathbf{H}$ in the general form, we can now consider the oblique incidence of a uniform plane wave at a plane boundary as illustrated in Figure 10.15(a). The plane defined by the propagation vector $\mathbf{k}$ and a unit normal vector $\mathbf{a}_n$ to the boundary is called the *plane of incidence*. The angle θ_i between $\mathbf{k}$ and $\mathbf{a}_n$ is the *angle of incidence*.

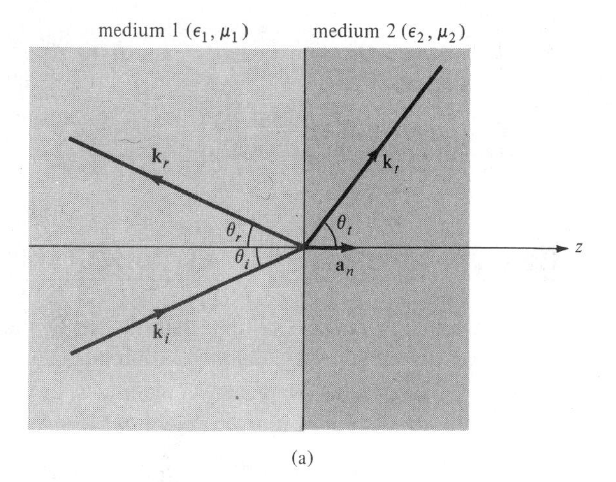

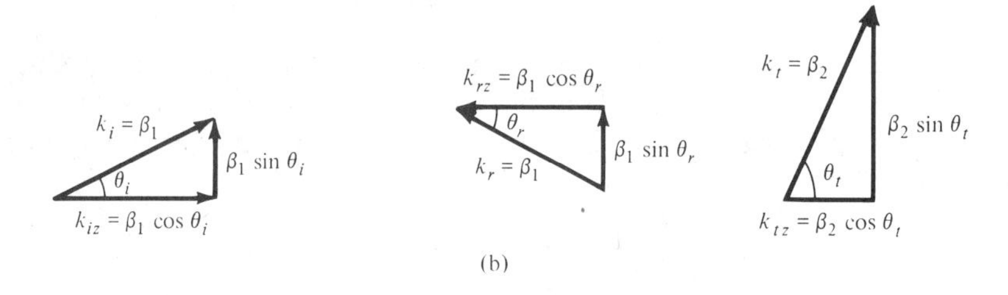

Figure 10.15 Oblique incidence of a plane wave: **(a)** illustration of θ_i, θ_r, and θ_t; **(b)** illustration of the normal and tangential components of $\mathbf{k}$.

Again, both the incident and the reflected waves are in medium 1 while the transmitted (or refracted wave) is in medium 2. Let

$$\mathbf{E}_i = \mathbf{E}_{io} \cos (k_{ix}x + k_{iy}y + k_{iz}z - \omega_i t) \quad [10.97a]$$

$$\mathbf{E}_r = \mathbf{E}_{ro} \cos (k_{rx}x + k_{ry}y + k_{rz}z - \omega_r t) \quad [10.97b]$$

$$\mathbf{E}_t = \mathbf{E}_{to} \cos (k_{tx}x + k_{ty}y + k_{tz}z - \omega_t t) \quad [10.97c]$$

where k_i, k_r, and k_t with their normal and tangential components are shown in Figure 10.15(b). Since the tangential component of $\mathbf{E}$ must be continuous at the boundary $z = 0$,

$$\mathbf{E}_i(z = 0) + \mathbf{E}_r(z = 0) = \mathbf{E}_t(z = 0) \quad [10.98]$$

The only way this boundary condition will be satisfied by the waves in eq. (10.97) for all x and y is that

1. $\omega_i = \omega_r = \omega_t = \omega$
2. $k_{ix} = k_{rx} = k_{tx} = k_x$
3. $k_{iy} = k_{ry} = k_{ty} = k_y$

Condition 1 implies that the frequency is unchanged. Conditions 2 and 3 require that the tangential components of the propagation vectors be continuous (called the *phase matching conditions*). This means that the propagation vectors $\mathbf{k}_i$, $\mathbf{k}_t$, and $\mathbf{k}_r$ must all lie in the plane of incidence. Thus, by conditions 2 and 3,

$$k_i \sin \theta_i = k_r \sin \theta_r \tag{10.99}$$

$$k_i \sin \theta_i = k_t \sin \theta_t \tag{10.100}$$

where θ_r is *the angle of reflection* and θ_t is the *angle of transmission*. But for lossless media,

$$k_i = k_r = \beta_1 = \omega\sqrt{\mu_1\varepsilon_1} \tag{10.101a}$$

$$k_t = \beta_2 = \omega\sqrt{\mu_2\varepsilon_2} \tag{10.101b}$$

From eqs. (10.99) and (10.101a), it is clear that

$$\boxed{\theta_r = \theta_i} \tag{10.102}$$

so that the angle of reflection θ_r equals the angle of incidence θ_i as in optics. Also from eqs. (10.100) and (10.101),

$$\frac{\sin \theta_t}{\sin \theta_i} = \frac{k_i}{k_t} = \frac{u_2}{u_1} = \sqrt{\frac{\mu_1\varepsilon_1}{\mu_2\varepsilon_2}} \tag{10.103}$$

where $u = \omega/k$ is the phase velocity. Equation (10.103) is the well-known *Snell's law*, which can be written as

$$\boxed{n_1 \sin \theta_i = n_2 \sin \theta_t} \tag{10.104}$$

where $n_1 = c\sqrt{\mu_1\varepsilon_1} = c/u_1$ and $n_2 = c\sqrt{\mu_2\varepsilon_2} = c/u_2$ are the *refractive indices* of the media.

Based on these general preliminaries on oblique incidence, we will now specifically consider two special cases: one with the **E** field perpendicular to the plane of incidence, the other with the **E** field parallel to it. Any other polarization may be considered as a linear combination of these two cases.

A. Parallel Polarization

This case is illustrated in Figure 10.16 where the **E** field lies in the xz-plane, the plane of incidence. In medium 1, we have both incident and reflected fields given by

$$\mathbf{E}_{is} = E_{io}(\cos \theta_i \, \mathbf{a}_x - \sin \theta_i \, \mathbf{a}_z) \, e^{-j\beta_1(x \sin \theta_i + z \cos \theta_i)} \tag{10.105a}$$

$$\mathbf{H}_{is} = \frac{E_{io}}{\eta_1} e^{-j\beta_1(x \sin \theta_i + z \cos \theta_i)} \, \mathbf{a}_y \tag{10.105b}$$

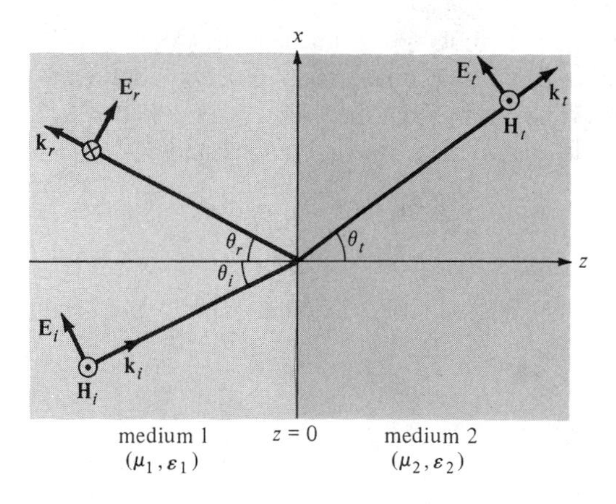

Figure 10.16 Oblique incidence with **E** parallel to the plane of incidence.

$$\mathbf{E}_{rs} = E_{ro}(\cos\theta_r\,\mathbf{a}_x + \sin\theta_r\,\mathbf{a}_z)\,e^{-j\beta_1(x\sin\theta_r - z\cos\theta_r)} \tag{10.106a}$$

$$\mathbf{H}_{rs} = -\frac{E_{ro}}{\eta_1}\,e^{-j\beta_1(x\sin\theta_i - z\cos\theta_i)}\,\mathbf{a}_y \tag{10.106b}$$

where $\beta_1 = \omega\sqrt{\mu_1\varepsilon_1}$. Notice carefully how we arrive at each field component. The trick in deriving the components is to first get the polarization vector **k** as shown in Figure 10.15(b) for incident, reflected, and transmitted waves. Once **k** is known, we define $\mathbf{E}_s$ such that $\nabla \cdot \mathbf{E}_s = 0$ or $\mathbf{k} \cdot \mathbf{E}_s = 0$ and then $\mathbf{H}_s$ is obtained from $\mathbf{H}_s = \dfrac{\mathbf{k}}{\omega\mu} \times \mathbf{E}_s = \mathbf{a}_k \times \dfrac{\mathbf{E}}{\eta}$.

The transmitted fields exist in medium 2 and are given by

$$\mathbf{E}_{ts} = E_{to}(\cos\theta_t\,\mathbf{a}_x - \sin\theta_t\,\mathbf{a}_z)\,e^{-j\beta_2(x\sin\theta_t + z\cos\theta_t)} \tag{10.107a}$$

$$\mathbf{H}_{ts} = \frac{E_{to}}{\eta_2}\,e^{-j\beta_2(x\sin\theta_t + z\cos\theta_t)}\,\mathbf{a}_y \tag{10.107b}$$

where $\beta_2 = \omega\sqrt{\mu_2\varepsilon_2}$. Should our assumption about the relative directions in eqs. (10.105) to (10.107) be wrong, the final result will show us by means of its sign.

Requiring that $\theta_r = \theta_i$ and that the tangential components of **E** and **H** be continuous at the boundary $z = 0$, we obtain

$$(E_{io} + E_{ro})\cos\theta_i = E_{to}\cos\theta_t \tag{10.108a}$$

$$\frac{1}{\eta_1}(E_{io} - E_{ro}) = \frac{1}{\eta_2}E_{to} \tag{10.108b}$$

Expressing E_{ro} and E_{to} in terms of E_{io}, we obtain

$$\boxed{\Gamma_{\parallel} = \frac{E_{ro}}{E_{io}} = \frac{\eta_2 \cos \theta_t - \eta_1 \cos \theta_i}{\eta_2 \cos \theta_t + \eta_1 \cos \theta_i}} \quad [10.109]$$

and

$$\boxed{\tau_{\parallel} = \frac{E_{to}}{E_{io}} = \frac{2\eta_2 \cos \theta_i}{\eta_2 \cos \theta_t + \eta_1 \cos \theta_i}} \quad [10.110]$$

Equations (10.109) and (10.110) are called *Fresnel's equations*. Note that the equations reduce to eqs. (10.81) and (10.82) when $\theta_i = \theta_t = 0$ as expected. Since θ_i and θ_t are related according to Snell's law of eq. (10.103), eqs. (10.109) and (10.110) can be written in terms of θ_i by substituting

$$\cos \theta_t = \sqrt{1 - \sin^2\theta_t} = \sqrt{1 - (u_2/u_1)^2 \sin^2\theta_i} \quad [10.111]$$

From eqs. (10.109) and (10.110), it is easily shown that

$$1 + \Gamma_{\parallel} = \tau_{\parallel}\left(\frac{\cos \theta_t}{\cos \theta_i}\right) \quad [10.112]$$

From eq. (10.109), it is evident that it is possible that $\Gamma_{\parallel} = 0$ because the numerator is the difference of two terms. Under this condition, there is no reflection ($E_{ro} = 0$) and the incident angle at which this takes place is called the *Brewster angle* $\theta_{B_\parallel}$. The Brewster angle is also known as the *polarizing angle* because an arbitrarily polarized incident wave will be reflected with only the component of **E** perpendicular to the plane of incidence. The Brewster effect is utilized in a laser tube where quartz windows are set at the Brewster angle to control polarization of emitted light. The Brewster angle is obtained by setting $\theta_i = \theta_{B_\parallel}$ when $\Gamma_{\parallel} = 0$ in eq. (10.109), that is,

$$\eta_2 \cos \theta_t = \eta_1 \cos \theta_{B_\parallel}$$

or

$$\eta_2^2 (1 - \sin^2\theta_t) = \eta_1^2 (1 - \sin^2\theta_{B_\parallel})$$

Introducing eq. (10.103) or (10.104) gives

$$\boxed{\sin^2\theta_{B_\parallel} = \frac{1 - \mu_2\varepsilon_1/\mu_1\varepsilon_2}{1 - (\varepsilon_1/\varepsilon_2)^2}} \quad [10.113]$$

It is of practical value to consider the case when the dielectric media are not only lossless but nonmagnetic as well—that is, $\mu_1 = \mu_2 = \mu_o$. For this situation, eq. (10.113) becomes

$$\sin^2\theta_{B_\parallel} = \frac{1}{1 + \varepsilon_1/\varepsilon_2} \rightarrow \sin\theta_{B_\parallel} = \sqrt{\frac{\varepsilon_2}{\varepsilon_1 + \varepsilon_2}}$$

or

$$\tan\theta_{B_\parallel} = \sqrt{\frac{\varepsilon_2}{\varepsilon_1}} = \frac{n_2}{n_1} \qquad [10.114]$$

showing that there is a Brewster angle for any combination of ε_1 and ε_2.

B. Perpendicular Polarization

In this case, the **E** field is perpendicular to the plane of incidence (the *xz*-plane) as shown in Figure 10.17. This may also be viewed as the case where **H** field is parallel to the plane of incidence. The incident and reflected fields in medium 1 are given by

$$\mathbf{E}_{is} = E_{io}e^{-j\beta_1(x\sin\theta_i + z\cos\theta_i)}\,\mathbf{a}_y \qquad [10.115a]$$

$$\mathbf{H}_{is} = \frac{E_{io}}{\eta_1}(-\cos\theta_i\,\mathbf{a}_x + \sin\theta_i\,\mathbf{a}_z)\,e^{-j\beta_1(x\sin\theta_i + z\cos\theta_i)} \qquad [10.115b]$$

$$\mathbf{E}_{rs} = E_{ro}e^{-j\beta_1(x\sin\theta_r - z\cos\theta_r)}\,\mathbf{a}_y \qquad [10.116a]$$

$$\mathbf{H}_{rs} = \frac{E_{ro}}{\eta_1}(\cos\theta_r\,\mathbf{a}_x + \sin\theta_r\,\mathbf{a}_z)\,e^{-j\beta_1(x\sin\theta_r - z\cos\theta_r)} \qquad [10.116b]$$

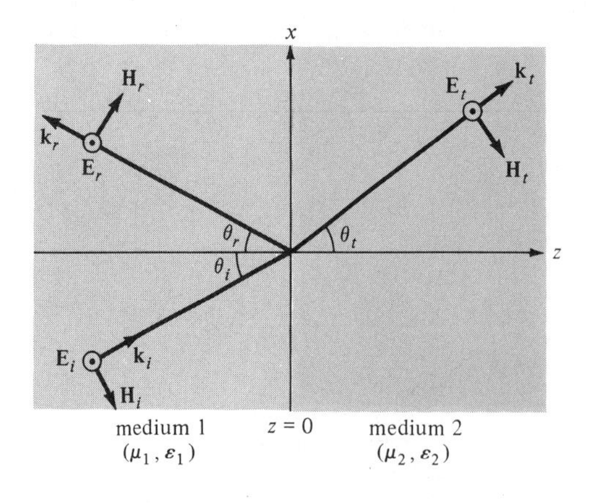

Figure 10.17 Oblique incidence with **E** perpendicular to the plane of incidence.

while the transmitted fields in medium 2 are given by

$$\mathbf{E}_{ts} = E_{to} e^{-j\beta_2(x \sin \theta_t + z \cos \theta_t)} \, \mathbf{a}_y \qquad [10.117a]$$

$$\mathbf{H}_{ts} = \frac{E_{to}}{\eta_2} (-\cos \theta_t \, \mathbf{a}_x + \sin \theta_t \, \mathbf{a}_z) \, e^{-j\beta_2(x \sin \theta_t + z \cos \theta_t)} \qquad [10.117b]$$

Notice that in defining the field components in eqs. (10.115) to (10.117), Maxwell's equations (10.95) are always satisfied. Again, requiring that the tangential components of **E** and **H** be continuous at $z = 0$ and setting θ_r equal to θ_i, we get

$$E_{io} + E_{ro} = E_{to} \qquad [10.118a]$$

$$\frac{1}{\eta_1} (E_{io} - E_{ro}) \cos \theta_i = \frac{1}{\eta_2} E_{to} \cos \theta_t \qquad [10.118b]$$

Expressing E_{ro} and E_{to} in terms of E_{io} leads to

$$\boxed{\Gamma_\perp = \frac{E_{ro}}{E_{io}} = \frac{\eta_2 \cos \theta_i - \eta_1 \cos \theta_t}{\eta_2 \cos \theta_i + \eta_1 \cos \theta_t}} \qquad [10.119]$$

and

$$\boxed{\tau_\perp = \frac{E_{to}}{E_{io}} = \frac{2\eta_2 \cos \theta_i}{\eta_2 \cos \theta_i + \eta_1 \cos \theta_t}} \qquad [10.120]$$

which are the *Fresnel's equations* for perpendicular polarization. From eqs. (10.119) and (10.120), it is easy to show that

$$1 + \Gamma_\perp = \tau_\perp \qquad [10.121]$$

which is similar to eq. (10.83) for normal incidence. Also, when $\theta_i = \theta_t = 0$, eqs. (10.119) and (10.120) become eqs. (10.81) and (10.82) as they should.

For no reflection, $\Gamma_\perp = 0$ (or $E_r = 0$). This is the same as the case of total transmission ($\tau_\perp = 1$). By replacing θ_i with the corresponding Brewster angle $\theta_{B\perp}$, we obtain

$$\eta_2 \cos \theta_{B\perp} = \eta_1 \cos \theta_t$$

or

$$\eta_2^2 \, (1 - \sin^2\theta_{B\perp}) = \eta_1^2 \, (1 - \sin^2\theta_t)$$

Incorporating eq. (10.104) yields

$$\sin^2\theta_{B\perp} = \frac{1 - \mu_1\varepsilon_2/\mu_2\varepsilon_1}{1 - (\mu_1/\mu_2)^2} \qquad [10.122]$$

Note that for nonmagnetic media ($\mu_1 = \mu_2 = \mu_o$), $\sin^2\theta_{B\perp} \rightarrow \infty$ in eq. (10.122), so $\theta_{B\perp}$ does not exist because the sine of an angle is never greater than unity. Also if $\mu_1 \neq \mu_2$ and $\varepsilon_1 = \varepsilon_2$, eq. (10.122) reduces to

$$\sin \theta_{B\perp} = \sqrt{\frac{\mu_2}{\mu_1 + \mu_2}}$$

or

$$\tan \theta_{B\perp} = \sqrt{\frac{\mu_2}{\mu_1}} \tag{10.123}$$

Although this situation is theoretically possible, it is rare in practice.

EXAMPLE 10.10

An EM wave travels in free space with the electric field component

$$\mathbf{E}_s = 100\, e^{j(0.866y + 0.5z)}\, \mathbf{a}_x \text{ V/m}$$

Determine

(a) ω and λ

(b) The magnetic field component

(c) The time average power in the wave

SOLUTION

(a) Comparing the given **E** with

$$\mathbf{E}_s = \mathbf{E}_o\, e^{j\mathbf{k} \cdot \mathbf{r}} = E_o e^{j(k_x x + k_y y + k_z z)}\, \mathbf{a}_x$$

it is clear that

$$k_x = 0, \qquad k_y = 0.866, \qquad k_z = 0.5$$

Thus

$$k = \sqrt{k_x^2 + k_y^2 + k_z^2} = \sqrt{(0.866)^2 + (0.5)^2} = 1$$

But in free space,

$$k = \beta = \omega \sqrt{\mu_o \varepsilon_o} = \frac{\omega}{c} = \frac{2\pi}{\lambda}$$

Hence,

$$\omega = kc = 3 \times 10^8 \text{ rad/s}$$

$$\lambda = \frac{2\pi}{k} = 2\pi = 6.283\ m$$

(b) From eq. (10.96), the corresponding magnetic field is given by

$$\mathbf{H}_s = \frac{1}{\mu\omega}\,\mathbf{k} \times \mathbf{E}_s$$

$$= \frac{(0.866\mathbf{a}_y + 0.5\mathbf{a}_z)}{4\pi \times 10^{-7} \times 3 \times 10^8} \times 100\ \mathbf{a}_x e^{j\mathbf{k}\cdot\mathbf{r}}$$

or

$$\mathbf{H}_s = (1.33\ \mathbf{a}_y - 2.3\ \mathbf{a}_z)\ e^{j(0.866y + 0.5z)}\ \text{mA/m}$$

(c) The time average power is

$$\mathcal{P}_{\text{ave}} = \frac{1}{2}\,\text{Re}\,(\mathbf{E}_s \times \mathbf{H}_s^*) = \frac{E_o^2}{2\,\eta}\,\mathbf{a}_k$$

$$= \frac{(100)^2}{2(120\pi)}\,(0.866\ \mathbf{a}_y + 0.5\ \mathbf{a}_z)$$

$$= 11.49\ \mathbf{a}_y + 6.631\ \mathbf{a}_z\ \text{W/m}^2$$

■

PRACTICE EXERCISE 10.10

Rework Example 10.10 if

$\mathbf{E} = (10\ \mathbf{a}_y + 5\mathbf{a}_z)\cos(\omega t + 2y - 4z)$ V/m

in free space.

ANSWER (a) 1.342×10^9 rad/s, 1.405 m, (b) $-29.66 \cos(1.342 \times 10^9 t + 2y - 4z)\ \mathbf{a}_x$ mA/m, (c) $-0.07415\ \mathbf{a}_y + 0.1489\ \mathbf{a}_z$ W/m^2.

EXAMPLE 10.11

A uniform plane wave in air with

$$\mathbf{E} = 8\cos(\omega t - 4x - 3z)\ \mathbf{a}_y\ \text{V/m}$$

is incident on a dielectric slab ($z \geq 0$) with $\mu_r = 1.0$, $\varepsilon_r = 2.5$, $\sigma = 0$. Find

(a) The polarization of the wave

(b) The angle of incidence

(c) The reflected **E** field

(d) The transmitted **H** field

SOLUTION (a) From the incident **E** field, it is evident that the propagation vector is

$$\mathbf{k}_i = 4\mathbf{a}_x + 3\mathbf{a}_z \rightarrow k_i = 5 = \omega\sqrt{\mu_o\varepsilon_o} = \frac{\omega}{c}$$

Hence,

$$\omega = 5c = 15 \times 10^8 \text{ rad/s.}$$

A unit vector normal to the interface ($z = 0$) is $\mathbf{a}_z$. The plane containing **k** and $\mathbf{a}_z$ is y = constant, which is the xz-plane, the plane of incidence. Since $\mathbf{E}_i$ is normal to this plane, we have perpendicular polarization (similar to Fig. 10.17).

(b) The propagation vectors are illustrated in Figure 10.18 where it is clear that

$$\tan\theta_i = \frac{k_{ix}}{k_{iz}} = \frac{4}{3} \rightarrow \theta_i = 53.13°$$

Alternatively, without Figure 10.18, we can obtain θ_i from the fact that θ_i is the angle between **k** and $\mathbf{a}_n$, that is,

$$\cos\theta_i = \mathbf{a}_k \cdot \mathbf{a}_n = \left(\frac{4\mathbf{a}_x + 3\mathbf{a}_z}{5}\right) \cdot \mathbf{a}_z = \frac{3}{5}$$

or

$$\theta_i = 53.13°$$

(c) An easy way to find $\mathbf{E}_r$ is to use eq. (10.116a) because we have noticed that this problem is similar to that considered in Section 10.9(b). Suppose we are not aware of this. Let

$$\mathbf{E}_r = E_{ro}\cos(\omega t - \mathbf{k}_r \cdot \mathbf{r})\,\mathbf{a}_y$$

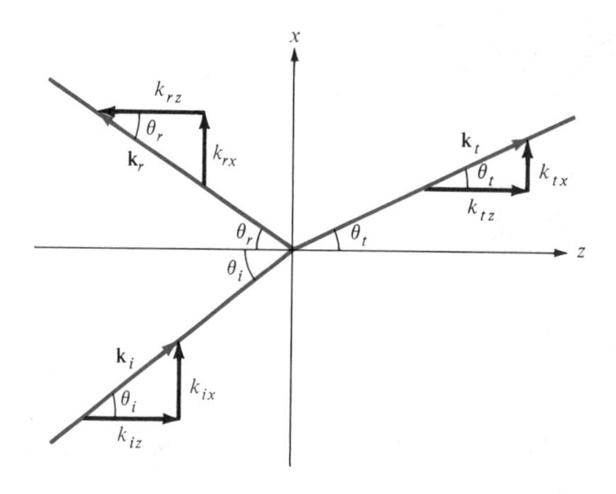

Figure 10.18 Propagation vectors of Example 10.11.

which is similar in form to the given $\mathbf{E}_i$. The unit vector $\mathbf{a}_y$ is chosen in view of the fact that the tangential component of $\mathbf{E}$ must be continuous at the interface. From Figure 10.18,

$$\mathbf{k}_r = k_{rx}\,\mathbf{a}_x - k_{rz}\,\mathbf{a}_z$$

where

$$k_{rx} = k_r \sin\theta_r, \qquad k_{rz} = k_r \cos\theta_r$$

But $\theta_r = \theta_i$ and $k_r = k_i = 5$ because both k_r and k_i are in the same medium. Hence,

$$k_r = 4\mathbf{a}_x - 3\mathbf{a}_z$$

To find E_{ro}, we need θ_t. From Snell's law

$$\sin\theta_t = \frac{n_1}{n_2}\sin\theta_i = \frac{c\sqrt{\mu_1\varepsilon_1}}{c\sqrt{\mu_2\varepsilon_2}}\sin\theta_i$$

$$= \frac{\sin 53.13^\circ}{\sqrt{2.5}}$$

or

$$\theta_t = 30.39^\circ$$

$$\Gamma_\perp = \frac{E_{ro}}{E_{io}}$$

$$= \frac{\eta_2\cos\theta_i - \eta_1\cos\theta_t}{\eta_2\cos\theta_i + \eta_1\cos\theta_t}$$

where $\eta_1 = \eta_o = 377$, $\quad \eta_2 = \sqrt{\dfrac{\mu_o\mu_{r_2}}{\varepsilon_o\varepsilon_{r_2}}} = \dfrac{377}{\sqrt{2.5}} = 238.4$

$$\Gamma_\perp = \frac{238.4\cos 35.13^\circ - 377\cos 30.39^\circ}{238.4\cos 53.13^\circ + 377\cos 30.39^\circ} = -0.389$$

Hence,

$$E_{ro} = \Gamma_\perp E_{io} = -0.389\,(8) = -3.112$$

and

$$\mathbf{E}_r = -\,3.112\cos(15\times 10^8 t - 4x + 3z)\,\mathbf{a}_y \text{ V/m}$$

(d) Similarly, let the transmitted electric field be

$$\mathbf{E}_t = E_{to}\cos(\omega t - \mathbf{k}_t\cdot\mathbf{r})\,\mathbf{a}_y$$

where

$$k_t = \beta_2 = \omega \sqrt{\mu_2 \varepsilon_2} = \frac{\omega}{c} \sqrt{\mu_{r_2} \varepsilon_{r_2}}$$

$$= \frac{15 \times 10^8}{3 \times 10^8} \sqrt{1 \times 2.5} = 7.906$$

From Figure 10.18,

$$k_{tx} = k_t \sin \theta_t = 4$$

$$k_{tz} = k_t \cos \theta_t = 6.819$$

or

$$\mathbf{k}_t = 4\mathbf{a}_x + 6.819\ \mathbf{a}_z$$

Notice that $k_{ix} = k_{rx} = k_{tx}$ as expected.

$$\tau_\perp = \frac{E_{to}}{E_{io}} = \frac{2\ \eta_2 \cos \theta_i}{\eta_2 \cos \theta_i + \eta_1 \cos \theta_t}$$

$$= \frac{2 \times 238.4 \cos 53.13°}{238.4 \cos 53.13° + 377 \cos 30.39°}$$

$$= 0.611.$$

The same result could be obtained from the relation $\tau_\perp = 1 + \Gamma_\perp$. Hence,

$$E_{to} = \tau_\perp\ E_{io} = 0.611 \times 8 = 4.888$$

$$\mathbf{E}_t = 4.888 \cos (15 \times 10^8 t - 4x - 6.819z)\ \mathbf{a}_y$$

From $\mathbf{E}_t$, $\mathbf{H}_t$ is easily obtained as

$$\mathbf{H}_t = \frac{1}{\mu_2 \omega} \mathbf{k}_t \times \mathbf{E}_t = \frac{\mathbf{a}_{k_t} \times \mathbf{E}_t}{\eta_2}$$

$$= \frac{4\mathbf{a}_x + 6.819\ \mathbf{a}_z}{7.906\ (238.4)} \times 4.888\ \mathbf{a}_y \cos (\omega t - \mathbf{k} \cdot \mathbf{r})$$

$$\mathbf{H}_t = (-17.69\ \mathbf{a}_x + 10.37\ \mathbf{a}_z) \cos (15 \times 10^8 t - 4x - 6.819z) \text{ mA/m.}$$ ■

PRACTICE EXERCISE 10.11

If the plane wave of Practice Exercise 10.10 is incident on a dielectric medium having $\sigma = 0$, $\varepsilon = 4\varepsilon_o$, $\mu = \mu_o$ and occupying $z \geq 0$, calculate

(a) The angles of incidence, reflection, and transmission
(b) The reflection and transmission coefficients
(c) The total **E** field in free space
(d) The total **E** field in the dielectric
(e) The Brewster angle.

ANSWER (a) 26.56°, 26.56°, 12.92°, (b) −0.295, 0.647, (c) $(10\,\mathbf{a}_y + 5\mathbf{a}_z)\cos(\omega t + 2y - 4z) + (-2.946\mathbf{a}_y + 1.473\mathbf{a}_z)\cos(\omega t + 2y + 4z)$ V/m, (d) $(7.055\mathbf{a}_y + 1.618\mathbf{a}_z)\cos(\omega t + 2y - 8.718z)$ V/m, (e) 63.43°.

SUMMARY

1. The wave equation is of the form

$$\frac{\partial^2 \Phi}{\partial t^2} - u^2 \frac{\partial^2 \Phi}{\partial z^2} = 0$$

with the solution

$$\Phi = A \sin(\omega t - \beta z)$$

where u = wave velocity, A = wave amplitude, ω = angular frequency (= $2\pi f$), and β = phase constant. Also, $\beta = \omega/u = 2\pi/\lambda$ or $u = f\lambda = \lambda/T$, where λ = wavelength and T = period.

2. In a lossy, charge-free medium, the wave equation based on Maxwell's equations is of the form

$$\nabla^2 \mathbf{A}_s - \gamma^2 \mathbf{A}_s = 0$$

where $\mathbf{A}_s$ is either $\mathbf{E}_s$ or $\mathbf{H}_s$ and $\gamma = \alpha + j\beta$ is the propagation constant. If we assume $\mathbf{E}_s = E_{xs}(z)\,\mathbf{a}_x$, we obtain EM waves of the form

$$\mathbf{E}(z, t) = E_o e^{-\alpha z} \cos(\omega t - \beta z)\,\mathbf{a}_x$$

$$\mathbf{H}(z, t) = H_o e^{-\alpha z} \cos(\omega t - \beta z - \theta_\eta)\,\mathbf{a}_y$$

where α = attenuation constant, β = phase constant, $\eta = |\eta| \underline{/\theta_\eta}$ = intrinsic impedance of the medium. The reciprocal of α is the skin depth ($\delta = 1/\alpha$). The relationships between β, ω, and λ as stated above remain valid for EM waves.

3. Wave propagation in other types of media can be derived from that for lossy media as special cases. For free space, set $\sigma = 0$, $\varepsilon = \varepsilon_o$, $\mu = \mu_o$; for lossless dielectric media, set $\sigma = 0$, $\varepsilon = \varepsilon_o\varepsilon_r$, and $\mu = \mu_o\mu_r$; and for good conductors, set $\sigma \simeq \infty$, $\varepsilon = \varepsilon_o$, $\mu = \mu_o$, or $\sigma/\omega\varepsilon \to 0$.

4. A medium is classified as lossy dielectric, lossless dielectric or good conductor depending on its loss tangent given by

$$\tan\theta = \frac{|\mathbf{J}_s|}{|\mathbf{J}_{d_s}|} = \frac{\sigma}{\omega\varepsilon} = \frac{\varepsilon''}{\varepsilon'}$$

where $\varepsilon_c = \varepsilon' - j\varepsilon''$ is the complex permittivity of the medium. For lossless dielectrics $\tan\theta \ll 1$, for good conductors $\tan\theta \gg 1$, and for lossy dielectrics $\tan\theta$ is of the order of unity.

5. In a good conductor, the fields tend to concentrate within the initial distance δ from the conductor surface. This phenomenon is called skin effect. For a conductor of width w and length ℓ, the effective or ac resistance is

$$R_{ac} = \frac{\ell}{\sigma w \delta}$$

where δ is the skin depth.

6. The Poynting vector, $\mathcal{P}$, is the power-flow vector whose direction is the same as the direction of wave propagation and magnitude the same as the amount of power flowing through a unit area normal to its direction.

$$\mathcal{P} = \mathbf{E} \times \mathbf{H}, \qquad \mathcal{P}_{ave} = 1/2 \text{ Re } (\mathbf{E}_s \times \mathbf{H}_s^*)$$

7. If a plane wave is incident normally from medium 1 to medium 2, the reflection coefficient Γ and transmission coefficient τ are given by

$$\Gamma = \frac{E_{ro}}{E_{io}} = \frac{\eta_2 - \eta_1}{\eta_2 + \eta_1}, \qquad \tau = \frac{E_{to}}{E_{io}} = 1 + \Gamma$$

The standing wave ratio, s, is defined as

$$s = \frac{1 + |\Gamma|}{1 - |\Gamma|}$$

8. For oblique incidence from lossless medium 1 to lossless medium 2, we have the Fresnel coefficients as

$$\Gamma_{\|} = \frac{\eta_2 \cos \theta_t - \eta_1 \cos \theta_i}{\eta_2 \cos \theta_t + \eta_1 \cos \theta_i}, \qquad \tau_{\|} = \frac{2\eta_2 \cos \theta_i}{\eta_2 \cos \theta_t + \eta_1 \cos \theta_i}$$

for parallel polarization and

$$\Gamma_{\perp} = \frac{\eta_2 \cos \theta_i - \eta_1 \cos \theta_t}{\eta_2 \cos \theta_i + \eta_1 \cos \theta_t}, \qquad \tau_{\perp} = \frac{2\eta_2 \cos \theta_i}{\eta_2 \cos \theta_i + \eta_1 \cos \theta_t}$$

for perpendicular polarization. As in optics,

$$\theta_r = \theta_i$$

$$\frac{\sin \theta_t}{\sin \theta_i} = \frac{\beta_1}{\beta_2} = \sqrt{\frac{\mu_1 \varepsilon_1}{\mu_2 \varepsilon_2}}$$

Total transmission or no reflection ($\Gamma = 0$) occurs when the angle of incidence θ_i is equal to the Brewster angle.

REVIEW QUESTIONS

10.1 Which of these is not a correct form of the wave $E_x = \cos(\omega t - \beta z)$?

(a) $\cos(\beta z - \omega t)$

(b) $\sin(\beta z - \omega t - \pi/2)$

(c) $\cos\left(\frac{2\pi t}{T} - \frac{2\pi z}{\lambda}\right)$

(d) $\text{Re}\,(e^{j(\omega t - \beta z)})$

(e) $\cos \beta(z - ut)$

10.2 Identify which of these functions do not satisfy the wave equation:

(a) $50e^{j\omega(t - 3z)}$

(b) $\sin \omega\,(10z + 5t)$

(c) $(x + 2t)^2$

(d) $\cos^2(y + 5t)$

(e) $\sin x \cos t$

(f) $\cos\,(5y + 2x)$

10.3 Which of the following statements is not true of waves in general?

(a) It may be a function of time only.

(b) It may be sinusoidal or cosinusoidal.

(c) It must be a function of time and space.

(d) For practical reasons, it must be finite in extent.

10.4 The electric field component of a wave in free space is given by $\mathbf{E} = 10 \cos (10^7 t + kz)\,\mathbf{a}_y$ V/m. It can be inferred that

(a) The wave propagates along $\mathbf{a}_y$.

(b) The wavelength $\lambda = 188.5$ m.

(c) The wave amplitude is 10 V/m.

(d) The wave number $k = 0.33$ rad/m.

(e) The wave attenuates as it travels.

10.5 Given that $\mathbf{H} = 0.5\,e^{-0.1x} \sin\,(10^6 t - 2x)\,\mathbf{a}_z$ A/m, which of these statements are incorrect?

(a) $\alpha = 0.1$ Np/m

(b) $\beta = -2$ rad/m

(c) $\omega = 10^6$ rad/s

(d) The wave travels along $\mathbf{a}_x$.

(e) The wave is polarized in the z-direction.

(f) The period of the wave is 1 μs.

10.6 What is the major factor for determining whether a medium is free space, lossless dielectric, lossy dielectric, or good conductor?

(a) Attenuation constant

(b) Constitutive parameters (σ, ε, μ)

(c) Loss tangent

(d) Reflection coefficient

10.7 In a certain medium, $\mathbf{E} = 10 \cos (10^8 t - 3y)\, \mathbf{a}_x$ V/m. What type of medium is it?

(a) Free space

(b) Perfect dielectric

(c) Lossless dielectric

(d) Perfect conductor

10.8 Electromagnetic waves travel faster in conductors than in dielectrics.

(a) True

(b) False

10.9 In a good conductor, **E** and **H** are in time phase.

(a) True

(b) False

10.10 The Poynting vector physically denotes the power density leaving or entering a given volume in a time-varying field.

(a) True

(b) False

Answers: 10.1b, 10.2d,f, 10.3a, 10.4b,c, 10.5b,f, 10.6c, 10.7c, 10.8b, 10.9b, 10.10a.

PROBLEMS

10.1 In air, a uniform plane wave is polarized in the z-direction and propagates along $\mathbf{a}_x$. Assuming that **E** is sinusoidal with adjacent minimum and maximum values of ± 0.01 V/m occurring at $x = 20$ m and $x = 170$ m respectively when $t = 0$ and that $\mathbf{E}(0, 0) = 0$,

(a) Find the instantaneous expression for **E**.

(b) Calculate **E** at $x = 100$ m, $t = 2$ μs.

10.2 (a) Derive eqs. (10.23) and (10.24) from eqs. (10.18) and (10.20).

(b) Using eq. (10.29) in conjunction with Maxwell's equations, show that

$$\eta = \frac{j\omega\mu}{\gamma}$$

(c) From part (b), derive eqs. (10.32) and (10.33).

10.3 The magnetic field component of a wave is given by

$$\mathbf{H} = 30 \cos (10^8 t - 6x) \, \mathbf{a}_x \text{ mA/m}$$

Determine (a) the direction of wave propagation, (b) the wavelength, and (c) the wave velocity.

10.4 By expressing eq. (10.17) in terms of field components E_{xs}, E_{ys}, and E_{zs}, obtain three equivalent scalar wave equations.

10.5 A lossy material has $\mu = 5\mu_o$, $\varepsilon = 2\varepsilon_o$. If at 5 MHz, the phase constant is 10 rad/m, calculate

(a) The loss tangent

(b) The conductivity of the material

(c) The complex permittivity

(d) The attenuation constant

(e) The intrinsic impedance

***10.6** A nonmagnetic medium has an intrinsic impedance $240 \angle 30° \ \Omega$. Find its

(a) Loss tangent

(b) Dielectric constant

(c) Complex permittivity

(d) Attenuation constant at 1 MHz.

***10.7** At a particular frequency, a medium has $\alpha = 0.1$ Np/m, $\eta = 250 \angle 35.26° \ \Omega$. Calculate the loss tangent, loss angle, and wavelength.

10.8 Seawater plays a vital role in the study of submarine communications. Assuming that for seawater, $\sigma = 4$ mhos/m, $\varepsilon_r = 80$, $\mu_r = 1$, and $f = 100$ MHz, calculate: (a) the phase velocity, (b) the wavelength, (c) the skin depth, (d) the intrinsic impedance.

10.9 In free space, the electric field component of a TEM wave is

$$\mathbf{E} = 5 \sin(3 \times 10^8 t + y) \, \mathbf{a}_z \text{ V/m}$$

(a) Determine its polarization.

(b) Find λ, T, and u.

(c) Sketch the wave at $t = 0$, $T/4$, and $T/2$.

(d) Calculate the corresponding $\mathbf{H}$.

10.10 A uniform wave in air has

$$\mathbf{E} = 10 \cos (2\pi \times 10^6 t - \beta z)\, \mathbf{a}_y$$

(a) Calculate β and λ.

(b) Sketch the wave at $z = 0, \lambda/4$.

(c) Find $\mathbf{H}$.

10.11 A medium has the following constitutive parameters:

$$\mu = \mu_o, \qquad \varepsilon = 9\varepsilon_o, \qquad \sigma = 5 \times 10^{-9} \text{ mhos/m}$$

Calculate the wavelength of a wave at 1 GHz propagating through the medium. Should the medium be regarded as free space, lossy dielectric, lossless dielectric, or good conductor?

10.12 If the $\mathbf{H}$ and $\mathbf{E}$ fields in air are given by $\mathbf{H} = -0.3 \sin (4\pi \times 10^6 t - \beta z)\, \mathbf{a}_x$ A/m, and $\mathbf{E} = \mathbf{A} \sin (4\pi \times 10^6 t - \beta z)$ V/m, find β, $\mathbf{a}_k$, and $\mathbf{A}$.

***10.13** By assuming the time-dependent fields $\mathbf{E} = \mathbf{E}_o e^{j(\mathbf{k} \cdot \mathbf{r} - \omega t)}$ and $\mathbf{H} = \mathbf{H}_o e^{j(\mathbf{k} \cdot \mathbf{r} - \omega t)}$ where $\mathbf{k} = k_x \mathbf{a}_x + k_y \mathbf{a}_y + k_z \mathbf{a}_z$ is the wave number vector and $\mathbf{r} = x\mathbf{a}_x + y\mathbf{a}_y + z\mathbf{a}_z$ is the radius vector, show that $\nabla \times \mathbf{E} = -\partial \mathbf{B}/\partial t$ can be expressed as $\mathbf{k} \times \mathbf{E} = \mu\omega\mathbf{H}$ and deduce $\mathbf{a}_k \times \mathbf{a}_E = \mathbf{a}_H$.

10.14 Assume the same fields as in Problem 10.13 and show that Maxwell's equations in a source-free region can be written as

$$\mathbf{k} \cdot \mathbf{E} = 0$$

$$\mathbf{k} \cdot \mathbf{H} = 0$$

$$\mathbf{k} \times \mathbf{E} = \omega\mu\mathbf{H}$$

$$\mathbf{k} \times \mathbf{H} = -\omega\varepsilon\mathbf{E}$$

From these equations deduce

$$\mathbf{a}_k \times \mathbf{a}_E = \mathbf{a}_H \qquad \text{and} \qquad \mathbf{a}_k \times \mathbf{a}_H = -\mathbf{a}_E$$

10.15 The magnetic field component of a plane wave in a lossless dielectric is

$$\mathbf{H} = 30 \sin (2\pi \times 10^8 t - 5x)\, \mathbf{a}_z \text{ mA/m}$$

(a) If $\mu_r = 1$, find ε_r.

(b) Calculate the wavelength and wave velocity.

(c) Determine the wave impedance.

(d) Determine the polarization of the wave.

(e) Find the corresponding electric field component.

(f) Find the displacement current density.

10.16 In a certain medium

$$\mathbf{E} = 10 \cos (2\pi \times 10^7 t - \beta x)(\mathbf{a}_y + \mathbf{a}_z) \text{ V/m}$$

If $\mu = 50\mu_o$, $\varepsilon = 2\varepsilon_o$, and $\sigma = 0$, find β and $\mathbf{H}$.

10.17 At 1 MHz, which of the following materials should be regarded as conducting media?

(a) Seawater ($\varepsilon = 80\varepsilon_o$, $\mu = \mu_o$, $\sigma = 4$ mhos/m)

(b) Freshwater ($\varepsilon = 80\varepsilon_o$, $\mu = \mu_o$, $\sigma = 10^{-3}$ mhos/m)

(c) Wet earth ($\varepsilon = 10\varepsilon_o$, $\mu = \mu_o$, $\sigma = 10^{-3}$ mhos/m)

(d) Dry earth ($\varepsilon = 3\varepsilon_o$, $\mu = \mu_o$, $\sigma = 10^{-5}$ mhos/m)

10.18 For silver ($\sigma = 6.1 \times 10^7$ mhos/m), find the frequency at which the skin depth is 2 mm. Determine the wavelength and wave velocity at that frequency.

10.19 Calculate the skin depth and the velocity of propagation for a uniform plane wave at frequency 6 MHz traveling in polyvinylchloride ($\mu_r = 1$, $\varepsilon_r = 4$, tan $\theta_\eta = 7 \times 10^{-2}$).

10.20 A uniform plane wave in a lossy medium has a phase constant of 1.6 rad/m at 10^7 Hz and its magnitude is reduced by 60% for every 2 m traveled. Find the skin depth and speed of the wave.

10.21 (a) Determine the dc resistance of a round copper wire ($\sigma = 5.8 \times 10^7$ mhos/m, $\mu_r = 1$, $\varepsilon_r = 1$) of radius 1.2 mm and length 600 m.

(b) Find the ac resistance at 100 MHz.

(c) Calculate the approximate frequency where dc and ac resistances are equal.

10.22 A 40-m-long aluminum ($\sigma = 3.5 \times 10^7$ mhos/m, $\mu_r = 1$, $\varepsilon_r = 1$) pipe with inner and outer radii 9 mm and 12 mm carries a total current of 6 sin $10^6\pi t$ A. Find the skin depth and the effective resistance of the pipe.

10.23 Given that in air, $\mathbf{H} = 0.1 \sin (\pi \times 10^8 t + \beta y)\, \mathbf{a}_x$ A/m, find the time average power density in the wavefront. In what direction is the Poynting vector?

10.24 In a material with $\sigma = 0$, $\mu = 2.25\mu_o$, and $\varepsilon = \varepsilon_o$, the phasor of the electric field is

$$\mathbf{E}_s = (40\underline{/20^\circ})e^{-j8y}\, \mathbf{a}_z \text{ V/m}$$

find: (a) the instantaneous form of the field, (b) the magnetic field strength in time domain, (c) λ and η, (d) the time-average Poynting vector.

10.25 Calculate the Poynting vector for the wave in

(a) Problem 10.10

(b) Practice Exercise 10.2

*10.26 Show that eqs. (10.67) and (10.68) are equivalent.

*10.27 The magnetic field component of a spherical wave in free space is $\mathbf{H} = \dfrac{\sin\theta}{r}$ $\cos(10^7 t - \beta r)\ \mathbf{a}_\theta$ A/m. Find

(a) β and $\mathbf{E}$

(b) The time-average power

(c) The total power crossing the spherical cap $r = 100$ m, $0 \le \theta \le \pi/3$, $0 \le \phi \le 2\pi$

10.28 Using eqs. (10.71) to (10.82), show that

$$\frac{H_{to}}{H_{io}} = \tau \frac{\eta_1}{\eta_2} = \frac{2\eta_1}{\eta_1 + \eta_2}$$

and

$$-\frac{H_{ro}}{H_{io}} = \Gamma = \frac{\eta_2 - \eta_1}{\eta_2 + \eta_1}$$

10.29 (a) For a normal incidence upon an interface between two lossless media, we define the ratios of the time-average incident, reflected, and transmitted powers as

$$R = \frac{(\mathcal{P}_r)_{\text{ave}}}{(\mathcal{P}_i)_{\text{ave}}} \quad \text{and} \quad T = \frac{(\mathcal{P}_t)_{\text{ave}}}{(\mathcal{P}_i)_{\text{ave}}}$$

Prove that $R + T = 1$; that is, energy conservation holds for reflection and transmission.

(b) Express R and T in terms of η_1 and η_2.

10.30 Region 1 is a lossless medium for which $y \ge 0$, $\mu = \mu_o$, $\varepsilon = 4\varepsilon_o$, whereas region 2 is free space, $y \le 0$. If a plane wave $\mathbf{E} = 5\cos(10^8 t + \beta y)\ \mathbf{a}_z$ V/m exists in region 1, find: (a) the total electric field component of the wave in region 2, (b) the time-average Poynting vector in region 1, (c) the time-average Poynting vector in region 2.

10.31 A plane wave in free space ($z \le 0$) is incident normally on a large block of material with $\varepsilon_r = 12$, $\mu_r = 3$, $\sigma = 0$ which occupies $z \ge 0$. If the incident electric field is

$$\mathbf{E} = 30\cos(\omega t - z)\ \mathbf{a}_y \text{ V/m}$$

find: (a) ω, (b) the standing wave ratio, (c) the reflected magnetic field, (d) the average power density of the transmitted wave.

10.32 A 30-MHz uniform plane wave with

$$\mathbf{H} = 10 \sin (\omega t + \beta x)\, \mathbf{a}_z \text{ mA/m}$$

exists in region $x \geq 0$ having $\sigma = 0$, $\varepsilon = 9\varepsilon_o$, $\mu = 4\mu_o$. At $x = 0$, the wave encounters free space. Determine (a) the polarization of the wave, (b) the phase constant β, (c) the displacement current density in region $x \geq 0$, (d) the reflected and transmitted magnetic field, and (e) the average power density in each region.

10.33 A uniform plane wave in air is normally incident on an infinite lossless dielectric material having $\varepsilon = 3\varepsilon_o$ and $\mu = \mu_o$. If the incident wave is $\mathbf{E}_i = 10 \cos (\omega t - z)\, \mathbf{a}_y$ V/m, find:

(a) λ and ω of the wave in air and the transmitted wave in the dielectric medium

(b) The incident $\mathbf{H}_i$ field

(c) Γ and τ

(d) The total electric field and the time-average power in both regions

***10.34** A signal in air ($z \geq 0$) with the electric field component

$$\mathbf{E} = 10 \sin (\omega t + 3z)\, \mathbf{a}_x \text{ V/m}$$

hits normally the ocean surface at $z = 0$ as in Figure 10.19. Assuming that the ocean surface is smooth and that $\varepsilon = 80\varepsilon_o$, $\mu = \mu_o$, $\sigma = 4$ mhos/m in ocean, determine

(a) ω

(b) The wavelength of the signal in air

(c) The loss tangent and intrinsic impedance of the ocean

(d) The reflected and transmitted **E** field

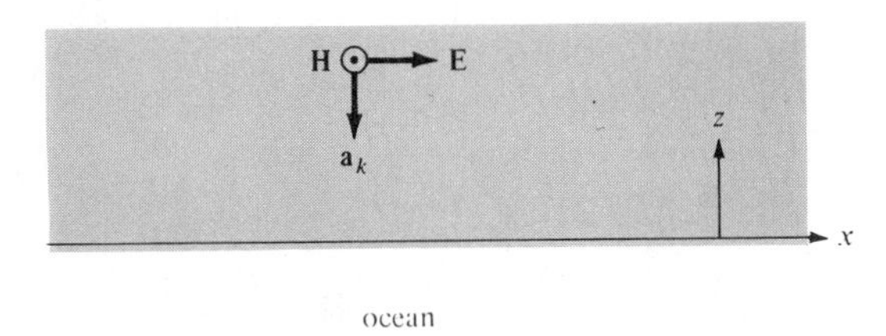

Figure 10.19 For Problem 10.34.

*10.35 An electromagnetic plane wave in free space with a power density of 3 W/m^2 impinges normally on a lossless dielectric boundary, causing a standing wave ratio of 2.2. What is the power density of the wave transmitted into the dielectric?

10.36 A plane wave in air is reflected at normal incidence from a lossless medium ($\varepsilon = \varepsilon_o$, $\mu = 9\mu_o$). If the amplitude of the incident wave is 2 V/m, find the time-average power/m^2 of the transmitted wave.

*10.37 If the conducting plate of Example 10.9 is replaced by a dielectric material ($\varepsilon = \varepsilon_o$, $\mu = \mu_o$, $\sigma = 0.02$ mhos/m), find the reflected and transmitted fields. Take $\omega = 10^9$ rad/s.

10.38 In air, a standing wave pattern exists due to a reflection of a lossless dielectric material. If two adjacent maxima in the wave are separated by 4 m and the standing wave ratio is 2, calculate

(a) The wavelength and frequency of the wave

(b) The reflection coefficient.

10.39 Show that the total fields in Example 10.9 form standing waves.

10.40 Sketch the standing wave in eq. (10.87) at $t = 0$, $T/8$, $T/4$, $3T/8$, $T/2$, and so on, where $T = 2\pi/\omega$.

10.41 A uniform plane wave is incident at an angle $\theta_i = 45°$ on a pair of dielectric slabs joined together as shown in Figure 10.20. Determine the angles of transmission θ_{t1} and θ_{t2} in the slabs.

10.42 Show that the field

$$\mathbf{E}_s = 20 \sin (k_x x) \cos (k_y y) \, \mathbf{a}_z$$

where $k_x^2 + k_y^2 = \omega^2\mu_o\varepsilon_o$, can be represented as the superposition of four propagating plane waves. Find the corresponding $\mathbf{H}_s$.

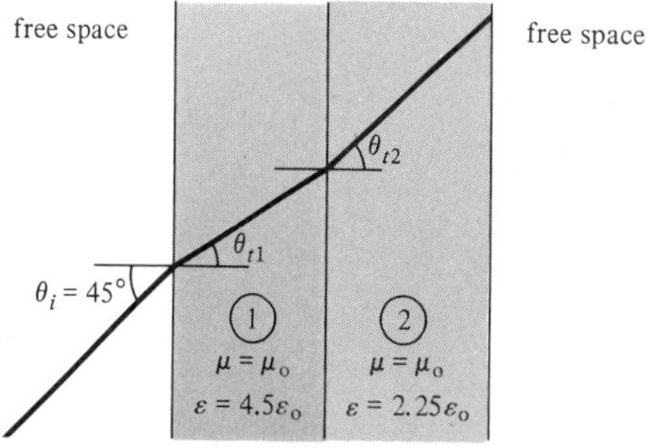

Figure 10.20 For Problem 10.41.

10.43 Show that for nonmagnetic dielectric media, the reflection and transmission coefficients for oblique incidence become

$$\Gamma_{\|} = \frac{\tan(\theta_t - \theta_i)}{\tan(\theta_t + \theta_i)}, \qquad \tau_{\|} = \frac{2\cos\theta_i \sin\theta_t}{\sin(\theta_t + \theta_i)\cos(\theta_t - \theta_i)}$$

$$\Gamma_{\perp} = \frac{\sin(\theta_t - \theta_i)}{\sin(\theta_t + \theta_i)}, \qquad \tau_{\perp} = \frac{2\cos\theta_i \sin\theta_t}{\sin(\theta_t + \theta_i)}$$

10.44 The electric field component of an EM wave propagating in air is

$$\mathbf{E} = 50 \cos(\omega t - \beta_1 x \sin 45° - \beta_1 z \cos 45°)\, \mathbf{a}_y \text{ V/m}$$

If the wave is incident on a lossless medium ($\varepsilon = 2.25\varepsilon_o$, $\mu = \mu_o$) in $z \geq 0$, determine the magnetic field component and the transmission coefficient.

10.45 In a dielectric medium ($\varepsilon = 9\varepsilon_o$, $\mu = \mu_o$), a plane wave with

$$\mathbf{H} = 0.2 \cos(10^9 t - kx - k\sqrt{8}z)\, \mathbf{a}_y \text{ A/m}$$

is incident on an air boundary at $z = 0$, find

(a) θ_r and θ_t

(b) k

(c) The wavelength in the dielectric and air

(d) The incident **E**

(e) The transmitted and reflected **E**

(f) The Brewster angle.

***10.46** A plane wave from free space to a lossless dielectric with $\mu = \mu_o$, $\varepsilon = 4\varepsilon_o$ is totally transmitted. Find θ_i and θ_t. What is the state of polarization of the wave?

***10.47** A plane wave in air with

$$\mathbf{E} = (8\mathbf{a}_x + 6\mathbf{a}_y + 5\mathbf{a}_z) \sin(\omega t + 3x - 4y) \text{ V/m}$$

is incident on a copper slab in $y \geq 0$. Find ω and the reflected wave. Assume copper is a perfect conductor. (*Hint:* Write down the field components in both media and match the boundary conditions.)

CHAPTER

Transmission Lines

11

Dictionary is the only place that success comes before work. Hard work is the price we must pay for success. I think you can accomplish almost anything if you're willing to pay the price.

—VINCE LOMBARDI

11.1 INTRODUCTION

Our discussion in the previous chapter was essentially on wave propagation in unbounded media, media of infinite extent. Such wave propagation is said to be unguided in that the uniform plane wave exists throughout all space and EM energy associated with the wave spreads over a wide area. However, wave propagation in unbounded media is used in radio or TV broadcasting, where the information being transmitted is meant for everyone who may be interested. Such means of wave propagation will not help in a situation like telephone conversation, where the information is received privately by one person.

Another means of transmitting power or information is by guided structures. Guided structures serve to guide (or direct) the propagation of energy from the source to the load. Typical examples of such structures are transmission lines and waveguides. Waveguides are discussed in the next chapter; transmission lines are considered in this chapter.

Transmission lines are commonly used in power distribution (at low frequencies) and in communications (at high frequencies). A transmission line basically consists of two or more parallel conductors used to connect a source to a load. The source may be a hydroelectric generator, a transmitter, or an oscillator; the load may be a factory, an antenna, or an oscilloscope, respectively. Typical transmission lines include coaxial cable, a two-wire line, a parallel-plate or planar line, a wire above the conducting plane, and a microstrip line. Cross-sectional views of these lines are portrayed in Figure 11.1. Notice that each of these lines consists of two conductors in parallel. Coaxial cables are routinely used in electrical laboratories and in connecting TV sets

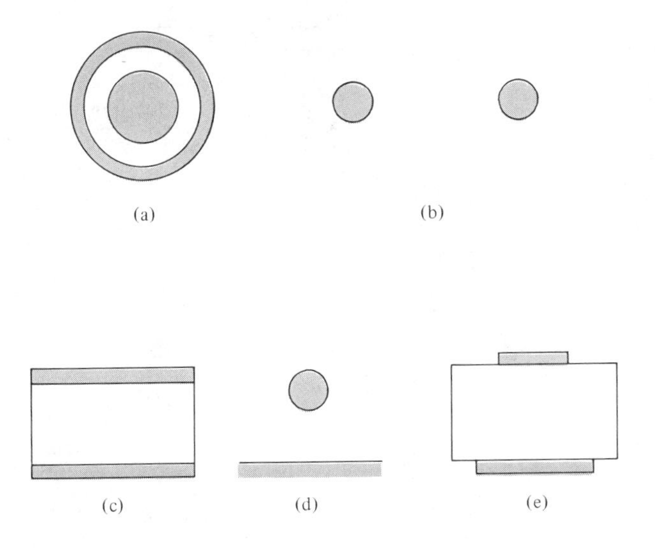

Figure 11.1 Cross-sectional view of typical transmission lines: **(a)** coaxial line, **(b)** two-wire line, **(c)** planar line, **(d)** wire above conducting plane, **(e)** microstrip line.

to TV antennas. Microstrip lines (similar to that in Figure 11.1e) are particularly important in integrated circuits where metallic strips connecting electronic elements are deposited on dielectric substrates.

Transmission line problems are usually solved using EM field theory and electric circuit theory, the two major theories on which electrical engineering is based. In this chapter, we use circuit theory because it is easier to deal with mathematically. The basic concepts of wave propagation (such as propagation constant, reflection coefficient, and standing wave ratio) covered in the previous chapter apply here.

Our analysis of transmission lines will include the derivation of the transmission-line equations and characteristic quantities, the use of the Smith chart, various practical applications of transmission lines, and transients on transmission lines.

11.2 TRANSMISSION LINE PARAMETERS

It is customary and convenient to describe a transmission line in terms of its line parameters, which are its resistance per unit length R, inductance per unit length L, conductance per unit length G, and capacitance per unit length C. Each of the lines shown in Figure 11.1 has specific formulas for finding R, L, G, and C. For coaxial, two-wire, and planar lines, the formulas for calculating the values of R, L, G, and C

Table 11.1 **Distributed Line Parameters at High Frequencies***

Parameters	Coaxial Line	Two-Wire Line	Planar Line
R (Ω/m)	$\dfrac{1}{2\pi\delta\sigma_c}\left[\dfrac{1}{a}+\dfrac{1}{b}\right]$ $(\delta \ll a, c-b)$	$\dfrac{1}{\pi a\delta\sigma_c}$ $(\delta \ll a)$	$\dfrac{2}{w\delta\sigma_c}$ $(\delta \ll t)$
L (H/m)	$\dfrac{\mu}{2\pi}\ln\dfrac{b}{a}$	$\dfrac{\mu}{\pi}\cosh^{-1}\dfrac{d}{2a}$	$\dfrac{\mu d}{w}$
G (℧/m)	$\dfrac{2\pi\sigma}{\ln\dfrac{b}{a}}$	$\dfrac{\pi\sigma}{\cosh^{-1}\dfrac{d}{2a}}$	$\dfrac{\sigma w}{d}$
C (F/m)	$\dfrac{2\pi\varepsilon}{\ln\dfrac{b}{a}}$	$\dfrac{\pi\varepsilon}{\cosh^{-1}\dfrac{d}{2a}}$	$\dfrac{\varepsilon w}{d}$ $(w \gg d)$

$^*\delta = \dfrac{1}{\sqrt{\pi f \mu_c \sigma_c}}$ = skin depth of the conductor; $\cosh^{-1}\dfrac{d}{2a} \simeq \ln\dfrac{d}{a}$ if $\left[\dfrac{d}{2a}\right]^2 \gg 1$.

are provided in Table 11.1. The dimensions of the lines are as shown in Figure 11.2. Some of the formulas[1] in Table 11.1 were derived in Chapters 6 and 8.

It should be noted that

1. The line parameters R, L, G, and C are not discrete or lumped but distributed as shown in Figure 11.3. By this we mean that the parameters are uniformly distributed along the entire length of the line.
2. For each line, the conductors are characterized by σ_c, μ_c, $\varepsilon_c = \varepsilon_o$, and the homogeneous dielectric separating the conductors is characterized by σ, μ, ε.
3. $G \neq 1/R$; R is the ac resistance per unit length of the conductors comprising the line and G is the conductance per unit length due to the dielectric medium separating the conductors.
4. The value of L shown in Table 11.1 is the external inductance per unit length; that is, $L = L_{ext}$. The effects of internal inductance L_{in} ($= R/\omega$) are negligible at high frequencies at which most communication systems operate.

[1]Similar formulas for other transmission lines can be obtained from engineering handbooks or data books—e.g., M. A. R. Guston, *Microwave Transmission-line Impedance Data*. London: Van Nostrand Reinhold, 1972.

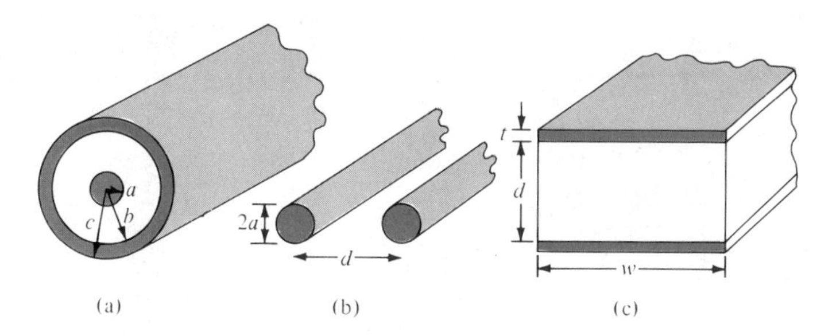

Figure 11.2 Common transmission lines: **(a)** coaxial line, **(b)** two-wire line, **(c)** planar line. For each line, the conductors are characterized by σ_c, μ_c, $\varepsilon_c = \varepsilon_0$ and the dielectric separating them by σ, μ, ε.

5. For each line,

$$LC = \mu\varepsilon \qquad \text{and} \qquad \frac{G}{C} = \frac{\sigma}{\varepsilon} \tag{11.1}$$

As a way of preparing for the next section let us consider how an EM wave propagates through a two-conductor transmission line. For example, consider the coaxial line connecting the generator or source to the load as in Figure 11.4(a). When switch S is closed, the inner conductor is made positive with respect to the outer one so that the **E** field is radially outward as in Figure 11.4(b). According to Ampere's law, the **H** field encircles the current carrying conductor as in Figure 11.4(b). The Poynting vector (**E** × **H**) points along the transmission line. Thus, closing the switch simply establishes a disturbance, which appears as a transverse electromagnetic (TEM) wave propagating along the line. This wave is a nonuniform plane wave and by means of it power is transmitted through the line.

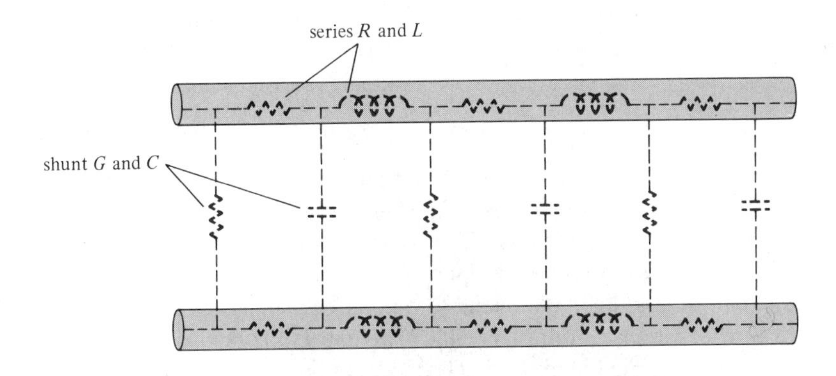

Figure 11.3 Distributed parameters of a two-conductor transmission line.

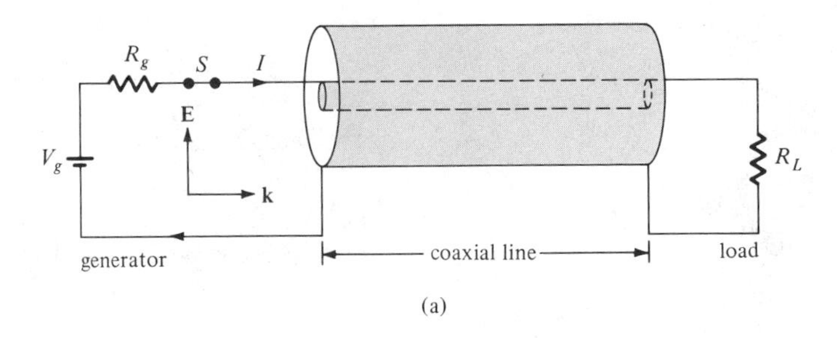

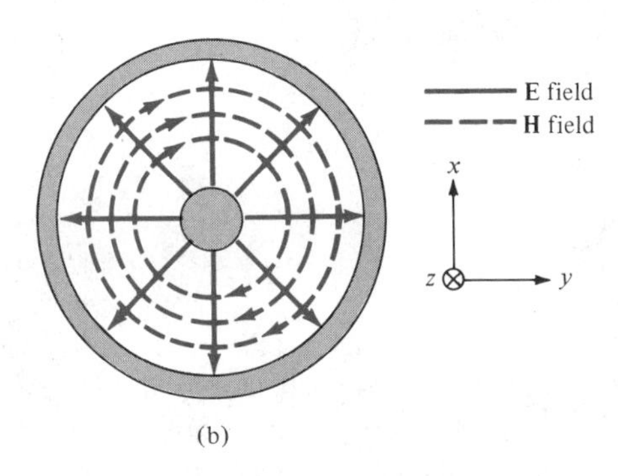

Figure 11.4 (a) Coaxial line connecting the generator to the load; (b) **E** and **H** fields on the coaxial line.

11.3 TRANSMISSION LINE EQUATIONS

As mentioned in the previous section, a two-conductor transmission line supports a TEM wave; that is, the electric and magnetic fields on the line are transverse to the direction of wave propagation. An important property of TEM waves is that the fields **E** and **H** are uniquely related to voltage V and current I respectively:

$$V = -\int \mathbf{E} \cdot d\mathbf{l}, \qquad I = \oint \mathbf{H} \cdot d\mathbf{l} \tag{11.2}$$

In view of this, we will use circuit quantities V and I in solving the transmission line problem instead of solving field quantities **E** and **H** (i.e., solving Maxwell's equations and boundary conditions). The circuit model is simpler and more convenient.

Let us examine an incremental portion of length Δz of a two-conductor transmission line. We intend to find an equivalent circuit for this line and derive the line equations. From Figure 11.3, we expect the equivalent circuit of a portion of the line to be as in Figure 11.5. The model in Figure 11.5 is in terms of the line parameters R, L, G, and C, and may represent any of the two-conductor lines of Figure 11.2. The

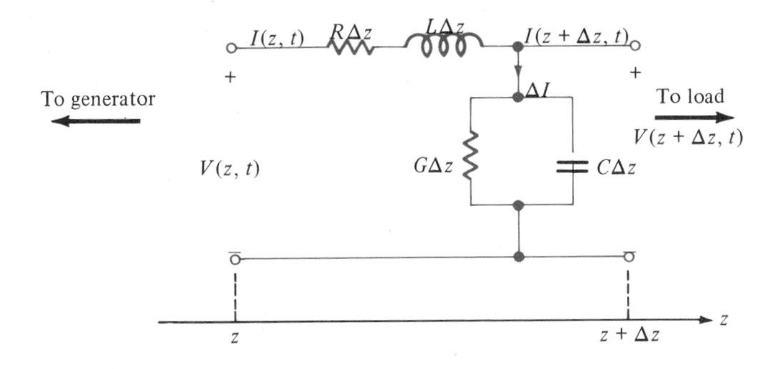

Figure 11.5 L-type equivalent circuit model of a differential length Δz of a two-conductor transmission line.

model is called the L-type equivalent circuit; there are other possible types (see Problem 11.1). In the model of Figure 11.5, we assume that the wave propagates along the $+z$ direction, from the generator to the load.

By applying Kirchhoff's voltage law to the outer loop of the circuit in Figure 11.5, we obtain

$$V(z, t) = R\,\Delta z\,I(z, t) + L\,\Delta z\,\frac{\partial I(z, t)}{\partial t} + V(z + \Delta z, t)$$

or

$$-\frac{V(z + \Delta z, t) - V(z, t)}{\Delta z} = R\,I(z, t) + L\,\frac{\partial I(z, t)}{\partial t} \tag{11.3}$$

Taking the limit of eq. (11.3) as $\Delta z \to 0$ leads to

$$\boxed{-\frac{\partial V(z, t)}{\partial z} = RI(z, t) + L\,\frac{\partial I(z, t)}{\partial t}} \tag{11.4}$$

Similarly, applying Kirchhoff's current law to the main node of the circuit in Figure 11.5 gives

$$\begin{aligned} I(z,t) &= I(z + \Delta z, t) + \Delta I \\ &= I(z + \Delta z, t) + G\,\Delta z\,V(z + \Delta z, t) + C\,\Delta z\,\frac{\partial V(z + \Delta z, t)}{\partial t} \end{aligned}$$

or

$$-\frac{I(z + \Delta z, t) - I(z, t)}{\Delta z} = G\,V(z + \Delta z, t) + C\,\frac{\partial V(z + \Delta z, t)}{\partial t} \tag{11.5}$$

As $\Delta z \rightarrow 0$, eq. (11.5) becomes

$$-\frac{\partial I(z, t)}{\partial z} = G\, V(z, t) + C\frac{\partial V(z, t)}{\partial t} \qquad [11.6]$$

If we assume harmonic time dependence so that

$$V(z, t) = \text{Re}\,(V_s(z)\, e^{j\omega t}) \qquad [11.7a]$$

$$I(z, t) = \text{Re}\,(I_s(z)\, e^{j\omega t}) \qquad [11.7b]$$

where $V_s(z)$ and $I_s(z)$ are the phasor forms of $V(z, t)$ and $I(z, t)$ respectively, eqs. (11.4) and (11.6) become

$$-\frac{dV_s}{dz} = (R + j\omega L)\, I_s \qquad [11.8]$$

$$-\frac{dI_s}{dz} = (G + j\omega C)\, V_s \qquad [11.9]$$

In the differential eqs. (11.8) and (11.9), V_s and I_s are coupled. To separate them, we take the second derivative of V_s in eq. (11.8) and employ eq. (11.9) so that we obtain

$$\frac{d^2V_s}{dz^2} = (R + j\omega L)\,(G + j\omega C)\, V_s$$

or

$$\frac{d^2V_s}{dz^2} - \gamma^2 V_s = 0 \qquad [11.10]$$

where

$$\gamma = \alpha + j\beta = \sqrt{(R + j\omega L)(G + j\omega C)} \qquad [11.11]$$

By taking the second derivative of I_s in eq. (11.9) and employing eq. (11.8), we get

$$\frac{d^2I_s}{dz^2} - \gamma^2 I_s = 0 \qquad [11.12]$$

We notice that eqs. (11.10) and (11.12) are respectively the wave equations for voltage and current similar in form to the wave equations obtained for plane waves in eqs. (10.17) and (10.19). Thus, in our usual notations, γ in eq. (11.11) is the propagation constant (in per meter), α is the attenuation constant (in nepers per meter or decibels per meter), and β is the phase constant (in radians per meter). The wavelength λ and wave velocity u are respectively given by

$$\lambda = \frac{2\pi}{\beta} \tag{11.13}$$

$$\boxed{u = \frac{\omega}{\beta} = f\lambda} \tag{11.14}$$

The solutions of the linear homogeneous differential equations (11.10) and (11.12) are similar to Case 2 of Example 6.5, namely,

$$V_s(z) = \underset{\longrightarrow +z}{V_o^+ e^{-\gamma z}} + \underset{-z \longleftarrow}{V_o^- e^{\gamma z}} \tag{11.15}$$

and

$$I_s(z) = \underset{\longrightarrow +z}{I_o^+ e^{-\gamma z}} + \underset{-z \longleftarrow}{I_o^- e^{\gamma z}} \tag{11.16}$$

where V_o^+, V_o^-, I_o^+, and I_o^- are wave amplitudes; the + and − signs respectively denote wave traveling along $+z$ and $-z$ directions, as is also indicated by the arrows. Thus, we obtain the instantaneous expression for voltage as

$$\begin{aligned} V(z, t) &= \text{Re}\,(V_s(z)\, e^{j\omega t}) \\ &= V_o^+ e^{-\alpha z} \cos(\omega t - \beta z) + V_o^- e^{\alpha z} \cos(\omega t + \beta z) \end{aligned} \tag{11.17}$$

We define the *characteristic impedance* Z_o of the line as the ratio of positively traveling voltage wave to current wave at any point on the line. Z_o is analogous to η, the intrinsic impedance of the medium of wave propagation. By substituting eqs. (11.15) and (11.16) into eqs. (11.8) and (11.9) and equating coefficients of terms $e^{\gamma z}$ and $e^{-\gamma z}$, we obtain

$$Z_o = \frac{V_o^+}{I_o^+} = -\frac{V_o^-}{I_o^-} = \frac{R + j\omega L}{\gamma} = \frac{\gamma}{G + j\omega C} \tag{11.18}$$

or

$$\boxed{Z_o = \sqrt{\frac{R + j\omega L}{G + j\omega C}} = R_o + jX_o} \tag{11.19}$$

where R_o and X_o are the real and imaginary parts of Z_o. R_o should not be mistaken for R—while R is in ohms per meter; R_o is in ohms. The propagation constant γ and the characteristic impedance Z_o are important properties of the line because they both depend on the line parameters R, L, G, and C and the frequency of operation. The reciprocal of Z_o is the characteristic admittance Y_o, that is, $Y_o = 1/Z_o$.

The transmission line considered thus far in this section is the *lossy* type in that the conductors comprising the line are imperfect ($\sigma_c \neq \infty$) and the dielectric in which the conductors are embedded is lossy ($\sigma \neq 0$). Having considered this general case, we may now consider two special cases of lossless transmission line and distortionless line.

A. Lossless Line ($R = 0 = G$)

A transmission line is said to be *lossless* if the conductors of the line are perfect ($\sigma_c \approx \infty$) and the dielectric medium separating them is lossless ($\sigma \simeq 0$). For such a line, it is evident from Table 11.1 that when $\sigma_c \simeq \infty$ and $\sigma \simeq 0$,

$$\boxed{R = 0 = G} \tag{11.20}$$

This is a necessary condition for a line to be lossless. Thus for such a line, eq. (11.20) forces eqs. (11.11), (11.14), and (11.19) to become

$$\alpha = 0, \qquad \gamma = j\beta = j\omega\sqrt{LC} \tag{11.21a}$$

$$u = \frac{\omega}{\beta} = \frac{1}{\sqrt{LC}} = f\lambda \tag{11.21b}$$

$$X_o = 0, \qquad Z_o = R_o = \sqrt{\frac{L}{C}} \tag{11.21c}$$

B. Distortionless Line ($R/L = G/C$)

A signal normally consists of a band of frequencies; wave amplitudes of different frequency components will be attenuated differently in a lossy line as α is frequency dependent. This results in distortion. A *distortionless line* is one in which the attenuation constant α is frequency independent while the phase constant β is linearly dependent on frequency. From the general expression for α and β (see Problem 11.6), it is evident that a distortionless line results if the line parameters are such that

$$\boxed{\frac{R}{L} = \frac{G}{C}} \tag{11.22}$$

Thus, for a distortionless line,

$$\gamma = \sqrt{RG\left(1 + \frac{j\omega L}{R}\right)\left(1 + \frac{j\omega C}{G}\right)}$$
$$= \sqrt{RG}\left(1 + \frac{j\omega C}{G}\right) = \alpha + j\beta$$

or

$$\alpha = \sqrt{RG}, \qquad \beta = \omega\sqrt{LC} \qquad \text{[11.23a]}$$

showing that α does not depend on frequency whereas β is a linear function of frequency. Also

$$Z_o = \sqrt{\frac{R\,(1 + j\omega L/R)}{G\,(1 + j\omega C/G)}} = \sqrt{\frac{R}{G}} = \sqrt{\frac{L}{C}} = R_o + jX_o$$

or

$$R_o = \sqrt{\frac{R}{G}} = \sqrt{\frac{L}{C}} \qquad X_o = 0 \qquad \text{[11.23b]}$$

and

$$u = \frac{\omega}{\beta} = \frac{1}{\sqrt{LC}} = f\lambda \qquad \text{[11.23c]}$$

Note that

1. The phase velocity is independent of frequency because the phase constant β linearly depends on frequency. We have shape distortion of signals unless α and u are independent of frequency.
2. u and Z_o remain the same as for lossless lines.
3. A lossless line is also a distortionless line, but a distortionless line is not necessarily lossless. Although lossless lines are desirable in power transmission, telephone lines are required to be distortionless.

A summary of our discussion is in Table 11.2. For the greater part of our analysis, we shall restrict our discussion to lossless transmission lines.

EXAMPLE 11.1

An air line has characteristic impedance of 70 Ω and phase constant of 3 rad/m at 100 MHz. Calculate the inductance per meter and the capacitance per meter of the line.

Table 11.2 **Transmission Line Characteristics**

Case	Propagation constant $\gamma = \alpha + j\beta$	Characteristic impedance $Z_o = R_o + jX_o$
General	$\sqrt{(R + j\omega L)(G + j\omega C)}$	$\sqrt{\dfrac{R + j\omega L}{G + j\omega C}}$
Lossless	$0 + j\omega\sqrt{LC}$	$\sqrt{\dfrac{L}{C}} + j0$
Distortionless	$\sqrt{RG} + j\omega\sqrt{LC}$	$\sqrt{\dfrac{L}{C}} + j0$

SOLUTION An air line can be regarded as a lossless line since $\sigma \simeq 0$. Hence

$$R = 0 = G \qquad \text{and} \qquad \alpha = 0$$

$$Z_o = R_o = \sqrt{\frac{L}{C}} \tag{11.1.1}$$

$$\beta = \omega\sqrt{LC} \tag{11.1.2}$$

Dividing eq. (11.1.1) by eq. (11.1.2) yields

$$\frac{R_o}{\beta} = \frac{1}{\omega C}$$

or

$$C = \frac{\beta}{\omega R_o} = \frac{3}{2\pi \times 100 \times 10^6 (70)} = 68.2 \text{ pF/m}$$

From eq. (11.1.1),

$$L = R_o^2 C = (70)^2 (68.2 \times 10^{-12}) = 334.2 \text{ nH/m}$$ ■

PRACTICE EXERCISE 11.1

A transmission line operating at 500 MHz has $Z_o = 80\ \Omega$, $\alpha = 0.04$ Np/m, $\beta = 1.5$ rad/m. Find the line parameters R, L, G, and C.

ANSWER 3.2 Ω/m, 38.2 nH/m, 5×10^{-4} ℧/m, 5.97 pF/m.

EXAMPLE 11.2

A distortionless line has $Z_o = 60\ \Omega$, $\alpha = 20$ mNp/m, $u = 0.6c$ where c is the speed of light in a vacuum. Find R, L, G, C, and λ at 100 MHz.

SOLUTION For a distortionless line,

$$RC = GL \qquad \text{or} \qquad G = \frac{RC}{L}$$

and hence

$$Z_o = \sqrt{\frac{L}{C}} \tag{11.2.1}$$

$$\alpha = \sqrt{RG} = R\sqrt{\frac{C}{L}} = \frac{R}{Z_o} \tag{11.2.2a}$$

or

$$R = \alpha\, Z_o \tag{11.2.2b}$$

But

$$u = \frac{\omega}{\beta} = \frac{1}{\sqrt{LC}} \tag{11.2.3}$$

From eq. (11.2.2b),

$$R = \alpha\, Z_o = (20 \times 10^{-3})(60) = 1.2\ \Omega/\text{m}$$

Dividing eq. (11.2.1) by eq. (11.2.3) results in

$$L = \frac{Z_o}{u} = \frac{60}{0.6\,(3 \times 10^8)} = 333\ \text{nH/m}$$

From eq. (11.2.2a),

$$G = \frac{\alpha^2}{R} = \frac{400 \times 10^{-6}}{1.2} = 333\ \mu\mho/\text{m}$$

Multiplying eqs. (11.2.1) and (11.2.3) together gives

$$uZ_o = \frac{1}{C}$$

or

$$C = \frac{1}{uZ_o} = \frac{1}{0.6\,(3 \times 10^8)\,60} = 92.59\ \text{pF/m}$$

$$\lambda = \frac{u}{f} = \frac{0.6\,(3 \times 10^8)}{10^8} = 1.8\ \text{m}$$

■

PRACTICE EXERCISE 11.2

A telephone line has $R = 30\ \Omega$/km, $L = 100$ mH/km, $G = 0$, and $C = 20\ \mu$F/km. At $f = 1$ kHz, obtain

(a) The characteristic impedance of the line

(b) The propagation constant

(c) The phase velocity

ANSWER (a) $70.75 \angle{-1.367°}\ \Omega$, (b) $2.121 \times 10^{-4} + j8.888 \times 10^{-3}$/m, (c) 7.069×105 m/s.

11.4 INPUT IMPEDANCE, SWR, AND POWER

Consider a transmission line of length ℓ, characterized by γ and Z_o, connected to a load Z_L as shown in Figure 11.6. Looking into the line, the generator sees the line with the load as an input impedance Z_{in}. It is our intention in this section to determine the input impedance, the standing wave ratio (SWR), and the power flow on the line.

Let the transmission line extend from $z = 0$ at the generator to $z = \ell$ at the load. First of all, we need V_o^+ and I_o^+ in eqs. (11.15) and (11.16), that is

$$V_s(z) = V_o^+ e^{-\gamma z} + V_o^- e^{\gamma z} \tag{11.24}$$

$$I_s(z) = \frac{V_o^+}{Z_o} e^{-\gamma z} - \frac{V_o^-}{Z_o} e^{\gamma z} \tag{11.25}$$

where eq. (11.18) has been incorporated. To find V_o^+ and V_o^-, the terminal conditions must be given. For example, if we are given the conditions at the input, say

$$V_o = V(z = 0), \qquad I_o = I(z = 0), \tag{11.26}$$

substituting these into eqs. (11.24) and (11.25) results in

$$V_o^+ = \frac{1}{2}(V_o + Z_o I_o) \tag{11.27a}$$

$$V_o^- = \frac{1}{2}(V_o - Z_o I_o) \tag{11.27b}$$

If the input impedance at the input terminals is Z_{in}, the input voltage V_o and the input current I_o are easily obtained from Figure 11.6 as

$$V_o = \frac{Z_{in}}{Z_{in} + Z_g} V_g, \qquad I_o = \frac{V_g}{Z_{in} + Z_g} \tag{11.28}$$

On the other hand, if we are given the conditions at the load, say

$$V_L = V(z = \ell), \qquad I_L = I(z = \ell) \tag{11.29}$$

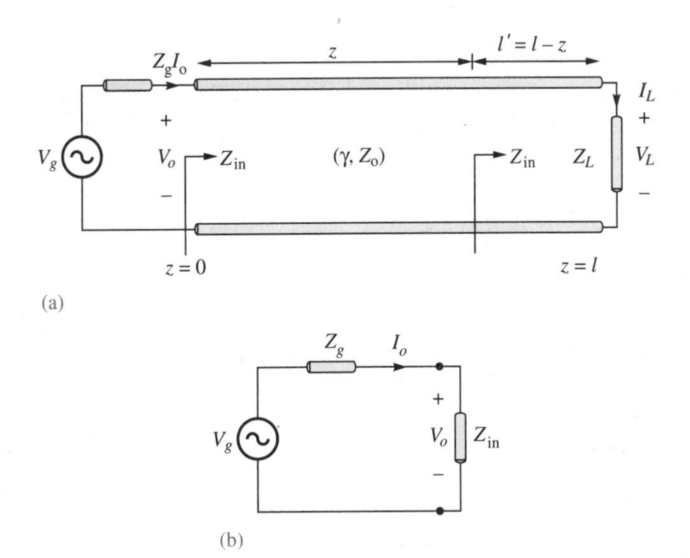

Figure 11.6 **(a)** Input impedance due to a line terminated by a load; **(b)** equivalent circuit for finding V_o and I_o in terms of Z_{in} at the input.

Substituting these into eqs. (11.24) and (11.25) gives

$$V_o^+ = \frac{1}{2}(V_L + Z_oI_L)e^{\gamma\ell} \quad [11.30a]$$

$$V_o^- = \frac{1}{2}(V_L - Z_oI_L)e^{-\gamma\ell} \quad [11.30b]$$

Next, we determine the input impedance $Z_{in} = V_s(z)/I_s(z)$ at any point on the line. At the generator, for example,

$$Z_{in} = \frac{V_s(z)}{I_s(z)} = \frac{Z_o(V_o^+ + V_o^-)}{V_o^+ - V_o^-} \quad [11.31]$$

Substituting eq. (11.30) into (11.31) and utilizing the fact that

$$\frac{e^{\gamma\ell} + e^{-\gamma\ell}}{2} = \cosh\gamma\ell, \qquad \frac{e^{\gamma\ell} - e^{-\gamma\ell}}{2} = \sinh\gamma\ell \quad [11.32a]$$

or

$$\tanh\gamma\ell = \frac{\sinh\gamma\ell}{\cosh\gamma\ell} = \frac{e^{\gamma\ell} - e^{-\gamma\ell}}{e^{\gamma\ell} + e^{-\gamma\ell}} \quad [11.32b]$$

we get

$$Z_{in} = Z_o \left[\frac{Z_L + Z_o \tanh \gamma \ell}{Z_o + Z_L \tanh \gamma \ell} \right] \quad \text{(lossy)} \tag{11.33}$$

Although eq. (11.33) has been derived for the input impedance Z_{in} at the generation end, it is a general expression for finding Z_{in} at any point on the line. To find Z_{in} at a distance ℓ' from the load as in Fig. 11.6(a), we replace ℓ by ℓ'. A formula for calculating the hyperbolic tangent of a complex number, required in eq. (11.33), is found in Appendix A.3.

For a lossless line, $\gamma = j\beta$, $\tanh j\beta\ell = j \tan \beta\ell$, and $Z_o = R_o$, so eq. (11.33) becomes

$$Z_{in} = Z_o \left[\frac{Z_L + jZ_o \tan \beta\ell}{Z_o + jZ_L \tan \beta\ell} \right] \quad \text{(lossless)} \tag{11.34}$$

showing that the input impedance varies periodically with distance ℓ from the load. The quantity $\beta\ell$ in eq. (11.34) is usually referred to as the *electrical length* of the line and can be expressed in degrees or radians.

We now define Γ_L as the *voltage reflection coefficient* (at the load). Γ_L is the ratio of the voltage reflection wave to the incident wave at the load, that is,

$$\Gamma_L = \frac{V_o^- e^{\gamma\ell}}{V_o^+ e^{-\gamma\ell}} \tag{11.35}$$

Substituting V_o^- and V_o^+ in eq. (11.30) into eq. (11.35) and incorporating $V_L = Z_L I_L$ gives

$$\Gamma_L = \frac{Z_L - Z_o}{Z_L + Z_o} \tag{11.36}$$

In general, the voltage reflection coefficient at any point on the line can be defined as the ratio of the magnitude of the reflected voltage wave to that of the incident wave, that is,

$$\Gamma(z) = \frac{V_o^- e^{\gamma z}}{V_o^+ e^{-\gamma z}} = \frac{V_o^-}{V_o^+} e^{2\gamma z}$$

But $z = \ell - \ell'$. Substituting and combining with eq. (11.35), we get

$$\Gamma(z) = \frac{V_o^-}{V_o^+} e^{2\gamma\ell} e^{-2\gamma\ell'} = \Gamma_L e^{-2\gamma\ell'} \tag{11.37}$$

The *current reflection coefficient* at any point on the line is negative of the voltage reflection coefficient at that point. Thus, the current reflection coefficient at the load is $I_o^- e^{\gamma\ell}/I_o^+ e^{-\gamma\ell} = -\Gamma_L$.

Just as we did for plane waves, we define the *standing wave ratio s* (otherwise denoted by SWR) as

$$s = \frac{V_{max}}{V_{min}} = \frac{I_{max}}{I_{min}} = \frac{1 + |\Gamma_L|}{1 - |\Gamma_L|} \tag{11.38}$$

It is easy to show that $I_{max} = V_{max}/Z_o$ and $I_{min} = V_{min}/Z_o$. The input impedance Z_{in} in eq. (11.34) has maxima and minima that occur respectively at the maxima and minima of the voltage and current standing wave. It is easily shown that

$$|Z_{in}|_{max} = \frac{V_{max}}{I_{min}} = sZ_o \tag{11.39a}$$

and

$$|Z_{in}|_{min} = \frac{V_{min}}{I_{max}} = \frac{Z_o}{s} \tag{11.39b}$$

As a way of demonstrating these concepts, consider a lossless line with characteristic impedance of $Z_o = 50\ \Omega$. For the sake of simplicity, we assume that the line is terminated in a pure resistive load $Z_L = 100\ \Omega$ and the voltage at the load is 100 V (rms). The conditions on the line are displayed in Figure 11.7. Note from the figure that conditions on the line repeat themselves every half wavelength.

The average input power at a distance z from the load is given by an equation similar to eq. (10.68); that is,

$$P_{ave} = \frac{1}{2} \text{Re}\left[V_s(z)\, I_s^*(z)\right] \tag{11.40}$$

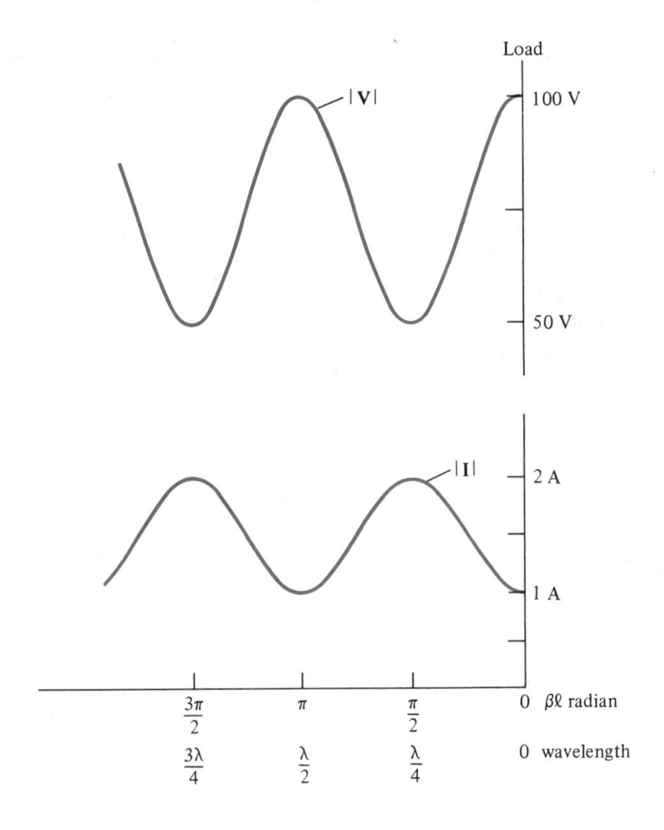

Figure 11.7 Voltage and current wave patterns on a lossless line terminated by a resistive load.

We now consider special cases when the line is connected to load $Z_L = 0$, $Z_L = \infty$, and $Z_L = Z_o$. These special cases can easily be derived from the general case.

A. Shorted Line ($Z_L = 0$)

For this case, eq. (11.34) becomes

$$Z_{sc} = Z_{in}\Big|_{Z_L = 0} = jZ_o \tan \beta\ell \qquad [11.41a]$$

Also,

$$\Gamma_L = -1, \qquad s = \infty \qquad [11.41b]$$

We notice from eq. (11.41a) that Z_{in} is a pure reactance, which could be capacitive or inductive depending on the value of ℓ. The variation of Z_{in} with ℓ is shown in Figure 11.8(a).

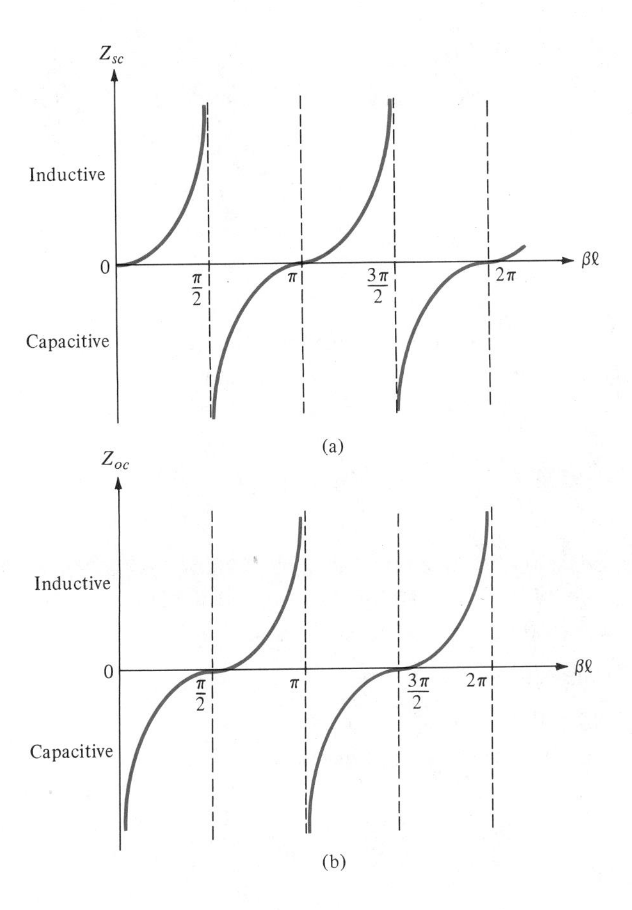

Figure 11.8 Input impedance of a lossless line: **(a)** when shorted, **(b)** when open.

B. Open-Circuited Line ($Z_L = \infty$)

In this case, eq. (11.34) becomes

$$Z_{oc} = \lim_{Z_L \to \infty} Z_{in} = \frac{Z_o}{j \tan \beta\ell} = -jZ_o \cot \beta\ell \qquad \text{[11.42a]}$$

and

$$\Gamma_L = 1, \qquad s = \infty \qquad \text{[11.42b]}$$

The variation of Z_{in} with ℓ is shown in Figure 11.8(b). Notice from eqs. (11.41a) and (11.42a) that

$$Z_{sc}Z_{oc} = Z_o^2 \qquad \text{[11.43]}$$

C. Matched Line ($Z_L = Z_o$)

This is the most desired case from the practical point of view. For this case, eq. (11.34) reduces to

$$Z_{in} = Z_o \quad [11.44a]$$

and

$$\Gamma_L = 0, \qquad s = 1 \quad [11.44b]$$

that is, $V_o^- = 0$, the whole wave is transmitted and there is no reflection. The incident power is fully absorbed by the load. Thus maximum power transfer is possible when a transmission line is matched to the load.

EXAMPLE 11.3

A certain transmission line operating at $\omega = 10^6$ rad/s has $\alpha = 8$ dB/m, $\beta = 1$ rad/m, and $Z_o = 60 + j40\ \Omega$, and is 2 m long. If the line is connected to a source of $10\underline{/0^\circ}$ V, $Z_g = 40\ \Omega$ and terminated by a load of $20 + j50\ \Omega$, determine

(a) The input impedance

(b) The sending-end current

(c) The current at the middle of the line

SOLUTION:

(a) Since 1 Np = 8.686 dB,

$$\alpha = \frac{8}{8.686} = 0.921 \text{ Np/m}$$

$$\gamma = \alpha + j\beta = 0.921 + j1 \text{ /m}$$

$$\gamma\ell = 2(0.921 + j1) = 1.84 + j2$$

Using the formula for $\tanh(x + jy)$ in Appendix A.3, we obtain

$$\tanh \gamma\ell = 1.033 - j0.03929$$

$$Z_{in} = Z_o \left(\frac{Z_L + Z_o \tanh \gamma\ell}{Z_o + Z_L \tanh \gamma\ell} \right)$$

$$= (60 + j40) \left(\frac{20 + j50 + (60 + j40)(1.033 - j0.03929)}{60 + j40 + (20 + j50)(1.033 - j0.03929)} \right)$$

$$Z_{in} = 60.25 + j38.79\ \Omega$$

(b) The sending-end current is $I(z = 0) = I_o$. From eq. (11.28),

$$I(z = 0) = \frac{V_g}{Z_{in} + Z_g} = \frac{10}{60.25 + j38.79 + 40}$$

$$= 93.03 \underline{/-21.15°} \text{ mA}$$

(c) To find the current at any point, we need V_o^+ and V_o^-. But

$$I_o = I(z = 0) = 93.03\underline{/-21.15°} \text{ mA}$$

$$V_o = Z_{in}I_o = (71.66\underline{/32.77°})(0.09303 \underline{/-21.15°}) = 6.667\underline{/11.62°} \text{ V}$$

From eq. (11.27),

$$V_o^+ = \frac{1}{2}(V_o + Z_oI_o)$$

$$= \frac{1}{2}[6.667\underline{/11.62°} + (60 + j40)(0.09303\underline{/-21.15°})] = 6.687\underline{/12.08°}$$

$$V_o^- = \frac{1}{2}(V_o - Z_oI_o) = 0.0518\underline{/260°}$$

At the middle of the line, $z = \ell/2$, $\gamma z = 0.921 + j1$. Hence, the current at this point is

$$I_s(z = \ell/2) = \frac{V_o^+}{Z_o} e^{-\gamma z} - \frac{V_o^-}{Z_o} e^{\gamma z}$$

$$= \frac{(6.687e^{j12.08°})e^{-0.921 - j1}}{60 + j40} - \frac{(0.0518e^{j260°})e^{0.921 + j1}}{60 + j40}$$

Note that $j1$ is in radians and is equivalent to $j57.3°$. Thus,

$$I_s(z = \ell/2) = \frac{6.687e^{j12.08°}e^{-0.921}e^{-j57.3°}}{72.1e^{j33.69°}} - \frac{0.0518e^{j260°}e^{0.921}e^{j57.3°}}{72.1e^{33.69°}}$$

$$= 0.0369e^{-j78.91°} - 0.001805e^{j283.61°}$$

$$= 6.673 - j34.456 \text{ mA}$$

$$= 35.10\underline{/281°} \text{ mA.}$$

■

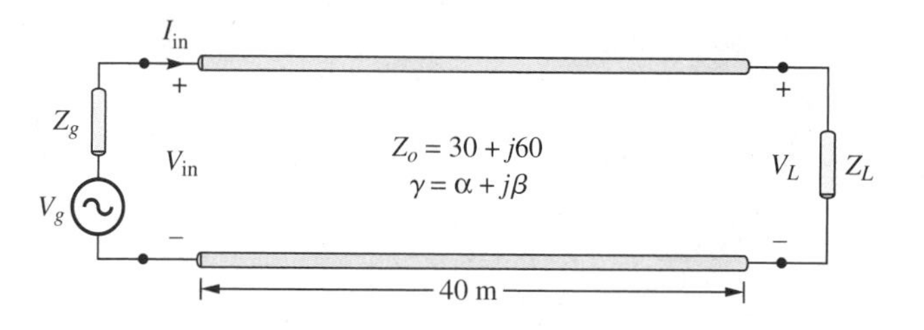

Figure 11.9 For Practice Exercise 11.3.

PRACTICE EXERCISE 11.3

A 40-m-long transmission line shown in Figure 11.9 has $V_g = 15\underline{/0°}$ V_{rms}, $Z_o = 30 + j60\ \Omega$, and $V_L = 5\underline{/-48°}$ V_{rms}. If the line is matched to the load, calculate:

(a) The input impedance Z_{in}

(b) The sending-end current I_{in} and voltage V_{in}

(c) The propagation constant γ

ANSWER (a) $30 + j60\ \Omega$, (b) $0.112\underline{/-63.43°}$ A, $7.5\underline{/0°}$ V_{rms}, (c) $0.0101 + j0.2094$ /m.

11.5 THE SMITH CHART

Prior to the advent of digital computers and calculators, engineers developed all sorts of aids (tables, charts, graphs, etc.) to facilitate their calculations for design and analysis. To reduce the tedious manipulations involved in calculating the characteristics of transmission lines, graphical means have been developed. The Smith chart[2] is the most commonly used of the graphical techniques. It is basically a graphical indication of the impedance of a transmission line as one moves along the line. It becomes very easy to use after a small amount of experience. We will first examine how the Smith chart is constructed and later employ it in our calculations of transmission line characteristics such as Γ_L, s, and Z_{in}. We will assume that the transmission line to which the Smith chart will be applied is lossless ($Z_o = R_o$) although this is not fundamentally required.

[2]Devised by Phillip H. Smith in 1939. See P. H. Smith, "Transmission line calculator." *Electronics*, vol. 12, pp. 29–31, 1939 and P. H. Smith, "An improved transmission line calculator." *Electronics*, vol. 17, pp. 130–133, 318–325, 1944.

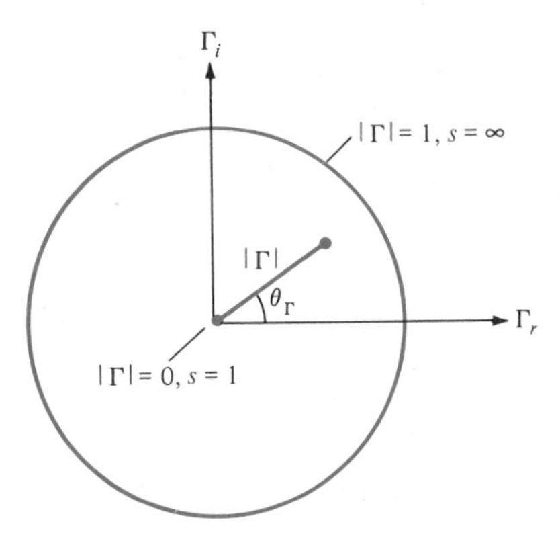

Figure 11.10 Unit circle on which the Smith chart is constructed.

The Smith chart is constructed within a circle of unit radius ($|\Gamma| \leq 1$) as shown in Figure 11.10. The construction of the chart is based on the relation in eq. (11.36)[3]; that is,

$$\Gamma = \frac{Z_L - Z_o}{Z_L + Z_o} \tag{11.45}$$

or

$$\Gamma = |\Gamma| \underline{/\theta_\Gamma} = \Gamma_r + j\Gamma_i \tag{11.46}$$

where Γ_r and Γ_i are the real and imaginary parts of the reflection coefficient Γ.

Instead of having separate Smith charts for transmission lines with different characteristic impedances such as $Z_o = 60$, 100, and 120 Ω, we prefer to have just one that can be used for any line. We achieve this by using a normalized chart in which all impedances are normalized with respect to the characteristic impedance Z_o of the particular line under consideration. For the load impedance Z_L, for example, the *normalized impedance* z_L is given by

$$z_L = \frac{Z_L}{Z_o} = r + jx \tag{11.47}$$

Substituting eq. (11.47) into eqs. (11.45) and (11.46) gives

$$\Gamma = \Gamma_r + j\Gamma_i = \frac{z_L - 1}{z_L + 1} \tag{11.48a}$$

....

[3]Whenever a subscript is not attached to Γ, we simply mean voltage reflection coefficient at the load ($\Gamma_L = \Gamma$).

or

$$z_L = r + jx = \frac{(1 + \Gamma_r) + j\Gamma_i}{(1 - \Gamma_r) - j\Gamma_i} \quad [11.48b]$$

Normalizing and equating components, we obtain

$$r = \frac{1 - \Gamma_r^2 - \Gamma_i^2}{(1 - \Gamma_r)^2 + \Gamma_i^2} \quad [11.49a]$$

$$x = \frac{2\,\Gamma_i}{(1 - \Gamma_r)^2 + \Gamma_i^2} \quad [11.49b]$$

Rearranging terms in eq. (11.49) leads to

$$\boxed{\left[\Gamma_r - \frac{r}{1 + r}\right]^2 + \Gamma_i^2 = \left[\frac{1}{1 + r}\right]^2} \quad [11.50]$$

and

$$\boxed{[\Gamma_r - 1]^2 + \left[\Gamma_i - \frac{1}{x}\right]^2 = \left[\frac{1}{x}\right]^2} \quad [11.51]$$

Each of eqs. (11.50) and (11.51) is similar to

$$(x - h)^2 + (y - k)^2 = a^2 \quad [11.52]$$

which is the general equation of a circle of radius a, centered at (h, k). Thus eq. (11.50) is an *r-circle (resistance circle)* with

$$\text{center at } (\Gamma_r, \Gamma_i) = \left(\frac{r}{1 + r}, 0\right) \quad [11.53a]$$

$$\text{radius} = \frac{1}{1 + r} \quad [11.53b]$$

For typical values of the normalized resistance r, the corresponding centers and radii of the r-circles are presented in Table 11.3. Typical examples of the r-circles based on the data in Table 11.3 are shown in Figure 11.11. Similarly, eq. (11.51) is an *x-circle (reactance circle)* with

$$\text{center at } (\Gamma_r, \Gamma_i) = \left(1, \frac{1}{x}\right) \quad [11.54a]$$

$$\text{radius} = \frac{1}{x} \quad [11.54b]$$

Table 11.3 Radii and Centers of r-Circles for Typical Values of r

Normalized resistance (r)	Radius $\left(\frac{1}{1+r}\right)$	Center $\left(\frac{r}{1+r}, 0\right)$
0	1	(0, 0)
1/2	2/3	(1/3, 0)
1	1/2	(1/2, 0)
2	1/3	(2/3, 0)
5	1/6	(5/6, 0)
∞	0	(1, 0)

Table 11.4 presents centers and radii of the x-circles for typical values of x, and Figure 11.12 shows the corresponding plots. Notice that while r is always positive, x can be positive (for inductive impedance) or negative (for capacitive impedance).

If we superpose the r-circles and x-circles, what we have is the Smith chart shown in Figure 11.13. On the chart, we locate a normalized impedance $z = 2 + j$, for example, as the point of intersection of the $r = 2$ circle and the $x = 1$ circle. This is point P_1 in Figure 11.13. Similarly, $z = 1 - j\,0.5$ is located at P_2, where the $r = 1$ circle and the $x = -0.5$ circle intersect.

Apart from the r- and x-circles (shown on the Smith chart), we can draw the *s-circles* or *constant standing-wave-ratio circles* (always not shown on the Smith chart) which are centered at the origin with s varying from 1 to ∞. The value of the standing wave ratio s is determined by locating where an s-circle crosses the Γ_r axis. Typical examples of s-circles for $s = 1, 2, 3$ and ∞ are shown in Figure 11.13. Since $|\Gamma|$ and

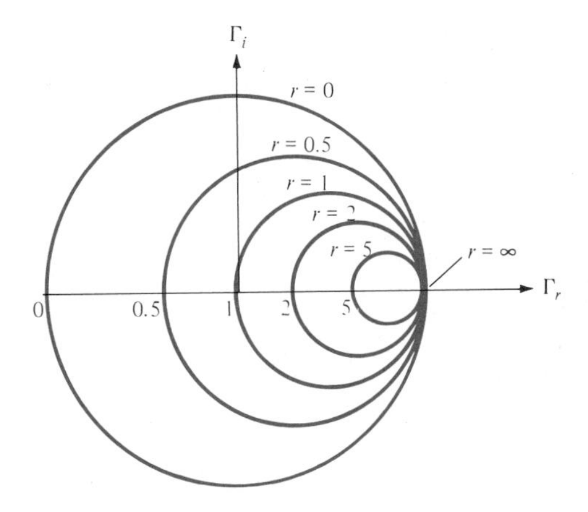

Figure 11.11 Typical r-circles for $r = 0, 0.5, 1, 2, 5, \infty$.

Table 11.4 Radii and Centers of x-Circles for Typical Values of x

Normalized reactance (x)	Radius $\left(\frac{1}{x}\right)$	Center $\left(1, \frac{1}{x}\right)$
0	∞	$(1, \infty)$
$\pm 1/2$	2	$(1, \pm 2)$
± 1	1	$(1, \pm 1)$
± 2	1/2	$(1, \pm 1/2)$
± 5	1/5	$(1, \pm 1/5)$
$\pm \infty$	0	$(1, 0)$

s are related according to eq. (11.38), the s-circles are sometimes referred to as $|\Gamma|$-circles with $|\Gamma|$ varying linearly from 0 to 1 as we move away from the center O toward the periphery of the chart while s varies nonlinearly from 1 to ∞.

The following points should be noted about the Smith chart:

1. At point P_{sc} on the chart, $r = 0$, $x = 0$; that is, $Z_L = 0 + j0$ showing that P_{sc} represents a short circuit on the transmission line. At point P_{oc}, $r = \infty$ and $x = \infty$, or $Z_L = \infty + j\infty$, which implies that P_{oc} corresponds to an open circuit on the line. Also at P_{oc}, $r = 0$ and $x = 0$, showing that P_{oc} is another location of a short circuit on the line.
2. A complete revolution (360°) around the Smith chart represents a distance of $\lambda/2$ on the line. Clockwise movement on the chart is regarded as moving toward the generator (or away from the load) as shown by the arrow G in

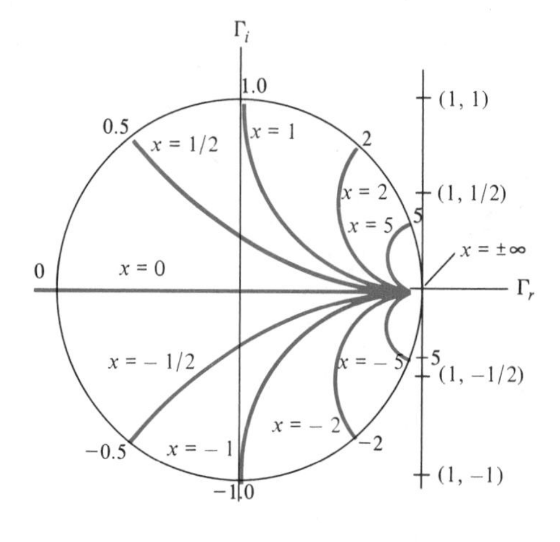

Figure 11.12 Typical x-circles for $x = 0, \pm 1/2, \pm 1, \pm 2, \pm 5, \pm\infty$.

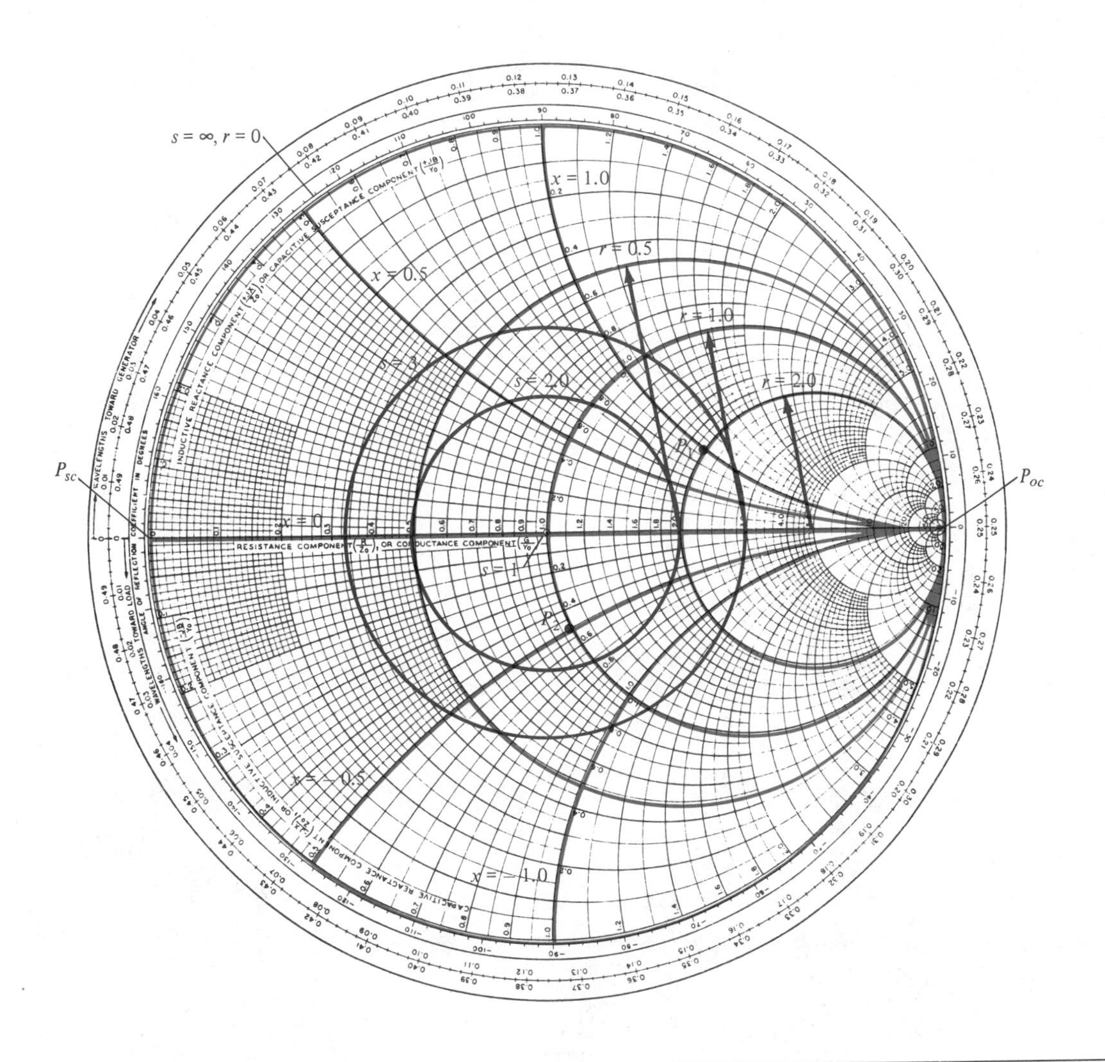

Figure 11.13 Illustration of the r-, x-, and s-circles on the Smith chart.

Figure 11.14(a) and (b). Similarly, counterclockwise movement on the chart corresponds to moving toward the load (or away from the generator) as indicated by the arrow L in Figure 11.14. Notice from Figure 11.14(b) that at the load, moving toward the load does not make sense (because we are already at the load). The same can be said of the case when we are at the generator end.

3. There are three scales around the periphery of the Smith chart as illustrated in Figure 11.14(a). The three scales are included for the sake of convenience but they are actually meant to serve the same purpose; one scale should be sufficient. The scales are used in determining the distance from the load or

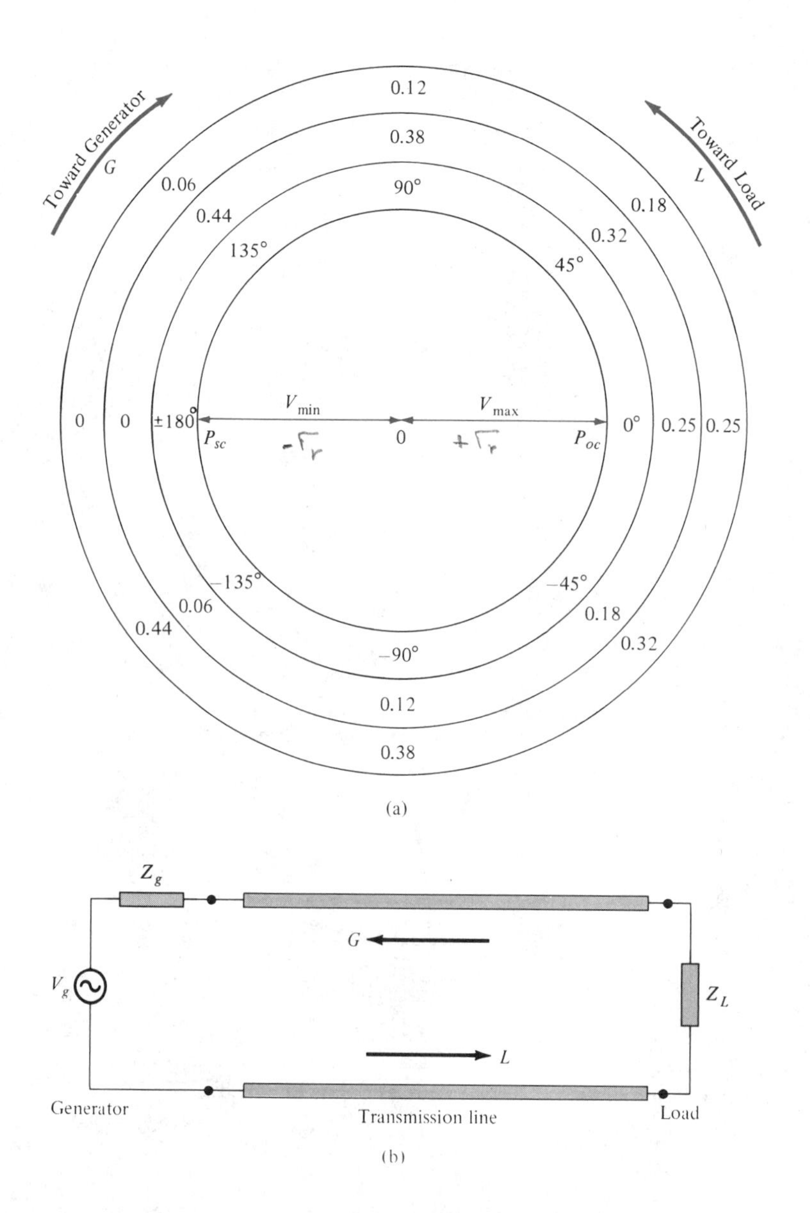

Figure 11.14 **(a)** Smith chart illustrating scales around the periphery and movements around the chart, **(b)** corresponding movements along the transmission line.

generator in degrees or wavelengths. The outermost scale is used to determine the distance on the line from the generator end in terms of wavelengths, and the next scale determines the distance from the load end in terms of wavelengths. The innermost scale is a protractor (in degrees) and is primarily used in determining θ_Γ; it can also be used to determine the distance from the load or generator. Since a $\lambda/2$ distance on the line corresponds to a movement of 360° on the chart, *λ distance on the line corresponds to a 720° movement on the chart.*

$$\boxed{\lambda \rightarrow 720} \qquad [11.55]$$

Thus we may ignore the other outer scales and use the protractor (the innermost scale) for all our θ_Γ and distance calculations.

4. V_{max} occurs where $Z_{in,\,max}$ is located on the chart (see eq. (11.39a)), and that is on the positive Γ_r axis or on OP_{oc} in Figure 11.14(a). V_{min} is located at the same point where we have $Z_{in,min}$ on the chart; that is, on the negative Γ_r axis or on OP_{sc} in Fig. 11.14(a). Notice that V_{max} and V_{min} (or $Z_{in,max}$ and $Z_{in,min}$) are $\lambda/4$ (or 180°) apart.
5. The Smith chart is used both as impedance chart and admittance chart ($Y = 1/Z$). As admittance chart (normalized impedance $y = Y/Y_o = g + jb$), the g- and b-circles correspond to r- and x-circles respectively.

Based on these important properties, the Smith chart may be used to determine, among other things, (a) $\Gamma = |\Gamma| \underline{/\theta_\Gamma}$ and s; (b) Z_{in} or Y_{in}; and (c) the locations of V_{max} and V_{min} provided that we are given Z_o, Z_L, and the length of the line. Some examples will clearly show how we can do all these and much more with the aid of the Smith chart, a compass, and a plain straightedge.

EXAMPLE 11.4

A 30-m-long lossless transmission line with $Z_o = 50\ \Omega$ operating at 2 MHz is terminated with a load $Z_L = 60 + j40\ \Omega$. If $u = 0.6c$ on the line, find

(a) The reflection coefficient Γ

(b) The standing wave ratio s

(c) The input impedance

SOLUTION

This problem will be solved with and without using the Smith chart.

Method 1: (Without the Smith chart)

(a) $$\Gamma = \frac{Z_L - Z_o}{Z_L + Z_o} = \frac{60 + j40 - 50}{60 + j40 + 50} = \frac{10 + j40}{110 + j40}$$

$$= 0.3523 \underline{/56°}$$

(b) $$s = \frac{1 + |\Gamma|}{1 - |\Gamma|} = \frac{1 + 0.3523}{1 - 0.3523} = 2.088$$

(c) Since $u = \omega/\beta$, or $\beta = \omega/u$,

$$\beta\ell = \frac{\omega\ell}{u} = \frac{2\pi\,(2 \times 10^6)(30)}{0.6\,(3 \times 10^8)} = \frac{2\pi}{3} = 120°$$

Note that $\beta\ell$ is the electrical length of the line.

$$Z_{\text{in}} = Z_o \left[\frac{Z_L + jZ_o \tan \beta\ell}{Z_o + jZ_L \tan \beta\ell}\right]$$

$$= \frac{50\,(60 + j40 + j50 \tan 120°)}{[50 + j(60 + j40) \tan 120°]}$$

$$= \frac{50\,(6 + j4 - j5\sqrt{3})}{(5 + 4\sqrt{3} - j6\sqrt{3})} = 24.01 \underline{/3.22°}$$

$$= 23.97 + j1.35\Omega$$

Method 2: (Using the Smith chart).

(a) Calculate the normalized load impedance

$$z_L = \frac{Z_L}{Z_o} = \frac{60 + j40}{50}$$

$$= 1.2 + j0.8.$$

Locate z_L on the Smith chart of Figure 11.15 at point P where the $r = 1.2$ circle and the $x = 0.8$ circle meet. To get Γ at z_L, extend OP to meet the $r = 0$ circle at Q and measure OP and OQ. Since OQ corresponds to $|\Gamma| = 1$, then at P,

$$|\Gamma| = \frac{OP}{OQ} = \frac{3.2 \text{ cm}}{9.1 \text{ cm}} = 0.3516$$

Note that $OP = 3.2$ cm and $OQ = 9.1$ cm were taken from the Smith chart used by the author; the Smith chart in Figure 11.15 is reduced but the ratio of OP/OQ remains the same.

Angle θ_Γ is read directly on the chart as the angle between OS and OP; that is

$$\theta_\Gamma = \text{angle } POS = 56°$$

Thus

$$\Gamma = 0.3516 \underline{/56°}$$

(b) To obtain the standing wave ratio s, draw a circle with radius OP and center at O. This is the constant s or $|\Gamma|$ circle. Locate point S where the s-circle meets the Γ_r-axis. (This is easily shown by setting $\Gamma_i = 0$ in eq. (11.49a).) The value of r at this point is s; that is

$$s = r \text{ (for } r \geq 1)$$

$$= 2.1$$

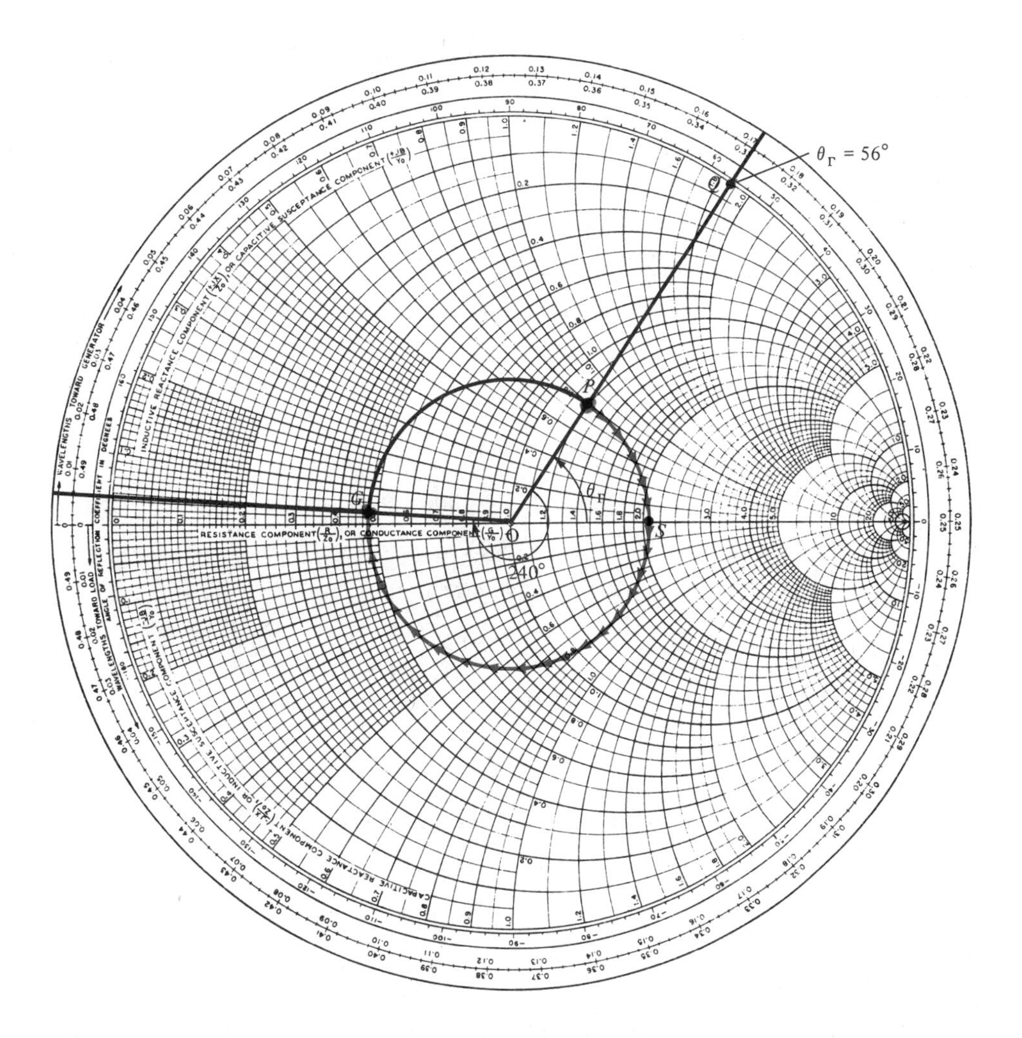

Figure 11.15 For Example 11.4.

(c) To obtain Z_{in}, first express ℓ in terms of λ or in degrees.

$$\lambda = \frac{u}{f} = \frac{0.6\,(3 \times 10^8)}{2 \times 10^6} = 90 \text{ m}$$

$$\ell = 30 \text{ m} = \frac{30}{90}\lambda = \frac{\lambda}{3} \rightarrow \frac{720^\circ}{3} = 240^\circ$$

Since λ corresponds to an angular movement of 720° on the chart, the length of the line corresponds to an angular movement of 240°. That means, we move toward the generator (or away from the load, in the clockwise direction) 240° on the s-circle from point P to point G. At G, we obtain

$$z_{in} = 0.47 + j0.035$$

Hence

$$Z_{in} = Z_o z_{in} = 50(0.47 + j0.035) = 23.5 + j1.75\ \Omega$$

Although the results obtained using the Smith chart are only approximate, for engineering purposes they are close enough to the exact ones obtained in Method 1. ■

PRACTICE EXERCISE 11.4

A 70-Ω lossless line has $s = 1.6$ and $\theta_\Gamma = 300°$. If the line is 0.6λ long, obtain

(a) Γ, Z_L, Z_{in}

(b) The distance of the first minimum voltage from the load

ANSWER (a) $0.228\ \underline{/300°}$, $80.5 - j33.6\ \Omega$, $47.6 - j17.5\ \Omega$, (b) $\lambda/6$.

EXAMPLE 11.5

A $100 + j150$–Ω load is connected to a 75-Ω lossless line. Find:

(a) Γ

(b) s

(c) The load admittance Y_L

(d) Z_{in} at 0.4λ from the load

(e) The locations of V_{max} and V_{min} with respect to the load if the line is 0.6λ long

(f) Z_{in} at the generator.

SOLUTION (a) We can use the Smith chart to solve this problem. The normalized load impedance is

$$z_L = \frac{Z_L}{Z_o} = \frac{100 + j150}{75} = 1.33 + j2$$

We locate this at point P on the Smith chart of Figure 11.16. At P, we obtain

$$|\Gamma| = \frac{OP}{OQ} = \frac{6\text{ cm}}{9.1\text{ cm}} = 0.659$$

$$\theta_\Gamma = \text{angle } POS = 40°$$

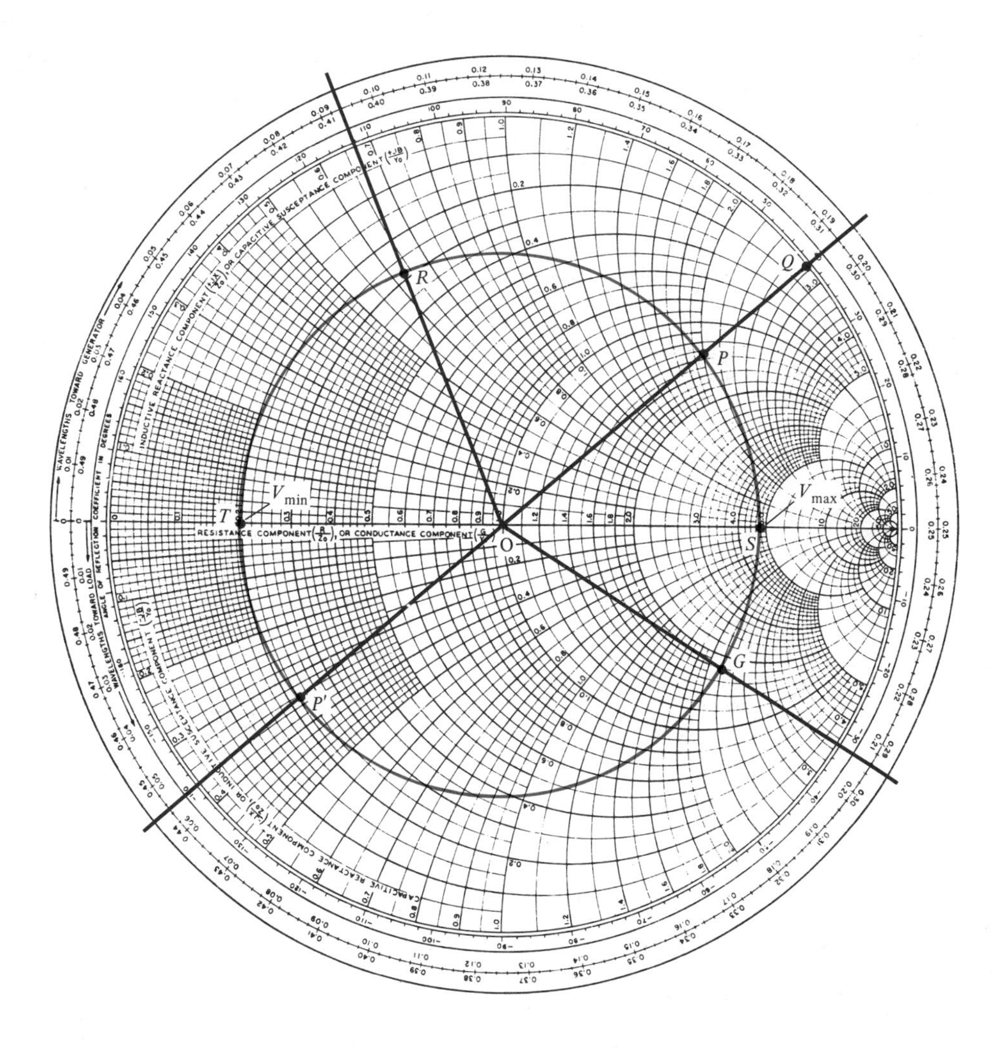

Figure 11.16 For Example 11.5.

Hence,

$$\Gamma = 0.659 \angle 40^\circ$$

Check:

$$\Gamma = \frac{Z_L - Z_o}{Z_L + Z_o} = \frac{100 + j150 - 75}{100 + j150 + 75}$$

$$= 0.659 \angle 40^\circ$$

(b) Draw the constant s-circle passing through P and obtain

$$s = 4.82$$

Check:

$$s = \frac{1 + |\Gamma|}{1 - |\Gamma|} = \frac{1 + 0.659}{1 - 0.659} = 4.865$$

(c) To obtain Y_L, extend PO to POP' and note point P' where the constant s-circle meets POP'. At P', obtain

$$y_L = 0.228 - j0.35$$

The load admittance is

$$Y_L = Y_o y_L = \frac{1}{75}(0.228 - j0.35) = 3.04 - j4.67 \text{ m℧}$$

Check:

$$Y_L = \frac{1}{Z_L} = \frac{1}{100 + j150} = 3.07 - j4.62 \text{ m℧}$$

(d) 0.4λ corresponds to an angular movement of $0.4 \times 720° = 288°$ on the constant s-circle. From P, we move 288° toward the generator (clockwise) on the s-circle to reach point R. At R,

$$z_{in} = 0.3 + j0.63$$

Hence

$$Z_{in} = Z_o z_{in} = 75\,(0.3 + j0.63)$$
$$= 22.5 + j47.25\ \Omega$$

Check:

$$\beta\ell = \frac{2\pi}{\lambda}(0.4\lambda) = 360°\,(0.4) = 144°$$

$$Z_{in} = Z_o\left[\frac{Z_L + jZ_o \tan \beta\ell}{Z_o + jZ_L \tan \beta\ell}\right]$$
$$= \frac{75\,(100 + j150 + j75 \tan 144°)}{[75 + j(100 + j150) \tan 144°]}$$
$$= 54.41\,\underline{/65.25°}$$

or

$$Z_{in} = 21.9 + j47.6 \text{ ℧}$$

(e) 0.6λ corresponds to an angular movement of

$$0.6 \times 720° = 432° = 1 \text{ revolution} + 72°$$

Thus, we start from P (load end), move along the s-circle 432°, or one revolution plus 72°, and reach the generator at point G. Note that to reach G from P, we have passed through point T (location of V_{min}) once and point S (location of V_{max}) twice. Thus, from the load,

$$\text{1st } V_{max} \text{ is located at } \frac{40°}{720°}\lambda = 0.055\lambda$$

$$\text{2nd } V_{max} \text{ is located at } 0.0555\lambda + \frac{\lambda}{2} = 0.555\lambda$$

and the only V_{min} is located at $0.055\lambda + \lambda/4 = 0.3055\lambda$

(f) At G (generator end),

$$z_{in} = 1.8 - j2.2$$

$$Z_{in} = 75(1.8 - j2.2) = 135 - j165\ \Omega.$$

This can be checked by using eq. (11.34), where $\beta\ell = \frac{2\pi}{\lambda}(0.6\lambda) = 216°$.

We can see how much time and effort is saved using the Smith chart. ■

PRACTICE EXERCISE 11.5

A lossless 60-Ω line is terminated by a 60 + j60–Ω load.

(a) Find Γ and s. If $Z_{in} = 120 - j60\ \Omega$, how far (in terms of wavelengths) is the load from the generator? Solve this without using the Smith chart.

(b) Solve the problem in (a) using the Smith chart. Calculate Z_{max} and $Z_{in,min}$. How far (in terms of λ) is the first maximum voltage from the load?

ANSWER (a) $0.4472\,\underline{/63.43°}$, 2.618, $\frac{\lambda}{8}(1 + 4n)$, $n = 0, 1, 2, \ldots$, (b) $0.4457\,\underline{/62°}$, 2.612, $\frac{\lambda}{8}(1 + 4n)$, 157.1 Ω, 22.92 Ω, 0.0861 λ.

11.6 SOME APPLICATIONS OF TRANSMISSION LINES

Transmission lines are used to serve different purposes. Here we consider how transmission lines are used for load matching and impedance measurements.

A. Quarter-Wave Transformer (Matching)

When $Z_o \neq Z_L$, we say that the load is *mismatched* and a reflected wave exists on the line. However, for maximum power transfer, it is desired that the load be matched to the transmission line ($Z_o = Z_L$) so that there is no reflection ($|\Gamma| = 0$ cr $s = 1$). The matching is achieved by using shorted sections of transmission lines.

We recall from eq. (11.34) that when $\ell = \lambda/4$ or $\beta\ell = (2\pi/\lambda)(\lambda/4) = \pi/2$,

$$Z_{in} = Z_o \left[\frac{Z_L + jZ_o \tan \pi/2}{Z_o + jZ_L \tan \pi/2}\right] = \frac{Z_o^2}{Z_L} \tag{11.56}$$

that is

$$\frac{Z_{in}}{Z_o} = \frac{Z_o}{Z_L}$$

or

$$z_{in} = \frac{1}{z_L} \rightarrow y_{in} = z_L \tag{11.57}$$

Thus by adding a $\lambda/4$ line on the Smith chart, we obtain the input admittance corresponding to a given impedance.

Also, a mismatched load Z_L can be properly matched to a line (with characteristic impedance Z_o) by inserting prior to the load a transmission line $\lambda/4$ long (with characteristic impedance Z_o') as shown in Figure 11.17. The $\lambda/4$ section of transmission line is called a *quarter-wave transformer* because it is used for impedance matching like an ordinary transformer. From eq. (11.56), Z_o' is selected such that ($Z_{in} = Z_o$)

$$Z_o' = \sqrt{Z_o Z_L} \tag{11.58}$$

where Z_o', Z_o, and Z_L are all real. If, for example, a 120-Ω load is to be matched to a 75-Ω line, the quarter-wave transformer must have a characteristic impedance of $\sqrt{(75)(120)} \simeq 95\ \Omega$. This 95-$\Omega$ quarter-wave transformer will also match a 75-Ω load to a 120-Ω line. The voltage standing wave patterns with and without the $\lambda/4$ transformer are shown in Figure 11.18(a) and (b), respectively. From Figure 11.18, we observe that although a standing wave still exists between the transformer and the load,

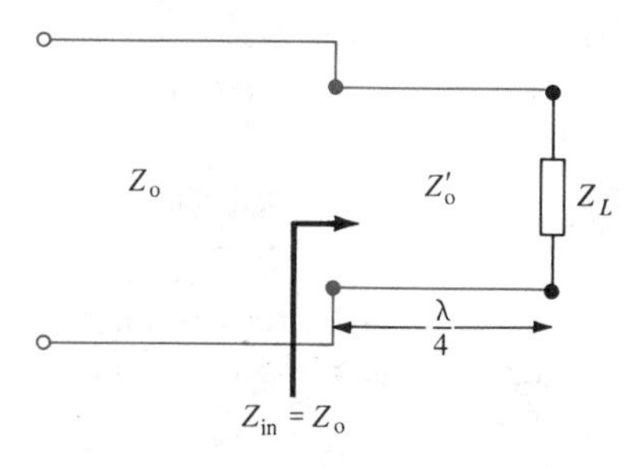

Figure 11.17 Load matching using a $\lambda/4$ transformer.

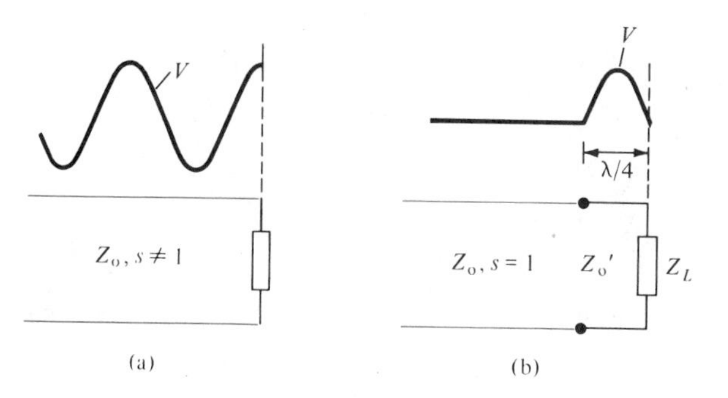

Figure 11.18 Voltage standing wave pattern of mismatched load: **(a)** without a $\lambda/4$ transformer, **(b)** with a $\lambda/4$ transformer.

there is no standing wave to the left of the transformer due to the matching. However, the reflected wave (or standing wave) is eliminated only at the desired wavelength (or frequency f); there will be reflection at a slightly different wavelength. Thus, the main disadvantage of the quarter-wave transformer is that it is a narrow-band or frequency-sensitive device.

B. Single-Stub Tuner (Matching)

The major drawback of using a quarter-wave transformer as a line-matching device is eliminated by using a *single-stub* tuner. The tuner consists of an open or shorted section of transmission line of length d connected in parallel with the main line at some distance ℓ from the load as in Figure 11.19. Notice that the stub has the same characteristic impedance as the main line. It is more difficult to use a series stub although it is theoretically feasible. An open-circuited stub radiates some energy at high frequencies. Consequently, shunt short-circuited stubs are preferred.

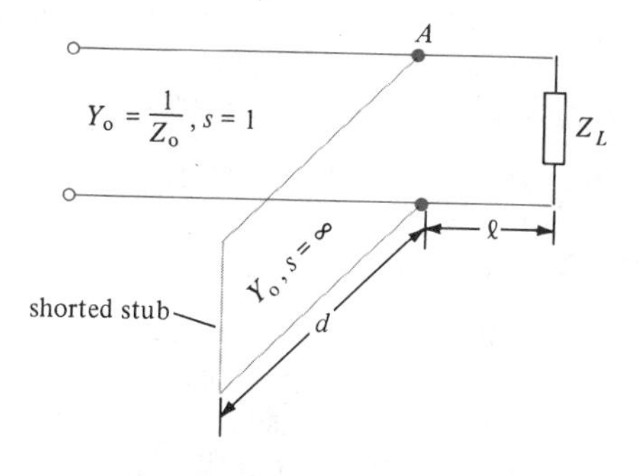

Figure 11.19 Matching with a single-stub tuner.

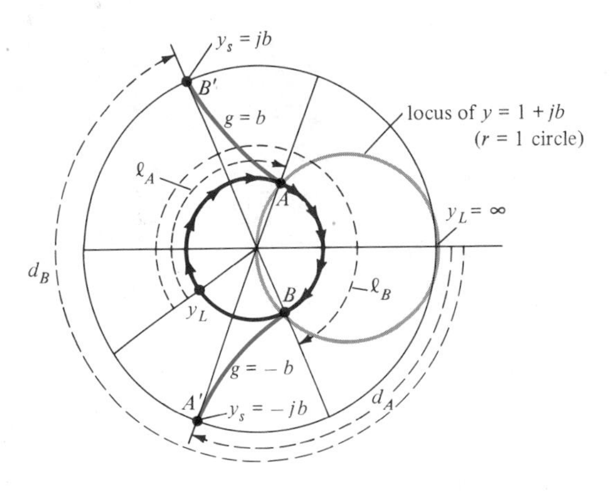

Figure 11.20 Using the Smith chart to determine ℓ and d of a shunt-shorted single-stub tuner.

As we intend that $Z_{in} = Z_o$; that is, $z_{in} = 1$ or $y_{in} = 1$ at point A on the line, we first draw the locus $y = 1 + jb$ ($r = 1$ circle) on the Smith chart as shown in Figure 11.20. If a shunt stub of admittance $y_s = -jb$ is introduced at A, then

$$y_{in} = 1 + jb + y_s = 1 + jb - jb = 1 + j0 \tag{11.59}$$

as desired. Since b could be positive or negative, two possible values of ℓ ($< \lambda/2$) can be found on the line. At A, $y_s = -jb$, $\ell = \ell_A$ and at B, $y_s = jb$, $\ell = \ell_B$ as in Figure 11.20. Due to the fact that the stub is shorted ($y_L' = \infty$), we determine the length d of the stub by finding the distance from P_{sc} (at which $z_L' = 0 + j0$) to the required stub admittance y_s. For the stub at A, we obtain $d = d_A$ as the distance from P to A' where A' corresponds to $y_s = -jb$ located on the periphery of the chart as in Figure 11.20. Similarly, we obtain $d = d_B$ as the distance from P_{sc} to B' ($y_s = jb$).

Thus we obtain $d = d_A$ and $d = d_B$, corresponding to A and B respectively, as shown in Figure 11.20. Note that $d_A + d_B = \lambda/2$ always. Since we have two possible shunted stubs, we normally choose to match the shorter stub or one at a position closer to the load. Instead of having a single stub shunted across the line, we may have two stubs. This is called *double-stub matching* and allows for the adjustment of the load impedance.

C. Slotted Line (Impedance Measurement)

At high frequencies, it is very difficult to measure current and voltage because measuring devices become significant in size and every circuit becomes a transmission line. The slotted line is a simple device used in determining the impedance of an unknown load at high frequencies up into the region of gigahertz. It consists of a section of an air (lossless) line with a slot in the outer conductor as shown in Figure 11.21. The line has a probe, along the **E** field (see Figure 11.4), which samples the **E**

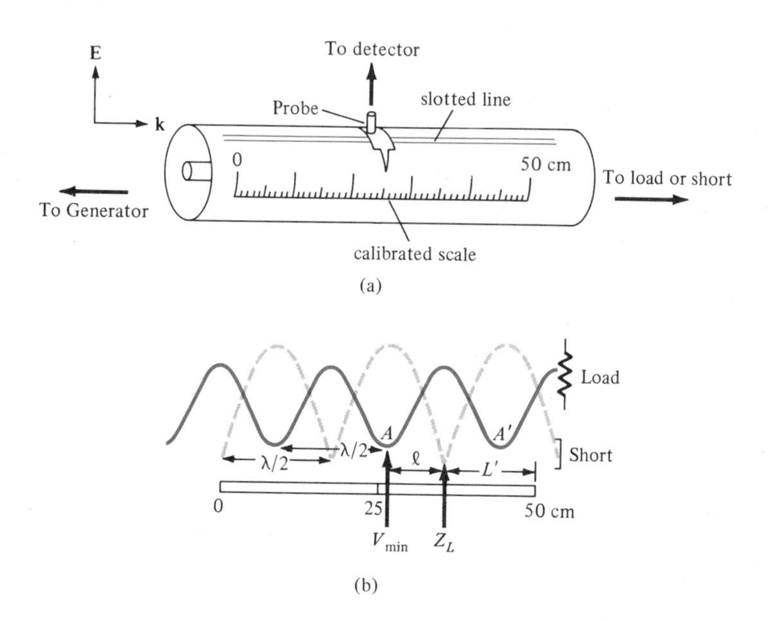

Figure 11.21 **(a)** Typical slotted line; **(b)** determining the location of the load Z_L and V_{min} on the line.

field and consequently measures the potential difference between the probe and its outer shield.

The slotted line is primarily used in conjunction with the Smith chart to determine the standing wave ratio s (the ratio of maximum voltage to the minimum voltage) and the load impedance Z_L. The value of s is read directly on the detection meter when the load is connected. To determine Z_L, we first find the position of the load by replacing the load by a short circuit and noting the locations of voltage minima (which are more accurately determined than the maxima because of the sharpness of the turning point) on the scale. Since impedances repeat every half wavelength, any of the minima may be selected as the load reference point. We now determine the distance from the selected reference point to the load by replacing the short circuit by the load and noting the locations of voltage minima. The distance ℓ (distance of V_{min} toward the load) expressed in terms of λ is used to locate the position of the load of an s-circle on the chart as shown in Figure 11.22. We could also locate the load by using ℓ', which is the distance of V_{min} toward the generator. Either ℓ or ℓ' may be used to locate Z_L.

It will be noted that the procedure involved in using the slotted line can be summarized as follows:

1. With the load connected, read s on the detection meter. With the value of s, draw the s-circle on the Smith chart.
2. With the load replaced by a short circuit, locate a reference position for z_L at a voltage minimum point.

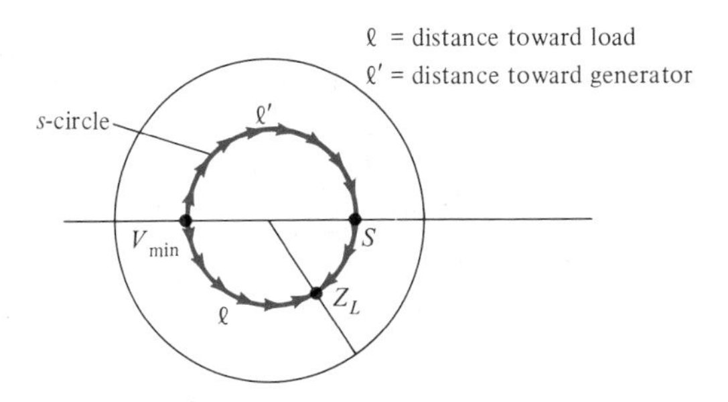

Figure 11.22 Determining the load impedance from the Smith chart using the data obtained from the slotted line.

3. With the load on the line, note the position of V_{min} and determine ℓ.
4. On the Smith chart, move toward the load a distance ℓ from the location of V_{min}. Find Z_L at that point.

EXAMPLE 11.6

With an unknown load connected to a slotted air line, $s = 2$ is recorded by a standing wave indicator and minima are found at 11 cm, 19 cm, . . . on the scale. When the load is replaced by a short circuit, the minima are at 16 cm, 24 cm, ... If $Z_o = 50\ \Omega$, calculate λ, f, and Z_L.

SOLUTION Consider the standing wave patterns as in Figure 11.23(a). From this, we observe that

$$\frac{\lambda}{2} = 19 - 11 = 8 \text{ cm} \qquad \text{or} \qquad \lambda = 16 \text{ cm}$$

$$f = \frac{u}{\lambda} = \frac{3 \times 10^8}{16 \times 10^{-2}} = 1.875 \text{ GHz}$$

Electrically speaking, the load can be located at 16 cm or 24 cm. If we assume that the load is at 24 cm, the load is at a distance ℓ from V_{min}, where

$$\ell = 24 - 19 = 5 \text{ cm} = \frac{5}{16}\lambda = 0.3125\ \lambda$$

This corresponds to an angular movement of $0.3125 \times 720° = 225°$ on the $s = 2$ circle. By starting at the location of V_{min} and moving 225° toward the load

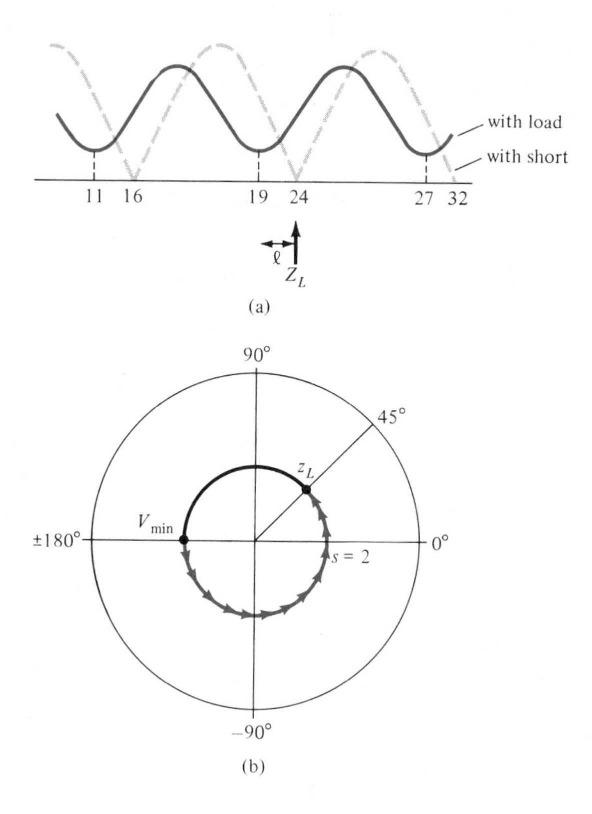

Figure 11.23 Determining Z_L using the slotted line: **(a)** wave pattern, **(b)** Smith chart for Example 11.6.

(counterclockwise), we reach the location of z_L as illustrated in Figure 11.23(b). Thus

$$z_L = 1.4 + j0.75$$

and

$$Z_L = Z_o z_L = 50\,(1.4 + j0.75) = 70 + j37.5\ \Omega$$

PRACTICE EXERCISE 11.6

The following measurements were taken using the slotted line technique: with load, $s = 1.8$, V_{max} occurred at 23 cm, 35.5 cm, . . .; with short, $s = \infty$, V_{max} occurred at 25 cm, 37.5 cm, ... If $Z_o = 50\ \Omega$, determine Z_L.

ANSWER $32.5 - j17.5\ \Omega$.

EXAMPLE 11.7

An antenna with impedance $40 + j30\ \Omega$ is to be matched to a 100-Ω lossless line with a shorted stub. Determine

(a) The required stub admittance

(b) The distance between the stub and the antenna

(c) The stub length

(d) The standing wave ratio on each section of the system

SOLUTION

(a) $z_L = \dfrac{Z_L}{Z_o} = \dfrac{40 + j30}{100} = 0.4 + j0.3$

Locate z_L on the Smith chart as in Figure 11.24 and from this draw the s-circle so that y_L can be located diametrically opposite z_L. Thus $y_L = 1.6 - j1.2$. Alternatively, we may find y_L using

$$y_L = \frac{Z_o}{Z_L} = \frac{100}{40 + j30} = 1.6 - j1.2$$

Locate points A and B where the s-circle intersects the $g = 1$ circle. At A, $y_s = -j1.04$ and at B, $y_s = +j1.04$. Thus the required stub admittance is

$$Y_s = Y_o\, y_s = \pm j1.04 \frac{1}{100} = \pm j10.4 \text{ m℧}$$

Both $j10.4$ m℧ and $-j10.4$ m℧ are possible values.

(b) From Figure 11.24, we determine the distance between the load (antenna in this case) y_L and the stub. At A,

$$\ell_A = \frac{\lambda}{2} - \frac{(62° - -39°)\lambda}{720°} = 0.36\lambda$$

At B:

$$\ell_B = \frac{(62° - 39°)}{720°} = 0.032\lambda$$

(c) Locate points A' and B' corresponding to stub admittance $-j1.04$ and $j1.04$ respectively. Determine the stub length (distance from P_{sc} to A' and B'):

$$d_A = \frac{88°}{720°}\lambda = 0.1222\lambda$$

$$d_B = \frac{272°\lambda}{720°} = 0.3778\lambda$$

Notice that $d_A + d_B = 0.5\lambda$ as expected.

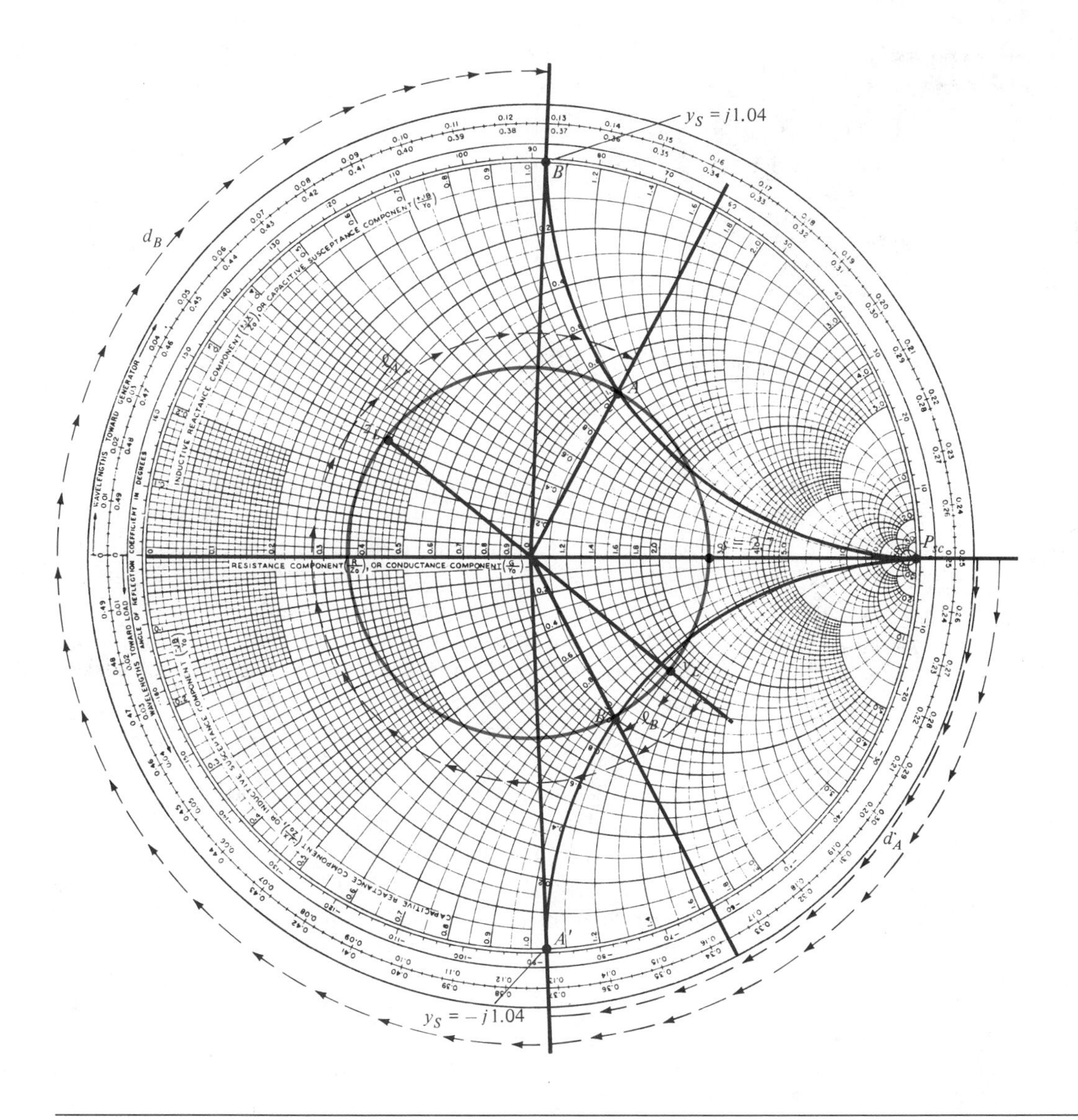

Figure 11.24 For Example 11.7.

(d) From Figure 11.24, $s = 2.7$. This is the standing wave ratio on the line segment between the stub and the load (see Figure 11.18); $s = 1$ to the left of the stub because the line is matched, and $s = \infty$ along the stub because the stub is shorted. ■

PRACTICE EXERCISE 11.7

A 75-Ω lossless line is to be matched to a $100 - j80$–Ω load with a shorted stub. Calculate the stub length, its distance from the load, and the necessary stub admittance.

ANSWER $\ell_A = 0.093\lambda$, $\ell_B = 0.272\lambda$, $d_A = 0.126\lambda$, $d_B = 0.374\lambda$, $\pm j12.67$ m℧.

†11.7 TRANSIENTS ON TRANSMISSION LINES

In our discussion so far, we have assumed that a transmission line operates at a single frequency. In some practical applications, such as in computer networks, pulsed signals may be sent through the line. From Fourier analysis, a pulse can be regarded as a superposition of waves of many frequencies. Thus, sending a pulsed signal on the line may be regarded as the same as simultaneously sending waves of different frequencies.

As in circuit analysis, when a pulse generator or battery connected to a transmission line is switched on, it takes some time for the current and voltage on the line to reach steady values. This transitional period is called the *transient*. The transient behavior just after closing the switch (or due to lightning strokes) is usually analyzed in the frequency domain using Laplace transform. For the sake of convenience, we treat the problem in the time domain.

Consider a lossless line of length ℓ and characteristic impedance Z_o as shown in Figure 11.25(a). Suppose that the line is driven by a pulse generator of voltage V_g with internal impedance Z_g at $z = 0$ and terminated with a purely resistive load Z_L. At the instant $t = 0$ that the switch is closed, the starting current "sees" only Z_g and Z_o, so the initial situation can be described by the equivalent circuit of Figure 11.25(b). From the figure, the starting current at $z = 0$, $t = 0^+$ is given by

$$I(0, 0^+) = I_o = \frac{V_g}{Z_g + Z_o} \qquad [11.60]$$

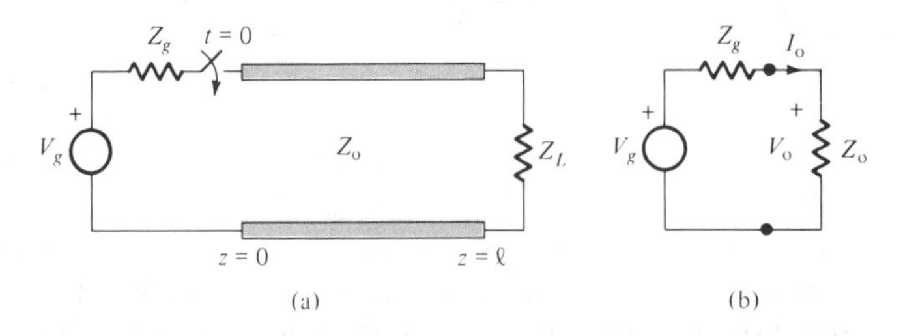

Figure 11.25 Transients on a transmission line: **(a)** a line driven by a pulse generator, **(b)** the equivalent circuit at $z = 0$, $t = 0^+$.

and the initial voltage is

$$V(0, 0^+) = V_o = I_o Z_o = \frac{Z_o}{Z_g + Z_o} V_g \tag{11.61}$$

After the switch is closed, waves $I^+ = I_o$ and $V^+ = V_o$ propagate toward the load at the speed

$$u = \frac{1}{\sqrt{LC}} \tag{11.62}$$

Since this speed is finite, it takes some time for the positively traveling waves to reach the load and interact with it. The presence of the load has no effect on the waves before the transit time given by

$$t_1 = \frac{\ell}{u} \tag{11.63}$$

After t_1 seconds, the waves reach the load. The voltage (or current) at the load is the sum of the incident and reflected voltages (or currents). Thus

$$V(\ell, t_1) = V^+ + V^- = V_o + \Gamma_L V_o = (1 + \Gamma_L)V_o \tag{11.64}$$

and

$$I(\ell, t_1) = I^+ + I^- = I_o - \Gamma_L I_o = (1 - \Gamma_L)I_o \tag{11.65}$$

where Γ_L is the load reflection coefficient given in eq. (11.36); that is,

$$\Gamma_L = \frac{Z_L - Z_o}{Z_L + Z_o} \tag{11.66}$$

The reflected waves $V^- = \Gamma_L V_o$ and $I^- = -\Gamma_L I_o$ travel back toward the generator in addition to the waves V_o and I_o already on the line. At time $t = 2t_1$, the reflected waves have reached the generator, so

$$V(0, 2t_1) = V^+ + V^- = \Gamma_G \Gamma_L V_o + (1 + \Gamma_L)V_o$$

or

$$V(0, 2t_1) = (1 + \Gamma_L + \Gamma_G \Gamma_L)\, V_o \tag{11.67}$$

and

$$I(0, 2t_1) = I^+ + I^- = -\Gamma_G(-\Gamma_L I_o) + (1 - \Gamma_L)I_o$$

or

$$I(0, 2t_1) = (1 - \Gamma_L + \Gamma_L \Gamma_G)I_o \tag{11.68}$$

where Γ_G is the generator reflection coefficient given by

$$\Gamma_G = \frac{Z_g - Z_o}{Z_g + Z_o} \tag{11.69}$$

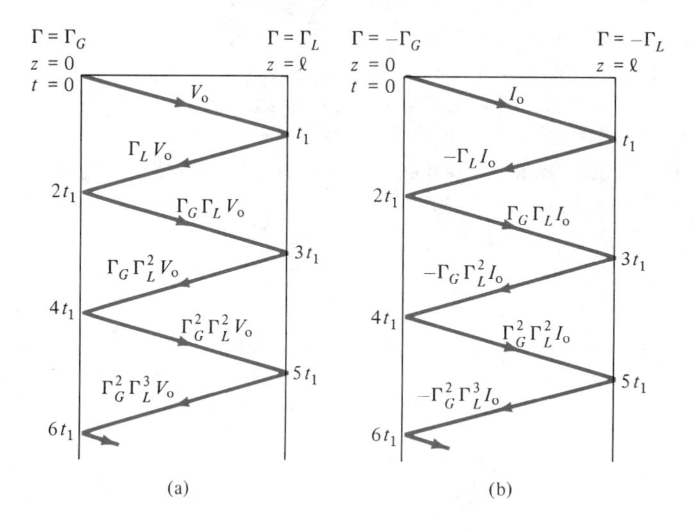

Figure 11.26 Bounce diagram for **(a)** a voltage wave, and **(b)** a current wave.

Again the reflected waves (from the generator end) $V^+ = \Gamma_G \Gamma_L V_o$ and $I^+ = \Gamma_G \Gamma_L I_o$ propagate toward the load and the process continues until the energy of the pulse is actually absorbed by the resistors Z_g and Z_L.

Instead of tracing the voltage and current waves back and forth, it is easier to keep track of the reflections using a *bounce diagram*, otherwise known as a *lattice diagram*. The bounce diagram consists of a zigzag line indicating the position of the voltage (or current) wave with respect to the generator end as shown in Figure 11.26. On the bounce diagram, the voltage (or current) at any time may be determined by adding those values that appear on the diagram above that time.

EXAMPLE 11.8

For the transmission line of Figure 11.27, calculate and sketch

(a) The voltage at the load and generator ends for $0 < t < 6\ \mu s$

(b) The current at the load and generator ends for $0 < t < 6\ \mu s$

SOLUTION

(a) We first calculate the voltage reflection coefficients at the generator and load ends.

$$\Gamma_G = \frac{Z_g - Z_o}{Z_g + Z_o} = \frac{100 - 50}{100 + 50} = \frac{1}{3}$$

$$\Gamma_L = \frac{Z_L - Z_o}{Z_L + Z_o} = \frac{200 - 50}{200 + 50} = \frac{3}{5}$$

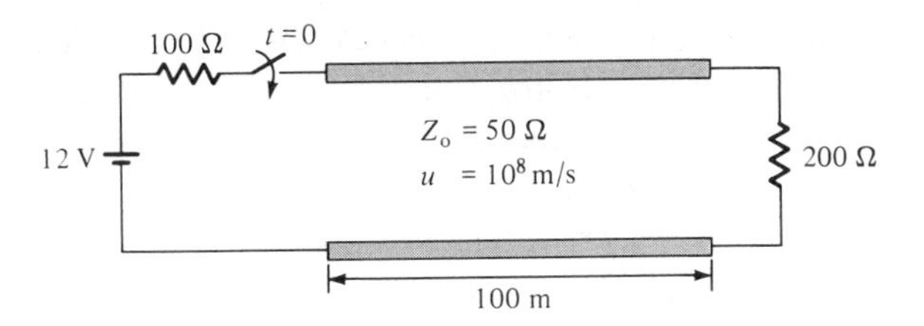

Figure 11.27 For Example 11.8.

The transit time $t_1 = \frac{\ell}{u} = \frac{100}{10^8} = 1\ \mu s$.

The initial voltage at the generator end is

$$V_o = \frac{Z_o}{Z_o + Z_g} V_g = \frac{50}{150}(12) = 4 \text{ V}$$

The 4 V is sent out to the load. The leading edge of the pulse arrives at the load at $t = t_1 = 1\ \mu s$. A portion of it, 4(3/5) = 2.4 V, is reflected back and reaches the generator at $t = 2t_1 = 2\ \mu s$. At the generator, 2.4(1/3) = 0.8 is reflected and the process continues. The whole process is best illustrated in the voltage bounce diagram of Figure 11.28.

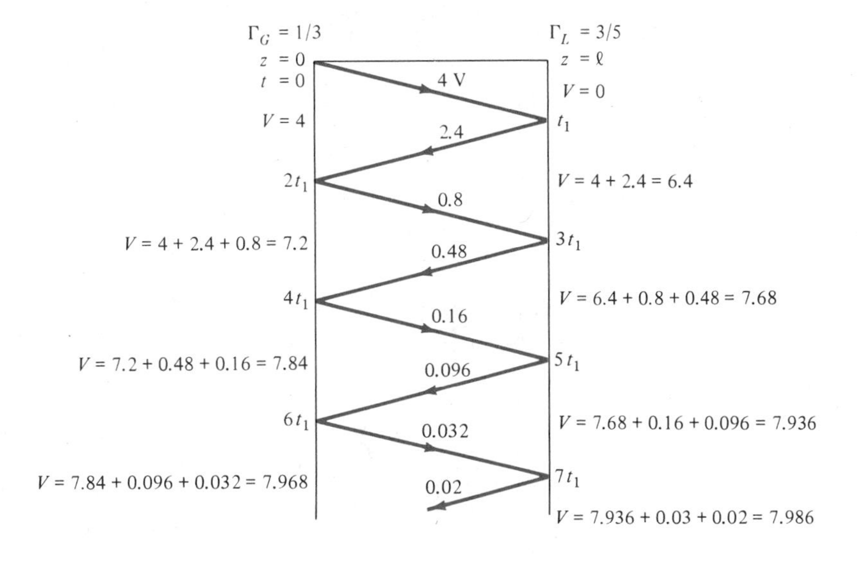

Figure 11.28 Voltage bounce diagram for Example 11.8.

From the bounce diagram, we can sketch $V(0, t)$ and $V(\ell, t)$ as functions of time as shown in Figure 11.29. Notice from Figure 11.29 that as $t \rightarrow \infty$, the voltages approach an asymptotic value of

$$V_\infty = \frac{Z_L}{Z_L + Z_g} V_g = \frac{200}{300}(12) = 8 \text{ V}$$

This should be expected because the equivalent circuits at $t = 0$ and $t = \infty$ are as shown in Figure 11.30 (see Problem 11.39 for proof).

(b) The current reflection coefficients at the generator and load ends are $-\Gamma_G = -1/3$ and $-\Gamma_L = -3/5$ respectively. The initial current is

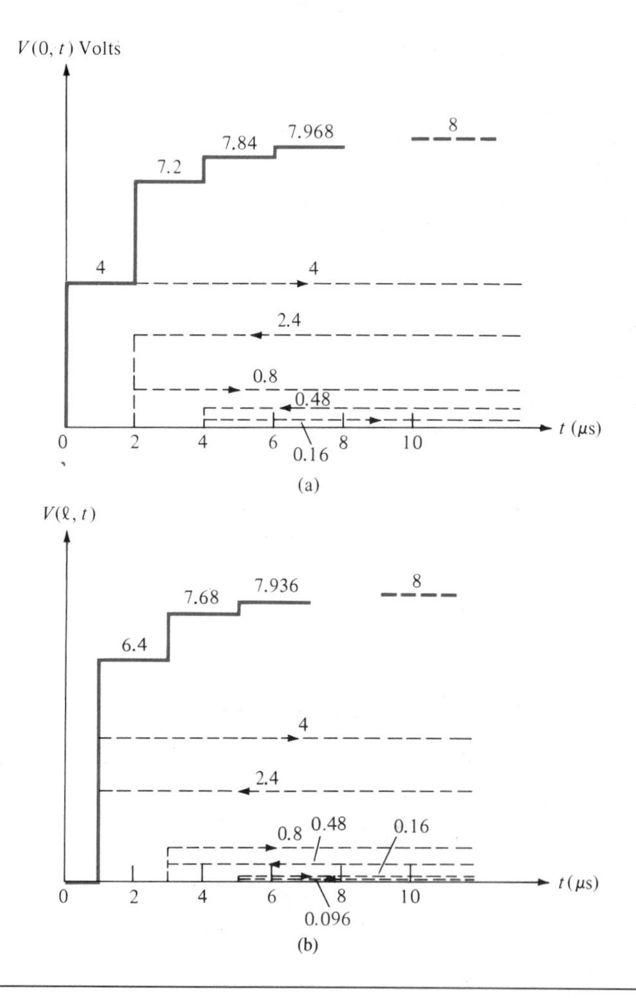

Figure 11.29 Voltage (not to scale): **(a)** at the generator end, **(b)** at the load end.

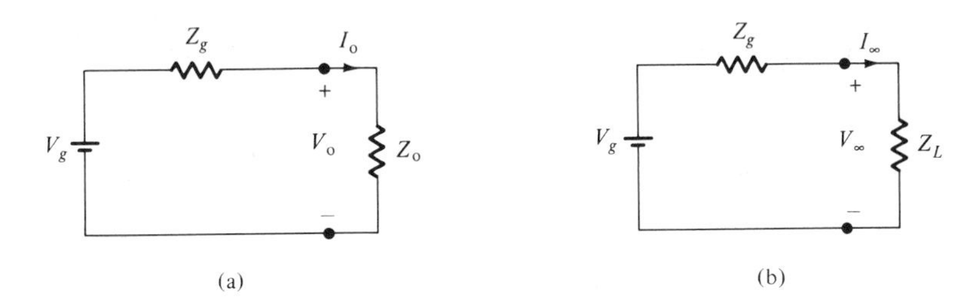

Figure 11.30 Equivalent circuits for the line in Figure 11.27 for **(a)** $t = 0$, and **(b)** $t = \infty$.

$$I_o = \frac{V_o}{Z_o} = \frac{4}{50} = 80 \text{ mA}$$

Again, $I(0, t)$ and $I(\ell, t)$ are easily obtained from the current bounce diagram shown in Figure 11.31. These currents are sketched in Figure 11.32. Note that $I(\ell, t) = V(\ell, t)/Z_L$. Hence, Figure 11.32(b) can be obtained either from the current bounce diagram of Figure 11.31 or by scaling Figure 11.29(b) by a factor of $1/Z_L = 1/200$. Notice from Figures 11.30(b) and 11.32 that the currents approach an asymptotic value of

$$I_\infty = \frac{V_g}{Z_g + Z_L} = \frac{12}{300} = 40 \text{ mA}$$

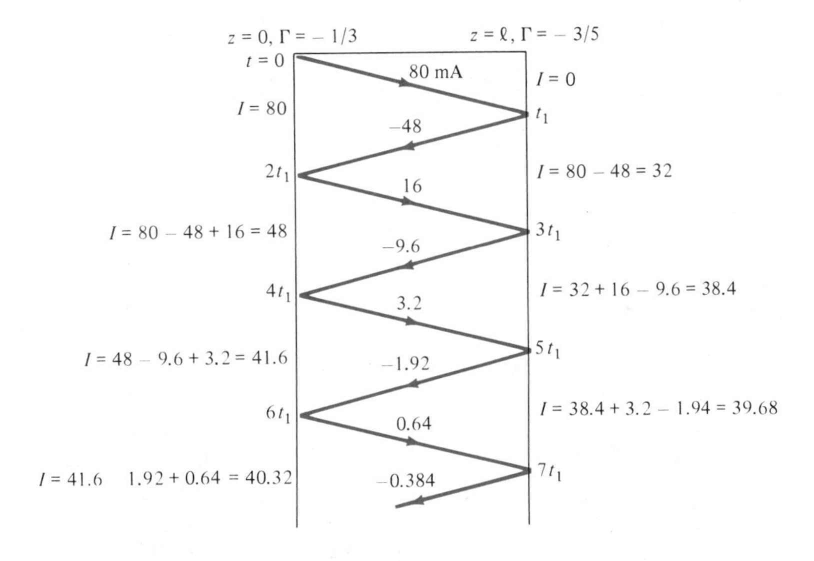

Figure 11.31 Current bounce diagram for Example 11.8.

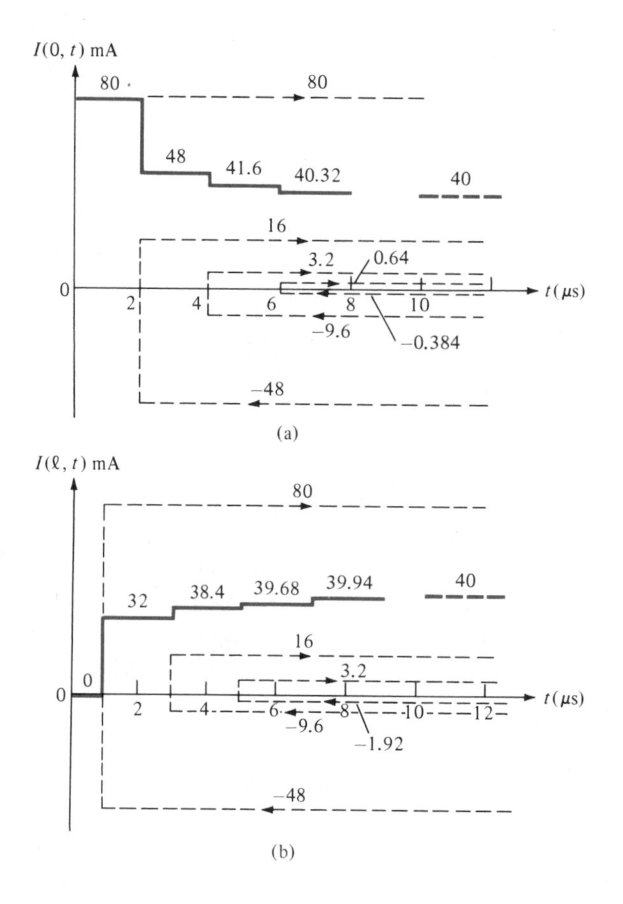

Figure 11.32 Current (not to scale): **(a)** at the generator end, **(b)** at the load end, for Example 11.8.

PRACTICE EXERCISE 11.8

Repeat Example 11.8 if the transmission line is

(a) Short-circuited

(b) Open-circuited

ANSWER (a) see Figure 11.33

(b) see Figure 11.34

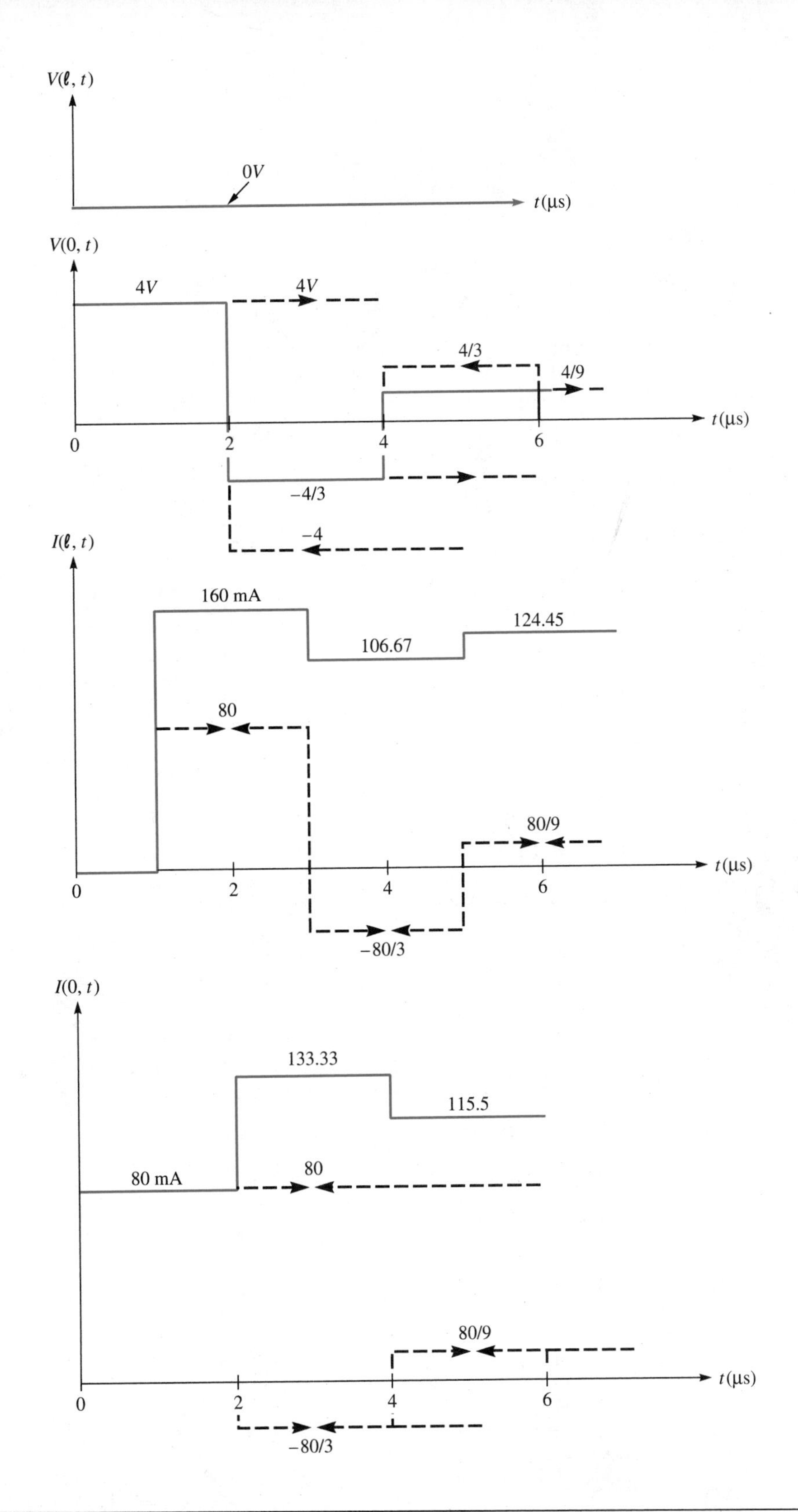

Figure 11.33 For Practice Exercise 11.8(a).

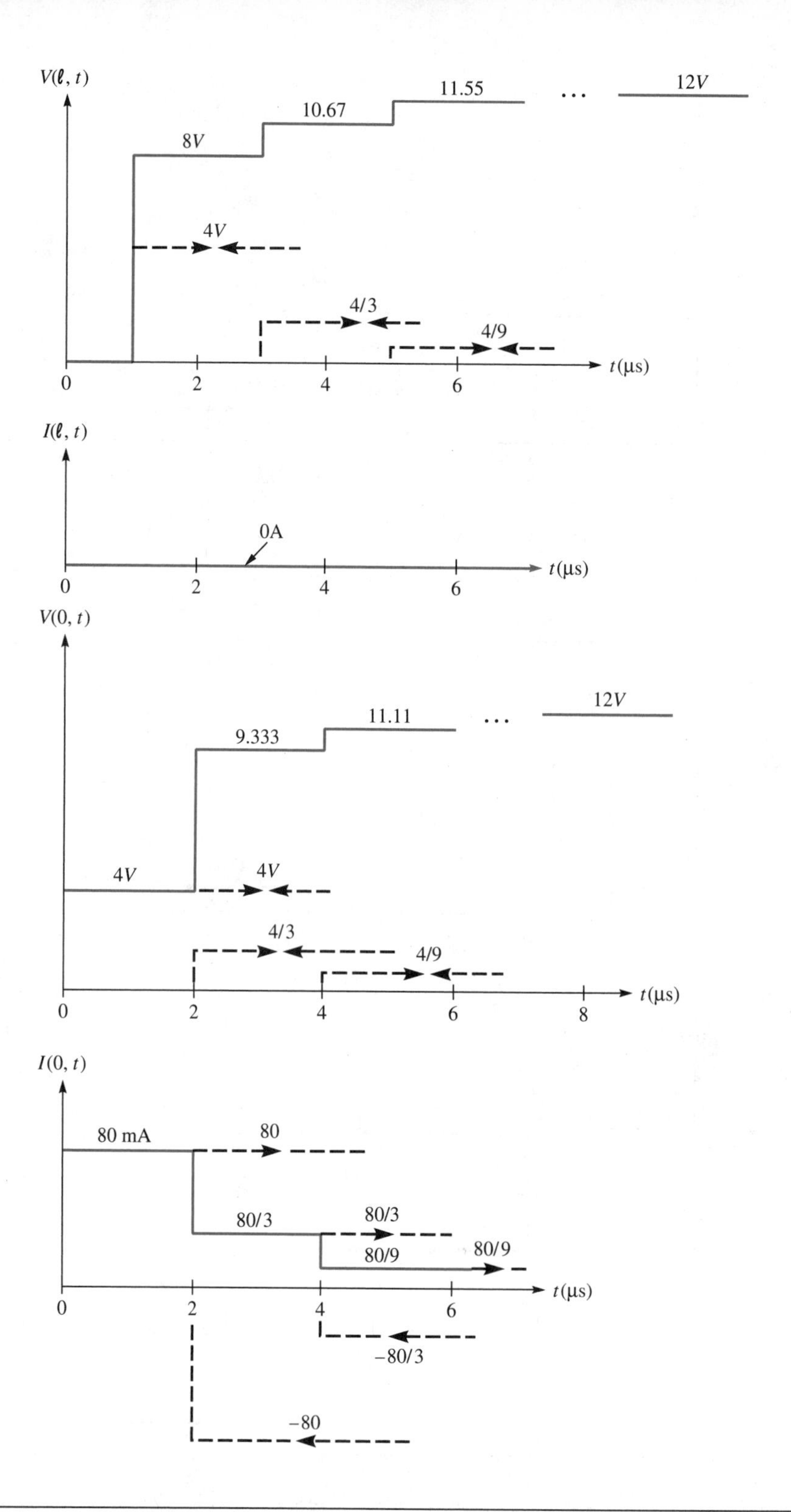

Figure 11.34 For Practice Exercise 11.8(b).

EXAMPLE 11.9

A 75-Ω transmission line of length 60 m is terminated by a 100-Ω load. If a rectangular pulse of width 5 μs and magnitude 4 V is sent out by the generator connected to the line, sketch $I(0, t)$ and $I(\ell, t)$ for $0 < t < 15$ μs. Take $Z_g = 25\ \Omega$ and $u = 0.1c$.

SOLUTION

In the previous example, the switching on of a battery created a step function, a pulse of infinite width. In this example, the pulse is of finite width of 5 μs. We first calculate the voltage reflection coefficients:

$$\Gamma_G = \frac{Z_g - Z_o}{Z_g + Z_o} = -\frac{1}{2}$$

$$\Gamma_L = \frac{Z_L - Z_o}{Z_L + Z_o} = \frac{1}{7}$$

The initial voltage and transit time are given by

$$V_o = \frac{Z_o}{Z_o + Z_g} V_g = \frac{75}{100}(4) = 3\ V$$

$$t_1 = \frac{\ell}{u} = \frac{60}{0.1\,(3 \times 10^8)} = 2\ \mu s$$

The time taken by V_o to go forth and back is $2t_1 = 4$ μs, which is less than the pulse duration of 5 μs. Hence, there will be overlapping.

The current reflection coefficients are

$$-\Gamma_L = -\frac{1}{7} \quad \text{and} \quad -\Gamma_G = \frac{1}{2}$$

The initial current $I_o = \dfrac{V_g}{Z_g + Z_o} = \dfrac{4}{100} = 40$ mA.

Let i and r denote incident and reflected pulses respectively. At the generator end:

$$0 < t < 5\ \mu s \qquad I_r = I_o = 40 \text{ mA}$$

$$4 < t < 9 \qquad I_i = -\frac{1}{7}(40) = -5.714$$

$$I_r = \frac{1}{2}(-5.714) = -2.857$$

$$8 < t < 13 \qquad I_i = -\frac{1}{7}(-2.857) = 0.4082$$

$$I_r = \frac{1}{2}(0.4082) = 0.2041$$

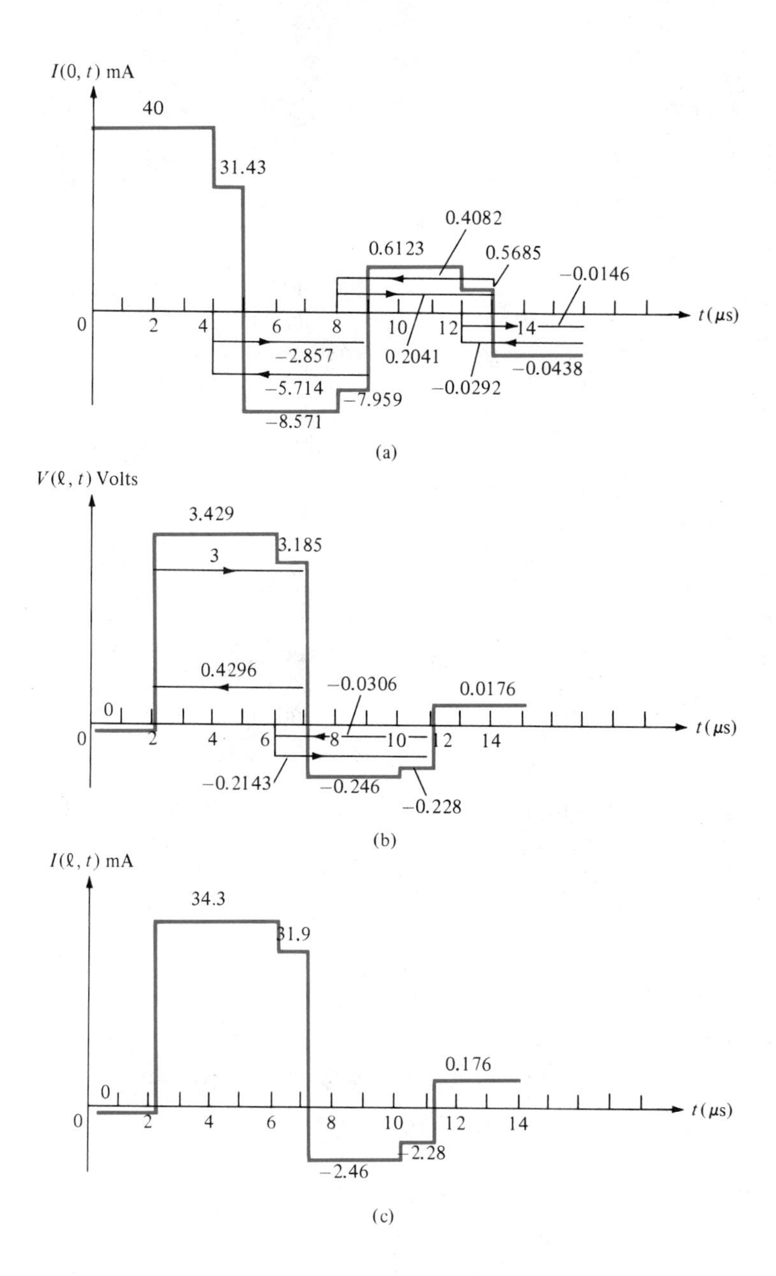

Figure 11.35 For Example 11.9 (not to scale).

$$12 < t < 17 \qquad I_i = -\frac{1}{7}(0.2041) = -0.0292$$

$$I_r = \frac{1}{2}(-0.0292) = -0.0146$$

and so on. Hence, the plot of $I(0, t)$ versus t is as shown in Figure 11.35(a).

At the load end:

$$0 < t < 2\ \mu s \qquad V = 0$$

$$2 < t < 7 \qquad V_i = 3$$

$$V_r = \frac{1}{7}(3) = 0.4296$$

$$6 < t < 11 \qquad V_i = -\frac{1}{2}(0.4296) = -0.2143$$

$$V_r = \frac{1}{7}(-0.2143) = -0.0306$$

$$10 < t < 14 \qquad V_i = -\frac{1}{2}(-0.0306) = 0.0154$$

$$V_r = \frac{1}{7}(0.0154) = 0.0022$$

and so on. From $V(\ell, t)$, we can obtain $I(\ell, t)$ as

$$I(\ell, t) = \frac{V(\ell, t)}{Z_o} = \frac{V(\ell, t)}{100}$$

The plots of $V(\ell, t)$ and $I(\ell, t)$ are shown in Figure 11.35(b) and (c). ■

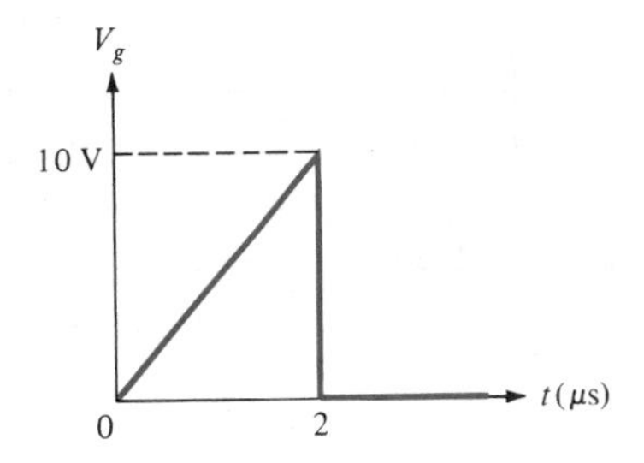

Figure 11.36 Triangular pulse of Practice Exercise 11.9.

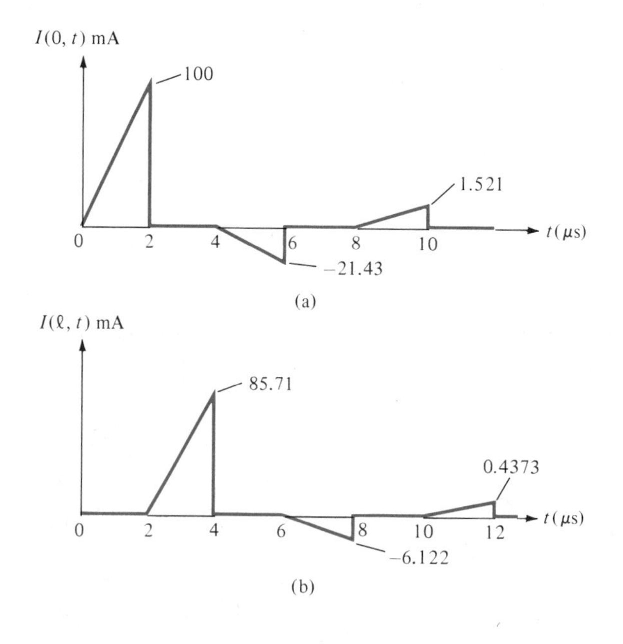

Figure 11.37 Current waves for Practice Exercise 11.9.

PRACTICE EXERCISE 11.9

Repeat Example 11.9 if the rectangular pulse is replaced by a triangular pulse of Figure 11.36.

ANSWER $(I_o)_{max} = 100$ mA. See Figure 11.37 for the current waveforms.

†11.8 MICROSTRIP TRANSMISSION LINES

Microstrip lines belong to a group of lines known as parallel-plate transmission lines. They are widely used in present-day electronics. Apart from being the most commonly used form of transmission lines for microwave integrated circuits, microstrips are used for circuit components such as filters, couplers, resonators, antennas, and so on. In comparison with the coaxial line, the microstrip line allows for greater flexibility and compactness of design.

A microstrip line consists of a single ground plane and an open strip conductor separated by dielectric substrate as shown in Figure 11.38. It is constructed by the photographic processes used for integrated circuits. Analytical derivation of the

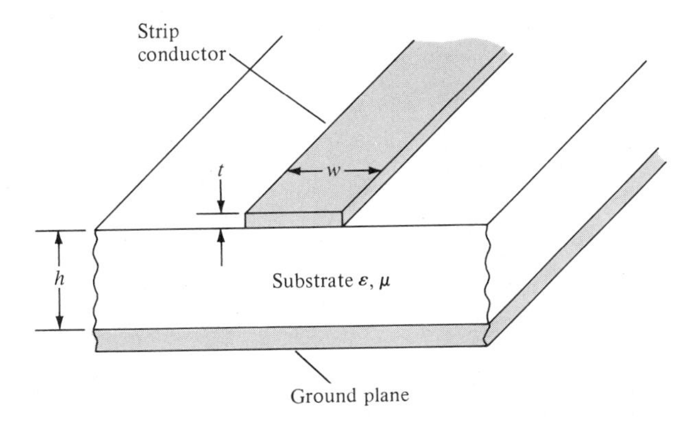

Figure 11.38 Microstrip line.

characteristic properties of the line is cumbersome.[4] We will only consider some basic, valid empirical formulas necessary for calculating the phase velocity, impedance, and losses of the line.

Due to the open structure of the microstrip line, the EM field is not confined to the dielectric, but is partly in the surrounding air as in Figure 11.39. Provided the frequency is not too high, the microstrip line will propagate a wave that, for all practical purposes, is a TEM wave. Because of the fringing, the *effective relative permittivity* ε_{eff} is less than the relative permittivity ε_r of the substrate. If w is the line width and h is the substrate thickness, we define the following parameters:

$$A = \frac{120}{\sqrt{2(\varepsilon_r + 1)}} \qquad [11.70a]$$

$$B = \frac{1}{2}\left[\frac{\varepsilon_r - 1}{\varepsilon + 1}\right]\left[\ln\frac{\pi}{2} + \frac{1}{\varepsilon_r}\ln\frac{4}{\pi}\right] \qquad [11.70b]$$

$$C = \ln\frac{8h}{w} + \frac{1}{32}\left[\frac{w}{h}\right]^2 \qquad [11.70c]$$

$$D = \frac{60\pi}{\sqrt{\varepsilon_r}} \qquad [11.70d]$$

$$E = \frac{w}{2h} + 0.4413 + 0.08226\left[\frac{\varepsilon_r - 1}{\varepsilon_r^2}\right] + \frac{\varepsilon_r + 1}{2\pi\varepsilon_r}\left[1.452 + \ln\left(\frac{w}{2h} + 0.94\right)\right] \qquad [11.70e]$$

[4]See H. A. Wheeler, "Transmission-line properties of parallel strips separated by a dielectric sheet," *IEEE Trans. Microwave Theory and Techniques*, MTT-3, no. 3, March 1965, pp. 172–185.

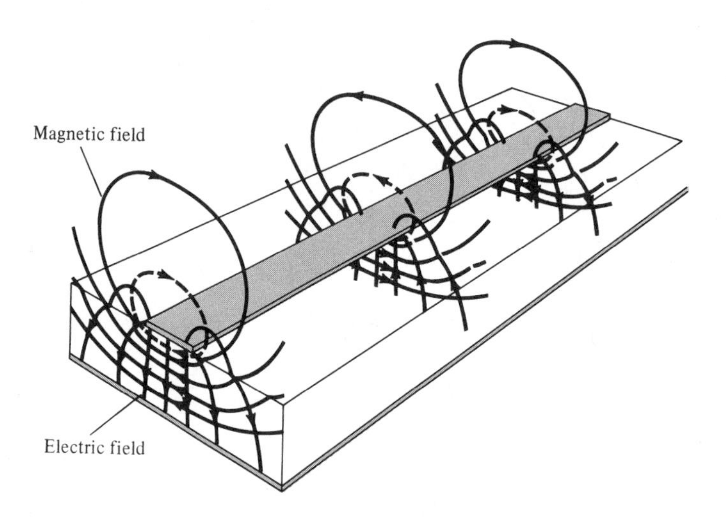

Figure 11.39 Pattern of the EM field of a microstrip line. Source: From D. Roddy, *Microwave Technology*, 1986, by permission of Prentice-Hall.

Note that A, B, and D depend on ε_r, C depends on the ratio w/h, and E depends on both ε_r and w/h. We use these parameters in two sets of formulas for ε_{eff} and Z_o covering narrow and wide strips with close agreement in the transition region.

For *narrow strips,*

$$\varepsilon_{\text{eff}} = \frac{\varepsilon_r + 1}{2\left[1 - \dfrac{B}{C}\right]^2} \qquad \left(\frac{w}{h} < 1.3\right) \tag{11.71}$$

$$Z_o = A(C - B), \qquad \left(\frac{w}{h} < 3.3\right) \tag{11.72}$$

for given ε_r and w/h. If ε_r and Z_o are known, we determine the ratio w/h using

$$\frac{h}{w} = \frac{e^C}{8} - \frac{e^{-C}}{4}, \qquad Z_o > (44 - 2\varepsilon_r) \tag{11.73}$$

where C is obtained from eq. (11.72) as

$$C = \frac{Z_o}{A} + B \tag{11.74}$$

For *wide strips,*

$$\varepsilon_{\text{eff}} = \frac{\varepsilon_r + 1}{2} + \left[\frac{\varepsilon_r - 1}{2}\right]\left[1 + \frac{10h}{w}\right]^{-1/2}, \qquad (w/h > 1.3) \tag{11.75}$$

$$Z_o = \frac{D}{E}, \qquad (w/h > 3.3) \tag{11.76}$$

given ε_r and w/h. If ε_r and Z_o are known,

$$\frac{w}{h} = \frac{2}{\pi}\left[\frac{\pi D}{Z_o} - 1 - \ln\left(\frac{2\pi D}{Z_o} - 1\right)\right] + \frac{\varepsilon_r - 1}{\pi\varepsilon_r}\left[\ln\left(\frac{\pi D}{Z_o} - 1\right) 0.293 - \frac{0.517}{\varepsilon_r}\right] \quad [11.77]$$

for $Z_o < (44 - 2\varepsilon_r)$. The characteristic impedance for a wide strip line is often low while that for a narrow strip line is high.

From the knowledge of ε_{eff} and Z_o, the phase velocity of a wave propagating on the microstrip is given by

$$u = \frac{c}{\sqrt{\varepsilon_{\text{eff}}}} \quad [11.78]$$

where c is the speed of light in a vacuum. The attenuation due to conduction (or ohmic) loss is (in dB/m)

$$\alpha_c \simeq 8.686 \frac{R_s}{wZ_o} \quad [11.79]$$

where $R_s = \dfrac{1}{\sigma_c \delta}$ is the skin resistance of the conductor. The attenuation due to dielectric loss is (in dB/m)

$$\alpha_d \simeq 27.3 \frac{(\varepsilon_{\text{eff}} - 1)\,\varepsilon_r}{(\varepsilon_r - 1)\,\varepsilon_{\text{eff}}} \frac{\tan\theta}{\lambda} \quad [11.80]$$

where $\lambda = u/f$ is the line wavelength and $\tan\theta = \sigma/\omega\varepsilon$ is the loss tangent of the substrate. The total attenuation constant is the sum of the ohmic attenuation constant α_c and the dielectric attenuation constant α_d, that is,

$$\alpha = \alpha_c + \alpha_d \quad [11.81]$$

Sometimes α_d is negligible in comparison with α_c. Although they offer an advantage of flexibility and compactness, the microstrip lines are not useful for long transmission due to excessive attenuation.

EXAMPLE 11.10

A certain microstrip line has fused quartz ($\varepsilon_r = 3.8$) as a substrate. If the ratio of line width to substrate thickness is $w/h = 4.5$, determine

(a) The effective relative permittivity of the substrate

(b) The characteristic impedance of the line

(c) The wavelength of the line at 10 GHz

SOLUTION (a) For $w/h = 4.5$, we have a wide strip. From eq. (11.75),

$$\varepsilon_{\text{eff}} = \frac{4.8}{2} + \frac{2.8}{2}\left[1 + \frac{10}{4.5}\right]^{-1/2} = 3.18$$

(b) From eqs. (11.70) and (11.76),

$$D = \frac{60\pi}{\sqrt{3.8}} = 96.7$$

$$E = \frac{4.5}{2} + 0.4413 + 0.08226 \times \frac{2.8}{(3.8)^2} + \frac{4.8}{2\pi(3.8)}$$
$$\left[1.452 + \ln\left(\frac{4.5}{2} + 0.94\right)\right]$$

$$E = 3.232$$

$$Z_o = \frac{D}{E} = \frac{96.7}{3.232} = 29.92\ \Omega$$

(c) $$\lambda = \frac{u}{f} = \frac{c}{f\sqrt{\varepsilon_{\text{eff}}}} = \frac{3 \times 10^8}{10^{10}\sqrt{3.18}}$$
$$= 1.68 \times 10^{-2}\ \text{m} = 16.8\ \text{mm}$$

■

PRACTICE EXERCISE 11.10

Repeat Example 11.10 for $w/h = 1.1$.

ANSWER (a) 2.693, (b) 73.93 Ω, (c) 18.28 mm.

EXAMPLE 11.11

At 10 GHz, a microstrip line has the following parameters:

$$h = 0.8\ \text{mm}$$
$$w = 1\ \text{mm}$$
$$\varepsilon_r = 6.6$$
$$\tan\theta = 10^{-4}$$
$$\sigma_c = 5.8 \times 10^7\ \text{mhos/m}$$

Calculate the attenuation due to conduction loss and dielectric loss.

SOLUTION The ratio $w/h = 1$ mm/0.8 mm $= 1.25$ shows that the microstrip line is narrow. Hence, from eqs. (11.70) to (11.72), we obtain the effective relative permittivity and characteristic impedance as

$$A = \frac{120}{\sqrt{2 \times 7.6}} = 30.78$$

$$B = \frac{1}{2}\left(\frac{5.6}{7.6}\right)\left(\ln\frac{\pi}{2} + \frac{1}{6.6}\ln\frac{4}{\pi}\right) = 0.1529$$

$$C = \ln\frac{8}{1.25} + \frac{1}{32}(1.25)^2 = 1.905$$

$$\varepsilon_{\text{eff}} = \frac{7.6}{2\left[1 - \dfrac{0.1529}{1.905}\right]^2} = 4.492$$

$$Z_o = 30.78(1.905 - 0.1529) = 53.93\ \Omega$$

The skin resistance of the conductor is

$$R_s = \frac{1}{\sigma_c \delta} = \sqrt{\frac{\pi f \mu_o}{\sigma_c}} = \sqrt{\frac{\pi \times 10 \times 10^9 \times 4\pi \times 10^{-7}}{5.8 \times 10^7}}$$
$$= 2.609 \times 10^{-2}\ \Omega/\text{m}^2$$

Using eq. (11.79), we obtain the conduction attenuation constant as

$$\alpha_c = 8.686 \times \frac{2.609 \times 10^{-2}}{1 \times 10^{-3} \times 53.93}$$
$$= 4.202\ \text{dB/m}$$

To find the dielectric attenuation constant, we need λ.

$$\lambda = \frac{u}{f} = \frac{c}{f\sqrt{\varepsilon_{\text{eff}}}} = \frac{3 \times 10^8}{10 \times 10^9\sqrt{4.492}}$$
$$= 1.415 \times 10^{-2}\ \text{m}$$

Applying eq. (11.80), we have

$$\alpha_d = 27.3 \times \frac{3.492 \times 6.6 \times 10^{-4}}{5.6 \times 4.492 \times 1.415 \times 10^{-2}}$$
$$= 0.1768\ \text{dB/m}$$

■

PRACTICE EXERCISE 11.11

Calculate the attenuation due to ohmic losses at 20 GHz for a microstrip line constructed of copper conductor having a width of 2.5 mm on an alumina substrate. Take the characteristic impedance of the line as 50 Ω.

ANSWER 2.564 dB/m.

SUMMARY

1. A transmission line is commonly described by its distributed parameters R (in Ω/m), L (in H/m), G (in ℧/m), and C (in F/m). Formulas for calculating R, L, G, and C are provided in Table 11.1 for coaxial, two-wire, and planar lines.
2. The distributed parameters are used in an equivalent circuit model to represent a differential length of the line. The transmission-line equations are obtained by applying Kirchhoff's laws and allowing the length of the line to approach zero. The voltage and current waves on the line are

$$V(z, t) = V_o^+ e^{-\alpha z} \cos(\omega t - \beta z) + V_o^- e^{\alpha z} \cos(\omega t + \beta z)$$

$$I(z, t) = \frac{V_o^+}{Z_o} e^{-\alpha z} \cos(\omega t - \beta z) - \frac{V_o^-}{Z_o} e^{\alpha z} \cos(\omega t + \beta z)$$

showing that there are two waves traveling in opposite directions on the line.
3. The characteristic impedance Z_o (analogous to the intrinsic impedance η of plane waves in a medium) of a line is given by

$$Z_o = \sqrt{\frac{R + j\omega L}{G + j\omega C}}$$

and the propagation constant γ (in per meter) is given by

$$\gamma = \alpha + j\beta = \sqrt{(R + j\omega L)(G + j\omega C)}$$

The wavelength and wave velocity are

$$\lambda = \frac{2\pi}{\beta}, \qquad u = \frac{\omega}{\beta} = f\lambda$$

4. The general case is that of the lossy transmission line ($G \neq 0 \neq R$) considered earlier. For a lossless line, $R = 0 = G$; for a distortionless line, $R/L = G/C$. It is desirable that power lines be lossless and telephone lines be distortionless.
5. The voltage reflection coefficient at the load end is defined as

$$\Gamma_L = \frac{V_o^-}{V_o^+} = \frac{Z_L - Z_o}{Z_L + Z_o}$$

and the standing wave ratio is

$$s = \frac{1 + |\Gamma_L|}{1 - |\Gamma_L|}$$

where Z_L is the load impedance.

6. At any point on the line, the ratio of the phasor voltage to phasor current is the impedance at that point looking towards the load and would be the input impedance to the line if the line were that long. For a lossy line,

$$Z(z) = \frac{V_s(z)}{I_s(z)} = Z_{\text{in}} = Z_o \left[\frac{Z_L + Z_o \tanh \gamma\ell}{Z_o + Z_L \tanh \gamma\ell}\right]$$

where ℓ is the distance from load to the point. For a lossless line ($\alpha = 0$), $\tanh \gamma\ell = j \tan \beta\ell$; for a shorted line, $Z_L = 0$; for an open-circuited line, $Z_L = \infty$; and for a matched line, $Z_L = Z_o$.

7. The Smith chart is a graphical means of obtaining line characteristics such as Γ, s, and Z_{in}. It is constructed within a circle of unit radius and based on the formula for Γ_L given above. For each r and x, it has two explicit circles (the resistance and reactance circles) and one implicit circle (the constant s-circle). It is conveniently used in determining the location of a stub tuner and its length. It is also used with the slotted line to determine the value of the unknown load impedance.

8. When a dc voltage is suddenly applied at the sending end of a line, a pulse moves forth and back on the line. The transient behavior is conveniently analyzed using the bounce diagrams.

9. Microstrip transmission lines are useful in microwave integrated circuits. Useful formulas for constructing microstrip lines and determining losses on the lines have been presented.

REVIEW QUESTIONS

11.1 Which of the following statements are not true of the line parameters R, L, G, and C?

(a) R and L are series elements.

(b) G and C are shunt elements.

(c) $G = \dfrac{1}{R}$

(d) $LC = \mu\varepsilon$ and $RG = \sigma\varepsilon$.

(e) Both R and G depend on the conductivity of the conductors forming the line.

(f) Only R depends explicitly on frequency.

(g) The parameters are not lumped but distributed.

11.2 For a lossy transmission line, the characteristic impedance does not depend on

(a) The operating frequency of the line

(b) The length of the line

(c) The load terminating the line

(d) The conductivity of the conductors

(e) The conductivity of the dielectric separating the conductors

11.3 Which of the following conditions will not guarantee a distortionless transmission line?

(a) $R = 0 = G$

(b) $RC = GL$

(c) Very low frequency range ($R \gg \omega L$, $G \gg \omega C$)

(d) Very high frequency range ($R \ll \omega L$, $G \ll \omega C$)

11.4 Which of these is not true of a lossless line?

(a) $Z_{in} = -jZ_o$ for a shorted line with $\ell = \lambda/8$.

(b) $Z_{in} = j\infty$ for a shorted line with $\ell = \lambda/4$.

(c) $Z_{in} = jZ_o$ for an open line with $\ell = \lambda/2$.

(d) $Z_{in} = Z_o$ for a matched line.

(e) At a half-wavelength from a load, $Z_{in} = Z_L$ and repeats for every half-wavelength thereafter.

11.5 A lossless transmission line of length 50 cm with $L = 10\ \mu$H/m, $C = 40$ pF/m is operated at 30 MHz. Its electrical length is

(a) 20λ

(b) 0.2λ

(c) $108°$

(d) 40π

(e) None of the above

11.6 Match the following normalized impedances with points A, B, C, D, and E on the Smith chart of Figure 11.40.

(i) $0 + j0$

(ii) $1 + j0$

(iii) $0 - j1$

(iv) $0 + j1$

(v) $\infty + j\infty$

(vi) $\left[\dfrac{Z_{in}}{Z_o}\right]_{min}$

(vii) $\left[\dfrac{Z_{in}}{Z_o}\right]_{max}$

(viii) Matched load ($\Gamma = 0$)

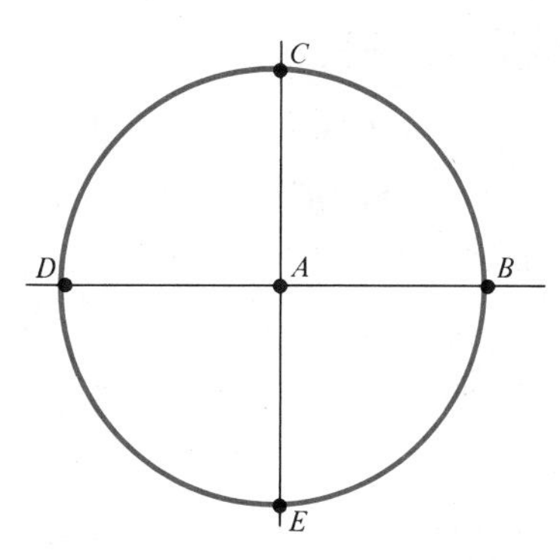

Figure 11.40 For Review Question 11.6.

11.7 A 500-m lossless transmission line is terminated by a load which is located at P on the Smith chart of Figure 11.41. If $\lambda = 150$ m, how many voltage maxima exist on the line?

(a) 7

(b) 6

(c) 5

(d) 3

(e) None

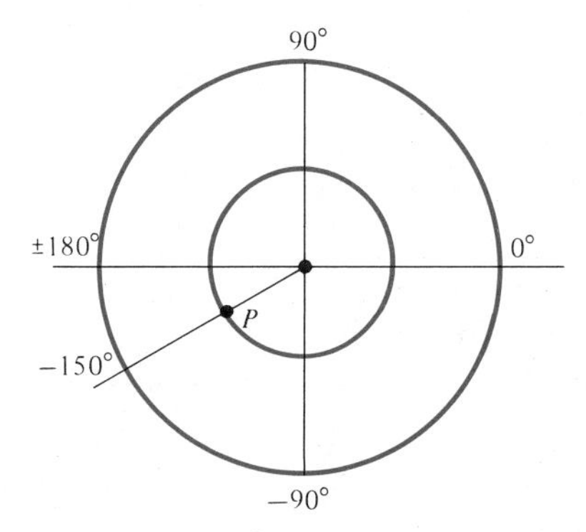

Figure 11.41 For Review Question 11.7.

11.8 Write true (T) or false (F) against each of the following statements.

(a) All r- and x-circles pass through point $(\Gamma_r, \Gamma_i) = (1, 0)$.

(b) Any impedance repeats itself every $\lambda/4$ on the Smith chart.

(c) An $s = 2$ circle is the same as $|\Gamma| = 0.5$ circle on the Smith chart.

(d) The basic principle of any matching scheme is to eliminate the reflected wave between the source and the matching device.

(e) The slotted line is used to determine Z_L only.

(f) At any point on a transmission line, the current reflection coefficient is the reciprocal of the voltage reflection coefficient at that point.

11.9 In an air line, adjacent maxima are found at 12.5 cm and 37.5 cm. The operating frequency is

(a) 1.5 GHz

(b) 600 MHz

(c) 300 MHz

(d) 1.2 GHz

11.10 Two identical pulses each of magnitude 12 V and width 2 μs are incident at $t = 0$ on a lossless transmission line of length 400 m terminated with a load. If the two pulses are separated 3 μs (similar to the case of Figure 11.53) and $u = 2 \times 10^8$ m/s, when does the contribution to $V_L(\ell, t)$ by the second pulse start overlapping that of the first?

(a) $t = 0.5$ μs

(b) $t = 2$ μs

(c) $t = 5$ μs

(d) $t = 5.5$ μs

(e) $t = 6$ μs

Answers: 11.1c,d,e, 11.2b,c, 11.3c, 11.4a,c, 11.5c, 11.6 (i) D,B, (ii) A, (iii) E, (iv) C, (v) B, (vi) D, (vii) B, (viii) A, 11.7a, 11.8 (a) T, (b) F, (c) F, (d) T, (e) F, (f) F, 11.9b, 11.10e.

PROBLEMS

***11.1** In Section 11.3, it was mentioned that the equivalent circuit of Figure 11.5 is not the only possible one. Show that eqs. (11.4) and (11.6) would remain the same if the Π-type and T-type equivalent circuits shown in Figure 11.42 were used.

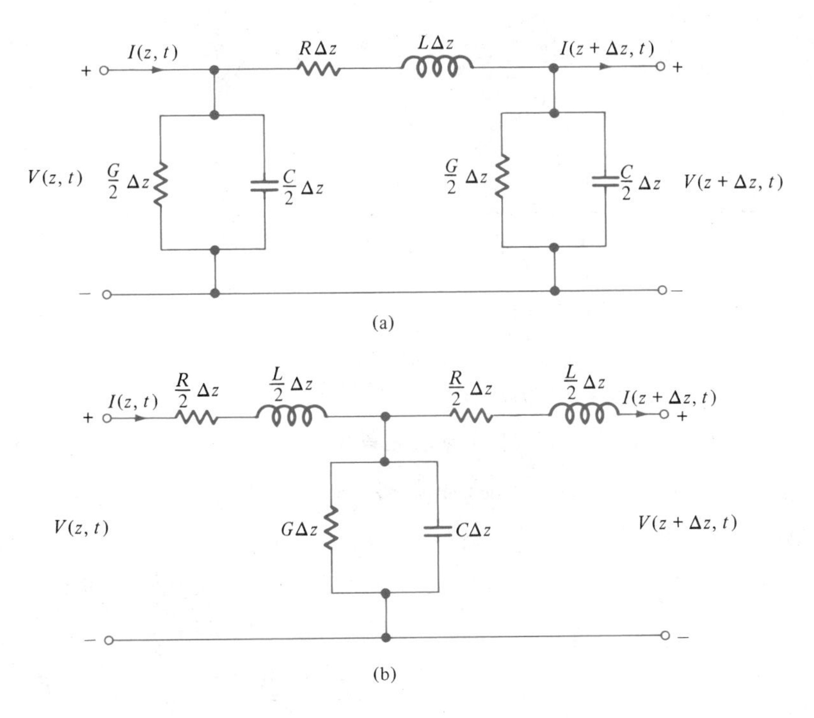

Figure 11.42 For Problem 11.1: **(a)** Π-type equivalent circuit, **(b)** T-type equivalent circuit.

11.2 For a lossless two-wire transmission line, show that

(a) The phase velocity $u = c = \dfrac{1}{\sqrt{LC}}$

(b) The characteristic impedance $Z_o = \dfrac{1}{cC} = \dfrac{\eta_o \varepsilon_o}{C}$

Is part (a) true of other lossless lines?

11.3 A lossless line has a voltage wave

$$V(z, t) = V_o \sin(\omega t - \beta z)$$

Find the corresponding current wave.

11.4 The copper leads of a diode are 16 mm in length and have a radius of 0.3 mm. They are separated by a distance of 2 mm as shown in Figure 11.43. Find the capacitance between the leads and the ac resistance at 10 MHz.

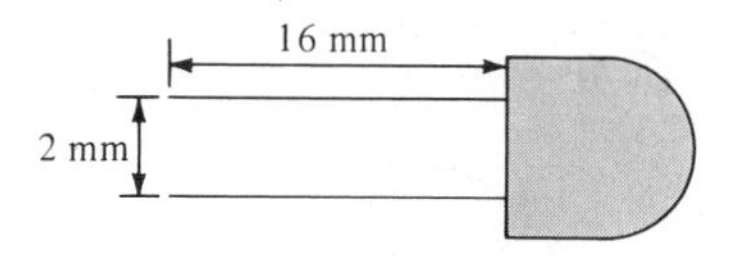

Figure 11.43 The diode of Problem 11.4.

11.5 Calculate the line parameters R, G, L, and C for

(a) A lossy line with $\alpha = 0.25$ Np/m, $\beta = 4.2$ rad/m, $Z_o = 100 - j5\ \Omega$, $f = 60$ MHz

(b) A lossless line with $\beta = 3$ rad/m, $Z_o = 50\ \Omega$, $f = 10$ MHz

(c) A distortionless line with $\gamma = 0.04 + j1.5$/m, $Z_o = 80\ \Omega$, $f = 500$ MHz

***11.6** (a) From eq. (11.11), show that

$$\alpha = \left[\frac{RG - \omega^2 LC + [(R^2 + \omega^2 L^2)(G^2 + \omega^2 C^2)]^{1/2}}{2}\right]^{1/2}$$

$$\beta = \left[\frac{-RG + \omega^2 LC + [(R^2 + \omega^2 L^2)(G^2 + \omega^2 C^2)]^{1/2}}{2}\right]^{1/2}$$

(b) Using eqs. (11.15) and (11.16) in conjunction with eqs. (11.8) and (11.9), show that eq. (11.18) is valid.

(c) Prove that the input admittance of a lossless line at a distance ℓ from the load is

$$Y_{in} = Y_o \left[\frac{Y_L + jY_o \tan \beta\ell}{Y_o + jY_L \tan \beta\ell}\right]$$

where $Y_o = 1/Z_o$ and $Y_L = 1/Z_L$.

11.7 A coaxial line 5.6 m long has distributed parameters $R = 6.5\ \Omega$/m, $L = 3.4$ μH/m, $G = 8.4$ m℧/m, and $C = 21.5$ pF/m. If the line operates at 2 MHz, calculate the characteristic impedance and the end-to-end propagation time delay

11.8 A lossless transmission line operating at 4.5 GHz has $L = 2.4\ \mu$H/m and $Z_o = 85\ \Omega$. Calculate the phase constant β and the phase velocity u.

11.9 A 300-m-long line has the following constants for the total length: 4.5 kΩ, 0.15 mH, 60 m℧, 12 nF. The line is terminated by a $30 + j60$–Ω load and operates at 6 MHz. Find γ, u, Γ, and s.

11.10 Given that $\gamma = 0.02 + j0.05$/m and $Z_o = 400 - j150\ \Omega$ for a lossy transmission line terminated by a load of $300 + j250\ \Omega$, find Z_{in} at 2 m and 10 m from the load.

11.11 Show that a lossy transmission line of length ℓ has an input impedance $Z_{sc} = Z_o \tanh \gamma\ell$ when shorted and $Z_{oc} = Z_o \coth \gamma\ell$ when open. Confirm eqs. (11.37) and (11.39).

11.12 Find the input impedance of a short-circuited coaxial transmission line of Figure 11.44 if $Z_o = 65 + j38\ \Omega$, $\gamma = 0.7 + j2.5$ /m, $\ell = 0.8$ m.

11.13 Determine the shortest length of a 42-Ω air line required to produce a reactance of $j75\ \Omega$ at 1 MHz if the line is

(a) Shorted

(b) Open

11.14 A short-circuited coaxial transmission line has $Z_o = 60\ \Omega$ and $\gamma = j8.5$ /m. Calculate the input impedance if the length of the line is

(a) 15 cm

(b) 1.5m

(c) $3\lambda/4$

(d) $\lambda/8$

11.15 A 500-Ω lossless line has $V_L = 10e^{j25°}$ V, $Z_L = 50e^{j30°}$. Find the current at $\lambda/8$ from the load.

11.16 A 60-Ω lossless line is connected to a source with $V_g = 10\underline{/0°}\ V_{rms}$ and $Z_g = 50 - j40\ \Omega$ and terminated with a load $j40\ \Omega$. If the line is 100 m long and $\beta = 0.25$ rad/m, calculate Z_{in} and V at

(a) The sending end

(b) The receiving end

(c) 4 m from the load

(d) 3 m from the source

11.17 A lossless transmission line with a characteristic impedance of 75 Ω is terminated by a load of 120 Ω. The length of the line is 1.25λ. If the line is energized by a source of 100 V (rms) with an internal impedance of 50 Ω, determine (a) the input impedance, and (b) the magnitude of the load voltage.

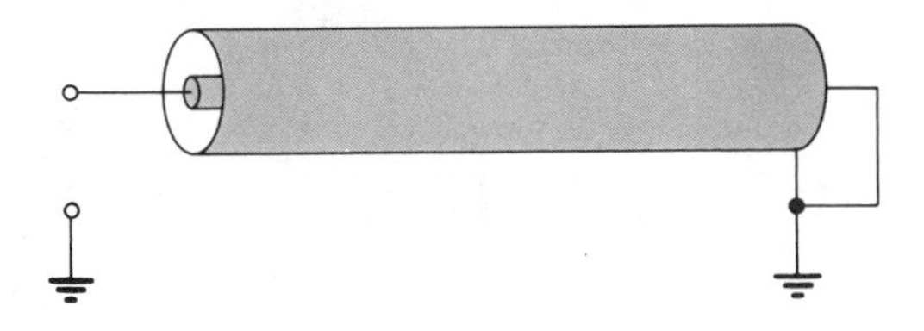

Figure 11.44 For Problem 11.12.

***11.18** Three lossless lines are connected as shown in Figure 11.45. Determine Z_{in}.

***11.19** Consider the two-port network shown in Figure 11.46(a). The relation between the input and output variables can be written in matrix form as

$$\begin{bmatrix} V_1 \\ I_1 \end{bmatrix} = \begin{bmatrix} A & B \\ C & D \end{bmatrix} \begin{bmatrix} V_2 \\ -I_2 \end{bmatrix}$$

For the lossy line in Figure 11.46(b), show that the *ABCD* matrix is

$$\begin{bmatrix} \cosh \gamma \ell & Z_o \sinh \gamma \ell \\ \dfrac{1}{Z_o} \sinh \gamma \ell & \cosh \gamma \ell \end{bmatrix}$$

11.20 A 50-Ω lossless line is 4.2 m long. At the operating frequency of 300 MHz, the input impedance at the middle of the line is $80 - j60\ \Omega$. Find the input impedance at the generator and the voltage reflection coefficient at the load. Take $u = 0.8c$.

11.21 A 50-Ω air line operating at 500 MHz is terminated by the series combination of a 60-Ω resistor and a variable capacitor. Find the value of the capacitance that will produce a standing wave ratio of 9 on the line. Calculate the reflection coefficient.

11.22 For a lossless line with $Z_o = 60\ \Omega$, the maximum Z_{in} of 180 Ω occurs at $\lambda/24 \simeq 0.042\ \lambda$ from the load. If the line is $0.3\ \lambda$ long, determine

(a) s

(b) Z_L

(c) Z_{in} at the generator

11.23 An 80 Ω transmission line operating at 12 MHz is terminated by a load Z_L. At 22 m from the load, the input impedance is $100 - j120\ \Omega$. If $u = 0.8c$,

(a) Calculate Γ_L, $Z_{in,max}$, and $Z_{in,min}$.

(b) Find Z_L, s, and the input impedance at 28 m from the load.

(c) How many $Z_{in,max}$ and $Z_{in,min}$ are there between the load and the $100 - j120\ \Omega$ input impedance?

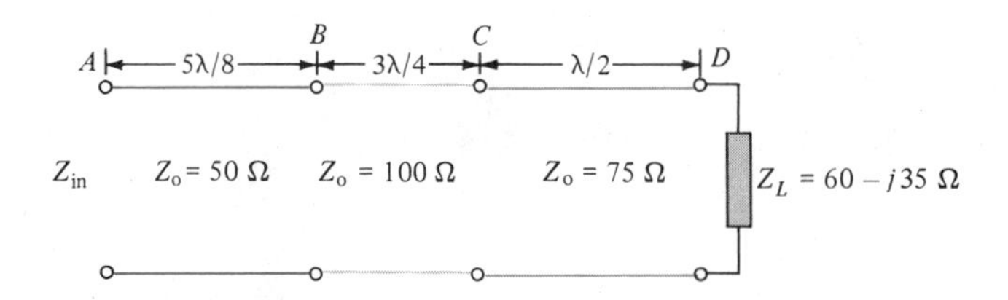

Figure 11.45 For Problem 11.18.

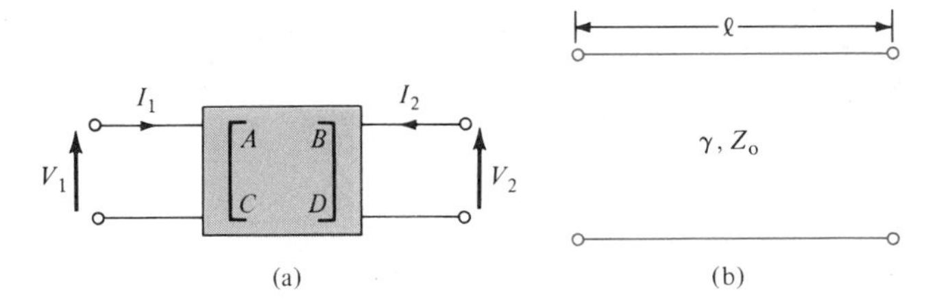

Figure 11.46 For Problem 11.19.

11.24 An antenna, connected to a 150-Ω lossless line, produces a standing wave ratio of 2.6. If measurements indicate that voltage maxima are 120 cm apart and that the last maximum is 40 cm from the antenna, calculate

(a) The operating frequency

(b) The antenna impedance

(c) The reflection coefficient. Assume that $u = c$.

11.25 The observed standing-wave ratio on a 100-Ω lossless line is 8. If the first maximum voltage occurs at 0.3λ from the load, calculate the load impedance and the voltage reflection coefficient at the load.

11.26 Two lossless transmission lines are connected together and terminated by a load as in Figure 11.47. If $Z_o = 75\ \Omega$, $Z_o' = 50\ \Omega$, $Z_L = 120 + j60\ \Omega$, $\lambda = 10$ m, calculate Z_{in} and Z_{in}' for

(a) $\ell_1 = 6$ m, $\ell_2 = 8$ m

(b) $\ell_1 = \lambda/4$, $\ell_2 = \lambda/2$.

Note that a $\lambda/2$ section of transmission line is regarded as a half-wave transformer.

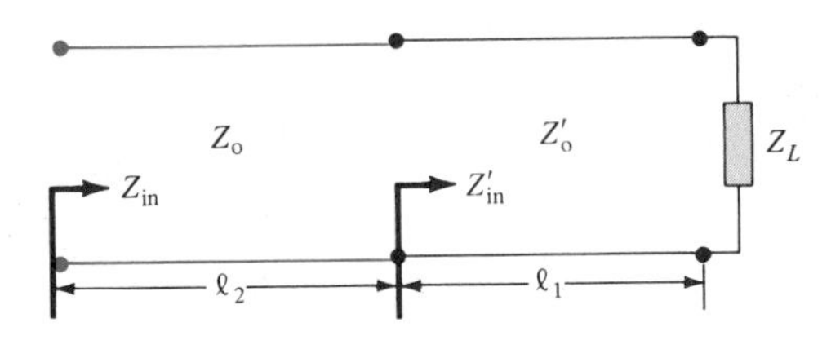

Figure 11.47 For Problem 11.26.

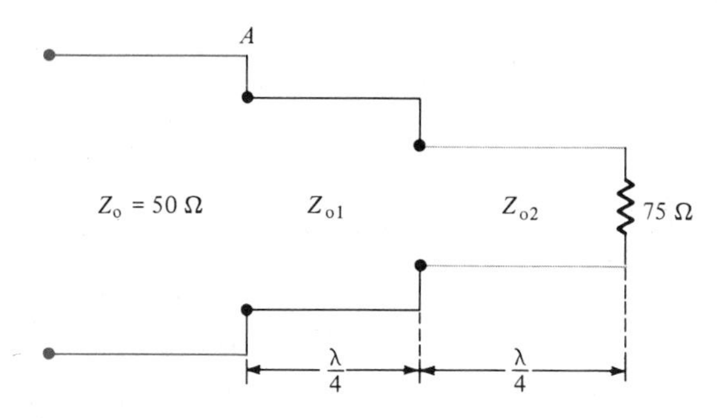

Figure 11.48 Double section transformer of Problem 11.27.

11.27 Two λ/4 transformers in tandem are to connect a 50-Ω line to a 75-Ω load as in Figure 11.48.

(a) Determine the characteristic impedance Z_{o1} if $Z_{o2} = 30\ \Omega$ and there is no reflected wave to the left of A.

(b) If the best results are obtained when

$$\left[\frac{Z_o}{Z_{o1}}\right]^2 = \frac{Z_{o1}}{Z_{o2}} = \left[\frac{Z_{o2}}{Z_L}\right]^2$$

determine Z_{o1} and Z_{o2} for this case.

11.28 A half-wavelength transformer repeats the load and acts like a 1 : 1 transformer. Explain.

11.29 Two identical antennas, each with input impedance 74 Ω are fed with three identical 50-Ω quarter-wave lossless transmission lines as shown in Figure 11.49. Calculate the input impedance at the source end.

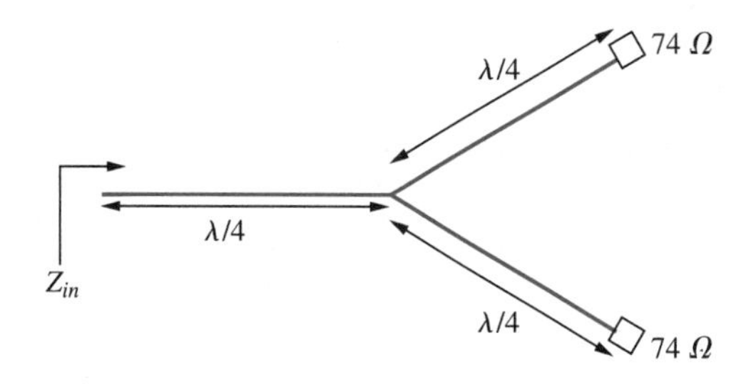

Figure 11.49 For Problems 11.29 and 11.30.

11.30 If the line in the previous problem is connected to a voltage source 120 V with internal impedance 80 Ω, calculate the average power delivered to either antenna.

11.31 Calculate the input impedance of a $\lambda/8$ section of a 50-Ω transmission line terminated with

(a) 40 Ω (inductive reactance)

(b) $30 - j60\ \Omega$

11.32 A section of lossless transmission line is shunted across the main line as in Figure 11.50. If $\ell_1 = \lambda/4$, $\ell_2 = \lambda/8$, and $\ell_3 = 7\lambda/8$, find y_{in_1}, y_{in_2} and y_{in_3} given that $Z_o = 100\ \Omega$, $Z_L = 200 + j150\ \Omega$. Repeat the calculations if the shorted section were open.

11.33 Design a shorted stub that will match a load $Z_L = 60 + j40\ \Omega$ to a 50-Ω transmission line. The stub should be located as close as possible to the load. Also find the required stub admittance.

11.34 A stub of length 0.12λ is used to match a 60-Ω lossless line to a load. If the stub is located at 0.3λ from the load, calculate

(a) The load impedance Z_L

(b) The length of an alternative stub and its location with respect to the load

(c) The standing wave ratio between the stub and the load

11.35 A 60-Ω lossless line terminated by load Z_L has a voltage wave as shown in Figure 11.51. Find s, Γ, and Z_L.

11.36 When a lossless line is connected to a $100 + j80\ \Omega$ load, $s = 2.8$ and the first voltage minimum occurs at 0.306λ from the load end. Determine Z_o and Γ.

11.37 When a 50-Ω air line is terminated with an unknown load Z_L, the standing wave ratio is measured to be 2.5 and the distance between adjacent minima is 10 cm.

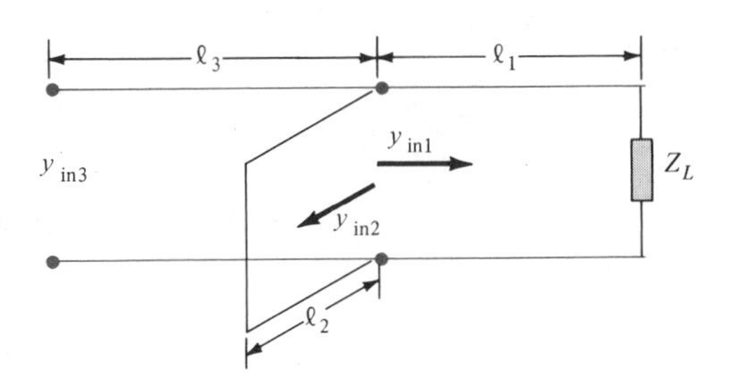

Figure 11.50 For Problem 11.32.

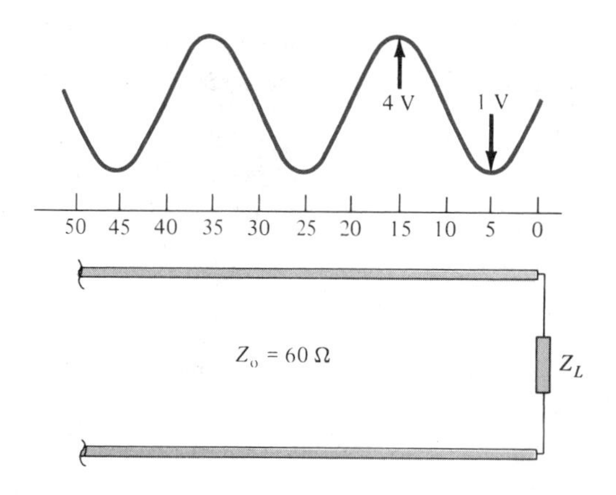

Figure 11.51 For Problem 11.35.

Calculate Z_L given that when the load is replaced by a short, the positions of the minima are shifted 4 cm

(a) Toward the load

(b) Toward the generator

***11.38** An unknown load attached to a 50-Ω air line produces $V_{max} = 0.8$ V, $V_{min} = 0.5$ V, and adjacent minima at 14 cm and 23.5 cm from the load. When the load is temporarily replaced with a short, the minima are (shifted toward the generator) at 9.5 and 19 cm from the load. Calculate s, f, Γ, and Z_L.

****11.39** Show that for a dc voltage V_g turned on at $t = 0$ (see Figure 11.30), the asymptotic values ($t \ll \ell/u$) of $V(\ell, t)$ and $I(\ell, t)$ are

$$V_\infty = \frac{V_g Z_L}{Z_g + Z_L} \quad \text{and} \quad I_\infty = \frac{V_g}{Z_g + Z_L}$$

11.40 A 60-Ω lossless line is connected to a 40-Ω pulse generator. The line is 6 m long and is terminated by a load of 100-Ω. If a rectangular pulse of width 5μ and magnitude 20 V is sent down the line, find $V(0, t)$ and $I(\ell, t)$ for $0 \leq t \leq 10\mu$s. Take $u = 3 \times 10^8$ m/s.

11.41 Refer to the circuit of Figure 11.52 with $Z_o = 75\ \Omega$, $Z_g = 25\ \Omega$, $Z_L = 125\ \Omega$, $V_g = 100$ V, $\ell = 300$ m, and $u = 10^8$ m/s.

(a) Draw the voltage and current bounce diagrams.

(b) Plot $V(0, t)$, $I(0, t)$, $V(\ell, t)$, and $I(\ell, t)$ for $0 < t < 15\ \mu$s.

(c) What are the final values of $V(0, t)$, $I(0, t)$, $V(\ell, t)$, and $I(\ell, t)$?

(d) Plot $V(\ell/2, t)$ and $I(\ell/2, t)$ for $0 < t < 15\ \mu$s.

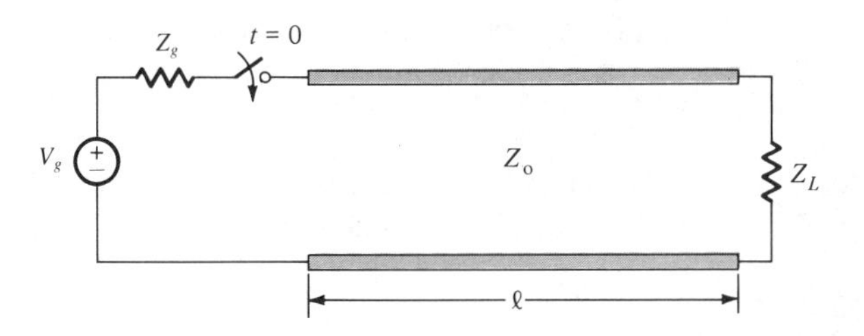

Figure 11.52 For Problems 11.41 and 11.42.

***11.42** Refer to Figure 11.52, where $Z_g = 25\ \Omega$, $Z_o = 50\ \Omega$, $Z_L = 150\ \Omega$, $\ell = 150$ m, $u = c$. If at $t = 0$, the pulse shown in Figure 11.53 is incident on the line

(a) Draw the voltage and current bounce diagrams.

(b) Determine $V(0, t)$, $V(\ell, t)$, $I(0, t)$, and $I(\ell, t)$ for $0 < t < 8\ \mu$s.

11.43 A 30-V battery in series with a 75-Ω resistor is switched into a 50-Ω air dielectric cable at $t = 0$. If the cable is 600 m long and is terminated in a load of 30 Ω, determine

(a) The voltage and current bounce diagrams

(b) $V(\ell/2, t)$ and $I(\ell/2, t)$

11.44 A lossless line of length $\ell = 300$ m is terminated by a load $Z_L = 100\ \Omega$. At the generator end the line is connected to a battery of 200 V, a resistance of 80 Ω,

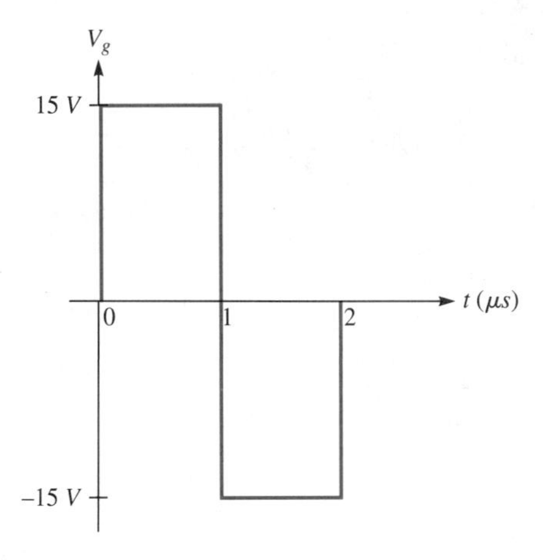

Figure 11.53 Two rectangular pulses of Problem 11.42.

and a switch. The switch is originally open but is closed at $t = 0$. Sketch and dimension $V(\ell/2, t)$ and $I(\ell/2, t)$ for $0 \leq t \leq 4\ell/u$. Take $Z_o = 50\ \Omega$ and $u = 0.25c$.

***11.45** For the previous problem, the switch has been closed for a long time and at $t = 0$, Z_L is short-circuited. Sketch and dimension $V(\ell/2, t)$ and $I(\ell/2, t)$.

11.46 A microstrip line is made of a copper conductor 4 mm wide on an alumina substrate ($\varepsilon_r = 9$, $\tan\theta = 6 \times 10^{-4}$ at 10 GHz) of thickness 1 mm. At 10 GHz, calculate

(a) The characteristic impedance of the line

(b) The effective permittivity of the line

(c) The dielectric attenuation constant

(d) The conduction attenuation constant

(e) The total loss in decibels per centimeter

11.47 A 20-Ω microstrip transmission line is to be constructed on sapphire ($\varepsilon_r = 10$). Calculate the required ratio of w/h and the effective relative permittivity.

11.48 A lossless microstrip line has $\varepsilon_r = 5$ and $Z_o = 60\ \Omega$. Calculate the inductance and capacitance per unit length of the line. (*Hint*: See Problem 11.2.)

11.49 An alumina substrate ($\varepsilon = 9.6\varepsilon_o$) of thickness 2mm is used for the construction of a microstrip circuit. If the circuit designer has the choice of making the line width to be within 0.4 to 8.0 mm, what is the range of characteristic impedance of the line?

CHAPTER

Waveguides

12

If you want to be happy for an hour, get drunk. If you want to be happy for three days, get married. If you want to be happy for eight days, kill your pig and eat it. If you want to be happy forever, invent a machine useful to your fellowmen.

— OLD CHINESE PROVERB (REVISED).

12.1 INTRODUCTION

As mentioned in the preceding chapter, a transmission line can be used to guide EM energy from one point (generator) to another (load). A waveguide is another means of achieving the same goal. However, a waveguide differs from a transmission line in some respects, although we may regard the latter as a special case of the former. In the first place, a transmission line can only support a transverse electromagnetic (TEM) wave, whereas a waveguide can support many possible field configurations. Second, at microwave frequencies (roughly 3–300 GHz), transmission lines become inefficient due to skin effect and dielectric losses; waveguides are used at that range of frequencies to obtain larger bandwidth and lower signal attenuation. Moreover, a transmission line may operate from dc ($f = 0$) to a very high frequency; a waveguide can only operate above a certain frequency called the *cutoff frequency* and therefore acts as a high-pass filter. Thus, waveguides cannot transmit dc, and they become excessively large at frequencies below microwave frequencies.

Although a waveguide may assume any arbitrary but uniform cross section, common waveguides are either rectangular or circular. Typical waveguides[1] are shown in Figure 12.1. Analysis of circular waveguides is involved and requires familiarity

[1]For other types of waveguides, see J. A. Seeger, *Microwave Theory, Components and Devices*. Englewood Cliffs, NJ: Prentice-Hall, 1986, pp. 128–133.

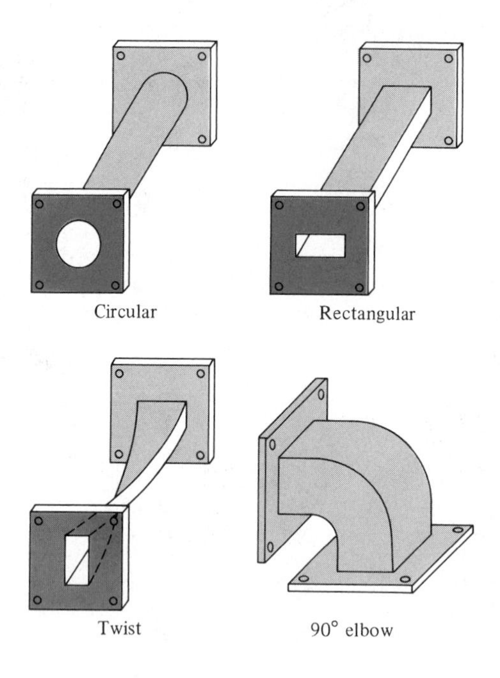

Figure 12.1 Typical waveguides.

with Bessel functions, which are beyond our scope.[2] We will only consider rectangular waveguides. By assuming lossless waveguides ($\sigma_c \simeq \infty$, $\sigma \approx 0$), we shall apply Maxwell's equations with the appropriate boundary conditions to obtain different modes of wave propagation and the corresponding **E** and **H** fields.

12.2 RECTANGULAR WAVEGUIDES

Consider the rectangular waveguide shown in Figure 12.2. We shall assume that the waveguide is filled with a source-free lossless dielectric material ($\sigma \simeq 0$) and its walls are perfectly conducting ($\sigma_c \simeq \infty$). From eqs. (10.17) and (10.19), we recall that for a lossless medium, Maxwell's equations in phasor form become

$$\nabla^2 \mathbf{E}_s + k^2 \mathbf{E}_s = 0 \quad [12.1]$$

$$\nabla^2 \mathbf{H}_s + k^2 \mathbf{H}_s = 0 \quad [12.2]$$

....

[2]Analysis of circular waveguides can be found in advanced EM or EM-related texts, e.g., S. Y. Liao, *Microwave Devices and Circuits*. Englewood Cliffs, NJ: Prentice-Hall, 3rd ed., 1990, pp. 119–141.

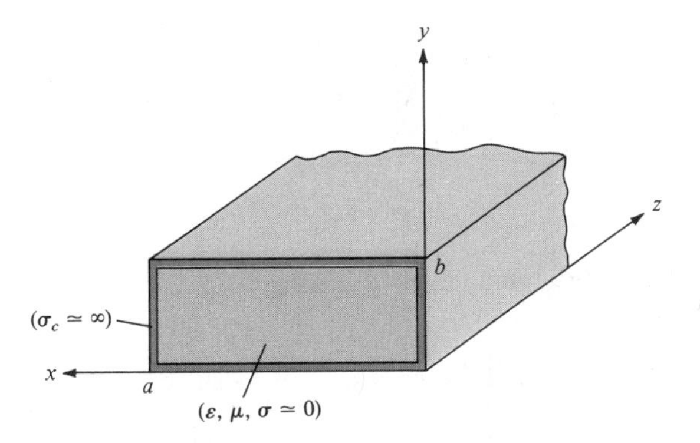

Figure 12.2 A rectangular waveguide with perfectly conducting walls, filled with a lossless material.

where

$$k = \omega\sqrt{\mu\varepsilon} \tag{12.3}$$

and the time factor $e^{j\omega t}$ is assumed. If we let

$$\mathbf{E}_s = (E_{xs}, E_{ys}, E_{zs}) \qquad \text{and} \qquad \mathbf{H}_s = (H_{xs}, H_{ys}, H_{zs})$$

each of eqs. (12.1) and (12.2) is comprised of three scalar Helmholtz equations (see Problem 10.4). In other words, to obtain **E** and **H** fields, we have to solve six scalar equations. For the z-component, for example, eq. (12.1) becomes

$$\frac{\partial^2 E_{zs}}{\partial x^2} + \frac{\partial^2 E_{zs}}{\partial y^2} + \frac{\partial^2 E_{zs}}{\partial z^2} + k^2 E_{zs} = 0 \tag{12.4}$$

which is a partial differential equation. From Example 6.5, we know that eq. (12.4) can be solved by separation of variables (product solution). So we let

$$E_{zs}(x, y, z) = X(x)\, Y(y)\, Z(z) \tag{12.5}$$

where $X(x)$, $Y(y)$, and $Z(z)$ are functions of x, y, and z respectively. Substituting eq. (12.5) into eq. (12.4) and dividing by XYZ gives

$$\frac{X''}{X} + \frac{Y''}{Y} + \frac{Z''}{Z} = -k^2 \tag{12.6}$$

Since the variables are independent, each term in eq. (12.6) must be constant, so the equation can be written as

$$-k_x^2 - k_y^2 + \gamma^2 = -k^2 \tag{12.7}$$

where $-k_x^2$, $-k_y^2$, and γ^2 are separation constants. Thus, eq. (12.6) is separated as

$$X'' + k_x^2 X = 0 \qquad [12.8a]$$

$$Y'' + k_y^2 Y = 0 \qquad [12.8b]$$

$$Z'' - \gamma^2 Z = 0 \qquad [12.8c]$$

By following the same argument as in Example 6.5, we obtain the solution to eq. (12.8) as

$$X(x) = c_1 \cos k_x x + c_2 \sin k_x x \qquad [12.9a]$$

$$Y(y) = c_3 \cos k_y y + c_4 \sin k_y y \qquad [12.9b]$$

$$Z(z) = c_5 e^{\gamma z} + c_6 e^{-\gamma z} \qquad [12.9c]$$

Substituting eq. (12.9) into eq. (12.5) gives

$$E_{zs}(x, y, z) = (c_1 \cos k_x x + c_2 \sin k_x x)(c_3 \cos k_y y + c_4 \sin k_y y) \cdot (c_5 e^{\gamma z} + c_6 e^{-\gamma z}) \qquad [12.10]$$

As usual, if we assume that the wave propagates along the waveguide in the $+z$ direction, the multiplicative constant $c_5 = 0$ because the wave has to be finite at infinity (i.e., $E_{zs}(x, y, z = \infty) = 0$). Hence eq. (12.10) is reduced to

$$E_{zs}(x, y, z) = (A_1 \cos k_x x + A_2 \sin k_x x)(A_3 \cos k_y y + A_4 \sin k_y y)e^{-\gamma z} \qquad [12.11]$$

where $A_1 = c_1 c_6$, $A_2 = c_2 c_6$, and so on. By taking similar steps, we get the solution of the z-component of eq. (12.2) as

$$H_{zs}(x, y, z) = (B_1 \cos k_x x + B_2 \sin k_x x)(B_3 \cos k_y y + B_4 \sin k_y y)e^{-\gamma z} \qquad [12.12]$$

Instead of solving for other field component E_{xs}, E_{ys}, H_{xs}, and H_{ys} in eqs. (12.1) and (12.2) in the same manner, we simply use Maxwell's equations to determine them from E_{zs} and H_{zs}. From

$$\nabla \times \mathbf{E}_s = -j\omega\mu \mathbf{H}_s$$

and

$$\nabla \times \mathbf{H}_s = j\omega\varepsilon \mathbf{E}_s$$

we obtain

$$\frac{\partial E_{zs}}{\partial y} - \frac{\partial E_{ys}}{\partial z} = -j\omega\mu H_{xs} \qquad [12.13a]$$

$$\frac{\partial H_{zs}}{\partial y} - \frac{\partial H_{ys}}{\partial z} = j\omega\varepsilon E_{xs} \qquad [12.13b]$$

$$\frac{\partial E_{xs}}{\partial z} - \frac{\partial E_{zs}}{\partial x} = -j\omega\mu H_{ys} \tag{12.13c}$$

$$\frac{\partial H_{xs}}{\partial z} - \frac{\partial H_{zs}}{\partial x} = j\omega\varepsilon E_{ys} \tag{12.13d}$$

$$\frac{\partial E_{ys}}{\partial x} - \frac{\partial E_{xs}}{\partial y} = -j\omega\mu H_{zs} \tag{12.13e}$$

$$\frac{\partial H_{ys}}{\partial x} - \frac{\partial H_{xs}}{\partial y} = j\omega\varepsilon E_{zs} \tag{12.13f}$$

We will now express E_{xs}, E_{ys}, H_{xs}, and H_{ys} in terms of E_{zs} and H_{zs}. For E_{xs}, for example, we combine eqs. (12.13b) and (12.13c) and obtain

$$j\omega\varepsilon E_{xs} = \frac{\partial H_{zs}}{\partial y} + \frac{1}{j\omega\mu}\left(\frac{\partial^2 E_{xs}}{\partial z^2} - \frac{\partial^2 E_{zs}}{\partial x \partial z}\right) \tag{12.14}$$

From eqs. (12.11) and (12.12), it is clear that all field components vary with z according to $e^{-\gamma z}$, that is,

$$E_{zs} \sim e^{-\gamma z}, \qquad E_{xs} \sim e^{-\gamma z}$$

Hence,

$$\frac{\partial E_{zs}}{\partial z} = -\gamma E_{zs}, \qquad \frac{\partial^2 E_{xx}}{\partial z^2} = \gamma^2 E_{xs}$$

and eq. (12.14) becomes

$$j\omega\varepsilon E_{xs} = \frac{\partial H_{zs}}{\partial y} + \frac{1}{j\omega\mu}\left(\gamma^2 E_{xs} + \gamma \frac{\partial E_{zs}}{\partial x}\right)$$

or

$$-\frac{1}{j\omega\mu}(\gamma^2 + \omega^2\mu\varepsilon)\, E_{xs} = \frac{\gamma}{j\omega\mu}\frac{\partial E_{zs}}{\partial x} + \frac{\partial H_{zs}}{\partial y}$$

Thus, if we let $h^2 = \gamma^2 + \omega^2\mu\varepsilon = \gamma^2 + k^2$,

$$E_{xs} = -\frac{\gamma}{h^2}\frac{\partial E_{zs}}{\partial x} - \frac{j\omega\mu}{h^2}\frac{\partial H_{zs}}{\partial y}$$

Similar manipulations of eq. (12.13) yield expressions for E_{ys}, H_{xs}, and H_{ys} in terms of E_{zs} and H_{zs}. Thus,

$$E_{xs} = -\frac{\gamma}{h^2}\frac{\partial E_{zs}}{\partial x} - \frac{j\omega\mu}{h^2}\frac{\partial H_{zs}}{\partial y} \qquad [12.15a]$$

$$E_{ys} = -\frac{\gamma}{h^2}\frac{\partial E_{zs}}{\partial y} + \frac{j\omega\mu}{h^2}\frac{\partial H_{zs}}{\partial x} \qquad [12.15b]$$

$$H_{xs} = \frac{j\omega\varepsilon}{h^2}\frac{\partial E_{zs}}{\partial y} - \frac{\gamma}{h^2}\frac{\partial H_{zs}}{\partial x} \qquad [12.15c]$$

$$H_{ys} = -\frac{j\omega\varepsilon}{h^2}\frac{\partial E_{zs}}{\partial x} - \frac{\gamma}{h^2}\frac{\partial H_{zs}}{\partial y} \qquad [12.15d]$$

where

$$h^2 = \gamma^2 + k^2 = k_x^2 + k_y^2 \qquad [12.16]$$

Thus we can use eq. (12.15) in conjunction with eqs. (12.11) and (12.12) to obtain E_{xs}, E_{ys}, H_{xs}, and H_{ys}.

From eqs. (12.11), (12.12), and (12.15), we notice that there are different types of field patterns or configurations. Each of these distinct field patterns is called a *mode*. Four different mode categories can exist, namely:

1. $E_{zs} = 0 = H_{zs}$ (TEM mode): This is the *transverse electromagnetic* (TEM) mode, in which both the **E** and **H** fields are transverse to the direction of wave propagation. From eq. (12.15), all field components vanish for $E_{zs} = 0 = H_{zs}$. Consequently, we conclude that a rectangular waveguide cannot support TEM mode.

2. $E_{zs} = 0$, $H_{zs} \neq 0$ (TE modes): For this case, the remaining components (E_{xs} and E_{ys}) of the electric field are transverse to the direction of propagation $\mathbf{a}_z$. Under this condition, fields are said to be in *transverse electric* (TE) modes. See Figure 12.3(a).

3. $E_{zs} \neq 0$, $H_{zs} = 0$ (TM modes): In this case, the **H** field is transverse to the direction of wave propagation. Thus we have *transverse magnetic* (TM) modes. See Figure 12.3(b).

4. $E_{zs} \neq 0$, $H_{zs} \neq 0$ (HE modes): This is the case when neither **E** nor **H** field is transverse to the direction of wave propagation. They are sometimes referred to as *hybrid* modes.

We should note the relationship between k in eq. (12.3) and β of eq. (10.43a). The phase constant β in eq. (10.43a) was derived for TEM mode. For the TEM mode, $h = 0$, so from eq. (12.16), $\gamma^2 = -k^2 \rightarrow \gamma = \alpha + j\beta = jk$; that is, $\beta = k$. For other modes, $\beta \neq k$. In the subsequent sections, we shall examine the TM and TE modes of propagation separately.

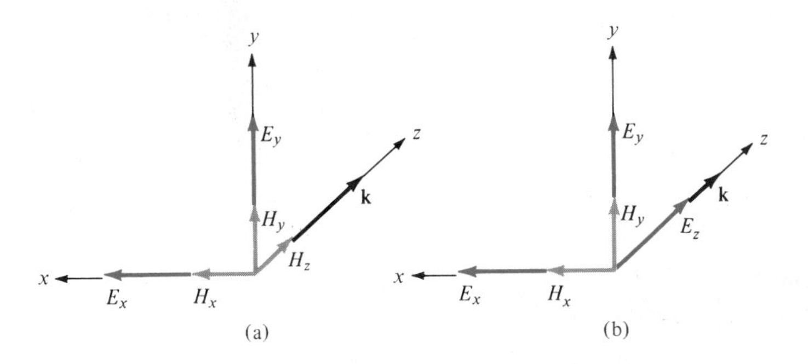

Figure 12.3 Components of EM fields in a rectangular waveguide: **(a)** TE mode $E_z = 0$, **(b)** TM mode, $H_z = 0$.

12.3 TRANSVERSE MAGNETIC (TM) MODES

For this case, the magnetic field has its components transverse (or normal) to the direction of wave propagation. This implies that we set $H_z = 0$ and determine E_x, E_y, E_z, H_x, and H_y using eqs. (12.11) and (12.15) and the boundary conditions. We shall solve for E_z and later determine other field components from E_z. At the walls of the waveguide, the tangential components of the **E** field must be continuous; that is,

$$E_{zs} = 0 \quad \text{at} \quad y = 0 \tag{12.17a}$$

$$E_{zs} = 0 \quad \text{at} \quad y = b \tag{12.17b}$$

$$E_{zs} = 0 \quad \text{at} \quad x = 0 \tag{12.17c}$$

$$E_{zs} = 0 \quad \text{at} \quad x = a \tag{12.17d}$$

Equations (12.17a) and (12.17c) require that $A_1 = 0 = A_3$ in eq. (12.11), so eq. (12.11) becomes

$$E_{zs} = E_o \sin k_x x \sin k_y y \, e^{-\gamma z} \tag{12.18}$$

where $E_o = A_2 A_4$. Also eqs. (12.17b) and (12.17d) when applied to eq. (12.18) require that

$$\sin k_x a = 0, \qquad \sin k_y b = 0 \tag{12.19}$$

This implies that

$$k_x a = m\pi, \qquad m = 1, 2, 3, \ldots \tag{12.20a}$$

$$k_y b = n\pi, \qquad n = 1, 2, 3, \ldots \tag{12.20b}$$

or

$$k_x = \frac{m\pi}{a}, \qquad k_y = \frac{n\pi}{b} \tag{12.21}$$

The negative integers are not chosen for m and n in eq. (12.20a) for the reason given in Example 6.5. Substituting eq. (12.21) into eq. (12.18) gives

$$E_{zs} = E_o \sin\left(\frac{m\pi x}{a}\right) \sin\left(\frac{n\pi y}{b}\right) e^{-\gamma z} \tag{12.22}$$

We obtain other field components from eqs. (12.22) and (12.15) bearing in mind that $H_{zs} = 0$. Thus

$$E_{xs} = -\frac{\gamma}{h^2}\left(\frac{m\pi}{a}\right) E_o \cos\left(\frac{m\pi x}{a}\right) \sin\left(\frac{n\pi y}{b}\right) e^{-\gamma z} \tag{12.23a}$$

$$E_{ys} = -\frac{\gamma}{h^2}\left(\frac{n\pi}{b}\right) E_o \sin\left(\frac{m\pi x}{a}\right) \cos\left(\frac{n\pi y}{b}\right) e^{-\gamma z} \tag{12.23b}$$

$$H_{xs} = \frac{j\omega\varepsilon}{h^2}\left(\frac{n\pi}{b}\right) E_o \sin\left(\frac{m\pi x}{a}\right) \cos\left(\frac{n\pi y}{b}\right) e^{-\gamma z} \tag{12.23c}$$

$$H_{ys} = -\frac{j\omega\varepsilon}{h^2}\left(\frac{m\pi}{a}\right) E_o \cos\left(\frac{m\pi x}{a}\right) \sin\left(\frac{n\pi y}{b}\right) e^{-\gamma z} \tag{12.23d}$$

where

$$h^2 = k_x^2 + k_y^2 = \left[\frac{m\pi}{a}\right]^2 + \left[\frac{n\pi}{b}\right]^2 \tag{12.24}$$

which is obtained from eqs. (12.16) and (12.21). Notice from eqs. (12.22) and (12.23) that each set of integers m and n gives a different field pattern or mode, referred to as TM_{mn} mode, in the waveguide. Integer m equals the number of half-cycle variations in the x-direction, and integer n is the number of half-cycle variations in the y-direction. We also notice from eqs. (12.22) and (12.23) that if (m, n) is $(0, 0)$, $(0, n)$, or $(m, 0)$, all field components vanish. Thus neither m nor n can be zero. Consequently, TM_{11} is the lowest-order mode of all the TM_{mn} modes.

By subtituting eq. (12.21) into eq. (12.16), we obtain the propagation constant

$$\gamma = \sqrt{\left[\frac{m\pi}{a}\right]^2 + \left[\frac{n\pi}{b}\right]^2 - k^2} \qquad [12.25]$$

where $k = \omega\sqrt{\mu\varepsilon}$ as in eq. (12.3). We recall that, in general, $\gamma = \alpha + j\beta$. In the case of eq. (12.25), we have three possibilities depending on k (or ω), m, and n:
Case 1 (cutoff): If

$$k^2 = \omega^2\mu\varepsilon = \left[\frac{m\pi}{a}\right]^2 + \left[\frac{n\pi}{b}\right]^2$$

$$\gamma = 0 \qquad \text{or} \qquad \alpha = 0 = \beta$$

The value of ω that causes this is called the *cutoff angular frequency* ω_c; that is,

$$\omega_c = \frac{1}{\sqrt{\mu\varepsilon}}\sqrt{\left[\frac{m\pi}{a}\right]^2 + \left[\frac{n\pi}{b}\right]^2} \qquad [12.26]$$

Case 2 (evanescent): If

$$k^2 = \omega^2\mu\varepsilon < \left[\frac{m\pi}{a}\right]^2 + \left[\frac{n\pi}{b}\right]^2$$

$$\gamma = \alpha, \qquad \beta = 0$$

In this case, we have no wave propagation at all. These nonpropagating or attenuating modes are said to be *evanescent*.
Case 3 (propagation): If

$$k^2 = \omega^2\mu\varepsilon > \left[\frac{m\pi}{a}\right]^2 + \left[\frac{n\pi}{b}\right]^2$$

$$\gamma = j\beta, \qquad \alpha = 0$$

that is, from eq. (12.25) the phase constant β becomes

$$\beta = \sqrt{k^2 - \left[\frac{m\pi}{a}\right]^2 - \left[\frac{n\pi}{b}\right]^2} \qquad [12.27]$$

This is the only case when propagation takes place because all field components will have the factor $e^{-\gamma z} = e^{-j\beta z}$.

Thus for each mode, characterized by a set of integers m and n, there is a corresponding *cutoff frequency* f_c below which attenuation occurs and above which propagation takes place. The waveguide therefore operates as a high-pass filter. The cutoff frequency is obtained from eq. (12.26) as

$$f_c = \frac{\omega_c}{2\pi} = \frac{1}{2\pi\sqrt{\mu\varepsilon}}\sqrt{\left[\frac{m\pi}{a}\right]^2 + \left[\frac{n\pi}{b}\right]^2}$$

or

$$f_c = \frac{u'}{2}\sqrt{\left(\frac{m}{a}\right)^2 + \left(\frac{n}{b}\right)^2} \qquad [12.28]$$

where $u' = \dfrac{1}{\sqrt{\mu\varepsilon}}$ = phase velocity of uniform plane wave in the lossless dielectric medium ($\sigma = 0, \mu, \varepsilon$) filling the waveguide. The *cutoff wavelength* λ_c is given by

$$\lambda_c = \frac{u'}{f_c}$$

or

$$\lambda_c = \frac{2}{\sqrt{\left(\frac{m}{a}\right)^2 + \left(\frac{n}{b}\right)^2}} \qquad [12.29]$$

It is evident from eqs. (12.28) and (12.29) that TM_{11} has the lowest cutoff frequency (or the longest cutoff wavelength) of all the TM modes. The phase constant β in eq. (12.27) can be written in terms of f_c as

$$\beta = \omega\sqrt{\mu\varepsilon}\sqrt{1 - \left[\frac{f_c}{f}\right]^2}$$

or

$$\beta = \beta'\sqrt{1 - \left[\frac{f_c}{f}\right]^2} \qquad [12.30]$$

where $\beta' = \omega/u' = \omega\sqrt{\mu\varepsilon}$ = phase constant of uniform plane wave in the dielectric medium. It should be noted that γ for evanescent mode can be expressed in terms of f_c, namely,

$$\gamma = \alpha = \beta'\sqrt{\left(\frac{f_c}{f}\right)^2 - 1} \qquad [12.30a]$$

The phase velocity u_p and the wavelength in the guide are respectively given by

$$u_p = \frac{\omega}{\beta}, \qquad \lambda = \frac{2\pi}{\beta} = \frac{u_p}{f} \qquad [12.31]$$

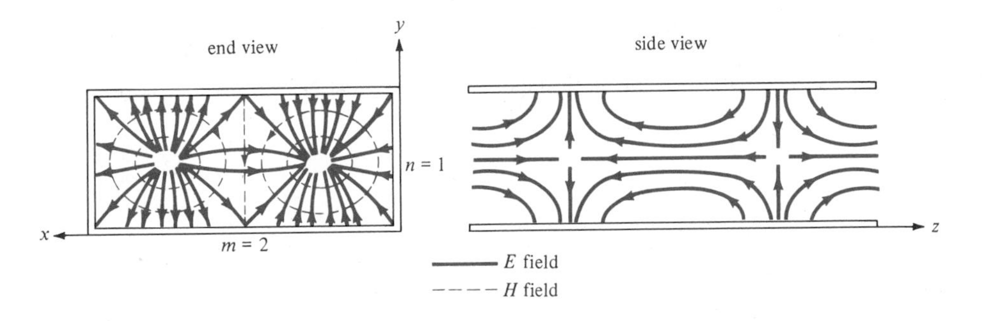

Figure 12.4 Field configuration for TM_{21} mode.

The intrinsic wave impedance of the mode is obtained from eq. (12.23) as ($\gamma = j\beta$)

$$\eta_{TM} = \frac{E_x}{H_y} = -\frac{E_y}{H_x}$$

$$= \frac{\beta}{\omega\varepsilon} = \sqrt{\frac{\mu}{\varepsilon}}\sqrt{1 - \left[\frac{f_c}{f}\right]^2}$$

or

$$\boxed{\eta_{TM} = \eta' \sqrt{1 - \left[\frac{f_c}{f}\right]^2}} \qquad [12.32]$$

where $\eta' = \sqrt{\mu/\varepsilon}$ =intrinsic impedance of uniform plane wave in the medium. Note the difference between u', β', and η', and u, β, and η. The quantities with prime are wave characteristics of the dielectric medium unbounded by the waveguide as discussed in Chapter 10 (i.e., for TEM mode). For example, u' would be the velocity of the wave if the waveguide were removed and the entire space were filled with the dielectric. The quantities without prime are the wave characteristics of the medium bounded by the waveguide.

As mentioned before, the integers m and n indicate the number of half-cycle variations in the $x-y$ cross section of the guide. Thus for a fixed time, the field configuration of Figure 12.4 results for TM_{21} mode, for example.

12.4 TRANSVERSE ELECTRIC (TE) MODES

In the TE modes, the electric field is transverse (or normal) to the direction of wave propagation. We set $E_z = 0$ and determine other field components E_x, E_y, H_x, H_y, and

H_z from eqs. (12.12) and (12.15) and the boundary conditions just as we did for the TM modes. The boundary conditions are obtained from the fact that the tangential components of the electric field must be continuous at the walls of the waveguide; that is,

$$E_{xs} = 0 \quad \text{at} \quad y = 0 \tag{12.33a}$$

$$E_{xs} = 0 \quad \text{at} \quad y = b \tag{12.33b}$$

$$E_{ys} = 0 \quad \text{at} \quad x = 0 \tag{12.33c}$$

$$E_{ys} = 0 \quad \text{at} \quad x = a \tag{12.33d}$$

From eqs. (12.15) and (12.33), the boundary conditions can be written as

$$\frac{\partial H_{zs}}{\partial y} = 0 \quad \text{at} \quad y = 0 \tag{12.34a}$$

$$\frac{\partial H_{zs}}{\partial y} = 0 \quad \text{at} \quad y = b \tag{12.34b}$$

$$\frac{\partial H_{zs}}{\partial x} = 0 \quad \text{at} \quad x = 0 \tag{12.34c}$$

$$\frac{\partial H_{zs}}{\partial x} = 0 \quad \text{at} \quad x = a \tag{12.34d}$$

Imposing these boundary conditions on eq. (12.12) yields

$$\boxed{H_{zs} = H_o \cos\left(\frac{m\pi x}{a}\right) \cos\left(\frac{n\pi y}{b}\right) e^{-\gamma z}} \tag{12.35}$$

where $H_o = B_1 B_3$. Other field components are easily obtained from eqs. (12.35) and (12.15) as

$$E_{xs} = \frac{j\omega\mu}{h^2}\left(\frac{n\pi}{b}\right) H_o \cos\left(\frac{m\pi x}{a}\right) \sin\left(\frac{n\pi y}{b}\right) e^{-\gamma z} \tag{12.36a}$$

$$E_{ys} = -\frac{j\omega\mu}{h^2}\left(\frac{m\pi}{a}\right) H_o \sin\left(\frac{m\pi x}{a}\right) \cos\left(\frac{n\pi y}{b}\right) e^{-\gamma z} \tag{12.36b}$$

$$H_{xs} = \frac{\gamma}{h^2}\left(\frac{m\pi}{a}\right) H_o \sin\left(\frac{m\pi x}{a}\right) \cos\left(\frac{n\pi y}{b}\right) e^{-\gamma z} \tag{12.36c}$$

$$H_{ys} = \frac{\gamma}{h^2}\left(\frac{n\pi}{b}\right) H_o \cos\left(\frac{m\pi x}{a}\right) \sin\left(\frac{n\pi y}{b}\right) e^{-\gamma z} \tag{12.36d}$$

where $m = 0, 1, 2, 3, \ldots$ and $n = 0, 1, 2, 3, \ldots$; h and γ remain as defined for the TM modes. Again, m and n denote the number of half-cycle variations in the $x-y$ cross section of the guide. For TE_{32} mode, for example, the field configuration is in Figure 12.5. The cutoff frequency f_c, the cutoff wavelength λ_c, the phase constant β, phase velocity u_p and wavelength λ for TE modes are the same as for TM modes (see eqs. (12.28) to (12.31)).

For TE modes, (m, n) may be $(0, 1)$ or $(1, 0)$ but not $(0, 0)$. Both m and n cannot be zero at the same time because this will force the field components in eq. (12.36) to vanish. This implies that the lowest mode can be TE_{10} or TE_{01} depending on the values of a and b, the dimensions of the guide. It is standard practice to have $a > b$ so that $1/a^2 < 1/b^2$ in eq. (12.28). Thus TE_{10} is the lowest mode. This mode is called the *dominant mode* of the waveguide and is of practical importance. The cutoff frequency for the TE_{10} mode is obtained from eq. (12.28) as $(m = 1, n = 0)$

$$f_{c_{10}} = \frac{u'}{2a} \tag{12.37}$$

and the cutoff wavelength for TE_{10} mode is obtained from eq. (12.29) as

$$\lambda_{c_{10}} = 2a \tag{12.38}$$

Note that from eq. (12.28) the cutoff frequency for TM_{11} is

$$\frac{u'[a^2 + b^2]^{1/2}}{2ab}$$

which is greater than the cutoff frequency for TE_{10}. Hence, TM_{11} cannot be regarded as the dominant mode. The dominant mode is the mode with the lowest cutoff frequency (or longest cutoff wavelength). Also note that any EM wave with frequency $f < f_{c_{10}}$ (or $\lambda > \lambda_{c_{10}}$) will not be propagated in the guide.

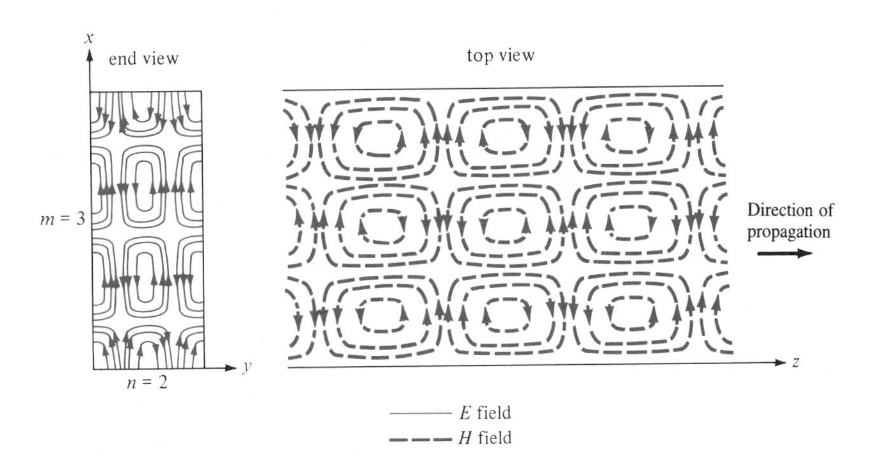

Figure 12.5 Field configuration for TE_{32} mode.

The intrinsic impedance for the TE mode is not the same as for TM modes. From eq. (12.36), it is evident that ($\gamma = j\beta$)

$$\eta_{\text{TE}} = \frac{E_x}{H_y} = -\frac{E_y}{H_x} = \frac{\omega\mu}{\beta}$$

$$= \sqrt{\frac{\mu}{\varepsilon}} \frac{1}{\sqrt{1 - \left[\frac{f_c}{f}\right]^2}}$$

or

$$\boxed{\eta_{TE} = \frac{\eta'}{\sqrt{1 - \left[\frac{f_c}{f}\right]^2}}} \quad [12.39]$$

Note from eqs. (12.32) and (12.39) that η_{TE} and η_{TM} are purely resistive and they vary with frequency as shown in Figure 12.6. Also note that

$$\eta_{\text{TE}}\, \eta_{\text{TM}} = \eta'^2 \quad [12.40]$$

Important equations for TM and TE modes are listed in Table 12.1 for convenience and quick reference.

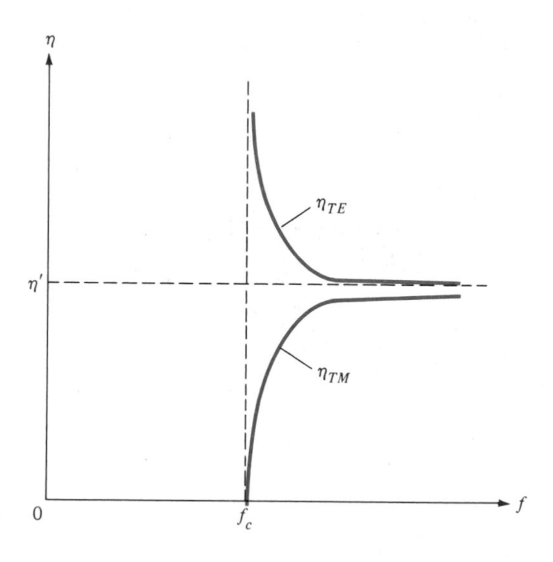

Figure 12.6 Variation of wave impedance with frequency for TE and TM modes.

Table 12.1 Important Equations for TM and TE Modes

TM modes	TE modes
$E_{xs} = -\frac{j\beta}{h^2}\left(\frac{m\pi}{a}\right) E_o \cos\left(\frac{m\pi x}{a}\right) \sin\left(\frac{n\pi y}{b}\right) e^{-\gamma z}$	$E_{xs} = \frac{j\omega\mu}{h^2}\left(\frac{n\pi}{b}\right) H_o \cos\left(\frac{m\pi x}{a}\right) \sin\left(\frac{n\pi y}{b}\right) e^{-\gamma z}$
$E_{ys} = -\frac{j\beta}{h^2}\left(\frac{n\pi}{b}\right) E_o \sin\left(\frac{m\pi x}{a}\right) \cos\left(\frac{n\pi y}{b}\right) e^{-\gamma z}$	$E_{ys} = -\frac{j\omega\mu}{h^2}\left(\frac{m\pi}{a}\right) H_o \sin\left(\frac{m\pi x}{a}\right) \cos\left(\frac{n\pi y}{b}\right) e^{-\gamma z}$
$E_{zs} = E_o \sin\left(\frac{m\pi x}{a}\right) \sin\left(\frac{n\pi y}{b}\right) e^{-\gamma z}$	$E_{zs} = 0$
$H_{xs} = \frac{j\omega\varepsilon}{h^2}\left(\frac{n\pi}{b}\right) E_o \sin\left(\frac{m\pi x}{a}\right) \cos\left(\frac{n\pi y}{b}\right) e^{-\gamma z}$	$H_{xs} = \frac{j\beta}{h^2}\left(\frac{m\pi}{a}\right) H_o \sin\left(\frac{m\pi x}{a}\right) \cos\left(\frac{n\pi y}{b}\right) e^{-\gamma z}$
$H_{ys} = -\frac{j\omega\varepsilon}{h^2}\left(\frac{m\pi}{a}\right) E_o \cos\left(\frac{m\pi x}{a}\right) \sin\left(\frac{n\pi y}{b}\right) e^{-\gamma z}$	$H_{ys} = \frac{j\beta}{h^2}\left(\frac{n\pi}{b}\right) H_o \cos\left(\frac{m\pi x}{a}\right) \sin\left(\frac{n\pi y}{b}\right) e^{-\gamma z}$
$H_{zs} = 0$	$H_{zs} = H_o \cos\left(\frac{m\pi x}{a}\right) \cos\left(\frac{n\pi y}{b}\right) e^{-\gamma z}$
$\eta = \eta' \sqrt{1 - \left(\frac{f_c}{f}\right)^2}$	$\eta = \frac{\eta'}{\sqrt{1 - \left(\frac{f_c}{f}\right)^2}}$

$$f_c = \frac{u'}{2}\sqrt{\left(\frac{m}{a}\right)^2 + \left(\frac{n}{b}\right)^2}$$

$$\lambda_c = \frac{u'}{f_c}$$

$$\beta = \beta' \sqrt{1 - \left(\frac{f_c}{f}\right)^2}$$

$$u_p = \frac{\omega}{\beta} = f\lambda$$

$$\text{where } h^2 = \left(\frac{m\pi}{a}\right)^2 + \left(\frac{n\pi}{b}\right)^2, \ u' = \frac{1}{\sqrt{\mu\varepsilon}}, \ \beta' = \frac{\omega}{u'}, \ \eta' = \sqrt{\frac{\mu}{\varepsilon}}$$

From eqs. (12.22), (12.23), (12.35), and (12.36), we obtain the field patterns for the TM and TE modes. For the dominant TE_{10} mode, $m = 1$ and $n = 0$, so eq. (12.35) becomes

$$H_{zs} = H_o \cos\left(\frac{\pi x}{a}\right) e^{-j\beta z} \tag{12.41}$$

In the time domain,

$$H_z = \text{Re}\,(H_{zs} e^{j\omega t})$$

or

$$H_z = H_o \cos\left(\frac{\pi x}{a}\right) \cos(\omega t - \beta z) \tag{12.42}$$

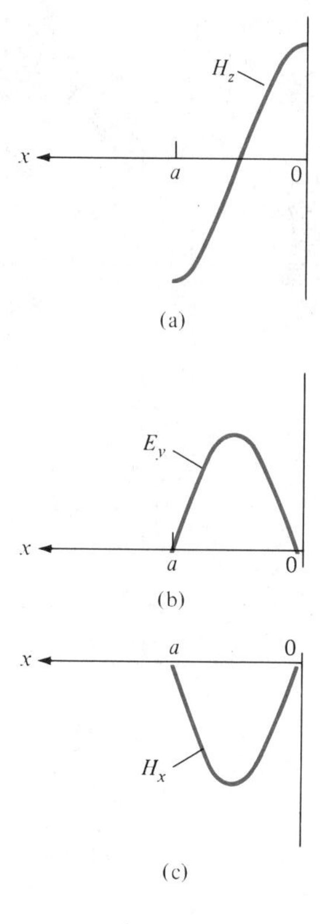

Figure 12.7 Variation of the field components with x for TE_{10} mode.

Similarly, from eq. (12.36),

$$E_y = \frac{\omega\mu a}{\pi} H_o \sin\left(\frac{\pi x}{a}\right) \sin(\omega t - \beta z) \quad \text{[12.43a]}$$

$$H_x = -\frac{\beta a}{\pi} H_o \sin\left(\frac{\pi x}{a}\right) \sin(\omega t - \beta z) \quad \text{[12.43b]}$$

$$E_z = E_x = H_y = 0 \quad \text{[12.43c]}$$

The variation of the **E** and **H** fields with x in an $x-y$ plane, say plane $\cos(\omega t - \beta z) = 1$ for H_z, and plane $\sin(\omega t - \beta z) = 1$ for E_y and H_x, is shown in Figure 12.7 for the TE_{10} mode. The corresponding field lines are shown in Figure 12.8.

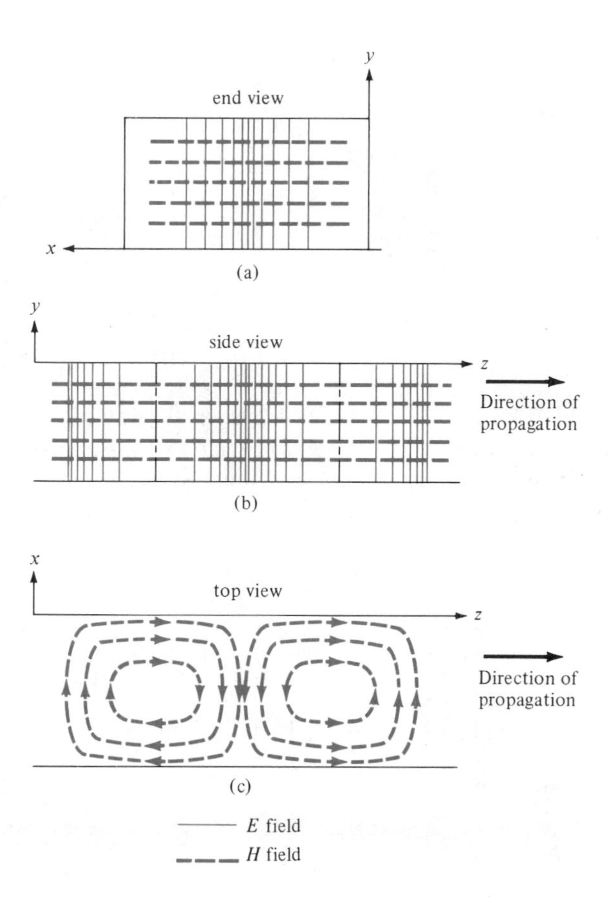

Figure 12.8 Field lines for TE_{10} mode.

EXAMPLE 12.1

A rectangular waveguide with dimensions $a = 2.5$ cm, $b = 1$ cm is to operate below 15.1 GHz. How many TE and TM modes can the waveguide transmit if the guide is filled with a medium characterized by $\sigma = 0$, $\varepsilon = 4\,\varepsilon_o$, $\mu_r = 1$? Calculate the cutoff frequencies of the modes.

SOLUTION The cutoff frequency is given by

$$f_{c_{mn}} = \frac{u'}{2}\sqrt{\frac{m^2}{a^2} + \frac{n^2}{b^2}}$$

where $\alpha = 2.5$b or $a/b = 2.5$, and

$$u' = \frac{1}{\sqrt{\mu\varepsilon}} = \frac{c}{\sqrt{\mu_r\varepsilon_r}} = \frac{c}{2}$$

Hence,

$$f_{c_{mn}} = \frac{c}{4a}\sqrt{m^2 + \frac{a^2}{b^2}n^2}$$

$$= \frac{3 \times 10^8}{4(2.5 \times 10^{-2})}\sqrt{m^2 + 6.25n^2}$$

or

$$f_{c_{mn}} = 3\sqrt{m^2 + 6.25n^2} \text{ GHz} \qquad [12.1.1]$$

We are looking for $f_{c_{mn}} < 15.1$ GHz. A systematic way of doing this is to fix m or n and increase the other until $f_{c_{mn}}$ is greater than 15.1 GHz. From eq. (12.1.1), it is evident that fixing m and increasing n will quickly give us an $f_{c_{mn}}$ that is greater than 15.1 GHz.

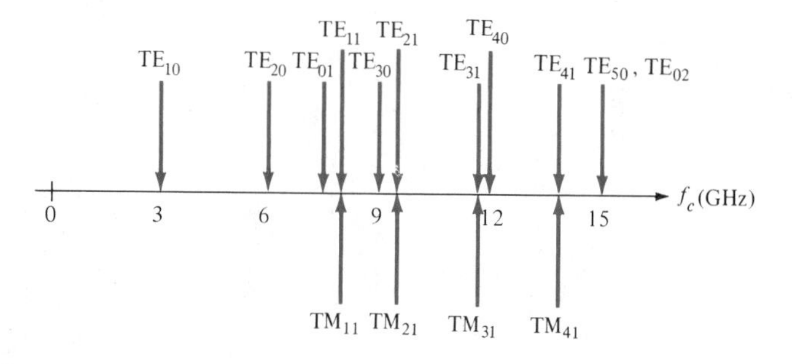

Figure 12.9 Cutoff frequencies of rectangular waveguide with $a = 2.5b$; for Example 12.1.

For TE_{01} mode ($m = 0, n = 1$), $f_{c_{01}} = 3(2.5) = 7.5$ GHz
TE_{02} mode ($m = 0, n = 2$), $f_{c_{02}} = 3(5) = 15$ GHz
TE_{03} mode, $f_{c_{03}} = 3(7.5) = 22.5$ GHz

Thus for $f_{c_{mn}} < 15.1$ GHz, the maximum $n = 2$. We now fix n and increase m until $f_{c_{mn}}$ is greater than 15.1 GHz.

For TE_{10} mode ($m = 1, n = 0$), $f_{c_{10}} = 3$ GHz
TE_{20} mode, $f_{c_{20}} = 6$ GHz
TE_{30} mode, $f_{c_{30}} = 9$ GHz
TE_{40} mode, $f_{c_{40}} = 12$ GHz
TE_{50} mode, $f_{c_{50}} = 15$ GHz (the same as for TE_{02})
TE_{60} mode, $f_{c_{60}} = 18$ GHz.

that is, for $f_{c_{mn}} < 15.1$ GHz, the maximum $m = 5$. Now that we know the maximum m and n, we try other possible combinations in between these maximum values.

For TE_{11}, TM_{11} (degenerate modes), $f_{c_{11}} = 3\sqrt{7.25} = 8.078$ GHz
TE_{21}, TM_{21}, $f_{c_{21}} = 3\sqrt{10.25} = 9.6$ GHz
TE_{31}, TM_{31}, $f_{c_{31}} = 3\sqrt{15.25} = 11.72$ GHz
TE_{41}, TM_{41}, $f_{c_{41}} = 3\sqrt{22.25} = 14.14$ GHz
TE_{12}, TM_{12}, $f_{c_{12}} = 3\sqrt{26} = 15.3$ GHz

Those modes whose cutoff frequencies are less or equal to 15.1 GHz will be transmitted—that is, eleven TE modes and four TM modes (all of the above modes except TE_{12}, TM_{12}, TE_{60}, and TE_{03}). The cutoff frequencies for the fifteen modes are illustrated in the line diagram of Figure 12.9. ■

PRACTICE EXERCISE 12.1

Consider the waveguide of Example 12.1. Calculate the phase constant, phase velocity and wave impedance for TE_{10} and TM_{11} modes at the operating frequency of 15 GHz.

ANSWER For TE_{10}, $\beta = 615.6$ rad/m, $u = 1.531 \times 10^8$ m/s, $\eta_{TE} = 192.4\ \Omega$.
For TM_{11}, $\beta = 529.4$ rad/m, $u = 1.78 \times 10^8$ m/s, $\eta_{TM} = 158.8\ \Omega$.

EXAMPLE 12.2

Write the general instantaneous field expressions for the TM and TE modes. Deduce those for TE_{01} and TM_{12} modes.

SOLUTION

The instantaneous field expressions are obtained from the phasor forms by using

$$\mathbf{E} = \text{Re}\,(\mathbf{E}_s e^{j\omega t}) \quad \text{and} \quad \mathbf{H} = \text{Re}\,(\mathbf{H}_s e^{j\omega t})$$

Applying these to eqs. (12.22) and (12.23) while replacing γ with $j\beta$ gives the following field components for the TM modes:

$$E_x = \frac{\beta}{h^2}\left[\frac{m\pi}{a}\right] E_o \cos\left(\frac{m\pi x}{a}\right) \sin\left(\frac{n\pi y}{b}\right) \sin(\omega t - \beta z)$$

$$E_y = \frac{\beta}{h^2}\left[\frac{n\pi}{b}\right] E_o \sin\left(\frac{m\pi x}{a}\right) \cos\left(\frac{n\pi y}{b}\right) \sin(\omega t - \beta z)$$

$$E_z = E_o \sin\left(\frac{m\pi x}{a}\right) \sin\left(\frac{n\pi y}{b}\right) \cos(\omega t - z)$$

$$H_x = -\frac{\omega\varepsilon}{h^2}\left[\frac{n\pi}{b}\right] E_o \sin\left(\frac{m\pi x}{a}\right) \cos\left(\frac{n\pi y}{b}\right) \sin(\omega t - \beta z)$$

$$H_y = \frac{\omega\varepsilon}{h^2}\left[\frac{m\pi}{a}\right] E_o \cos\left(\frac{m\pi x}{a}\right) \sin\left(\frac{n\pi y}{b}\right) \sin(\omega t - \beta z)$$

$$H_z = 0$$

Similarly, for the TE modes, eqs. (12.35) and (12.36) become

$$E_x = -\frac{\omega\mu}{h^2}\left[\frac{n\pi}{b}\right] H_o \cos\left(\frac{m\pi x}{a}\right) \sin\left(\frac{n\pi y}{b}\right) \sin(\omega t - \beta z)$$

$$E_y = \frac{\omega\mu}{h^2}\left[\frac{m\pi}{a}\right] H_o \sin\left(\frac{m\pi x}{a}\right) \cos\left(\frac{n\pi y}{b}\right) \sin(\omega t - \beta z)$$

$$E_z = 0$$

$$H_x = -\frac{\beta}{h^2}\left[\frac{m\pi}{a}\right] H_o \sin\left(\frac{m\pi x}{a}\right) \cos\left(\frac{n\pi y}{b}\right) \sin(\omega t - \beta z)$$

$$H_y = -\frac{\beta}{h^2}\left[\frac{n\pi}{b}\right] H_o \cos\left(\frac{m\pi x}{a}\right) \sin\left(\frac{n\pi y}{b}\right) \sin(\omega t - \beta z)$$

$$H_z = H_o \cos\left(\frac{m\pi x}{a}\right) \cos\left(\frac{n\pi y}{b}\right) \cos(\omega t - \beta z)$$

For the TE_{01} mode, we set $m = 0$, $n = 1$ to obtain

$$h^2 = \left[\frac{\pi}{b}\right]^2$$

$$E_x = -\frac{\omega\mu b}{\pi} H_o \sin\left(\frac{\pi y}{b}\right) \sin(\omega t - \beta z)$$

$$E_y = 0 = E_z = H_x$$

$$H_y = -\frac{\beta b}{\pi} H_o \sin\left(\frac{\pi y}{b}\right) \sin(\omega t - \beta z)$$

$$H_z = H_o \cos\left(\frac{\pi y}{b}\right) \cos(\omega t - \beta z)$$

For the TM_{12} mode, we set $m = 1$, $n = 2$ to obtain

$$E_x = \frac{\beta}{h^2}\left(\frac{\pi}{a}\right) E_o \cos\left(\frac{\pi x}{a}\right) \sin\left(\frac{2\pi y}{b}\right) \sin(\omega t - \beta z)$$

$$E_y = \frac{\beta}{h^2}\left(\frac{2\pi}{b}\right) E_o \sin\left(\frac{\pi x}{a}\right) \cos\left(\frac{2\pi y}{b}\right) \sin(\omega t - \beta z)$$

$$E_z = E_o \sin\left(\frac{\pi x}{a}\right) \sin\left(\frac{2\pi y}{b}\right) \cos(\omega t - \beta z)$$

$$H_x = -\frac{\omega\varepsilon}{h^2}\left(\frac{2\pi}{b}\right) E_o \sin\left(\frac{\pi x}{a}\right) \cos\left(\frac{2\pi y}{b}\right) \sin(\omega t - \beta z)$$

$$H_y = \frac{\omega\varepsilon}{h^2}\left(\frac{\pi}{a}\right) E_o \cos\left(\frac{\pi x}{a}\right) \sin\left(\frac{2\pi y}{b}\right) \sin(\omega t - \beta z)$$

$$H_z = 0$$

where

$$h^2 = \left[\frac{\pi}{a}\right]^2 + \left[\frac{2\pi}{b}\right]^2$$

■

PRACTICE EXERCISE 12.2

An air-filled 5- by 2-cm waveguide has

$$E_{zs} = 20 \sin 40\pi x \sin 50\pi y\, e^{-j\beta z} \text{ V/m}$$

at 15 GHz.

(a) What mode is being propagated?
(b) Find β.
(c) Determine E_y/E_x.

ANSWER (a) TM_{21}, (b) 241.3 rad/m, (c) 1.25 tan $40\pi x$ cot $50\pi y$.

EXAMPLE 12.3

In a rectangular waveguide for which $a = 1.5$ cm, $b = 0.8$ cm, $\sigma = 0$, $\mu = \mu_o$, and $\varepsilon = 4\varepsilon_o$,

$$H_x = 2 \sin\left(\frac{\pi x}{a}\right) \cos\left(\frac{3\pi y}{b}\right) \sin(\pi \times 10^{11}t - \beta z) \text{ A/m}$$

Determine

(a) The mode of operation
(b) The cutoff frequency
(c) The phase constant β
(d) The propagation constant γ
(e) The intrinsic wave impedance η.

SOLUTION (a) It is evident from the given expression for H_x and the field expressions of the last example that $m = 1$, $n = 3$; that is, the guide is operating at TM_{13} or TE_{13}. Suppose we choose TM_{13} mode (the possibility of having TE_{13} mode is left as an exercise in Practice Exercise 12.3).

(b)
$$f_{c_{mn}} = \frac{u'}{2}\sqrt{\frac{m^2}{a^2} + \frac{n^2}{b^2}}$$

$$u' = \frac{1}{\sqrt{\mu\varepsilon}} = \frac{c}{\sqrt{\mu_r\varepsilon_r}} = \frac{c}{2}$$

Hence

$$f_{c_{13}} = \frac{c}{4}\sqrt{\frac{1}{[1.5 \times 10^{-2}]^2} + \frac{9}{[0.8 \times 10^{-2}]^2}}$$

$$= \frac{3 \times 10^8}{4}(\sqrt{0.444 + 14.06}) \times 10^2 = 28.57 \text{ GHz}$$

(c)
$$\beta = \omega\sqrt{\mu\varepsilon}\sqrt{1 - \left[\frac{f_c}{f}\right]^2} = \frac{\omega\sqrt{\varepsilon_r}}{c}\sqrt{1 - \left[\frac{f_c}{f}\right]^2}$$

$$\omega\omega = 2\pi f = \pi \times 10^{11} \qquad \text{or} \qquad f = \frac{100}{2} = 50 \text{ GHz}$$

$$\beta = \frac{\pi \times 10^{11}(2)}{3 \times 10^8} \sqrt{1 - \left[\frac{28.57}{50}\right]^2} = 1718.81 \text{ rad/m}$$

(d) $\gamma = j\beta = j1718.81$ /m

(e)

$$\eta_{\text{TM}_{13}} = \eta' \sqrt{1 - \left[\frac{f_c}{f}\right]^2} = \frac{377}{\sqrt{\varepsilon_r}} \sqrt{1 - \left[\frac{28.57}{50}\right]^2}$$

$$= 154.7 \ \Omega$$

■

PRACTICE EXERCISE 12.3

Repeat Example 12.3 if TE_{13} mode is assumed. Determine other field components for this mode.

ANSWER

$$f_c = 28.57 \text{ GHz}, \ \beta = 1718.81 \text{ rad/m}, \ \mu = j\beta, \ \eta_{\text{TE}_{13}} = 229.69 \ \Omega$$

$$E_x = 2584.1 \cos\left(\frac{\pi x}{a}\right) \sin\left(\frac{3\pi y}{b}\right) \sin(\omega t - \beta z) \text{ V/m}$$

$$E_y = -459.4 \sin\left(\frac{\pi x}{a}\right) \cos\left(\frac{3\pi y}{b}\right) \sin(\omega t - \beta z) \text{ V/m} \qquad E_z = 0$$

$$H_y = 11.25 \cos\left(\frac{\pi x}{a}\right) \sin\left(\frac{3\pi y}{b}\right) \sin(\omega t - \beta z) \text{ A/m}$$

$$H_z = -7.96 \cos\left(\frac{\pi x}{a}\right) \cos\left(\frac{3\pi y}{b}\right) \cos(\omega t - \beta z) \text{ A/m.}$$

12.5 WAVE PROPAGATION IN THE GUIDE

Examination of eq. (12.36) shows that the field components all involve the terms sine or cosine of $(m\pi/a)x$ or $(n\pi/b)y$ times $e^{-\gamma z}$. Since

$$\sin \theta = \frac{1}{2j}(e^{j\theta} - e^{-j\theta}) \tag{12.44a}$$

$$\cos \theta = \frac{1}{2}(e^{j\theta} + e^{-j\theta}) \tag{12.44b}$$

a wave within the waveguide can be resolved into a combination of plane waves reflected from the waveguide walls. For the TE_{10} mode, for example,

$$\begin{aligned} E_{ys} &= -\frac{j\omega\mu a}{\pi}\sin\left(\frac{\pi x}{a}\right) e^{-j\beta z} \\ &= -\frac{\omega\mu a}{2\pi}\left(e^{j\pi x/a} - e^{-j\pi x/a}\right) e^{-j\beta z} \\ &= \frac{\omega\mu a}{2\pi}\left[e^{-j\beta(z + \pi x/\beta a)} - e^{-j\beta(z - \pi x/\beta a)}\right] \end{aligned} \quad [12.45]$$

The first term of eq. (12.45) represents a wave traveling in the positive z-direction, at an angle

$$\theta = \tan^{-1}\left(\frac{\pi}{\beta a}\right) \quad [12.46]$$

with the z-axis. The second term of eq. (12.45) represents a wave traveling in the positive z-direction at an angle $-\theta$. The field may be depicted as a sum of two plane TEM waves propagating along zigzag paths between the guide walls at $x = 0$ and $x = a$ as illustrated in Figure 12.10(a). The decomposition of the TE_{10} mode into two plane waves can be extended to any TE and TM mode. When n and m are both different from zero, four plane waves result from the decomposition.

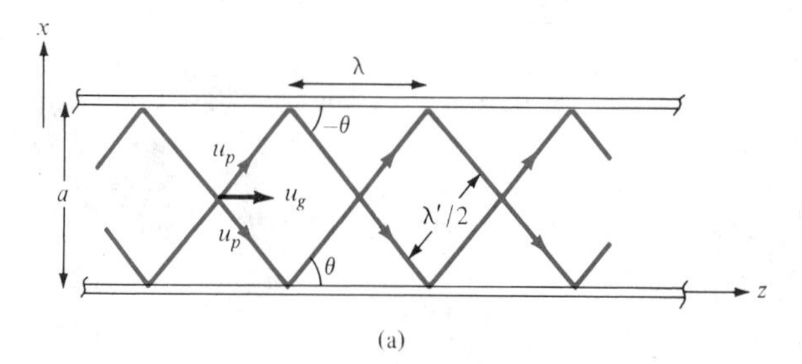

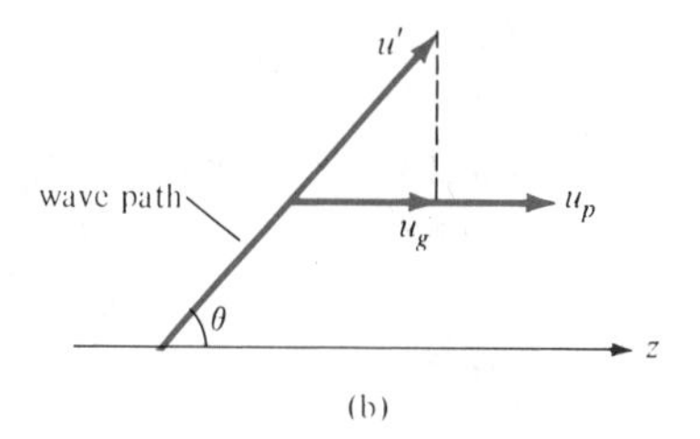

Figure 12.10 **(a)** Decomposition of TE_{10} mode into two plane waves; **(b)** relationship between u', u_p, and u_g.

The wave component in the z-direction has a different wavelength from that of the plane waves. This wavelength along the axis of the guide is called the *waveguide wavelength* and is given by (see Problem 12.16)

$$\lambda = \frac{\lambda'}{\sqrt{1 - \left[\frac{f_c}{f}\right]^2}} \qquad [12.47]$$

where $\lambda' = u'/f$.

As a consequence of the zigzag paths, we have three types of velocity: the *medium velocity* u', the *phase velocity* u_p, and the *group velocity* u_g. Figure 12.10(b) illustrates the relationship between the three different velocities. The medium velocity $u' = 1/\sqrt{\mu\varepsilon}$ is as explained in the previous sections. The phase velocity u_p is the velocity at which loci of constant phase are propagated down the guide and is given by eq. (12.31), that is,

$$u_p = \frac{\omega}{\beta} \qquad [12.48a]$$

or

$$u_p = \frac{u'}{\cos\theta} = \frac{u'}{\sqrt{1 - \left[\frac{f_c}{f}\right]^2}} \qquad [12.48b]$$

This shows that $u_p \geq u'$. If $u' = c$, then u_p is greater than the speed of light in vacuum. Does this violate Einstein's relativity theory that messages cannot travel faster than the speed of light? Not really, because information (or energy) in a waveguide generally does not travel at the phase velocity. Information travels at the group velocity, which must be less than the speed of light. The group velocity u_g is the velocity with which the resultant repeated reflected waves are traveling down the guide and is given by

$$\boxed{u_g = \frac{1}{\partial\beta/\partial\omega}} \qquad [12.49a]$$

or

$$u_g = u'\cos\theta = u'\sqrt{1 - \left[\frac{f_c}{f}\right]^2} \qquad [12.49b]$$

Although the concept of group velocity is fairly complex and is beyond the scope of this chapter, a group velocity is essentially the velocity of propagation of the wave-packet envelope of a group of frequencies. It is the energy propagation velocity in the guide and is always less than or equal to u'. From eqs. (12.48) and (12.49), it is evident that

$$u_p u_g = u'^2 \qquad [12.50]$$

This relation is similar to eq. (12.40). Hence the variation of u_p and u_g with frequency is similar to that in Figure 12.6 for η_{TE} and η_{TM}.

EXAMPLE 12.4

A standard air-filled rectangular waveguide with dimensions $a = 8.636$ cm, $b = 4.318$ cm is fed by a 4-GHz carrier from a coaxial cable. Determine if a TE_{10} mode will be propagated. If so, calculate the phase velocity and the group velocity.

SOLUTION For the TE_{10} mode, $f_c = u'/2a$. Since the waveguide is air-filled, $u' = c = 3 \times 10^8$. Hence,

$$f_c = \frac{3 \times 10^8}{2 \times 8.636 \times 10^{-2}} = 1.737 \text{ GHz}$$

As $f = 4$ GHz $> f_c$, the TE_{10} mode will propagate.

$$u_p = \frac{u'}{\sqrt{1 - (f_c/f)^2}} = \frac{3 \times 10^8}{\sqrt{1 - (1.737/4)^2}}$$

$$= 3.33 \times 10^8 \text{ m/s}$$

$$u_g = \frac{u'^2}{u_p} = \frac{9 \times 10^{16}}{3.33 \times 10^8} = 2.702 \times 10^8 \text{ m/s}$$

■

PRACTICE EXERCISE 12.4

Repeat Example 12.4 for the TM_{11} mode.

ANSWER 12.5×10^8 m/s, 7.203×10^7 m/s.

12.6 POWER TRANSMISSION AND ATTENUATION

To determine power flow in the waveguide, we first find the average Poynting vector (from eq. (10.68)),

$$\mathcal{P}_{ave} = \frac{1}{2} \text{Re} (\mathbf{E}_s \times \mathbf{H}_s^*) \qquad [12.51]$$

In this case, the Poynting vector is along the z-direction so that

$$\mathcal{P}_{\text{ave}} = \frac{1}{2}\,\text{Re}\,(E_{xs}H_{ys}^* - E_{ys}H_{xs}^*)\mathbf{a}_z$$

$$= \frac{|E_{xs}|^2 + |E_{ys}|^2}{2\eta}\,\mathbf{a}_z \tag{12.52}$$

where $\eta = \eta_{\text{TE}}$ for TE modes or $\eta = \eta_{\text{TM}}$ for TM modes. The total average power transmitted across the cross section of the waveguide is

$$P_{\text{ave}} = \int \mathcal{P}_{\text{ave}} \cdot d\mathbf{S}$$

$$= \int_{x=0}^{a}\int_{y=0}^{b} \frac{|E_{xs}|^2 + |E_{ys}|^2}{2\eta}\,dy\,dx \tag{12.53}$$

Of practical importance is the attenuation in a lossy waveguide. In our analysis thus far, we have assumed lossless waveguides ($\sigma = 0$, $\sigma_c \simeq \infty$) for which $\alpha = 0$, $\gamma = j\beta$. When the dielectric medium is lossy ($\sigma \neq 0$) and the guide walls are not perfectly conducting ($\sigma_c \neq \infty$), there is a continuous loss of power as a wave propagates along the guide. According to eqs. (10.69) and (10.70), the power flow in the guide is of the form

$$P_{\text{ave}} = P_o e^{-2\alpha z} \tag{12.54}$$

In order that energy be conserved, the rate of decrease in P_{ave} must equal the time average power loss P_L per unit length, that is,

$$P_L = -\frac{dP_{\text{ave}}}{dz} = 2\alpha P_{\text{ave}}$$

or

$$\alpha = \frac{P_L}{2P_{\text{ave}}} \tag{12.55}$$

In general,

$$\alpha = \alpha_c + \alpha_d \tag{12.56}$$

where α_c and α_d are attenuation constants due to ohmic or conduction losses ($\sigma_c \neq \infty$) and dielectric losses ($\sigma \neq 0$) respectively.

To determine α_d, recall that we started with eq. (12.1) assuming a lossless dielectric medium ($\sigma = 0$). For a lossy dielectric, we need to incorporate the fact that $\sigma \neq 0$. All our equations still hold except that $\gamma = j\beta$ needs to be modified. This is achieved by replacing ε in eq. (12.25) by the complex permittivity of eq. (10.40). Thus, we obtain

$$\gamma = \alpha_d + j\beta_d = \sqrt{\left(\frac{m\pi}{a}\right)^2 + \left(\frac{n\pi}{b}\right)^2 - \omega^2\mu\varepsilon_c} \tag{12.57}$$

where

$$\varepsilon_c = \varepsilon' - j\varepsilon'' = \varepsilon - j\frac{\sigma}{\omega} \qquad [12.58]$$

Substituting eq. (12.58) into eq. (12.57) and squaring both sides of the equation, we obtain

$$\gamma^2 = \alpha_d^2 - \beta_d^2 + 2j\alpha_d\beta_d = \left(\frac{m\pi}{a}\right)^2 + \left(\frac{n\pi}{b}\right)^2 - \omega^2\mu\varepsilon + j\omega\mu\sigma$$

Equating real and imaginary parts,

$$\alpha_d^2 - \beta_d^2 = \left(\frac{m\pi}{a}\right)^2 + \left(\frac{n\pi}{b}\right)^2 - \omega^2\mu\varepsilon \qquad [12.59a]$$

$$2\alpha_d\beta_d = \omega\mu\sigma \qquad \text{or} \qquad \alpha_d = \frac{\omega\mu\sigma}{2\beta_d} \qquad [12.59b]$$

Assuming that $\alpha_d^2 \ll \beta_d^2$, $\alpha_d^2 - \beta_d^2 \simeq -\beta_d^2$, so eq. (12.59a) gives

$$\beta_d = \sqrt{\omega^2\mu\varepsilon - \left(\frac{m\pi}{a}\right)^2 - \left(\frac{n\pi}{b}\right)^2}$$

$$= \omega\sqrt{\mu\varepsilon}\sqrt{1 - \left(\frac{f_c}{f}\right)^2} \qquad [12.60]$$

which is the same as β in eq. (12.30). Substituting eq. (12.60) into eq. (12.59b) gives

$$\boxed{\alpha_d = \frac{\sigma\eta'}{2\sqrt{1 - \left(\frac{f_c}{f}\right)^2}}} \qquad [12.61]$$

where $\eta' = \sqrt{\mu/\varepsilon}$.

The determination of α_c for TM_{mn} and TE_{mn} modes is time-consuming and tedious. We shall illustrate the procedure by finding α_c for the TE_{10} mode. For this mode, only E_y, H_x, and H_z exist. Substituting eq. (12.43a) into eq. (12.53) yields

$$P_{\mathrm{ave}} = \int_{x=0}^{a}\int_{y=0}^{b} \frac{|E_{ys}|^2}{2\eta}\,dx\,dy = \frac{\omega^2\mu^2a^2H_o^2b}{2\pi^2\eta}\int_0^b dy\int_0^a \sin^2\frac{\pi x}{a}\,dx$$

$$P_{\mathrm{ave}} = \frac{\omega^2\mu^2a^3H_o^2b}{4\pi^2\eta} \qquad [12.62]$$

The total power loss per unit length in the walls is

$$P_L = P_L \big|_{y=0} + P_L \big|_{y=b} + P_L \big|_{x=0} + P_L \big|_{x=a}$$
$$= 2\left(P_L \big|_{y=0} + P_L \big|_{x=0}\right) \quad [12.63]$$

since the same amount is dissipated in the walls $y = 0$ and $y = b$ or $x = 0$ and $x = a$. For the wall $y = 0$,

$$P_L \big|_{y=0} = \frac{1}{2} \text{Re}\left[\eta_c \int (|H_{xs}|^2 + |H_{zs}|^2)\, dx\right]\Bigg|_{y=0}$$
$$= \frac{1}{2} R_s \left[\int_0^a \frac{\beta^2 a^2}{\pi^2} H_o^2 \sin^2 \frac{\pi x}{a}\, dx + \int_o^a H_o^2 \cos^2 \frac{\pi x}{a}\, dx\right]$$
$$= \frac{R_s a H_o^2}{4}\left(1 + \frac{\beta^2 a^2}{\pi^2}\right) \quad [12.64]$$

where R_s is the real part of the intrinsic impedance η_c of the conducting wall. From eq. (10.56),

$$R_s = \frac{1}{\sigma_c \delta} \quad [12.65]$$

where δ is the skin depth. R_s is the skin resistance of the wall; it may be regarded as the resistance of 1 m by δ by 1 m of the conducting material. For the wall $x = 0$,

$$P_L \big|_{x=0} = \frac{1}{2} \text{Re}\left[\eta_c \int (|H_{zs}|^2)\, dy\right] \big|_{x=0} = \frac{1}{2} R_s \int_0^b H_o^2\, dy$$
$$= \frac{R_s b H_o^2}{2} \quad [12.66]$$

Substituting eqs. (12.64) and (12.66) into eq. (12.63) gives

$$P_L = R_s H_o^2 \left[b + \frac{a}{2}\left(1 + \frac{\beta^2 a^2}{\pi^2}\right)\right] \quad [12.67]$$

Finally, substituting eqs. (12.62) and (12.67) into eq. (12.55),

$$\alpha_c = \frac{R_s H_o^2 \left[b + \dfrac{a}{2}\left(1 + \dfrac{\beta^2 a^2}{\pi^2}\right)\right] 4\pi^2 \eta}{\omega^2 \mu^2 a^3 H_o^2\, b} \quad [12.68a]$$

It is convenient to express α_c in terms of f and f_c. After some manipulations, we obtain for the TE_{10} mode

$$\alpha_c = \frac{2R_s}{b\eta' \sqrt{1 - \left[\frac{f_c}{f}\right]^2}} \left(\frac{1}{2} + \frac{b}{a}\left[\frac{f_c}{f}\right]^2\right) \quad [12.68b]$$

By following the same procedure, the attenuation constant for the TE_{mn} modes ($n \neq 0$) can be obtained as

$$\alpha_c \big|_{TE} = \frac{2R_s}{b\eta' \sqrt{1 - \left[\frac{f_c}{f}\right]^2}} \left[\left(1 + \frac{b}{a}\right)\left[\frac{f_c}{f}\right]^2 + \frac{\frac{b}{a}\left(\frac{b}{a} m^2 + n^2\right)}{\frac{b^2}{a^2} m^2 + n^2}\left(1 - \left[\frac{f_c}{f}\right]^2\right)\right] \quad [12.69]$$

and for the TM_{mn} modes as

$$\alpha_c \big|_{TM} = \frac{2R_s}{b\eta' \sqrt{1 - \left[\frac{f_c}{f}\right]^2}} \frac{(b/a)^3 m^2 + n^2}{(b/a)^2 m^2 + n^2} \quad [12.70]$$

The total attenuation constant α is obtained by substituting eqs. (12.61) and (12.69) or (12.70) into eq. (12.56).

†12.7 WAVEGUIDE CURRENT AND MODE EXCITATION

For either TM or TE modes, the surface current density **K** on the walls of the waveguide may be found using

$$\mathbf{K} = \mathbf{a}_n \times \mathbf{H} \quad [12.71]$$

where $\mathbf{a}_n$ is the unit outward normal to the wall and **H** is the field intensity evaluated on the wall. The current flow on the guide walls for TE_{10} mode propagation can be found using eq. (12.71) with eqs. (12.42) and (12.43). The result is sketched in Figure 12.11.

The surface charge density ρ_S on the walls is given by

$$\rho_S = \mathbf{a}_n \cdot \mathbf{D} = \mathbf{a}_n \cdot \varepsilon\mathbf{E} \quad [12.72]$$

where **E** is the electric field intensity evaluated on the guide wall.

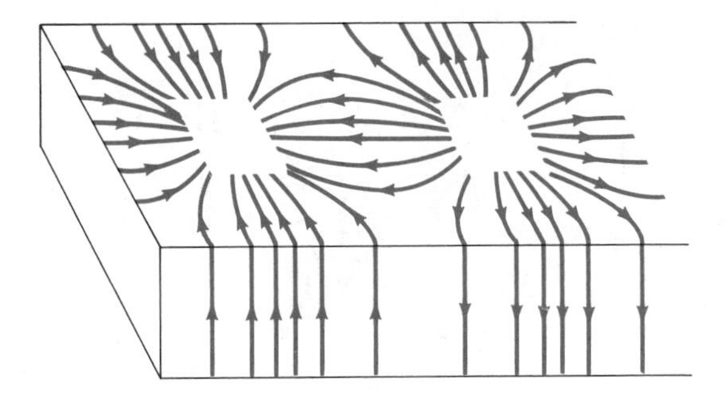

Figure 12.11 Surface current on guide walls for TE_{10} mode.

A waveguide is usually fed or excited by a coaxial line or another waveguide. Most often, a probe (central conductor of a coaxial line) is used to establish the field intensities of the desired mode and achieve a maximum power transfer. The probe is located so as to produce **E** and **H** fields that are roughly parallel to the lines of **E** and **H** fields of the desired mode. To excite the TE_{10} mode, for example, we notice from eq. (12.43a) that E_y has maximum value at $x = a/2$. Hence, the probe is located at $x = a/2$ to excite the TE_{10} mode as shown in Figure 12.12(a) where the field lines are similar to those of Figure 12.8. Similarly, the TM_{11} mode is launched by placing the probe along the z-direction as in Figure 12.12(b).

EXAMPLE

12.5

An air-filled rectangular waveguide of dimensions $a = 4$ cm, $b = 2$ cm transports energy in the dominant mode at a rate of 2 mW. If the frequency of operation is 10 GHz, determine the peak value of the electric field in the waveguide.

SOLUTION The dominant mode for $a > b$ is TE_{10} mode. The field expressions corresponding to this mode ($m = 1$, $n = 0$) are in eq. (12.36) or (12.43), namely

$$E_{xs} = 0, \qquad E_{ys} = -jE_o \sin\left(\frac{\pi x}{a}\right) e^{-j\beta z}, \qquad \text{where } E_o = \frac{\omega\mu a}{\pi} H_o$$

$$f_c = \frac{u'}{2a} = \frac{3 \times 10^8}{2\,(4 \times 10^{-2})} = 3.75 \text{ GHz}$$

$$\eta = \eta_{TE} = \frac{\eta'}{\sqrt{1 - \left[\dfrac{f_c}{f}\right]^2}} = \frac{377}{\sqrt{1 - \left[\dfrac{3.75}{10}\right]^2}} = 406.7\ \Omega$$

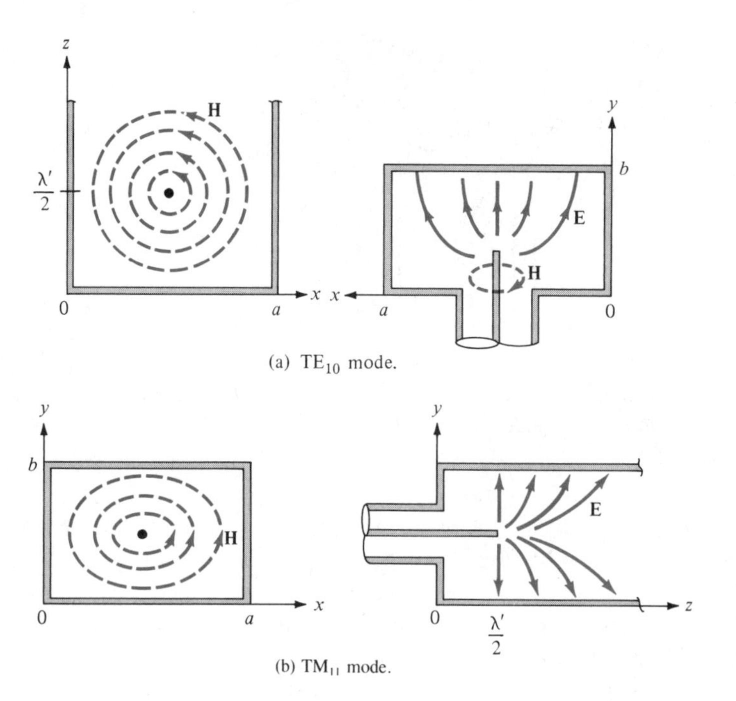

Figure 12.12 Excitation of modes in a rectangular waveguide.

From eq. (12.53), the average power transmitted is

$$P_{ave} = \int_{y=0}^{b} \int_{x=0}^{a} \frac{|E_{ys}|^2}{2\eta} \, dx \, dy = \frac{E_o^2}{2\eta} \int_0^b dy \int_0^a \sin^2\left(\frac{\pi x}{a}\right) dx$$

$$= \frac{E_o^2 \, ab}{4\eta}$$

Hence,

$$E_o^2 = \frac{4\eta P_{ave}}{ab} = \frac{4\,(406.7) \times 2 \times 10^{-3}}{8 \times 10^{-4}} = 4067$$

$$E_o = 63.77 \text{ V/m}$$

■

PRACTICE EXERCISE 12.5

In Example 12.5, calculate the peak value H_o of the magnetic field in the guide if $a = 2$ cm, $b = 4$ cm while other things remain the same.

ANSWER 63.34 mA/m.

EXAMPLE 12.6

A copper-plated waveguide ($\sigma_c = 5.8 \times 10^7$ mhos/m) operating at 4.8 GHz is supposed to deliver a minimum power of 1.2 kW to an antenna. If the guide is filled with polystyrene ($\sigma = 10^{-17}$ mhos/m, $\varepsilon = 2.55\varepsilon_o$) and its dimensions are $a = 4.2$ cm, $b = 2.6$ cm, calculate the power dissipated in a length 60 cm of the guide in the TE_{10} mode.

SOLUTION Let

P_d = power loss or dissipated
P_a = power delivered to the antenna
P_o = input power to the guide

so that $P_o = P_d + P_a$
From eq. (12.54),

$$P_a = P_o e^{-2\alpha z}$$

Hence,

$$P_a = (P_d + P_a)\, e^{-2\alpha z}$$

or

$$P_d = P_a(e^{2\alpha z} - 1)$$

Now we need to determine α from

$$\alpha = \alpha_d + \alpha_c$$

From eq. (12.61),

$$\alpha_d = \frac{\sigma\eta'}{2\sqrt{1 - \left[\frac{f_c}{f}\right]^2}}$$

Since the loss tangent

$$\frac{\sigma}{\omega\varepsilon} = \frac{10^{-17}}{2\pi \times 4.8 \times 10^9 \times \frac{10^{-9}}{36\pi} \times 2.55}$$

$$= 1.47 \times 10^{-17} \ll 1 \qquad \text{(lossless dielectric medium)}$$

then

$$\eta' \simeq \sqrt{\frac{\mu}{\varepsilon}} = \frac{377}{\sqrt{\varepsilon_r}} = 236.1$$

$$u' = \frac{1}{\sqrt{\mu\varepsilon}} = \frac{c}{\sqrt{\varepsilon_r}} = 1.879 \times 10^8 \text{ m/s}$$

$$f_c = \frac{u'}{2a} = \frac{1.879 \times 10^8}{2 \times 4.2 \times 10^{-2}} = 2.234 \text{ GHz}$$

$$\alpha_d = \frac{10^{-17} \times 236.1}{2\sqrt{1 - \left[\frac{2.234}{4.8}\right]^2}}$$

$$\alpha_d = 1.334 \times 10^{-15} \text{ Np/m}$$

For the TE_{10} mode, eq. (12.68b) gives

$$\alpha_c = \frac{2R_s}{b\eta'\sqrt{1 - \left[\frac{f_c}{f}\right]^2}}\left(0.5 + \frac{b}{a}\left[\frac{f_c}{f}\right]^2\right)$$

where

$$R_s = \frac{1}{\sigma_c\delta} = \sqrt{\frac{\pi f\mu}{\sigma_c}} = \sqrt{\frac{\pi \times 4.8 \times 10^9 \times 4\pi \times 10^{-7}}{5.8 \times 10^7}}$$

$$= 1.808 \times 10^{-2}\ \Omega$$

Hence

$$\alpha_c = \frac{2 \times 1.808 \times 10^{-2}\left(0.5 + \frac{2.6}{4.2}\left[\frac{2.234}{4.8}\right]^2\right)}{2.6 \times 10^{-2} \times 236.1\sqrt{1 - \left[\frac{2.234}{4.8}\right]^2}}$$

$$= 4.218 \times 10^{-3} \text{ Np/m}$$

Note that $\alpha_d \ll \alpha_c$, showing that the loss due to the finite conductivity of the guide walls is more important than the loss due to the dielectric medium. Thus

$$\alpha = \alpha_d + \alpha_c \simeq \alpha_c = 4.218 \times 10^{-3} \text{ Np/m}$$

and the power dissipated is

$$P_d = P_a\,(e^{2\alpha z} - 1) = 1.2 \times 10^3(e^{2 \times 4.218 \times 10^{-3} \times 0.6} - 1)$$

$$= 6.089 \text{ W}$$

■

PRACTICE EXERCISE 12.6

A brass waveguide ($\sigma_c = 1.1 \times 10^7$ mhos/m) of dimensions $a = 4.2$ cm, $b =$ 1.5 cm is filled with Teflon ($\varepsilon_r = 2.6$, $\sigma = 10^{-15}$ mhos/m). The operating frequency is 9 GHz. For the TE_{10} mode:

(a) Calculate α_d and α_c.

(b) What is the loss in decibels in the guide if it is 40 cm long?

ANSWER (a) 1.206×10^{-13} Np/m, 1.744×10^{-2} Np/m, (b) 0.0606 dB.

EXAMPLE 12.7

Sketch the field lines for the TM_{11} mode. Derive the instantaneous expressions for the surface current density of this mode.

SOLUTION From Example 12.2, we obtain the fields for TM_{11} mode ($m = 1$, $n = 1$) as

$$E_x = \frac{\beta}{h^2}\left(\frac{\pi}{a}\right) E_o \cos\left(\frac{\pi x}{a}\right) \sin\left(\frac{\pi y}{b}\right) \sin(\omega t - \beta z)$$

$$E_y = \frac{\beta}{h^2}\left(\frac{\pi}{b}\right) E_o \sin\left(\frac{\pi x}{a}\right) \cos\left(\frac{\pi y}{b}\right) \sin(\omega t - \beta z)$$

$$E_z = E_o \sin\left(\frac{\pi x}{a}\right) \sin\left(\frac{\pi y}{b}\right) \cos(\omega t - \beta z)$$

$$H_x = -\frac{\omega\varepsilon}{h^2}\left(\frac{\pi}{b}\right) E_o \sin\left(\frac{\pi x}{a}\right) \cos\left(\frac{\pi y}{b}\right) \sin(\omega t - \beta z)$$

$$H_y = \frac{\omega\varepsilon}{h^2}\left(\frac{\pi}{a}\right) E_o \cos\left(\frac{\pi x}{a}\right) \sin\left(\frac{\pi y}{b}\right) \sin(\omega t - \beta z)$$

$$H_z = 0$$

For the electric field lines, applying eq. (4.84),

$$\frac{dy}{dx} = \frac{E_y}{E_x} = \frac{a}{b} \tan\left(\frac{\pi x}{a}\right) \cot\left(\frac{\pi y}{b}\right)$$

For the magnetic field lines,

$$\frac{dy}{dx} = \frac{H_y}{H_x} = -\frac{b}{a} \cot\left(\frac{\pi x}{a}\right) \tan\left(\frac{\pi y}{b}\right)$$

Notice that $(E_y/E_x)(H_y/H_x) = -1$, showing that electric and magnetic field lines are mutually orthogonal. This should also be observed in Figure 12.13 where the field lines are sketched.

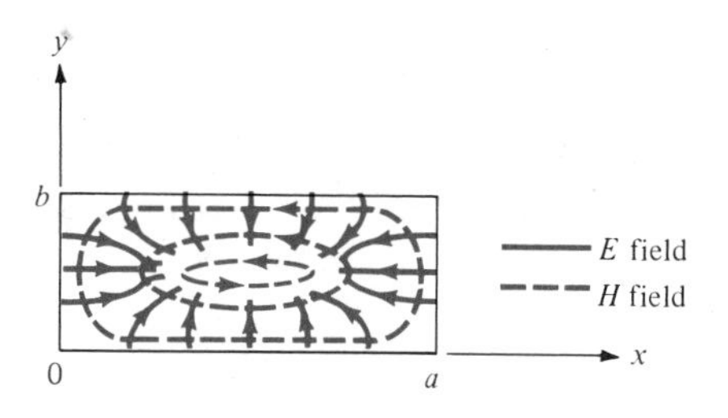

Figure 12.13 Field lines for TM_{11} mode; for Example 12.7.

The surface current density on the walls of the waveguide is given by

$$\mathbf{K} = \mathbf{a}_n \times \mathbf{H} = \mathbf{a}_n \times (H_x, H_y, 0).$$

At $x = 0$, $\mathbf{a}_n = \mathbf{a}_x$, $\mathbf{K} = H_y(0, y, z, t)\,\mathbf{a}_z$, that is,

$$\mathbf{K} = \frac{\omega\varepsilon}{h^2}\left(\frac{\pi}{a}\right) E_o \sin\left(\frac{\pi y}{b}\right) \sin(\omega t - \beta z)\,\mathbf{a}_z$$

At $x = a$, $\mathbf{a}_n = -\mathbf{a}_x$, $\mathbf{K} = -H_y(a, y, z, t)\,\mathbf{a}_z$
or

$$\mathbf{K} = \frac{\omega\varepsilon}{h^2}\left(\frac{\pi}{a}\right) E_o \sin\left(\frac{\pi y}{b}\right) \sin(\omega t - \beta z)\,\mathbf{a}_z$$

At $y = 0$, $\mathbf{a}_n = \mathbf{a}_y$, $\mathbf{K} = -H_x(x, 0, z, t)\,\mathbf{a}_z$
or

$$\mathbf{K} = \frac{\omega\varepsilon}{h^2}\left(\frac{\pi}{b}\right) E_o \sin\left(\frac{\pi x}{a}\right) \sin(\omega t - \beta z)\,\mathbf{a}_z$$

At $y = b$, $\mathbf{a}_n = -\mathbf{a}_y$, $\mathbf{K} = H_x(x, b, z, t)\,\mathbf{a}_z$
or

$$\mathbf{K} = \frac{\omega\varepsilon}{h^2}\left(\frac{\pi}{b}\right) E_o \sin\left(\frac{\pi x}{a}\right) \sin(\omega t - \beta z)\,\mathbf{a}_z$$

■

PRACTICE EXERCISE 12.7

Sketch the field lines for the TE_{11} mode.

ANSWER See Figure 12.14. The strength of the field at any point is indicated by the density of the lines; the field is strongest (or weakest) where the lines are closest together (or farthest apart).

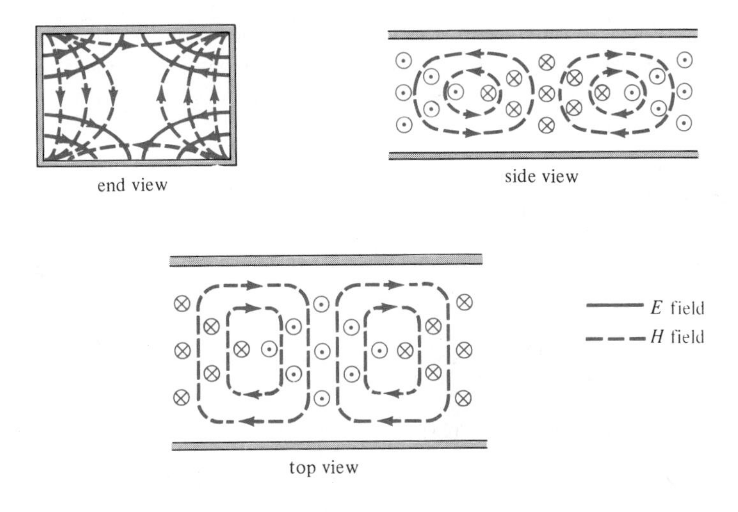

Figure 12.14 For Practice Exercise 12.7; for TE_{11} mode.

12.8 WAVEGUIDE RESONATORS

Resonators are primarily used for energy storage. At high frequencies (100 MHz and above) the *RLC* circuit elements are inefficient when used as resonators because the dimensions of the circuits are comparable with the operating wavelength, and consequently, unwanted radiation takes place. Therefore, at high frequencies the *RLC* resonant circuits are replaced by electromagnetic cavity resonators. Such resonator cavities are used in klystron tubes, bandpass filters, and wave meters.

Consider the rectangular cavity (or closed conducting box) shown in Figure 12.15. We notice that the cavity is simply a rectangular waveguide shorted at both ends. We therefore expect to have standing wave and also TM and TE modes of wave propagation. Depending on how the cavity is excited, the wave can propagate in the x-, y-, or z-direction. We will choose the $+z$-direction as the "direction of wave

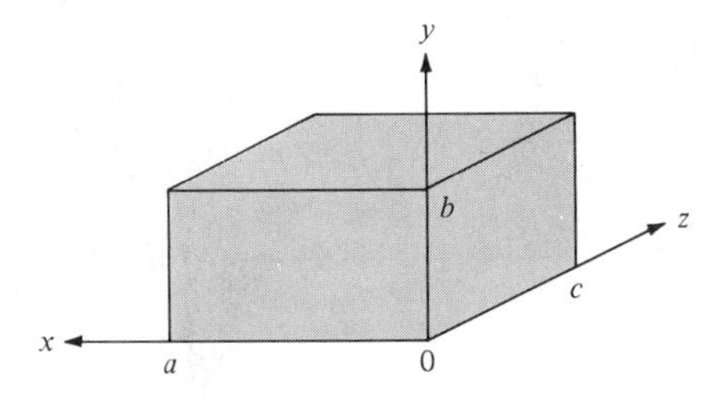

Figure 12.15 Rectangular cavity.

propagation." In fact, there is no wave propagation. Rather, there are standing waves. We recall from Section 10.8 that a standing wave is a combination of two waves traveling in opposite directions.

A. TM Mode to z

For this case, $H_z = 0$ and we let

$$E_{zs}(x, y, z) = X(x)\, Y(y)\, Z(z) \quad [12.73]$$

be the product solution of eq. (12.1). We follow the same procedure taken in Section 12.2 and obtain

$$X(x) = c_1 \cos k_x x + c_2 \sin k_x x \quad [12.74a]$$

$$Y(y) = c_3 \cos k_y y + c_4 \sin k_y y \quad [12.74b]$$

$$Z(z) = c_5 \cos k_z z + c_6 \sin k_z z \quad [12.74c]$$

where

$$k^2 = k_x^2 + k_y^2 + k_z^2 = \omega^2 \mu \varepsilon \quad [12.75]$$

The boundary conditions are:

$$E_z = 0 \quad \text{at} \quad x = 0, a \quad [12.76a]$$

$$E_z = 0 \quad \text{at} \quad y = 0, b \quad [12.76b]$$

$$E_y = 0, E_x = 0 \quad \text{at} \quad z = 0, c \quad [12.76c]$$

As shown in Section 12.3, the conditions in eqs. (12.76a, b) are satisfied when $c_1 = 0 = c_3$ and

$$k_x = \frac{m\pi}{a}, \qquad k_y = \frac{n\pi}{b} \quad [12.77]$$

where $m = 1, 2, 3, \ldots, n = 1, 2, 3, \ldots$. To invoke the conditions in eq. (12.76c), we notice that eq. (12.14) (with $H_{zs} = 0$) yields

$$j\omega\varepsilon E_{xs} = \frac{1}{j\omega\mu}\left(\frac{\partial^2 E_{xs}}{\partial z^2} - \frac{\partial^2 E_{zs}}{\partial z\, \partial x}\right) \quad [12.78]$$

Similarly, combining eqs. (12.13a) and (12.13d) (with $H_{zs} = 0$) results in

$$j\omega\varepsilon E_{ys} = \frac{1}{-j\omega\mu}\left(\frac{\partial^2 E_{zs}}{\partial y\, \partial z} - \frac{\partial^2 E_{ys}}{\partial z^2}\right) \quad [12.79]$$

From eqs. (12.78) and (12.79), it is evident that eq. (12.76c) is satisfied if

$$\frac{\partial E_{zs}}{\partial z} = 0 \quad \text{at} \quad z = 0, c \quad [12.80]$$

This implies that $c_6 = 0$ and $\sin k_z c = 0 = \sin p\pi$. Hence,

$$k_z = \frac{p\pi}{c} \quad [12.81]$$

where $p = 0, 1, 2, 3, \ldots$. Substituting eqs. (12.77) and (12.81) into eq. (12.74) yields

$$\boxed{E_{zs} = E_o \sin\left(\frac{m\pi x}{a}\right) \sin\left(\frac{n\pi y}{b}\right) \cos\left(\frac{p\pi z}{c}\right)} \quad [12.82]$$

where $E_o = c_2 c_4 c_5$. Other field components are obtained from eqs. (12.82) and (12.13). The phase constant β is obtained from eqs. (12.75), (12.77), and (12.81) as

$$\beta^2 = k^2 = \left[\frac{m\pi}{a}\right]^2 + \left[\frac{n\pi}{b}\right]^2 + \left[\frac{p\pi}{c}\right]^2 \quad [12.83]$$

Since $\beta^2 = \omega^2 \mu\varepsilon$, from eq. (12.83), we obtain the *resonant frequency* f_r

$$2\pi f_r = \omega_r = \frac{\beta}{\sqrt{\mu\varepsilon}} = \beta u'$$

or

$$\boxed{f_r = \frac{u'}{2} \sqrt{\left[\frac{m}{a}\right]^2 + \left[\frac{n}{b}\right]^2 + \left[\frac{p}{c}\right]^2}} \quad [12.84]$$

The corresponding resonant wavelength is

$$\boxed{\lambda_r = \frac{u'}{f_r} = \frac{2}{\sqrt{\left[\frac{m}{a}\right]^2 + \left[\frac{n}{b}\right]^2 + \left[\frac{p}{c}\right]^2}}} \quad [12.85]$$

From eq. (12.84), we notice that the lowest-order TM mode is TM_{110}.

B. TE Mode to z

In this case, $E_z = 0$ and

$$H_{zs} = (b_1 \cos k_x x + b_2 \sin k_x x)(b_3 \cos k_y y + b_4 \sin k_y y) \cdot (b_5 \cos k_z z + \sin k_z z) \quad [12.86]$$

The boundary conditions in eq. (12.76c) combined with eq. (12.13) yields

$$H_{zs} = 0 \quad \text{at} \quad z = 0, c \qquad [12.87a]$$

$$\frac{\partial H_{zs}}{\partial x} = 0 \quad \text{at} \quad x = 0, a \qquad [12.87b]$$

$$\frac{\partial H_{zs}}{\partial y} = 0 \quad \text{at} \quad y = 0, b \qquad [12.87c]$$

Imposing the conditions in eq. (12.87) on eq. (12.86) in the same manner as for TM mode to z leads to

$$H_{zs} = H_o \cos\left(\frac{m\pi x}{a}\right) \cos\left(\frac{n\pi y}{b}\right) \sin\left(\frac{p\pi z}{c}\right) \qquad [12.88]$$

where $m = 0, 1, 2, 3, \ldots, n = 0, 1, 2, 3, \ldots$, and $p = 1, 2, 3, \ldots$. Other field components can be obtained from eqs. (12.13) and (12.88). The resonant frequency is the same as that of eq. (12.84) except that m or n (but not both at the same time) can be zero for TE modes. The reason why m and n cannot be zero at the same time is that the field components will be zero if they are zero (see Problem 12.27). The mode that has the lowest resonant frequency for a given cavity size (a, b, c) is the *dominant mode*. If $a > b < c$, it implies that $1/a < 1/b > 1/c$ and hence the dominant mode is TE_{101}. Note that for $a > b < c$, the resonant frequency of TM_{110} mode is higher than that for TE_{101} mode; hence, TE_{101} is dominant. When different modes have the same resonant frequency, we say that the modes are *degenerate;* one mode will dominate others depending on how the cavity is excited.

A practical resonant cavity has walls with finite conductivity σ_c and is, therefore, capable of losing stored energy. The *quality factor* Q is a means of determining the loss. As used in *RLC* circuits, the quality factor is also a measure of the bandwidth of the cavity resonator. It may be defined as

$$Q = 2\pi \cdot \frac{\text{Time average energy stored}}{\text{Energy loss per cycle of oscillation}}$$

$$= 2\pi \cdot \frac{W}{P_L T} = \omega \frac{W}{P_L} \qquad [12.89]$$

where $T = 1/f$ = the period of oscillation, P_L is the time average power loss in the cavity, and W is the total time average energy stored in electric and magnetic fields in the cavity. Q is usually very high for a cavity resonator compared with that for an *RLC* resonant circuit. By following a procedure similar to that used in deriving α_c in Section 12.6, it can be shown that the quality factor for the dominant TE_{101} is given by[3]

[3]For the proof, see S. V. Marshall and G. G. Skitek, *Electromagnetic Concepts and Applications.* Englewood Cliffs, NJ: Prentice-Hall, 3rd ed., 1990, pp. 440–442.

$$\boxed{Q_{\text{TE}_{101}} = \frac{(a^2 + c^2)abc}{\delta[2b(a^3 + c^3) + ac(a^2 + c^2)]}} \qquad [12.90]$$

where $\delta = \dfrac{1}{\sqrt{\pi f_{101}\mu_o\sigma_c}}$ is the skin depth of the cavity walls.

EXAMPLE

12.8

An air-filled resonant cavity with dimensions $a = 5$ cm, $b = 4$ cm, and $c = 10$ cm is made of copper ($\sigma_c = 5.8 \times 10^7$ mhos/m). Find

(a) The five lowest order modes

(b) The quality factor for TE_{101} mode

SOLUTION

(a) The resonant frequency is given by

$$f_r = \frac{u'}{2}\sqrt{\left[\frac{m}{a}\right]^2 + \left[\frac{n}{b}\right]^2 + \left[\frac{p}{c}\right]^2}$$

where

$$u' = \frac{1}{\sqrt{\mu\varepsilon}} = c$$

Hence

$$f_r = \frac{3 \times 10^8}{2}\sqrt{\left[\frac{m}{5 \times 10^{-2}}\right]^2 + \left[\frac{n}{4 \times 10^{-2}}\right]^2 + \left[\frac{p}{10 \times 10^{-2}}\right]^2}$$
$$= 15\sqrt{0.04m^2 + 0.0625n^2 + 0.01p^2} \text{ GHz}$$

Since $c > a > b$ or $1/c < 1/a < 1/b$, the lowest order mode is TE_{101}. Notice that TM_{101} and TE_{100} do not exist because $m = 1, 2, 3, \ldots, n = 1, 2, 3, \ldots$, and $p = 0, 1, 2, 3, \ldots$ for the TM modes, and $m = 0, 1, 2, \ldots, n = 0, 1, 2, \ldots$, and $p = 1, 2, 3, \ldots$ for the TE modes. The resonant frequency for the TE_{101} mode is

$$f_{r_{101}} = 15\sqrt{0.04 + 0 + 0.01} = 3.35 \text{ GHz}$$

The next higher mode is TE_{011} (TM_{011} does not exist), with

$$f_{r_{011}} = 15\sqrt{0 + 0.0625 + 0.01} = 4.04 \text{ GHz}$$

The next mode is TE_{102} (TM_{102} does not exist), with

$$f_{r_{102}} = 15\sqrt{0.04 + 0 + 0.04} = 4.243 \text{ GHz}$$

The next mode is TM_{110} (TE_{110} does not exist), with

$$f_{r_{110}} = 15\sqrt{0.04 + 0.0625 + 0} = 4.8 \text{ GHz}$$

The next two modes are TE_{111} and TM_{111} (degenerate modes), with

$$f_{r_{111}} = 15\sqrt{0.04 + 0.0625 + 0.01} = 5.031 \text{ GHz}$$

The next mode is TM_{103} with

$$f_{r_{103}} = 15\sqrt{0.04 + 0 + 0.09} = 5.408 \text{ GHz}$$

Thus the five lowest order modes in ascending order are

TE_{101}	(3.35 GHz)
TE_{011}	(4.04 GHz)
TE_{102}	(4.243 GHz)
TM_{110}	(4.8 GHz)
TE_{111} or TM_{111}	(5.031 GHz)

(b) The quality factor for TE_{101} is given by

$$\begin{aligned} Q_{TE_{101}} &= \frac{(a^2 + c^2)\, abc}{\delta[2b(a^3 + c^3) + ac(a^2 + c^2)]} \\ &= \frac{(25 + 100)\, 200 \times 10^{-2}}{\delta[8(125 + 1000) + 50(25 + 100)]} \\ &= \frac{1}{61\delta} = \frac{\sqrt{\pi f_{101}\, \mu_o \sigma_c}}{61} \\ &= \frac{\sqrt{\pi (3.35 \times 10^9)\, 4\pi \times 10^{-7}\, (5.8 \times 10^7)}}{61} \\ &= 14{,}358 \end{aligned}$$

■

PRACTICE EXERCISE 12.8

If the resonant cavity of Example 12.8 is filled with a lossless material ($\mu_r = 1$, $\varepsilon_r = 3$), find the resonant frequency f_r and the quality factor for TE_{101} mode.

ANSWER 1.936 GHz, 1.093×10^4

SUMMARY

1. Waveguides are structures used in guiding EM waves at high frequencies. Assuming a lossless rectangular waveguide ($\sigma_c \simeq \infty$, $\sigma \simeq 0$), we apply Maxwell's equations in analyzing EM wave propagation through the guide. The resulting partial differential equation is solved using the method of separation of variables. On applying the boundary conditions on the walls of the guide, the basic formulas for the guide are obtained for different modes of operation.
2. Two modes of propagation (or field patterns) are the TM_{mn} and TE_{mn} where m and n are positive integers. For TM modes, $m = 1, 2, 3, \ldots$, and $n = 1, 2, 3, \ldots$ and for TE modes, $m = 0, 1, 2, \ldots$, and $n = 0, 1, 2, \ldots$, $n = m \neq 0$.
3. Each mode of propagation has associated propagation constant and cutoff frequency. The propagation constant $\gamma = \alpha + j\beta$ does not only depend on the constitutive parameters (ε, μ, σ) of the medium as in the case of plane waves in an unbounded space, it also depends on the cross-sectional dimensions (a, b) of the guide. The cutoff frequency is the frequency at which γ changes from being purely real (attenuation) to purely imaginary (propagation). The dominant mode of operation is the lowest mode possible. It is the mode with the lowest cutoff frequency. If $a > b$, the dominant mode is TE_{10}.
4. The basic equations for calculating the cutoff frequency f_c, phase constant β, and phase velocity u are summarized in Table 12.1. Formulas for calculating the attenuation constants due to lossy dielectric medium and imperfectly conducting walls are also provided.
5. The group velocity (or velocity of energy flow) u_g is related to the phase velocity u_p of the wave propagation by

$$u_p u_g = u'^2$$

where $u' = 1/\sqrt{\mu\varepsilon}$ is the medium velocity—i.e., the velocity of the wave in the dielectric medium unbounded by the guide. Although u_p is greater than u', u_p does not exceed u'.
6. The mode of operation for a given waveguide is dictated by the method of excitation.
7. A waveguide resonant cavity is used for energy storage at high frequencies. It is nothing but a waveguide shorted at both ends. Hence its analysis is similar to that of a waveguide. The resonant frequency for both the TE and TM modes to z is given by

$$f_r = \frac{u'}{2}\sqrt{\left[\frac{m}{a}\right]^2 + \left[\frac{n}{b}\right]^2 + \left[\frac{p}{c}\right]}$$

For TM modes, $m = 1, 2, 3, \ldots$, $n = 1, 2, 3, \ldots$, and $p = 0, 1, 2, 3, \ldots$, and for TE modes, $m = 0, 1, 2, 3, \ldots$, $n = 0, 1, 2, 3, \ldots$, and $p = 1, 2, 3, \ldots$, $m = n \neq 0$. If $a > b < c$, the dominant mode (one with the lowest resonant frequency) is TE_{101}.

8. The quality factor, a measure of the energy loss in the cavity, is given by

$$Q = \omega \frac{W}{P_L}$$

REVIEW QUESTIONS

12.1 At microwave frequencies, we prefer waveguides to transmission lines for transporting EM energy because of all the following *except* that

(a) Losses in transmission lines are prohibitively large.

(b) Waveguides have larger bandwidths and lower signal attenuation.

(c) Transmission lines are larger in size than waveguides.

(d) Transmission lines only support TEM mode.

12.2 An evanscent mode occurs when

(a) A wave is attenuated rather than propagated.

(b) The propagation constant is purely imaginary.

(c) $m = 0 = n$ so that all field components vanish.

(d) The wave frequency is the same as the cutoff frequency.

12.3 The dominant mode for rectangular waveguides is

(a) TE_{11}

(b) TM_{11}

(c) TE_{101}

(d) TE_{10}

12.4 The TM_{10} mode can exist in a rectangular waveguide.

(a) True

(b) False

12.5 For TE_{30} mode, which of the following field components exist?

(a) E_x

(b) E_y

(c) E_z

(d) H_x

(e) H_y

12.6 If in a rectangular waveguide for which $a = 2b$, the cutoff frequency for TE_{02} mode is 12 GHz, the cutoff frequency for TM_{11} mode is

(a) 3 GHz

(b) $3\sqrt{5}$ GHz

(c) 12 GHz

(d) $6\sqrt{5}$ GHz

(e) None of the above

12.7 If a tunnel is 4 by 7 m in cross section, a car in the tunnel will not receive an AM radio signal (e.g., $f = 10$ MHz).

(a) True

(b) False

12.8 When the electric field is at its maximum value, the magnetic energy of a cavity is

(a) At its maximum value

(b) At $\sqrt{2}$ of its maximum value

(c) At $\dfrac{1}{\sqrt{2}}$ of its maximum value

(d) At 1/2 of its maximum value

(e) Zero

12.9 Which of these modes does not exist in a rectangular resonant cavity?

(a) TE_{110}

(b) TE_{011}

(c) TM_{110}

(d) TM_{111}

12.10 How many degenerate dominant modes exist in a rectangular resonant cavity for which $a = b = c$?

(a) 0

(b) 2

(c) 3

(d) 5

(e) ∞

Answers: 12.1c, 12.2a, 12.3d, 12.4b, 12.5b,d, 12.6b, 12.7a, 12.8e, 12.9a, 12.10c.

PROBLEMS

12.1 (a) Show that a rectangular waveguide does not support TM_{10} and TM_{01} modes.

(b) Explain the difference between TE_{mn} and TM_{mn} modes.

12.2 An air-filled rectangular waveguide has dimensions $a = 2$ cm, $b = 1$ cm. Determine the cutoff frequencies for the following modes: TE_{01}, TE_{10}, TE_{11}, TE_{02}, TE_{20}, TE_{12}, TE_{21}, TM_{11}, TM_{12}, TM_{21}, TM_{22}, TM_{13}, and TM_{31}. Arrange the cutoff frequencies in ascending order.

12.3 The cutoff frequency of an air-filled rectangular waveguide is 2.4 GHz for the TE_{10} mode. What would be the cutoff frequency if the same guide were filled with a lossless nonmagnetic material whose dielectric permittivity is six times that of air?

12.4 Design a rectangular waveguide with an aspect ratio of 3 to 1 for use in the k band (18–26.5 GHz). Assume that the guide is air-filled.

12.5 A rectangular waveguide with dimensions $a = 2b = 5$ cm operates at 11 GHz. Assuming that the guide is filled with air, find the highest possible operating mode and the intrinsic wave impedance at that mode.

12.6 Assume the waveguide in Problem 12.5 is operated at 12.1 GHz and the average power transmitted down the guide is 45 mW. When the guide is operating at the highest possible mode, calculate the electric field in the guide.

12.7 In an air-filled rectangular waveguide, the cutoff frequency of a TE_{10} mode is 5 GHz, whereas that of TE_{01} mode is 12 GHz. Calculate

(a) The dimensions of the guide

(b) The cutoff frequencies of the next three higher TE modes

(c) The cutoff frequency for TE_{11} mode if the guide is filled with a lossless material having $\varepsilon_r = 2.25$ and $\mu_r = 1$.

12.8 An air-filled hollow rectangular waveguide is 150 m long and is capped at the end with a metal plate. If a short pulse of frequency 7.2 GHz is introduced into the input end of the guide, how long does it take the pulse to return to the input end? Assume that the cutoff frequency of the guide is 6.5 GHz.

12.9 Calculate the dimensions of an air-filled rectangular waveguide for which the cutoff frequencies for TM_{11} and TE_{03} modes are both equal to 12 GHz. At 8 GHz, determine whether the dominant mode will propagate or evanesce in the waveguide.

12.10 An air-filled rectangular waveguide has cross-sectional dimensions $a = 6$ cm and $b = 3$ cm. Given that

$$E_z = 5 \sin\left(\frac{2\pi x}{a}\right) \sin\left(\frac{3\pi y}{b}\right) \cos\,(10^{12}t - \beta z) \text{ V/m}$$

calculate the intrinsic impedance of this mode and the average power flow in the guide.

12.11 A TM mode operating at 5 GHz is propagated in an air-filled waveguide. If

$$E_z = \sin\left(\frac{2\pi x}{a}\right)\sin\left(\frac{\pi y}{b}\right)\cos(\omega t - 10z)\ \text{V/m}$$

find: (a) the cutoff frequency, (b) H_z and E_x.

12.12 In an air-filled square waveguide with $a = 1.2$ cm,

$$E_x = -10\sin\left(\frac{2\pi y}{a}\right)\sin(\omega t - 150z)\ \text{V/m}$$

(a) What is the mode of propagation?

(b) Find the cutoff wavelength λ_c.

(c) Calculate the frequency of operation f.

(d) Determine γ and η.

12.13 For Problem 12.12, determine

(a) Other field components E_y, E_z, H_x, H_y, and H_z

(b) The surface current density

(c) The surface charge density at $(x, y) = (0, a/2)$ and $(a/2, 0)$

***12.14** A square guide ($a = 2.5$ cm) filled with a lossless dielectric material ($\varepsilon_r = 4$) is excited at a frequency 75 percent of cutoff. At what distance along the guide is the signal strength reduced to 1 percent of the excitation if the guide operates in the

(a) TE_{10} mode

(b) TE_{12} mode

12.15 For the TM_{11} mode, derive a formula for the average power transmitted down the guide.

12.16 (a) Show that for a rectangular waveguide,

$$u_p = \frac{u'}{\sqrt{1 - \left[\frac{f_c}{f}\right]^2}} \qquad \lambda = \frac{\lambda'}{\sqrt{1 - \left[\frac{f_c}{f}\right]^2}}$$

(b) For an air-filled waveguide with $a = 2b = 2.5$ cm operating at 20 GHz, calculate u_p and λ for TE_{11} and TE_{21} modes.

12.17 Given that in a rectangular waveguide with dielectric of free space,

$$E_z = 50 \sin 40\pi x \sin 30\pi y \cos(\omega t - \beta z) \text{ V/m}$$

$$H_z = 0.1 \cos 40\pi x \cos 30\pi y \cos(\omega t - \beta z) \text{ A/m}$$

where $\omega = 20\pi \times 10^9$ rad/s, find E_x.

12.18 A 240-degree phase shift is produced by a 4-GHz signal when traveling along a dielectric-filled waveguide 3 cm long. If the cutoff frequency of the waveguide when air-filled is 10 GHz, calculate the relative permittivity of the dielectric.

12.19 An EM wave at $f = 10$ GHz is propagated down an air-filled rectangular waveguide of dimensions $a = 2.26$ cm, $b = 1.25$ cm. Determine

(a) The mode of propagation

(b) The group and phase velocities

12.20 A microwave transmitter is connected by an air-filled waveguide of cross section 2.5 cm × 1 cm to an antenna. For transmission at 11 GHz, find the ratio of (a) the phase velocity to the medium velocity, and (b) the group velocity to the medium velocity.

12.21 A rectangular waveguide is filled with polyethylene ($\varepsilon = 2.25\varepsilon_o$) and operates at 24 GHz. If the cutoff frequency of a certain TE mode is 16 GHz, find the group velocity and intrinsic impedance of the mode.

12.22 A rectangular waveguide with cross sections shown in Figure 12.16 has dielectric discontinuity. Calculate the standing wave ratio if the guide operates at 8 GHz in the dominant mode.

***12.23** Analysis of circular waveguide requires solution of the scalar Helmholtz equation in cylindrical coordinates, namely

$$\nabla^2 E_{zs} + k^2 E_{zs} = 0$$

or

$$\frac{1}{\rho}\frac{\partial}{\partial \rho}\left(\rho \frac{\partial E_{zs}}{\partial \rho}\right) + \frac{1}{\rho^2}\frac{\partial^2 E_{zs}}{\partial \phi^2} + \frac{\partial^2 E_{zs}}{\partial z^2} + k^2 E_{zs} = 0$$

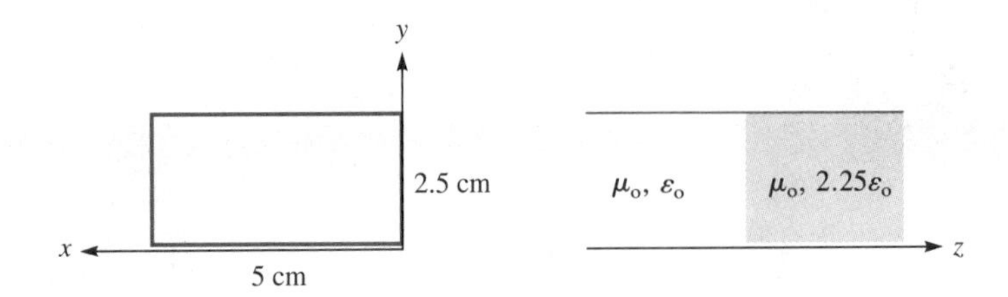

Figure 12.16 For Problem 12.22.

By assuming the product solution

$$E_{zs}(\rho, \phi, z) = R(\rho)\, \Phi(\phi)\, Z(z),$$

show that the separated equations are:

$$Z'' - k_z^2 Z = 0$$

$$\Phi'' + k_\phi^2 \Phi = 0$$

$$\rho^2 R'' + \rho R' + (k_\rho^2 \rho^2 - k_\phi^2) R = 0$$

where

$$k_\rho^2 = k^2 + k_z^2$$

12.24 A 1 cm × 2cm waveguide is made of copper ($\sigma_c = 5.8 \times 10^7$ mhos/m) and filled with a dielectric material for which $\varepsilon = 2.6\varepsilon_o$, $\mu = \mu_o$, $\sigma_d = 10^{-4}$ mhos/m. If the guide operates at 9 GHz, evaluate α_c and α_d for (a) TE_{10}, and (b) TM_{11}.

***12.25** A rectangular waveguide is filled with a certain material having $\sigma = 10^{-5}$ mhos/m, $\mu = \mu_o$, $\varepsilon = 5\varepsilon_o$ and operates at 4 GHz. If

$$H_{zs} = 5 \cos 30\pi x\, e^{-\gamma z}, \qquad E_{zs} = 0$$

calculate

(a) The cutoff frequency f_c and the cutoff wavelength λ_c

(b) α, β, and γ

(c) The phase velocity u_p.
Assume perfectly conducting waveguide walls.

12.26 An air-filled waveguide of dimensions $a = 2b = 7.5$ cm is operated at 10 GHz. If the guide is made of copper ($\sigma_c = 5.8 \times 10^7$ mhos/m), calculate the attenuation constant for TE_{10}, TE_{21}, and TM_{11} modes. For an input power of 5 kW, determine the power dissipated in a 2-m length of the guide in the TE_{10} mode.

***12.27** Show that for the rectangular cavity considered in section 12.8,

$$E_{xs} = \frac{j\omega\mu}{h^2}\left(\frac{n\pi}{b}\right) H_o \cos\left(\frac{m\pi x}{a}\right) \sin\left(\frac{n\pi y}{b}\right) \sin\left(\frac{p\pi z}{c}\right)$$

for the TE mode to z. Find other field components.

***12.28** Show that for TM mode to z in a rectangular cavity,

$$E_{xs} = -\frac{1}{h^2}\left(\frac{m\pi}{a}\right)\left(\frac{p\pi}{c}\right) E_o \cos\left(\frac{m\pi x}{a}\right) \sin\left(\frac{n\pi y}{b}\right) \sin\left(\frac{p\pi z}{c}\right)$$

Determine other field components.

12.29 In a rectangular resonant cavity, which mode is dominant when

(a) $a < b < c$

(b) $a > b > c$

(c) $a = c > b$

12.30 For an air-filled rectangular cavity with dimensions $a = 3$ cm, $b = 2$ cm, $c = 4$ cm, determine the resonant frequencies for the following modes: TE_{011}, TE_{101}, TM_{110}, and TM_{111}. List the resonant frequencies in ascending order.

12.31 A rectangular cavity with dimensions $a = 3.5$ cm, $b = 5$ cm, and $c = 10$ cm is filled with a dielectric material ($\varepsilon_r = 4.5$). Calculate the resonant frequency of the dominant mode.

12.32 A rectangular cavity filled with Teflon ($\varepsilon_r = 2.6$) has dimensions $a = 4$ cm, $b = 3$ cm, and $c = 1$ cm. Enumerate all modes for which the resonant frequency does not exceed 10 GHz.

12.33 (a) Show that for an air-filled cubical resonant cavity ($a = b = c$),

$$Q_{\text{TE}_{101}} = \frac{a}{3\delta}$$

(b) If the cavity is made of brass ($\sigma_c = 1.5 \times 10^7$ mhos/m), calculate the size of the cavity that would produce a quality factor of 6000 at the dominant resonant frequency.

(c) Determine the resonant frequency.

CHAPTER

Antennas 13

School seeks to get you ready for examination; life gives the finals.

— SAY

13.1 INTRODUCTION

Up till now, we have not asked ourselves how EM waves are produced. Recall that electric charges are the sources of EM fields. If the sources are time-varying, EM waves propagate away from the sources and radiation is said to have taken place. Radiation may be thought of as the process of transmitting electric energy. The radiation or launching of the waves into space is efficiently accomplished with the aid of conducting or dielectric structures called *antennas*. Theoretically, any structure can radiate EM waves but not all structures can serve as efficient radiation mechanisms.

An antenna may also be viewed as a transducer used in matching the transmission line or waveguide (used in guiding the wave to be launched) to the surrounding medium or vice versa. Figure 13.1 shows how an antenna is used to accomplish a match between the line or guide and the medium. The antenna is needed for two main reasons: efficient radiation and matching wave impedances in order to minimize reflection. The antenna uses voltage and current from the transmission line (or the EM fields from the waveguide) to launch an EM wave into the medium. An antenna may be used for either transmitting or receiving EM energy.

Typical antennas are illustrated in Figure 13.2. The dipole antenna in Figure 13.2(a) consists of two straight wires lying along the same axis. The loop antenna in Figure 13.2(b) consists of one or more turns of wire. The helical antenna in Figure 13.2(c) consists of a wire in the form of a helix backed by a ground plane. Antennas in Figure 13.2(a–c) are called *wire antennas;* they are used in automobiles, buildings, aircraft, ships, and so on. The horn antenna in Figure 13.2(d), an example of an *aperture antenna*, is a tapered section of waveguide providing a transition between a waveguide and the surroundings. Since it is conveniently flush-mounted, it is

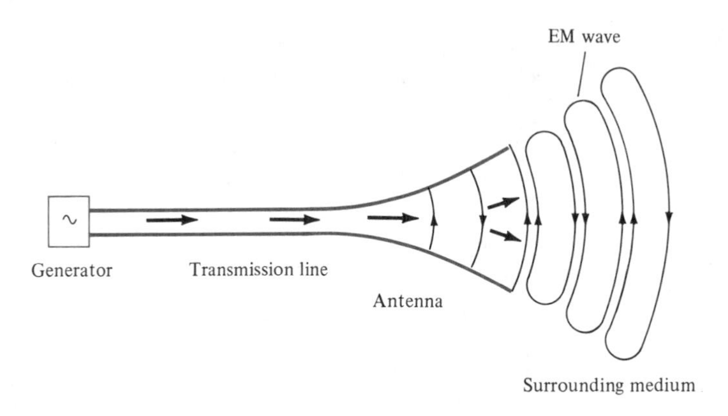

Figure 13.1 Antenna as a matching device between the guiding structure and the surrounding medium.

useful in various applications such as aircraft. The parabolic dish reflector in Figure 13.2(e) utilizes the fact that EM waves are reflected by a conducting sheet. When used as a transmitting antenna, a feed antenna such as a dipole or horn, is placed at the focal point. The radiation from the source is reflected by the dish (acting like a mirror) and a parallel beam results. Parabolic dish antennas are used in communications, radar, and astronomy.

The phenomenon of radiation is rather complicated, so we have intentionally delayed its discussion until this chapter. We will not attempt a broad coverage of antenna theory; our discussion will be limited to the basic types of antennas such as the Hertzian dipole, the half-wave dipole, the quarter-wave monopole, and the small loop. For each of these types, we will determine the radiation fields by taking the following steps:

1. Select an appropriate coordinate system and determine the magnetic vector potential $\mathbf{A}$.
2. Find $\mathbf{H}$ from $\mathbf{B} = \mu\mathbf{H} = \nabla \times \mathbf{A}$.
3. Determine $\mathbf{E}$ from $\nabla \times \mathbf{H} = \varepsilon \dfrac{\partial \mathbf{E}}{\partial t}$ or $\mathbf{E} = \eta \mathbf{H} \times \mathbf{a}_k$ assuming a lossless medium ($\sigma = 0$).
4. Find the far field and determine the time-average power radiated using

$$P_{\text{rad}} = \int \mathcal{P}_{\text{ave}} \cdot d\mathbf{S} \qquad \text{where} \qquad \mathcal{P}_{\text{ave}} = \frac{1}{2} \operatorname{Re} (\mathbf{E}_s \times \mathbf{H}_s^*)$$

Note that P_{rad} throughout this chapter is the same as P_{ave} in eq. (10.70).

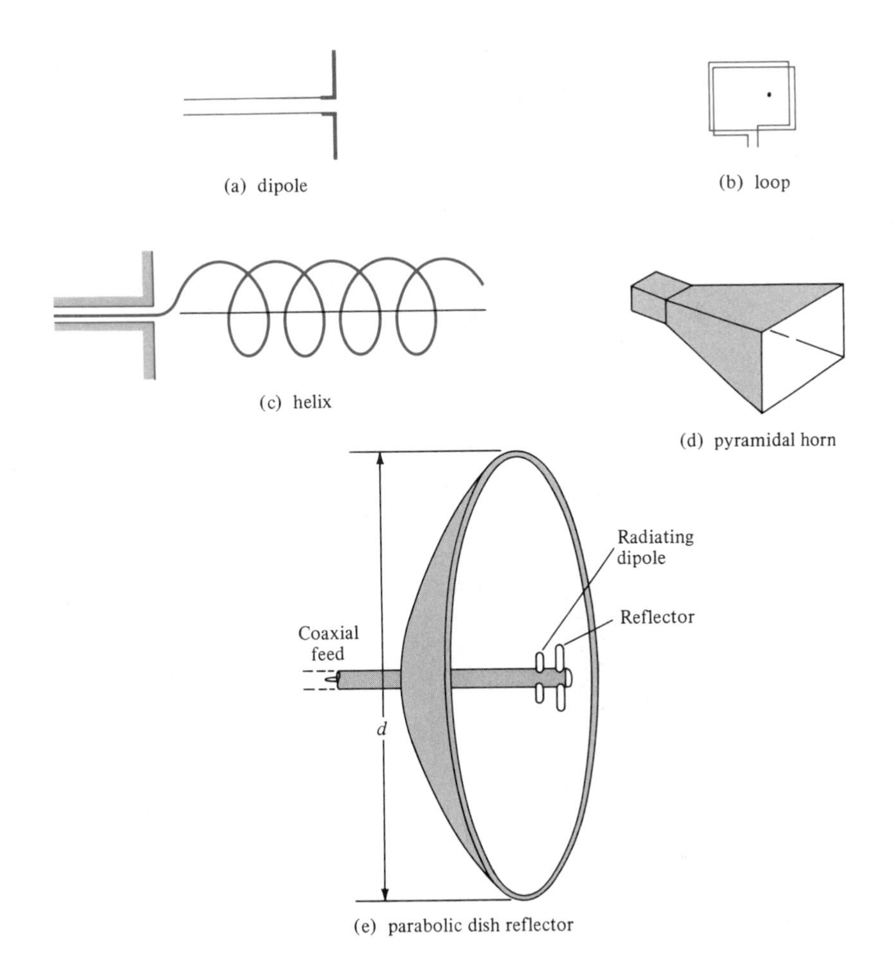

Figure 13.2 Typical antennas.

13.2 HERTZIAN DIPOLE

By a Hertzian dipole, we mean an infinitesimal current element $I\,dl$. Although such a current element does not exist in real life, it serves as a building block from which the field of a practical antenna can be calculated by integration.

Consider the Hertzian dipole shown in Figure 13.3. We assume that it is located at the origin of a coordinate system and that it carries a uniform current (independent of position on the dipole), $I = I_o \cos \omega t$. From eq. (9.54), the retarded magnetic vector potential at the field point P, due to the dipole, is given by

$$\mathbf{A} = \frac{\mu[I]\,dl}{4\pi r}\,\mathbf{a}_z \tag{13.1}$$

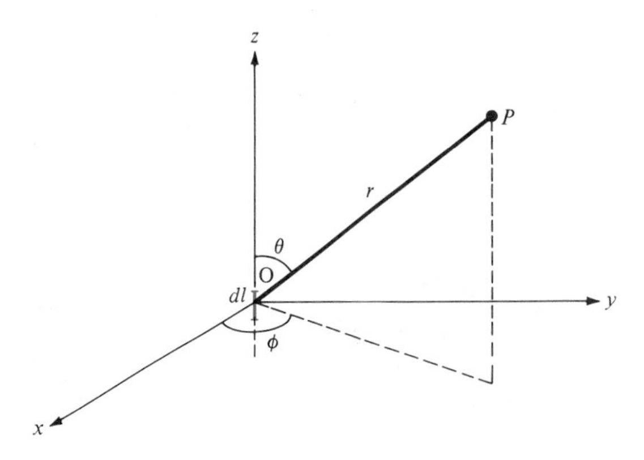

Figure 13.3 A Hertzian dipole carrying current $I = I_o \cos \omega t$.

where $[I]$ is the retarded current given by

$$[I] = I_o \cos \omega \left(t - \frac{r}{u}\right) = I_o \cos (\omega t - \beta r) \\ = \text{Re}\left[I_o e^{j(\omega t - \beta r)}\right] \tag{13.2}$$

where $\beta = \omega/u = 2\pi/\lambda$, and $u = 1/\sqrt{\mu\varepsilon}$. The current is said to be *retarded* at point P because there is a propagation time delay r/u or phase delay βr from O to P. By substituting eq. (13.2) into eq. (13.1), we may write **A** in phasor form as

$$A_{zs} = \frac{\mu I_o dl}{4\pi r} e^{-j\beta r} \tag{13.3}$$

Transforming this vector in Cartesian to spherical coordinates yields

$$\mathbf{A}_s = (A_{rs}, A_{\theta s}, A_{\phi s})$$

where

$$A_{rs} = A_{zs} \cos \theta, \qquad A_{\theta s} = -A_{zs} \sin \theta, \qquad A_{\phi s} = 0 \tag{13.4}$$

But $\mathbf{B} = \mu\mathbf{H} = \nabla \times \mathbf{A}$; hence, we obtain the **H** field as

$$H_{\phi s} = \frac{I_o dl}{4\pi} \sin \theta \left[\frac{j\beta}{r} + \frac{1}{r^2}\right] e^{-j\beta r} \tag{13.5a}$$

$$H_{rs} = 0 = H_{\theta s} \tag{13.5b}$$

We find the **E** field using $\nabla \times \mathbf{H} = \varepsilon\, \partial\mathbf{E}/\partial t$ or $\nabla \times \mathbf{H}_s = j\omega\varepsilon\mathbf{E}_s$,

$$E_{rs} = \frac{\eta I_o dl}{2\pi} \cos \theta \left[\frac{1}{r^2} - \frac{j}{\beta r^3}\right] e^{-j\beta r} \tag{13.6a}$$

$$E_{\theta s} = \frac{\eta I_o dl}{4\pi} \sin\theta \left[\frac{j\beta}{r} + \frac{1}{r^2} - \frac{j}{\beta r^3}\right] e^{-j\beta r} \qquad [13.6b]$$

$$E_{\phi s} = 0 \qquad [13.6c]$$

where

$$\eta = \frac{\beta}{\omega\varepsilon} = \sqrt{\frac{\mu}{\varepsilon}}$$

A close observation of the field equations in eqs. (13.5) and (13.6) reveals that we have terms varying as $1/r^3$, $1/r^2$, and $1/r$. The $1/r^3$ term is called the *electrostatic field* since it corresponds to the field of an electric dipole (see eq. (4.82)). This term dominates over other terms in a region very close to the Hertzian dipole. The $1/r^2$ term is called the *inductive field,* and it is predictable from the Biot-Savart law (see eq. (7.3)). The term is important only at near field, that is, at distances close to the current element. The $1/r$ term is called the *far field* or *radiation field* because it is the only term that remains at the far zone, that is, at a point very far from the current element. Here, we are mainly concerned with the far field or radiation zone ($\beta r \gg 1$ or $2\pi r \gg \lambda$), where the terms in $1/r^3$ and $1/r^2$ can be neglected in favor of the $1/r$ term. Thus at far field,

$$H_{\phi s} = \frac{jI_o\beta dl}{4\pi r} \sin\theta\, e^{-j\beta r} \qquad E_{\theta s} = \eta\, H_{\phi s} \qquad [13.7a]$$

$$H_{rs} = H_{\theta s} = E_{rs} = E_{\phi s} = 0 \qquad [13.7b]$$

Note from eq. (13.7a) that the radiation terms of $H_{\phi s}$ and $E_{\theta s}$ are in time phase and orthogonal just as the fields of a uniform plane wave. Also note that near-zone and far-zone fields are determined respectively to be the inequalities $\beta r \ll 1$ and $\beta r \gg 1$. More specifically, we define the boundary between the near and the far zones by the value of r given by

$$r = \frac{2d^2}{\lambda} \qquad [13.8]$$

where d is the largest dimension of the antenna.

The time-average power density is obtained as

$$\mathcal{P}_{\text{ave}} = \frac{1}{2}\,\text{Re}\,(\mathbf{E}_s \times \mathbf{H}_s^*) = \frac{1}{2}\,\text{Re}\,(E_{\theta s}\, H_{\phi s}^*\, \mathbf{a}_r)$$

$$= \frac{1}{2}\,\eta |H_{\phi s}|^2\, \mathbf{a}_r \qquad [13.9]$$

Substituting eq. (13.7) into eq. (13.9) yields the time-average radiated power as

$$\begin{aligned} P_{rad} &= \int \mathcal{P}_{ave} \cdot d\mathbf{S} \\ &= \int_{\phi=0}^{2\pi} \int_{\theta=0}^{\pi} \frac{I_o^2 \eta \beta^2 \, dl^2}{32\pi^2 r^2} \sin^2\theta \, r^2 \sin\theta \, d\theta \, d\phi \\ &= \frac{I_o^2 \eta \beta^2 \, dl^2}{32\pi^2} 2\pi \int_0^{\pi} \sin^3\theta \, d\theta \end{aligned} \tag{13.10}$$

But

$$\begin{aligned} \int_0^{\pi} \sin^3\theta \, d\theta &= \int_0^{\pi} (1 - \cos^2\theta) \, d(-\cos\theta) \\ &= \frac{\cos^3\theta}{3} - \cos\theta \Bigg|_0^{\pi} = \frac{4}{3} \end{aligned}$$

and $\beta^2 = 4\pi^2/\lambda^2$. Hence eq. (13.10) becomes

$$P_{rad} = \frac{I_o^2 \pi \eta}{3} \left[\frac{dl}{\lambda}\right]^2 \tag{13.11a}$$

If free space is the medium of propagation, $\eta = 120\pi$ and

$$P_{rad} = 40\pi^2 \left[\frac{dl}{\lambda}\right]^2 I_o^2 \tag{13.11b}$$

This power is equivalent to the power dissipated in a fictitious resistance R_{rad} by current I_o; that is

$$P_{rad} = I_{rms}^2 R_{rad}$$

or

$$\boxed{P_{rad} = \frac{1}{2} I_o^2 R_{rad}} \tag{13.12}$$

where I_{rms} is the root-mean-square value of I. From eqs. (13.11) and (13.12), we obtain

$$R_{rad} = \frac{2P_{rad}}{I_o^2} \tag{13.13a}$$

or

$$\boxed{R_{rad} = 80\pi^2 \left[\frac{dl}{\lambda}\right]^2} \tag{13.13b}$$

The resistance R_{rad} is a characteristic property of the Hertzian dipole antenna and is called its *radiation resistance*. For example, if $dl = \lambda/20$, $R_{\text{rad}} \simeq 2\ \Omega$, which is small in that it can deliver relatively small amounts of power. From eqs. (13.12) and (13.13), we observe that it requires antennas with large radiation resistances to deliver large amounts of power to space. It should be noted that R_{rad} in eq. (13.13b) is for a Hertzian dipole in free space. If the dipole is in a different, lossless medium, $\eta = \sqrt{\mu/\varepsilon}$ is substituted in eq. (13.10) and R_{rad} is determined using eq. (13.13a).

Note that the Hertzian dipole is assumed to be infinitesimally small ($\beta\, dl \ll 1$ or $dl \leq \lambda/10$). Consequently, its radiation resistance is very small and it is in practice difficult to match it with a real transmission line. We have also assumed that the dipole has a uniform current; this requires that the current be nonzero at the end points of the dipole. This is practically impossible because the surrounding medium is not conducting. However, our analysis will serve as a useful, valid approximation for an antenna with $dl \leq \lambda/10$. A more practical (and perhaps the most important) antenna is the half-wave dipole considered in the next section.

13.3 HALF-WAVE DIPOLE ANTENNA

The half-wave dipole derives its name from the fact that its length is half a wavelength ($\ell = \lambda/2$). As shown in Figure 13.4(a), it consists of a thin wire fed or excited at the midpoint by a voltage source connected to the antenna via a transmission line (e.g., a two-wire line). The field due to the dipole can be easily obtained if we consider it as consisting of a chain of Hertzian dipoles. The magnetic vector potential at P due to a differential length dl ($= dz$) of the dipole carrying a phasor current $I_s = I_o \cos \beta z$ is

$$dA_{zs} = \frac{\mu I_o \cos \beta z\, dz}{4\pi r'} e^{-j\beta r'} \qquad [13.14]$$

Notice that to obtain eq. (13.14), we have assumed a sinusoidal current distribution because the current must vanish at the ends of the dipole; a triangular current distribution is also possible (see Problem 13.7) but would give less accurate results. The actual current distribution on the antenna is not precisely known. It is determined by solving Maxwell's equations subject to the boundary conditions on the antenna, but the procedure is mathematically complex. However, the sinusoidal current assumption approximates the distribution obtained by solving the boundary-value problem and is commonly used in antenna theory.

If $r \gg \ell$, as explained in Section 4.10 on electric dipoles (see Figure 4.24), then

$$r - r' = z \cos \theta \qquad \text{or} \qquad r' = r - z \cos \theta$$

Thus we may substitute $r' \simeq r$ in the denominator of eq. (13.14) where the magnitude of the distance is needed. For the phase term in the numerator of eq. (13.14), the difference between βr and $\beta r'$ is significant, so we replace r' by $r - z \cos \theta$ and not r. In other words, we maintain the cosine term in the exponent while neglecting it in the denominator because the exponent involves the phase constant while the denominator does not. Thus,

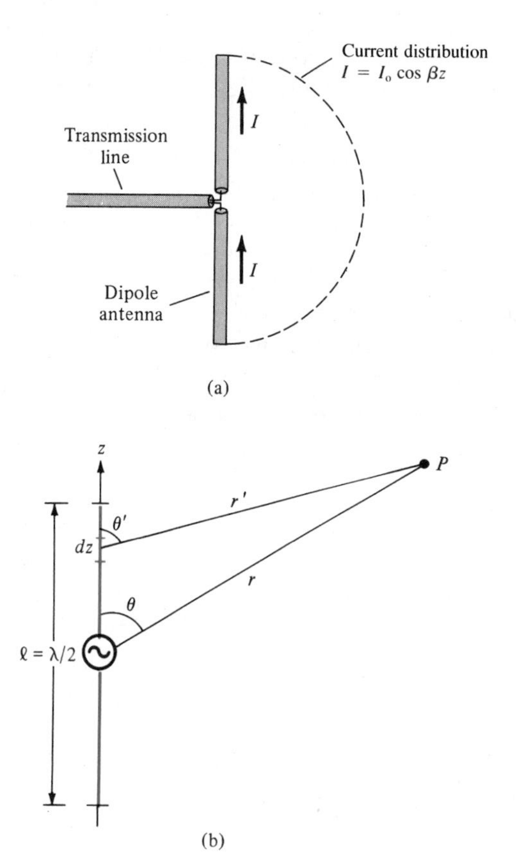

Figure 13.4 A half-wave dipole.

$$A_{zs} = \frac{\mu I_o}{4\pi r} \int_{-\lambda/4}^{\lambda/4} e^{-j\beta(r - z\cos\theta)} \cos \beta z \, dz$$

$$= \frac{\mu I_o}{4\pi r} e^{-j\beta r} \int_{-\lambda/4}^{\lambda/4} e^{j\beta z \cos\theta} \cos \beta z \, dz \qquad [13.15]$$

From the integral tables of Appendix A.8,

$$\int e^{az} \cos bz \, dz = \frac{e^{az}(a \cos bz + b \sin bz)}{a^2 + b^2}$$

Applying this to eq. (13.15) gives

$$A_{zs} = \frac{\mu I_o e^{-j\beta r} e^{j\beta z \cos\theta}}{4\pi r} \frac{(j\beta \cos\theta \cos \beta z + \beta \sin \beta z)}{-\beta^2 \cos^2\theta + \beta^2} \Bigg|_{-\lambda/4}^{\lambda/4} \qquad [13.16]$$

Since $\beta = 2\pi/\lambda$ or $\beta\ \lambda/4 = \pi/2$ and $-\cos^2\theta + 1 = \sin^2\theta$, eq. (13.16) becomes

$$A_{zs} = \frac{\mu I_o e^{-j\beta r}}{4\pi r\beta^2\sin^2\theta}\left[e^{j(\pi/2)\cos\theta}(0 + \beta) - e^{-j(\pi/2)\cos\theta}(0 - \beta)\right] \tag{13.17}$$

Using the identity $e^{jx} + e^{-jx} = 2\cos x$, we obtain

$$A_{zs} = \frac{\mu I_o e^{-j\beta r}\cos\left(\frac{\pi}{2}\cos\theta\right)}{2\pi r\beta\sin^2\theta} \tag{13.18}$$

We use eq. (13.4) in conjunction with the fact that $\mathbf{B}_s = \mu\mathbf{H}_s = \nabla \times \mathbf{A}_s$ and $\nabla \times \mathbf{H}_s = j\omega\varepsilon\mathbf{E}_s$ to obtain the magnetic and electric fields at far zone (discarding the $1/r^3$ and $1/r^2$ terms) as

$$\boxed{H_{\phi s} = \frac{jI_o e^{-j\beta r}\cos\left(\frac{\pi}{2}\cos\theta\right)}{2\pi r\sin\theta}, \qquad E_{\theta s} = \eta H_{\phi s}} \tag{13.19}$$

Notice again that the radiation term of $H_{\phi s}$ and $E_{\theta s}$ are in time phase and orthogonal.

Using eqs. (13.9) and (13.19), we obtain the time-average power density as

$$\begin{aligned}\mathcal{P}_{\text{ave}} &= \frac{1}{2}\eta\,|H_{\phi s}|^2\,\mathbf{a}_r \\ &= \frac{\eta I_o^2\cos^2\left(\frac{\pi}{2}\cos\theta\right)}{8\pi^2r^2\sin^2\theta}\,\mathbf{a}_r\end{aligned} \tag{13.20}$$

The time-average radiated power can be determined as

$$\begin{aligned}P_{\text{rad}} &= \int \mathcal{P}_{\text{ave}}\cdot d\mathbf{S} \\ &= \int_{\phi=0}^{2\pi}\int_{\theta=0}^{\pi}\frac{\eta I_o^2\cos^2\left(\frac{\pi}{2}\cos\theta\right)}{8\pi^2r^2\sin^2\theta}\,r^2\sin\theta\,d\theta\,d\phi \\ &= \frac{\eta I_o^2}{8\pi^2}\,2\pi\int_0^{\pi}\frac{\cos^2\left(\frac{\pi}{2}\cos\theta\right)}{\sin\theta}\,d\theta \\ &= 30\,I_o^2\int_0^{\pi}\frac{\cos^2\left(\frac{\pi}{2}\cos\theta\right)}{\sin\theta}\,d\theta\end{aligned} \tag{13.21}$$

where $\eta = 120\pi$ has been substituted assuming free space as the medium of propagation. Due to the nature of the integrand in eq. (13.21),

$$\int_0^{\pi/2} \frac{\cos^2\left(\dfrac{\pi}{2}\cos\theta\right)}{\sin\theta}\, d\theta = \int_{\pi/2}^{\pi} \frac{\cos^2\left(\dfrac{\pi}{2}\cos\theta\right)}{\sin\theta}\, d\theta$$

This is easily illustrated by a rough sketch of the variation of the integrand with θ. Hence

$$P_{\text{rad}} = 60 I_o^2 \int_0^{\pi/2} \frac{\cos^2\left(\dfrac{\pi}{2}\cos\theta\right) d\theta}{\sin\theta} \tag{13.22}$$

Changing variables, $u = \cos\theta$, and using partial fraction reduces eq. (13.22) to

$$\begin{aligned} P_{\text{rad}} &= 60 I_o^2 \int_0^1 \frac{\cos^2 \frac{1}{2}\pi u}{1 - u^2}\, du \\ &= 30 I_o^2 \left[\int_0^1 \frac{\cos^2 \frac{1}{2}\pi u}{1 + u}\, du + \int_0^1 \frac{\cos^2 \frac{1}{2}\pi u}{1 - u}\, du \right] \end{aligned} \tag{13.23}$$

Replacing $1 + u$ with v in the first integrand and $1 - u$ with v in the second results in

$$\begin{aligned} P_{\text{rad}} &= 30 I_o^2 \left[\int_0^1 \frac{\sin^2 \frac{1}{2}\pi v}{v}\, dv + \int_1^2 \frac{\sin^2 \frac{1}{2}\pi v}{v}\, dv \right] \\ &= 30 I_o^2 \int_0^2 \frac{\sin^2 \frac{1}{2}\pi v}{v}\, dv \end{aligned} \tag{13.24}$$

Changing variables, $w = \pi v$, yields

$$\begin{aligned} P_{\text{rad}} &= 30 I_o^2 \int_0^{2\pi} \frac{\sin^2 \frac{1}{2} w}{w}\, dw \\ &= 15 I_o^2 \int_0^{2\pi} \frac{(1 - \cos w)}{w}\, dw \\ &= 15 I_o^2 \int_0^{2\pi} \left[\frac{w}{2!} - \frac{w^3}{4!} + \frac{w^5}{6!} - \frac{w^7}{8!} + \cdots \right] dw \end{aligned} \tag{13.25}$$

since $\cos w = 1 - \frac{w^2}{2!} + \frac{w^4}{4!} - \frac{w^6}{6!} + \frac{w^8}{8!} - \cdots$. Integrating eq. (13.25) term by term and evaluating at the limit leads to

$$P_{\text{rad}} = 15 I_o^2 \left[\frac{(2\pi)^2}{2(2!)} - \frac{(2\pi)^4}{4(4!)} + \frac{(2\pi)^6}{6(6!)} - \frac{(2\pi)^8}{8(8!)} + \cdots \right]$$

$$\simeq 36.56\, I_o^2 \qquad [13.26]$$

The radiation resistance R_{rad} for the half-wave dipole antenna is readily obtained from eqs. (13.12) and (13.26) as

$$\boxed{R_{\text{rad}} = \frac{2P_{\text{rad}}}{I_o^2} = 73\ \Omega} \qquad [13.27]$$

Note the significant increase in the radiation resistance of the half-wave dipole over that of the Hertzian dipole. Thus the half-wave dipole is capable of delivering greater amounts of power to space than the Hertzian dipole.

The total input impedance Z_{in} of the antenna is the impedance seen at the terminals of the antenna and is given by

$$Z_{\text{in}} = R_{\text{in}} + jX_{\text{in}} \qquad [13.28]$$

where $R_{\text{in}} = R_{\text{rad}}$ for lossless antenna. Deriving the value of the reactance X_{in} involves a complicated procedure beyond the scope of this text. It is found that $X_{\text{in}} = 42.5\ \Omega$, so $Z_{\text{in}} = 73 + j42.5\ \Omega$ for a dipole length $\ell = \lambda/2$. The inductive reactance drops rapidly to zero as the length of the dipole is slightly reduced. For $\ell = 0.485\ \lambda$, the dipole is resonant, with $X_{\text{in}} = 0$. Thus in practice, a $\lambda/2$ dipole is designed such that X_{in} approaches zero and $Z_{\text{in}} \simeq 73\ \Omega$. This value of the radiation resistance of the $\lambda/2$ dipole antenna is the reason for the standard 75-Ω coaxial cable. Also, the value is easy to match to transmission lines. These factors in addition to the resonance property are the reasons for the dipole antenna's popularity and its extensive use.

13.4 QUARTER-WAVE MONOPOLE ANTENNA

Basically, the quarter-wave monopole antenna consists of one half of a half-wave dipole antenna located on a conducting ground plane as in Figure 13.5. The monopole antenna is perpendicular to the plane, which is usually assumed to be infinite and perfectly conducting. It is fed by a coaxial cable connected to its base.

Using image theory of Section 6.6, we replace the infinite, perfectly conducting ground plane with the image of the monopole. The field produced in the region above the ground plane due to the $\lambda/4$ monopole with its image is the same as the field due

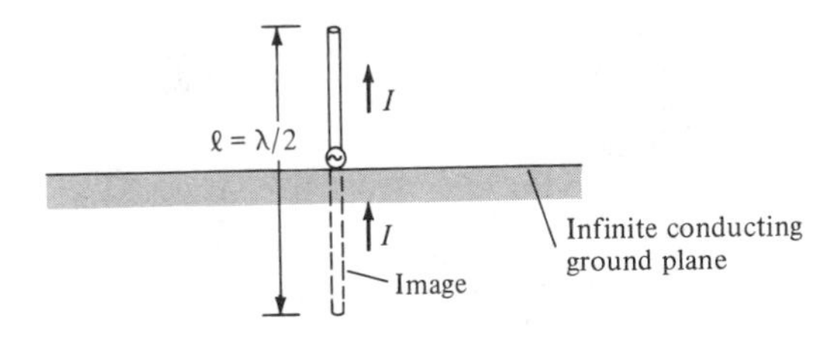

Figure 13.5 The monopole antenna.

to a λ/2 wave dipole. Thus eq. (13.19) holds for the λ/4 monopole. However, the integration in eq. (13.21) is only over the hemispherical surface above the ground plane (i.e., $0 \leq \theta \leq \pi/2$) because the monopole radiates only through that surface. Hence, the monopole radiates only half as much power as the dipole with the same current. Thus for a λ/4 monopole,

$$P_{\text{rad}} \simeq 18.28\, I_o^2 \qquad [13.29]$$

and

$$R_{\text{rad}} = \frac{2P_{\text{rad}}}{I_o^2}$$

or

$$\boxed{R_{\text{rad}} = 36.5\ \Omega} \qquad [13.30]$$

By the same token, the total input impedance for a λ/4 monopole is $Z_{\text{in}} = 36.5 + j21.25\ \Omega$.

13.5 SMALL LOOP ANTENNA

The loop antenna is of practical importance. It is used as a directional finder (or search loop) in radiation detection and as a TV antenna for ultra-high frequencies. The term "small" implies that the dimensions (such as ρ_o) of the loop are much smaller than λ.

Consider a small filamentary circular loop of radius ρ_o carrying a uniform current, $I_o \cos \omega t$, as in Figure 13.6. The loop may be regarded as an elemental magnetic dipole. The magnetic vector potential at the field point P due to the loop is

$$\mathbf{A} = \oint_L \frac{\mu[I]\, d\mathbf{l}}{4\pi r'} \qquad [13.31]$$

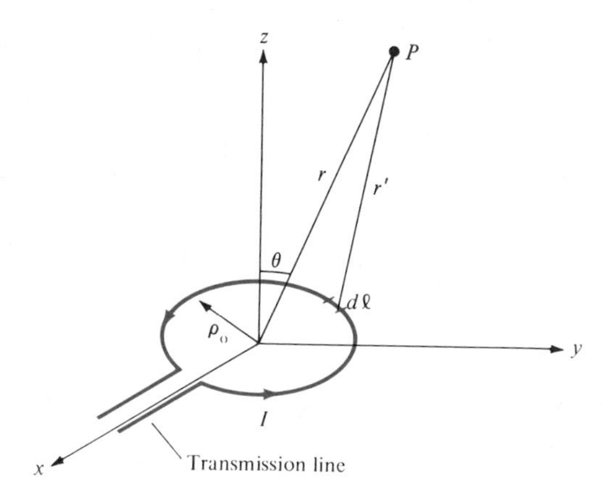

Figure 13.6 The small loop antenna.

where $[I] = I_o \cos(\omega t - \beta r') = \text{Re}\,[I_o e^{j(\omega t - \beta r')}]$. Substituting $[I]$ into eq. (13.31), we obtain **A** in phasor form as

$$\mathbf{A}_s = \frac{\mu I_o}{4\pi} \oint_L \frac{e^{-j\beta r'}}{r'} \, d\mathbf{l} \tag{13.32}$$

Evaluating this integral requires a lengthy procedure. It can be shown that for a small loop ($\rho_o \ll \lambda$), r' can be replaced by r in the denominator of eq. (13.32) and $\mathbf{A}_s$ has only ϕ-component given by

$$A_{\phi s} = \frac{\mu I_o S}{4\pi r^2}(1 + j\beta r)e^{-j\beta r} \sin\theta \tag{13.33}$$

where $S = \pi\rho_o^2$ = loop area for a loop with N turns, $S = N\pi\rho_o^2$. Using the fact that $\mathbf{B}_s = \mu\mathbf{H}_s = \nabla \times \mathbf{A}_s$ and $\nabla \times \mathbf{H}_s = j\omega\varepsilon\mathbf{E}_s$, we obtain the electric and magnetic fields from eq. (13.33) as

$$E_{\phi s} = \frac{-j\omega\mu I_o S}{4\pi} \sin\theta \left[\frac{j\beta}{r} + \frac{1}{r^2}\right] e^{-j\beta r} \tag{13.34a}$$

$$H_{rs} = \frac{j\omega\mu I_o S}{2\pi\eta} \cos\theta \left[\frac{1}{r^2} - \frac{j}{\beta r^3}\right] e^{-j\beta r} \tag{13.34b}$$

$$H_{\theta s} = \frac{j\omega\mu I_o S}{4\pi\eta} \sin\theta \left[\frac{j\beta}{r} + \frac{1}{r^2} - \frac{j}{\beta r^3}\right] e^{-j\beta r} \tag{13.34c}$$

$$E_{rs} = E_{\theta s} = H_{\phi s} = 0 \tag{13.34d}$$

Comparing eqs. (13.5) and (13.6) with eq. (13.34), we observe the dual nature of the field due to an electric dipole of Figure 13.3 and the magnetic dipole of Figure 13.6

(see Table 8.2 also). At far field, only the $1/r$ term (the radiation term) in eq. (13.34) remains. Thus at far field,

$$E_{\phi s} = \frac{\omega \mu I_o S}{4\pi r} \beta \sin\theta \, e^{-j\beta r}$$

$$= \frac{\eta \pi I_o S}{r\lambda^2} \sin\theta \, e^{-j\beta r}$$

or

$$E_{\phi s} = \frac{120\pi^2 I_o}{r} \frac{S}{\lambda^2} \sin\theta \, e^{-j\beta r}, \qquad H_{\theta s} = -\frac{E_{\phi s}}{\eta} \tag{13.35a}$$

$$E_{rs} = E_{\theta s} = H_{rs} = H_{\phi s} = 0 \tag{13.35b}$$

where $\eta = 120\pi$ for free space has been assumed. Though the far field expressions in eq. (13.35) are obtained for a small circular loop, they can be used for a small square loop with one turn ($S = a^2$) or any small loop provided that the loop dimensions are small ($d \leq \lambda/10$, where d is the largest dimension of the loop). It is left as an exercise to show that using eqs. (13.13a) and (13.35) gives the radiation resistance of a small loop antenna as

$$R_{\text{rad}} = \frac{320\,\pi^4 S^2}{\lambda^4} \tag{13.36}$$

EXAMPLE 13.1

A magnetic field strength of 5 μA/m is required at a point on $\theta = \pi/2$, 2 km from an antenna in air. Neglecting ohmic loss, how much power must the antenna transmit if it is

(a) A Hertzian dipole of length $\lambda/25$?

(b) A half-wave dipole?

(c) A quarter-wave monopole?

(d) A 10-turn loop antenna of radius $\rho_o = \lambda/20$?

SOLUTION

(a) For a Hertzian dipole,

$$|H_{\phi s}| = \frac{I_o \beta \, dl \sin\theta}{4\pi r}$$

where $dl = \lambda/25$ or $\beta\, dl = \frac{2\pi}{\lambda} \cdot \frac{\lambda}{25} = \frac{2\pi}{25}$. Hence,

$$5 \times 10^{-6} = \frac{I_o \cdot \frac{2\pi}{25}(1)}{4\pi\,(2 \times 10^3)} = \frac{I_o}{10^5}$$

or

$$I_o = 0.5 \text{ A}$$

$$P_{rad} = 40\pi^2 \left[\frac{dl}{\lambda}\right]^2 I_o^2 = \frac{40\pi^2(0.5)^2}{(25)^2}$$

$$= 158 \text{ mW}$$

(b) For a $\lambda/2$ dipole,

$$|H_{\phi s}| = \frac{I_o \cos\left(\frac{\pi}{2}\cos\theta\right)}{2\pi r \sin\theta}$$

$$5 \times 10^{-6} = \frac{I_o \cdot 1}{2\pi\,(2 \times 10^3) \cdot (1)}$$

or

$$I_o = 20\pi \text{ mA}$$

$$P_{rad} = 1/2 I_o^2 R_{rad} = 1/2(20\pi)^2 \times 10^{-6}\,(73)$$

$$= 144 \text{ mW}$$

(c) For a $\lambda/4$ monopole,

$$I_o = 20\pi \text{ mA}$$

as in part (b).

$$P_{rad} = 1/2 I_o^2 R_{rad} = 1/2(20\pi)^2 \times 10^{-6}\,(36.56)$$

$$= 72 \text{ mW.}$$

(d) For a loop antenna,

$$|H_{\theta s}| = \frac{\pi I_o}{r}\frac{S}{\lambda^2}\sin\theta$$

For a single turn, $S = \pi\rho_o^2$. For N-turn, $S = N\pi\rho_o^2$. Hence,

$$5 \times 10^{-6} = \frac{\pi I_o 10\pi}{2 \times 10^3}\left[\frac{\rho_o}{\lambda}\right]^2$$

or

$$I_o = \frac{10}{10\pi^2}\left[\frac{\lambda}{\rho_o}\right]^2 \times 10^{-3} = \frac{20^2}{\pi^2} \times 10^{-3}$$

$$= 40.53 \text{ mA}$$

$$R_{rad} = \frac{320\ \pi^4\ S^2}{\lambda^4} = 320\ \pi^6\ N^2\left[\frac{\rho_o}{\lambda}\right]^4$$

$$= 320\ \pi^6 \times 100\left[\frac{1}{20}\right]^4 = 192.3\ \Omega.$$

$$P_{rad} = \frac{1}{2} I_o^2\ R_{rad} = \frac{1}{2}(40.53)^2 \times 10^{-6}\ (192.3)$$

$$= 158 \text{ mW}$$

■

PRACTICE EXERCISE 13.1

A Hertzian dipole of length $\lambda/100$ is located at the origin and fed with a current of $0.25 \sin 10^8 t$ A. Determine the magnetic field at

(a) $r = \lambda/5$, $\theta = 30°$

(b) $r = 200\lambda$, $\theta = 60°$

ANSWER (a) $0.2119 \sin (10^8 t - 20.5°)\ \mathbf{a}_\phi$ mA/m, (b) $0.2871 \sin (10^8 t + 90°)\ \mathbf{a}_\phi$ μA/m.

EXAMPLE 13.2

An electric field strength of 10 μV/m is to be measured at an observation point $\theta = \pi/2$, 500 km from a half-wave (resonant) dipole antenna operating in air at 50 MHz.

(a) What is the length of the dipole?

(b) Calculate the current that must be fed to the antenna.

(c) Find the average power radiated by the antenna.

(d) If a transmission line with $Z_o = 75\ \Omega$ is connected to the antenna, determine the standing wave ratio.

SOLUTION (a) The wavelength $\lambda = \dfrac{c}{f} = \dfrac{3 \times 10^8}{50 \times 10^6} = 6$ m.

Hence, the length of the half-dipole is $\ell = \dfrac{\lambda}{2} = 3$ m.

(b) From eq. (13.19),

$$|E_{\theta s}| = \frac{\eta_o I_o \cos\left(\dfrac{\pi}{2}\cos\theta\right)}{2\pi r \sin\theta}$$

or

$$I_o = \frac{|E_{\theta s}|\, 2\pi r \sin\theta}{\eta_o \cos\left(\dfrac{\pi}{2}\cos\theta\right)}$$

$$= \frac{10 \times 10^{-6}\, 2\pi\, (500 \times 10^3) \cdot (1)}{120\pi\, (1)}$$

$$= 83.33 \text{ mA.}$$

(c)

$$R_{rad} = 73\ \Omega$$

$$P_{rad} = \frac{1}{2} I_o^2 R_{rad} = \frac{1}{2}(83.33)^2 \times 10^{-6} \times 73$$

$$= 253.5 \text{ mW.}$$

(d)

$$\Gamma = \frac{Z_L - Z_o}{Z_L + Z_o} \quad (Z_L = Z_{in} \text{ in this case})$$

$$= \frac{73 + j42.5 - 75}{73 + j42.5 + 75} = \frac{-2 + j42.5}{148 + j42.5}$$

$$= \frac{42.55\underline{/92.69^\circ}}{153.98\underline{/16.02^\circ}} = 0.2763\underline{/76.67^\circ}$$

$$s = \frac{1 + |\Gamma|}{1 - |\Gamma|} = \frac{1 + 0.2763}{1 - 0.2763} = 1.763$$

■

PRACTICE EXERCISE 13.2

Repeat Example 13.2 if the dipole antenna is replaced by a $\lambda/4$ monopole.

ANSWER (a) 1.5m, (b) 83.33 mA, (c) 126.8 mW, (d) 2.265.

13.6 ANTENNA CHARACTERISTICS

Having considered the basic elementary antenna types, we now discuss some important characteristics of an antenna as a radiator of electromagnetic energy. These characteristics include: (a) antenna pattern, (b) radiation intensity, (c) directive gain, (d) power gain.

A. Antenna Patterns

An antenna pattern or radiation pattern is a three-dimensional plot of its radiation at far field. When the amplitude of a specified component of the **E** field is plotted, it is called the *field pattern* or *voltage pattern*. When the square of the amplitude of **E** is plotted, it is called the *power pattern*. A three-dimensional plot of an antenna pattern is avoided by plotting separately the normalized $|E_s|$ versus θ for a constant ϕ (this is called an *E-plane pattern* or *vertical pattern*) and the normalized $|E_s|$ versus ϕ for $\theta = \pi/2$ (called the *H-plane pattern* or *horizontal pattern*). The normalization of $|E_s|$ is with respect to the maximum value of the $|E_s|$ so that the maximum value of the normalized $|E_s|$ is unity.

For the Hertzian dipole, for example, the normalized $|E_s|$ is obtained from eq. (13.7) as

$$f(\theta) = |\sin \theta| \qquad [13.37]$$

which is independent of ϕ. From eq. (13.37), we obtain the E-plane pattern as the polar plot of $f(\theta)$ with θ varying from 0° to 180°. The result is shown in Figure 13.7(a). Note that the plot is symmetric about the z-axis ($\theta = 0$). For the H-plane pattern, we set $\theta = \pi/2$ so that $f(\theta) = 1$ which is circle of radius 1 as shown in Figure 13.7(b). When the two plots of Figures 13.7(a) and (b) are combined, we have a three-dimensional field pattern of Figure 13.7(c), which has the shape of a doughnut.

A plot of the time-average power, $|\mathcal{P}_{\text{ave}}| = \mathcal{P}_{\text{ave}}$, for a fixed distance r is the power pattern of the antenna. It is obtained by plotting separately $\mathcal{P}_{\text{ave}}$ versus θ for constant ϕ and $\mathcal{P}_{\text{ave}}$ versus ϕ for constant θ.

For the Hertzian dipole, the normalized power pattern is easily obtained from eqs. (13.37) or (13.9) as

$$f^2(\theta) = \sin^2\theta \qquad [13.38]$$

which is sketched in Figure 13.8. Notice that Figures 13.7(b) and 13.8(b) show circles because $f(\theta)$ is independent of ϕ and that the value of OP in Figure 13.8(a) is the relative average power for that particular θ. Thus, at point Q ($\theta = 45°$), the average power is one-half the maximum average power (the maximum average power is at $\theta = \pi/2$).

B. Radiation Intensity

The radiation intensity of an antenna is defined as

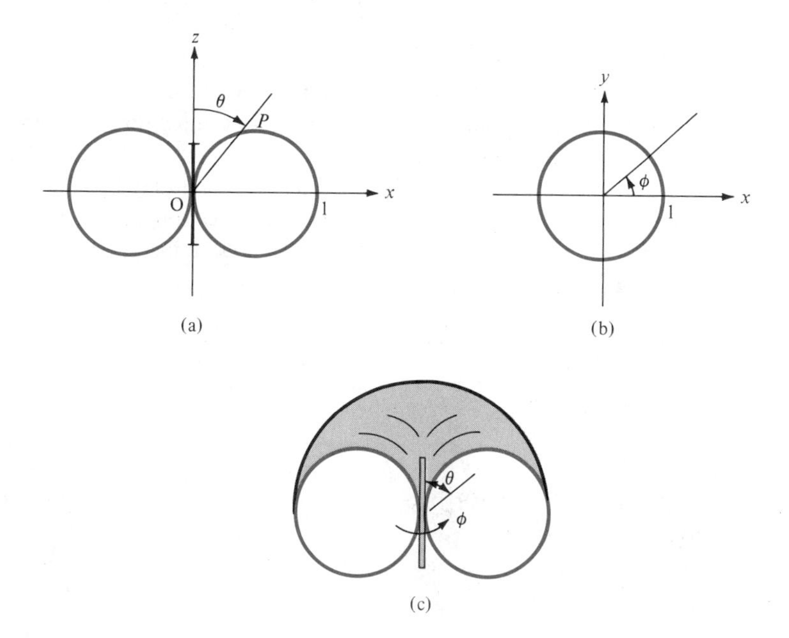

Figure 13.7 Field patterns of the Hertzian dipole: **(a)** normalized E-plane or vertical pattern (ϕ = constant = 0), **(b)** normalized H-plane or horizontal pattern ($\theta = \pi/2$), **(c)** three-dimensional pattern.

$$\boxed{U(\theta, \phi) = r^2\, \mathcal{P}_{\text{ave}}} \qquad [13.39]$$

From eq. (13.39), the total average power radiated can be expressed as

$$\begin{aligned} P_{\text{rad}} &= \oint_S \mathcal{P}_{\text{ave}}\, dS = \oint_S \mathcal{P}_{\text{ave}}\, r^2 \sin\theta\, d\theta\, d\phi \\ &= \int_S U(\theta, \phi) \sin\theta\, d\theta\, d\phi \\ &= \int_{\phi=0}^{2\pi} \int_{\theta=0}^{\pi} U(\theta, \phi)\, d\Omega \end{aligned} \qquad [13.40]$$

where $d\Omega = \sin\theta\, d\theta\, d\phi$ is the *differential solid angle* in steradian (sr). Hence the radiation intensity $U(\theta, \phi)$ is measured in watts per steradian (W/sr). The average value of $U(\theta, \phi)$ is the total radiated power divided by 4π sr; that is,

$$\boxed{U_{\text{ave}} = \frac{P_{\text{rad}}}{4\pi}} \qquad [13.41]$$

C. Directive Gain

Besides the antenna patterns described above, we are often interested in measurable quantities such as gain and directivity to determine the radiation characteristics of an antenna. The *directive gain* $G_d(\theta, \phi)$ of an antenna is a measure of the concentration of the radiated power in a particular direction (θ, ϕ). It may be regarded as the ability of the antenna to direct radiated power in a given direction. It is usually obtained as the ratio of radiation intensity in a given direction (θ, ϕ) to the average radiation intensity, that is

$$G_d(\theta, \phi) = \frac{U(\theta, \phi)}{U_{\text{ave}}} = \frac{4\pi\, U(\theta, \phi)}{P_{\text{rad}}} \tag{13.42}$$

By substituting eq. (13.39) into eq. (13.42), $\mathcal{P}_{\text{ave}}$ may be expressed in terms of directive gain as

$$\mathcal{P}_{\text{ave}} = \frac{G_d}{4\pi r^2} P_{\text{rad}} \tag{13.43}$$

The directive gain $G_d(\theta, \phi)$ depends on antenna pattern. For the Hertzian dipole (as well as for $\lambda/2$ dipole and $\lambda/4$ monopole), we notice from Figure 13.8 that $\mathcal{P}_{\text{ave}}$ is maximum at $\theta = \pi/2$ and minimum (zero) at $\theta = 0$ or π. Thus the Hertzian dipole

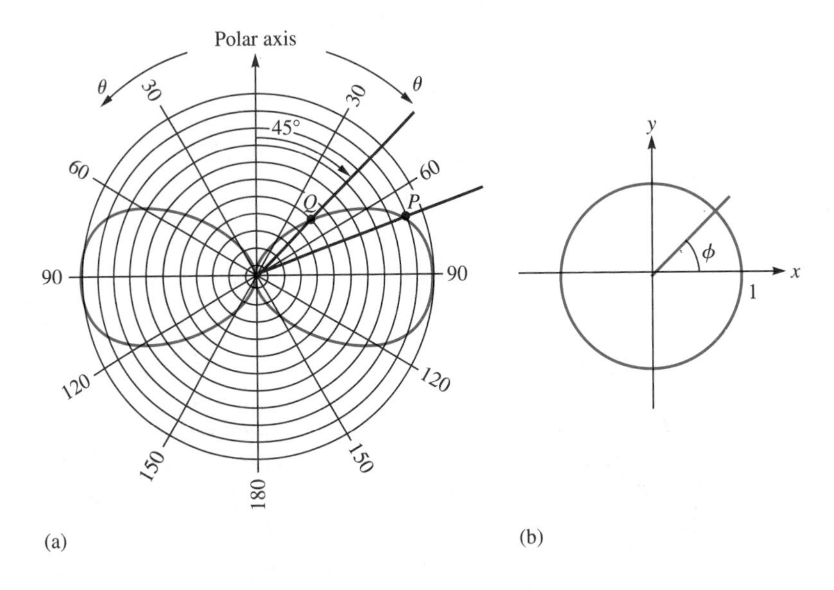

Figure 13.8 Power pattern of the Hertzian dipole: **(a)** ϕ = constant = 0; **(b)** θ = constant = $\pi/2$).

radiates power in a direction broadside to its length. For an *isotropic* antenna (one that radiates equally in all directions), $G_d = 1$. However, such an antenna is not a physicality but an ideality.

The *directivity* D of an antenna is defined as the ratio of the maximum radiation intensity to the average radiation intensity. Obviously, D is the maximum directive gain G_d, max. Thus

$$D = \frac{U_{max}}{U_{ave}} = G_d, \text{max} \quad [13.44a]$$

or

$$D = \frac{4\pi\, U_{max}}{P_{rad}} \quad [13.44b]$$

$D = 1$ for an isotropic antenna; this is the smallest value D can have. For the Hertzian dipole,

$$G_d(\theta, \phi) = 1.5 \sin^2\theta, \qquad D = 1.5. \quad [13.45]$$

For the $\lambda/2$ dipole,

$$G_d(\theta, \phi) = \frac{\eta}{\pi R_{rad}} f^2(\theta), \qquad D = 1.64 \quad [13.46]$$

where $\eta = 120\pi$, $R_{rad} = 73\ \Omega$, and

$$f(\theta) = \frac{\cos\left(\dfrac{\pi}{2}\cos\theta\right)}{\sin\theta} \quad [13.47]$$

D. Power Gain

Our definition of the directive gain in eq. (13.42) does not account for the ohmic power loss P_ℓ of the antenna. P_ℓ is due to the fact that the antenna is made of a conductor with finite conductivity. As illustrated in Figure 13.9, if P_{in} is the total input power to the antenna,

$$\begin{aligned} P_{in} &= P_\ell + P_{rad} \\ &= \frac{1}{2}|I_{in}|^2 (R_\ell + R_{rad}) \end{aligned} \quad [13.48]$$

where I_{in} is the current at the input terminals and R_ℓ is the *loss* or *ohmic resistance* of the antenna. In other words, P_{in} is the power accepted by the antenna at its terminals during the radiation process, and P_{rad} is the power radiated by the antenna; the difference between the two powers is P_ℓ, the power dissipated within the antenna.

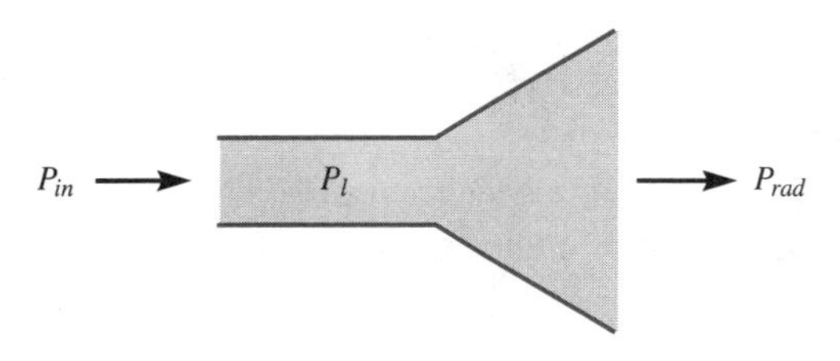

Figure 13.9 Relating P_{in}, P_ℓ, and P_{rad}.

We define the *power gain* $G_p(\theta, \phi)$ of the antenna as

$$G_p(\theta, \phi) = \frac{4\pi\, U(\theta, \phi)}{P_{\text{in}}} \tag{13.49}$$

The ratio of the power gain in any specified direction to the directive gain in that direction is referred to as the *radiation efficiency* η_r of the antennas, that is

$$\eta_r = \frac{G_p}{G_d} = \frac{P_{\text{rad}}}{P_{\text{in}}}$$

Introducing eq. (13.48) leads to

$$\eta_r = \frac{P_{\text{rad}}}{P_{\text{in}}} = \frac{R_{\text{rad}}}{R_{\text{rad}} + R_\ell} \tag{13.50}$$

For many antennas, η_r is close to 100 percent so that $G_P \simeq G_d$. It is customary to express directivity and gain in decibels (dB). Thus

$$D \text{ (dB)} = 10 \log_{10} D \tag{13.51a}$$

$$G \text{ (dB)} = 10 \log_{10} G \tag{13.51b}$$

It should be mentioned at this point that the radiation patterns of an antenna are usually measured in the far field region. The far field region of an antenna is commonly taken to exist at distances $r \geq r_{\text{min}}$ where

$$r_{\text{min}} = \frac{2d^2}{\lambda} \tag{13.52}$$

and d is the largest dimension of the antenna. For example, $d = \ell$ for the electric dipole antenna and $d = 2\rho_o$ for the small loop antenna.

EXAMPLE 13.3

Show that the directive gain of the Hertzian dipole is

$$G_d(\theta, \phi) = 1.5 \sin^2\theta$$

and that of the half-wave dipole is

$$G_d(\theta, \phi) = 1.64 \frac{\cos^2\left(\frac{\pi}{2}\cos\theta\right)}{\sin^2\theta}$$

SOLUTION From eq. (13.42),

$$G_d(\theta, \phi) = \frac{4\pi f^2(\theta)}{\int f^2(\theta)\, d\Omega}$$

(a) For the Hertzian dipole,

$$G_d(\theta, \phi) = \frac{4\pi \sin^2\theta}{\int_{\phi=0}^{2\pi}\int_{\theta=0}^{\pi} \sin^3\theta\, d\theta\, d\phi} = \frac{4\pi \sin^2\theta}{2\pi\,(4/3)}$$

$$= 1.5 \sin^2\theta$$

as required.

(b) For the half-wave dipole,

$$G_d(\theta, \phi) = \frac{\dfrac{4\pi\cos^2\left(\frac{\pi}{2}\cos\theta\right)}{\sin^2\theta}}{\displaystyle\int_{\phi=0}^{2\pi}\int_{\theta=0}^{\pi} \frac{\cos^2\left(\frac{\pi}{2}\cos\theta\right) d\theta\, d\phi}{\sin\theta}}$$

From eq. (13.26), the integral in the denominator gives $2\pi(1.2188)$. Hence,

$$G_d(\theta, \phi) = \frac{4\pi\cos^2\left(\frac{\pi}{2}\cos\theta\right)}{\sin^2\theta} \cdot \frac{1}{2\pi\,(1.2188)}$$

$$= 1.64 \frac{\cos^2\left(\frac{\pi}{2}\cos\theta\right)}{\sin^2\theta}$$

as required. ■

PRACTICE EXERCISE 13.3

Calculate the directivity of

(a) The Hertzian monopole

(b) The quarter-wave monopole

ANSWER (a) 3, (b) 3.28.

EXAMPLE 13.4

Determine the electric field intensity at a distance of 10 km from an antenna having a directive gain of 5 dB and radiating a total power of 20 kW.

SOLUTION

$$5 = G_d \text{ (dB)} = 10 \log_{10} G_d$$

or

$$0.5 = \log_{10} G_d \rightarrow G_d = 10^{0.5} = 3.162$$

From eq. (13.43),

$$\mathcal{P}_{\text{ave}} = \frac{G_d P_{\text{rad}}}{4\pi r^2}$$

But

$$\mathcal{P}_{\text{ave}} = \frac{|E_s|^2}{2\eta}$$

Hence,

$$|E_s|^2 = \frac{\eta G_d P_{\text{rad}}}{2\pi r^2} = \frac{120\pi(3.162)(20 \times 10^3)}{2\pi\,[10 \times 10^3]^2}$$

$$|E_s| = 0.1948 \text{ V/m}$$

■

PRACTICE EXERCISE 13.4

A certain antenna with an efficiency of 95% has maximum radiation intensity of 0.5 W/sr. Calculate its directivity when

(a) The input power is 0.4 W

(b) The radiated power is 0.3 W

ANSWER (a) 16.53, (b) 20.94.

EXAMPLE 13.5

The radiation intensity of a certain antenna is

$$U(\theta, \phi) = \begin{cases} 2 \sin \theta \sin^3\phi, & 0 \leq \theta \leq \pi, 0 \leq \phi \leq \pi \\ 0, & \text{elsewhere} \end{cases}$$

Determine the directivity of the antenna.

SOLUTION The directivity is defined as

$$D = \frac{U_{max}}{U_{ave}}$$

From the given U,

$$U_{max} = 2$$

$$\begin{aligned} U_{ave} &= \frac{1}{4\pi} \int U \, d\Omega \; (=P_{rad}/4\pi) \\ &= \frac{1}{4\pi} \int_{\phi=0}^{\pi} \int_{\theta=0}^{\pi} 2 \sin \theta \sin^3\phi \sin \theta \, d\theta \, d\phi \\ &= \frac{1}{2\pi} \int_0^{\pi} \sin^2\theta \, d\theta \int_0^{\pi} \sin^3\phi \, d\phi \\ &= \frac{1}{2\pi} \int_0^{\pi} \frac{1}{2}(1 - \cos 2\theta) \, d\theta \int_0^{\pi} (1 - \cos^2\phi) \, d(-\cos \phi) \\ &= \frac{1}{2\pi} \frac{1}{2} \left(\theta - \frac{\sin 2\theta}{2} \right) \bigg|_0^{\pi} \left(\frac{\cos^3\phi}{3} - \cos \phi \right) \bigg|_0^{\pi} \\ &= \frac{1}{2\pi} \left(\frac{\pi}{2} \right) \left(\frac{4}{3} \right) = \frac{1}{3} \end{aligned}$$

Hence

$$D = \frac{2}{(1/3)} = 6$$

■

PRACTICE EXERCISE 13.5

Evaluate the directivity of an antenna with normalized radiation intensity

$$U(\theta, \phi) = \begin{cases} \sin \theta, \; 0 \leq \theta \leq \pi/2, \; 0 \leq \phi \leq 2\pi \\ 0, \; \text{otherwise} \end{cases}$$

ANSWER 2.546.

13.7 ANTENNA ARRAYS

In many practical applications (e.g., in an AM broadcast station), it is necessary to design antennas with more energy radiated in some particular directions and less in other directions. This is tantamount to requiring that the radiation pattern be concentrated in the direction of interest. This is hardly achievable with a single antenna element. An antenna array is used to obtain greater directivity than can be obtained with a single antenna element. An *antenna array* is a group of radiating elements arranged so as to produce some particular radiation characteristics. It is practical and convenient that the array consists of identical elements but this is not fundamentally required. We shall consider the simplest case of a two-element array and extend our results to the more complicated, general case of an N-element array.

Consider an antenna consisting of two Hertzian dipoles placed in free space along the z-axis but oriented parallel to the x-axis as depicted in Figure 13.10. We assume that the dipole at $(0, 0, d/2)$ carries current $I_{1s} = I_o \underline{/\alpha}$ and the one at $(0, 0, -d/2)$ carries current $I_{2s} = I_o \underline{/0}$, where α is the phase difference between the two currents. By varying the spacing d and phase difference α, the fields from the array can be made to interfere constructively (add) in certain directions of interest and interfere destructively (cancel) in other directions. The total electric field at point P is the vector sum of the fields due to the individual elements. If P is in the far field zone, we obtain the total electric field at P from eq. (13.7a) as

$$\mathbf{E}_s = \mathbf{E}_{1s} + \mathbf{E}_{2s}$$

$$= \frac{j\eta\beta I_o dl}{4\pi}\left[\cos\theta_1 \frac{e^{-j\beta r_1}}{r_1} e^{j\alpha}\,\mathbf{a}_{\theta_1} + \cos\theta_2 \frac{e^{-j\beta r_2}}{r_2}\,\mathbf{a}_{\theta_2}\right] \qquad [13.53]$$

Note that $\sin\theta$ in eq. (13.7a) has been replaced by $\cos\theta$ since the element of Figure 13.3 is z-directed whereas those in Figure 13.10 are x-directed. Since P is far from the array, $\theta_1 \simeq \theta \simeq \theta_2$ and $\mathbf{a}_{\theta_1} \simeq \mathbf{a}_\theta \simeq \mathbf{a}_{\theta_2}$. In the amplitude, we can set $r_1 \simeq r \approx r_2$ but in the phase, we use

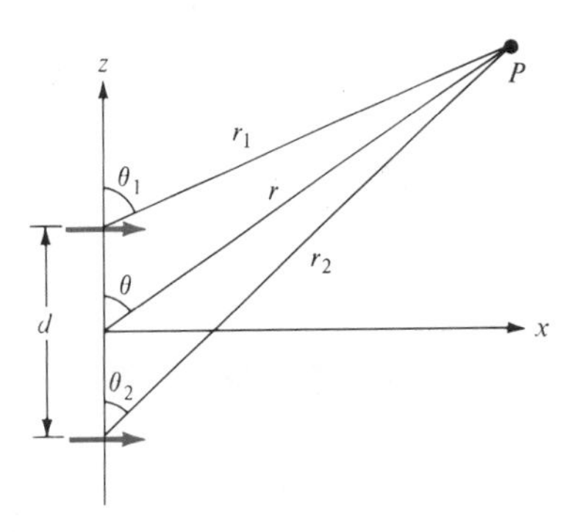

Figure 13.10 A two-element array.

$$r_1 \simeq r - \frac{d}{2}\cos\theta \tag{13.54a}$$

$$r_2 \simeq r + \frac{d}{2}\cos\theta \tag{13.54b}$$

Thus eq. (13.53) becomes

$$\begin{aligned}\mathbf{E}_s &= \frac{j\eta\beta I_o\, dl}{4\pi\, r}\cos\theta\; e^{-j\beta r} e^{j\alpha/2}\left[e^{j(\beta d\cos\theta)/2} e^{j\alpha/2} + e^{-j(\beta d\cos\theta)/2} e^{-j\alpha/2}\right]\mathbf{a}_\theta \\ &= \frac{j\eta\beta I_o\, dl}{4\pi\, r}\cos\theta\; e^{-j\beta r} e^{j\alpha/2} 2\cos\left[\frac{1}{2}(\beta d\cos\theta + \alpha)\right]\mathbf{a}_\theta\end{aligned} \tag{13.55}$$

Comparing this with eq. (13.7a) shows that the total field of an array is equal to the field of single element located at the origin multiplied by an *array factor* given by

$$AF = 2\cos\left[\frac{1}{2}(\beta d\cos\theta + \alpha)\right] e^{j\alpha/2} \tag{13.56}$$

Thus, in general, the far field due to a two-element array is given by

$$\mathbf{E}\text{ (total)} = (\mathbf{E}\text{ due to single element at origin}) \times (\text{array factor}) \tag{13.57}$$

Also, from eq. (13.55), note that $|\cos\theta|$ is the radiation pattern due to a single element whereas the normalized array factor, $|\cos[1/2(\beta d\cos\theta + \alpha)]|$, is the radiation pattern of the array if the elements were isotropic. These may be regarded as "unit pattern" and "group pattern" respectively. Thus the "resultant pattern" is the product of the unit pattern and the group pattern, that is,

$$\text{Resultant pattern} = \text{Unit pattern} \times \text{Group pattern} \tag{13.58}$$

This is known as *pattern multiplication*. It is possible to sketch, almost by inspection, the pattern of an array by pattern multiplication. It is, therefore, a useful tool in the design of an array. We should note that while the unit pattern depends on the type of elements the array is comprised of, the group pattern is independent of the element type so long as the spacing d and phase difference α, and the orientation of the elements remain the same.

Let us now extend the results on the two-element array to the general case of an N-element array shown in Figure 13.11. We assume that the array is *linear* in that the elements are spaced equally along a straight line and lie along the z-axis. Also, we

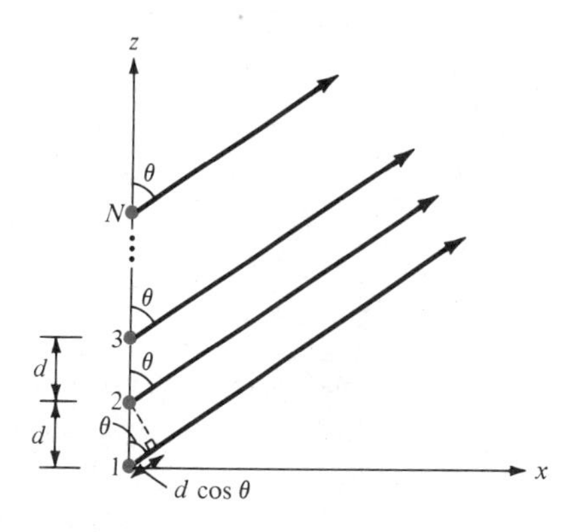

Figure 13.11 An N-element uniform linear array.

assume that the array is *uniform* so that each element is fed with current of the same magnitude but of progressive phase shift α, that is, $I_{1s} = I_o\underline{/0}$, $I_{2s} = I_o\underline{/\alpha}$, $I_{3s} = I_o\underline{/2\alpha}$, and so on. We are mainly interested in finding the array factor; the far field can easily be found from eq. (13.57) once the array factor is known. For the uniform linear array, the array factor is the sum of the contributions by all the elements. Thus,

$$AF = 1 + e^{j\psi} + e^{j2\psi} + e^{j3\psi} + \cdots + e^{j(N-1)\psi} \tag{13.59}$$

where

$$\psi = \beta d \cos \theta + \alpha \tag{13.60}$$

In eq. (13.60), $\beta = 2\pi/\lambda$, d and α are respectively the spacing and interelement phase shift. Notice that the right-hand side of eq. (13.59) is a geometric series of the form

$$1 + x + x^2 + x^3 + \cdots + x^{N-1} = \frac{1 - x^N}{1 - x} \tag{13.61}$$

Hence eq. (13.59) becomes

$$AF = \frac{1 - e^{jN\psi}}{1 - e^{j\psi}} \tag{13.62}$$

which can be written as

$$\begin{aligned} AF &= \frac{e^{jN\psi} - 1}{e^{j\psi} - 1} = \frac{e^{jN\psi/2}}{e^{j\psi/2}} \frac{e^{jN\psi/2} - e^{-jN\psi/2}}{e^{j\psi/2} - e^{-j\psi/2}} \\ &= e^{j(N-1)\psi/2} \frac{\sin (N\psi/2)}{\sin (\psi/2)} \end{aligned} \tag{13.63}$$

The phase factor $e^{j(N-1)\psi/2}$ would not be present if the array were centered about the origin. Neglecting this unimportant term,

$$AF = \frac{\sin \dfrac{N\psi}{2}}{\sin \dfrac{\psi}{2}}, \qquad \psi = \beta d \cos\theta + \alpha \tag{13.64}$$

Note that this equation reduces to eq. (13.56) when $N = 2$ as expected (see Problem 13.25). Also, note the following:

1. AF has the maximum value of N; thus the normalized AF is obtained by dividing AF by N. The principal maximum occurs when $\psi = 0$, that is

$$0 = \beta d \cos\theta + \alpha \quad \text{or} \quad \cos\theta = -\frac{\alpha}{\beta d} \tag{13.65}$$

2. AF has *nulls* (or *zeros*) when $AF = 0$, that is

$$\frac{N\psi}{2} = \pm k\pi, \qquad k = 1, 2, 3, \ldots \tag{13.66}$$

   where k is not a multiple of N.
3. A *broadside* array has its maximum radiation directed normal to the axis of the array, that is, $\psi = 0$, $\theta = 90°$ so that $\alpha = 0$.
4. An *end-fire* array has its maximum radiation directed along the axis of the array, that is, $\psi = 0$, $\theta = \begin{cases} 0 \\ \pi \end{cases}$ so that $\alpha = \begin{cases} -\beta d \\ \beta d \end{cases}$

These points are helpful in plotting AF. For $N = 2$, 3, and 4, the plots of AF are sketched in Figure 13.12.

EXAMPLE 13.6

For the two-element antenna array of Figure 13.10, sketch the normalized field pattern when the currents are:

(a) Fed in phase ($\alpha = 0$), $d = \lambda/2$

(b) Fed 90° out of phase ($\alpha = \pi/2$), $d = \lambda/4$

SOLUTION The normalized field of the array is obtained from eqs. (13.55) to (13.57) as

$$f(\theta) = \left| \cos\theta \cos\left[\frac{1}{2}(\beta d \cos\theta + \alpha)\right] \right|$$

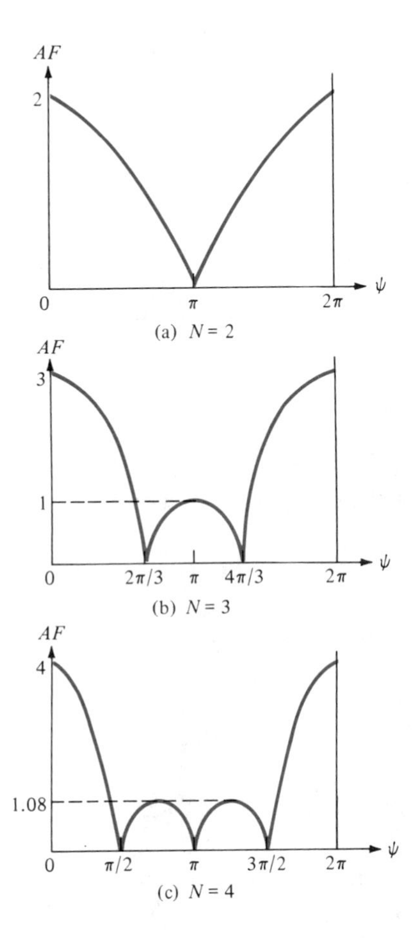

Figure 13.12 Array factor for uniform linear array.

(a) If $\alpha = 0$, $d = \lambda/2$, $\beta d = \dfrac{2\pi}{\lambda}\dfrac{\lambda}{2} = \pi$. Hence,

$$f(\theta) = |\cos\theta| \quad \left|\cos\left(\frac{\pi}{2}\cos\theta\right)\right|$$

$$\downarrow \qquad \downarrow \qquad \downarrow$$

$$\text{resultant pattern} = \text{unit pattern} \times \text{group pattern}$$

The sketch of the unit pattern is straightforward. It is merely a rotated version of that in Figure 13.7(a) for the Hertzian dipole and is shown in Figure 13.13(a). To

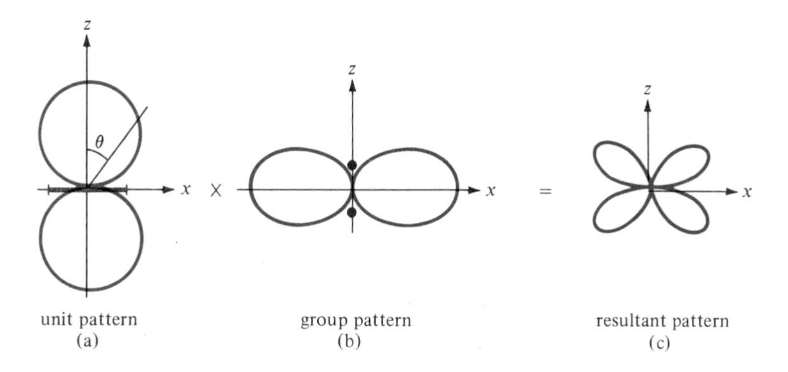

Figure 13.13 For Example 13.6(a); field patterns in the plane containing the axes of the elements.

sketch a group pattern requires that we first determine its nulls and maxima. For the nulls (or zeros),

$$\cos\left(\frac{\pi}{2}\cos\theta\right) = 0 \rightarrow \frac{\pi}{2}\cos\theta = \pm\frac{\pi}{2}, \pm\frac{3\pi}{2}, \ldots$$

or

$$\theta = 0°, 180°$$

For the maxima,

$$\cos\left(\frac{\pi}{2}\cos\theta\right) = 1 \rightarrow \cos\theta = 0$$

or

$$\theta = 90°$$

The group pattern is as shown in Figure 13.12(b). It is the polar plot obtained by sketching $\left|\cos\left(\frac{\pi}{2}\cos\theta\right)\right|$ for $\theta = 0°, 5°, 10°, 15°, \ldots, 360°$ and incorporating the nulls and maxima at $\theta = 0°, 180°$ and $\theta = 90°$ respectively. Multiplying Figure 13.13(a) with Figure 13.13(b) gives the resultant pattern in Figure 13.13(c). It should be observed that the field patterns in Figure 13.13 are in the plane containing the axes of the elements. Note that: (1) In the *yz*-plane, which is normal to the axes of the elements, the unit pattern (= 1) is a circle (see Figure 13.7(b)) while the group pattern remains as in Figure 13.13(b); therefore, the resultant pattern is the same as the group pattern in this case. (2) In the *xy*-plane, $\theta = \pi/2$, so the unit pattern vanishes while the group pattern (= 1) is a circle.

(b) If $\alpha = \pi/2$, $d = \lambda/4$, and $\beta d = \dfrac{2\pi}{\lambda}\dfrac{\lambda}{4} = \dfrac{\pi}{2}$

$$
\begin{array}{ccccc}
f(\theta) & = & |\cos\theta| & & \left|\cos\left(\dfrac{\pi}{4}\cos\theta + 1\right)\right| \\
\downarrow & & \downarrow & & \downarrow \\
\text{resultant pattern} & = & \text{unit pattern} & \times & \text{group pattern}
\end{array}
$$

The unit pattern remains as in Figure 13.13(a). For the group pattern, the null occurs when

$$\cos\frac{\pi}{4}(1 + \cos\theta) = 0 \rightarrow \frac{\pi}{4}(1 + \cos\theta) = \pm\frac{\pi}{2}, \pm\frac{3\pi}{2}, \ldots$$

or

$$\cos\theta = 1 \rightarrow \theta = 0$$

The maxima and minima occur when

$$\frac{d}{d\theta}\left[\cos\frac{\pi}{4}(1 + \cos\theta)\right] = 0 \rightarrow \sin\theta\sin\frac{\pi}{4}(1 + \cos\theta) = 0$$

$$\sin\theta = 0 \rightarrow \theta = 0^\circ, 180^\circ$$

and

$$\sin\frac{\pi}{4}(1 + \cos\theta) = 0 \rightarrow \cos\theta = -1 \quad \text{or} \quad \theta = 180^\circ$$

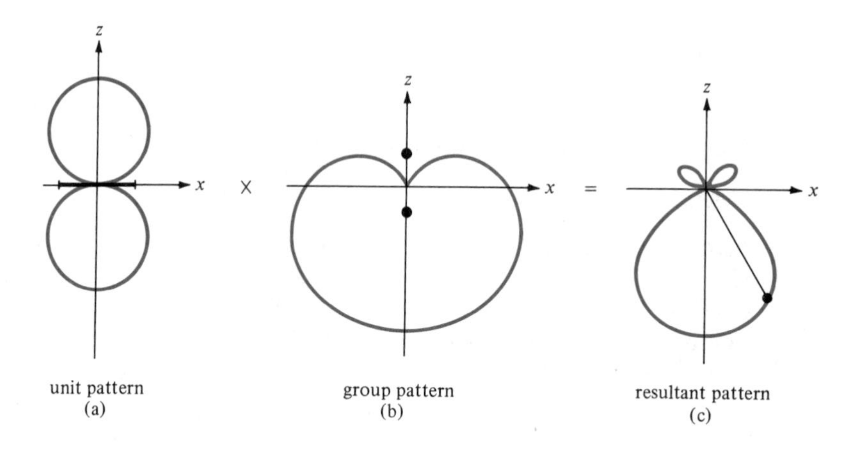

Figure 13.14 For Example 13.6(b); field patterns in the plane containing the axes of the elements.

Note that $\theta = 180°$ corresponds to the maximum value of AF whereas $\theta = 0°$ corresponds to the null. Thus the unit, group, and resultant patterns in the plane containing the axes of the elements are shown in Figure 13.14. Observe from the group patterns that the broadside array ($\alpha = 0$) in Figure 13.13 is bidirectional while the end-fire array ($\alpha = \beta d$) in Figure 13.14 is unidirectional. ■

PRACTICE EXERCISE 13.6

Repeat Example 13.6 for cases when:

(a) $\alpha = \pi$, $d = \lambda/2$, (b) $\alpha = -\pi/2$, $d = \lambda/4$.

ANSWER See Figure 13.15.

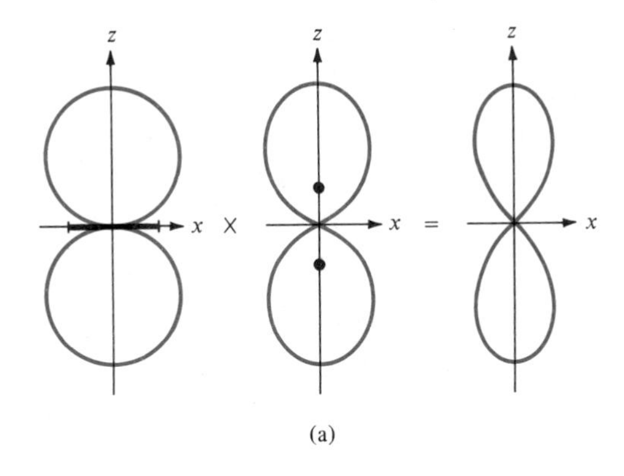

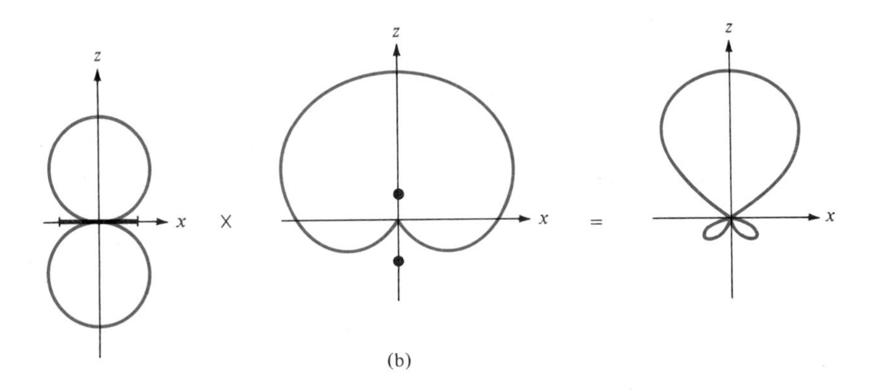

Figure 13.15 For Practice Exercise 13.6.

EXAMPLE 13.7

Consider a three-element array that has current ratios 1:2:1 as in Figure 13.16(a). Sketch the group pattern in the plane containing the axes of the elements.

SOLUTION For the purpose of analysis, we split the middle element in Figure 13.16(a) carrying current $2I \underline{/0°}$ into two elements each carrying current $I \underline{/0°}$. This results in four elements instead of three as shown in Figure 13.16(b). If we consider elements 1 and 2 as a group and elements 3 and 4 as another group, we have a two-element array of Figure 13.16(c). Each group is a two-element array with $d = \lambda/2$, $\alpha = 0$, that the group pattern of the two-element array (or the unit pattern for the three-element array) is as shown in Figure 13.13(b). The two groups form a two-element array similar to Example 13.6(a) with $d = \lambda/2$, $\alpha = 0$, so the group pattern is the same as that in Figure 13.13(b). Thus, in this case, both the unit and group patterns are the same pattern in Figure 13.13(b). The resultant group pattern is obtained in Figure 13.17(c). We should note that the pattern in Figure 13.17(c) is not the resultant pattern but the group pattern of the three-element array. The resultant group pattern of the array is Figure 13.17(c) multiplied by the field pattern of the element type.

An alternative method of obtaining the resultant group pattern of the three-element array of Figure 13.16 is following similar steps taken to obtain eq. (13.59). We obtain the normalized array factor (or the group pattern) as

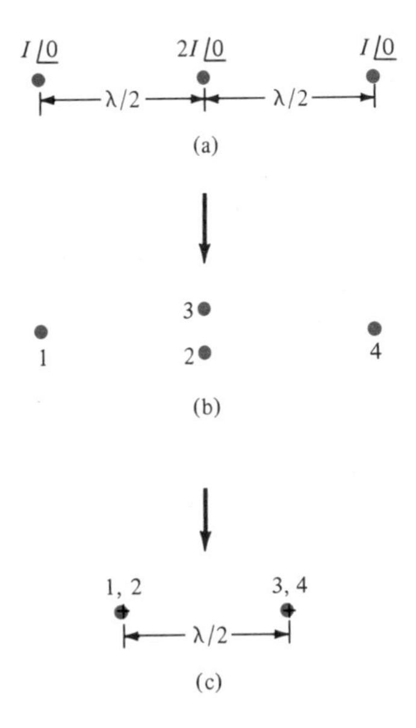

Figure 13.16 For Example 13.7: **(a)** a three-element array with current ratios 1:2:1; **(b)** and **(c)** equivalent two-element arrays.

$$(AF)_n = \frac{1}{4}|1 + 2e^{j\psi} + e^{j2\psi}|$$

$$= \frac{1}{4}|e^{j\psi}||2 + e^{-j\psi} + e^{j\psi}|$$

$$= \frac{1}{2}|1 + \cos\psi| = \left|\cos\frac{\psi}{2}\right|^2$$

where $\psi = \beta d \cos\theta + \alpha$ if the elements are placed along the z-axis but oriented parallel to the x-axis. Since $\alpha = 0$, $d = \lambda/2$, $\beta d = \frac{2\pi}{\lambda} \cdot \frac{\lambda}{2} = \pi$,

$$(AF)_n = \left|\cos\left(\frac{\pi}{2}\cos\theta\right)\right|^2$$

$$(AF)_n = \left|\cos\left(\frac{\pi}{2}\cos\theta\right)\right| \quad \left|\cos\left(\frac{\pi}{2}\cos\theta\right)\right|$$

↓	↓		↓
resultant group pattern	unit pattern	×	group pattern

The sketch of these patterns is exactly what is in Figure 13.17.

If two three-element arrays in Figure 13.16(a) are displaced by $\lambda/2$, we obtain a four-element array with current ratios 1:3:3:1 as in Figure 13.18. Two of such four-element arrays, displaced by $\lambda/2$, give a five-element array with current ratios 1:4:6:4:1. Continuing this process results in an N-element array, spaced $\lambda/2$ and $(N-1)\lambda/2$ long, whose current ratios are the binomial coefficients. Such an array is called a linear *binomial array*. ■

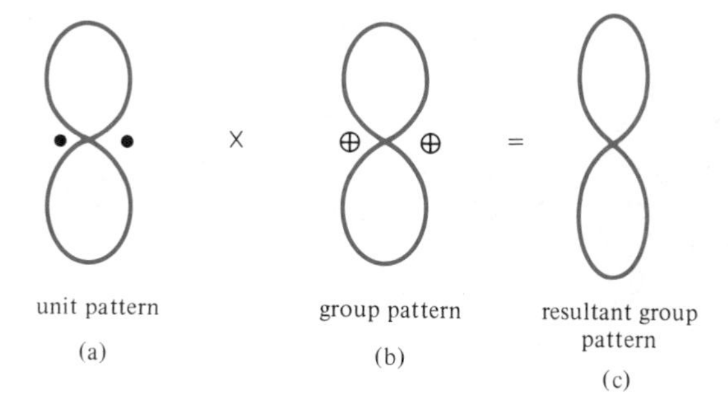

Figure 13.17 For Example 13.7; obtaining the resultant group pattern of the three-element array of Figure 13.16(a).

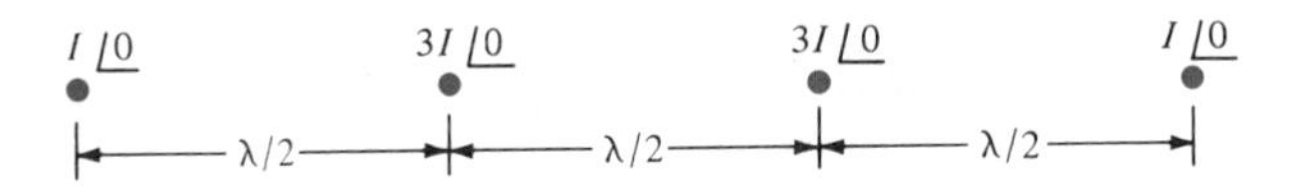

Figure 13.18 A four-element array with current ratios 1:3:3:1; for Practice Exercise 13.7.

PRACTICE EXERCISE 13.7

(a) Sketch the resultant group pattern for the four-element array with current ratios 1:3:3:1 shown in Figure 13.18.
(b) Derive an expression for the group pattern of a linear binomial array of N elements. Assume that the elements are placed along the z-axis, oriented parallel to the x-axis with spacing d and inter-element phase shift α.

ANSWER (a) See Figure 13.19, (b) $\left|\cos \frac{\psi}{2}\right|^{N-1}$, where $\psi = \beta d \cos \theta + \alpha$.

†13.8 EFFECTIVE AREA AND THE FRIIS EQUATION

In a situation where the incoming EM wave is normal to the entire surface of a receiving antenna, the power received is

$$P_r = \int \mathcal{P}_{\text{ave}} \cdot d\mathbf{S} = \mathcal{P}_{\text{ave}} S \qquad [13.67]$$

But in most cases, the incoming EM wave is not normal to the entire surface of the antenna. This necessitates the idea of the effective area of a receiving antenna.

The concept of effective area or effective aperture (receiving cross section of an antenna) is usually employed in the analysis of receiving antennas. We define the *effective area* A_e of a receiving antenna as the ratio of the time-average power received

Figure 13.19 For Practice Exercise 13.7(a).

P_r (or delivered to the load, to be strict) to the time-average power density $\mathcal{P}_{\text{ave}}$ of the incident wave at the antenna, that is

$$A_e = \frac{P_r}{\mathcal{P}_{\text{ave}}} \tag{13.68}$$

From eq. (13.67), we notice that the effective area is a measure of the ability of the antenna to extract energy from a passing EM wave.

Let us derive the formula for calculating the effective area of the Hertzian dipole acting as a receiving antenna. The Thevenin equivalent circuit for the receiving antenna is shown in Figure 13.20 where V_{oc} is the open-circuit voltage induced on the antenna terminals, $Z_{\text{in}} = R_{\text{rad}} + jX_{\text{in}}$ is the antenna impedance, and $Z_L = R_L + jX_L$ is the external load impedance, which might be the input impedance to the transmission line feeding the antenna. For maximum power transfer, $Z_L = Z_{\text{in}}^*$ and $X_L = -X_{\text{in}}$. The time-average power delivered to the matched load is therefore

$$P_r = \frac{1}{2}\left[\frac{|V_{\text{oc}}|}{2R_{\text{rad}}}\right]^2 R_{\text{rad}} = \frac{|V_{\text{oc}}|^2}{8\,R_{\text{rad}}} \tag{13.69}$$

For the Hertzian dipole, $R_{\text{rad}} = 80\pi^2(dl/\lambda)^2$ and $V_{\text{oc}} = E\,dl$ where E is the effective field strength parallel to the dipole axis. Hence, eq. (13.69) becomes

$$P_r = \frac{E^2\lambda^2}{640\pi^2} \tag{13.70}$$

The time-average power at the antenna is

$$\mathcal{P}_{\text{ave}} = \frac{E^2}{2\eta_o} = \frac{E^2}{240\pi} \tag{13.71}$$

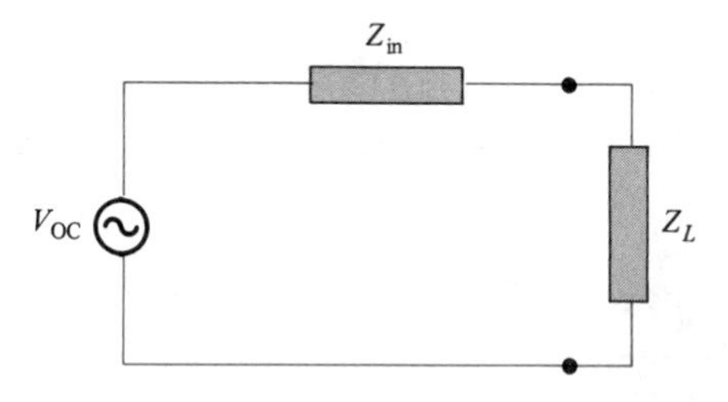

Figure 13.20 Thevenin equivalent of a receiving antenna.

Inserting eqs. (13.70) and (13.71) in eq. (13.68) gives

$$A_e = \frac{3\lambda^2}{8\pi} = 1.5\,\frac{\lambda^2}{4\pi}$$

or

$$A_e = \frac{\lambda^2}{4\pi} D \quad [13.72]$$

where $D = 1.5$ is the directivity of the Hertzian dipole. Although eq. (13.72) was derived for the Hertzian dipole, it holds for any antenna if D is replaced by $G_d(\theta, \phi)$. Thus, in general

$$\boxed{A_e = \frac{\lambda^2}{4\pi} G_d(\theta, \phi)} \quad [13.73]$$

Now suppose we have two antennas separated by distance r in free space as shown in Figure 13.21. The transmitting antenna has effective area A_{et} and directive gain G_{dt}, and transmits a total power P_t ($= P_{\text{rad}}$). The receiving antenna has effective area of A_{er} and directive gain G_{dr}, and receives a total power of P_r. Assuming that the antennas are lossless, the time-average power density at the receiving antenna is given by eq. (13.43) as

$$\mathcal{P}_{\text{ave}} = \frac{|E_s|^2}{2\eta_o} = \frac{P_t}{4\pi r^2} G_{dt} \quad [13.74]$$

By applying eqs. (13.68) and (13.73), we obtain the time-average power received as

$$P_r = \mathcal{P}_{\text{ave}} A_{er} = \frac{\lambda^2}{4\pi} G_{dr} \mathcal{P}_{\text{ave}} \quad [13.75]$$

Substituting eq. (13.74) into eq. (13.75) results in

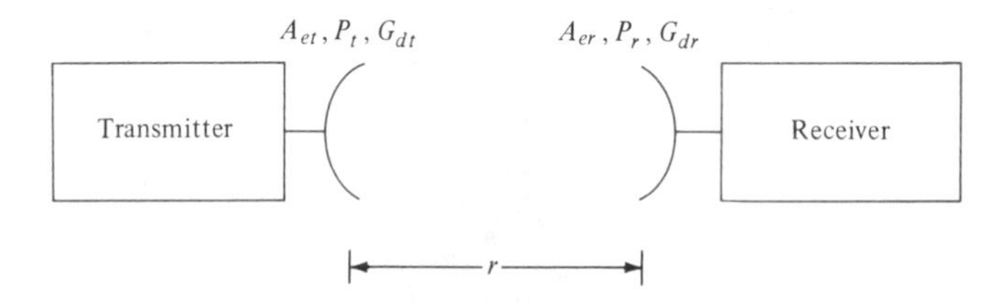

Figure 13.21 Transmitting and receiving antennas in free space.

$$P_r = G_{dr}G_{dt}\left[\frac{\lambda}{4\pi r}\right]^2 P_t \qquad [13.76]$$

This is referred to as the *Friis transmission formula*. It relates the power received by one antenna to the power transmitted by the other, provided that the two antennas are separated by $r > 2d^2/\lambda$, where d is the largest dimension of either antenna (see eq. (13.52)). Therefore, in order to apply the Friis equation, we must make sure that the two antennas are in the far field of each other.

†13.9 THE RADAR EQUATION

Radars are electromagnetic devices used for detection and location of objects. The term *radar* is derived from the phrase *ra*dio *d*etection *a*nd *r*anging. In a typical radar system shown in Figure 13.22(a), pulses of EM energy are transmitted to a distant object. The same antenna is used for transmitting and receiving, so the time interval between the transmitted and reflected pulses is used to determine the distance of the target. If r is the distance between the radar and target and c is the speed of light, the elapsed time between the transmitted and received pulse is $2r/c$. By measuring the elapsed time, r is determined.

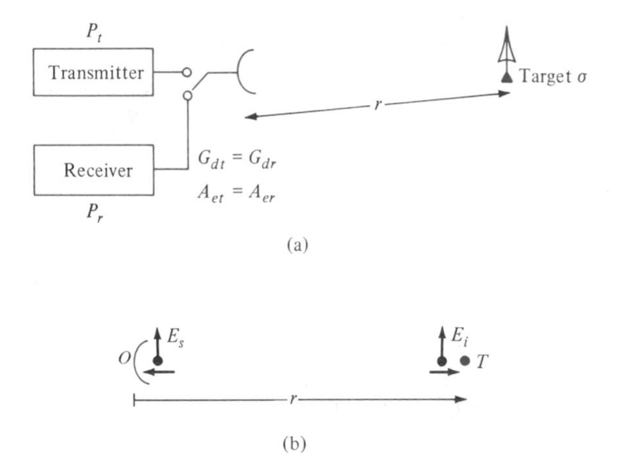

Figure 13.22 **(a)** Typical radar system, **(b)** simplification of the system in (a) for calculating the target cross section σ.

The ability of the target to scatter (or reflect) energy is characterized by the *scattering cross section* σ (also called the *radar cross section*) of the target. The scattering cross section has the units of area and can be measured experimentally. It is defined as the equivalent area intercepting that amount of power which, when scattering isotropically, produces at the radar a power density which is equal to that scattered (or reflected) by the actual target. That is,

$$\mathcal{P}_s = \lim_{r \to \infty} \left[\frac{\sigma \mathcal{P}_i}{4\pi r^2} \right]$$

or

$$\sigma = \lim_{r \to \infty} 4\pi r^2 \frac{\mathcal{P}_s}{\mathcal{P}_i} \tag{13.77}$$

where $\mathcal{P}_i$ is the incident power density at the target T while $\mathcal{P}_s$ is the scattered power density at the transreceiver O as in Figure 13.22(b).

From eq. (13.43), the incident power density $\mathcal{P}_i$ at the target T is

$$\mathcal{P}_i = \mathcal{P}_{\text{ave}} = \frac{G_d}{4\pi r^2} P_{\text{rad}} \tag{13.78}$$

The power received at transreceiver O is

$$P_r = A_{er}\, \mathcal{P}_s$$

or

$$\mathcal{P}_s = \frac{P_r}{A_{er}} \tag{13.79}$$

Note that $\mathcal{P}_i$ and $\mathcal{P}_s$ are the time-average power densities in watts/m^2 and P_{rad} and P_r are the total time-average powers in watts. Since $G_{dr} = G_{dt} = G_d$ and $A_{er} = A_{et} = A_e$, substituting eqs. (13.78) and (13.79) into eq. (13.77) gives

$$\sigma = (4\pi r^2)^2 \frac{P_r}{P_{\text{rad}}} \frac{1}{A_{er}\, G_d} \tag{13.80a}$$

or

$$P_r = \frac{A_e \sigma G_d P_{\text{rad}}}{(4\pi r^2)^2} \tag{13.80b}$$

From eq. (13.73), $A_e = \lambda^2 G_d / 4\pi$. Hence,

$$\boxed{P_r = \frac{(\lambda G_d)^2 \sigma P_{\text{rad}}}{(4\pi)^3 r^4}} \tag{13.81}$$

This is the *radar transmission equation* for free space. It is the basis for measurement of scattering cross section of a target. Solving for r in eq. (13.81) results in

$$r = \left[\frac{\lambda^2 G_d^2 \sigma}{(4\pi)^3} \cdot \frac{P_{\text{rad}}}{P_r} \right]^{1/4} \qquad [13.82]$$

Equation (13.82) is called the *radar range equation*. Given the minimum detectable power of the receiver, the equation determines the maximum range for a radar. It is also useful for obtaining engineering information concerning the effects of the various parameters on the performance of a radar system.

The radar considered so far is the *monostatic* type because of the predominance of this type of radar in practical applications. A *bistatic radar* is one in which the transmitter and receiver are separated. If the transmitting and receiving antennas are at distances r_1 and r_2 from the target and $G_{dr} \neq G_{dt}$, eq. (13.81) for bistatic radar becomes

$$P_r = \frac{G_{dt} G_{dr}}{4\pi} \left[\frac{\lambda}{4\pi r_1 r_2} \right]^2 \sigma P_{\text{rad}} \qquad [13.83]$$

Radar transmission frequencies range from 25 to 70,000 MHz. Table 13.1 shows radar frequencies and their designations as commonly used by radar engineers.

Table 13.1 **Designations of Radar Frequencies**

Designation	Frequency
UHF	300–1000 MHz
L	1000–2000 MHz
S	2000–4000 MHz
C	4000–8000 MHz
X	8000–12,500 MHz
Ku	12.5–18 GHz
K	18–26.5 GHz
Millimeter	>35 GHz

EXAMPLE 13.8

The transmitting and receiving antennas are separated by a distance of 200 λ and have directive gains of 25 and 18 dB respectively. If 5 mW of power is to be received, calculate the minimum transmitted power.

SOLUTION Given that G_{dt} (dB) $= 25$ dB $= 10 \log_{10} G_{dt}$,

$$G_{dt} = 10^{2.5} = 316.23$$

Similarly,

$$G_{dr}\ (dB) = 18 \text{ db} \qquad \text{or} \qquad G_{dr} = 10^{1.8} = 63.1$$

Using the Friis equation, we have

$$P_r = G_{dr} G_{dt} \left[\frac{\lambda}{4\pi r} \right]^2 P_t$$

or

$$P_t = P_r \left[\frac{4\pi r}{\lambda} \right]^2 \frac{1}{G_{dr} G_{dt}}$$

$$= 5 \times 10^{-3} \left[\frac{4\pi \times 200\ \lambda}{\lambda} \right]^2 \frac{1}{(63.1)(316.23)}$$

$$= 1.583 \text{ W}$$

PRACTICE EXERCISE 13.8

An antenna in air radiates a total power of 100 kW so that a maximum radiated electric field strength of 12 mV/m is measured 20 km from the antenna. Find: (a) its directivity in dB, (b) its maximum power gain if $\eta_r = 98$ percent.

ANSWER (a) 3.34 dB, (b) 2.117.

EXAMPLE 13.9

An S-band radar transmitting at 3 GHz radiates 200 kW. Determine the signal power density at ranges 100 and 400 nautical miles if the effective area of the radar antenna is 9 m^2. With a 20-m^2 target at 300 nautical miles, calculate the power of the reflected signal at the radar.

SOLUTION The nautical mile is a common unit in radar communications.

$$1 \text{ nautical mile (nm)} = 1852 \text{ m}$$

$$\lambda = \frac{c}{f} = \frac{3 \times 10^8}{3 \times 10^9} = 0.1 \text{ m}$$

$$G_{dt} = \frac{4\pi}{\lambda^2} A_{et} = \frac{4\pi}{(0.1)^2} 9 = 3600\pi$$

For $r = 100 \text{ nm} = 1.852 \times 10^5$ m

$$\mathcal{P} = \frac{G_{dt} P_{\text{rad}}}{4\pi r^2} = \frac{3600\pi \times 200 \times 10^3}{4\pi\,(1.852)^2 \times 10^{10}}$$

$$= 5.248 \text{ mW/m}^2$$

For $r = 400 \text{ nm} = 4\,(1.852 \times 10^5)$ m

$$\mathcal{P} = \frac{5.248}{(4)^2} = 0.328 \text{ mW/m}^2$$

Using eq. (13.80b)

$$P_r = \frac{A_e \sigma\, G_d\, P_{\text{rad}}}{[4\pi r^2]^2}$$

where $r = 300 \text{ nm} = 5.556 \times 10^5$ m

$$P_r = \frac{9 \times 20 \times 3600\pi \times 200 \times 10^3}{[4\pi \times 5.556^2]^2 \times 10^{20}} = 2.706 \times 10^{-14} \text{ W}$$

The same result can be obtained using eq. (13.81). ■

PRACTICE EXERCISE 13.9

A C-band radar with an antenna 1.8 m in radius transmits 60 kW at a frequency of 6000 MHz. If the minimum detectable power is 0.26 mW, for a target cross section of 5 m^2, calculate the maximum range in nautical miles and the signal power density at half this range. Assume unity efficiency and that the effective area of the antenna is 70 percent of the actual area.

ANSWER 0.6309 nm, 500.90 W/m^2.

SUMMARY

1. We have discussed the fundamental ideas and definitions in antenna theory. The basic types of antenna considered include the Hertzian (or differential length) dipole, the half-wave dipole, the quarter-wave monopole, and the small loop.

2. Theoretically, if we know the current distribution on an antenna, we can find the retarded magnetic vector potential **A**, and from it we can find the retarded electromagnetic fields **H** and **E** using

$$\mathbf{H} = \nabla \times \frac{\mathbf{A}}{\mu}, \qquad \mathbf{E} = \eta\, \mathbf{H} \times \mathbf{a}_k$$

The far-zone fields are obtained by retaining only $1/r$ terms.
3. The analysis of the Hertzian dipole serves as a stepping-stone for other antennas. The radiation resistance of the dipole is very small. This limits the practical usefulness of the Hertzian dipole.
4. The half-wave dipole has a length equal to $\lambda/2$. It is more popular and of more practical use than the Hertzian dipole. Its input impedance is $73 + j42.5\ \Omega$.
5. The quarter-wave monopole is essentially half a half-wave dipole placed on a conducting plane.
6. The radiation patterns commonly used are the field intensity, power intensity, and radiation intensity patterns. The field pattern is usually a plot of $|E_s|$ or its normalized form $f(\theta)$. The power pattern is the plot of $\mathcal{P}_{\text{ave}}$ or its normalized form $f^2(\theta)$.
7. The directive gain is the ratio of $U(\theta, \phi)$ to its average value. The directivity is the maximum value of the directive gain.
8. An antenna array is a group of radiating elements arranged so as to produce some particular radiation characteristics. Its radiation pattern is obtained by multiplying the unit pattern (due to a single element in the group) with the group pattern, which is the plot of the normalized array factor. For an N-element linear uniform array,

$$AF = \left| \frac{\sin(N\psi/2)}{\sin(\psi/2)} \right|$$

where $\psi = \beta d \cos\theta + \alpha$, $\beta = 2\pi/\lambda$, d = spacing between the elements, and α = inter-element phase shift.
9. The Friis transmission formula characterizes the coupling between two antennas in terms of their directive gains, separation distance, and frequency of operation.
10. For a bistatic radar (one in which the transmitting and receiving antennas are separated), the power received is given by

$$P_r = \frac{G_{dt}G_{dr}}{4\pi}\left[\frac{\lambda}{4\pi r_1 r_2}\right]^2 \sigma\, P_{\text{rad}}$$

For a monostatic radar, $r_1 = r_2 = r$ and $G_{dt} = G_{dr}$.

REVIEW QUESTIONS

13.1 An antenna located in a city is a source of radio waves. How much time does it take the wave to reach a town 12,000 km away from the city?

(a) 36 s

(b) 20 μs

(c) 20 ms

(d) 40 ms

(e) None of the above

13.2 In eq. (13.34), which term is the radiation term?

(a) $1/r$ term

(b) $1/r^2$ term

(c) $1/r^3$ term

(d) All of the above

13.3 A very small thin wire of length $\lambda/100$ has a radiation resistance of

(a) $\simeq 0\ \Omega$

(b) $0.08\ \Omega$

(c) $7.9\ \Omega$

(d) $790\ \Omega$

13.4 A quarter-wave monopole antenna operating in air at frequency 1 MHz must have an overall length of

(a) $\ell \gg \lambda$

(b) 300 m

(c) 150 m

(d) 75 m

(e) $\ell \ll \lambda$

13.5 If a small single-turn loop antenna has a radiation resistance of $0.04\ \Omega$, how many turns are needed to produce a radiation resistance of $1\ \Omega$?

(a) 150

(b) 125

(c) 50

(d) 25

(e) 5

13.6 At a distance of 8 km from a differential antenna, the field strength is 12 μV/m. The field strength at a location 20 km from the antenna is

(a) 75 μV/m

(b) 30 μV/m

(c) 4.8 μV/m

(d) 1.92 μV/m

13.7 An antenna has $U_{max} = 10$ W/sr, $U_{ave} = 4.5$ W/sr, and $\eta_r = 95$ percent. The input power to the antenna is

(a) 2.222 W

(b) 12.11 W

(c) 55.55 W

(d) 59.52 W

13.8 A receiving antenna in an airport has a maximum dimension of 3 m and operates at 100 MHz. An aircraft approaching the airport is 1/2 km from the antenna. The aircraft is in the far field region of the antenna.

(a) True

(b) False

13.9 A receiving antenna is located 100 m away from the transmitting antenna. If the effective area of the receiving antenna is 500 cm^2 and the power density at the receiving location is 2 mW/m^2, the total power received is:

(a) 10 nW

(b) 100 nW

(c) 1 μW

(d) 10 μW

(e) 100 μW

13.10 Let R be the maximum range of a monostatic radar. If a target with radar cross section of 5 m^2 exists at $R/2$, what should be the target cross section at $3R/2$ to result in an equal signal strength at the radar?

(a) 0.0617 m^2

(b) 0.555 m^2

(c) 15 m^2

(d) 45 m^2

(e) 405 m^2

Answers: 13.1d, 13.2a, 13.3b, 13.4d, 13.5e, 13.6c, 13.7d, 13.8a, 13.9e, 13.10e.

PROBLEMS

13.1 The magnetic vector potential at point $P(r, \theta, \phi)$ due to a small antenna located at the origin is given by

$$\mathbf{A}_s = \frac{50\, e^{-j\beta r}}{r}\, \mathbf{a}_x$$

where $r^2 = x^2 + y^2 + z^2$. Find $\mathbf{E}(r, \theta\ \phi, t)$ and $\mathbf{H}(r, \theta, \phi, t)$ at the far field.

13.2 Show that eq. (13.3) is a solution of the wave equation $\nabla^2 A_{zs} + k^2 A_{zs} = 0$. Determine the value of k.

13.3 An electric dipole of length $\lambda/15$ in free space is to operate under lake water, a nonmagnetic medium with $\varepsilon = 81\varepsilon_o$. Assuming that the lake water is lossless and infinite in extent, calculate the radiation resistance of the antenna in the medium.

13.4 An electric dipole of length 1 m is operated in air. Calculate the time-average power radiated by the antenna assuming that it is excited at its center by a sinusoidal current of amplitude 200 mA at a frequency

(a) In the AM broadcast band, $f = 1.5$ MHz

(b) In the FM/TV band, $f = 150$ MHz

13.5 A half-wave dipole antenna operating at 2 MHz in free space has a current of 60 A at its center. On a plane perpendicular to the antenna ($\theta = \pi/2$), calculate

(a) The E-field 100 m from the antenna

(b) The H-field 300 m from it

13.6 At a distance 10 km from a half-wave dipole antenna in air, the maximum electric field strength is 5 V/m. Calculate

(a) The maximum electric field strength 20 km from the antenna

(b) The maximum magnetic field strength 30 km from it

(c) The maximum average power density 40 km from it

13.7 (a) Instead of a constant current distribution assumed for the short dipole of Section 13.2, assume a triangular current distribution $I_s = I_o \left(1 - \frac{2|z|}{\ell}\right)$ shown in Figure 13.23. Show that

$$R_{\text{rad}} = 20\, \pi^2 \left[\frac{\ell}{\lambda}\right]^2$$

which is one-fourth of that in eq. (13.13). Thus R_{rad} depends on the current distribution.

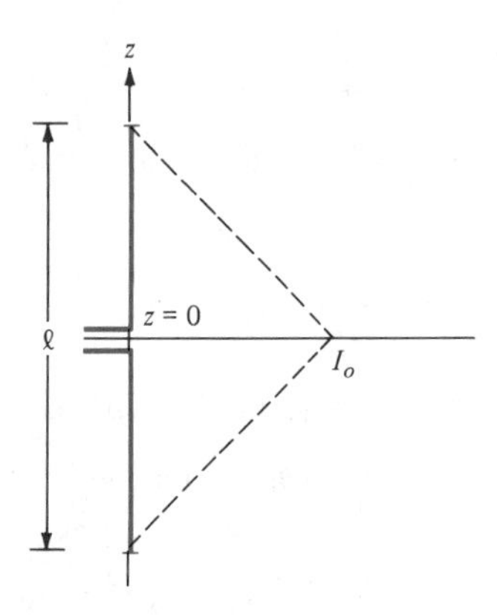

Figure 13.23 Short dipole antenna with triangular current distribution; for Problem 13.7.

(b) Calculate the length of the dipole that will result in a radiation resistance of 0.5 Ω.

***13.8** (a) Show that the generated far field expressions for a thin dipole of length ℓ carrying sinusoidal current $I_o \cos \beta z$ are

$$H_{\phi s} = \frac{jI_o e^{-\beta r}}{2\pi r} \frac{\cos\left(\dfrac{\beta\ell}{2}\cos\theta\right) - \cos\dfrac{\beta\ell}{2}}{\sin\theta}, \qquad E_{\theta s} = \eta H_{\phi s}$$

(*Hint*: Use Figure 13.4 and start with eq. (13.14).)

(b) On a polar coordinate sheet, plot $f(\theta)$ in part (a) for $\ell = \lambda$, $3\lambda/2$ and 2λ

13.9 A dipole antenna is located at the origin and fed with a current $5 \cos 8\pi \times 10^8 t$. If the antenna radiates in free space and is one-quarter wavelength long, calculate $E_{\theta s}$ at

(a) (10 m, 30°, 60°)

(b) (100 m, 60°, 30°)

***13.10** For Problem 13.7:

(a) Determine $\mathbf{E}_s$ and $\mathbf{H}_s$ at the far field

(b) Calculate the directivity of the dipole

***13.11** An antenna located on the surface of a flat earth transmits an average power of 200 kW. Assuming that all the power is radiated uniformly over the surface of a hemisphere with the antenna at the center, calculate (a) the time-average Poynting vector at 50 km, and (b) the maximum electric field at that location.

13.12 A circular loop antenna of radius 10 cm is made of thin copper wire. If the antenna is operating at 100 MHz, find its radiation resistance assuming that

(a) It has a single turn

(b) It has 20 turns

13.13 A 100-turn loop antenna of radius 20 cm operating at 100 MHz in air is to give a 50 mV/m field strength at a distance 3 m from the loop. Determine

(a) The current that must be fed to the antenna

(b) The average power radiated by the antenna

13.14 A small circular loop antenna of radius $\rho_o = 0.025\ \lambda$ is to be designed to provide a radiation resistance of 36 Ω. How many turns are needed?

13.15 Sketch the normalized E-field and H-field patterns for

(a) A half-wave dipole

(b) A quarter-wave monopole

13.16 Based on the result of Problem 13.8, plot the vertical field patterns of monopole antennas of lengths $\ell = 3\lambda/2, \lambda, 5\lambda/8$. Note that a $5\lambda/8$ monopole is often used in practice.

13.17 At the far field, the electric field produced by an antenna is

$$\mathbf{E}_s = \frac{10}{r} e^{-j\beta r} \cos\theta \cos\phi\ \mathbf{a}_z$$

Sketch the vertical pattern of the antenna. Your plot should include as many points as possible.

13.18 For an Hertzian dipole, show that the time-average power density is related to the radiation power according to

$$P_{\text{ave}} = \frac{1.5 \sin^2\theta}{4\pi r^2} P_{\text{rad}}$$

13.19 At $\theta = 40°$ and $\theta = 60°$, determine the directive gain of

(a) A Hertzian dipole

(b) A half-wave dipole

(c) A small loop antenna

13.20 At the far field, an antenna produces

$$P_{\text{ave}} = \frac{2 \sin\theta \cos\phi}{r^2} \mathbf{a}_r \text{ W/m}^2, \qquad 0 < \theta < \pi,\ 0 < \phi < \pi/2$$

Calculate the directive gain and the directivity of the antenna.

13.21 Determine the directivity of an antenna whose normalized intensity is given by

(a) $U(\theta, \phi) = \cos\theta,\ 0 \leq \theta \leq \pi/2,\ 0 \leq \phi \leq 2\pi$

(b) $U(\theta, \phi) = \sin\theta \sin^2\phi,\ 0 \leq \theta \leq \pi,\ 0 \leq \phi \leq \pi$

(c) $U(\theta, \phi) = \sin^2\theta \sin\phi,\ 0 \leq \theta \leq \pi,\ 0 \leq \phi \leq \pi$

13.22 Find U_{max} and U_{ave} if

(a) $U(\theta, \phi) = \sin^2\theta \sin^2\phi,\ 0 \leq \theta \leq \pi,\ 0 \leq \phi \leq 2\pi$

(b) $U(\phi, \phi) = 10 \sin^2\theta \sin^3\phi,\ 0 \leq \theta \leq \pi,\ 0 \leq \phi \leq \pi$

***13.23** If the half-wave dipole of Figure 13.4 has a current distribution $I(z) = I_o \cos^2 \beta z$, show that

$$H_{\phi s} = \frac{jI_o \sin\theta \sin\left(\frac{\pi}{2}\cos\theta\right) e^{-j\beta r}}{4\pi r \cos\theta\,(1 - 0.25\cos^2\theta)}$$

Find the radiation intensity.

(*Hint*: $\int e^{ax}\cos^2 bx\, dx = a^{ax}\left[\frac{\cos bx}{a^2 + 4b^2}(a\cos bx + 2b\sin bx) + \frac{2b^2/a}{a^2 + 4b^2}\right]$.)

13.24 The loss resistance of an antenna carrying a uniform current is given by

$$R_\ell = \frac{\ell}{2\pi a} R_s$$

where a is the wire radius, ℓ is the length of the wire, and R_s is the skin resistance (see eq. (10.56)). Calculate the radiation efficiency of a $\lambda/80$ Hertzian dipole operating at 3 MHz. Assume that the dipole is made of copper wire of radius 1.024 mm.

13.25 Show that eq. (13.64) reduces to eq. (13.56) when $N = 2$.

13.26 Derive $\mathbf{E}_s$ at far field due to the two-element array shown in Figure 13.24. Assume that the Hertzian dipole elements are fed in phase with uniform current $I_o \cos\omega t$.

13.27 An array comprises two dipoles that are separated by one wavelength. If the dipoles are fed by currents of the same magnitude and phase,

(a) Find the array factor.

(b) Calculate the angles where the nulls of the pattern occur.

(c) Determine the angles where the maxima of the pattern occur.

(d) Sketch the group pattern in the plane containing the elements.

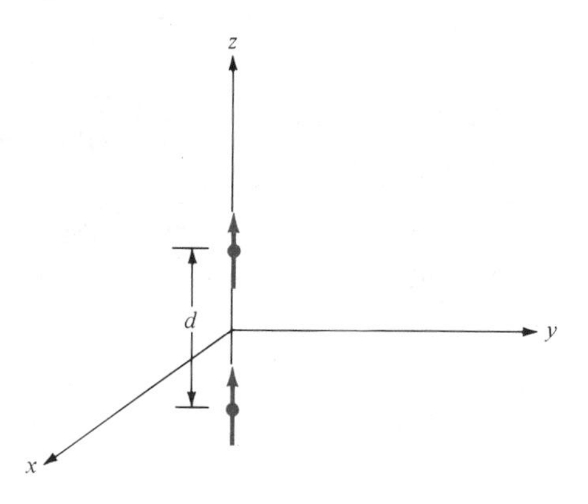

Figure 13.24 Two-element array of Problem 13.26.

13.28 Sketch the group pattern in the xz-plane of the two-element array of Figure 13.10 with

(a) $d = \lambda,\ \alpha = \pi/2$

(b) $d = \lambda/4,\ \alpha = 3\pi/4$

(c) $d = 3\lambda/4,\ \alpha = 0$

13.29 Calculate the approximate value of the directivity of the two-element array of Figure 13.10 when $\alpha = \pi$, $d \ll \lambda$. Express your result in decibels.

13.30 Sketch the resultant group patterns for the four-element arrays shown in Figure 13.25.

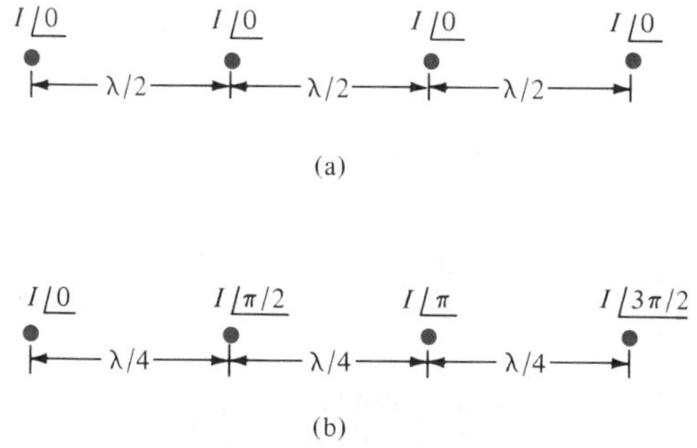

$I\angle 0$ $\quad I\angle 0$ $\quad I\angle 0$ $\quad I\angle 0$

$\lambda/2$ $\quad \lambda/2$ $\quad \lambda/2$

(a)

$I\angle 0$ $\quad I\angle \pi/2$ $\quad I\angle \pi$ $\quad I\angle 3\pi/2$

$\lambda/4$ $\quad \lambda/4$ $\quad \lambda/4$

(b)

Figure 13.25 For Problem 13.30.

13.31 (a) Using the identity

$$\left|\frac{\sin(N\psi/2)}{N\sin(\psi/2)}\right|^2 = \frac{1}{N} + \frac{2}{N^2}\sum_{m=1}^{N-1}(N-m)\cos m\psi$$

show that for an N-element linear uniform array, spaced a distance d apart with inter-element phase shift α, the directivity is given by

$$D = \frac{1}{\dfrac{1}{N} + \dfrac{2}{N^2}\displaystyle\sum_{m=1}^{N-1}\frac{N-m}{m\beta d}\sin m\beta d\cos m\alpha}$$

where $\beta = 2\pi/\lambda$.

(b) Derive an expression for the directivity of a two-element broadside array ($\alpha = 0$).

(c) What is the directivity when $N = 8$, $d = \lambda/2$, $\alpha = 0$?

13.32 For a 10-turn loop antenna of radius 15 cm operating at 100 MHz, calculate the effective area at $\theta = 30°$, $\phi = 90°$.

13.33 An antenna receives a power of 2 μW from a radio station. Calculate its effective area if the antenna is located in the far zone of the station where E = 50 mV/m.

13.34 (a) Show that the Friis transmission equation can be written as

$$\frac{P_r}{P_t} = \frac{A_{er}A_{et}}{\lambda^2 r^2}$$

(b) Two half-wave dipole antennas are operated at 100 MHz and separated by 1 km. If 80 W is transmitted by one, how much power is received by the other?

13.35 Two 100 percent efficient antennas spaced 5 km apart operate at 300 MHz in air. If 50 W is delivered to the transmitting antenna, determine the maximum power that will be obtained at the receiving terminals if

(a) Both antennas are $\lambda/2$ dipoles

(b) Both are $\lambda/4$ monopoles

(c) One is $\lambda/2$ dipole and the other is $\lambda/4$ monopole

In all cases, assume $G_d = D$.

13.36 The power transmitted by a synchronous orbit satellite antenna is 320 W. If the antenna has a gain of 40 dB at 15 GHz, calculate the power received by another antenna with a gain of 32 dB at the range of 24,567 km.

13.37 The directive gain of an antenna is 34 dB. If the antenna radiates 7.5 kW at a distance of 40 km, find the time-average power density at that distance.

13.38 Two identical antennas in an anechoic chamber are separated by 12 m and are oriented for maximum directive gain. At a frequency of 5 GHz, the power received by one is 30 dB down from that transmitted by the other. Calculate the gain of the antennas in dB.

13.39 What is the maximum power that can be received over a distance of 1.5 km in free space with a 1.5-GHz circuit consisting of a transmitting antenna with a gain of 25 dB and a receiving antenna with a gain of 30 dB? The transmitted power is 200 W.

13.40 An L-band pulse radar with a common transmitting and receiving antenna having a directive gain of 3500 operates at 1500 MHz and transmits 200 kW. If the object is 120 km from the radar and its scattering cross section 8 m^2, find

(a) The magnitude of the incident electric field intensity of the object

(b) The magnitude of the scattered electric field intensity at the radar

(c) The amount of power captured by the object

(d) The power absorbed by the antenna from the scattered wave

13.41 In the bistatic radar system of Figure 13.26, the ground-based antennas are separated by 4 km and the 2.4 m^2 target is at a height of 3 km. The system operates at 5 GHz. For G_{dt} of 36 dB and G_{dr} of 20 dB, determine the minimum necessary radiated power to obtain a return power of 8×10^{-12} W.

***13.42** An antenna with effective area of 25 cm^2 at 1 GHz is used to transmit and receive energy scattered from a perfectly conducting sphere of radius 2λ. If the scattering cross-section of the sphere is equal to the geometrical cross section πa^2, calculate the maximum distance between the antenna and the sphere. Take the minimum detectable power as $-$ 136 dBm/m^2 and the transmitted power as 80 dBm/m^2. (*Note*: P in mW/m^2 = $10 \log_{10} P$ dBm/m^2.)

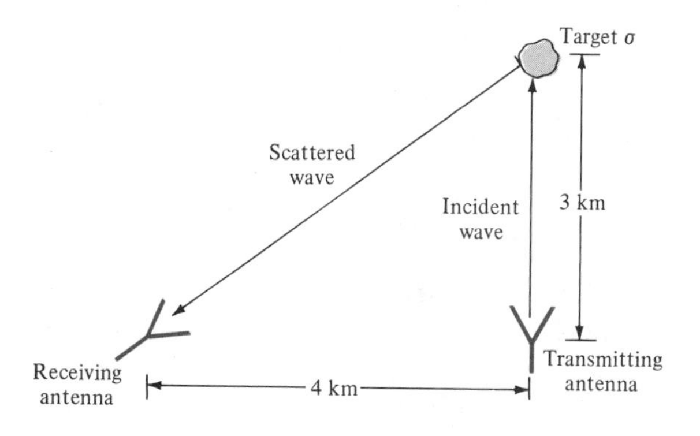

Figure 13.26 For Problem 13.41.

CHAPTER

Numerical Methods

14

The man who graduates today and stops learning tomorrow is uneducated the day after.

—NEWTON D. BAKER

14.1 INTRODUCTION

In the preceding chapters we considered various analytic techniques for solving EM problems and obtaining solutions in closed form. A *closed form solution* is one in the form of an explicit, algebraic equation in which values of the problem parameters can be substituted. Some of these analytic solutions were obtained assuming certain situations, thereby making the solutions applicable to those idealized situations. For example, in deriving the formula for calculating the capacitance of a parallel-plate capacitor, we assumed that the fringing effect was negligible and that the separation distance was very small compared with the width and length of the plates. Also, our application of Laplace's equation in Chapter 6 was restricted to problems with boundaries coinciding with coordinate surfaces. Analytic solutions have an inherent advantage of being exact. They also make it easy to observe the behavior of the solution for variation in the problem parameters. However, analytic solutions are available only for problems with simple configurations.

When the complexities of theoretical formulas make analytic solution intractable, we resort to nonanalytic methods, which include (1) graphical methods, (2) experimental methods, (3) analog methods, and (4) numerical methods. Graphical, experimental, and analog methods are applicable to solving relatively few problems. Numerical methods have come into prominence and become more attractive with the advent of fast digital computers. The most commonly used simple numerical techniques in EM are moment method, finite difference method, and finite element method. Most EM problems involve either partial differential equations or integral equations. Partial differential equations are usually solved using the finite difference method or the finite element method; integral equations are solved conveniently using the moment method. Although numerical methods give approximate solutions, the solutions are sufficiently accurate for engineering purposes. We should not get the impression that analytic techniques are outdated by numerical methods; rather they are complementary. As will be observed later, every numerical method involves an analytic simplification to the point where it is easy to apply the method.

The codes developed for computer implementation of the concepts developed in this chapter are simplified and self-explanatory for instructional purposes. The notations used in the programs are as close as possible to those used in the main text; some are defined wherever necessary. These programs are by no means unique; there are several ways of writing a computer program. Therefore, users may decide to modify the programs to suit their objectives.

†14.2 FIELD PLOTTING

In Section 4.10, we used field lines and equipotential surfaces for visualizing an electrostatic field. However, the graphical representations in Figure 4.24 for electrostatic fields and in Figures 7.8(b) and 7.16 for magnetostatic fields are very simple, trivial, and qualitative. Accurate pictures of more complicated charge distributions would be more helpful. In this section, a numerical technique which may be developed into an interactive computer program is presented. It generates data points for electric field lines and equipotential lines for arbitrary configuration of point sources.

Electric field lines and equipotential lines can be plotted for coplanar point sources with simple programs. Suppose we have N point charges located at position vectors $\mathbf{r}_1$, $\mathbf{r}_2, \ldots, \mathbf{r}_N$, the electric field intensity $\mathbf{E}$ and potential V at position vector $\mathbf{r}$ are given respectively by

$$\mathbf{E} = \sum_{k=1}^{N} \frac{Q_k (\mathbf{r} - \mathbf{r}_k)}{4\pi\varepsilon |\mathbf{r} - \mathbf{r}_k|^3} \qquad [14.1]$$

and

$$V = \sum_{k=1}^{N} \frac{Q_k}{4\pi\varepsilon |\mathbf{r} - \mathbf{r}_k|} \qquad [14.2]$$

If the charges are on the same plane (z = constant), eqs. (14.1) and (14.2) become

$$\mathbf{E} = \sum_{k=1}^{N} \frac{Q_k[(x - x_k)\mathbf{a}_x + (y - y_k)\mathbf{a}_y]}{4\pi\varepsilon\,[(x - x_k)^2 + (y - y_k)^2]^{3/2}} \qquad [14.3]$$

$$V = \sum_{k=1}^{N} \frac{Q_k}{4\pi\varepsilon\,[(x - x_k)^2 + (y - y_k)^2]^{1/2}} \qquad [14.4]$$

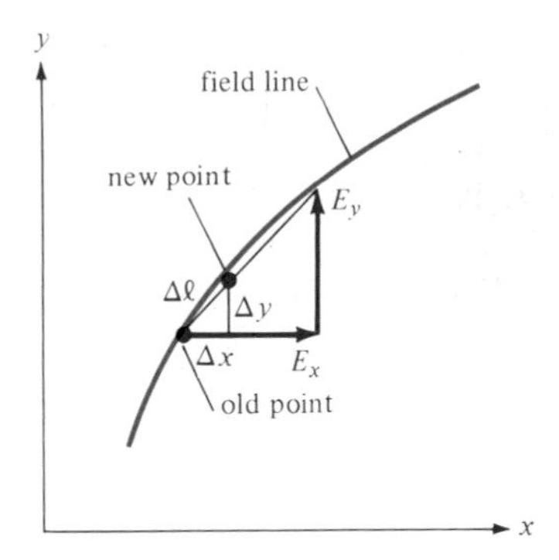

Figure 14.1 A small displacement on a field line.

To plot the electric field lines, follow these steps:

1. Choose a starting point on the field line.
2. Calculate E_x and E_y at that point using eq. (14.3).
3. Take a small step along the field line to a new point in the plane. As shown in Figure 14.1, a movement $\Delta\ell$ along the field line corresponds to movements Δx and Δy along x- and y-directions respectively. From the figure, it is evident that

$$\frac{\Delta x}{\Delta\ell} = \frac{E_x}{E} = \frac{E_x}{[E_x^2 + E_y^2]^{1/2}}$$

   or

$$\Delta x = \frac{\Delta\ell \cdot E_x}{[E_x^2 + E_y^2]^{1/2}} \qquad [14.5]$$

   Similarly,

$$\Delta y = \frac{\Delta\ell \cdot E_y}{[E_x^2 + E_y^2]^{1/2}} \qquad [14.6]$$

   Move along the field line from the old point (x, y) to a new point $x' = x + \Delta x$, $y' = y + \Delta y$.

4. Go back to step 2 and repeat the calculations. Continue to generate new points until a line is completed within a given range of coordinates. On completing the line, go back to step 1 and choose another starting point. Note that since there are an infinite number of field lines, any starting point is likely to be on a field line. The points generated can be plotted by hand or by a plotter as illustrated in Figure 14.2.

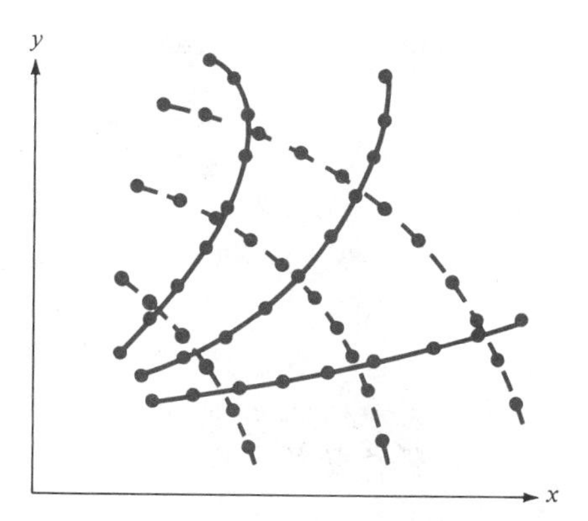

Figure 14.2 Generated points on E-field lines (shown thick) and equipotential lines (shown dotted).

To plot the equipotential lines, follow these steps:

1. Choose a starting point.
2. Calculate the electric field (E_x, E_y) at that point using eq. (14.3).
3. Move a small step along the line perpendicular to E-field line at that point. Utilize the fact that if a line has slope m, a perpendicular line must have slope $-1/m$. Since an E-field line and an equipotential line meeting at a given point are mutually orthogonal there,

$$\Delta x = \frac{-\Delta \ell \cdot E_y}{[E_x^2 + E_y^2]^{1/2}} \tag{14.7}$$

$$\Delta y = \frac{\Delta \ell \cdot E_x}{[E_x^2 + E_y^2]^{1/2}} \tag{14.8}$$

   Move along the equipotential line from the old point (x, y) to a new point $(x + \Delta x, y + \Delta y)$. As a way of checking the new point, calculate the potential at the new and old points using eq. (14.4); they must be equal because the points are on the same equipotential line.
4. Go back to step 2 and repeat the calculations. Continue to generate new points until a line is completed within the given range of x and y. After completing the line, go back to step 1 and choose another starting point. Join the points generated by hand or by a plotter as illustrated in Figure 14.2.

The magnetic field line due to various current distributions can be plotted using Biot-Savart law. Programs for determining the magnetic field line due to line current, a current loop, a Helmholtz pair, and a solenoid can be developed. Programs for drawing the electric and magnetic field lines inside a rectangular waveguide or the power radiation pattern produced by a linear array of vertical half-wave electric dipole antennas can also be written.

EXAMPLE 14.1

Write a program to plot the electric field and equipotential lines due to:

(a) Two point charges Q and $-4Q$ located at $(x, y) = (-1, 0)$ and $(1, 0)$ respectively

(b) Four point charges Q, $-Q$, Q, and $-Q$ located at $(x, y) = (-1, -1)$, $(1, -1)$, $(1, 1)$, and $(-1, 1)$ respectively. Take $Q/4\pi\varepsilon = 1$ and $\Delta\ell = 0.1$. Consider the range $-5 \leq x, y, \leq 5$.

SOLUTION

Based on the steps given in Section 14.2, the program in Figure 14.3 was developed. Enough comments are inserted to make the program as self-explanatory as possible. Further explanation of the program is provided in the following paragraphs.

Since the E-field lines emanate from positive charges and terminate on negative charges, it seems reasonable to generate starting points (x_s, y_s) for the E-field lines on small circles centered at charge locations (x_Q, y_Q); that is,

$$x_s = x_Q + r \cos \theta \qquad \text{[14.1.1a]}$$

$$y_s = y_Q + r \sin \theta \qquad \text{[14.1.1b]}$$

where r is the radius of the small circle (e.g., $r = 0.1$ or 0.05), and θ is a prescribed angle chosen for each E-field line. The starting points for the equipotential lines can be generated in different ways: along the x- and y-axes, along line $y = x$, and so on. However, to make the program as general as possible, the starting points should depend on the charge locations like those for the E-field lines. They could be chosen using eq. (14.1.1) but with fixed θ (e.g., 45°) and variable r (e.g., 0.5, 1.0, 2.0, . . .).

```
0001   C   PROGRAM FOR PLOTTING THE ELECTRIC FIELD LINES
0002   C   AND EQUIPOTENTIAL LINES DUE TO COPLANAR POINT CHARGES
0003   C   THE PLOT IS TO BE WITHIN THE RANGE -5 < X,Y < 5
0004   C
0005   C   DLE OR DLV = THE INCREMENT ALONG E & V LINES
0006   C   NLE = NO. OF  E-FIELD LINES PER CHARGE
0007   C   NLV = NO. OF EQUIPOTENTIAL LINES PER CHARGE
0008   C   NOTE THAT CONSTANT  Q/4*PIE*ER IS SET EQUAL TO 1.0
0009
0010       DIMENSION Q(50),XQ(50),YQ(50)
0011       DATA DLE,DLV,NLE,NLV/0.1,0.01,8,2/
0012
0013   C   ENTER INFORMATION ABOUT CHARGE DISTRIBUTION
0014
0015       PRINT *,'ENTER NO. OF CHARGES, NQ'
```

Figure 14.3 Computer program for Example 14.1.

```
0016              READ(5,*) NQ
0017              DO 10 K=1,NQ
0018              PRINT *,'ENTER THE VALUE OF THE CHARGE, Q, ONE AT A TIME'
0019              READ(5,*) Q(K)
0020              PRINT *,'ENTER CHARGE LOCATION (X,Y)'
0021              READ(5,*) XQ(K), YQ(K)
0022      10      CONTINUE
0023      C       NOW DETERMINE THE E-FIELD LINES
0024      C       FOR CONVENIENCE, THE STARTING POINTS (XS,YS) ARE RADIALLY
0025      C       DISTRIBUTED ABOUT CHARGE LOCATIONS
0026              JJ=1
0027              DO 80 K=1,NQ
0028              DO 70 I=1,NLE
0029              THETA = 360*FLOAT(I-1)/FLOAT(NLE)
0030              XS=XQ(K) + 0.1*COSD(THETA)
0031              YS=YQ(K) + 0.1*SIND(THETA)
0032              N=1
0033              XE=XS
0034              YE=YS
0035              WRITE(6,20) N,XS,YS
0036      20      FORMAT(I5,2F6.2)
0037      C       FIND INCREMENT AND NEW POINT  ( X,Y )
0038      30      EX=0.0
0039              EY=0.0
0040              DO 40 J=1,NQ
0041              R=SQRT( (XE-XQ(J))**2 + (YE-YQ(J))**2 )
0042              EX=EX + Q(J)*(XE-XQ(J))/(R**3)
0043              EY=EY + Q(J)*(YE-YQ(J))/(R**3)
0044      40      CONTINUE
0045              E=SQRT(EX**2 + EY**2)
0046      C       CHECK FOR SINGULAR POINT
0047              IF(E.LE.0.00005) GO TO 70
0048              DX=DLE*EX/E
0049              DY=DLE*EY/E
0050      C       FOR NEGATIVE CHARGE, NEGATE DX & DY SO THAT INCREMENT
0051      C       IS AWAY FROM THE CHARGE
0052              IF( Q(K).LT.0 )  DX= -DX
0053              IF( Q(K).LT.0 )  DY= -DY
0054              XE=XE + DX
0055              YE=YE + DY
0056      C       CHECK WHETHER NEW POINT IS WITHIN THE GIVEN RANGE OR TO
0057      C       CLOSE TO ANY OF THE POINT CHARGES - TO AVOID SINGULAR POINT
0058              IF( ABS(XE).GE.5) GO TO 70
0059              IF( ABS(YE).GE.5) GO TO 70
0060              DO 50 L=1,NQ
0061              IF( ABS(XE-XQ(L)).LT.0.05.AND.
0062           1      ABS(YE-YQ(L)).LT.0.05) GO TO 70
0063      50     CONTINUE
0064      C      SELECT EVERY 5TH POINT FOR PLOTTING
0065             IF(JJ.EQ.5) GO TO 60
0066             JJ=JJ + 1
0067             GO TO 30
0068      60     JJ=1
0069              N=N+1
0070              WRITE(6,20) N,XE,YE
0071              GO TO 30
0072      70     CONTINUE
0073      80     CONTINUE
0074
0075      C       NEXT, DETERMINE THE EQUIPOTENTIAL LINES
0076      C       FOR CONVENIENCE, THE STARTING POINTS (XS,YS) ARE
```

Figure 14.3 (Continued)

```
0077      C       CHOSEN LIKE THOSE FOR THE E-FIELD LINES
0078              JJ=1
0079              ANGLE=45.0
0080              DO 180 K=1,NQ
0081              FACTOR=0.5
0082              DO 170 KK=1,NLV
0083              XS=XQ(K) + FACTOR*COSD(ANGLE)
0084              YS=YQ(K) + FACTOR*SIND(ANGLE)
0085              IF( ABS(XS).GE.5.OR.ABS(YS).GE.5) GO TO 180
0086              PRINT *,K,XS,YS
0087              N=1
0088              DIR=1.0
0089              XV=XS
0090              YV=YS
0091              WRITE(6,20) N,XV,YV
0092      C       FIND INCREMENT AND NEW POINT ( XV,YV )
0093      90      EX=0.0
0094              EY=0.0
0095              DO 100 J=1,NQ
0096              R=SQRT( (XV-XQ(J))**2 + (YV-YQ(J))**2 )
0097              EX=EX + Q(J)*(XV-XQ(J))/(R**3)
0098              EY=EY + Q(J)*(YV-YQ(J))/(R**3)
0099      100     CONTINUE
0100              E=SQRT(EX**2 + EY**2)
0101              IF(E.LE.0.00005) GO TO 160
0102              DX= -DLV*EY/E
0103              DY= DLV*EX/E
0104              XV=XV + DIR*DX
0105              YV=YV + DIR*DY
0106      C       CHECK IF THE EQUIPOTENTIAL LINE LOOPS BACK TO (XS,YS)
0107              RO=SQRT( (XV - XS)**2  +  (YV - YS)**2 )
0108              IF( RO.LT.0.2.AND.N.GT.50 ) GO TO 160
0109      C       CHECK WHETHER NEW POINT IS WITHIN THE GIVEN RANGE
0110      C       IF FOUND OUT OF RANGE, GO BACK TO THE STARTING POINT
0111      C       (XS,YS) BUT INCREMENT IN THE OPPOSITE DIRECTION
0112              IF( ABS(XV).GT.5.OR.ABS(YV).GT.5) GO TO 110
0113              GO TO 120
0114      110     DIR=DIR - 2.0
0115              N=1
0116              XV=XS
0117              YV=YS
0118              WRITE(6,20) N,XV,YV
0119              IF( ABS(DIR).GT.1 ) GO TO 160
0120              GO TO 90
0121      120     DO 130 L=1,NQ
0122              IF( ABS(XV-XQ(L)).LT.0.005.AND.
0123            1      ABS(YV-YQ(L)).LT.0.005) GO TO 140
0124      130     CONTINUE
0125      C       SELECT EVERY 5TH POINT FOR PLOTTING
0126      140     IF(JJ.EQ.5) GO TO 150
0127              JJ=JJ + 1
0128              GO TO 90
0129      150     JJ=1
0130              N=N+1
0131              WRITE(6,20) N,XV,YV
0132              GO TO 90
0133      160     FACTOR=2.0*FACTOR
0134      170     CONTINUE
0135      180     CONTINUE
0136              STOP
0137              END
```

Figure 14.3 (Continued)

The value of incremental length $\Delta\ell$ is crucial for accurate plots. Although the smaller the value of $\Delta\ell$, the more accurate the plots, we must keep in mind that the smaller the value of $\Delta\ell$, the more points we generate and memory storage may be a problem. For example, a line may consist of more than 1000 generated points. In view of the large number of points to be plotted, the points are usually stored in a data file and a graphics routine is used to plot the data.

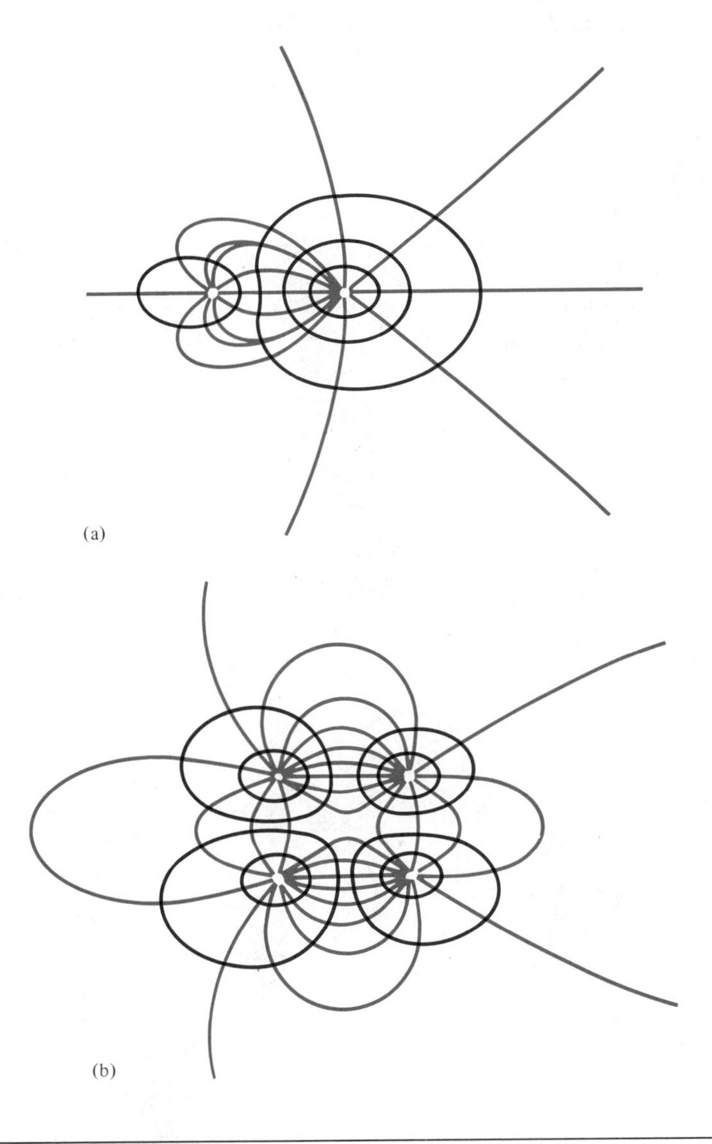

Figure 14.4 For Example 14.1; plots of E-field lines and equipotential lines due to **(a)** two point charges, and **(b)** four point charges (a two-dimensional quadrupole).

For both the E-field and equipotential lines, different checks are inserted in the program in Figure 14.3:

(a) Check for singular point ($\mathbf{E} = 0$?).

(b) Check whether the point generated is too close to a charge location.

(c) Check whether the point is within the given range of $-5 < x, y < 5$.

(d) Check whether the (equipotential) line loops back to the starting point.

The plot of the points generated for the cases of two point charges and four point charges are shown in Figure 14.4(a) and (b) respectively. ■

PRACTICE EXERCISE 14.1

Write a complete program for plotting the electric field lines and equipotential lines due to coplanar point charges. Run the program for $N = 3$; that is, there are three point charges $-Q$, $+Q$, and $-Q$ located at $(x, y) = (-1, 0)$, $(0, 1)$, and $(1, 0)$ respectively. Take $Q/4\pi\varepsilon = 1$, $\Delta\ell = 0.1$ or 0.01 for greater accuracy and limit your plot to $-5 \leq x, y, \leq 5$.

ANSWER See Figure 14.5.

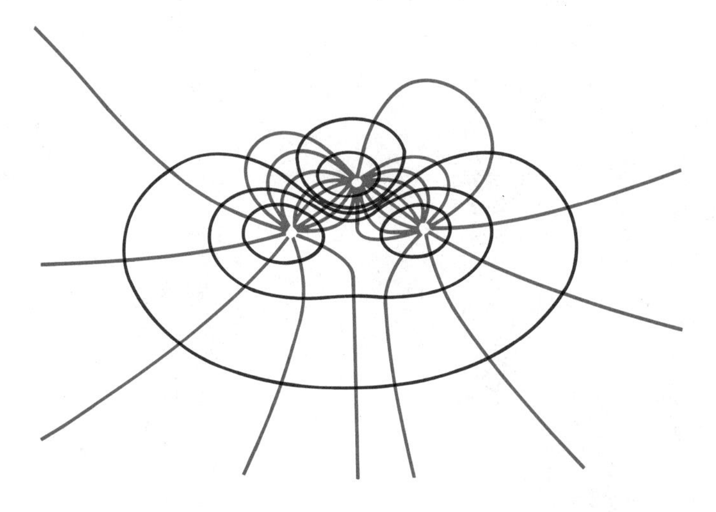

Figure 14.5 For Practice Exercise 14.1.

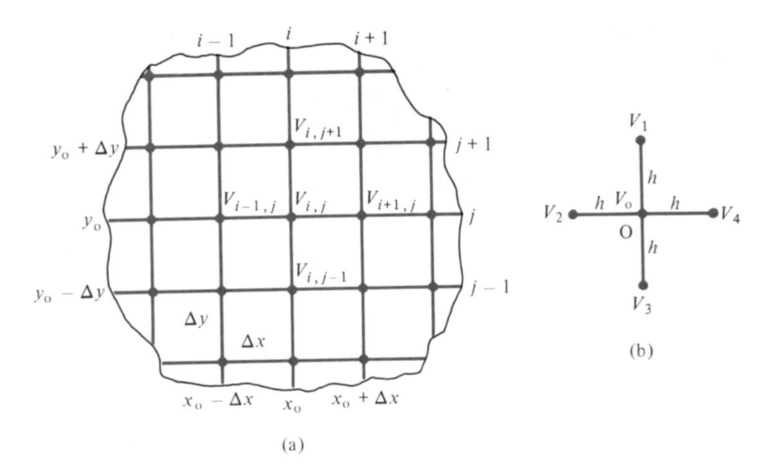

Figure 14.6 Finite difference solution pattern: **(a)** division of the solution into grid points, **(b)** finite difference five-node molecule.

14.3 THE FINITE DIFFERENCE METHOD

The finite difference method[1] is a simple numerical technique used in solving partial differential equations like those solved analytically in Chapter 6. A finite difference solution to Poisson's or Laplace's equation, for example, proceeds in two steps: (1) approximating the differential equation and boundary conditions by a set of linear algebraic equations (called difference equations) on grid points within the solution region, and (2) solving this set of algebraic equations.

Suppose we intend to apply the finite difference method to determine the electric potential in a region shown in Figure 14.6(a). The solution region is divided into rectangular meshes with *grid points* or *nodes* as in Figure 14.6(a). A node on the boundary of the region where the potential is specified is called a *fixed node* (fixed by the problem) and interior points in the region are called *free points* (free in that the potential is unknown). Our objective is to obtain the finite difference approximation to Poisson's equation and use this to determine the potentials at all the free points. We recall that Poisson's equation is given by

$$\nabla^2 V = -\frac{\rho_v}{\varepsilon} \tag{14.9a}$$

[1]For an extensive treatment of the finite difference method, see G. D. Smith, *Numerical Solution of Partial Differential Equations: Finite Difference Methods,* Oxford: Clarendon, 2nd edition, 1978.

For two-dimensional solution region such as in Figure 14.6(a), ρ_v is replaced by ρ_S, $\dfrac{\partial^2 V}{\partial z^2} = 0$, so

$$\frac{\partial^2 V}{\partial x^2} + \frac{\partial^2 V}{\partial y^2} = -\frac{\rho_S}{\varepsilon} \tag{14.9b}$$

From the definition of the derivative of $V(x, y)$ at point (x_o, y_o),

$$V' = \left.\frac{\partial V}{\partial x}\right|_{x = x_o} \simeq \frac{V(x_o + \Delta x, y_o) - V(x_o - \Delta x, y_o)}{2\Delta x}$$

$$= \frac{V_{i+1,j} - V_{i-1,j}}{2\,\Delta x} \tag{14.10}$$

where Δx is sufficiently small. For the second derivative, which is the derivative of the first derivative V',

$$V'' = \left.\frac{\partial^2 V}{\partial x^2}\right|_{x = x_o} = \frac{\partial V'}{\partial x} \simeq \frac{V'(x_o + \Delta x/2, y_o) - V'(x_o - \Delta x/2, y_o)}{\Delta x}$$

$$= \frac{V(x_o + \Delta x, y_o) - 2V(x_o, y_o) + V(x_o - \Delta x, y_o)}{(\Delta x)^2} \tag{14.11}$$

$$= \frac{V_{i+1,j} - 2V_{i,j} + V_{i-1,j}}{(\Delta x)^2}$$

Equations (14.10) and (14.11) are the finite difference approximations for the first and second partial derivatives of V with respect to x, evaluated at $x = x_o$. The approximation in eq. (14.10) is associated with an error of the order of Δx while that of eq. (14.11) has an associated error on the order of $(\Delta x)^2$. Similarly,

$$\left.\frac{\partial^2 V}{\partial y^2}\right|_{y = y_o} \simeq \frac{V(x_o, y_o + \Delta y) - 2V(x_o, y_o) + V(x_o, y_o - \Delta y)}{(\Delta y)^2}$$

$$= \frac{V_{i,j+1} - 2V_{i,j} + V_{i,j-1}}{(\Delta y)^2} \tag{14.12}$$

Substituting eqs. (14.11) and (14.12) into eq. (14.9b) and letting $\Delta x = \Delta y = h$ gives

$$V_{i+1,j} + V_{i-1,j} + V_{i,j+1} + V_{i,j-1} - 4V_{i,j} = -\frac{h^2 \rho_S}{\varepsilon}$$

or

$$\boxed{V_{i,j} = \frac{1}{4}\left(V_{i+1,j} + V_{i-1,j} + V_{i,j+1} + V_{i,j-1} + \frac{h^2 \rho_S}{\varepsilon}\right)} \tag{14.13}$$

where h is called the *mesh size*. Equation (14.13) is the finite difference approximation to Poisson's equation. If the solution region is charge-free ($\rho_S = 0$), eq. (14.9) becomes Laplace's equation:

$$\nabla^2 V = \frac{\partial^2 V}{\partial x^2} + \frac{\partial^2 V}{\partial y^2} = 0 \qquad [14.14]$$

The finite difference approximation to this equation is obtained from eq. (14.13) by setting $\rho_S = 0$; that is

$$V_{i,j} = \frac{1}{4}(V_{i+1,j} + V_{i-1,j} + V_{i,j+1} + V_{i,j-1}) \qquad [14.15]$$

This equation is essentially a five-node finite difference approximation for the potential at the central point of a square mesh. Figure 14.6(b) illustrates what is called the finite difference *five-node molecule*. The molecule in Figure 14.6(b) is taken out of Figure 14.6(a). Thus eq. (14.15) applied to the molecule becomes

$$\boxed{V_o = \frac{1}{4}(V_1 + V_2 + V_3 + V_4)} \qquad [14.16]$$

This equation clearly shows the average-value property of Laplace's equation. In other words, Laplace's equation can be interpreted as a differential means of stating the fact that the potential at a specific point is the average of the potentials at the surrounding points.

To apply eq. (14.16) (or eq. (14.13)) to a given problem, one of the following three methods is commonly used:

A. Iteration Method

We start by setting initial values of the potentials at the free nodes equal to zero or to any reasonable guessed value. Keeping the potentials at the fixed nodes unchanged at all times, we apply eq. (14.16) to every free node in turn until the potentials at all free nodes are calculated. The potentials obtained at the end of this first iteration are not accurate but just approximate. To increase the accuracy of the potentials, we repeat the calculation at every free node using old values to determine new ones. The iterative or repeated modification of the potential at each free node is continued until a prescribed degree of accuracy is achieved or until the old and the new values at each node are satisfactorily close.

B. Band Matrix Method

Equation (14.16) applied to all free nodes results in a set of simultaneous equations of the form

$$[A]\,[V] = [B] \qquad [14.17]$$

where $[A]$ is a *sparse* matrix (i.e., one having many zero terms), $[V]$ consists of the unknown potentials at the free nodes, and $[B]$ is another column matrix formed by the known potentials at the fixed nodes. Matrix $[A]$ is also *banded* in that its nonzero terms appear clustered near the main diagonal because only nearest neighboring nodes affect the potential at each node. The sparse, band matrix is easily inverted to determine $[V]$. Thus we obtain the potentials at the free nodes from matrix $[V]$ as

$$[V] = [A]^{-1}[B] \tag{14.18}$$

C. Relaxation Method

A convenient initial guess for the potential at the free nodes is made. The initialization is done by setting all values equal to zero or guessing based on analytic solution of a similar but less difficult problem. The degree of accuracy of the values is measured by the *residuals* at the nodes. The residuals R_o at node free O in Figure 14.6(b)is the amount by which eq. (14.16) fails to hold and is given by

$$R_o = V_1 + V_2 + V_3 + V_4 - 4V_o \tag{14.19}$$

We modify the potential at the free node so as to relax (or reduce) the residuals at that node. The relaxation process is repeated at each free node and the whole grid is gone over several times until the residuals at the free nodes are reduced to zero or to a sufficiently small value.

The finite difference method can be applied to solve time-varying problems. For example, consider the one-dimensional wave equation of eq. (10.1), namely

$$u^2 \frac{\partial^2 \Phi}{\partial x^2} = \frac{\partial^2 \Phi}{\partial t^2} \tag{14.20}$$

where u is the wave velocity and Φ is the E- or H-field component of the EM wave. The difference approximation of the derivatives at (x_o, t_o) or (i, j)th node shown in Figure 14.7 are

$$\left.\frac{\partial^2 \Phi}{\partial x^2}\right|_{x = x_o} \simeq \frac{\Phi_{i-1,j} - 2\Phi_{i,j} + \Phi_{i+1,j}}{(\Delta x)^2} \tag{14.21a}$$

$$\left.\frac{\partial^2 \Phi}{\partial t^2}\right|_{t = t_o} \simeq \frac{\Phi_{i,j-1} - 2\Phi_{i,j} + \Phi_{i,j+1}}{(\Delta t)^2} \tag{14.21b}$$

Inserting eq. (14.21) in eq. (14.20) and solving for $\Phi_{i,j+1}$ gives

$$\boxed{\Phi_{i,j+1} \simeq \alpha\,(\Phi_{i-1,j} + \Phi_{i+1,j}) + 2(1 - \alpha)\,\Phi_{i,j} - \Phi_{i,j-1}} \tag{14.22}$$

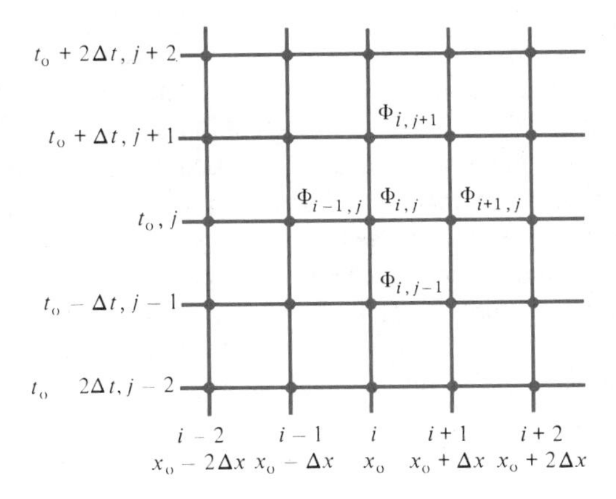

Figure 14.7 Finite difference solution pattern for wave equation.

where

$$\alpha = \left[\frac{u\,\Delta t}{\Delta x}\right]^2 \qquad [14.23]$$

It can be shown that for the solution in eq. (14.22) to be stable, $\alpha \leq 1$. To start the finite difference algorithm in eq. (14.22), we use the initial condition. We assume that at $t = 0$, $\partial\Phi_{i,0}/\partial t = 0$ and use (central) difference approximation (see Review Question 14.2) to get

$$\frac{\partial\Phi_{i,0}}{\partial t} \simeq \frac{\Phi_{i,1} - \Phi_{i,-1}}{2\Delta t} = 0$$

or

$$\Phi_{i,1} = \Phi_{i,-1}. \qquad [14.24]$$

Substituting eq. (14.24) into eq. (14.22) and taking $j = 0$ ($t = 0$), we obtain

$$\Phi_{i,1} \simeq \alpha(\Phi_{i-1,0} + \Phi_{i+1,0}) + 2(1-\alpha)\Phi_{i,0} - \Phi_{i,1}$$

or

$$\Phi_{i,1} \simeq \frac{1}{2}\left[\alpha(\Phi_{i-1,0} + \Phi_{i+1,0}) + 2(1-\alpha)\Phi_{i,0}\right] \qquad [14.25]$$

With eq. (14.25) as the ''starting'' formula, the value of Φ at any point on the grid can be obtained directly from eq. (14.22). Note that the three methods discussed for solving eq. (14.16) do not apply to eq. (14.22) because eq. (14.22) can be used directly with eq. (14.25) as the starting formula. In other words, we do not have a set of simultaneous equations; eq. (14.22) is an explicit formula.

The concept of finite difference method can be extended to Poisson's, Laplace's, or wave equations in other coordinate systems. The accuracy of the method depends on the fineness of the grid and the amount of time spent in refining the potentials. We can reduce computer time and increase the accuracy and convergence rate by the method of successive overrelaxation, by making reasonable guesses at initial values, by taking advantage of symmetry if possible, by making the mesh size as small as possible, and by using more complex finite difference molecules (see Figure 14.41). One limitation of the finite difference method is that interpolation of some kind must be used to determine solutions at points not on the grid. One obvious way to overcome this is to use a finer grid, but this would require a greater number of computations and a larger amount of computer storage.

EXAMPLE 14.2

Solve the one-dimensional boundary-value problem $-\Phi'' = x^2$, $0 \leq x \leq 1$ subject to $\Phi(0) = 0 = \Phi(1)$. Use the finite difference method.

SOLUTION

First, we obtain the finite difference approximation to the differential equation $\Phi'' = -x^2$, which is Poisson's equation in one dimension. Next, we divide the entire domain $0 \leq x \leq 1$ into N equal segments each of length h $(= 1/N)$ as in Figure 14.8(a) so that there are $(N + 1)$ nodes.

$$-x_o^2 = \left.\frac{d^2\Phi}{dx^2}\right|_{x = x_o} \simeq \frac{\Phi(x_o + h) - 2\Phi(x_o) + \Phi(x_o - h)}{h^2}$$

or

$$-x_j^2 = \frac{\Phi_{j+1} - 2\Phi_j + \Phi_{j-1}}{h^2}$$

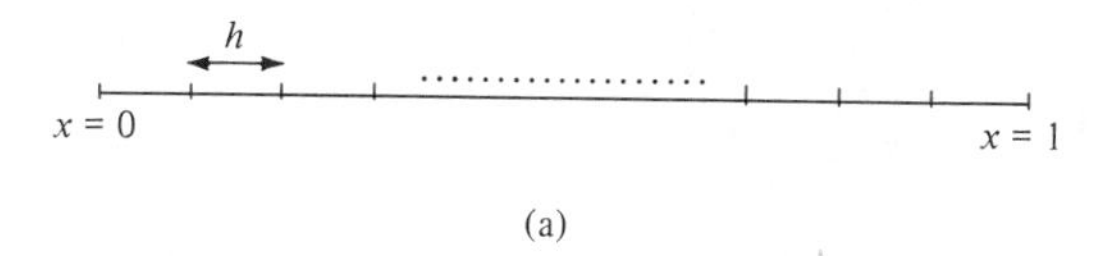

(a)

Φ_{j-1} Φ_j Φ_{j+1}

x_{j-1} x_j x_{j+1}

(b)

Figure 14.8 For Example 14.2.

Thus

$$-2\Phi_j = -x_j^2h^2 - \Phi_{j+1} - \Phi_{j-1}$$

or

$$\Phi_j = \frac{1}{2}(\Phi_{j+1} + \Phi_{j-1} + x_j^2\, h^2)$$

Using this finite difference scheme, we obtain an approximate solution for various values of N. The code is shown in Figure 14.9. The number of iterations NI depends on the degree of accuracy desired. For a one-dimensional problem such as this, $NI = 50$ may suffice. For two- or three-dimensional problems, larger values of NI would be required (see Table 14.1). It should be noted that the values of Φ at end points (fixed nodes) are held fixed. The solutions for $N = 4$ and 10 are shown in Figure 14.10.

We may compare this with the exact solution obtained as follows. Given that $d^2\Phi/dx^2 = -x^2$, integrating twice gives

$$\Phi = -\frac{x^4}{12} + Ax + B$$

```
0001   C     ONE-DIMENSIONAL PROBLEM OF EXAMPLE 14.2
0002   C     SOLVED USING FINITE DIFFERENCE METHOD
0003   C
0004   C     H = MESH SIZE
0005   C     NI = NO. OF ITERATIONS DESIRED
0006
0007         REAL L
0008         DIMENSION PHI (0:50)
0009
0010         N=20
0011         NI=1000
0012         L=1.0
0013         H=L/N
0014         DO 10 I=0,N
0015         PHI(I)=0.0
0016   10    CONTINUE
0017         DO 30 K=1,NI
0018         DO 20 J=1,N-1
0019         XO=H*FLOAT(J)
0020         PHI (J)=0.5* ( (XO*H) **2 + PHI (J+1) + PHI (J-1) )
0021   20    CONTINUE
0022   30    CONTINUE
0023   C     CALCULATE THE EXACT VALUE ALSO
0024         DO 50 J=0,N
0025         XO=H*FLOAT(J)
0026         PHIEX=XO* (1.0 - XO**3) /12.0
0027         WRITE (6,40) J,PHI (J) ,PHIEX
0028   40    FORMAT (2X, 'J=',I3, 2X, 'PHI=',F10.4, 2X, 'PHIEX=',F10.4,/)
0029   50    CONTINUE
0030         STOP
0031         END
```

Figure 14.9 Computer program for Example 14.2.

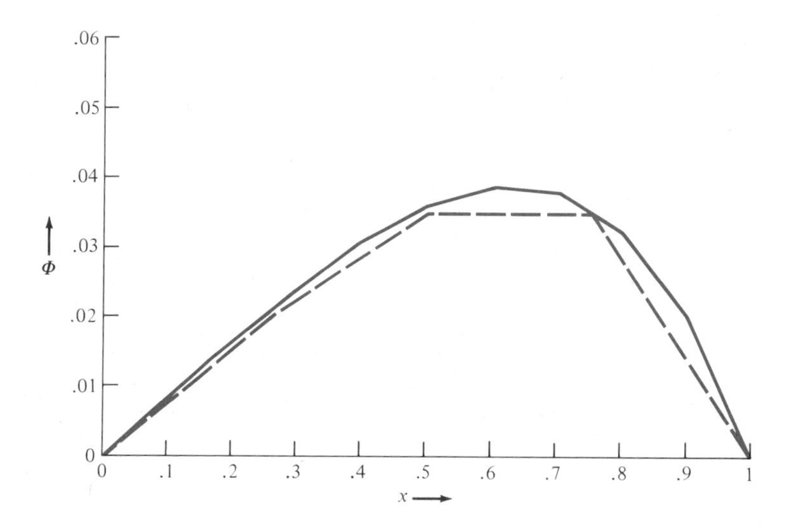

Figure 14.10 For Example 14.2: plot of $\Phi(x)$. Continuous curve is for $N = 10$; dashed curve is for $N = 4$.

where A and B are integration constants. From the boundary conditions,

$$\Phi(0) = 0 \rightarrow B = 0$$

$$\Phi(1) = 0 \rightarrow 0 = -\frac{1}{12} + A \qquad \text{or} \qquad A = \frac{1}{12}$$

Hence, the exact solution is $\Phi = x(1 - x^3)/12$, which is calculated in Figure 14.9 and found to be very close to case $N = 10$. ■

PRACTICE EXERCISE 14.2

Solve the differential equation $d^2y/dx^2 + y = 0$ with the boundary conditions $y(0) = 0$, $y(1) = 1$ using the finite difference method. Take $\Delta x = 1/4$.

ANSWER Compare your result with the exact solution $y(x) = \dfrac{\sin(x)}{\sin(1)}$.

EXAMPLE 14.3

Determine the potential at the free nodes in the potential system of Figure 14.11 using the finite difference method.

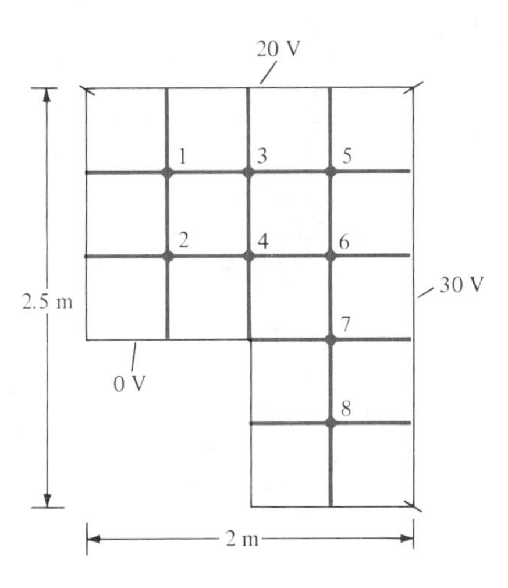

Figure 14.11 For Example 14.3.

SOLUTION This problem will be solved using the iteration method and band matrix method.

Method 1 (Iteration Method): We first set the initial values of the potential at the free nodes equal to zero. We apply eq. (14.16) to each free node using the newest surrounding potentials each time the potential at that node is calculated. For the first iteration:

$$V_1 = 1/4(0 + 20 + 0 + 0) = 5$$

$$V_2 = 1/4(5 + 0 + 0 + 0) = 1.25$$

$$V_3 = 1/4(5 + 20 + 0 + 0) = 6.25$$

$$V_4 = 1/4(1.25 + 6.25 + 0 + 0) = 1.875$$

and so on. To avoid confusion, each time a new value at a free node is calculated, we cross out the old value as shown in Figure 14.12. After V_8 is calculated, we start the second iteration at node 1:

$$V_1 = 1/4(0 + 20 + 1.25 + 6.25) = 6.875$$

$$V_2 = 1/4(6.875 + 0 + 0 + 1.875) = 2.187$$

and so on. If this process is continued, we obtain the uncrossed values shown in Figure 14.12 after five iterations. After 10 iterations (not shown in Figure 14.12), we obtain

$$V_1 = 10.04, \quad V_2 = 4.956, \quad V_3 = 15.22, \quad V_4 = 9.786$$

$$V_5 = 21.05, \quad V_6 = 18.97, \quad V_7 = 15.06, \quad V_8 = 11.26$$

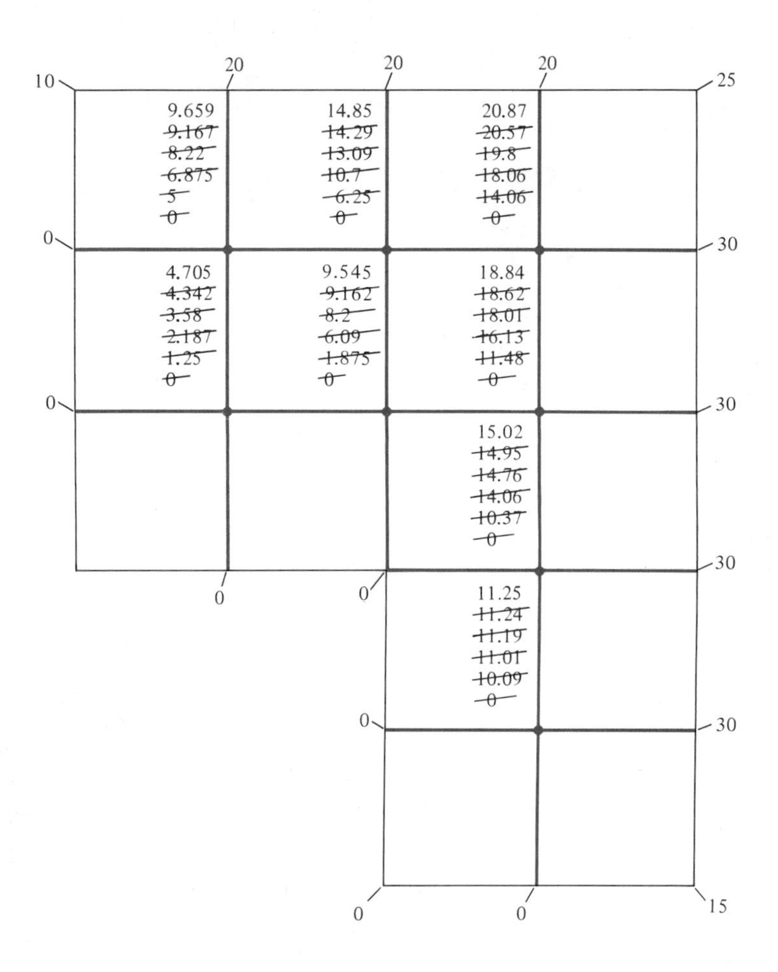

Figure 14.12 For Example 14.3; the values not crossed out are the solutions after five iterations.

Method 2 (Band Matrix Method): This method reveals the sparse structure of the problem. We apply eq. (14.16) to each free node and keep the known terms (prescribed potentials at the fixed nodes) on the right side while the unknown terms (potentials at free nodes) are on the left side of the resulting system of simultaneous equations which will be expressed in matrix form as $[A]\ [V] = [B]$.
For node 1,

$$-4V_1 + V_2 + V_3 = -20 - 0$$

For node 2,

$$V_1 + 4V_2 + V_4 = -0 - 0$$

For node 3,

$$V_1 - 4V_3 + V_4 + V_5 = -20$$

For node 4,

$$V_2 + V_3 - 4V_4 + V_6 = -0$$

For node 5,

$$V_3 - 4V_5 + V_6 = -20 - 30$$

For node 6,

$$V_4 + V_5 - 4V_6 + V_7 = -30$$

For node 7,

$$V_6 - 4V_7 + V_8 = -30 - 0$$

For node 8,

$$V_7 - 4V_8 = -0 - 0 - 30$$

Note that we have five terms at each node since we are using a five-node molecule. The eight equations obtained are put in matrix form as:

$$\begin{bmatrix} -4 & 1 & 1 & 0 & 0 & 0 & 0 & 0 \\ 1 & -4 & 0 & 1 & 0 & 0 & 0 & 0 \\ 1 & 0 & -4 & 1 & 1 & 0 & 0 & 0 \\ 0 & 1 & 1 & -4 & 0 & 1 & 0 & 0 \\ 0 & 0 & 1 & 0 & -4 & 1 & 0 & 0 \\ 0 & 0 & 0 & 1 & 1 & -4 & 1 & 0 \\ 0 & 0 & 0 & 0 & 0 & 1 & -4 & 1 \\ 0 & 0 & 0 & 0 & 0 & 0 & 1 & -4 \end{bmatrix} \begin{bmatrix} V_1 \\ V_2 \\ V_3 \\ V_4 \\ V_5 \\ V_6 \\ V_7 \\ V_8 \end{bmatrix} = \begin{bmatrix} -20 \\ 0 \\ -20 \\ 0 \\ -50 \\ -30 \\ -30 \\ -30 \end{bmatrix}$$

or

$$[A]\,[V] = [B]$$

where $[A]$ is the band, sparse matrix, $[V]$ is the column matrix consisting of the unknown potentials at the free nodes, and $[B]$ is the column matrix formed by the potential at the fixed nodes. The ''band'' nature of $[A]$ is shown by the dotted loop.

Notice that matrix $[A]$ could have been obtained directly from Figure 14.11 without writing down eq. (14.16) at each free node. To do this, we simply set the diagonal (or self) terms $A_{ii} = -4$ and set $A_{ij} = 1$ if i and j nodes are connected or $A_{ij} = 0$ if i and j nodes are not directly connected. For example, $A_{23} = A_{32} = 0$ because nodes 2 and 3 are not connected whereas $A_{46} = A_{64} = 1$ because nodes 4 and 6 are connected. Similarly, matrix [*B*] is obtained directly from Figure 14.11 by setting B_i equal to minus the sum of the potentials at fixed nodes connected to node i. For example, $B_5 = -(20 + 30)$ because node 5 is connected to two fixed nodes with potentials 20 V and 30 V. If node i is not connected to any fixed node, $B_i = 0$.

By inverting matrix [*A*] (using the subroutine INVERSE of Appendix C.1), we obtain

$$[V] = [A]^{-1} [B]$$

or

$$V_1 = 10.04, \quad V_2 = 4.958, \quad V_3 = 15.22, \quad V_4 = 9.788$$

$$V_5 = 21.05, \quad V_6 = 18.97, \quad V_7 = 15.06, \quad V_8 = 11.26$$

which compares well with the result obtained using the iteration method. ■

PRACTICE EXERCISE 14.3

Use the iteration method to find the finite difference approximation to the potentials at points *a* and *b* of the system in Figure 14.13.

ANSWER $V_a = 10.01$ V, $V_b = 28.3$ V.

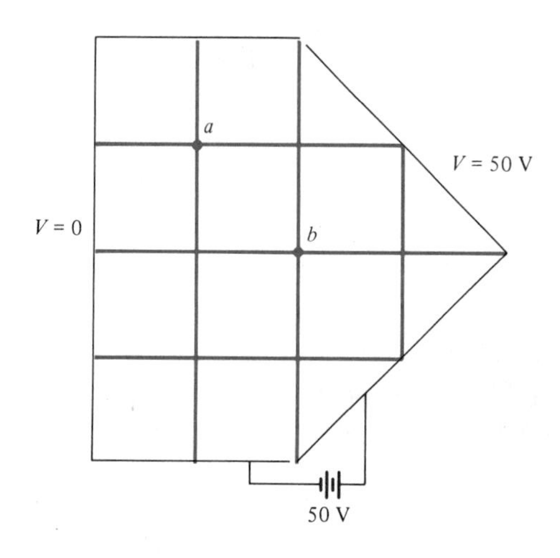

Figure 14.13 For Practice Exercise 14.3.

EXAMPLE 14.4

Obtain the solution of Laplace's equation for an infinitely long trough whose rectangular cross section is shown in Figure 14.14. Let $V_1 = 10$ V, $V_2 = 100$ V, $V_3 = 40$ V, and $V_4 = 0$ V.

SOLUTION

We shall solve this problem using the iteration method and the relaxation method.

Method 1 (Iteration Method): In this case, the solution region has a regular boundary. We can easily write a program to determine the potentials at the grid points within the trough. We divide the region into square meshes. If we decide to use a 15×10 grid, the number of grid points along x is $15 + 1 = 16$ and the number of grid points along y is $10 + 1 = 11$. The mesh size $h = 1.5/15 = 0.1$ m. The 15×10 grid is illustrated in Figure 14.15. The grid points are numbered (i, j) starting from the lower left-hand corner of the trough. Applying eq. (14.15) and using the iteration method, the computer program in Figure 14.16 was developed to determine the potential at the free nodes. At points $(x, y) = (0.5, 0.5)$, $(0.8, 0.8)$, $(1.0, 0.5)$, and $(0.8, 0.2)$ corresponding to $(i, j) = (5, 5)$, $(8, 8)$, $(10, 5)$, and $(8, 2)$ respectively, the potentials after 50, 100, and 200 iterations are shown in Table 14.1. The exact values (see Problem 6.5), obtained using the method of separation of variables and a program similar to that of Figure 6.11, are also shown. It should be noted that the degree of accuracy depends on the mesh size h. It is always desirable to make h as small as possible. Also note that the potentials at the fixed nodes are held constant throughout the calculations.

Method 2 (Relaxation Method): In addition to applying eq. (14.15) to each free node, we determine the residue using eq. (14.19) and check whether the residue is close to a predetermined value. If not, we repeat calculations at the free nodes. A simple program to do this is shown in Figure 14.17. It is essentially a modified version of that in Figure 14.16. A loop counter NCOUNT is inserted to determine

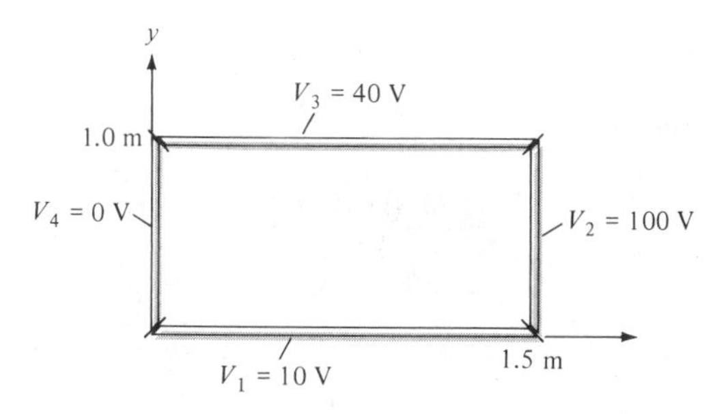

Figure 14.14 For Example 14.4.

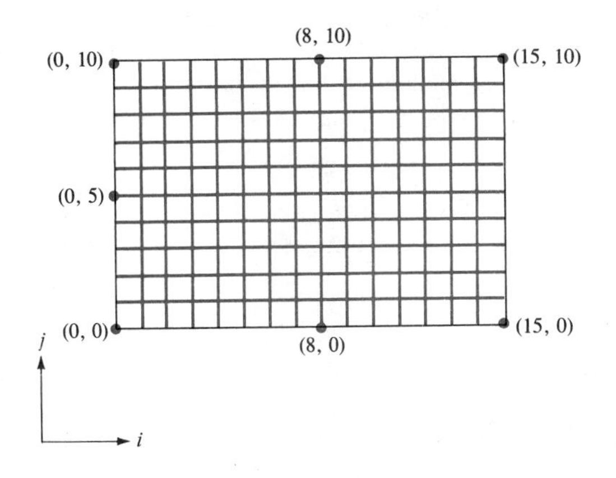

Figure 14.15 For Example 14.4; a 15 × 10 grid.

how many times the calculations are made for the residue to reach a tolerance value of 10^{-4}.

By making $V = 0.0$ the initial potential at every free node, we obtain NCOUNT = 150 and

$$V(5, 5) = 22.48, \qquad V(8, 8) = 38.59$$

$$V(10, 5) = 43.22, \qquad V(8, 2) = 20.97$$

A considerable saving in computer time (and hence less value of NCOUNT) results when the initial guess is made more intelligently. If we make initial potentials at the free nodes equal the average value of the potentials on the boundary; that is, $V = 1/4(V_1 + V_2 + V_3 + V_4)$ as in Figure 14.17, we obtain NCOUNT = 86. The final values of the potentials at the free nodes remain the same since the same tolerance is used. To increase the accuracy of the calculations, we may reduce tolerance or reduce the mesh size or both. ■

Table 14.1 **Solution of Example 14.4 (Iteration Method) at Selected Points**

Coordinates	Number of iterations			
(x, y)	*50*	*100*	*200*	**Exact value**
(0.5, 0.5)	20.91	22.44	22.49	22.44
(0.8, 0.8)	37.7	38.56	38.59	38.55
(1.0, 0.5)	41.83	43.18	43.2	43.22
(0.8, 0.2)	19.87	20.94	20.97	20.89

```
0001    C     USING FINITE DIFFERENCE (ITERATION) METHOD
0002    C     THIS PROGRAM SOLVES THE TWO-DIMENSIONAL BOUNDARY-VALUE
0003    C     PROBLEM (LAPLACE'S EQUATION) SHOWN IN FIG.14.14
0004    C
0005    C     NI = NO. OF ITERATIONS
0006    C     NX = NO. OF X GRID POINTS
0007    C     NY = NO. OF Y GRID POINTS
0008    C     V(I,J) = POTENTIAL AT GRID POINT (I,J) OR (X,Y) WITH
0009    C     NODE NUMBERING STARTING FROM THE LOWER LEFT-HAND
0010    C     CORNER OF THE TROUGH
0011
0012          DIMENSION V(0:100,0:100)
0013
0014    C     SPECIFY BOUNDARY VALUES AND NECESSARY PARAMETERS
0015          V1=10.0
0016          V2=100.0
0017          V3=40.0
0018          V4=0.0
0019          NI=200
0020          NX=15
0021          NY=10
0022    C     SET INITIAL VALUES EQUAL TO ZEROS
0023          DO 10 I=0,NX
0024          DO 10 J=0,NY
0025    10    V(I,J)=0.0
0026    C     FIX POTENTIALS AT FIXED NODES
0027          DO 20 I=1,NX-1
0028          V(I,0)=V1
0029          V(I,NY)=V3
0030    20    CONTINUE
0031          DO 30 J=1,NY-1
0032          V(0,J)=V4
0033          V(NX,J)=V2
0034    30    CONTINUE
0035          V(0,0)=(V1 + V4)/2.0
0036          V(NX,0)=(V1 + V2)/2.0
0037          V(0,NY)=(V3 + V4)/2.0
0038          V(NX,NY)=(V2 + V3)/2.0
0039    C     FIND V(I,J) USING EQ.(14.15) AFTER NI ITERATIONS
0040          DO 50 K=1,NI
0041          DO 40 I=1,NX-1
0042          DO 40 J=1,NY-1
0043          V(I,J)=( V(I+1,J) + V(I-1,J) + V(I,J+1) + V(I,J-1) )/4.
0044    40    CONTINUE
0045    50    CONTINUE
0046          PRINT *,V(5,5),V(8,8),V(10,5),V(8,2)
0047          WRITE(6,60) NI
0048    60    FORMAT(2X,'NO. OF ITERATIONS =',I4,/)
0049          DO 80 J=0,NY
0050          DO 80 I=0,NX
0051          WRITE(6,70) J,I,V(I,J)
0052    70    FORMAT(2X,'J=',I3,3X,'I=',I3,3X,'V(I,J)=',F10.4,/)
0053    80    CONTINUE
0054          STOP
0055          END
```

Figure 14.16 Computer program for Example 14.4.

```
0001    C       USING FINITE DIFFERENCE (RELAXATION) METHOD
0002    C       THIS PROGRAM SOLVES THE TWO-DIMENSIONAL BOUNDARY-VALUE
0003    C       PROBLEM (LAPLACE'S EQUATION) SHOWN IN FIG.14.14
0004    C
0005    C       NX = NO. OF X GRID POINTS
0006    C       NY = NO. OF Y GRID POINTS
0007    C       RMIN IS TOLERANCE VALUE OF THE RESIDUE RES
0008    C       V(I,J) = POTENTIAL AT GRID POINT (I,J) OR (X,Y) WITH
0009    C       NODE NUMBERING STARTING FROM THE LOWER LEFT-HAND
0010    C       CORNER OF THE TROUGH
0011
0012            DIMENSION V(0:100,0:100)
0013
0014    C       SPECIFY BOUNDARY VALUES AND NECESSARY PARAMETERS
0015            V1=10.0
0016            V2=100.0
0017            V3=40.0
0018            V4=0.0
0019            NX=15
0020            NY=10
0021    C       SET INITIAL VALUES TO ZEROS OR TO GUESSED VALUES
0022            DO 10 I=0,NX
0023            DO 10 J=0,NY
0024    10      V(I,J)=(V1 + V2 + V3 + V4)/4.0
0025            DO 20 I=1,NX-1
0026    C       FIX POTENTIALS AT FIXED NODES
0027            V(I,0)=V1
0028            V(I,NY)=V3
0029    20      CONTINUE
0030            DO 30 J=1,NY-1
0031            V(0,J)=V4
0032            V(NX,J)=V2
0033    30      CONTINUE
0034            V(0,0)=(V1 + V4)/2.0
0035            V(NX,0)=(V1 + V2)/2.0
0036            V(0,NY)=(V3 + V4)/2.0
0037            V(NX,NY)=(V2 + V3)/2.0
0038            RMIN=0.0001
0039            NCOUNT=0
0040    40      RES=0.0
0041            NCOUNT=NCOUNT + 1
0042    C       APPLY EQ.(14.15) AND FIND THE LARGEST RESIDUE
0043            DO 50 I=1,NX-1
0044            DO 50 J=1,NY-1
0045            VNEW=0.25*( V(I+1,J) + V(I-1,J) + V(I,J+1) + V(I,J-1) )
0046            R=ABS( V(I,J) - VNEW )
0047            IF(R.GT.RES) RES=R
0048    C       REPLACE OLD VALUE WITH NEWLY COMPUTED VALUE
0049    50      V(I,J)=VNEW
0050    C       TEST FOR CONVERGENCE - RESIDUE RES SHOULD DECREASE
0051    C       CONSISTENTLY OR THE PROCESS IS DIVERGING
0052            PRINT *,NCOUNT,RES,V(5,5),V(8,8),V(10,5),V(8,2)
0053           IF(RES.GE.RMIN) GO TO 40
0054            STOP
0055            END
```

Figure 14.17 Computer program for Example 14.4.

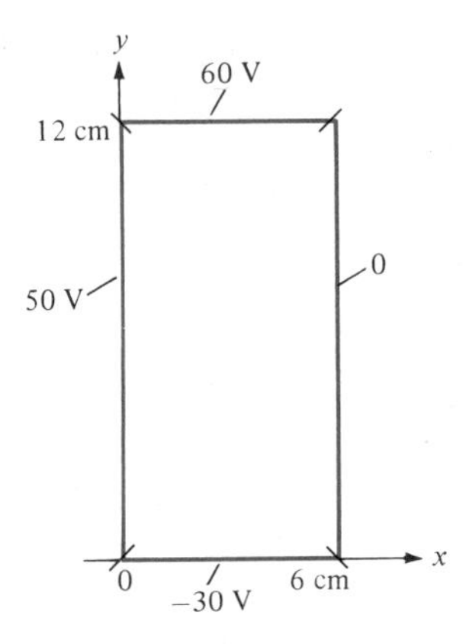

Figure 14.18 For Practice Exercise 14.4.

PRACTICE EXERCISE 14.4

Consider the trough of Figure 14.18. Use a five-node finite difference scheme to find the potential at the center of the trough using (a) a 4×8 grid, and (b) a 12×24 grid.

ANSWER (a) 23.8 V, (b) 23.89 V.

14.4 THE MOMENT METHOD

Like the finite difference method, the moment method[2] has the advantage of being conceptually simple. While the finite difference method is used in solving differential equations, the moment method is commonly used in solving integral equations.

For example, suppose we want to apply the moment method to solve Poisson's equation in eq. (14.9a). It can be shown that an integral solution to Poisson's equation is

$$V = \int \frac{\rho_v \, dv}{4\pi\varepsilon r} \tag{14.26}$$

We recall from Chapter 4 that eq. (14.26) can be derived from Coulomb's law. We also recall that given the charge distribution $\rho_v(x, y, z)$, we can always find the potential $V(x, y, z)$, the electric field $\mathbf{E}(x, y, z)$, and the total charge Q. If, on the other hand, the potential V is known and the charge distribution is unknown, how do we determine

[2]The term "moment method" was first used in Western literature by Harrington. For further exposition on the method, see R. F. Harrington, *Field Computation by Moment Methods*, Malabar, FL: Krieger, 1968.

ρ_v from eq. (14.26)? In that situation, eq. (14.26) becomes what is called an *integral equation*. In general, an integral equation is one involving the unknown function under the integral sign. It has the general form of

$$V(x) = \int_a^b K(x, t)\,\rho(t)\,dt \tag{14.27}$$

where the functions $K(x, t)$ and $V(t)$ and the limits a and b are known. The unknown function $\rho(t)$ is to be determined; the function $K(x, t)$ is called the *kernel* of the equation. The moment method is a common numerical technique used in solving integral equations such as in eq. (14.27). The method is probably best explained with an example.

Consider a thin conducting wire of radius a, length L ($L \gg a$) located in free space as shown in Figure 14.19. Let the wire be maintained at a potential of V_o. Our goal is to determine the charge density ρ_L along the wire using the moment method. Once we determine ρ_L, related field quantities can be found. At any point on the wire, eq. (14.26) reduces to an integral equation of the form

$$V_o = \int_0^L \frac{\rho_L\,dl}{4\pi\varepsilon_o r} \tag{14.28}$$

Since eq. (14.28) applies for observation points everywhere on the wire, at a fixed point y_k known as the *match point,*

$$V_o = \frac{1}{4\pi\varepsilon_o}\int_0^L \frac{\rho_L(y)\,dy}{|y_k - y|} \tag{14.29}$$

We recall from calculus that integration is essentially finding the area under a curve. If Δy is small, the integration of $f(y)$ over $0 < y < L$ is given by

$$\begin{aligned}\int_0^L f(y)\,dy &\simeq f(y_1)\,\Delta y + f(y_2)\,\Delta y + \cdots + f(y_N)\,\Delta y \\ &= \sum_{k=1}^{N} f(y_k)\,\Delta y\end{aligned} \tag{14.30}$$

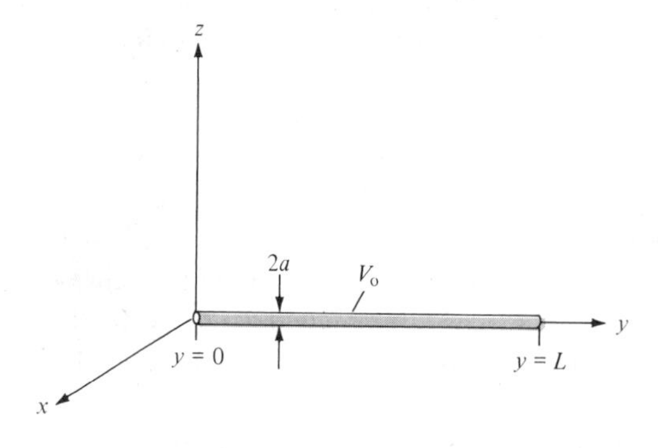

Figure 14.19 Thin conducting wire held at a constant potential.

where the interval L has been divided into N units of each length Δy. With the wire divided into N segments of equal length Δ as shown in Figure 14.20, eq. (14.29) becomes

$$4\pi\varepsilon_o V_o \simeq \frac{\rho_1 \Delta}{|y_k - y_1|} + \frac{\rho_2 \Delta}{|y_k - y_2|} + \cdots + \frac{\rho_N \Delta}{|y_k - y_N|} \tag{14.31}$$

where $\Delta = L/N = \Delta y$. The assumption in eq. (14.31) is that the unknown charge density ρ_k on the kth segment is constant. Thus in eq. (14.31), we have unknown constants $\rho_1, \rho_2, \ldots, \rho_N$. Since eq. (14.31) must hold at all points on the wire, we obtain N similar equations by choosing N match points at $y_1, y_2, \ldots, y_k, \ldots y_N$ on the wire. Thus we obtain

$$4\pi\varepsilon_o V_o = \frac{\rho_1 \Delta}{|y_1 - y_1|} + \frac{\rho_2 \Delta}{|y_1 - y_2|} + \cdots + \frac{\rho_N \Delta}{|y_1 - y_N|} \tag{14.32a}$$

$$4\pi\varepsilon_o V_o = \frac{\rho_1 \Delta}{|y_2 - y_1|} + \frac{\rho_2 \Delta}{|y_2 - y_2|} + \cdots + \frac{\rho_N \Delta}{|y_2 - y_N|} \tag{14.32b}$$

$$\vdots$$

$$4\pi\varepsilon_o V_o = \frac{\rho_1 \Delta}{|y_N - y_1|} + \frac{\rho_2 \Delta}{|y_N - y_2|} + \cdots + \frac{\rho_N \Delta}{|y_N - y_N|} \tag{14.32c}$$

The idea of matching the left-hand side of eq. (14.29) with the right-hand side of the equation at the match points is similar to the concept of taking moments in mechanics. Here lies the reason why this technique is called moment method. Notice from Figure

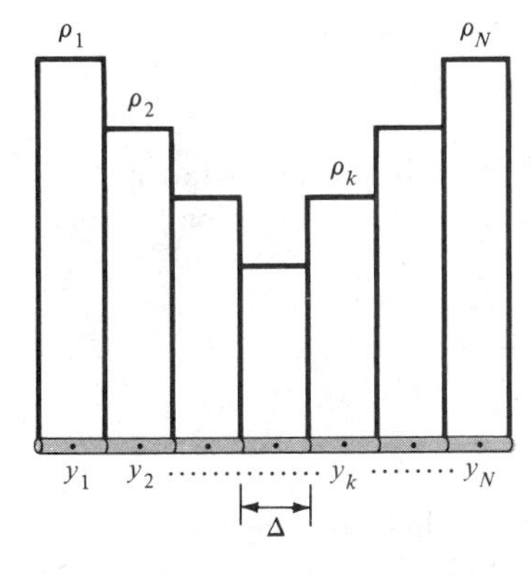

Figure 14.20 Division of the wire into N segments.

14.20 that the match points $y_1, y_2, \ldots, y_N$ are placed at the center of each segment. Equation (14.32) can be put in matrix form as

$$[B] = [A]\,[\rho] \tag{14.33}$$

where

$$[B] = 4\pi\varepsilon_o V_o \begin{bmatrix} 1 \\ 1 \\ \cdot \\ \cdot \\ \cdot \\ 1 \end{bmatrix} \tag{14.34}$$

$$[A] = \begin{bmatrix} A_{11} & A_{12} & \cdots & A_{1N} \\ A_{21} & A_{22} & \cdots & A_{2N} \\ \cdot & & & \cdot \\ \cdot & & & \cdot \\ \cdot & & & \cdot \\ A_{N1} & A_{N2} & \cdots & A_{NN} \end{bmatrix} \tag{14.35a}$$

$$A_{mn} = \frac{\Delta}{|y_m - y_n|}, \qquad m \neq n \tag{14.35b}$$

$$[\rho] = \begin{bmatrix} \rho_1 \\ \rho_2 \\ \cdot \\ \cdot \\ \cdot \\ \rho_N \end{bmatrix} \tag{14.36}$$

In eq. (14.33), $[\rho]$ is the matrix whose elements are unknown. We can determine $[\rho]$ from eq. (14.33) using Cramer's rule, matrix inversion, or Gaussian elimination technique. Using matrix inversion,

$$\boxed{[\rho] = [A]^{-1}\,[B]} \tag{14.37}$$

where $[A]^{-1}$ is the inverse of matrix $[A]$. In evaluating the diagonal elements (or self terms) of matrix $[A]$ in eq. (14.32) or (14.35), caution must be exercised. Since the wire is conducting, a surface charge density ρ_S is expected over the wire surface. Hence at the center of each segment,

$$V\ (\text{center}) = \frac{1}{4\pi\varepsilon_o} \int_0^{2\pi} \int_{-\Delta/2}^{\Delta/2} \frac{\rho_S a\, d\phi\, dy}{[a^2 + y^2]^{1/2}}$$

$$= \frac{2\pi a \rho_S}{4\pi\varepsilon_o} \ln \left\{ \frac{\Delta/2 + [(\Delta/2)^2 + a^2]^{1/2}}{-\Delta/2 + [(\Delta/2)^2 + a^2]^{1/2}} \right\}$$

Assuming $\Delta \gg a$,

$$V\ (\text{center}) = \frac{2\pi a \rho_S}{4\pi\varepsilon_o}\, 2 \ln\left(\frac{\Delta}{a}\right) \quad [14.38]$$

$$= \frac{2\rho_L}{4\pi\varepsilon_o} \ln\left(\frac{\Delta}{a}\right)$$

where $\rho_L = 2\pi\ a\rho_S$. Thus, the self terms ($m = n$) are

$$A_{nn} = 2 \ln\left(\frac{\Delta}{a}\right) \quad [14.39]$$

Eq. (14.33) now becomes

$$\begin{bmatrix} 2\ln\left(\frac{\Delta}{a}\right) & \frac{\Delta}{|y_1 - y_2|} & \cdots & \frac{\Delta}{|y_1 - y_N|} \\ \frac{\Delta}{|y_2 - y_1|} & 2\ln\left(\frac{\Delta}{a}\right) & \cdots & \frac{\Delta}{|y_2 - y_N|} \\ \cdot & & & \cdot \\ \cdot & & & \cdot \\ \cdot & & & \cdot \\ \frac{\Delta}{|y_N - y_1|} & \frac{\Delta}{|y_N - y_2|} & \cdots & 2\ln\left(\frac{\Delta}{a}\right) \end{bmatrix} \begin{bmatrix} \rho_1 \\ \rho_2 \\ \cdot \\ \cdot \\ \cdot \\ \rho_N \end{bmatrix} = 4\pi\varepsilon_o V_o \begin{bmatrix} 1 \\ 1 \\ \cdot \\ \cdot \\ \cdot \\ 1 \end{bmatrix} \quad [14.40]$$

Using eq. (14.37) with eq. (14.40) and letting $V_o = 1\ V$, $L = 1$ m, $a = 1$ mm, and $N = 10$ ($\Delta = L/N$), a FORTRAN code such as in Figure 14.21 can be developed. The program in Figure 14.21 is self-explanatory. It calls subroutine INVERSE in Appendix C.1 to invert matrix [A] and uses subroutine PLOT in Appendix C.2 to plot ρ_L against y. The plot is shown in Figure 14.22. The program also determines the total charge on the wire using

$$Q = \int \rho_L\, dl \quad [14.41]$$

which can be written in discrete form as

$$Q = \sum_{k=1}^{N} \rho_k\, \Delta \quad [14.42]$$

```
0001    C     THIS PROGRAM DETERMINES THE CHARGE DISTRIBUTION
0002    C     ON A CONDUCTING THIN WIRE, OF RADIUS AA AND
0003    C     LENGTH L, MAINTAINED AT  VO VOLT
0004    C     THE WIRE IS LOCATED AT  0 < Y < L
0005    C     ALL DIMENSIONS ARE IN S.I. UNITS
0006    C
0007    C     MOMENT METHOD IS USED
0008    C     N IS THE NO. OF SEGMENTS INTO WHICH THE WIRE IS DIVIDED
0009    C     RO IS THE LINE CHARGE DENSITY
0010    C     RO = (1/A).B
0011    C
0012
0013          REAL L
0014          DIMENSION A(100,100),B(100),RO(100),YY(100)
0015
0016    C     FIRST, SPECIFY PROBLEM PARAMETERS
0017
0018          PIE=3.14159
0019          ER=1.0
0020          EO=8.8541E-12
0021          VO=1.0
0022          AA=0.001
0023          L=1.0
0024          N=20
0025          DELTA=L/N
0026    C     SECOND, CALCULATE THE ELEMENTS OF THE COEFFICIENT
0027    C     MATRIX  A
0028          DO 30 I=1,N
0029          Y=DELTA*( FLOAT(I) - 0.5)
0030          YY(I)=Y
0031          DO 20 J=1,N
0032          YP=DELTA*( FLOAT(J) - 0.5)
0033          IF(I-J) 11,10,11
0034    10    A(I,J)=2.0*ALOG(DELTA/AA)
0035          GO TO 12
0036    11    A(I,J)=DELTA/ABS(Y - YP)
0037    12    CONTINUE
0038    20    CONTINUE
0039    30    CONTINUE
0040    C     NOW DETERMINE THE MATRIX OF CONSTANT VECTOR B
0041          DO 40 K=1,N
0042          B(K)=4.0*PIE*EO*ER*VO
0043    40    CONTINUE
0044    C
0045    C     INVERT MATRIX  A(I,J) AND CALCULATE MATRIX RO(N)
0046    C     CONSISTING OF THE UNKNOWN ELEMENTS
0047    C     ALSO CALCULATE THE TOTAL CHARGE Q
0048          NIV=N
0049          NMAX=100
0050    C     NIV IS THE SIZE OF THE MATRIX A TO BE INVERTED
0051    C     NMAX IS THE DIMENSION OF A
0052          CALL INVERSE(A,NIV,NMAX)
0053          SUM=0.0
0054          DO 60 I=1,N
0055          RO(I)=0.0
0056          DO 50 M=1,N
0057          RO(I)=RO(I) + A(I,M)*B(M)
0058    50    CONTINUE
0059          SUM=SUM + RO(I)
0060    60    CONTINUE
```

Figure 14.21 FORTRAN code for the conducting thin wire of Figure 14.19.

```
0061            Q=SUM*DELTA
0062            WRITE(6,70) Q
0063     70     FORMAT(2X,'Q=',E14.6,/)
0064            DO 90 I=1,N
0065            WRITE(6,80) YY(I),RO(I)
0066     80     FORMAT(2X,'Y=',F10.4,3X,'RO=',2X,E14.6,/)
0067     90     CONTINUE
0068     C      FINALLY PLOT RO AGAINST Y
0069            CALL PLOT(RO,N,DELTA,DELTA,1)
0070            STOP
0071            END
```

Figure 14.21 (Continued)

With the chosen parameters, the value of the total charge was found to be $Q = 8.536$ pC. If desired, the electric field at any point can be calculated using

$$\mathbf{E} = \int \frac{\rho_L \, dl}{4\pi\varepsilon_o R^2} \mathbf{a}_R \tag{14.43}$$

which can be written as

$$\mathbf{E} = \sum_{k=1}^{N} \frac{\rho_k \, \Delta \mathbf{R}}{4\pi\varepsilon_o R^3} \tag{14.44}$$

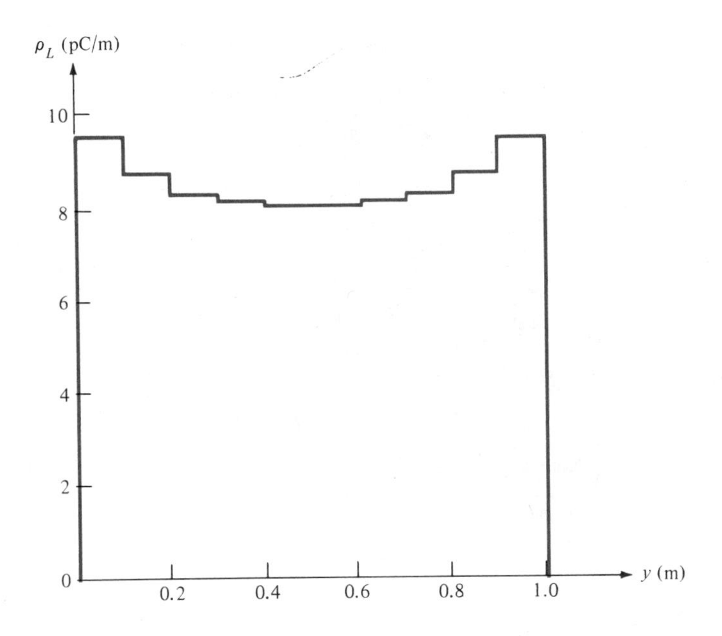

Figure 14.22 Plot of ρ_L against y.

where $R = |\mathbf{R}|$ and

$$\mathbf{R} = \mathbf{r} - \mathbf{r}_k = (x - x_k)\mathbf{a}_x + (y - y_k)\mathbf{a}_y + (z - z_k)\mathbf{a}_z$$

$\mathbf{r} = (x, y, z)$ is the position vector of the observation point, and $\mathbf{r}_k = (x_k, y_k, z_k)$ is that of the source point.

Notice that to obtain the charge distribution in Figure 14.22, we have taken $N = 10$. It should be expected that a smaller value of N would give a less accurate result and a larger value of N would yield a more accurate result. However, if N is too large, we may have the computation problem of inverting the square matrix $[A]$. The capacity of the computing facilities at our disposal can limit the accuracy of the numerical experiment.

EXAMPLE 14.5

Use the moment method to find the capacitance of the parallel-plate capacitor of Figure 14.23. Take $a = 1$ m, $b = 1$ m, $d = 1$ m, and $\varepsilon_r = 1.0$.

SOLUTION Let the potential difference between the plates be $V_o = 2$ V so that the top plate P_1 is maintained at $+1$ V while the bottom plate P_2 is at -1 V. We would like to determine the surface charge density ρ_s on the plates so that the total charge on each plate can be found as

$$Q = \int \rho_S \, dS$$

Once Q is known, we can calculate the capacitance as

$$C = \frac{Q}{V_o} = \frac{Q}{2}$$

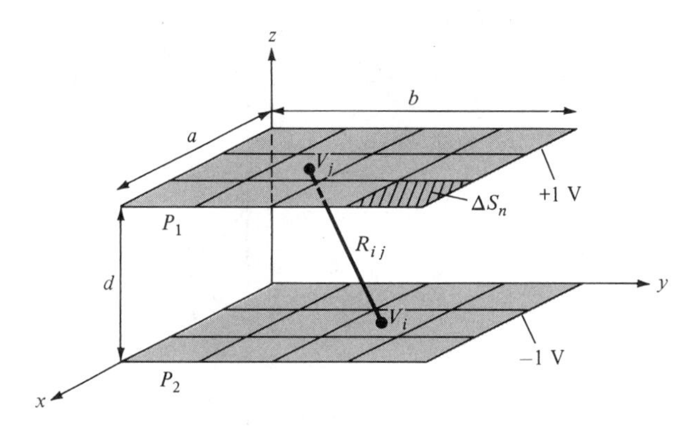

Figure 14.23 Parallel-plate capacitor; for Example 14.5.

To determine ρ_S using the moment method, we divide P_1 into n subsections: ΔS_1, $\Delta S_2, \ldots, \Delta S_n$ and P_2 into n subsections: $\Delta S_{n+1}, \Delta S_{n+2}, \ldots, \Delta S_{2n}$. The potential V_i at the center of a typical subsection ΔS_i is

$$V_i = \int_S \frac{\rho_S \, dS}{4\pi\varepsilon_o R} \simeq \sum_{j=1}^{2n} \frac{1}{4\pi\varepsilon_o} \int_{\Delta S_i} \frac{\rho_j \, dS}{R_{ij}}$$

$$= \sum_{j=1}^{2n} \rho_j \frac{1}{4\pi\varepsilon_o} \int_{\Delta S_j} \frac{dS}{R_{ij}}$$

It has been assumed that there is uniform charge distribution on each subsection. The last equation can be written as

$$V_i = \sum_{j=1}^{2n} \rho_j A_{ij}$$

where

$$A_{ij} = \frac{1}{4\pi\varepsilon_o} \int_{\Delta S_i} \frac{dS}{R_{ij}}$$

Thus

$$V_1 = \sum_{j=1}^{2n} \rho_j A_{1j} = 1$$

$$V_2 = \sum_{j=1}^{2n} \rho_j A_{2j} = 1$$

$$\vdots$$

$$V_n = \sum_{j=1}^{2n} \rho_j A_{nj} = 1$$

$$V_{n+1} = \sum_{j=1}^{2n} \rho_j A_{n+1,j} = -1$$

$$\vdots$$

$$V_{2n} = \sum_{j=1}^{2n} \rho_j A_{2n,j} = -1$$

yielding a set of $2n$ simultaneous equations with $2n$ unknown charge densities ρ_j. In matrix form,

$$\begin{bmatrix} A_{11} & A_{12} & \cdots & A_{1,2n} \\ A_{21} & A_{22} & \cdots & A_{2,2n} \\ \vdots & & & \vdots \\ & & & \\ A_{2n,1} & A_{2n,2} & \cdots & A_{2n,2n} \end{bmatrix} \begin{bmatrix} \rho_1 \\ \rho_2 \\ \vdots \\ \\ \rho_{2n} \end{bmatrix} = \begin{bmatrix} 1 \\ 1 \\ \vdots \\ -1 \\ -1 \end{bmatrix}$$

or

$$[A]\,[\rho] = [B]$$

Hence,

$$[\rho] = [A]^{-1}\,[B]$$

where $[B]$ is the column matrix defining the potentials and $[A]$ is a square matrix containing elements A_{ij}. To determine A_{ij}, consider the two subsections i and j shown in Figure 14.24 where the subsections could be on different plates or on the same plate.

$$A_{ij} = \frac{1}{4\pi\varepsilon_o} \int_{y=y_1}^{y_2} \int_{x=x_1}^{x_2} \frac{dx\,dy}{R_{ij}}$$

where

$$R_{ij} = [(x_j - x_i)^2 + (y_j - y_i)^2 + (z_j - z_i)^2]^{1/2}$$

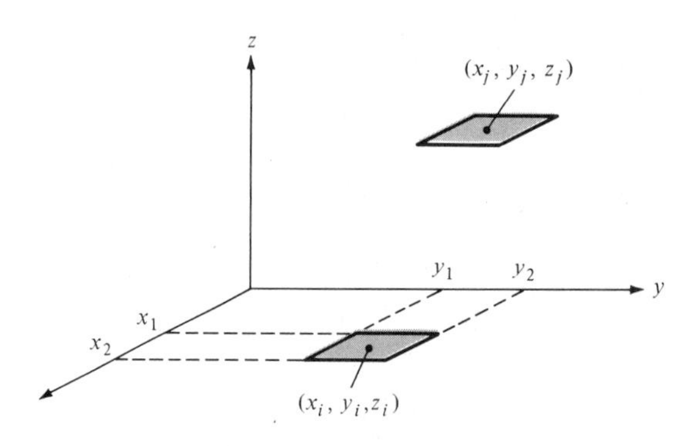

Figure 14.24 Subsections i and j; for Example 14.5.

```
0001    C       USING MOMENT METHOD,
0002    C       THIS PROGRAM DETERMINES THE CAPACITANCE OF A
0003    C       PARALLEL-PLATE CAPACITOR CONSISTING OF TWO CONDUCTING
0004    C       PLATES, EACH OF DIMENSION  AA X BB, SEPARATED BY
0005    C       A DISTANCE D  MAINTAINED AT 1 VOLT AND - 1 VOLT.
0006    C
0007    C       ONE PLATE IS LOCATED ON THE Z=0 PLANE WHILE THE OTHER
0008    C       IS LOCATED ON THE  Z=D  PLANE.
0009    C
0010    C       ALL DIMENSIONS ARE IN S.I. UNITS
0011    C
0012    C       N IS THE NUMBER OF SUBSECTIONS INTO WHICH EACH PLATE IS DIVIDED
0013
0014            DIMENSION  A(100,100), B(100), X(100), Y(100), Z(100), RO(100)
0015
0016    C       FIRST, SPECIFY THE PARAMETERS
0017
0018            PIE=3.14159
0019            ER=1.0
0020            EO=8.8541E-12
0021            AA=1.0
0022            BB=1.0
0023            D=1.0
0024            N=9
0025            NT=2*N
0026            M=SQRT( FLOAT(N) )
0027            DX=AA/M
0028            DY=BB/M
0029            DL=DX
0030    C
0031    C       SECOND, CALCULATE THE ELEMENTS OF THE COEFFICIENT
0032    C       MATRIX  A
0033            DO 25  K1=1,2
0034            DO 20  K2=1,M
0035            DO 10  K3=1,M
0036            K=K + 1
0037            X(K)= DX*( FLOAT(K2)  - 0.5 )
0038            Y(K)= DY*( FLOAT(K3)  - 0.5 )
0039    10      CONTINUE
0040    20      CONTINUE
0041    25      CONTINUE
0042            DO 30 K1=1,N
0043            Z(K1)=0
0044            Z(K1 + N)=D
0045    30      CONTINUE
0046            DO 14 I=1,NT
0047            DO 14 J=1,NT
0048            IF(I-J)  12,11,12
0049    11      A(I,J)= DL*0.8814/(PIE*EO)
0050            GO TO 13
0051    12      R=SQRT( (X(I)  - X(J))**2 + (Y(I)  - Y(J))**2 + (Z(I)  - Z(J))**2)
0052            A(I,J)= DL**2/(4.*PIE*EO*R)
0053    13      CONTINUE
0054    14      CONTINUE
0055    C       NOW DETERMINE THE MATRIX OF CONSTANT VECTOR B
0056            DO 40 K=1,N
0057            B(K)=1.0
0058            B(K+N)= -1.0
0059    40      CONTINUE
0060    C
0061    C       INVERT MATRIX  A(I,J) AND CALCULATE MATRIX RO(N)
```

Figure 14.25 Computer program for Example 14.5 (continued on next page).

```
0062    C      CONSISTING OF THE UNKNOWN ELEMENTS
0063    C      ALSO CALCULATE THE TOTAL CHARGE Q
0064    C      AND THE CAPACITANCE C
0065           NIV=NT
0066           NMAX=100
0067           CALL INVERSE(A,NIV,NMAX)
0068           DO 60 I=1,NT
0069           RO(I)=0.0
0070           DO 50 J=1,NT
0071           WRITE(6,51) I,J,A(I,J)
0072    51     FORMAT(2X,'I=',I5,2X,'J=',I5,2X,'A=',E14.6,/)
0073           RO(I)=RO(I) + A(I,J)*B(J)
0074    50     CONTINUE
0075    60     CONTINUE
0076           SUM=0.0
0077           DO 65 I=1,N
0078           SUM= SUM + RO(I)
0079    65     CONTINUE
0080           Q=SUM*(DL**2)
0081           VO=2.0
0082           C=ABS(Q)/VO
0083           PRINT *,C
0084           WRITE(6,70) C
0085    70     FORMAT(2X,'C=',E14.6,/)
0086           DO 90 I=1,NT
0087           WRITE(6,80) X(I),Y(I),Z(I),RO(I)
0088    80     FORMAT(2X,'X=',F10.4,2X,'Y=',F10.4,2X,'Z=',F10.4,3X,
0089          1    'RO=',E14.6,/)
0090    90     CONTINUE
0091           STOP
0092           END
```

Figure 14.25 (Continued)

For the sake of convenience, if we assume that the subsections are squares,

$$x_2 - x_1 = \Delta\ell = y_2 - y_1$$

it can be shown that

$$A_{ij} = \frac{\Delta S_i}{4\pi\varepsilon_o R_{ij}} = \frac{(\Delta\ell)^2}{4\pi\varepsilon_o R_{ij}} \qquad i \neq j$$

and

$$A_{ii} = \frac{\Delta\ell}{\pi\varepsilon_o}\ln(1 + \sqrt{2}) = \frac{\Delta\ell}{\pi\varepsilon_o}(0.8814)$$

With these formulas, the code in Figure 14.25 was developed. With $n = 9$, $C = 26.51$ pF, with $n = 16$, $C = 27.27$ pF, and with $n = 25$, $C = 27.74$ pF. ■

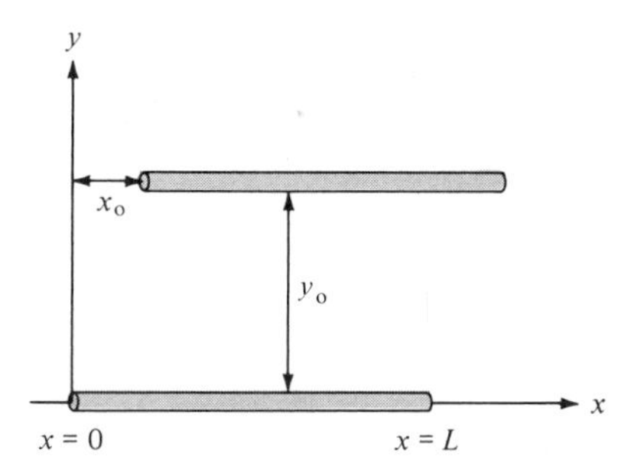

Figure 14.26 Parallel conducting wires of Practice Exercise 14.5.

Table 14.2 Capacitance for Practice Exercise 14.5

x_o (m)	C (pF)
0.0	4.91
0.2	4.891
0.4	4.853
0.6	4.789
0.8	4.71
1.0	4.643

PRACTICE EXERCISE 14.5

Using the moment method, write a program to determine the capacitance of two identical parallel conducting wires separated at a distance y_o and displaced by x_o as shown in Figure 14.26. If each wire is of length L and radius a, find the capacitance for cases $x_o = 0, 0.2, 0.4, \ldots, 1.0$ m. Take $y_o = 0.5$ m, $L = 1$ m, $a = 1$ mm, $\varepsilon_r = 1$.

ANSWER For $N = 10$ = number of segments per wire, see Table 14.2.

14.5 THE FINITE ELEMENT METHOD

The finite element method has its origin in the field of structural analysis. The method was not applied to EM problems until 1968.[3] Like the finite difference method, the finite element method is useful in solving differential equations. As noticed in Section 14.3, the finite difference method represents the solution region by an array of grid points; its application becomes difficult with problems having irregularly shaped

[3]See P. P. Silvester and R. L. Ferrari, *Finite Elements for Electrical Engineers*. Cambridge, England: Cambridge Univ. Press, 1983.

boundaries. Such problems can be handled more easily using the finite element method.

The finite element analysis of any problem involves basically four steps: (a) discretizing the solution region into a finite number of subregions or *elements,* (b) deriving governing equations for a typical element, (c) assembling of all elements in the solution region, and (d) solving the system of equations obtained.

A. Finite Element Discretization

We divide the solution region into a number of *finite elements* as illustrated in Figure 14.27 where the region is subdivided into four nonoverlapping elements (two triangular and two quadrilateral) and seven nodes. We seek an approximation for the potential V_e within an element e and then interrelate the potential distributions in various elements such that the potential is continuous across interelement boundaries. The approximate solution for the whole region is

$$V(x, y) \simeq \sum_{e=1}^{N} V_e(x, y) \qquad [14.45]$$

where N is the number of triangular elements into which the solution region is divided.

The most common form of approximation for V within an element is polynomial approximation, namely

$$V_e(x, y) = a + bx + cy \qquad [14.46]$$

for a triangular element and

$$V_e(x, y) = a + bx + cy + dxy \qquad [14.47]$$

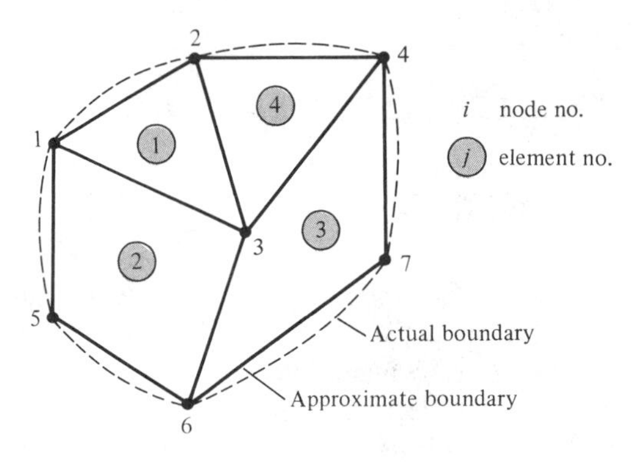

Figure 14.27 A typical finite element subdivision of an irregular domain.

for a quadrilateral element. The potential V_e in general is nonzero within element e but zero outside e. It is difficult to approximate the boundary of the solution region with quadrilateral elements; such elements are useful for problems whose boundaries are sufficiently regular. In view of this, we prefer to use triangular elements throughout our analysis in this section. Notice that our assumption of linear variation of potential within the triangular element as in eq. (14.46) is the same as assuming that the electric field is uniform within the element; that is,

$$\mathbf{E}_e = -\nabla V_e = -(b\mathbf{a}_x + c\mathbf{a}_y) \tag{14.48}$$

B. Element Governing Equations

Consider a typical triangular element shown in Figure 14.28. The potential V_{e1}, V_{e2}, and V_{e3} at nodes 1, 2, and 3 respectively are obtained using eq. (14.46); that is,

$$\begin{bmatrix} V_{e1} \\ V_{e2} \\ V_{e3} \end{bmatrix} = \begin{bmatrix} 1 & x_1 & y_1 \\ 1 & x_2 & y_2 \\ 1 & x_3 & y_3 \end{bmatrix} \begin{bmatrix} a \\ b \\ c \end{bmatrix} \tag{14.49}$$

The coefficients a, b, and c are determined from eq. (14.49) as

$$\begin{bmatrix} a \\ b \\ c \end{bmatrix} = \begin{bmatrix} 1 & x_1 & y_1 \\ 1 & x_2 & y_2 \\ 1 & x_3 & y_3 \end{bmatrix}^{-1} \begin{bmatrix} V_{e1} \\ V_{e2} \\ V_{e3} \end{bmatrix} \tag{14.50}$$

Substituting this into eq. (14.46) gives

$$V_e = \begin{bmatrix} 1 & x & y \end{bmatrix} \frac{1}{2A} \begin{bmatrix} (x_2y_3 - x_3y_2) & (x_3y_1 - x_1y_3) & (x_1y_2 - x_2y_1) \\ (y_2 - y_3) & (y_3 - y_1) & (y_1 - y_2) \\ (x_3 - x_2) & (x_1 - x_3) & (x_2 - x_1) \end{bmatrix} \begin{bmatrix} V_{e1} \\ V_{e2} \\ V_{e3} \end{bmatrix}$$

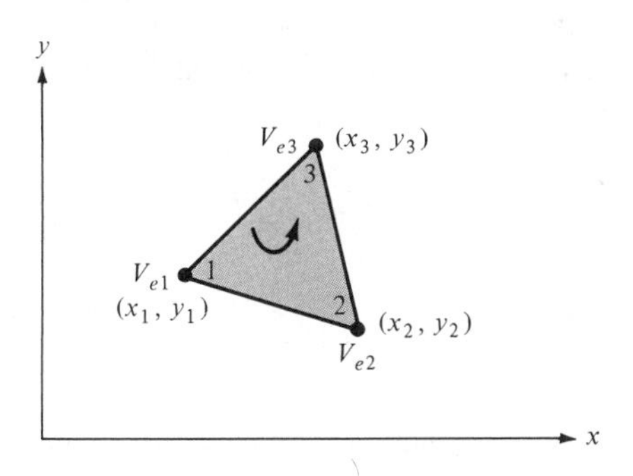

Figure 14.28 Typical triangular element; the local node numbering 1-2-3 must be counterclockwise as indicated by the arrow.

or

$$\boxed{V_e = \sum_{i=1}^{3} \alpha_i(x, y)\, V_{ei}} \qquad [14.51]$$

where

$$\alpha_1 = \frac{1}{2A}\left[(x_2y_3 - x_3y_2) + (y_2 - y_3)\,x + (x_3 - x_2)\,y\right] \qquad [14.52a]$$

$$\alpha_2 = \frac{1}{2A}\left[(x_3y_1 - x_1y_3) + (y_3 - y_1)\,x + (x_1 - x_3)\,y\right] \qquad [14.52b]$$

$$\alpha_3 = \frac{1}{2A}\left[(x_1y_2 - x_2y_1) + (y_1 - y_2)\,x + (x_2 - x_1)\,y\right] \qquad [14.52c]$$

and A is the area of the element e; that is,

$$2A = \begin{vmatrix} 1 & x_1 & y_1 \\ 1 & x_2 & y_2 \\ 1 & x_3 & y_3 \end{vmatrix}$$

$$= (x_1y_2 - x_2y_1) + (x_3y_1 - x_1y_3) + (x_2y_3 - x_3y_2)$$

or

$$A = 1/2\left[(x_2 - x_1)(y_3 - y_1) - (x_3 - x_1)(y_2 - y_1)\right] \qquad [14.53]$$

The value of A is positive if the nodes are numbered counterclockwise (starting from any node) as shown by the arrow in Figure 14.28. Note that eq. (14.51) gives the potential at any point (x, y) within the element provided that the potentials at the vertices are known. This is unlike the situation in finite difference analysis where the potential is known at the grid points only. Also note that α_i are linear interpolation functions. They are called the *element shape functions* and they have the following properties:

$$\alpha_i(x_j, y_j) = \begin{bmatrix} 1, & i = j \\ 0, & i \neq j \end{bmatrix} \qquad [14.54a]$$

$$\sum_{i=1}^{3} \alpha_i(x, y) = 1 \qquad [14.54b]$$

The shape functions α_1 and α_2, for example, are illustrated in Figure 14.29.

The energy per unit length associated with the element e is given by eq. (4.103); that is,

$$W_e = \frac{1}{2}\int \varepsilon\,|\mathbf{E}|^2\,dS = \frac{1}{2}\int \varepsilon\,|\nabla V_e|^2\,dS \qquad [14.55]$$

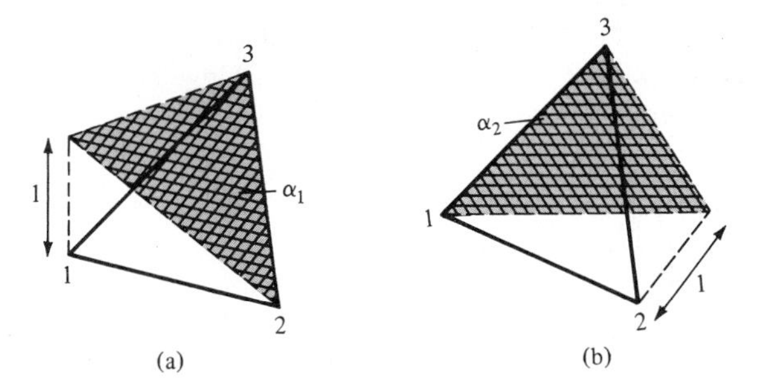

Figure 14.29 Shape functions α_1 and α_2 for a triangular element.

where a two-dimensional solution region free of charge ($\rho_S = 0$) is assumed. But from eq. (14.51),

$$\nabla V_e = \sum_{i=1}^{3} V_{ei}\, \nabla \alpha_i \tag{14.56}$$

Substituting eq. (14.56) into eq. (14.55) gives

$$W_e = \frac{1}{2} \sum_{i=1}^{3} \sum_{j=1}^{3} \varepsilon V_{ei} \left[\int \nabla \alpha_i \cdot \nabla \alpha_j \, dS \right] V_{ej} \tag{14.57}$$

If we define the term in brackets as

$$C_{ij}^{(e)} = \int \nabla \alpha_i \cdot \nabla \alpha_j \, dS \tag{14.58}$$

we may write eq. (14.57) in matrix form as

$$W_e = \frac{1}{2} \varepsilon \, [V_e]^T \, [C^{(e)}] \, [V_e] \tag{14.59}$$

where the supercript T denotes the transpose of the matrix,

$$[V_e] = \begin{bmatrix} V_{e1} \\ V_{e2} \\ V_{e3} \end{bmatrix} \tag{14.60a}$$

and

$$[C^{(e)}] = \begin{bmatrix} C_{11}^{(e)} & C_{12}^{(e)} & C_{13}^{(e)} \\ C_{21}^{(e)} & C_{22}^{(e)} & C_{23}^{(e)} \\ C_{31}^{(e)} & C_{32}^{(e)} & C_{33}^{(e)} \end{bmatrix} \tag{14.60b}$$

The matrix $[C^{(e)}]$ is usually called the *element coefficient matrix* (or "stiffness matrix" in structural analysis). The matrix element $C_{ij}^{(e)}$ of the coefficient matrix may be regarded as the coupling between nodes i and j; its value is obtained from eqs. (14.52) and (14.58). For example,

$$C_{12}^{(e)} = \int \nabla\alpha_1 \cdot \nabla\alpha_2 \, dS$$

$$= \frac{1}{4A^2}[(y_2 - y_3)(y_3 - y_1) + (x_3 - x_2)(x_1 - x_3)] \int dS$$

$$= \frac{1}{4A}[(y_2 - y_3)(y_3 - y_1) + (x_3 - x_2)(x_1 - x_3)] \tag{14.61a}$$

Similarly:

$$C_{13}^{(e)} = \frac{1}{4A}[(y_2 - y_3)(y_1 - y_2) + (x_3 - x_2)(x_2 - x_1)] \tag{14.61b}$$

$$C_{23}^{(e)} = \frac{1}{4A}[(y_3 - y_1)(y_1 - y_2) + (x_1 - x_3)(x_2 - x_1)] \tag{14.61c}$$

$$C_{11}^{(e)} = \frac{1}{4A}[(y_2 - y_3)^2 + (x_3 - x_2)^2] \tag{14.61d}$$

$$C_{22}^{(e)} = \frac{1}{4A}[(y_3 - y_1)^2 + (x_1 - x_3)^2] \tag{14.61e}$$

$$C_{33}^{(e)} = \frac{1}{4A}[(y_1 - y_2)^2 + (x_2 - x_1)^2] \tag{14.61f}$$

Also

$$C_{21}^{(e)} = C_{12}^{(e)}, \qquad C_{31}^{(e)} = C_{13}^{(e)}, \qquad C_{32}^{(e)} = C_{23}^{(e)} \tag{14.61g}$$

However, our calculations will be easier if we define

$$P_1 = (y_2 - y_3), \qquad P_2 = (y_3 - y_1), \qquad P_3 = (y_1 - y_2) \tag{14.62a}$$
$$Q_1 = (x_3 - x_2), \qquad Q_2 = (x_1 - x_3), \qquad Q_3 = (x_2 - x_1)$$

With P_i and Q_i ($i = 1, 2, 3$ are the local node numbers), each term in the element coefficient matrix is found as

$$\boxed{C_{ij}^{(e)} = \frac{1}{4A}[P_iP_j + Q_iQ_j]} \tag{14.62b}$$

where

$$A = \frac{1}{2}(P_2Q_3 - P_3Q_2) \tag{14.62c}$$

Note that $P_1 + P_2 + P_3 = 0 = Q_1 + Q_2 + Q_3$ and hence $\sum_{i=1}^{3} C_{ij}^{(e)} = 0 = \sum_{j=1}^{3} C_{ij}^{(e)}$.

C. Assembling of All Elements

Having considered a typical element, the next step is to assemble all such elements in the solution region. The energy associated with the assemblage of all elements in the mesh is

$$W = \sum_{e=1}^{N} W_e = \frac{1}{2}\varepsilon\,[V]^T\,[C]\,[V] \tag{14.63}$$

where

$$[V] = \begin{bmatrix} V_1 \\ V_2 \\ \cdot \\ \cdot \\ \cdot \\ V_n \end{bmatrix} \tag{14.64}$$

n is the number of nodes, N is the number of elements, and $[C]$ is called the *overall* or *global coefficient matrix,* which is the assemblage of individual element coefficient matrices. The major problem now is obtaining $[C]$ from $[C^{(e)}]$.

The process by which individual element coefficient matrices are assembled to obtain the global coefficient matrix is best illustrated with an example. Consider the finite element mesh consisting of three finite elements as shown in Figure 14.30.

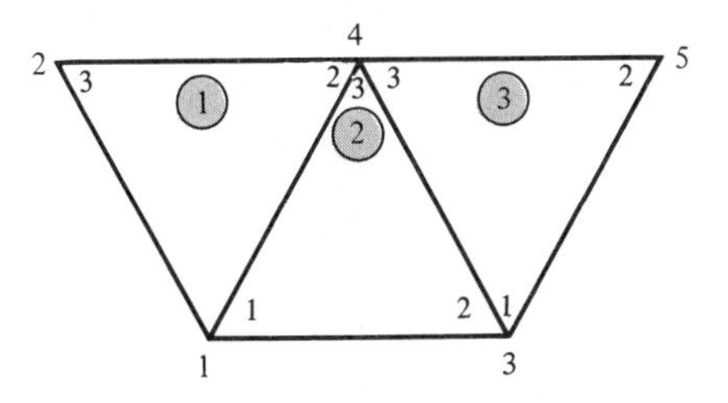

Figure 14.30 Assembly of three elements; *i-j-k* corresponds to local numbering 1-2-3 of the element in Figure 14.28.

Observe the numberings of the nodes. The numbering of nodes as 1, 2, 3, 4, and 5 is called *global* numbering. The numbering *i-j-k* is called *local* numbering and it corresponds with 1-2-3 of the element in Figure 14.28. For example, for element 3 in Figure 14.30, the global numbering 3-5-4 corresponds with local numbering 1-2-3 of the element in Figure 14.28. Note that the local numbering must be in counterclockwise sequence starting from any node of the element. For element 3, for example, we could choose 4-3-5 or 5-4-3 instead of 3-5-4 to correspond with 1-2-3 of the element in Figure 14.28. Thus the numbering in Figure 14.30 is not unique. However, we obtain the same [C] whichever numbering is used. Assuming the particular numbering in Figure 14.30, the global coefficient matrix is expected to have the form

$$[C] = \begin{bmatrix} C_{11} & C_{12} & C_{13} & C_{14} & C_{15} \\ C_{21} & C_{22} & C_{23} & C_{24} & C_{25} \\ C_{31} & C_{32} & C_{33} & C_{34} & C_{35} \\ C_{41} & C_{42} & C_{43} & C_{44} & C_{45} \\ C_{51} & C_{52} & C_{53} & C_{54} & C_{55} \end{bmatrix} \quad [14.65]$$

which is a 5×5 matrix since five nodes ($n = 5$) are involved. Again, C_{ij} is the coupling between nodes i and j. We obtain C_{ij} by utilizing the fact that the potential distribution must be continuous across interelement boundaries. The contribution to the i, j position in [C] comes from all elements containing nodes i and j. To find C_{11}, for example, we observe from Figure 14.30 that global node 1 belongs to elements 1 and 2 and it is local node 1 in both; hence,

$$C_{11} = C_{11}^{(1)} + C_{11}^{(2)} \quad [14.66a]$$

For C_{22}, global node 2 belongs to element 1 only and is the same as local node 3; hence,

$$C_{22} = C_{33}^{(1)} \quad [14.66b]$$

For C_{44}, global node 4 is the same as local node 2, 3, and 3 in elements 1, 2, and 3 respectively; hence,

$$C_{44} = C_{22}^{(1)} + C_{33}^{(2)} + C_{33}^{(3)} \quad [14.66c]$$

For C_{14}, global link 14 is the same as the local link 12 and 13 in elements 1 and 2 respectively; hence,

$$C_{14} = C_{12}^{(1)} + C_{13}^{(2)} \quad [14.66d]$$

Since there is no coupling (or direct link) between nodes 2 and 3,

$$C_{23} = C_{32} = 0 \quad [14.66e]$$

Continuing in this manner, we obtain all the terms in the global coefficient matrix by inspection of Fig. 14.30 as

$$[C] = \begin{bmatrix} C_{11}^{(1)} + C_{11}^{(2)} & C_{13}^{(1)} & C_{12}^{(2)} & C_{12}^{(1)} + C_{13}^{(2)} & 0 \\ C_{31}^{(1)} & C_{33}^{(1)} & 0 & C_{32}^{(1)} & 0 \\ C_{21}^{(2)} & 0 & C_{22}^{(2)} + C_{11}^{(3)} & C_{23}^{(2)} + C_{13}^{(3)} & C_{12}^{(3)} \\ C_{21}^{(1)} + C_{31}^{(2)} & C_{23}^{(1)} & C_{32}^{(2)} + C_{31}^{(3)} & C_{22}^{(1)} + C_{33}^{(2)} + C_{33}^{(3)} & C_{32}^{(3)} \\ 0 & 0 & C_{21}^{(3)} & C_{23}^{(3)} & C_{22}^{(3)} \end{bmatrix} \qquad [14.67]$$

Note that element coefficient matrices overlap at nodes shared by elements and that there are 27 terms (nine for each of the three elements) in the global coefficient matrix $[C]$. Also note the following properties of the matrix $[C]$:

1. It is symmetric ($C_{ij} = C_{ji}$) just as the element coefficient matrix.
2. Since $C_{ij} = 0$ if no coupling exists between nodes i and j, it is evident that for a large number of elements $[C]$ becomes sparse and banded.
3. It is singular. Although this is not so obvious, it can be shown using the element coefficient matrix of eq. (14.60b). Adding columns 2 and 3 to column 1 yields zeros in column 1 in eq. (14.60b).

D. Solving the Resulting Equations

From variational calculus, it is known that Laplace's (or Poisson's) equation is satisfied when the total energy in the solution region is minimum. Thus we require that the partial derivatives of W with respect to each nodal value of the potential be zero; that is,

$$\frac{\partial W}{\partial V_1} = \frac{\partial W}{\partial V_2} = \cdots = \frac{\partial W}{\partial V_n} = 0$$

or

$$\frac{\partial W}{\partial V_k} = 0 \qquad k = 1, 2, \ldots, n \qquad [14.68]$$

For example, to get $\partial W/\partial V_1 = 0$ for the finite element mesh of Figure 14.30, we substitute eq. (14.65) into eq. (14.63) and take the partial derivative of W with respect to V_1. We obtain

$$0 = \frac{\partial W}{\partial V_1} = 2V_1C_{11} + V_2C_{12} + V_3C_{13} + V_4C_{14} + V_5C_{15}$$
$$+ V_2C_{21} + V_3C_{31} + V_4C_{41} + V_5C_{51}$$

or

$$0 = V_1C_{11} + V_2C_{12} + V_3C_{13} + V_4C_{14} + V_5C_{15} \qquad [14.69]$$

In general, $\partial W/\partial V_k = 0$ leads to

$$0 = \sum_{i=1}^{n} V_i C_{ik} \qquad [14.70]$$

where n is the number of nodes in the mesh. By writing eq. (14.70) for all nodes $k = 1, 2, \ldots, n$, we obtain a set of simultaneous equations from which the solution of $[V]^T = [V_1, V_2, \ldots, V_n]$ can be found. This can be done in two ways similar to those used in solving finite difference equations obtained from Laplace's (or Poisson's) equation.

Iteration Method: This approach is similar to that used in finite difference method. Let us assume that node 1 in Figure 14.30, for example, is a free node. The potential at node 1 can be obtained from eq. (14.69) as

$$V_1 = -\frac{1}{C_{11}} \sum_{i=2}^{5} V_i C_{1i} \qquad [14.71]$$

In general, the potential at a free node k is obtained from eq. (14.70) as

$$\boxed{V_k = -\frac{1}{C_{kk}} \sum_{i=1, i \neq k}^{n} V_i C_{ik}} \qquad [14.72]$$

This is applied iteratively to all the free nodes in the mesh with n nodes. Since $C_{ki} = 0$ if node k is not directly connected to node i, only nodes that are directly linked to node k contribute to V_k in eq. (14.72).

Thus if the potentials at nodes connected to node k are known, we can determine V_k using eq. (14.72). The iteration process begins by setting the potentials at the free nodes equal to zero or to the average potential

$$V_{\text{ave}} = 1/2\,(V_{\min} + V_{\max}) \qquad [14.73]$$

where $V_{\min}$ and $V_{\max}$ are the minimum and maximum values of the prescribed potentials at the fixed nodes. With those initial values, the potentials at the free nodes are calculated using eq. (14.72). At the end of the first iteration, when the new values have been calculated for all the free nodes, the values become the old values for the second iteration. The procedure is repeated until the change between subsequent iterations becomes negligible.

Band Matrix Method: If all free nodes are numbered first and the fixed nodes last, eq. (14.63) can be written such that

$$W = \frac{1}{2}\varepsilon \begin{bmatrix} V_f & V_p \end{bmatrix} \begin{bmatrix} C_{ff} & C_{fp} \\ C_{pf} & C_{pp} \end{bmatrix} \begin{bmatrix} V_f \\ V_p \end{bmatrix} \qquad [14.74]$$

where subscripts f and p respectively refer to nodes with free and fixed (or prescribed) potentials. Since V_p is constant (it consists of known, fixed values), we only differentiate with respect to V_f so that applying eq. (14.68) to eq. (14.74) yields

$$C_{ff}V_p + C_{fp}V_p = 0$$

or

$$[C_{ff}]\,[V_f] = -[C_{fp}]\,[V_p] \qquad \textbf{[14.75]}$$

This equation can be written as

$$[A]\,[V] = [B] \qquad \textbf{[14.76a]}$$

or

$$\boxed{[V] = [A]^{-1}\,[B]} \qquad \textbf{[14.76b]}$$

where $[V] = [V_f]$, $[A] = [C_{ff}]$, and $[B] = -[C_{fp}]\,[V_p]$. Since $[A]$ is, in general, nonsingular, the potential at the free nodes can be found using eq. (14.75). We can solve for $[V]$ in eq. (14.76a) using Gaussian elimination technique. We can also solve for $[V]$ in eq. (14.76b) using matrix inversion if the size of the matrix to be inverted is not large.

Notice that as from eq. (14.55) onward, our solution has been restricted to a two-dimensional problem involving Laplace's equation, $\nabla^2 V = 0$. The basic concepts developed in this section can be extended to finite element analysis of problems involving Poisson's equation ($\nabla^2 V = -\rho_v/\varepsilon$, $\nabla^2 \mathbf{A} = -\mu\mathbf{J}$) or wave equation ($\nabla^2\phi - \gamma^2\phi = 0$). A major problem associated with finite element analysis is the relatively large amount of computer memory required in storing the matrix elements and the associated computational time. However, several algorithms have been developed to alleviate the problem to some degree.

The finite element method (FEM) has a number of advantages over the finite difference method (FDM) and the method of moments (MOM). First, the FEM can easily handle complex solution region. Second, the generality of FEM makes it possible to construct a general-purpose program for solving a wide range of problems. A single program can be used to solve different problems (described by the same partial differential equations) with different solution regions and different boundary conditions; only the input data to the problem needs be changed. However, FEM has its own drawbacks. It is harder to understand and program than FDM and MOM. It also requires preparing input data, a process which could be tedious.

EXAMPLE 14.6

Consider the two-element mesh shown in Figure 14.31a. Using the finite element method, determine the potentials within the mesh.

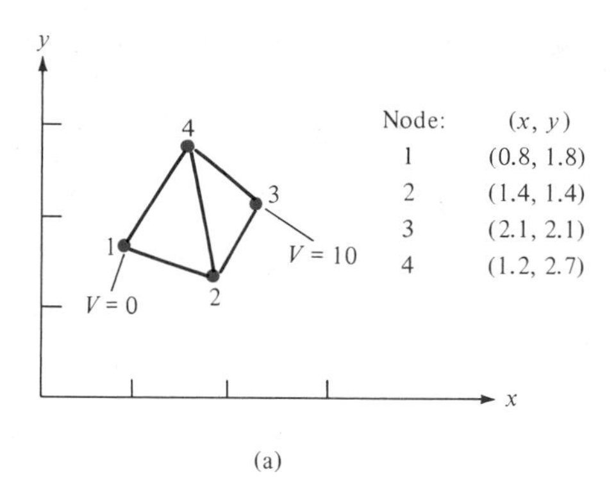

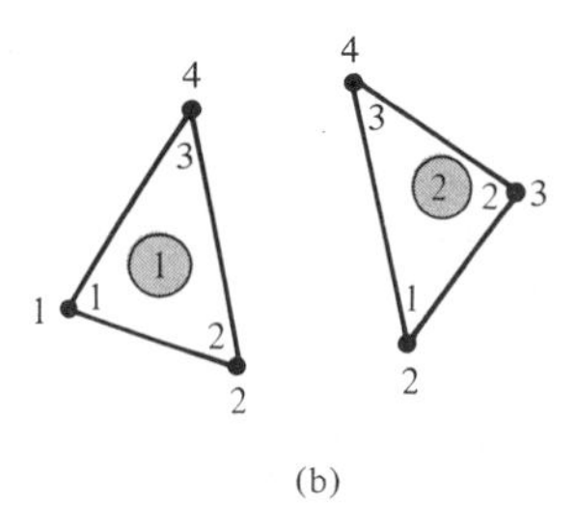

Figure 14.31 For Example 14.6: **(a)** two-element mesh, **(b)** local and global numbering of the elements.

SOLUTION The element coefficient matrices can be calculted using eq. (14.62). For element 1, consisting of nodes 1-2-4 corresponding to the local numbering 1-2-3 as in Figure 14.31b,

$$P_1 = -1.3, \qquad P_2 = 0.9, \qquad P_3 = 0.4$$

$$Q_1 = -0.2, \qquad Q_2 = -0.4, \qquad Q_3 = 0.6$$

$$A = 1/2\ (0.54 + 0.16) = 0.35$$

Substituting all these into eq. (14.62b) gives

$$[C^{(1)}] = \begin{bmatrix} 1.236 & -0.7786 & -0.4571 \\ -0.7786 & 0.6929 & 0.0857 \\ -0.4571 & 0.0857 & 0.3714 \end{bmatrix} \tag{14.6.1}$$

Similarly, for element 2 consisting of nodes 2-3-4 corresponding to local numbering 1-2-3 as in Figure 14.31b,

$$P_1 = -0.6, \qquad P_2 = 1.3, \qquad P_3 = -0.7$$

$$Q_1 = -0.9, \qquad Q_2 = 0.2, \qquad Q_3 = 0.7$$

$$A = 1/2\ (0.91 + 0.14) = 0.525$$

Hence,

$$[C^{(2)}] = \begin{bmatrix} 0.5571 & -0.4571 & -0.1 \\ -0.4571 & 0.8238 & -0.3667 \\ -0.1 & -0.3667 & 0.4667 \end{bmatrix} \quad \textbf{[14.6.2]}$$

Applying eq. (14.75) gives

$$\begin{bmatrix} C_{22} & C_{24} \\ C_{42} & C_{44} \end{bmatrix} \begin{bmatrix} V_2 \\ V_4 \end{bmatrix} = -\begin{bmatrix} C_{21} & C_{23} \\ C_{41} & C_{43} \end{bmatrix} \begin{bmatrix} V_1 \\ V_3 \end{bmatrix} \quad \textbf{[14.6.3]}$$

This can be written in a more convenient form as

$$\begin{bmatrix} 1 & 0 & 0 & 0 \\ 0 & C_{22} & 0 & C_{24} \\ 0 & 0 & 1 & 0 \\ 0 & C_{42} & 0 & C_{44} \end{bmatrix} \begin{bmatrix} V_1 \\ V_2 \\ V_3 \\ V_4 \end{bmatrix} = \begin{bmatrix} 1 & 0 \\ -C_{21} & -C_{23} \\ 0 & 1 \\ -C_{41} & -C_{43} \end{bmatrix} \begin{bmatrix} V_1 \\ V_3 \end{bmatrix} \quad \textbf{[14.6.4a]}$$

or

$$[C]\,[V] = [B] \quad \textbf{[14.6.4b]}$$

The terms of the global coefficient matrix are obtained as follows:

$$C_{22} = C_{22}^{(1)} + C_{11}^{(2)} = 0.6929 + 0.5571 = 1.25$$

$$C_{24} = C_{23}^{(1)} + C_{13}^{(2)} = 0.0857 - 0.1 = -0.0143$$

$$C_{44} = C_{33}^{(1)} + C_{33}^{(2)} = 0.3714 + 0.4667 = 0.8381$$

$$C_{21} = C_{21}^{(1)} = -0.7786$$

$$C_{23} = C_{12}^{(2)} = -0.4571$$

$$C_{41} = C_{31}^{(1)} = -0.4571$$

$$C_{43} = C_{32}^{(2)} = -0.3667$$

Note that we follow local numbering for the element coefficient matrix and global numbering for the global coefficient matrix. Thus the square matrix $[C]$ is obtained as

$$[C] = \begin{bmatrix} 1 & 0 & 0 & 0 \\ 0 & 1.25 & 0 & -0.0143 \\ 0 & 0 & 1 & 0 \\ 0 & -0.0143 & 0 & 0.8381 \end{bmatrix} \quad \textbf{[14.6.5]}$$

and the matrix $[B]$ on the right-hand side of eq. (14.6.4a) is obtained as

$$[B] = \begin{bmatrix} 0 \\ 4.571 \\ 10.0 \\ 3.667 \end{bmatrix} \quad [14.6.6]$$

By inverting matrix $[C]$ in eq. (14.6.5),we obtain

$$[V] = [C]^{-1}[B] = \begin{bmatrix} 0 \\ 3.708 \\ 10.0 \\ 4.438 \end{bmatrix}$$

Thus $V_1 = 0$, $V_2 = 3.708$, $V_3 = 10$, and $V_4 = 4.438$. Once the values of the potentials at the nodes are known, the potential at any point within the mesh can be determined using eq. (14.51). ■

PRACTICE EXERCISE 14.6

Calculate the global coefficient matrix for the two-element mesh shown in Figure 14.32 when: (a) node 1 is linked with node 3 and the local numbering (*i-j-k*) is as indicated in Figure 14.32(a), (b) node 2 is linked with node 4 with local numbering as in Figure 14.32 (b).

ANSWER (a) $\begin{bmatrix} 0.9964 & 0.05 & -0.2464 & -0.8 \\ 0.05 & 0.7 & -0.75 & 0.0 \\ -0.2464 & -0.75 & 1.5964 & -0.6 \\ -0.8 & 0.0 & -0.6 & 1.4 \end{bmatrix}$

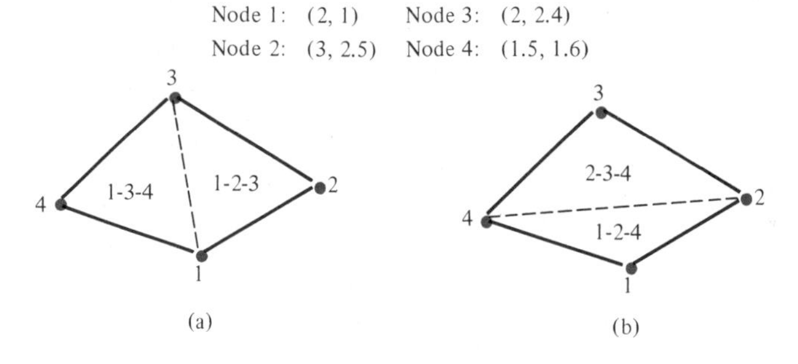

Figure 14.32 For Practice Exercise 14.6.

$$(b)\quad \begin{bmatrix} 1.333 & -0.7777 & 0.0 & -1.056 \\ -0.0777 & 0.8192 & -0.98 & 0.2386 \\ 0.0 & -0.98 & 2.04 & -1.06 \\ -1.056 & 0.2386 & -1.06 & 1.877 \end{bmatrix}$$

EXAMPLE 14.7

Write a FORTRAN program to solve Laplace's equation using the finite element method. Apply the program to the two-dimensional problem shown in Figure 14.33(a).

SOLUTION The solution region is divided into 25 three-node triangular elements with the total number of nodes being 21 as shown in Figure 14.33(b). This is a necessary step in order to have input data defining the geometry of the problem. Based on our discussions in Section 14.5, a general FORTRAN program for solving problems involving Laplace's equation using three-node triangular elements was developed as in Figure 14.34. The development of the program basically involves four steps indicated in the program and explained as follows.

Step 1: This involves inputting the necessary data defining the problem. This is the only step that depends on the geometry of the problem at hand. Through a data file, we input the number of elements, the number of nodes, the number of fixed nodes, the prescribed values of the potentials at the free nodes, the x and y coordinates of all nodes, and a list identifying the nodes belonging to each element in the order of the local numbering 1-2-3. For the problem in Figure 14.33, the three sets of data for coordinates, element-node relationship, and prescribed potentials at fixed nodes are shown in Tables 14.3, 14.4, and 14.5 respectively.

Step 2: This step entails finding the element coefficient matrix $[C^{(e)}]$ for each element and the global matrix $[C]$. The procedure explained in the previous example is applied. The matrix $[B]$ on the right-hand side of eq. (14.6.4) is also obtained at this stage.

Step 3: The global matrix obtained in the previous step is inverted by calling subroutine INVERSE of Appendix C.1. The values of the potentials at all nodes are obtained by matrix multiplication as in eq. (14.76b). Instead of inverting the global matrix, it is also possible to solve for the potentials at the nodes using Gaussian elimination technique.

Step 4: This involves outputting the result of the computation.

The input and output data are presented in Tables 14.6 and 14.7 respectively. ■

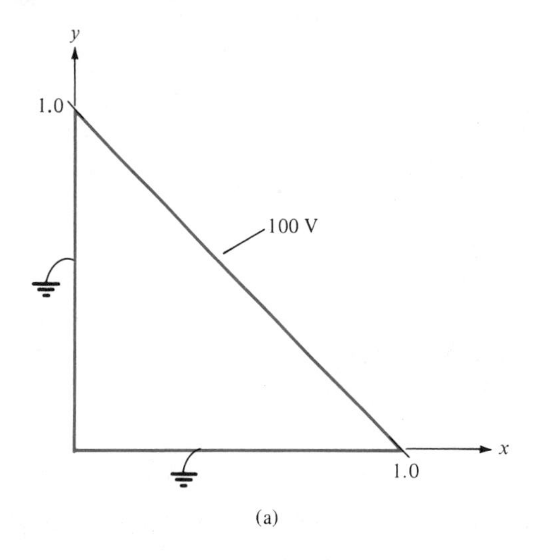

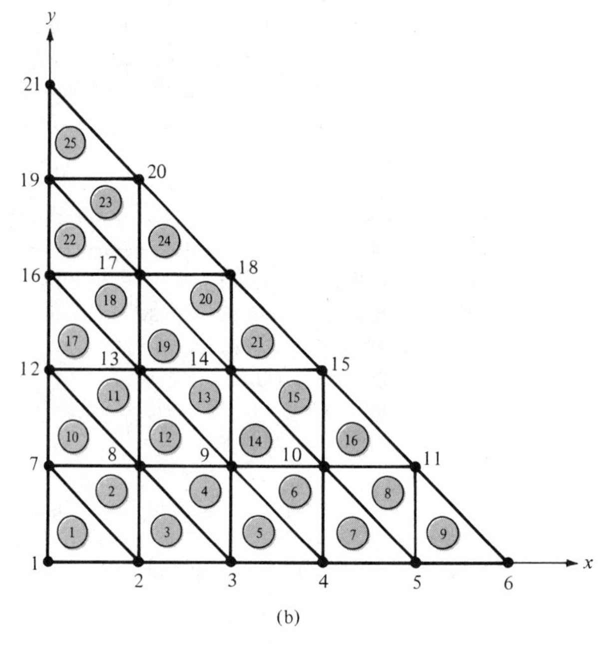

Figure 14.33 For Example 14.7: **(a)** two-dimensional electrostatic problem, **(b)** solution region divided into 25 triangular elements.

```
0001   C      FINITE ELEMENT SOLUTION OF LAPLACE'S EQUATION FOR
0002   C      TWO-DIMENSIONAL PROBLEMS
0003   C      TRIANGULAR ELEMENTS ARE USED
0004   C
0005   C      ND = NO. OF NODES
0006   C      NE = NO. OF ELEMENTS
0007   C      NP = NO. OF FIXED NODES (WHERE POTENTIAL IS PRESCRIBED)
0008   C      NDP(I) = NODE NO. OF PRESCRIBED POTENTIAL, I = 1,2,...NP
0009   C      VAL(I)) = VALUE OF PRESCRIBED POTENTIAL AT NODE NDP(I)
0010   C      NL(I,J) = LIST OF NODES FOR EACH ELEMENT I, WHERE
0011   C      J = 1, 2, 3  IS THE LOCAL NODE NUMBER
0012   C      CE(I,J) = ELEMENT COEFFICIENT MATRIX
0013   C      C(I,J)  = GLOBAL COEFFICIENT MATRIX
0014   C      B(I) = RIGHT-HAND SIDE MATRIX IN THE SYSTEM OF
0015   C      SIMULTANEOUS EQUATIONS; SEE EQ.(14.6.4)
0016   C      X(I), Y(I) = GLOBAL COORDINATES OF NODE I
0017   C      XL(J), YL(J) = LOCAL COORDINATES OF NODE J = 1,2,3
0018   C      V(I) = POTENTIAL AT NODE I
0019   C      MATRICES P(I) and Q(I) ARE DEFINED IN EQ.(14.6.2)
0020
0021          DIMENSION  X(100), Y(100), C(100,100), CE(100,100)
0022          DIMENSION  B(100), NL(100,3), NDP(100), VAL(100)
0023          DIMENSION  V(100), P(3), Q(3), XL(3), YL(3)
0024
0025   C     ***************************************************
0026   C     FIRST STEP - INPUT DATA DEFINING GEOMETRY AND
0027   C                  BOUNDARY CONDITIONS
0028   C     ***************************************************
0029         READ(5,10) NE,ND, NP
0030   10    FORMAT(3I3)
0031         READ(5,20)( I, ( NL(I,J), J=1,3),I=1,NE)
0032   20    FORMAT(4I3)
0033         READ(5,30)  (I, X(I), Y(I), I=1,ND)
0034   30    FORMAT(I3, 2F6.2)
0035         READ(5,40)  ( NDP(I), VAL(I), I=1,NP)
0036   40    FORMAT(I3,F6.2)
0037   C     ***************************************************
0038   C     SECOND STEP - EVALUATE COEFFICIENT MATRIX FOR EACH
0039   C                   ELEMENT AND ASSEMBLE GLOBALLY
0040   C     ***************************************************
0041         DO 50 M =1, ND
0042         B(M)=0.0
0043         DO 50 N=1,ND
0044         C(M,N) = 0.0
0045   50    CONTINUE
0046         DO 140  I = 1, NE
0047   C     FIND LOCAL COORDINATES XL(J), YL(J) FOR ELEMENT I
0048         DO 60 J=1,3
0049         K=NL(I,J)
0050         XL(J) = X(K)
0051         YL(J) = Y(K)
0052   60    CONTINUE
0053         P(1) = YL(2) - YL(3)
0054         P(2) = YL(3) - YL(1)
0055         P(3) = YL(1) - YL(2)
0056         Q(1) = XL(3) - XL(2)
0057         Q(2) = XL(1) - XL(3)
```

Figure 14.34 Computer program for Example 14.7 (continued on next page).

```
0058            Q(3) = XL(2) - XL(1)
0059            AREA = 0.5*ABS( P(2)*Q(3) - Q(2)*P(3) )
0060      C     DETERMINE COEFFICIENT MATRIX FOR ELEMENT I
0061            DO 70 M=1,3
0062            DO 70 N=1,3
0063            CE(M,N) = ( P(M)*P(N) + Q(M)*Q(N) )/(4.0*AREA)
0064      70    CONTINUE
0065      C     ASSEMBLE GLOBALLY - FIND C(I,J) AND B(I)
0066            DO 130 J=1,3
0067            IR=NL(I,J)
0068      C     CHECK IF ROW CORRESPONDS TO FIXED NODE
0069            DO 80 K=1,NP
0070            IF(IR.EQ.NDP(K)) GO TO 120
0071      80    CONTINUE!IR IS A FREE NODE
0072            DO 110 L=1,3
0073            IC = NL(I,L)
0074      C     CHECK IF COLUMN CORRESPONDS TO FIXED NODE
0075            DO 90 K=1,NP
0076            IF( IC.EQ.NDP(K) ) GO TO 100
0077      90    CONTINUE!IC IS A FREE NODE
0078            C(IR,IC) = C(IR,IC) + CE(J,L)
0079            GO TO 110
0080      100   B(IR) = B(IR) - CE(J,L)*VAL(K)
0081      110   CONTINUE
0082            GO TO 130
0083      120   CONTINUE
0084            C(IR,IR) = 1.0
0085            B(IR) = VAL(K)
0086      130   CONTINUE
0087      140   CONTINUE
0088      C     ********************************************
0089      C     THIRD STEP - SOLVE THE RESULTING SYSTEM OF
0090      C                  SIMULTANEOUS EQUATIONS
0091      C     ********************************************
0092            NMAX=100
0093            CALL INVERSE(C,ND,NMAX)
0094            DO 150  I=1,ND
0095            V(I) = 0.0
0096            DO 150 J=1,ND
0097            V(I) = V(I) + C(I,J)*B(J)
0098      150   CONTINUE
0099      C     ********************************************
0100      C     FOURTH STEP - OUTPUT THE RESULTS, THE POTENTIAL
0101      C                   V(I) AT NODE I,   I = 1,2,....ND
0102      C     *********************************************
0103            WRITE(6,160) ND,NE,NP
0104      160   FORMAT(2X,'NO. OF NODES = ',I3,2X,'NO. OF ELEMENTS =',
0105           1   I3,2X,'NO. OF FIXED NODES = ',I3,/)
0106            WRITE(6,170)
0107      170   FORMAT(2X,'NODE',5X,'X         Y',7X,'POTENTIAL',/)
0108            WRITE(6,180)  (I, X(I), Y(I), V(I), I=1,ND )
0109      180   FORMAT(2X,I3,2X,F6.2,2X,F6.2,2X,F10.4,/)
0110            STOP
0111            END
```

Figure 14.34 (Continued)

Table 14.3 **Nodal Coordinates of the Finite Element Mesh of Figure 14.33**

Node	x	y	Node	x	y
1	0.0	0.0	12	0.0	0.4
2	0.2	0.0	13	0.2	0.4
3	0.4	0.0	14	0.4	0.4
4	0.6	0.0	15	0.6	0.4
5	0.8	0.0	16	0.0	0.6
6	1.0	0.0	17	0.2	0.6
7	0.0	0.2	18	0.4	0.6
8	0.2	0.2	19	0.0	0.8
9	0.4	0.2	20	0.2	0.8
10	0.6	0.2	21	0.0	1.0
11	0.8	0.2			

Table 14.4 **Element-Node Identification**

	Local node no.				Local node no.		
Element no.	*1*	*2*	*3*	**Element no.**	*1*	*2*	*3*
1	1	2	7	14	9	10	14
2	2	8	7	15	10	15	14
3	2	3	8	16	10	11	15
4	3	9	8	17	12	13	16
5	3	4	9	18	13	17	16
6	4	10	9	19	13	14	17
7	4	5	10	20	14	18	17
8	5	11	10	21	14	15	18
9	5	6	11	22	16	17	19
10	7	8	12	23	17	20	19
11	8	13	12	24	17	18	20
12	8	9	13	25	19	20	21
13	9	14	13				

Table 14.5 **Prescribed Potentials at Fixed Nodes**

Node no.	Prescribed potential	Node no.	Prescribed potential
1	0.0	18	100.0
2	0.0	20	100.0
3	0.0	21	50.0
4	0.0	19	0.0
5	0.0	16	0.0
6	50.0	12	0.0
11	100.0	7	0.0
15	100.0		

Table 14.6 **Input Data for the Finite Element Program in Figure 14.34**

25	21	15	
1	1	2	7
2	2	8	7
3	2	3	8
4	3	9	8
5	3	4	9
6	4	10	9
7	4	5	10
8	5	11	10
9	5	6	11
10	7	8	12
11	8	13	12
12	8	9	13
13	9	14	13
14	9	10	14
15	10	15	14
16	10	11	15
17	12	13	16
18	13	17	16
19	13	14	17
20	14	18	17
21	14	15	18

Table 14.6 **Input Data for the Finite Element Program in Figure 14.34** (Continued)

25	21	15	
22	16	17	19
23	17	20	19
24	17	18	20
25	19	20	21
1	0.0	0.0	
2	0.2	0.0	
3	0.4	0.0	
4	0.6	0.0	
5	0.8	0.0	
6	1.0	0.0	
7	0.0	0.2	
8	0.2	0.2	
9	0.4	0.2	
10	0.6	0.2	
11	0.8	0.2	
12	0.0	0.4	
13	0.2	0.4	
14	0.4	0.4	
15	0.6	0.4	
16	0.0	0.6	
17	0.2	0.6	
18	0.4	0.6	
19	0.0	0.8	
20	0.2	0.8	
21	0.0	1.0	
1	0.0		
2	0.0		
3	0.0		
4	0.0		
5	0.0		
6	50.0		
11	100.0		
15	100.0		
18	100.0		
20	100.0		
21	50.0		
19	0.0		
16	0.0		
12	0.0		
7	0.0		

Table 14.7 Output Data of the Program in Figure 14.34

Node *No. of nodes = 21*	X *No. of elements = 25*	Y	Potential *No. of fixed nodes = 15*
1	0.00	0.00	0.000
2	0.20	0.00	0.000
3	0.40	0.00	0.000
4	0.60	0.00	0.000
5	0.80	0.00	0.000
6	1.00	0.00	50.000
7	0.00	0.20	0.000
8	0.20	0.20	18.182
9	0.40	0.20	36.364
10	0.60	0.20	59.091
11	0.80	0.20	100.000
12	0.00	0.40	0.000
13	0.20	0.40	36.364
14	0.40	0.40	68.182
15	0.60	0.40	100.000
16	0.00	0.60	0.000
17	0.20	0.60	59.091
18	0.40	0.60	100.000
19	0.00	0.80	0.000
20	0.20	0.80	100.000
21	0.00	1.00	50.000

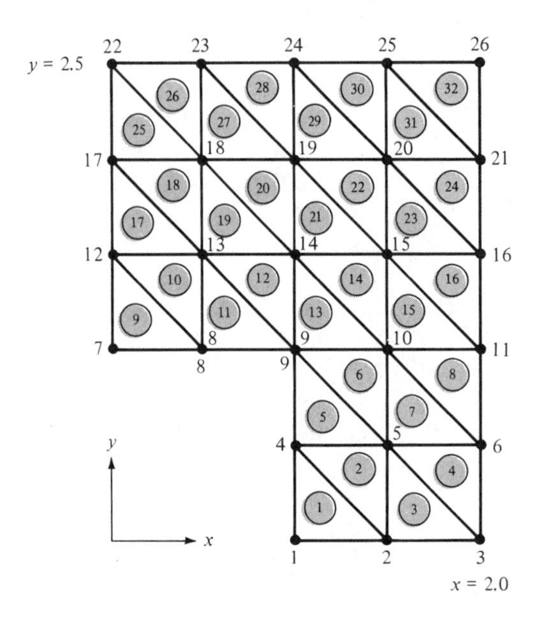

Figure 14.35 For Practice Exercise 14.7.

PRACTICE EXERCISE 14.7

Rework Example 14.3 using the finite element method. Divide the solution region into triangular elements as shown in Figure 14.35. Compare the solution with that obtained in Example 14.3 using the finite difference method.

ANSWER See Example 14.3.

SUMMARY

1. Electric field lines and equipotential lines due to coplanar point sources can be plotted using the numerical technique presented in this chapter. The basic concept can be extended to plotting magnetic field lines.
2. An EM problem in the form of a partial differential equation can be solved using the finite difference method. The finite difference equation that approximates the differential equation is applied at grid points spaced in an ordered manner over the whole solution region. The field quantity at the free points are determined using a suitable method.
3. An EM problem in the form of an integral equation is conveniently solved using the moment method. The unknown quantity under the integral sign is determined by matching both sides of the integral equation at a finite number of points in the domain of the quantity.
4. While the finite difference method is restricted to problems with regularly shaped solution regions, the finite element method can handle problems with complex geometries. This method involves dividing the solution region into finite elements, deriving equations for a typical element, assembling all elements in the region, and solving the resulting system of equations.

Typical examples on how to apply each method to some practical problems have been shown. Computer programs for solving the problems are provided wherever needed.

REVIEW QUESTIONS

14.1 At the point (1, 2, 0) in an electric field due to coplanar point charges, $\mathbf{E} = 0.3\,\mathbf{a}_x - 0.4\,\mathbf{a}_y$ V/m. A differential displacement of 0.05 m on an equipotential line at that point will lead to point

(a) (1.04, 2.03, 0)

(b) (0.96, 1.97, 0)

(c) (1.04, 1.97, 0)

(d) (0.96, 2.03, 0)

14.2 Which of the following is *not* a correct finite difference approximation to dV/dx at x_o if $h = \Delta x$?

(a) $\dfrac{V(x_o + h) - V(x_o)}{h}$

(b) $\dfrac{V(x_o) - V(x_o - h)}{h}$

(c) $\dfrac{V(x_o + h) - V(x_o - h)}{h}$

(d) $\dfrac{V(x_o + h) - V(x_o - h)}{2h}$

(e) $\dfrac{V(x_o + h/2) - V(x_o - h/2)}{h}$

14.3 The triangular element of Figure 14.36 is in free space. The approximate value of the potential at the center of the triangle is

(a) 10 V

(b) 7.5 V

(c) 5 V

(d) 0 V

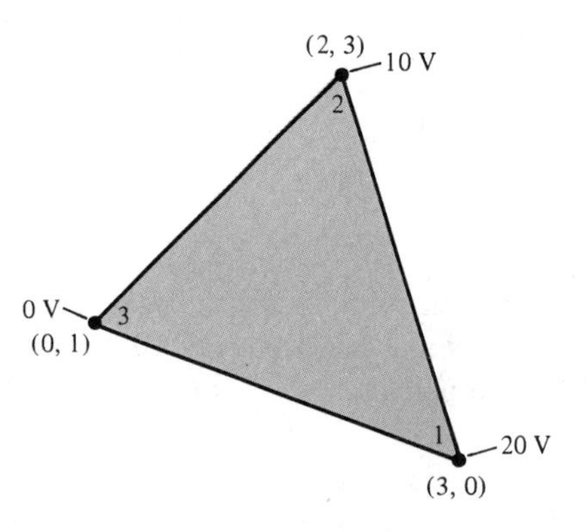

Figure 14.36 For Review Questions 14.3 and 14.10.

14.4 For finite difference analysis, a rectangular plate measuring 10 by 20 cm is divided into eight subregions by lines 5 cm apart parallel to the edges of the plates. How many free nodes are there if the edges are connected to some source?

(a) 15

(b) 12

(c) 9

(d) 6

(e) 3

14.5 Using the difference equation $V_n = V_{n-1} + V_{n+1}$ with $V_o = V_5 = 1$ and starting with initial values $V_n = 0$ for $1 \leq n \leq 4$, the value of V_2 after the third iteration is

(a) 1

(b) 3

(c) 9

(d) 15

(e) 25

14.6 The coefficient matrix $[A]$ obtained in the moment method does *not* have one of these properties:

(a) It is dense (i.e., has many nonzero terms).

(b) It is banded.

(c) It is square and symmetric.

(d) It depends on the geometry of the given problem.

14.7 A major difference between the finite difference and the finite element methods is that

(a) Using one, a sparse matrix results in the solution.

(b) In one, the solution is known at all points in the domain.

(c) One applies to solving partial differential equation.

(d) One is limited to time-invariant problems.

14.8 If the plate of Question 14.4 is to be discretized for finite element analysis such that we have the same number of grid points, how many triangular elements are there?

(a) 32

(b) 16

(c) 12

(d) 9

14.9 Which of these statements is *not* true about shape functions?

(a) They are interpolatory in nature.

(b) They must be continuous across the elements.

(c) Their sum is identically equal to unity at every point within the element.

(d) The shape function associated with a given node vanishes at any other node.

(e) The shape function associated with a node is zero at that node.

14.10 The area of the element in Figure 14.36 is

(a) 14

(b) 8

(c) 7

(d) 4

Answers: 14.1a, 14.2c[4], 14.3a, 14.4e, 14.5c, 14.6b, 14.7a, 14.8b, 14.9e, 14.10d.

PROBLEMS

14.1 Run the program developed in Practice Example 14.1 for these cases:

(a) $N = 2$ with charges $-Q$ and $+Q$ located at $(x, y) = (-1, 0)$ and $(1, 0)$ respectively.

(b) $N = 4$ with identical charges Q located at $(1, 1)$, $(-1, 1)$, $(-1, -1)$, and $(1, -1)$ respectively.

[4]The formula in (a) is known as a forward-difference formula, that in (b) as a backward-difference formula, and that in (d) or (e) is a central-difference formula.

14.2 Given the one-dimensional differential equation

$$\frac{d^2 y}{dx^2} = 0, \qquad 0 \le x \le 1$$

subject to $y(0) = 0$, $y(1) = 10$, use the finite difference (iterative) method to find $y(0.25)$. You may take $\Delta = 0.25$ and perform 5 iterations.

14.3 Find dy/dx and d^2y/dx^2 at $x = 0.75$ given that

x	0.65	0.7	0.75	0.8	0.85
y	0.1615	0.2417	0.3263	0.4153	0.5088

14.4 Another way of deriving eq. (14.15) is using the Taylor series expansion. For example, the Taylor series expansion of $V(x, y)$ at points $(x \pm a, y)$ is

$$V(x \pm a, y) = V(x, y) \pm a \frac{\partial V}{\partial x}(x, y) + \frac{a^2}{2!}\frac{\partial^2 V}{\partial x^2}(x, y) \pm \frac{a^3}{3!}\frac{\partial^3 V}{\partial x^3}(x, y) + \cdots$$

Use Taylor series expansion to derive eq. (14.15).

14.5 Show that the finite difference equation for Laplace's equation in cylindrical coordinates, $V = V(\rho, z)$, is

$$V(\rho_o, z_o) = \frac{1}{4}\left[V(\rho_o, z_o + h) + V(\rho_o, z_o - h) + \left(1 + \frac{h}{2\rho_o}\right) V(\rho_o + h, z_o) + \left(1 - \frac{h}{2\rho_o}\right) V(\rho_o - h, z_o)\right]$$

where $h = \Delta z = \Delta \rho$.

14.6 Using the finite difference representation in cylindrical coordinates (ρ, ϕ) at a grid point P shown in Figure 14.37, let $\rho = m\,\Delta\rho$ and $\phi = n\,\Delta\phi$ so that $V(\rho, \phi)|_P = V(m\Delta\rho, n\Delta\phi) = V_m^n$. Show that

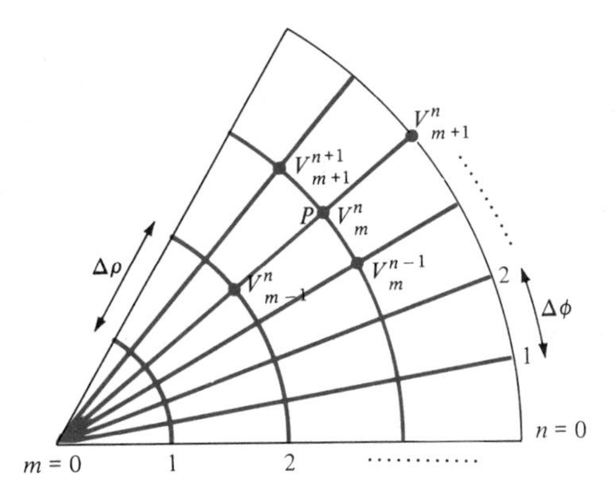

Figure 14.37 Finite difference grid in cylindrical coordinates; for Problem 14.6.

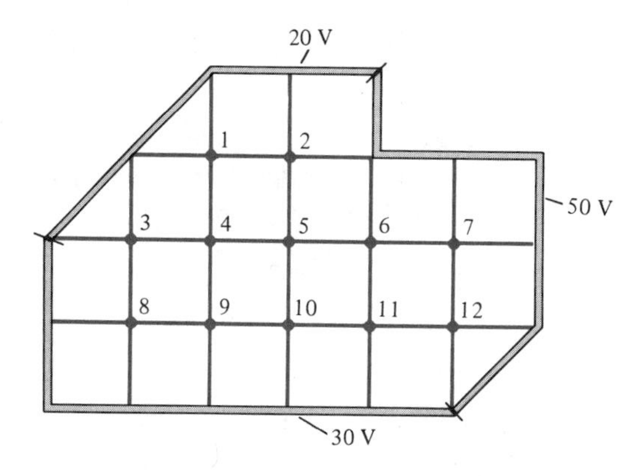

Figure 14.38 For Problem 14.8.

$$\nabla^2 V|_{m,n} = \frac{1}{\Delta\rho^2}\left[\left(1 - \frac{1}{2m}\right) V^n_{m-1} - 2V^n_m + \left(1 + \frac{1}{2m}\right) V^n_{m+1} + \frac{1}{(m\,\Delta\phi)^2}\left(V^{n-1}_m - 2\,V^n_m + V^{n+1}_m\right)\right]$$

14.7 A square conducting trough has its four sides held at potentials −10, 0, 30, and 60 V. Determine the potential at the center of the trough.

14.8 Estimate the potential at the grid points due to the potential system shown in Figure 14.38.

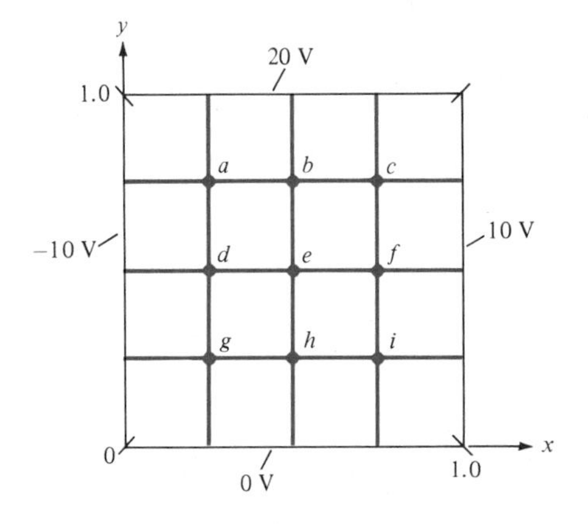

Figure 14.39 For Problem 14.9.

14.9 (a) How would you modify matrices [*A*] and [*B*] of Example 14.3 if the solution region had charge density ρ_S?

(b) Write a program to solve for the potentials at the grid points shown in Figure 14.39 assuming a charge density $\rho_S = x(y - 1)$ nC/m^2. Use the iterative finite difference method and take $\varepsilon_r = 1.0$.

14.10 Apply the band matrix technique to set up a system of simultaneous difference equations for each of the problems in Figure 14.40. Obtain matrices [*A*] and [*B*].

14.11 (a) The two-dimensional wave equation is given by

$$\frac{1}{c^2}\frac{\partial^2 \Phi}{\partial t^2} = \frac{\partial^2 \Phi}{\partial x^2} + \frac{\partial^2 \Phi}{\partial z^2}$$

By letting $\Phi^j_{m,n}$ denote the finite difference approximation of $\Phi(x_m, z_n, t_j)$, show that the finite difference scheme for the wave equation is

$$\Phi^{j+1}_{m,n} = 2\,\Phi^j_{m,n} - \Phi^{j-1}_{m,n} + \alpha\,(\Phi^j_{m+1,n} + \Phi^j_{m-1,n} - 2\,\Phi^j_{m,n}) + \alpha\,(\Phi^j_{m,n+1} + \Phi^j_{m,n-1} - 2\,\Phi^j_{m,n})$$

where $h = \Delta x = \Delta z$ and $\alpha = (c\Delta t/h)^2$.

(b) Write a program that uses the finite difference scheme to solve the one-dimensional wave equation

$$\frac{\partial^2 V}{\partial x^2} = \frac{\partial^2 V}{\partial t^2}, \qquad 0 \leq x \leq 1, \qquad t > 0$$

given boundary conditions $V(0, t) = 0$, $V(1, t) = 0$, $t > 0$ and the initial condition $\partial V/\partial t\,(x, 0) = 0$, $V(x, 0) = \sin \pi x$, $0 < x < 1$. Take $\Delta x = \Delta t = 0.1$. Compare your solution with the exact solution $V(x, t) = \sin \pi x \cos \pi t$ for $0 < t < 4$.

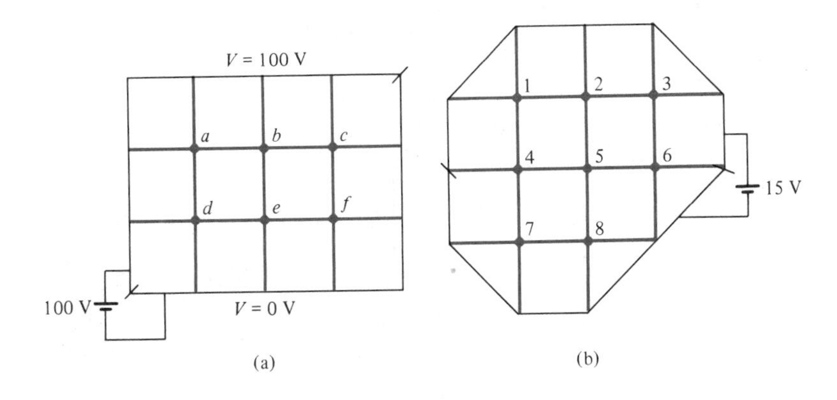

Figure 14.40 For Problem 14.10.

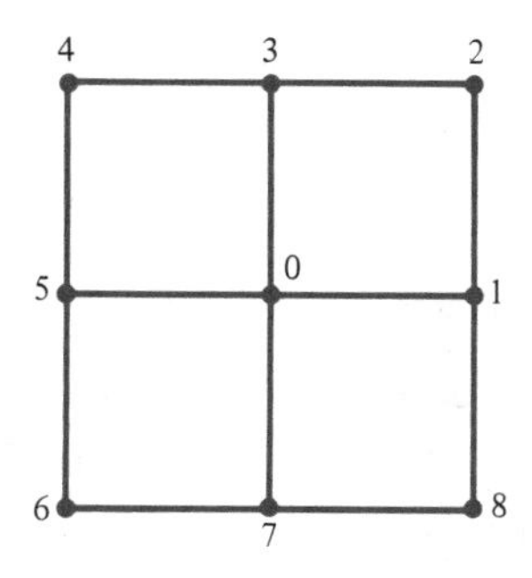

Figure 14.41 Nine-node molecule of Problem 14.12.

14.12 (a) Show that the finite difference representation of Laplace's equation using the nine-node molecule of Figure 14.41 is

$$V_o = 1/8\ (V_1 + V_2 + V_3 + V_4 + V_5 + V_6 + V_7 + V_8)$$

(b) Using this scheme, rework Example 14.4.

14.13 A thin conducting wire of length L and radius a is maintained at a potential V_o at a height H parallel to an infinitely large perfectly conducting ground plane as shown in Figure 14.42. Code a FORTRAN program and determine the charge distribution on the line. Calculate the induced charge on the ground plane. Take $L = 1$ m, $a = 1$ mm, $V_o = 1$ V, and $H = 0.5$ m.
(*Hint*: use image theory in conjunction with the moment method.)

14.14 Two conducting wires of equal length L and radius a are separated by a small gap and inclined at an angle θ as shown in Figure 14.43. Find the capacitance between the wires using the method of moments for cases $\theta = 10°, 20°, \ldots, 180°$. Take the gap as 2 mm, $a = 1$ mm, $L = 2$ m, $\varepsilon_r = 1$.

14.15 Determine the potential and electric field at point $(-1, 4, 5)$ due to the thin conducting wire of Figure 14.19. Take $V_o = 1$ V, $L = 1$ m, $a = 1$ mm.

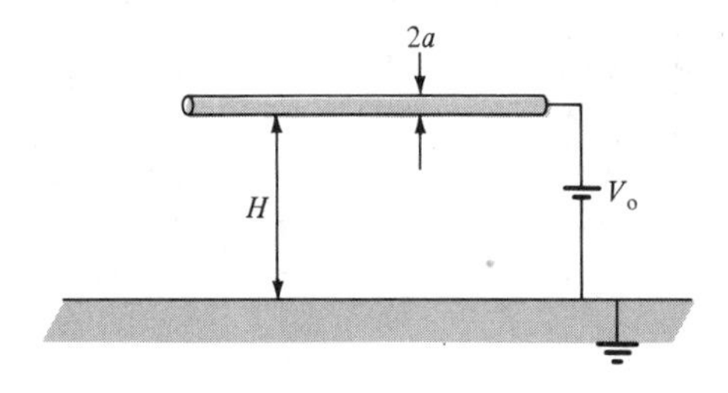

Figure 14.42 Conducting wire over a ground plane; for Problem 14.13.

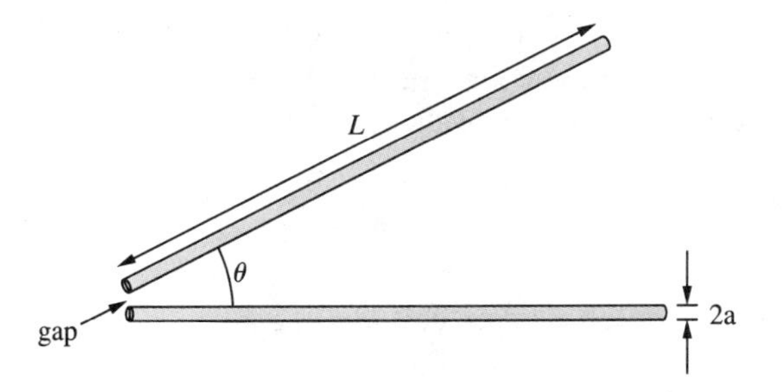

Figure 14.43 For Problem 14.14.

14.16 Given an infinitely long thin strip transmission line shown in Figure 14.44(a), we want to determine the characteristic impedance of the line using the moment method. We divide each strip into N subareas as in Figure 14.44(b) so that on subarea i,

$$V_i = \sum_{j=1}^{2N} A_{ij} \rho_j$$

where

$$A_{ij} = \begin{cases} \dfrac{-\Delta\ell}{2\pi\varepsilon_o} \ln R_{ij}, & i \neq j \\ \dfrac{-\Delta\ell}{2\pi\varepsilon_o} [\ln \Delta\ell - 1.5], & i = j \end{cases}$$

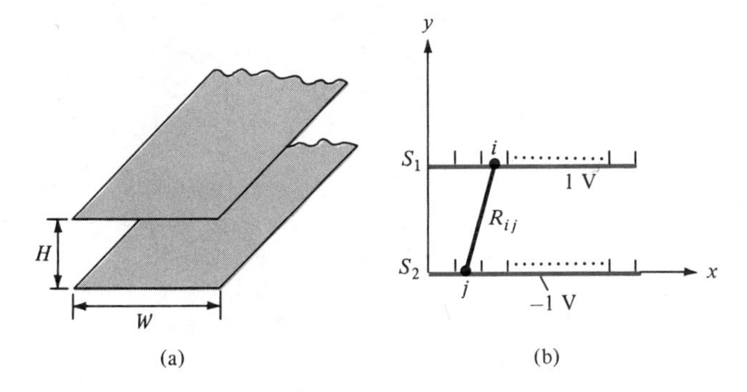

Figure 14.44 Analysis of strip transmission line using moment method; for Problem 14.16.

R_{ij} is the distance between the ith and jth subareas, and $V_i = 1$ or -1 depending on whether the ith subarea is on strip 1 or 2 respectively. Write a program to find the characteristic impedance of the line using the fact that

$$Z_o = \frac{\sqrt{\mu_o \varepsilon_o}}{C}$$

where C is the capacitance per unit length and

$$C = \frac{Q}{V_d} = \frac{\sum_{i=1}^{N} \rho_i \, \Delta \ell}{V_d}$$

and $V_d = 2$ V is the potential difference between strips. Take $H = 2$ m, $W = 5$ m, and $N = 20$.

14.17 Consider the coaxial line of arbitrary cross section shown in Figure 14.45(a). Using the moment method to find the capacitance C per length involves dividing each conductor into N strips so that the potential on the jth strip is given by

$$V_j = \sum_{i=1}^{2N} \rho_i \, A_{ij}$$

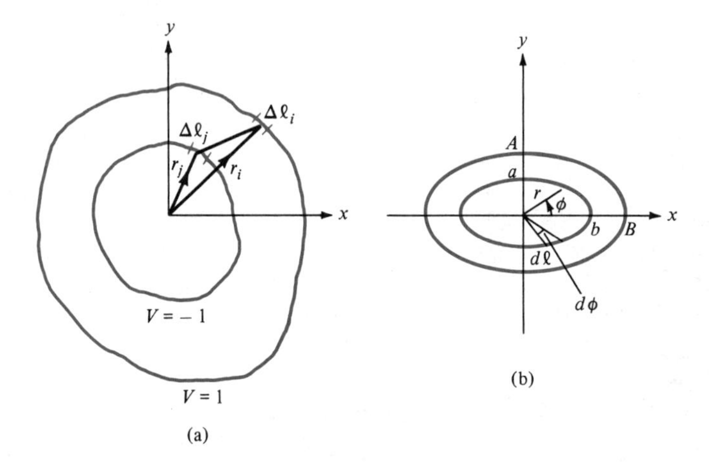

Figure 14.45 For Problem 14.17; coaxial line of **(a)** arbitrary cross section, and **(b)** elliptical cylindrical cross section.

where

$$A_{ij} = \begin{cases} \dfrac{-\Delta\ell}{2\pi\varepsilon} \ln \dfrac{R_{ij}}{r_o}, & i \neq j \\ \dfrac{-\Delta\ell}{2\pi\varepsilon} \left[\ln \dfrac{\Delta\ell_i}{r_o} - 1.5 \right], & i = j \end{cases}$$

and $V_j = -1$ or 1 depending on whether $\Delta\ell_i$ lies on the inner or outer conductor respectively. Code a FORTRAN program to determine the total charge per length on a coaxial cable of elliptical cylindrical cross section shown in Figure 14.45(b) using

$$Q = \sum_{i=1}^{N} \rho_i$$

and the capacitance per unit length using $C = Q/2$.

(a) As a way of checking your program, take $A = B = 2$ cm and $a = b = 1$ cm (coaxial line with circular cross section), and compare your result with the exact value of $C = 2\pi\varepsilon/\ln(A/a)$.

(b) Take $A = 2$ cm, $B = 4$ cm, $a = 1$ cm, and $b = 2$ cm.
(*Hint*: For the inner ellipse of Figure 14.45(b), for example,

$$r = \frac{a}{\sqrt{\sin^2\phi + v^2\cos^2\phi}}$$

where $v = a/b$, $d\ell = r\, d\phi$. Take $r_o = 1$ cm.)

14.18 A coaxial line in which the outer conductor has a step is shown in Figure 14.46(a). To determine the charge density and the capacitance of length L of the line (a portion of an infinite coaxial line), given that the line is uniform outside L, we truncate the infinite line at some distance T beyond L at each end of length L and divide each of surfaces S_1 and S_2 into N subelements. The potential on the jth subelement is

$$V_j = \sum_{j=1}^{2N} \rho_i A_{ij}$$

where

$$A_{ij} = \begin{cases} \dfrac{\Delta S_i}{4\pi\varepsilon_o R_{ij}}, & i \neq j \\ \dfrac{r_i}{\varepsilon_o} \ln \left(\dfrac{\Delta\ell}{r_i} \right), & i = j \end{cases}$$

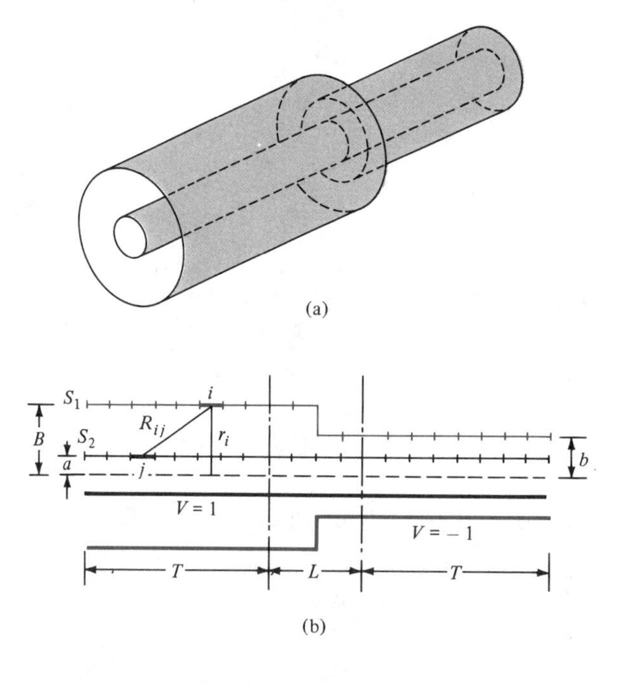

Figure 14.46 Coaxial line with a step; for Problem 14.18.

and ΔS_i is the surface area of the subelement ($\Delta S_i = 2\pi r_i \, \Delta \ell$). Assuming T/L is sufficiently large, we can find S_1 and S_2 using the moment method and obtain the capacitance of L as

$$C^- = -\sum_{LS_1} \rho_i \, \Delta S_i$$

$$C^+ = \sum_{LS_2} \rho_i \, \Delta S_i$$

where LS_1 and LS_2 respectively are the surfaces of the inner and outer conductors within the length L. Write a program to find C^- and C^+ given that $L = 1.0$ cm, $T = 15.5$ cm, $N = 12$, $a = 0.635$ cm, $b = 0.714$ cm, and $B = 1.0$ cm.

14.19 For the one-dimensional finite element of Figure 14.47, show that the shape functions are

$$\alpha_1 = \frac{x_2 - x}{L}, \qquad \alpha_2 = \frac{x - x_1}{L}$$

so that at any point within the element, $\phi = \alpha_1 \phi_1 + \alpha_2 \phi_2$.

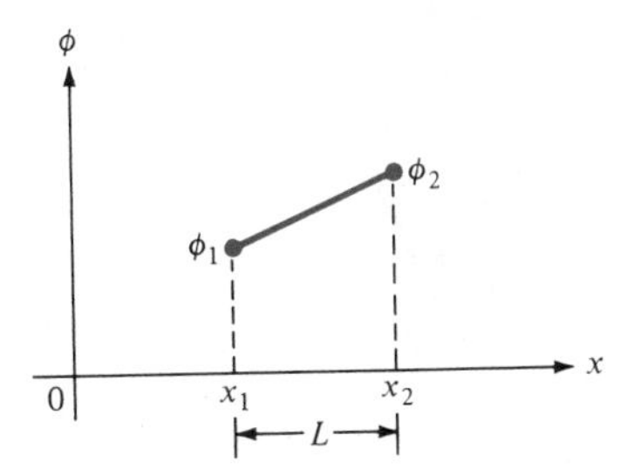

Figure 14.47 One-dimensional finite element of Problem 14.19.

14.20 Another way of defining the shape functions at an arbitrary point (x, y) in a finite element is using the areas A_1, A_2, and A_3 shown in Figure 14.48. Show that

$$\alpha_k = \frac{A_k}{A}, \qquad k = 1, 2, 3$$

where $A = A_1 + A_2 + A_3$ is the total area of the triangular element.

14.21 For each of the triangular elements of Figure 14.49,

(a) Calculate the shape functions.

(b) Determine the coefficient matrix.

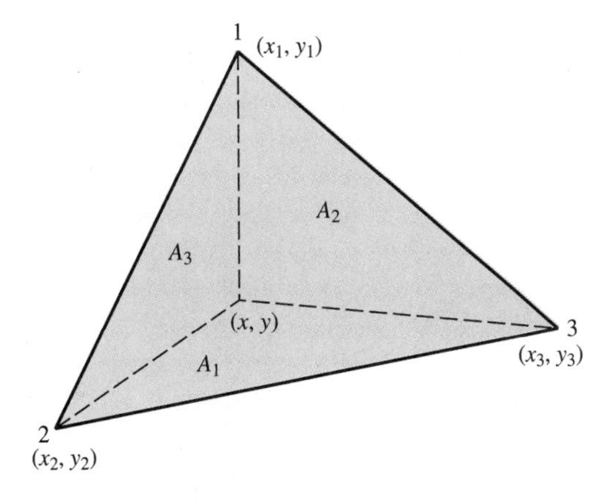

Figure 14.48 For Problem 14.20.

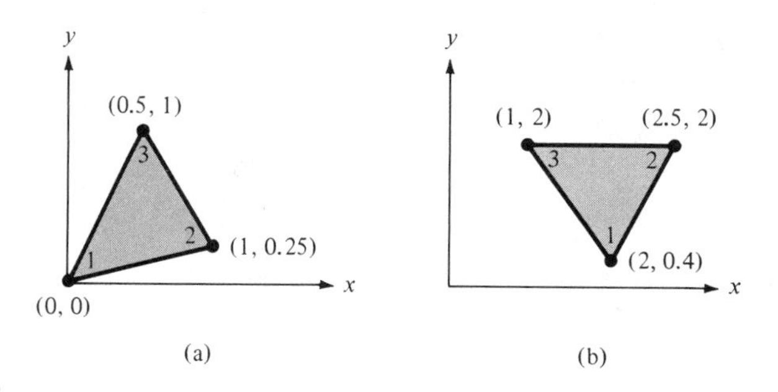

Figure 14.49 Triangular elements of Problem 14.21.

14.22 Show that

$$\sum_{i=1}^{3} C_{ij} = 0 = \sum_{j=1}^{3} C_{ij}$$

14.23 A point source of magnitude 60 W is located at (2, 3) in the element shown in Figure 14.50. Determine the contribution of the source to each node.

14.24 The nodal potential values for the triangular element of Figure 14.51 are $V_1 = 100$ V, $V_2 = 50$ V, and $V_3 = 30$ V. (a) Determine where the 80 V equipotential line intersects the boundaries of the element. (b) Calculate the potential of (2, 1).

14.25 The triangular element shown in Figure 14.52 is part of a finite element mesh. If $V_1 = 8$ V, $V_2 = 12$ V, and $V_3 = 10$ V, find the potential at (a) (1,2) and (b) the center of the element.

14.26 Determine the global coefficient matrix for the two-element region shown in Figure 14.53.

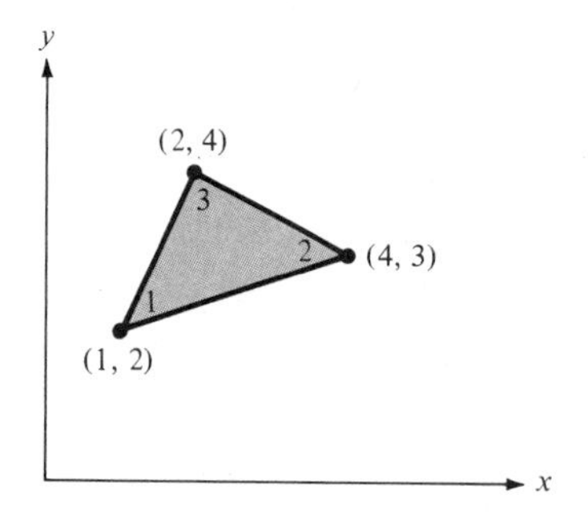

Figure 14.50 For Problem 14.23.

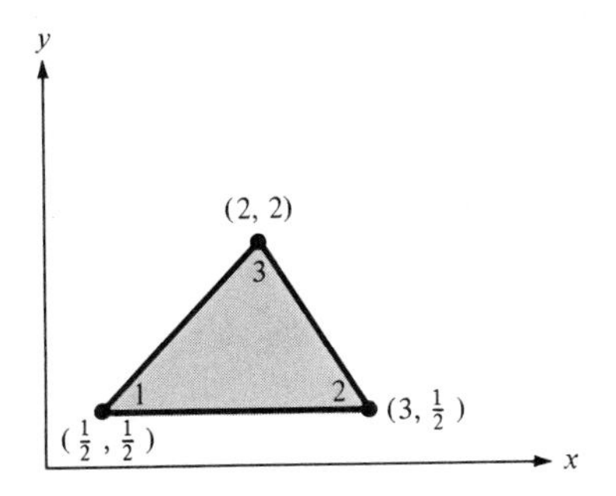

Figure 14.51 For Problem 14.24.

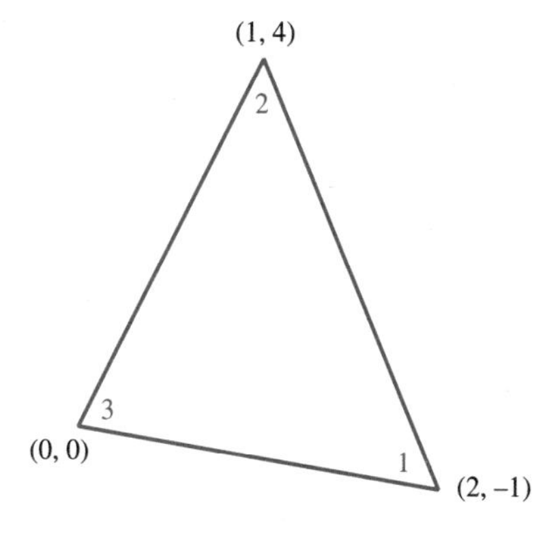

Figure 14.52 For Problem 14.25.

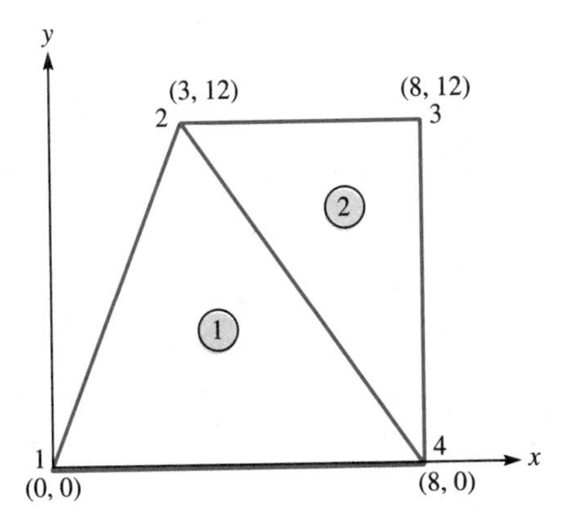

Figure 14.53 For Problem 14.26.

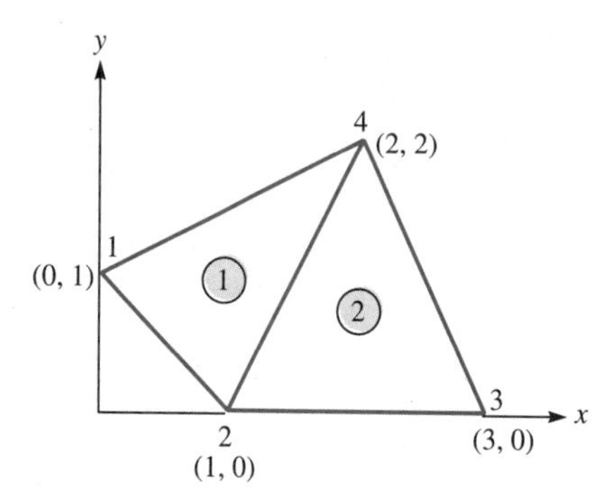

Figure 14.54 For Problem 14.27.

14.27 Find the global coefficient matrix of the two-element mesh of Figure 14.54.

14.28 In the mesh shown in Figure 14.55, calculate elements $C_{1,7}$ and $C_{1,1}$ of the global coefficient matrix. The shaded region is conducting and has no finite elements.

14.29 Use the program in Figure 14.34 to solve Laplace's equation in the problem shown in Figure 14.56 where $V_o = 100$ V. Compare the finite element solution to the exact solution in Example 6.5; that is,

$$V(x, y) = \frac{4V_o}{\pi} \sum_{k=0}^{\infty} \frac{\sin n\pi x \sinh n\pi y}{n \sinh n\pi}, \qquad n = 2k + 1$$

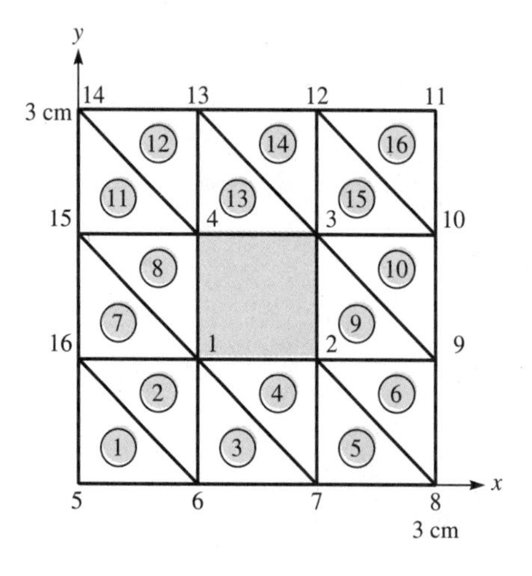

Figure 14.55 For Problem 14.28.

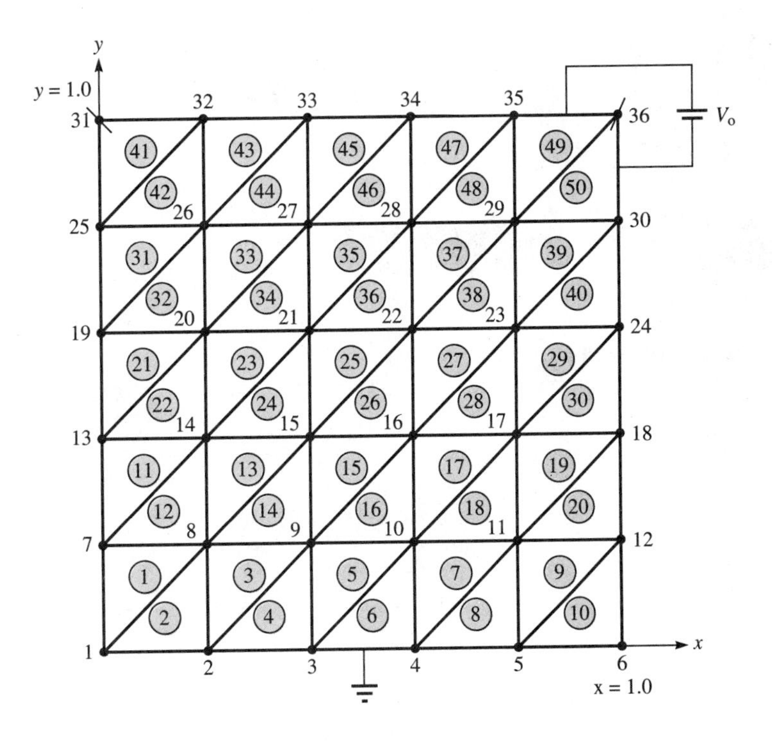

Figure 14.56 For Problem 14.29.

14.30 Repeat the preceding problem for $V_o = 100 \sin \pi x$. Compare the finite element solution with the theoretical solution (similar to Example 6.6(a)); that is,

$$V(x, y) = \frac{100 \sin \pi x \sinh \pi y}{\sinh \pi}$$

14.31 In steps 2 and 3 of the program in Figure 14.34, the element coefficient matrices are assembled by elements and the band matrix method is used in solving the resulting set of simultaneous equations. Rewrite the program so that the element coefficient matrices are assembled by nodes (see eq. (14.72)) and use the iteration method to determine the potentials at the free nodes. Test the program using the data in Example 14.7.

APPENDIX

Mathematical Formulas

A

A.1 TRIGONOMETRIC IDENTITIES

$$\tan A = \frac{\sin A}{\cos A}, \qquad \cot A = \frac{1}{\tan A}$$

$$\sec A = \frac{1}{\cos A}, \qquad \csc A = \frac{1}{\sin A}$$

$$\sin^2 A + \cos^2 A = 1, \qquad 1 + \tan^2 A = \sec^2 A$$

$$\sin (A \pm B) = \sin A \cos B \pm \cos A \sin B$$

$$\cos (A \pm B) = \cos A \cos B \mp \sin A \sin B$$

$$2 \sin A \sin B = \cos (A - B) - \cos (A + B)$$

$$2 \sin A \cos B = \sin (A + B) + \sin (A - B)$$

$$2 \cos A \cos B = \cos (A + B) + \cos (A - B)$$

$$\sin A + \sin B = 2 \sin \frac{A + B}{2} \cos \frac{A - B}{2}$$

$$\sin A - \sin B = 2 \cos \frac{A + B}{2} \sin \frac{A - B}{2}$$

$$\cos A + \cos B = 2 \cos \frac{A + B}{2} \cos \frac{A - B}{2}$$

$$\cos A - \cos B = -2 \sin \frac{A + B}{2} \sin \frac{A - B}{2}$$

$$\cos (A \pm 90^\circ) = \mp \sin A$$

$$\sin (A \pm 90^\circ) = \pm \cos A$$

$$\tan (A \pm 90^\circ) = -\cot A$$

$$\sin 2A = 2 \sin A \cos A$$

$$\cos 2A = \cos^2 A - \sin^2 A = 2\cos^2 A - 1 = 1 - 2\sin^2 A$$

$$\tan(A \pm B) = \frac{\tan A \pm B}{1 \mp \tan A \tan B}$$

$$\tan 2A = \frac{2\tan A}{1 - \tan^2 A}$$

$$\sin A = \frac{e^{jA} - e^{-jA}}{2j}, \qquad \cos A = \frac{e^{jA} + e^{-jA}}{2}$$

$$e^{jA} = \cos A + j\sin A \qquad \text{(Euler's identity)}$$

$$\pi = 3.1416$$

$$1 \text{ rad} = 57.296°$$

A.2 COMPLEX VARIABLES

A complex number may be represented as

$$z = x + jy = r\,\underline{/\theta} = re^{j\theta} = r(\cos\theta + j\sin\theta)$$

where $x = \text{Re } z = r\cos\theta, \qquad y = \text{Im } z = r\sin\theta$

$$r = |z| = \sqrt{x^2 + y^2}, \qquad \theta = \tan^{-1}\frac{y}{x}$$

$$j = \sqrt{-1}, \qquad \frac{1}{j} = -j, \qquad j^2 = -1$$

The complex conjugate of $z = z^* = x - jy = r\,\underline{/-\theta} = re^{-j\theta}$

$$= r(\cos\theta - j\sin\theta)$$

$$(e^{j\theta})^n = e^{jn\theta} = \cos n\theta + j\sin n\theta \qquad \text{(de Moivre's theorem)}$$

If $z_1 = x_1 + jy_1$ and $z_2 = x_2 + jy_2$, then $z_1 = z_2$ only if $x_1 = x_2$ and $y_1 = y_2$.

$$z_1 + z_2 = (x_1 + x_2) + j(y_1 + y_2)$$

$$z_1 z_2 = (x_1 x_2 - y_1 y_2) + j(x_1 y_2 + x_2 y_1)$$

or

$$z_1 z_2 = r_1 r_2\, e^{j(\theta_1 + \theta_2)} = r_1 r_2\,\underline{/\theta_1 + \theta_2}$$

$$\frac{z_1}{z_2} = \frac{(x_1 + jy_1)}{(x_2 + jy_2)} \cdot \frac{(x_2 - jy_2)}{(x_2 - jy_2)} = \frac{x_1 x_2 + y_1 y_2}{x_2^2 + y_2^2} + j\,\frac{x_2 y_1 - x_1 y_2}{x_2^2 + y_2^2}$$

or

$$\frac{z_1}{z_2} = \frac{r_1}{r_2} e^{j(\theta_1 - \theta_2)} = \frac{r_1}{r_2} \underline{/\theta_1 - \theta_2}$$

$$\sqrt{z} = \sqrt{x + jy} = \sqrt{r}\, e^{j\theta/2} = \sqrt{r}\, \underline{/\theta/2}$$

$$z^n = (x + jy)^n = r^n e^{jn\theta} = r^n \underline{/n\theta} \qquad (n = \text{integer})$$

$$z^{1/n} = (x + jy)^{1/n} = r^{1/n} e^{j\theta/n} = r^{1/n} \underline{/\theta/n + 2\pi k/n}\ (k = 0, 1, 2, \ldots, n - 1)$$

$$\ln (re^{j\theta}) = \ln r + \ln e^{j\theta} = \ln r + j\theta + j2k\pi \qquad (k = \text{integer})$$

A.3 HYPERBOLIC FUNCTIONS

$$\sinh x = \frac{e^x - e^{-x}}{2}, \qquad \cosh x = \frac{e^x + e^{-x}}{2}$$

$$\tanh x = \frac{\sinh x}{\cosh x}, \qquad \coth x = \frac{1}{\tanh x}$$

$$\operatorname{csch} x = \frac{1}{\sinh x}, \qquad \operatorname{sech} x = \frac{1}{\cosh x}$$

$$\sin jx = j \sinh x, \qquad \cos jx = \cosh x$$

$$\sinh jx = j \sin x, \qquad \cosh jx = \cos x$$

$$\sinh (x \pm y) = \sinh x \cosh y \pm \cosh x \sinh y$$

$$\cosh (x \pm y) = \cosh x \cosh y \pm \sinh x \sinh y$$

$$\sinh (x \pm jy) = \sinh x \cos y \pm j \cosh x \sin y$$

$$\cosh (x \pm jy) = \cosh x \cos y \pm j \sinh x \sin y$$

$$\tanh (x \pm jy) = \frac{\sinh 2x}{\cosh 2x + \cos 2y} \pm j \frac{\sin 2y}{\cosh 2x + \cos 2y}$$

$$\cosh^2 x - \sinh^2 x = 1$$

$$\operatorname{sech}^2 x + \tanh^2 x = 1$$

$$\sin (x \pm jy) = \sin x \cosh y \pm j \cos x \sinh y$$

$$\cos (x \pm jy) = \cos x \cosh y \mp j \sin x \sinh y$$

A.4 LOGARITHMIC IDENTITIES

$$\log xy = \log x + \log y$$

$$\log \frac{x}{y} = \log x - \log y$$

$$\log x^n = n \log x$$

$$\log_{10} x = \log x \text{ (common logarithm)}$$

$$\log_e x = \ln x \text{ (natural logarithm)}$$

If $|x| \ll 1$, $\ln (1 + x) \simeq x$

A.5 EXPONENTIAL IDENTITIES

$$e^x = 1 + x + \frac{x^2}{2!} + \frac{x^3}{3!} + \frac{x^4}{4!} + \cdots$$

where $e \simeq 2.7183$

$$e^x e^y = e^{x+y}$$

$$[e^x]^n = e^{nx}$$

$$\ln e^x = x$$

A.6 APPROXIMATIONS FOR SMALL QUANTITIES

If $|x| \ll 1$,

$$(1 \pm x)^n \simeq 1 \pm nx$$

$$e^x \simeq 1 + x$$

$$\ln (1 + x) \simeq x$$

$$\sin x \simeq x \quad \text{or} \quad \lim_{x \to 0} \frac{\sin x}{x} = 1$$

$$\cos \simeq 1$$

$$\tan x \simeq x$$

A.7 DERIVATIVES

If $U = U(x)$, $V = V(x)$, and $a =$ constant,

$$\frac{d}{dx}(aU) = a\frac{dU}{dx}$$

$$\frac{d}{dx}(UV) = U\frac{dV}{dx} + V\frac{dU}{dx}$$

$$\frac{d}{dx}\left[\frac{U}{V}\right] = \frac{V\frac{dU}{dx} - U\frac{dV}{dx}}{V^2}$$

$$\frac{d}{dx}(aU^n) = naU^{n-1}$$

$$\frac{d}{dx}\log_a U = \frac{\log_a e}{U}\frac{dU}{dx}$$

$$\frac{d}{dx}\ln U = \frac{1}{U}\frac{dU}{dx}$$

$$\frac{d}{dx}a^U = d^U \ln a\frac{dU}{dx}$$

$$\frac{d}{dx}e^U = e^U\frac{dU}{dx}$$

$$\frac{d}{dx}U^V = VU^{v-1}\frac{dU}{dx} + U^V \ln U\frac{dV}{dx}$$

$$\frac{d}{dx}\sin U = \cos U\frac{dU}{dx}$$

$$\frac{d}{dx}\cos U = -\sin U\frac{dU}{dx}$$

$$\frac{d}{dx}\tan U = \sec^2 U\frac{dU}{dx}$$

$$\frac{d}{dx}\sinh U = \cosh U\frac{dU}{dx}$$

$$\frac{d}{dx}\cosh U = \sinh U\frac{dU}{dx}$$

$$\frac{d}{dx}\tanh U = \text{sech}^2 U \frac{dU}{dx}$$

A.8 INDEFINITE INTEGRALS

If $U = U(x)$, $V = V(x)$, and a = constant,

$$\int a\,dx = ax + C$$

$$\int U\,dV = UV - \int V\,dU \qquad \text{(integration by parts)}$$

$$\int U^n\,dU = \frac{U^{n+1}}{n+1} + C, \qquad n \neq -1$$

$$\int \frac{dU}{U} = \ln U + C$$

$$\int a^U\,dU = \frac{a^U}{\ln a} + C, \qquad a > 0,\ a \neq 1$$

$$\int e^U\,dU = e^U + C$$

$$\int e^{ax}\,dx = \frac{1}{a}e^{ax} + C$$

$$\int xe^{ax}\,dx = \frac{e^{ax}}{a^2}(ax - 1) + C$$

$$\int x^2e^{ax}\,dx = \frac{e^{ax}}{a^3}(a^2x^2 - 2ax + 2) + C$$

$$\int \ln x\,dx = x\ln x - x + C$$

$$\int \sin ax\,dx = -\frac{1}{a}\cos ax + C$$

$$\int \cos ax\,dx = \frac{1}{a}\sin ax + C$$

$$\int \tan ax\,dx = \frac{1}{a}\ln\sec ax + C = -\frac{1}{a}\ln\cos ax + C$$

$$\int \sec ax\,dx = \frac{1}{a}\ln(\sec ax + \tan ax) + C$$

$$\int \sin^2 ax\, dx = \frac{x}{2} - \frac{\sin 2ax}{4a} + C$$

$$\int \cos^2 ax\, dx = \frac{x}{2} + \frac{\sin 2ax}{4a} + C$$

$$\int x \sin ax\, dx = \frac{1}{a^2}(\sin ax - ax \cos ax) + C$$

$$\int x \cos ax\, dx = \frac{1}{a^2}(\cos ax + ax \sin ax) + C$$

$$\int e^{ax} \sin bx\, dx = \frac{e^{ax}}{a^2 + b^2}(a \sin bx - b \cos bx) + C$$

$$\int e^{ax} \cos bx\, dx = \frac{e^{ax}}{a^2 + b^2}(a \cos bx + b \sin bx) + C$$

$$\int \sin ax \sin bx\, dx = \frac{\sin(a-b)x}{2(a-b)} - \frac{\sin(a+b)x}{2(a+b)} + C, \qquad a^2 \neq b^2$$

$$\int \sin ax \cos bx\, dx = -\frac{\cos(a-b)x}{2(a-b)} - \frac{\cos(a+b)x}{2(a+b)} + C, \qquad a^2 \neq b^2$$

$$\int \cos ax \cos bx\, dx = \frac{\sin(a-b)x}{2(a-b)} + \frac{\sin(a+b)x}{2(a+b)} + C, \qquad a^2 \neq b^2$$

$$\int \sinh ax\, dx = \frac{1}{a} \cosh ax + C$$

$$\int \cosh ax\, dx = \frac{1}{a} \sinh ax + C$$

$$\int \tanh ax\, dx = \frac{1}{a} \ln \cosh ax + C$$

$$\int \frac{dx}{x^2 + a^2} = \frac{1}{a} \tan^{-1} \frac{x}{a} + C$$

$$\int x^2 + a^2 = \frac{1}{2} \ln(x^2 + a^2) + C$$

$$\int \frac{x^2\, dx}{x^2 + a^2} = x - a \tan^{-1} \frac{x}{a}$$

$$\int \frac{dx}{x^2 - a^2} = \begin{cases} \dfrac{1}{2a} \ln \dfrac{x - a}{x + a} + C, & x^2 > a^2 \\ \dfrac{1}{2a} \ln \dfrac{a - x}{a + x} + C, & x^2 < a^2 \end{cases}$$

$$\int \frac{dx}{\sqrt{a^2 - x^2}} = \sin^{-1} \frac{x}{a} + C$$

$$\int \frac{dx}{\sqrt{x^2 \pm a^2}} = \ln \left(x + \sqrt{x^2 \pm a^2} \right) + C$$

A.9 DEFINITE INTEGRALS

$$\int_0^\pi \sin mx \sin nx \, dx = \int_0^\pi \cos mx \cos nx \, dx = \begin{cases} 0, & m \neq n \\ \pi/2, & m = n \end{cases}$$

$$\int_0^\pi \sin mx \cos nx \, dx = \begin{cases} 0 & m + n = \text{even} \\ \dfrac{2m}{m^2 - n^2} & m + n = \text{odd} \end{cases}$$

$$\int_0^{2\pi} \sin mx \sin nx \, dx = \int_{-\pi}^{\pi} \sin mx \sin nx \, dx = \begin{cases} 0, & m \neq n \\ \pi, & m = n \end{cases}$$

$$\int_0^\infty \frac{\sin ax}{x} dx = \begin{cases} \pi/2, & a > 0, \\ 0, & a = 0 \\ -\pi/2, & a < 0 \end{cases}$$

$$\int_0^\infty \frac{\sin^{2x}}{x} dx = \frac{\pi}{2}$$

$$\int_0^\infty \frac{\sin^2 ax}{x^2} dx = |a| \frac{\pi}{2}$$

$$\int_0^\infty x^n e^{-ax} \, dx = \frac{n!}{a^{n+1}}$$

$$\int_0^\infty e^{-ax^2} \, dx = \frac{1}{2} \sqrt{\frac{\pi}{a}}$$

$$\int_{-\infty}^\infty e^{-ax^2} \, dx = \sqrt{\frac{\pi}{a}}$$

$$\int_{-\infty}^\infty e^{-(ax^2 + bx + c)} \, dx = \sqrt{\frac{\pi}{a}} \, e^{(b^2 - 4ac)/4a}$$

$$\int_0^\infty e^{-ax} \cos bx \, dx = \frac{a}{a^2 + b^2}$$

$$\int_0^\infty e^{-ax} \sin bx \, dx = \frac{b}{a^2 + b^2}$$

A.10 VECTOR IDENTITIES

If **A** and **B** are vector fields while U and V are scalar fields, then

$$\nabla (U + V) = \nabla U + \nabla V$$

$$\nabla (UV) = U \nabla V + V \nabla U$$

$$\nabla \left[\frac{U}{V}\right] = \frac{V(\nabla U) - U(\nabla V)}{V^2}$$

$$\nabla V^n = n V^{n-1} \nabla V \qquad (n = \text{integer})$$

$$\nabla (\mathbf{A} \cdot \mathbf{B}) = (\mathbf{A} \cdot \nabla) \mathbf{B} + (\mathbf{B} \cdot \nabla) \mathbf{A} + \mathbf{A} \times (\nabla \times \mathbf{B}) + \mathbf{B} \times (\nabla \times \mathbf{A})$$

$$\nabla \cdot (\mathbf{A} + \mathbf{B}) = \nabla \cdot \mathbf{A} + \nabla \cdot \mathbf{B}$$

$$\nabla \cdot (\mathbf{A} \times \mathbf{B}) = \mathbf{B} \cdot (\nabla \times \mathbf{A}) - \mathbf{A} \cdot (\nabla \times \mathbf{B})$$

$$\nabla \cdot (V\mathbf{A}) = V \nabla \cdot \mathbf{A} + \mathbf{A} \cdot \nabla V$$

$$\nabla \cdot (\nabla V) = \nabla^2 V$$

$$\nabla \cdot (\nabla \times \mathbf{A}) = 0$$

$$\nabla \times (\mathbf{A} + \mathbf{B}) = \nabla \times \mathbf{A} + \nabla \times \mathbf{B}$$

$$\nabla \times (\mathbf{A} \times \mathbf{B}) = \mathbf{A} (\nabla \cdot \mathbf{B}) - \mathbf{B} (\nabla \cdot \mathbf{A}) + (\mathbf{B} \cdot \nabla)\mathbf{A} - (\mathbf{A} \cdot \nabla) \mathbf{B}$$

$$\nabla \times (V\mathbf{A}) = \nabla V \times \mathbf{A} + V(\nabla \times \mathbf{A})$$

$$\nabla \times (\nabla V) = 0$$

$$\nabla \times (\nabla \times \mathbf{A}) = \nabla(\nabla \cdot \mathbf{A}) - \nabla^2 \mathbf{A}$$

$$\oint_L \mathbf{A} \cdot d\mathbf{l} = \int_S \nabla \times \mathbf{A} \cdot d\mathbf{S}$$

$$\oint_L V \, d\mathbf{l} = -\int_S \nabla V \times d\mathbf{S}$$

$$\oint_S \mathbf{A} \cdot d\mathbf{S} = \int_v \nabla \cdot \mathbf{A} \, dv$$

$$\oint_S V \, d\mathbf{S} = \int_v \nabla V \, dv$$

$$\oint_S \mathbf{A} \times d\mathbf{S} = -\int_v \nabla \times \mathbf{A} \, dv$$

APPENDIX

Material Constants

B

Table B.1 **Approximate Conductivity* of Some Common Materials at 20°C**

Material	Conductivity (mhos/meter)
Conductors	
Silver	6.1×10^7
Copper (standard annealed)	5.8×10^7
Gold	4.1×10^7
Aluminum	3.5×10^7
Tungsten	1.8×10^7
Zinc	1.7×10^7
Brass	1.1×10^7
Iron (pure)	10^7
Lead	5×10^6
Mercury	10^6
Carbon	3×10^4
Water (sea)	4
Semiconductors	
Germanium (pure)	2.2
Silicon (pure)	4.4×10^{-4}

*The values vary from one published source to another due to the fact that there are many varieties of most materials and that conductivity is sensitive to temperature, moisture content, impurities, and the like.

Material	Conductivity (mhos/meter)
Insulators	
Water (distilled)	10^{-4}
Earth (dry)	10^{-5}
Bakelite	10^{-10}
Paper	10^{-11}
Glass	10^{-12}
Porcelain	10^{-12}
Mica	10^{-15}
Paraffin	10^{-15}
Rubber (hard)	10^{-15}
Quartz (fused)	10^{-17}
Wax	10^{-17}

Table B.2 **Approximate Dielectric Constant or Relative Permittivity (ε_r) and Strength of Some Common Materials***

Material	Dielectric Constant ε_r (dimensionless)	Dielectric Strength E (V/m)
Barium titanate	1200	7.5×10^6
Water (sea)	80	
Water (distilled)	81	
Nylon	8	
Paper	7	12×10^6
Glass	5–10	35×10^6
Mica	6	70×10^6
Porcelain	6	
Bakelite	5	20×10^6
Quartz (fused)	5	30×10^6
Rubber (hard)	3.1	25×10^6
Wood	2.5–8.0	
Polystyrene	2.55	
Polypropylene	2.25	
Paraffin	2.2	30×10^6
Petroleum oil	2.1	12×10^6
Air (1 atm.)	1	3×10^6

*The values given here are only typical; they vary from one published source to another due to different varieties of most materials and the dependence of ε_r on temperature, humidity, and the like.

Table B.3 **Relative Permeability (μ_r) of Some Materials***

Material	μ_r
Diamagnetic	
Bismuth	0.999833
Mercury	0.999968
Silver	0.9999736
Lead	0.9999831
Copper	0.9999906
Water	0.9999912
Hydrogen (s.t.p.)	≃ 1.0
Paramagnetic	
Oxygen (s.t.p.)	0.999998
Air	1.00000037
Aluminum	1.000021
Tungsten	1.00008
Platinum	1.0003
Manganese	1.001
Ferromagnetic	
Cobalt	250
Nickel	600
Soft Iron	5000
Silicon-Iron	7000

*The values given here are only typical; they vary from one published source to another due to different varieties of most materials.

APPENDIX

Computer Programs

C

C.1 SUBROUTINE INVERSE

```
0001   C     SUBROUTINE INVERSE
0002   C     SX IS THE MATRIX TO BE INVERTED; IT IS DESTROYED
0003   C     IN THE COMPUTATION AND REPLACED BY THE INVERSE
0004   C     N IS THE ORDER OF SX
0005   C     IDM IS THE DIMENSION OF SX
0006
0007         SUBROUTINE INVERSE (SX, N, IDM)
0008         DIMENSION SX(IDM, IDM)
0009
0010         ESP=1.0E-5
0011         DO 50 K=1, N
0012         DO 30 J=1, N
0013         IF(J.EQ.K) GO TO 30
0014         IF(SX(K, K)) 20, 10, 20
0015   10    SX(K, K)=ESP
0016   20    SX(K, J)=SX(K, J)/SX(K, K)
0017   30    CONTINUE
0018         SX(K, K)=1.0/SX(K, K)
0019         DO 40 I=1, N
0020         IF(I.EQ.K) GO TO 40
0021         DO 40 J=1, N
0022         IF(J.EQ.K) GO TO 40
0023         SX(I, J)=SX(I, J) - SX(K, J)*SX(I, K)
0024   40    CONTINUE
0025         DO 50 I=1, N
0026         IF(I.EQ.K) GO TO 50
0027         SX(I, K)= - SX(I, K)*SX(K, K)
0028   50    CONTINUE
0029         RETURN
0030         END
```

C.2 SUBROUTINE PLOT

```
0001   C     PLOTTING SUBROUTINE PLOT
0002   C     Y IS THE FUNCTION TO BE PLOTTED
0003   C     N IS THE NUMBER OF ARRAY ELEMENTS
0004   C     DX IS THE SAMPLING INCREMENT
0005   C     FIRST IS WHERE THE SAMPLING BEGINS
0006   C     JSKIP, IF NOT 1, IS USED TO INCREASE THE SAMPLING RATE
0007   C     i.e. THE MODIFIED Y(X)=Y(DX*JSKIP)
0008
0009         SUBROUTINE PLOT (Y, N, FIRST, DX, JSKIP)
0010         DIMENSION Y(400), A(26), SYM(4)
0011         DATA BLNK/4H    /, SYM(1)/4H*   /, SYM(2)/4H *  /,
0012        1SYM(3)/4H  * /, SYM(4)/4H   */, IPRINT/6/
0013         ISKIP=JSKIP
0014         IF (ISKIP.LE.0)  ISKIP=1
0015         DO 5 I=1, 26
0016   5     A(I)=BLNK
0017         ZERO=FIRST
0018         C=Y(1)
0019         D=C
0020         DO 10 I=1, N, ISKIP
0021         LOC=I
0022         E=Y(LOC)
0023         IF (E.GT.D) D=E
0024         IF (E.LT.C) C=E
0025   10    CONTINUE
0026         IGT=1
0027         IF (D.GT.9999.) IGT=2
0028         IF (C.LT.-9999.) IGT=2
0029         IF (D.LT.0.1.AND.C.GT. -.1) IGT=2
0030         DMC=D-C
0031         AA=(DMC)/103.
0032         BB=(D+C)/2.
0033         WRITE (IPRINT, 45) C, BB, D
0034         IF (ABS(DMC).GT.1.E -20) GO TO 15
0035         WRITE (IPRINT, 50)
0036         GO TO 40
0037   15    DO 35 I=1, N, ISKIP
0038         LOC=1
0039         XP=Y(LOC)
0040         J=(XP-C)/AA+.5
0041         J4=J/4
0042         JU=J-J4*4+1
0043         J4=J4+1
0044         A(J4)=SYM(JU)
0045         GO TO (20,25), IGT
0046   20    WRITE (IPRINT, 55) ZERO, XP, A
```

```
0047          GO TO 30
0048    25    WRITE (IPRINT, 60) ZERO, XP, A
0049    30    ZERO=ZERO + DX*ISKIP
0050          A(J4)=BLNK
0051    35    CONTINUE
0052    40    RETURN
0053    45    FORMAT(//17X, E10.2, 38X, E10.2, 38X, E10.2/3X, 'X=CX', 3X,
0054         1 'Y(X)',4X,'+',51('.'),'+',51('.'),'+')
0055    50    FORMAT (26HODATA ALL EQUAL...NO PLOT.,//)
0056    55    FORMAT (1X, F6.2, E10.3, 1X, 26A4)
0057    60    FORMAT (1X, F7.4, E11.4, 1X, 26A4)
0058          END
```

APPENDIX

List of Symbols

Symbol*	Quantity	Units	Dimensions
A. Regular Alphabet			
$\mathbf{A}$	Magnetic vector potential general vector function	weber/meter	ML/TQ
A_e	effective area	meter^2	L^2
$\mathbf{a}_A$	unit vector along $\mathbf{A}$	—	—
$\mathbf{B}$	magnetic flux density	weber/meter^2	M/TQ
C	capacitance	farad	T^2Q^2/ML^2
c	speed of light	meters/second	L/T
$\mathbf{D}$	electric flux density or electric displacement	coulomb/meter^2	Q/L^2
D	directivity	—	—
$d\mathbf{l}$	differential length element	meter	L
$d\mathbf{S}$	differential surface element	meter^2	L^2
dv	differential volume element	meter^3	L^3
$\mathbf{E}$	electric field intensity or electric field strength	volt/meter	ML/T^2Q
e	electron charge	coulomb	Q
$\mathbf{F}$	force	newton	ML/T^2
f	frequency	hertz	$1/T$
f_c	cutoff frequency	hertz	$1/T$
G	conductance	mho	TQ^2/ML^2

G_d	directive gain	—	—
G_p	power gain	—	—
$\mathbf{H}$	magnetic field intensity	ampere/meter	Q/TL
I	current	ampere	Q/T
$\mathbf{J}$	volume current density	ampere/meter^2	Q/L^2T
$\mathbf{J}_b$	bound current density	ampere/meter^2	Q/L^2T
$\mathbf{J}_d$	displacement current density	ampere/meter^2	Q/L^2T
j	$\sqrt{-1}$	—	—
$\mathbf{K}$	surface current density	ampere/meter	Q/TL
L	inductance	henry	ML^2/Q^2
L_{ext}	external inductance	henry	ML^2/Q^2
L_{int}	internal inductance	henry	ML^2/Q^2
$\mathbf{M}$	magnetization	ampere/meter	Q/LT
$\mathbf{m}$	magnetic moment	ampere-meter^2	QL^2/T
m	mass	kilogram	M
n	index of refraction	—	—
$\mathbf{P}$	polarization	coulomb/meter^2	Q/L^2
P	power	watt	ML^2/T^3
P_{ave}	time-average power	watt/meter^2	M/T^3
$\mathbf{p}$	dipole moment	coulomb-meter	QL
q,Q	charge	coulomb	Q
R	resistance	ohm	ML^2/TQ^2
R_{rad}	radiation resistance	ohm	ML^2/TQ^2
r,ϕ,θ	spherical coordinates	meter, radian, radian	$L,-,-$
S	surface area	meter^2	L^2
s	standing wave ratio	—	—
$\mathbf{T}$	torque	newton-meter	ML^2/T^2
t	time	second	T

u	velocity	meter/second	M/T
V	potential general scalar function	volt	ML^2T^2Q
v	volume	meter^3	L^3
W	energy, work	joule	ML^2/T^2
w	energy density	joule/meter^3	M/LT^2
X	reactance	ohm	ML^2/TQ^2
x,y,z	Cartesian coordinates	meter	L
Y	admittance	mho	TQ^2/ML^2
Y_o	characteristic admittance	mho	TQ^2/ML^2
Z	impedance	ohm	ML^2/TQ^2
	impedance per length	ohm/meter	ML/TQ^2
Z_o	characteristic impedance	ohm	ML^2/TQ^2

B. Greek Alphabet

α	attenuation constant	neper/meter	$1/L$
β	phase shift constant	radian/meter	$1/L$
Γ	reflection coefficient	—	—
δ	skin depth	meter	L
γ	propagation constant	1/meter	$1/L$
ε	permittivity	farad/meter	T^2Q^2/ML^3
ε_o	permittivity of free space	farad/meter	T^2Q^2/ML^3
ε_r	relative permittivity	—	—
η	intrinsic impedance	ohm	ML^2/TQ^2
Λ	flux linkage	weber	ML^2/TQ
λ	wavelength	meter	L
μ	permeability	henry/meter	ML/Q^2
μ_o	permeability of free space	henry/meter	ML/Q^2
μ_r	relative permeability	—	—

ρ,ϕ,z	cylindrical coordinates	meter, radian, radian	$L,-,-$
ρ_L	line charge density	coulomb/meter	Q/L
ρ_{ps}	polarization surface charge density	coulomb/meter^2	Q/L^2
ρ_{pv}	polarization volume charge density	coulomb/meter^3	Q/L^3
ρ_S	surface charge density	coulomb/meter^2	Q/L^2
ρ_v	volume charge density	coulomb/meter^3	Q/L^3
σ	conductivity	mho/meter	TQ^2/ML^3
τ	relaxation time	second	T
	transmission coefficient	—	—
Ψ	electric flux	coulomb	Q
	magnetic flux	weber	ML^2/TQ
	general flux		
χ_e	electric susceptibility	—	—
χ_m	magnetic susceptibility	—	—
ω	angular frequency	radian/second	$1/T$
C. Special Alphabet			
$\mathcal{F}$	magnetomotive force	ampere-turn	Q/T
ℓ	length	meter	L
$\mathcal{P}$	poynting vector	watt/meter^2	M/T^3
$\mathcal{P}$	permeance	weber/ampere-turn	ML^2/Q^2
$\mathcal{R}$	reluctance	ampere-turn/weber	Q^2/ML^2
∇	del operator	—	—

*Symbols used in Chapters 1 to 3 are arbitrary.

APPENDIX

E

Answers to Odd-Numbered Problems

Chapter 1

1.1 (a) $8\mathbf{a}_x - 3\mathbf{a}_y - 3\mathbf{a}_z$
(b) 8.544
(c) ± 0.4
(d) $0.8571\mathbf{a}_x + 0.6428\mathbf{a}_y + 1.642\mathbf{a}_z$

1.3 (a) 5
(b) 0.1569
(c) $4\mathbf{a}_x + 4\mathbf{a}_y - 5\mathbf{a}_z$
(d) 0.9913
(e) -31
(f) -31
(g) $13\mathbf{a}_x + 17\mathbf{a}_y + 24\mathbf{a}_z$
(h) $17\mathbf{a}_x + 17\mathbf{a}_y + 21\mathbf{a}_z$

1.5 $U_x = -\frac{69}{44}$, $V_y = -\frac{27}{22}$, $W_x = -\frac{97}{22}$

1.7 -14.43, $-8.33\mathbf{a}_x + 8.33\mathbf{a}_y - 8.33\mathbf{a}_z$

1.9 5

1.11 25.72

1.13 Proof

1.15 (c) 2.151

1.17 $7.5\mathbf{a}_x + 2.5\mathbf{a}_y$, $2.5\mathbf{a}_x - 7.5\mathbf{a}_y + 4\mathbf{a}_z$

1.19 (a) $0.9762\mathbf{a}_y + 0.2169\mathbf{a}_z$
(b) $-0.09523\mathbf{a}_x - 0.02381\mathbf{a}_y + 0.119\mathbf{a}_z$
(c) $2\mathbf{a}_x + 3\mathbf{a}_y - \mathbf{a}_z$

1.21 Proof

Chapter 2

2.1 (a) $P_1(-2.5, 4.33, 0)$
(b) $P_2(0.866, 0.5, -10)$

(c) $P_3(0, 7.071, -7.071)$
(d) $P_4(-0.75, -1.299, 2.598)$

2.3 (a) $(\rho \sin\phi + z)[\cos\phi \mathbf{a}_\rho - \sin\phi \mathbf{a}_\phi]$, $r(\sin\theta \sin\phi + \cos\theta)[\sin\theta \cos\phi \mathbf{a}_r + \cos\theta \cos\phi \mathbf{a}_\theta - \sin\phi \mathbf{a}_\phi]$
(b) $\frac{1}{2}\rho \sin 2\phi(1 + z)\mathbf{a}_\rho + \rho(z \cos^2\phi - \sin^2\phi)\mathbf{a}_\phi + \rho(\cos\phi + \sin\phi)\mathbf{a}_z$

$$\left[\frac{r}{2} \sin^2\theta \sin 2\phi + \frac{r^2}{2} \sin^2\theta \cos\theta \sin 2\phi + \frac{r}{2}(\cos\phi + \sin\phi)\cos 2\theta\right]\mathbf{a}_r$$
$$+ \left[\frac{r}{4} \sin 2\theta \sin 2\phi + \frac{r^2}{2} \cos^2\theta \sin\theta \sin 2\phi - r(\cos\phi + \sin\phi)\sin^2\theta\right]\mathbf{a}_\theta$$
$$+ \left[-r \sin\theta \sin^2\phi + \frac{r^2}{2} \sin 2\theta \cos^2\phi\right]\mathbf{a}_\phi$$

(c) $\cos\phi \mathbf{a}_\rho + \rho^2 \sin\phi \mathbf{a}_\phi + \mathbf{a}_z$, $(\sin\theta \cos\phi + \cos\theta)\mathbf{a}_r + (\cos\theta \cos\phi - \sin\theta)\mathbf{a}_\theta + r^2 \sin^2\theta \sin\phi \mathbf{a}_\phi$
(d) $-\frac{1}{\rho}\mathbf{a}_\rho + 10\mathbf{a}_z$, $10 \cos\theta \mathbf{a}_r - 10 \sin\theta \mathbf{a}_\theta - \frac{\mathbf{a}_\phi}{r \sin\theta}$

2.5 Proof

2.7 (a) $r \sin\theta[\sin\phi \cos\theta(r \sin\theta + \cos\phi)\mathbf{a}_r + \sin\phi(r \cos^2\theta - \sin\theta \cos\phi)\mathbf{a}_\theta + 3 \cos\phi \mathbf{a}_\phi]$, $5\mathbf{a}_\theta - 21.21\mathbf{a}_\phi$
(b) $\sqrt{\rho^2 + z^2}\left[\rho \mathbf{a}_\rho + \frac{\rho}{\rho^2 + z^2}\mathbf{a}_\phi + z\mathbf{a}_z\right]$, $4.472\mathbf{a}_\rho + 0.8944\mathbf{a}_\phi + 2.236\mathbf{a}_z$

2.9 (a) 5.385
(b) 10
(c) 9.956

2.11 (a) an infinite line parallel to the z-axis
(b) point $(2, -1, 10)$
(c) circle of radius 5
(d) an infinite line parallel to the z-axis
(e) a semi-infinite line parallel to the $x - y$ plane
(f) a circle of radius 5 in the $x - y$ plane

2.13 (a) $\mathbf{a}_x - \mathbf{a}_y + 7\mathbf{a}_z$
(b) $143.36°$
(c) -8.789

2.15 (a) $-\mathbf{a}_\theta = \mathbf{a}_z$
(b) $0.6931\mathbf{a}_\phi$
(c) $-\mathbf{a}_\theta + 0.6931\mathbf{a}_\phi$
(d) $0.6931\mathbf{a}_\phi$

2.17 (a) $\cos^{-1}(\sin\theta \cos\phi)$, $\cos^{-1}(\sin\theta \sin\phi)$, θ

2.19 $2 \cos\theta \mathbf{a}_r + \sin\theta \mathbf{a}_\theta$

Chapter 3

3.1 (a) 10.47
(b) 7.854
(c) 32.72

3.3 (a) 5.498
(b) 8.552
(c) 16

3.5 (a) −50
(b) −39.5

3.7 $2\pi a$

3.9 1.5

3.11 $\pi a^3/3$, $2\pi a^3/3$

3.13 1.257

3.15 (a) $4z^2\mathbf{a}_x + 3z\mathbf{a}_y + (8xz + 3y)\mathbf{a}_z$
(b) $2e^{(2x+3y)}\cos 5z\mathbf{a}_x + 3e^{(2x+3y)}\cos 5z\mathbf{a}_y - 5e^{(2x+3y)}\sin 5z\mathbf{a}_z$
(c) $2(z^2+1)\cos\phi\mathbf{a}_\rho - 2(z^2+1)\sin\phi\mathbf{a}_\phi + 4\rho z\cos\phi\mathbf{a}_z$
(d) $5e^{-2z}\sin\phi\mathbf{a}_\rho + 5e^{-2z}\cos\phi\mathbf{a}_\phi - 10e^{-2z}\sin\phi\mathbf{a}_z$
(e) $2r\cos\theta\cos\phi\mathbf{a}_r - r\sin\theta\cos\phi\mathbf{a}_\theta - r\cot\theta\sin\phi\mathbf{a}_\phi$
(f) $-\dfrac{3}{r^4}\sin\theta\sin\phi\mathbf{a}_r + \dfrac{\cos\theta\sin\phi}{r^4}\mathbf{a}_\theta + \dfrac{\cos\phi}{r^4}\mathbf{a}_\phi$

3.17 Along $2\mathbf{a}_x + 2\mathbf{a}_y - \mathbf{a}_z$

3.19 (a) $1 + 2y + yz$, $2x\mathbf{a}_x + (xy - 2x - y)\mathbf{a}_y + (2x - z - xz)\mathbf{a}_z$
(b) $z\cos\phi, -\left(\dfrac{\sin 2\phi}{\rho} + \rho\sin\phi\right)\mathbf{a}_\rho + \dfrac{1}{\rho}\mathbf{a}_\phi + \dfrac{\cos^2\phi}{\rho}\mathbf{a}_z$
(c) $4r + \dfrac{2\sin\theta\cos\phi}{r} + 2\cos\theta\cos\phi + \dfrac{\cos\phi}{r\sin\theta}, \left(\dfrac{\sec^2\theta}{r^2} + \dfrac{1}{r^2} + \dfrac{1}{r}\sin\phi\cot\theta + \sin\phi\right)\mathbf{a}_r - \dfrac{2\sin\phi}{r}\mathbf{a}_\theta + \left(2\sin\theta\cos\phi - \dfrac{\cos\theta\cos\phi}{r}\right)\mathbf{a}_\phi$

3.21 Proof.

3.23 (a) $6yz\mathbf{a}_x + 3xy^2\mathbf{a}_y + 3x^2yz\mathbf{a}_z$
(b) $4yz\mathbf{a}_x + 3xy^2\mathbf{a}_y + 4x^2yz\mathbf{a}_z$
(c) $6xyz + 3xy^3 + 3x^2yz^2$
(d) $2(x^2 + y^2 + z^2)$

3.25 (a) 0
(b) $\dfrac{1}{\rho}(6z - 1)z\sin\phi\mathbf{a}_\rho + \left[(2 - 3z)\dfrac{z}{\rho} + 6\rho\right]\cos\phi\mathbf{a}_\phi$, $\mathbf{a}_\rho$

(c) $-\left[\frac{4\cos\theta\cos\phi}{r} + \frac{2\cos\theta\cos\phi}{r\sin^2\theta}\right]\mathbf{a}_r - \frac{4\sin\theta\cos\phi}{r}\mathbf{a}_\theta - \left[\frac{1}{r^{3/2}\sin^2\theta}\right.$

$\left. + \frac{4\cos\theta\sin\phi}{r\sin\theta} - \frac{3}{4r^{3/2}}\right]\mathbf{a}_\phi$, $5.196\mathbf{a}_r - \mathbf{a}_\theta - 9.25\mathbf{a}_\phi$

3.27 (a) $\frac{7}{6}$

(b) $\frac{7}{6}$

(c) Yes

3.29 (a) 3

(b) 3

3.31 $\mathbf{A} = \left(\frac{x^3}{3} + c_1\right)\mathbf{a}_x + \left(\frac{y^3}{3} + c_2\right)\mathbf{a}_y$

3.33 $\frac{190\pi}{3}$

3.35 (a) 4π

(b) 4π

(c) -4π

(d) $4\pi\left(\frac{\pi}{3} - \frac{\sqrt{3}}{2}\right)$

(e) $\frac{4\pi\sqrt{3}}{3}$

(f) $\frac{4\pi}{3}\left(\pi - \frac{\sqrt{3}}{2}\right)$

3.37 Proof

3.39 -6

3.41 (a) Proof

(b) (i) $\frac{1}{6}$, 2, (ii) $\frac{1}{6}$, 2

Chapter 4

4.1 (b) $142\mathbf{a}_x - 364\mathbf{a}_y + 10\mathbf{a}_z$ N

4.3 (a) -29.46 nC

(b) 22.1 nC

4.5 (a) 0.5 C
(b) 1.206 μC
(c) 157.91 C

4.7 (a) $27.14\mathbf{a}_x$ V/m
(b) -68.23 nC

4.9 (a) $-0.3\mathbf{a}_x + 0.9\mathbf{a}_z$ V/m
(b) $-0.1611\mathbf{a}_x - 0.3889\mathbf{a}_y + 0.576\mathbf{a}_z$ V/m

4.11 $-0.0216\mathbf{a}_x + 18\mathbf{a}_y - 264.7\mathbf{a}_z$ V/m

4.13 $-16.36\mathbf{a}_x - 49.08\mathbf{a}_y$ nC/m^3

4.15 (a) 188.5 nC/m
(b) $7.5\mathbf{a}_z$ nC/m^2

4.17 (a) $8y$ C/m^3
(b) $4z$ C/m^3
(c) $2 \cos \theta \left(\frac{1}{r^4} - \frac{1}{r^3}\right)$

4.19 (a) $23.8\mathbf{a}_r$ V/m
(b) $23.35\mathbf{a}_r$ V/m

4.21 (a) 0
(b) 9 μC
(c) 14 μC

4.23 (a) 706.9 C
(b) -63.62C
(c) 63.62 C
(d) 706.9 C

4.25 251.3 nC

4.27 (a) -40μJ
(b) 20μJ
(c) -40μJ
(d) -60μJ

4.29 (a) 14.27 V
(b) 2.86 V

4.31 (a) 15 kV
(b) 15 kV
(c) 15 kV

4.33 Proof

4.35 (a) $e^{-x}[-\sinh y \sin z\mathbf{a}_x + \cosh y \sin z\mathbf{a}_y + \sinh y \cos z\mathbf{a}_z]$ V/m
(b) $\dfrac{3(x\mathbf{a}_x + y\mathbf{a}_y + z\mathbf{a}_z)}{[x^2 + y^2 + z^2]^{5/2}}$ V/m
(c) $e^{-z}[-\cos \phi\mathbf{a}_\rho + \sin \phi\mathbf{a}_\phi + \rho \cos \phi\mathbf{a}_z]$ V/m

(d) $\frac{1}{r^3}[2 \sin\theta \cos\phi \mathbf{a}_r - \cos\theta \cos\phi \mathbf{a}_\theta + \sin\phi \mathbf{a}_\phi]$ V/m

4.37 (b) 5.933×10^5
(c) 2.557 kV

4.39 (a) $\frac{2a^3\rho_o}{15\varepsilon_o r^2}\mathbf{a}_r, \frac{2a^3\rho_o}{15\varepsilon_o r}$
(b) $\frac{\rho_o}{\varepsilon_o}\left[\frac{r}{3} - \frac{r^3}{5a^2}\right]\mathbf{a}_r, \frac{\rho_o}{\varepsilon_o}\left[\frac{a^2}{4} + \frac{r^4}{20a^2} - \frac{r^2}{6}\right]$
(c) $0.745a$
(d) $r = 0, V_{\max} = \frac{\rho_o a^2}{4\varepsilon_o}$

4.41 $\mathbf{E} = \frac{Qd}{4\pi\varepsilon_o r^3}(2 \sin\theta \sin\phi \mathbf{a}_r - \cos\theta \sin\phi \mathbf{a}_\theta - \cos\phi \mathbf{a}_\phi)$ V/m

4.43 (a) 18 nJ
(b) −36 nJ

4.45 18.85 nJ

4.47 6.487 nJ

Chapter 5

5.1 5.026 A

5.3 −0.3 mA, −166 nA

5.5 (a) 10 μC/s
(b) 0.8 μC/m^2

5.7 1.26 W

5.9 (a) 0.27 mΩ
(b) 50.3 A (copper), 9.7 A (steel)
(c) 0.322 mΩ

5.11 1.000182

5.13 (a) $-12.73\, z\, \mathbf{a}_z$ nC/m^2, −12.73 nC/m^3
(b) $-7.427\, z\, \mathbf{a}_z$ nC/m^2 $\mathbf{a}_z$, 7.427 nC/m^3

5.15 (a) $P_r = \frac{Q}{4\pi r^2}\left(1 - \frac{1}{\varepsilon_r}\right)$
(b) 0
(c) $-\frac{Q}{4\pi a^2}\left(1 - \frac{1}{\varepsilon_r}\right), \frac{Q}{4\pi b^2}\left(1 - \frac{1}{\varepsilon_r}\right)$

5.17 (a) 428 μA
(b) -5.67 nC/m^3

5.19 10^{-12} mhos/m, glass or porcelain

5.21 (a) $-1.061\mathbf{a}_x + 1.768\mathbf{a}_y + 1.547\mathbf{a}_z$ nC/m^2
(b) $-0.7958\mathbf{a}_x + 1.326\mathbf{a}_y + 1.161\mathbf{a}_z$ nC/m^2
(c) 39.79°

5.23 (a) $26.53\mathbf{a}_\rho + 66.31\mathbf{a}_\phi - 53.05\mathbf{a}_z$ nC/m^2, $\dfrac{-26.53}{\rho}$ nC/m^3
(b) $5\mathbf{a}_\rho + 5\mathbf{a}_\phi - 4\mathbf{a}_z$ kV/m, $110.52\mathbf{a}_\rho + 110.52\mathbf{a}_\phi - 88.42\mathbf{a}_z$ nC/m^2

5.25 (a) 705.9 V/m, 0° (glass), 6000 V/m, 0° (air)
(b) 1940.5 V/m, 84.6° (glass), 2478.6 V/m, 51.2° (air)

5.27 (a) 381.97 nC/m^2
(b) $\dfrac{0.955\mathbf{a}_r}{r^2}$ nC/m^2
(c) 12.96 μJ

5.29 Proof

Chapter 6

6.1 Proof

6.3 (a) 79.67 V
(b) $-36.14\mathbf{a}_\rho$ V/m

6.5 Proof

6.7 Proof

6.9 252 kV, $-25\mathbf{a}_z$ kV/m, $-332\mathbf{a}_z$ nC/m^2, $\pm 332\mathbf{a}_z$ nC/m^2

6.11 $-\dfrac{10.53}{r} + 105.3V$, $-\dfrac{10.53}{r^2}\mathbf{a}_r$ V/m, $-\dfrac{93.2}{r^2}\mathbf{a}_r$ pC/m^2

6.13 11.7 V, $-17.86\mathbf{a}_\theta$ V/m

6.15 Proof

6.17 (a) $$\frac{4V_o}{\pi} \sum_{n=\text{odd}}^{\infty} \frac{\sin \dfrac{n\pi x}{b} \sinh \dfrac{n\pi}{b}(a - y)}{n \sinh \dfrac{n\pi a}{b}}$$

(b) $\dfrac{4V_o}{\pi} \displaystyle\sum_{n=\text{odd}}^{\infty} \dfrac{\sin \dfrac{n\pi y}{a} \sinh \dfrac{n\pi x}{a}}{n \sinh \dfrac{n\pi b}{a}}$

(c) $\dfrac{4V_o}{\pi} \displaystyle\sum_{n=\text{odd}}^{\infty} \dfrac{\sin \dfrac{n\pi y}{b} \sinh \dfrac{n\pi}{b}(a - x)}{n \sinh \dfrac{n\pi a}{b}}$

6.19 Proof

6.21 Proof

6.23 Proof

6.25 6 pF

6.27 Proof

6.29 (a) $-\dfrac{2V_o}{3d^2}(x + d)\mathbf{a}_x$ V/m

(b) $-\dfrac{2V_o}{3d^2}\, \varepsilon_o(d - x)\mathbf{a}_x$ C/m^2

(c) $\dfrac{2V_o\varepsilon_o}{3d}$, 0

(d) 94.31 pF

6.31 280.5 nF/km

6.33 21.85 pF

6.35 693.1 s

6.37 Proof

6.39 0.7078 mF

6.41 (a) 1 nC

(b) 5.25 nN

6.43 $-0.1891(\mathbf{a}_x + \mathbf{a}_y + \mathbf{a}_z)$ N

6.45 $-565.5\mathbf{a}_x$ V/m

Chapter 7

7.1 $14.85\mathbf{a}_x - 22.27\mathbf{a}_z$ nA/m

7.3 (b) $-0.255\mathbf{a}_x + 0.4775\mathbf{a}_y + 0.382\mathbf{a}_z$ A/m

7.5 (a) $-0.6792\mathbf{a}_z$ A/m
(b) $0.1989\mathbf{a}_z$ A/m
(c) $0.1989\mathbf{a}_x + 0.1989\mathbf{a}_y$ A/m

7.7 Proof

7.9 (a) $1.964\mathbf{a}_z$ A/m
(b) $1.78\mathbf{a}_z$ A/m
(c) $-0.1178\mathbf{a}_z$ A/m
(d) $-0.3457\mathbf{a}_x - 0.3165\mathbf{a}_y + 0.1798\mathbf{a}_z$ A/m

7.11 $102.32\mathbf{a}_z$ A/m

7.13 (a) $1.36\mathbf{a}_z$ A/m
(b) $0.884\mathbf{a}_z$ A/m

7.15 (b) $\mathbf{H} = \begin{cases} \dfrac{I\rho}{2\pi a^2}\,\mathbf{a}_\phi, & 0 < \rho < a \\ \dfrac{I}{2\pi\rho}\,\mathbf{a}_\phi, & \rho > a \end{cases}$

$|\mathbf{H}|$ is sketched in Fig. E.1

7.17 (a) Proof, $\dfrac{I}{\pi a^2}\,\mathbf{a}_z$
(b) $11.94\mathbf{a}_\phi$ A/m, $11.94\mathbf{a}_\phi$ A/m

7.19 (a) $9\mathbf{a}_y$ A/m^2
(b) 1.333 A

7.21 2.513 Wb

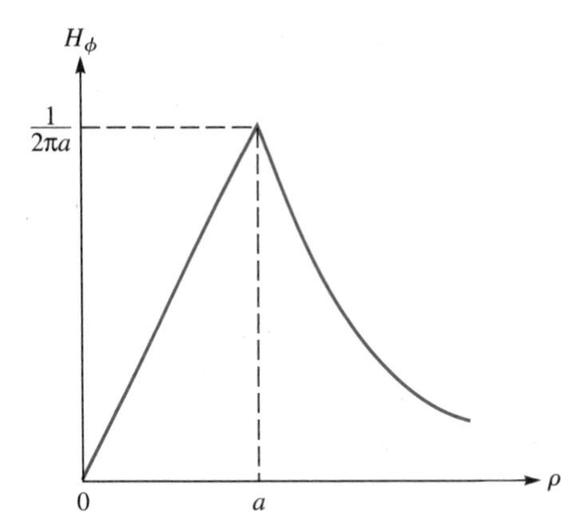

Figure E.1 For Problem 7.15.

7.23 (a) $31.43\mathbf{a}_y$ A/m
(b) $12.79\mathbf{a}_x + 6.366\mathbf{a}_y$ A/m

7.25 Proof

7.27 (a) **E** field in a charge-free region
(b) **H** field
(c) **E** field in a charge-free region
(d) **H** field
(e) neither

7.29 (a) $-(e^{-z} \cos y + \cos x)\mathbf{a}_y + e^{-z} \sin y\mathbf{a}_z$ Wb/m^2
(b) 6.283 Wb

7.31 $\mathbf{B} = \dfrac{\mu_o IL(-y\mathbf{a}_x + x\mathbf{a}_y)}{2\pi(x^2 + y^2 + z^2)^{3/2}}$

7.33 $\dfrac{I_o\rho}{2\pi a^2}\mathbf{a}_\phi$

7.35 Proof

7.37 $A_z = \dfrac{\mu_o I}{28\pi}\left(\dfrac{\rho^2}{a^2} - 9\right) - \dfrac{8\mu_o I}{7\pi} \ln \dfrac{\rho}{3a}$

7.39 (a) 50 A
(b) -250 A

7.41 Proof

Chapter 8

8.1 (a) $10\mathbf{a}_x - 30\mathbf{a}_y + 80\mathbf{a}_z$ N
(b) $4\mathbf{a}_x - 12\mathbf{a}_y + 2\mathbf{a}_z$ N
(c) $14\mathbf{a}_x - 42\mathbf{a}_y + 82\mathbf{a}_z$ N

8.3 (a) $18\mathbf{a}_x + 15\mathbf{a}_y$ m/s^2
(b) $22\mathbf{a}_x + 15\mathbf{a}_y + 3\mathbf{a}_z$ m/s
(c) 718 J
(d) (14, 5.5, 3)

8.5 (0.242, 1, 1.923), $4.071\mathbf{a}_x - 2.903\mathbf{a}_z$ m/s, $x^2 + (z - 19/12)^2 = (5/12)^2$, $y = 1$; i.e., a circle on plane $y = 1$.

8.7 (a) Proof
(b) (i) 5 μN/m attractive, (ii) 5 μN/m repulsive

8.9 -15.59 mJ

8.11 $1.949\mathbf{a}_x$ mN/m

8.13 $2.133\mathbf{a}_x - 0.2667\mathbf{a}_y$ Wb/m^2

8.15 104.2 mWb/m^2

8.17 (a) 0.03142$\mathbf{a}_z$ A·m
(b) 69.44$\mathbf{a}_r$ + 49.1$\mathbf{a}_\theta$ μA/m
(c) 1.571$\mathbf{a}_r$ + 0.785$\mathbf{a}_\theta$ pWb/m^2

8.19 (a) $-18.52\mathbf{a}_z$ mWb/m^2
(b) $-4\mathbf{a}_z$ mWb/m^2
(c) $-111\mathbf{a}_r + 78.6\mathbf{a}_\theta$ mWb/m^2

8.21 (a) 3
(b) 398$\mathbf{a}_x$ − 995$\mathbf{a}_y$ + 796$\mathbf{a}_z$ A/m
(c) 1.19$\mathbf{a}_x$ − 2.98$\mathbf{a}_y$ + 2.39$\mathbf{a}_z$ kA/m
(d) 4.475 J/m^3

8.23 (a) 1.273$\mathbf{a}_z$ MA/m^2, 1.592$\mathbf{a}_z$ MA/m^2, 159.15$\mathbf{a}_\phi$ A/m at ρ = 1 mm, 0, 0, 0 at ρ = 3 mm
(b) $-1.273\mathbf{a}_z$ kA/m^2

8.25 (a) 12$\mathbf{a}_\rho$ − 6$\mathbf{a}_\phi$ + 7$\mathbf{a}_z$ kA/m
(b) 15.1$\mathbf{a}_\rho$ − 22.6$\mathbf{a}_\phi$ + 26.4$\mathbf{a}_z$ mWb/m^2
(c) $\theta_2 = 66.55°$, $\theta_1 = 37.53°$

8.27 26.83$\mathbf{a}_x$ − 30$\mathbf{a}_y$ + 33.96$\mathbf{a}_z$ A/m

8.29 (a) $-5\mathbf{a}_y$ A/m, $-6.283\mathbf{a}_y$ μWb/m^2
(b) $-35\mathbf{a}_y$ A/m, $-110\mathbf{a}_y$ μWb/m^2
(c) 5$\mathbf{a}_y$ A/m, 6.283$\mathbf{a}_y$ μWb/m^2

8.31 24.674 mH

8.33 (a) 250 mH/m
(b) 2 μJ/m^3

8.35 5103 turns

8.37 Proof

8.39 (a) 5.25 × 10^5 A·t/Wb
(b) 420 A·t

8.41 (a) 1.15 × 10^6 A·t/m, 919 A·t, 0.8 mWb/m^2
(b) 2.387 × 10^7 A·t/m, 19,081 A·t, 0.8 mWb/m^2

8.43 53.05 kN

8.45 7694 kg

Chapter 9

9.1 0.4738 sin 377t V

9.3 (a) 0.75 mA
(b) 27.8 cos(120πt − 0.3) mA

9.5 9.888 μV, point A is at higher potential

9.7 3.89 mV

9.9 6A, counterclockwise

9.11 4.32 mV

9.13 $4 \cos 10^7 t$ nA/m^2, $-8.842 \sin 10^7 t$ nA/m^2

9.15 900 kHz

9.17 $\nabla \times \mathbf{E} = -\dfrac{\partial \mathbf{B}}{\partial t}, \nabla \times \mathbf{H} = \dfrac{\partial \mathbf{D}}{\partial t}$

9.19 $-\dfrac{60}{\mu_o \omega} \cos 2x \cos(kz - \omega t)\mathbf{a}_z - \dfrac{30k}{\mu_o \omega} \sin 2x \sin(kz - \omega t)\, \mathbf{a}_x$ A/m

$-30\omega\varepsilon_o \sin 2x \cos(kz - \omega t)\mathbf{a}_y$ A/m

9.21 Proof

9.23 $(2 - \rho)(1 + t)e^{-\rho - t}\mathbf{a}_z$ Wb/m^2, $\dfrac{(1 + t)(3 - \rho)}{4\pi} 10^7 e^{-\rho - t}\mathbf{a}_\phi$ A/m^2

9.25 Proof

9.27 (a) $0.354 \angle -202.38°$
(b) $3.14 \angle -49.26°$
(c) $2.77 \angle 33.7°$
(d) $2.76 \angle 72.4°$
(e) $6.325 \angle 127.5°$

9.29 (a) $5 \cos(\omega t + 70°)\mathbf{a}_x - 5x \cos(\omega t + 53.13°)\mathbf{a}_y$
(b) $10 \cos(\omega t - kz)\mathbf{a}_x - 5 \sin(\omega t + kz + \pi/4)\mathbf{a}_y$
(c) $2 \sin x \sin(\omega t - 3x) + e^{3x} \cos(\omega t - 4x)$

9.31 $20 \sin(k_x x) \sin(k_y y) \cos(\omega t)\mathbf{a}_z$ V/m

$-\dfrac{20}{\omega} [k_y \sin(k_x x) \cos(k_y y) \sin(\omega t)\mathbf{a}_x - k_x \cos(k_x x) \sin(k_y y) \sin(\omega t)\mathbf{a}_y]$ Wb/m^2

Chapter 10

10.1 (a) $0.01 \sin\left(2\pi \cdot 10^6 t - \dfrac{\pi x}{150}\right) \mathbf{a}_z$
(b) $-8.66\mathbf{a}_z$ mV/m

10.3 (a) $+\mathbf{a}_x$
(b) 1.0472 m
(c) 1.667×10^7 m/s

10.5
(a) 1823
(b) 1.013 mhos/m
(c) $1.768 \times 10^{-11} - j3.224 \times 10^{-8}$ F/m
(d) 9.995 Np/m
(e) $13.97 \angle 44.98° \ \Omega$

10.7 2.827, 70.52°, 44.43 m

10.9 (a) along z-direction
(b) 6.283 m, 20.94 ns, 3×10^8 m/s
(c) See Fig. E2
(d) $-13.26 \sin(3 \times 10^8 t + y)\mathbf{a}_x$ mA/m

10.11 0.1m, lossless medium

10.13 Proof

10.15 (a) 5.6993
(b) 1.2566 m, 1.257×10^8 m
(c) 157.91 Ω
(d) along y
(e) $4.737 \sin(2\pi \times 10^8 t - 5x)\mathbf{a}_y$ V/m
(f) $0.15 \cos(2\pi \times 10^8 t - 5x)\mathbf{a}_y$ A/m

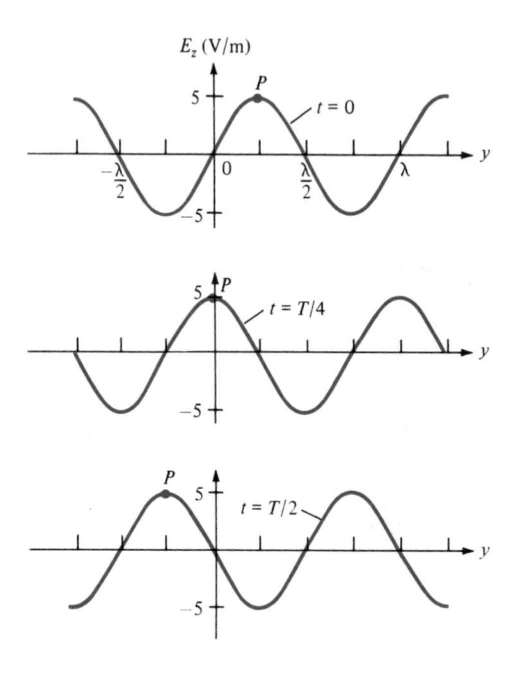

Figure E.2 For Problem 10.9(c).

10.17 (a) 900
(b) 0.225
(c) 1.8
(d) 0.06

10.19 113.75 m, 1.5×10^8 m/s

10.21 (a) 2.287 Ω
(b) 207.88 Ω
(c) 12.137 kHz

10.23 1.885 W/m^2 along $-\mathbf{a}_y$

10.25 (a) $0.2652 \cos(2\pi \times 10^6 t - \beta z)\mathbf{a}_z$ W/m^2
(b) $1.41 e^{-2z/3} \sin(10^8 t - \beta z)\sin(10^8 t - \beta z - 13.63°)\mathbf{a}_z$ mWb/m^2

10.27 (a) 0.0333 rad/m, $-377\dfrac{\sin\theta}{r}\cos(10^7 t - \beta r)\mathbf{a}_\phi$ V/m
(b) $188.5\dfrac{\sin^2\theta}{r^2}\mathbf{a}_r$ W/m^2
(c) 246.74 W

10.29 (a) Proof
(b) $\left[\dfrac{\eta_2 - \eta_1}{\eta_2 + \eta_1}\right]^2, \dfrac{4\eta_1\eta_2}{(\eta_1 + \eta_2)^2}$

10.31 (a) 0.5×10^8 rad/s
(b) 2
(c) $-26.53 \cos(0.5 \times 10^8 t + z)\mathbf{a}_x$ mA/m
(d) $1.061\mathbf{a}_z$ W/m^2

10.33 (a) 6.283 m, 3×10^8 rad/s
(b) $-0.0265 \cos(\omega t - z)\mathbf{a}_x$ A/m
(c) -0.268, 0.732
(d) $\mathbf{E}_1 = 10 \cos(\omega t - z)\mathbf{a}_y - 2.68 \cos(\omega t + z)\mathbf{a}_y$ V/m, $\mathbf{E}_2 = 7.32 \cos(\omega t - z)\mathbf{a}_y$ V/m, $\mathcal{P}_{1\text{ave}} = 0.1231\mathbf{a}_z$ W/m^2, $\mathcal{P}_{2\text{ave}} = 0.1231\mathbf{a}_z$ W/m^2

10.35 2.578 W/m^2

10.37 $\mathbf{E}_r = 14.84 \cos(10^9 t + 3.33z + 130.61°)\mathbf{a}_x + 11.13\sin(10^9 t + 3.33z + 130.61°)\mathbf{a}_y$ V/m
$\mathbf{H}_r = 29.52 \sin(10^9 t + 3.33z + 130.61°)\mathbf{a}_x - 39.36 \cos(10^9 t + 3.33z + 130.61°)\mathbf{a}_y$ mA/m
$\mathbf{E}_t = 32.4 \cos(10^9 t - 4.393z + 20.3°)\mathbf{a}_x + 24.3 \sin(10^9 t - 4.393z + 20.3°)\mathbf{a}_y$ V/m
$\mathbf{H}_t = -101.6 \sin(10^9 t - 4.393z - 12.78°)\mathbf{a}_x + 135.5\cos(10^9 t - 4.393z - 12.78°)\mathbf{a}_y$ mA/m

10.39 Proof

10.41 19.47°, 13.63°

10.43 Proof

10.45 (a) 19.47°, 90°
(b) 3.333 rad/m
(c) 0.6283 m, 1.885m
(d) $(-213.3\mathbf{a}_x + 75.4\mathbf{a}_z)\cos\left(10^9 t - kx - k\sqrt{8}z\right)$ V/m
(e) $1357\cos(10^9 t - 3.333x)\mathbf{a}_z$ V/m, $(213.3\mathbf{a}_x + 75.4\mathbf{a}_z)\cos\left(10^9 t - kx + k\sqrt{8}z\right)$ V/m
(f) 18.43°

10.47 (a) 15×10^8 rad/s
(b) $(-8\mathbf{a}_x + 6\mathbf{a}_y - 5\mathbf{a}_z)\sin(15 \times 10^8 t + 3x + 4y)$ V/m

Chapter 11

11.1 Proof

11.3 $\dfrac{V_o}{Z_o}\sin(\omega t - \beta z)$ A

11.5 (a) 46 Ω/m, 4.007×10^{-4} mhos/m, 1.111 μH/m, 111.5 pF/m
(b) 0, 0, 2.387 μH/m, 955 pF/m
(c) 3.2 Ω/m, 0.5 mmhos/m, 0.038 μH/m, 5.98 pF/m

11.7 $55.12 + j45.85$ Ω, 0.1738 μs

11.9 $0.075 + j0.1761$/m, 2.142×10^8 m/s, $0.841\,\underline{/79.25^\circ}$, 11.578

11.11 Proof

11.13 (a) 50 m
(b) 125 m

11.15 $0.2\,\underline{/40^\circ}$

11.17 (a) 46.875Ω
(b) 48.39 V

11.19 Proof

11.21 2.184 pF, $0.8\,\underline{/-33.12^\circ}$

11.23 (a) $0.5543\,\underline{/25^\circ}$, 296 Ω, 21.622 Ω
(b) $184 + j124$ Ω, 3.7, $38.4 + j60.8$ Ω
(c) Three $Z_{in,max}$ and two $Z_{in,min}$

11.25 $15 - j32$ Ω, $0.7742\,\underline{/216^\circ}$

11.27 (a) 24.5 Ω
(b) 55.33 Ω, 67.74 Ω

11.29 148 Ω

11.31 (a) $j450$ Ω
(b) $11.54 - j7.692$ Ω

11.33 0.231λ, $-j15$ mmhos

11.35 4, $0.6 \angle -90°$, $27.6 - j52.8\ \Omega$

11.37 (a) $83 - j49\ \Omega$
(b) $83 + j49\ \Omega$

11.39 Proof

11.41 (a) See Fig. E.3 (a)
(b) See Fig. E.3 (b) and (c)
(c) 83.33 V, 666.7 mA, 83.33 V, 666.7 mA
(d) See Fig. E.3 (d)

11.43 See Fig. E.4

11.45 See Fig. E.5

11.47 4.172, 7.941

11.49 $9.112\ \Omega < Z_o < 21.03\ \Omega$

Chapter 12

12.1 Proof

12.3 980 MHz

12.5 $TE_{31}/\ TM_{31}$, $2093\ \Omega/67.91\ \Omega$

12.7 (a) $a = 3$cm, $b = 1.25$ cm
(b) TE_{20} (10 GHz), TE_{11} (13 GHz), TE_{30} (15 GHz)
(c) 8.67 GHz

12.9 $a = 1.326$ m, $b = 3.75$ cm, will propagate

12.11 (a) 4.977 GHz
(b) $0,\ 9.203 \times 10^{-4}\left(\frac{2\pi}{a}\right) \cos\left(\frac{2\pi x}{a}\right) \sin\left(\frac{\pi y}{b}\right) \sin(\omega t - 10z)$ V/m

12.13 (a) $E_y = 0 = E_z = H_x$

$$H_y = -7.286 \sin\left(\frac{2\pi y}{a}\right) \sin(\omega t - 150z) \text{ mA/m}$$

$$H_z = 25.43 \cos\left(\frac{2\pi y}{a}\right) \cos(\omega t - 150z) \text{ mA/m}$$

(b) At $x = 0$, $-25.43 \cos\left(\frac{2\pi y}{a}\right) \cos(\omega t - 150z)\mathbf{a}_y - 7.286 \sin\left(\frac{2\pi y}{a}\right)$

$\sin(\omega t - 150z)\mathbf{a}_z$ mA/m

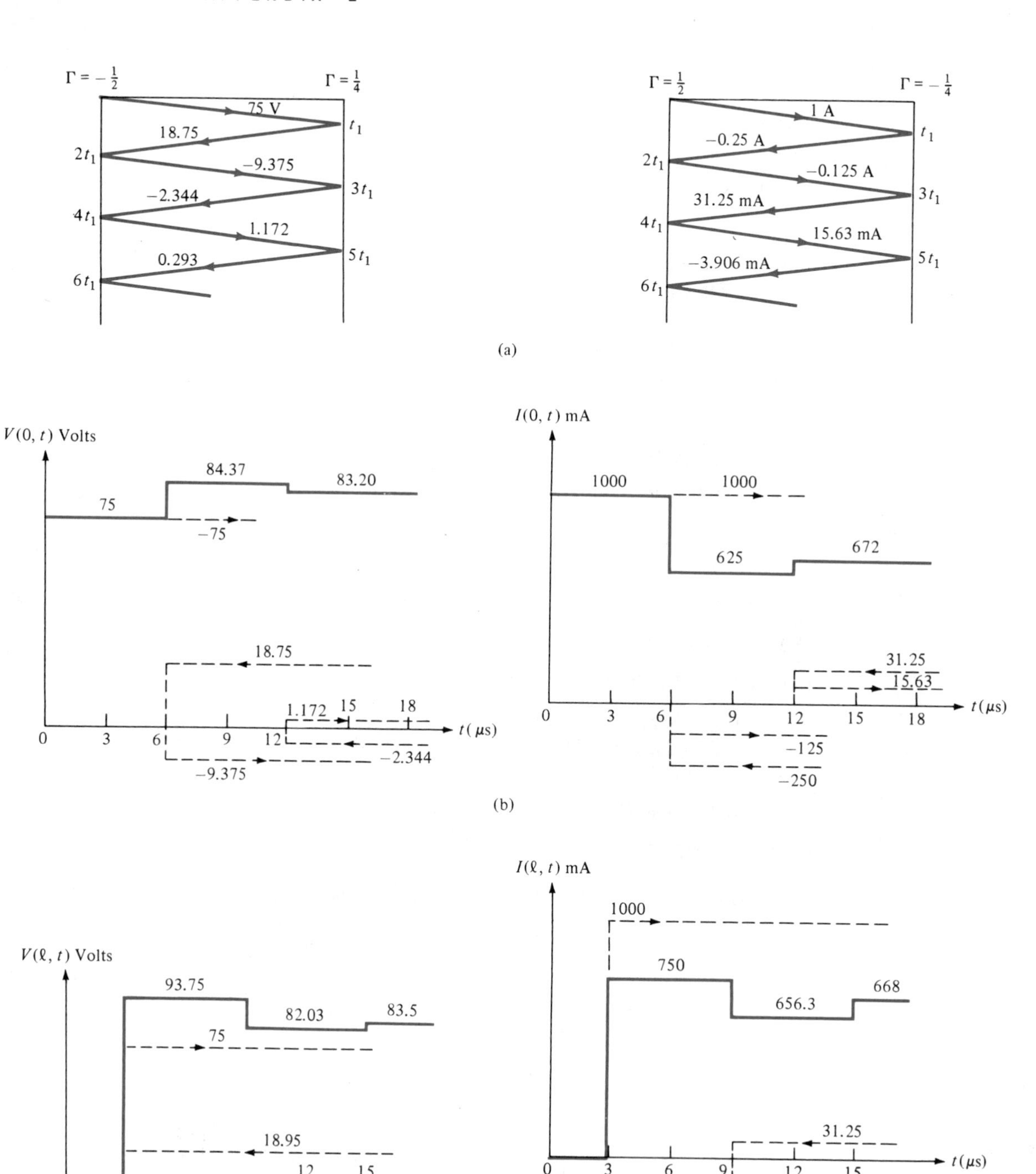

Figure E.3 For Problem 11.41: **(a)** voltage and current bounce diagrams ($t_1 = 3\mu s$, **(b)** plot of $V(0, t)$ and $I(0, t)$, **(c)** plot of $V(\ell, t)$ and $I(\ell, t)$, **(d)** plot of $V(\ell/2, t)$ and $I(\ell/2, t)$.

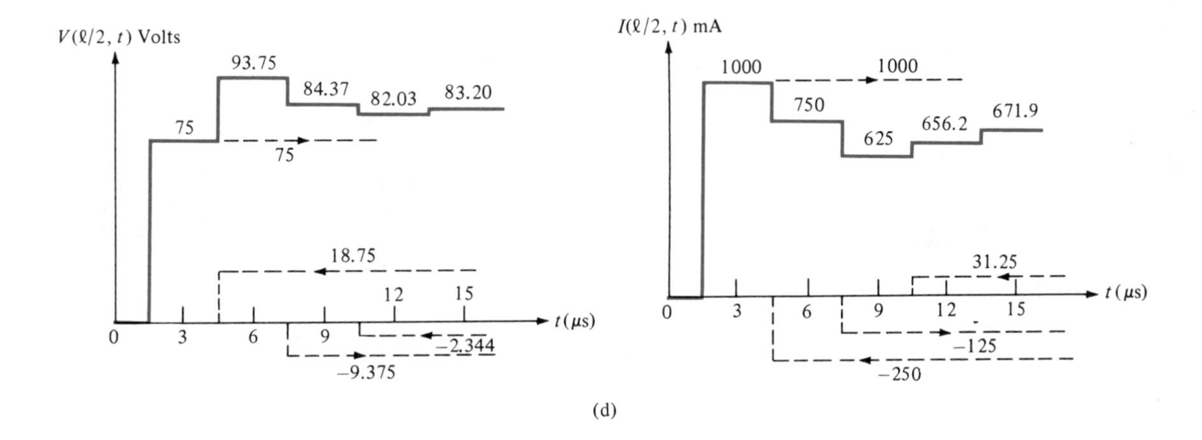

(d)

Figure E.3 (Continued)

at $x = a$, $25.43 \cos\left(\frac{2\pi y}{a}\right) \cos(\omega t - 150z)\mathbf{a}_y + 7.286 \sin\left(\frac{2\pi y}{a}\right)$

$\sin(\omega t - 150z)\mathbf{a}_z$ mA/m
at $y = 0$, $25.43 \cos(\omega t - 150z)\mathbf{a}_x$ mA/m
at $y = b$, $-25.43 \cos(\omega t - 150z)\mathbf{a}_x$ mA/m
(c) $-88.42 \sin(\omega t - 150z)$ pC/m², 0

12.15 $\dfrac{\beta^2 E_o^2}{8\pi^2 \eta_{TM11}} \dfrac{a^3 b^3}{a^2 + b^2}$

12.17 $5.12 \cos 40\pi x \sin 30\pi y \sin(\omega t - \beta z)$ V/m

12.19 (a) TE_{10} mode
(b) 4.01×10^8 m/s, 2.24×10^8 m/s

12.21 1.491×10^8 m/s, 337.2 Ω

12.23 Proof

12.25 (a) 2.012 GHz, 6.667 cm
(b) 9.754×10^{-4} Np/m, 161.9 rad/m, $9.754 \times 10^{-4} + j161.9$ /m
(c) 1.551×10^8 m/s

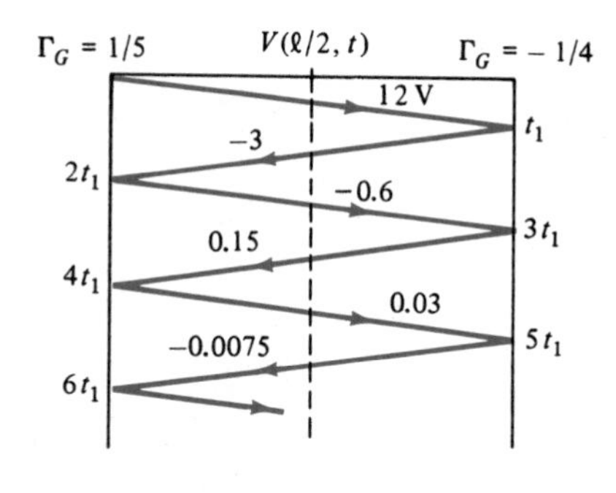

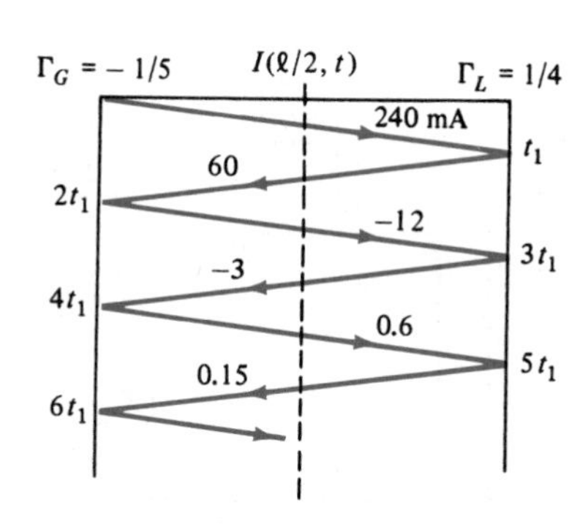

(a)

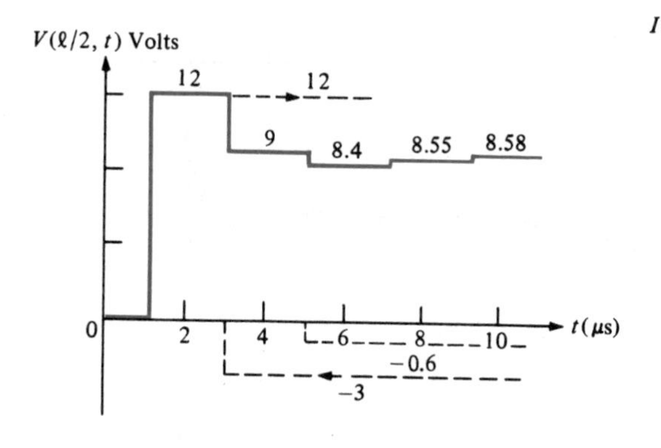

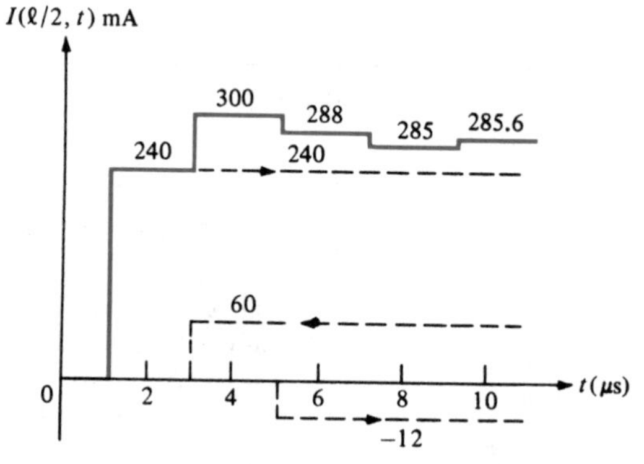

(b)

Figure E.4 For Problem 11.43: **(a)** voltage and current bounce diagrams ($t_1 = 2\mu s$), **(b)** plot of $V(\ell/2, t)$ and $I(\ell/2, t)$.

12.27 $E_{zs} = 0$

$$E_{xs} = \frac{j\omega\mu}{h^2}\left(\frac{n\pi}{b}\right) H_o \cos\left(\frac{m\pi x}{a}\right) \sin\left(\frac{n\pi y}{b}\right) \sin\left(\frac{p\pi z}{c}\right)$$

$$E_{ys} = -\frac{j\omega\mu}{h^2}\left(\frac{m\pi}{a}\right) H_o \sin\left(\frac{m\pi x}{a}\right) \cos\left(\frac{n\pi y}{b}\right) \sin\left(\frac{p\pi z}{c}\right)$$

$$H_{xs} = -\frac{1}{h^2}\left(\frac{m\pi}{a}\right)\left(\frac{p\pi}{c}\right) H_o \sin\left(\frac{m\pi x}{a}\right) \cos\left(\frac{n\pi y}{b}\right) \cos\left(\frac{p\pi z}{c}\right)$$

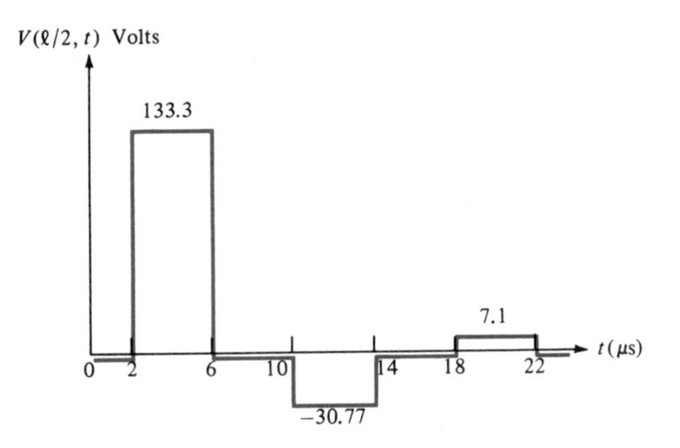

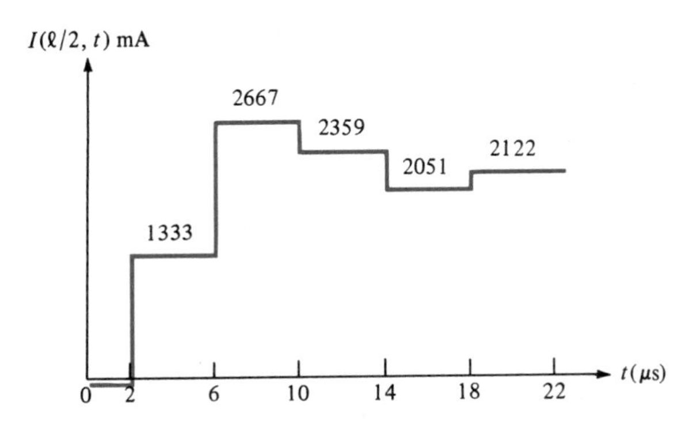

Figure E.5 For Problem 11.45.

$$H_{ys} = -\frac{1}{h^2}\left(\frac{n\pi}{b}\right)\left(\frac{p\pi}{c}\right) H_o \cos\left(\frac{m\pi x}{a}\right) \sin\left(\frac{n\pi y}{b}\right) \cos\left(\frac{p\pi z}{c}\right)$$

$$H_{zs} = H_o \cos\left(\frac{m\pi x}{a}\right) \cos\left(\frac{n\pi y}{b}\right) \sin\left(\frac{p\pi z}{c}\right)$$

$$\text{where } h^2 = \left(\frac{m\pi}{a}\right)^2 + \left(\frac{n\pi}{b}\right)^2$$

12.29 (a) TE_{011}
(b) TM_{110}
(c) TE_{101}

12.31 1.581 GHz

12.33 (a) Proof
(b) 2.579 cm
(c) 8.225 GHz

Chapter 13

13.1 $-\dfrac{50\eta\beta}{\mu r}\sin(\omega t - \beta r)(-\sin\phi\,\mathbf{a}_\phi + \cos\theta\cos\phi\,\mathbf{a}_\theta)$ V/m

$-\dfrac{50\beta}{\mu r}\sin(\omega t - \beta r)(\sin\phi\,\mathbf{a}_\theta + \cos\theta\cos\phi\,\mathbf{a}_\phi)$ A/m

13.3 31.58 Ω

13.5 (a) $E_{\theta s} = -31.2 - j18$ V/m
(b) $H_{\phi s} = -j31.83$ mA/m

13.7 (a) Proof (b) 0.05λ

13.9 (a) $4.224 \underline{/-30^\circ}$ V/m
(b) $0.751 \underline{/-30^\circ}$ V/m

13.11 (a) $12.73\mathbf{a}_r$ μW/m²
(b) 0.098 V/m

13.13 (a) 90.71 μA
(b) 0.25 mW

13.15 See Fig. E.6

13.17 See Fig. E.7

13.19 (a) 0.6198, 1.125
(b) 0.5124, 1.5462
(c) 0.6198, 1.125

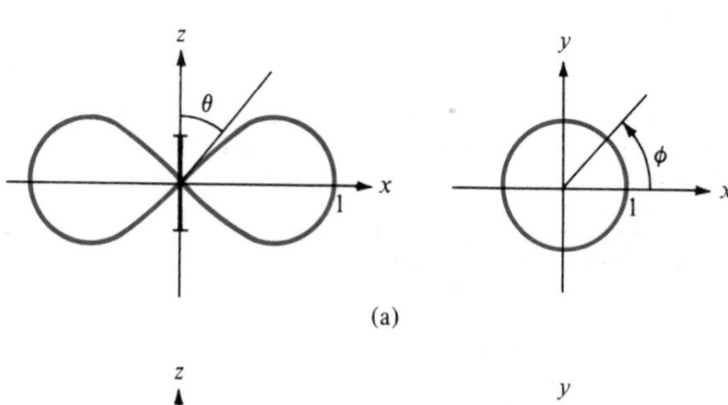

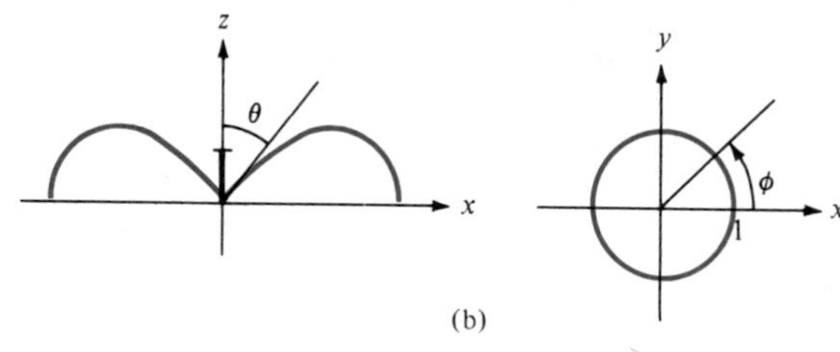

Figure E.6 For Problem 13.17: E-plane patterns ($\phi = 0$), and H-plane ($\theta = \pi/2$).

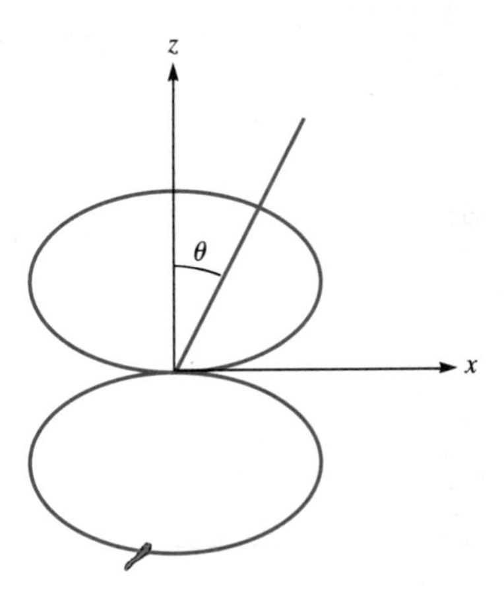

Figure E.7 For Problem 13.17.

13.21 (a) 4.0
(b) 5.093
(c) 4.712

13.23 $$\frac{\eta I_o^2 \sin^2\theta \sin^2\left(\frac{\pi}{2}\cos\theta\right)}{32\pi^2\cos^2\theta[1 - 0.25\cos^2\theta]^2}$$

13.25 Proof

13.27 (a) $2\cos(\pi\cos\theta)$
(b) 60°, 120°
(c) 0°, 90°, 180°
(d) see Fig. E.8

13.29 6.99 dB

13.31 (a) Proof
(b) $$\frac{2}{1 + \dfrac{\sin\beta d}{\beta d}}$$
(c) 8

13.33 0.6031

13.35 (a) 0.03406 μW
(b) 0.1362 μW
(c) 0.06812 μW

13.37 0.937 mW/m^2

13.39 7.12 mW

13.41 1.038 kW

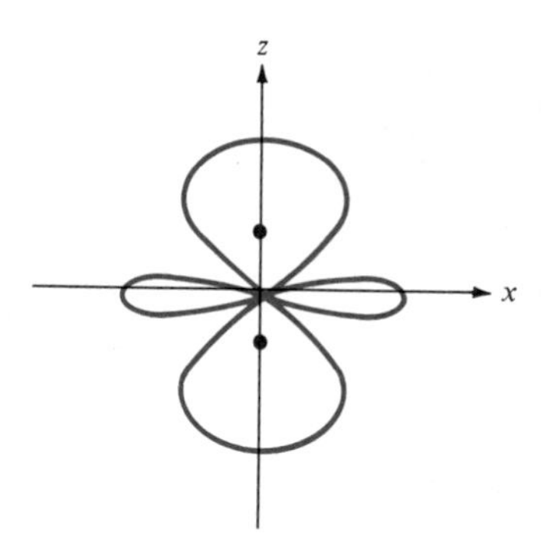

Figure E.8 For Problem 13.27.

Chapter 14

14.3 1.736, 1.76

14.5 Proof

14.7 20 V

14.9 (a) Matrix $[A]$ remains the same, but $-h^2\beta/\varepsilon$ must be added to each term of matrix $[B]$
(b) $V_a = 4.276$, $V_b = 9.577$, $V_c = 11.126$
$V_d = -2.013$, $V_e = 2.919$, $V_f = 6.069$
$V_g = -3.424$, $V_h = -0.109$, $V_i = 2.909$

14.11 (b) Numerical result agrees completely with the exact solution, e.g., for $t = 0$, $V(0,0) = 0$, $V(0.1,0) = 0.3090$, $V(0.2, 0) = 0.5878$, $V(0.3, 0) = 0.809$, $V(0.4, 0) = 0.9511$, $V(0.5, 0) = 1.0$, $V(0.6, 0) = 0.9511$, etc.

14.13 4.621 pF

14.15 12.47 mV, $-0.3266\mathbf{a}_x + 1.1353\mathbf{a}_y + 1.6331\mathbf{a}_z$ V/m

14.17 (a) Exact: $C = 80.26$ pF/m, $Z_o = 41.56\ \Omega$
Numerical:

N	C (pF/m)	$Z_o(\Omega)$
10	82.386	40.486
20	80.966	41.197
40	80.438	41.467
100	80.025	41.562

(b) Numerical:

N	C (pF/m)	$Z_o(\Omega)$
10	109.51	30.458
20	108.71	30.681
40	108.27	30.807
100	107.93	30.905

14.19 Proof

14.21 (a) $\alpha_1^{(1)} = 1.0 - 0.8571x - 0.5714y$, $\alpha_2^{(1)} = 1.1428x - 0.5714y$, $\alpha_3^{(1)} = -0.2857x + 1.1429y$, $\alpha_1^{(2)} = 1.25 - 0.625y$, $\alpha_2^{(2)} = -1.5 + 0.6667x + 0.4167y$, $\alpha_3^{(2)} = 1.25 - 0.6667x + 0.2083y$

(b) $$[C^{(1)}] = \begin{bmatrix} 0.4643 & -0.2857 & -0.1786 \\ -0.2857 & 0.7143 & -0.4286 \\ -0.1786 & -0.4286 & 0.6071 \end{bmatrix}$$

$$[C^{(2)}] = \begin{bmatrix} 0.4688 & -0.3125 & -0.1563 \\ -0.3125 & 0.7417 & -0.4292 \\ -0.1563 & -0.4292 & 0.5854 \end{bmatrix}$$

14.23 24 W, 12 W, and 24 W at nodes 1, 2, and 3 respectively

14.25 (a) 10.667 V
(b) 10 V

14.27 $$\begin{bmatrix} 0.8333 & -0.6667 & 0 & -0.1667 \\ -0.6667 & 1.4583 & -0.375 & -0.4167 \\ 0 & -0.375 & 0.625 & -0.25 \\ -0.1667 & -0.4167 & -0.25 & 0.833 \end{bmatrix}$$

14.29

Node no.	FEM	Exact
8	4.546	4.366
9	7.197	7.017
10	7.197	7.017
11	4.546	4.366
14	10.98	10.60
15	17.05	16.84
16	17.05	16.84
17	10.98	10.60
20	22.35	21.78
21	32.95	33.16
22	32.95	33.16
23	22.35	21.78
26	45.45	45.63
27	59.47	60.60
28	59.47	60.60
29	45.45	45.63

14.31 Compare results with those of Example 14.6

INDEX

PHYSICAL CONSTANTS

Quantity (Units)	Symbol	Best Experimental Value	Approximate Value for Problem Work
Permittivity of free space (F/m)	ε_o	8.854×10^{-12}	$\frac{10^{-9}}{36\pi}$
Permeability of free space (H/m)	μ_o	$4\pi \times 10^{-7}$	12.6×10^{-7}
Intrinsic impedance of free space (Ω)	η_o	376.6	120π
Speed of light in vacuum (m/s)	c	2.998×10^{8}	3×10^{8}
Electron charge (C)	e	-1.6030×10^{-19}	-1.6×10^{-19}
Electron mass (Kg)	m_e	9.1066×10^{-31}	9.1×10^{-31}
Proton mass (kg)	m_p	1.67248×10^{-27}	1.67×10^{-27}
Neutron mass (Kg)	m_n	1.6749×10^{-27}	1.67×10^{-27}
Boltzmann constant (J/K)	κ	1.38047×10^{-23}	1.38×10^{-23}
Avogadro's number (/Kg-mole)	N	6.0228×10^{26}	6×10^{26}
Planck's constant (J · s)	h	6.624×10^{-34}	6.62×10^{-34}
Acceleration due to gravity (m/s^2)	g	9.81	9.8
Universal contant of gravitation (m^2/Kg · s^2)	G	6.658×10^{-11}	6.66×10^{-11}
Electron-volt (J)	eV	1.6030×10^{-19}	1.6×10^{-19}

POWERS OF TEN

Power	Prefix	Symbol
10^{18}	Exa	E
10^{15}	Peta	P
10^{12}	Tera	T
10^{9}	Giga	G
10^{6}	Mega	M
10^{3}	kilo	k
10^{2}	hecto	h
10^{1}	deka	da
10^{-1}	deci	d
10^{-2}	centi	c
10^{-3}	milli	m
10^{-6}	micro	μ
10^{-9}	nano	n
10^{-12}	pico	p
10^{-15}	femto	f
10^{-18}	atto	a

THE GREEK ALPHABET

Upper Case	Lower Case	Name	Upper Case	Lower Case	Name
A	α	Alpha	N	ν	Nu
B	β	Beta	Ξ	ξ	Xi
Γ	γ	Gamma	O	o	Omicron
Δ	δ	Delta	Π	π	Pi
E	ε	Epsilon	P	ρ	Rho
Z	ζ	Zeta	Σ	σ,s	Sigma
H	η	Eta	T	τ	Tau
Θ	θ	Theta	Υ	υ	Upsilon
I	ι	Iota	Φ	ϕ	Phi
K	κ	Kappa	X	χ	Chi
Λ	λ	Lambda	Ψ	ψ	Psi
M	μ	Mu	Ω	ω	Omega

VECTOR DERIVATIVES

Cartesian Coordinates (x, y, z)

$$\mathbf{A} = A_x\mathbf{a}_x + A_y\mathbf{a}_y + A_z\mathbf{a}_z$$

$$\nabla V = \frac{\partial V}{\partial x}\mathbf{a}_x + \frac{\partial V}{\partial y}\mathbf{a}_y + \frac{\partial V}{\partial z}\mathbf{a}_z$$

$$\nabla \cdot \mathbf{A} = \frac{\partial A_x}{\partial x} + \frac{\partial A_y}{\partial y} + \frac{\partial A_z}{\partial z}$$

$$\nabla \times \mathbf{A} = \begin{vmatrix} \mathbf{a}_x & \mathbf{a}_y & \mathbf{a}_z \\ \dfrac{\partial}{\partial x} & \dfrac{\partial}{\partial y} & \dfrac{\partial}{\partial z} \\ A_x & A_y & A_z \end{vmatrix}$$

$$= \left[\frac{\partial A_z}{\partial y} - \frac{\partial A_y}{\partial z}\right]\mathbf{a}_x + \left[\frac{\partial A_x}{\partial z} - \frac{\partial A_z}{\partial x}\right]\mathbf{a}_y + \left[\frac{\partial A_y}{\partial x} - \frac{\partial A_x}{\partial y}\right]\mathbf{a}_z$$

$$\nabla^2 V = \frac{\partial^2 V}{\partial x^2} + \frac{\partial^2 V}{\partial y^2} + \frac{\partial^2 V}{\partial z^2}$$

Cylindrical Coordinates (ρ, ϕ, z)

$$\mathbf{A} = A_\rho\mathbf{a}_\rho + A_\phi\mathbf{a}_\phi + A_z\mathbf{a}_z$$

$$\nabla \mathbf{V} = \frac{\partial V}{\partial \rho}\mathbf{a}_\rho + \frac{1}{\rho}\frac{\partial V}{\partial \phi}\mathbf{a}_\phi + \frac{\partial V}{\partial z}\mathbf{a}_z$$

$$\nabla \cdot \mathbf{A} = \frac{1}{\rho}\frac{\partial}{\partial \rho}(\rho A_\rho) + \frac{1}{\rho}\frac{\partial A_\phi}{\partial \phi} + \frac{\partial A_z}{\partial z}$$

$$\nabla \times \mathbf{A} = \frac{1}{\rho}\begin{vmatrix} \mathbf{a}_\rho & \rho\mathbf{a}_\phi & \mathbf{a}_z \\ \dfrac{\partial}{\partial \rho} & \dfrac{\partial}{\partial \phi} & \dfrac{\partial}{\partial z} \\ A_\rho & \rho A_\phi & A_z \end{vmatrix}$$

$$= \left[\frac{1}{\rho}\frac{\partial A_z}{\partial \phi} - \frac{\partial A_\phi}{\partial z}\right]\mathbf{a}_\rho + \left[\frac{\partial A_\rho}{\partial z} - \frac{\partial A_z}{\partial \rho}\right]\mathbf{a}_\phi + \frac{1}{\rho}\left[\frac{\partial}{\partial \rho}(\rho A_\phi) - \frac{\partial A_\rho}{\partial \phi}\right]\mathbf{a}_z$$

$$\nabla^2 V = \frac{1}{\rho}\frac{\partial}{\partial \rho}\left(\rho\frac{\partial V}{\partial \rho}\right) + \frac{1}{\rho^2}\frac{\partial^2 V}{\partial \phi^2} + \frac{\partial^2 V}{\partial z^2}$$

VECTOR DERIVATIVES (cont.)

Spherical Coordinates (r, θ, ϕ)

$$\mathbf{A} = A_r\mathbf{a}_r + A_\theta\mathbf{a}_\theta + A_\phi\mathbf{a}_\phi$$

$$\nabla \mathbf{V} = \frac{\partial V}{\partial r}\mathbf{a}_r + \frac{1}{r}\frac{\partial V}{\partial \theta}\mathbf{a}_\theta + \frac{1}{r\sin\theta}\frac{\partial V}{\partial \phi}\mathbf{a}_\phi$$

$$\nabla \cdot \mathbf{A} = \frac{1}{r^2}\frac{\partial}{\partial r}(r^2 A_r) + \frac{1}{r\sin\theta}\frac{\partial}{\partial \theta}(A_\theta \sin\theta) + \frac{1}{r\sin\theta}\frac{\partial A_\phi}{\partial \phi}$$

$$\nabla \times \mathbf{A} = \frac{1}{r^2\sin\theta}\begin{vmatrix} \mathbf{a}_r & r\mathbf{a}_\theta & (r\sin\theta)\,\mathbf{a}_\phi \\ \dfrac{\partial}{\partial r} & \dfrac{\partial}{\partial \theta} & \dfrac{\partial}{\partial \phi} \\ A_r & rA_\theta & (r\sin\theta)\,A_\phi \end{vmatrix}$$

$$= \frac{1}{r\sin\theta}\left[\frac{\partial}{\partial\theta}(A_\phi\sin\theta) - \frac{\partial A_\theta}{\partial\phi}\right]\mathbf{a}_r + \frac{1}{r}\left[\frac{1}{\sin\theta}\frac{\partial A_r}{\partial\phi} - \frac{\partial}{\partial r}(rA_\phi)\right]\mathbf{a}_\theta$$

$$+ \frac{1}{r}\left[\frac{\partial}{\partial r}(rA_\theta) - \frac{\partial A_r}{\partial\theta}\right]\mathbf{a}_\phi$$

$$\nabla^2 V = \frac{1}{r^2}\frac{\partial}{\partial r}\left(r^2\frac{\partial V}{\partial r}\right) + \frac{1}{r^2\sin\theta}\frac{\partial}{\partial\theta}\left(\sin\theta\frac{\partial V}{\partial\theta}\right) + \frac{1}{r^2\sin^2\theta}\frac{\partial^2 V}{\partial\phi^2}$$

VECTOR DERIVATIVES

Cartesian Coordinates (x, y, z)

$$\mathbf{A} = A_x\mathbf{a}_x + A_y\mathbf{a}_y + A_z\mathbf{a}_z$$

$$\nabla V = \frac{\partial V}{\partial x}\mathbf{a}_x + \frac{\partial V}{\partial y}\mathbf{a}_y + \frac{\partial V}{\partial z}\mathbf{a}_z$$

$$\nabla \cdot \mathbf{A} = \frac{\partial A_x}{\partial x} + \frac{\partial A_y}{\partial y} + \frac{\partial A_z}{\partial z}$$

$$\nabla \times \mathbf{A} = \begin{vmatrix} \mathbf{a}_x & \mathbf{a}_y & \mathbf{a}_z \\ \dfrac{\partial}{\partial x} & \dfrac{\partial}{\partial y} & \dfrac{\partial}{\partial z} \\ A_x & A_y & A_z \end{vmatrix}$$

$$= \left[\frac{\partial A_z}{\partial y} - \frac{\partial A_y}{\partial z}\right]\mathbf{a}_x + \left[\frac{\partial A_x}{\partial z} - \frac{\partial A_z}{\partial x}\right]\mathbf{a}_y + \left[\frac{\partial A_y}{\partial x} - \frac{\partial A_x}{\partial y}\right]\mathbf{a}_z$$

$$\nabla^2 V = \frac{\partial^2 V}{\partial x^2} + \frac{\partial^2 V}{\partial y^2} + \frac{\partial^2 V}{\partial z^2}$$

Cylindrical Coordinates (ρ, ϕ, z)

$$\mathbf{A} = A_\rho\mathbf{a}_\rho + A_\phi\mathbf{a}_\phi + A_z\mathbf{a}_z$$

$$\nabla \mathbf{V} = \frac{\partial V}{\partial \rho}\mathbf{a}_\rho + \frac{1}{\rho}\frac{\partial V}{\partial \phi}\mathbf{a}_\phi + \frac{\partial V}{\partial z}\mathbf{a}_z$$

$$\nabla \cdot \mathbf{A} = \frac{1}{\rho}\frac{\partial}{\partial \rho}(\rho A_\rho) + \frac{1}{\rho}\frac{\partial A_\phi}{\partial \phi} + \frac{\partial A_z}{\partial z}$$

$$\nabla \times \mathbf{A} = \frac{1}{\rho}\begin{vmatrix} \mathbf{a}_\rho & \rho\mathbf{a}_\phi & \mathbf{a}_z \\ \dfrac{\partial}{\partial \rho} & \dfrac{\partial}{\partial \phi} & \dfrac{\partial}{\partial z} \\ A_\rho & \rho A_\phi & A_z \end{vmatrix}$$

$$= \left[\frac{1}{\rho}\frac{\partial A_z}{\partial \phi} - \frac{\partial A_\phi}{\partial z}\right]\mathbf{a}_\rho + \left[\frac{\partial A_\rho}{\partial z} - \frac{\partial A_z}{\partial \rho}\right]\mathbf{a}_\phi + \frac{1}{\rho}\left[\frac{\partial}{\partial \rho}(\rho A_\phi) - \frac{\partial A_\rho}{\partial \phi}\right]\mathbf{a}_z$$

$$\nabla^2 V = \frac{1}{\rho}\frac{\partial}{\partial \rho}\left(\rho\frac{\partial V}{\partial \rho}\right) + \frac{1}{\rho^2}\frac{\partial^2 V}{\partial \phi^2} + \frac{\partial^2 V}{\partial z^2}$$